National Fire Alarm Code® Handbook

National Fire Alarm Code® Handbook

THIRD EDITION

EDITED BY

Merton W. Bunker, Jr., P.E.   Wayne D. Moore, P.E.

With the complete text of the 1999 edition of NFPA 72, *National Fire Alarm Code*®

 National Fire Protection Association, Quincy, Massachusetts

Product Manager: Pam Powell
Project Editor: Joyce Grandy
Copy Editor: Susan Desrocher
Text Processing: Kathy Barber
Composition: Argosy

Art Coordinator: Nancy Maria
Illustrations: Todd K. Bowman and George Nichols
Cover Design: Greenwood Associates
Manufacturing Buyer: Ellen Glisker
Printer: Quebecor World/Taunton

Notice Concerning Liability: Publication of this handbook is for the purpose of circulating information and opinion among those concerned for fire and electrical safety and related subjects. While every effort has been made to achieve a work of high quality, neither the NFPA nor the contributors to this handbook guarantee the accuracy or completeness of or assume any liability in connection with the information and opinions contained in this handbook. The NFPA and the contributors shall in no event be liable for any personal injury, property, or other damages of any nature whatsoever, whether special, indirect, consequential, or compensatory, directly or indirectly resulting from the publication, use of, or reliance upon this handbook.

This handbook is published with the understanding that the NFPA and the contributors to this handbook are supplying information and opinion but are not attempting to render engineering or other professional services. If such services are required, the assistance of an appropriate professional should be sought.

Notice Concerning Code Interpretations: This third edition of *National Fire Alarm Code® Handbook* is based on the 1999 edition of NFPA 72 *National Fire Alarm Code®*. All NFPA codes, standards, and recommended practices, and guides are developed in accordance with the published procedures of the NFPA technical committees comprised of volunteers drawn from a broad array of relevant interests. The handbook contains the complete text of NFPA 72 and any applicable Formal Interpretations issued by the Association. These documents are accompanied by explanatory commentary and other supplementary materials.

The commentary and supplementary materials in this handbook are not a part of the Code and do not constitute Formal Interpretations of the NFPA (which can be obtained only through requests processed through the responsible technical committee in accordance with the published procedures of the NFPA). The commentary and supplementary materials, therefore, solely reflect the personal opinions of the editor or other contributors and do not necessarily represent the official position of the NFPA or its technical committees.

®Registered trademark National Fire Protection Association, Inc.

NFPA No.: F9-72HB99
ISBN: 0-87765-445-X
Library of Congress Card Catalog No.: 99-075955

Printed in the United States of America
04   03   02   01   00   5   4   3   2

The 1999 *National Fire Alarm Code Handbook* is dedicated to the authorities having jurisdiction who are responsible for enforcing the requirements contained herein and to the many users of the *National Fire Alarm Code*.

Contents

Preface

The 1999 edition of NFPA 72, *National Fire Alarm Code*®, represents the culmination of the first century of signaling standards. The first signaling standard, NFPA 71-D, *General Rules for the Installation of Wiring and Apparatus for Automatic Fire Alarms, Hatch Closers, Sprinkler Alarms, and Other Automatic Alarm Systems and Their Auxiliaries*, was written in 1899. That document was only fifteen pages in length, including the committee report! We are certain the original framers of that document would be astonished to see what their work looks like today.

Fire alarm signaling has come a long way since NFPA published that first signaling standard one hundred years ago. Many technologies related to fire alarm systems have evolved while others have changed little since the middle part of the nineteenth century. For example, fixed-temperature heat detectors and McCulloh loops have not changed significantly since they were invented in the late 1800s. Many technologies emerged in just the past thirty or forty years. More recent technologies, such as electronic addressable analog smoke detectors, continue to improve. Additionally, the computer age has ushered in an era of major changes in fire alarm system control units. Software-driven system designs have resulted in fire alarm systems that are more flexible, richer in features, and easier to test and maintain.

Certainly, the coming years will bring many more changes. As computer systems become more sophisticated, fire alarm system designers will integrate these systems with other building systems such as HVAC systems, security and access control systems, and energy management systems. Integration of these systems will require technicians from both the fire alarm and non-fire alarm system fields to possess a more detailed and functional knowledge of code requirements. Systems integration will also require a more complete understanding of the application and operation of the various building systems technologies and how they interact with fire alarm systems. Education will play a critical role to the understanding and application of fire alarm systems and their integration with other building systems.

This handbook contains many significant changes. Most notably, reordering of the chapters makes the code easier to use. Chapter 8, "Dwellings," has been rewritten to include performance-based requirements and is now a stand-alone chapter. A restructured Chapter 3 presents the material in a more orderly fashion. Additionally, a new Chapter 6 provides requirements for public fire alarm reporting systems, where formerly two separate chapters contained these requirements. A complete revision to Appendix B, "Engineering Guide for Automatic Fire Detector Spacing," makes this important material easier to use. Finally, four new supplements at the end of the handbook provide explanations, guidance, and tutorials that cover important and interesting subjects relating to fire alarm system design and application.

These changes to the 1999 code could not have occurred without the dedication and teamwork of nearly 240 individuals who serve on the technical committees that develop NFPA 72. These individuals collectively spend tens of thousands of hours during the development of, and processing of changes to, the *National Fire Alarm Code*. This code truly embodies a document of the people, by the people, and for the people. Future editions will surely contain increasing numbers of changes. Again and again, committee members will provide the necessary expertise to properly bring these changes into the code, and its users will take part in its development by submitting proposals for changes and comments on proposed changes.

Changes may seem inevitable, but some things never change. The NFPA continues to offer an open, consensus-based standards-making system. We remind our readers that anyone can submit a proposal or public comment to change this code. As with all of the codes and standards in the NFPA system, this code is subject to a rigorous public review process. And, in our opinion, this edition of the *National Fire Alarm Code Handbook* represents one of the best documents available in the world to detail the installation requirements for fire alarm systems.

The editors wish to thank Dean K. Wilson, P.E., of Hughes Associates, Inc.; Ronald H. Kirby of Simplex Time Recorder Co.; Guyléne Proulx of the National Research Council of Canada; John M. Cholin, P.E., of J.M. Cholin Consultants, Inc.; Robert P. Schifiliti, P.E., of R.P. Schifiliti Associates; Inc.; Christopher Marrion, P.E., of ARUP Fire Protection Engineering; and Bruce Fraser and A.J. Capowski

of Simplex Time Recorder Co. Their contributions made this handbook possible. Additionally, the editors wish to thank John Flaherty of John Flaherty Photographers, Charlie Beaulieu, Chuck Moran, and David Dionne of Mammoth Fire Alarms, and the many manufacturers whose expertise, generosity, and patience helped us to provide many new photographs for this edition. We also thank Pam Powell, our "guiding light," for her patience with missed deadlines and her attention to quality. She obviously understands how to keep authors and editors moving in the right direction.

We also wish to thank our wives, Maureen and Joan, for their patience, love, and understanding. Their support made the development of this handbook much easier for both of us.

Merton W. Bunker, Jr., P.E.
Wayne D. Moore, P.E.

NFPA 72,
National Fire Alarm Code®,
1999 Edition, with Commentary

Part One of this handbook includes the complete text and illustrations of the 1999 edition of NFPA 72, *National Fire Alarm Code*®. The text and illustrations from the code are printed in black and are the official requirements of NFPA 72. Line drawings and photographs from the code are labeled Figures.

Paragraphs that begin with the letter A are extracted from Appendix A of the code. Although printed in black ink, this nonmandatory material is purely explanatory in nature. For ease of use, this handbook places Appendix A material immediately after the code paragraph to which it refers.

Part One also includes Formal Interpretations of NFPA 72; these clarifications apply to previous and subsequent editions for which the requirements remain substantially unchanged. Formal Interpretations are not part of the code and are printed in shaded grey boxes.

In addition to code text, appendixes, and Formal Interpretations, Part One includes commentary that provides the history and other background information for specific paragraphs in the code. This insightful commentary takes the reader behind the scenes, into the reasons underlying the requirements.

To readily identify commentary material, commentary text, illustration captions, and tables are printed in red. So the reader may easily distinguish between line drawings and photographs in the code and the commentary, line drawings and photographs in the commentary are labeled Exhibits and each is enclosed in a red box. The distinction between figures in the code and exhibits in the commentary is new for this edition of the *National Fire Alarm Code Handbook*.

CHAPTER 1

Fundamentals of Fire Alarm Systems

Chapter 1 covers fundamentals of fire alarm systems, described as "the basic functions of a complete fire alarm system" (see 1-5.1.1). Separate sections address requirements for power supplies, compatibility of fire detection devices with the control unit, system functions, performance and limitations, fire alarm control units, zoning and annunciation, and documentation.

Chapter 1 also presents definitions used throughout NFPA 72.

1-1 Scope

NFPA 72 covers the application, installation, location, performance, and maintenance of fire alarm systems and their components.

The *National Fire Alarm Code®* provides the minimum installation, test, maintenance, and performance requirements for fire alarm systems. The code also provides requirements for the application, location, and limitations of fire alarm components such as manual fire alarm boxes, automatic fire detectors, and notification appliances.

Other codes and authorities, such as building codes, NFPA *101®*, *Life Safety Code®*, and insurance companies, require the installation of fire alarm systems. The *National Fire Alarm Code* provides the requirements for how to install fire alarm systems.

Partial systems or equipment not necessarily required by the code or other codes for use in fire alarm systems must also be installed to satisfy the requirements of the code; see 3-2.4. The requirement in 2-1.4.4 was also revised to allow detection devices installed for protection of a specific hazard; for example, a building owner wants to protect a specific hazard by installing automatic smoke detection. However, other applicable codes do not require these detectors. Nonetheless, the detection devices installed in that space must satisfy the spacing requirements described in Chapter 2.

1-2 Purpose

1-2.1* The purpose of this code is to define the means of signal initiation, transmission, notification, and annunciation; the levels of performance; and the reliability of the various types of fire alarm systems. This code defines the features associated with these systems and also provides the information necessary to modify or upgrade an existing system to meet the requirements of a particular system classification. It is the intent of this code to establish the required levels of performance, extent of redundancy, and quality of installation but not to establish the methods by which these requirements are to be achieved.

The code describes the various types of fire alarm and supervisory initiating devices, and the fire alarm, supervisory, and trouble audible and visible notification appliances. It provides requirements for how these devices and appliances must be used. The code also describes the types of systems, the methods of signal transmission, and the features that determine system reliability and performance. However, the code is not an installation specification, an approval guide, or a training manual.

Except in the case of fire warning equipment for dwelling units, the *National Fire Alarm Code* does not specify which occupancies are required to have a fire alarm system. The *Life Safety Code* or other local building codes determine what type of protection or system is required for a given occupancy. These codes reference NFPA 72, which provides the performance and installation requirements for fire alarm systems.

A-1-2.1 Fire alarm systems intended for life safety should be designed, installed, and maintained to provide indication and warning of abnormal fire conditions. The system should alert building occupants and summon appropriate aid in adequate time to allow for occupants to travel to a safe place and for rescue operations to occur. The fire alarm system

3

should be part of a life safety plan that also includes a combination of prevention, protection, egress, and other features particular to that occupancy.

Chapter 3 describes the system capabilities required for protected premises fire alarm systems. Chapter 5 contains descriptions of the three types of supervising station fire alarm systems. It also provides the requirements for various transmission technologies.

Whenever a system is modified or updated, it is vital for the system designer to have a thorough understanding of the existing equipment, including its capabilities and the system's wiring, i.e., circuit style, type, and configuration. Often, the existing equipment is too old to interface easily with the newer technology used in the planned additional equipment. This existing equipment may or may not be able to be modified to conform to current code requirements.

1-2.2 Any reference or implied reference to a particular type of hardware is for the purpose of clarity and shall not be interpreted as an endorsement.

NFPA does not manufacture, test, distribute, endorse, approve, or list services, products, or components. See Section 1-4 and A-1-4 for definitions of Approved and Listed.

1-2.3 Unless otherwise noted, it is not intended that the provisions of this document be applied to facilities, equipment, structures, or installations that were existing or approved for construction or installation prior to the effective date of the document.

Exception: Those cases where it is determined by the authority having jurisdiction that the existing situation involves a distinct hazard to life or property.

With the exception of Chapter 7, the code does not apply retroactively to existing installations. See the commentary following 7-1.1.4, which clearly indicates the retroactive application of Chapter 7 requirements.

1-3 General

1-3.1 This code classifies fire alarm systems as follows:

(1) Household fire warning systems

Fire warning equipment in dwelling units is installed to warn the occupants of a fire emergency so they may immediately evacuate the building. This system may be comprised of a control unit with connected initiating devices and notification appliances powered by the control unit. Alternatively, the system may consist of single-station smoke alarms or multiple-station smoke alarms located in specific areas of the family living unit. Single-station and multiple-station smoke

alarms may also be used in hotel rooms and apartment buildings. The requirements for fire warning equipment for dwelling units are detailed in Chapter 8 of the code.

(2) Protected premises fire alarm systems

The primary purpose of a protected premises fire alarm system is to warn building occupants to evacuate the premises. Other purposes of protected premises fire alarm systems include actuating the building fire protection features, providing property protection, assuring mission continuity, providing heritage preservation, and providing environmental protection. Chapter 3 of the code describes the requirements for protected premises fire alarm systems.

(3) Supervising station fire alarm systems

Supervising station fire alarm systems, described in Chapter 5 of the code, provide requirements for means of communication between the protected premises and a location called a supervising station. The types of supervising station fire alarm systems are enumerated in (a) through (e).

a. Auxiliary fire alarm systems

Auxiliary systems provide a direct means of communicating between the protected premises and the fire department using public fire alarm reporting systems. Public fire alarm reporting systems provide manual fire alarm boxes at strategic locations throughout a municipality. Citizens may initiate a fire alarm signal by actuating one of the street fire alarm boxes. An auxiliary system provides a connection from a protected premises fire alarm system to a public fire alarm reporting system master box. A master box is a public fire alarm reporting system manual fire alarm box that has been equipped for remote actuation. If a municipality does not have a public fire alarm reporting system, then an auxiliary fire alarm system cannot be provided. See Chapter 6 for requirements pertaining to auxiliary systems and public fire reporting systems.

1. Local energy type

When a fire alarm signal at a protected premises is actuated, contacts in the protected premises fire alarm system control unit actuate a circuit that, in turn, causes a municipal fire alarm master box to transmit a fire alarm signal to the public fire service communications center. The municipal fire alarm box may be mounted on the inside or outside of the protected building.

Power to operate the local energy interface circuit comes from the protected premises fire alarm system control unit. In addition, the protected premises fire alarm system control unit monitors the interface circuit for integrity. It also monitors the set or unset condition of the public fire alarm reporting system master box. A coded wired public fire alarm reporting system, a series telephone public fire alarm reporting system, and a coded radio public fire alarm

reporting system all use a local energy interface circuit to allow a protected premises fire alarm system control unit to actuate the master box. See Chapter 6 for requirements pertaining to local energy-type auxiliary systems.

2. Parallel telephone type

The parallel telephone public fire alarm reporting system uses telephone lines to connect the street boxes with the public fire service communications center. When a protected premises fire alarm system interfaces with a parallel telephone system, it does so without the need for a street box. Rather, the telephone circuit directly connects the protected premises fire alarm system control unit with the public fire service communications center. See Chapter 6 for requirements pertaining to parallel telephone-type auxiliary systems.

3. Shunt type

As an alternative to a local energy interface, some coded wired public fire alarm reporting system boxes offer a shunt connection. A closed contact at the protected premises is electrically connected to a circuit that is derived from the public fire alarm reporting telegraph circuit. When the closed contacts open the circuit and remove the shunt, the box trips and initiates a signal to the public fire service communications center.

A ground fault on the shunt circuit also becomes a ground fault on the public fire alarm reporting circuit. If an open fault occurs on the public fire alarm reporting circuit, a subsequent actuation of the shunt circuit will not cause the public fire alarm reporting box to initiate an alarm signal. Unfortunately, unless the fire department somehow notifies all of the building owners that the public fire alarm reporting circuit has an impairment, the owners will not know that their connection to the public circuit is also impaired.

The code limits the devices connected to a shunt circuit to manual fire alarm boxes and automatic sprinkler waterflow switches. Automatic fire detectors may not be connected to a shunt circuit. A shunt-type system has very specific requirements and is not allowed to be interconnected to a protected premises system unless the city circuits entering the protected premises are installed in rigid conduit. This helps to prevent faults in one premises from disabling the circuit. Faults in the circuit may prevent transmission from other protected premises, leaving them unprotected. See Chapter 6 for requirements pertaining to shunt-type auxiliary systems.

b. Remote supervising station fire alarm systems

If a building owner does not wish to use a central station fire alarm system or a proprietary supervising station fire alarm system, or if a public fire alarm reporting system is not available, the owner may choose to install a remote supervising station fire alarm system. These systems provide a means for transmitting alarm, supervisory, and trouble signals from the protected premises to a remote supervising station. Such systems normally transmit alarm signals to a public fire service communications center. They also transmit supervisory and trouble signals to a constantly attended location that is acceptable to the authority having jurisdiction. See Chapter 5 for requirements pertaining to remote supervising station systems.

c. Proprietary supervising station systems

Proprietary supervising station systems require trained, competent personnel in constant attendance at the supervising station. The station personnel monitor the protected premises and take appropriate action when necessary. A proprietary supervising station is located at the protected property or at another property of the same owner. It is owned and operated by the property owner and receives signals from one or more properties under the same ownership. The property may consist of a single building, such as a high-rise building, or several buildings, such as a college campus where the dormitories and other buildings report to a single proprietary supervising station at the campus police department or campus fire department. The property may be contiguous or noncontiguous. If noncontiguous, it may consist of protected properties at remote locations, such as across town or across the country. An example of a proprietary supervising station with contiguous property is a college campus. A proprietary supervising station with noncontiguous property would be a hotel chain with properties across the country that are monitored from a single location owned by the hotel chain. See Chapter 5 for requirements pertaining to proprietary supervising stations.

d. Central station fire alarm systems

The code requires that central station supervising fire alarm systems be controlled and operated by a person, firm, or corporation whose business is the furnishing of such systems. Such a company will have obtained specific listing from a nationally recognized testing laboratory as a provider of central station service. Normally, authorities having jurisdiction only require central station supervising station fire alarm systems where a facility has either a high risk of loss or a high value. For example, a facility that has a large number of nonambulatory people may benefit from central station service. Similarly, a high-hazard or high-value manufacturing facility may benefit from central station service. See Chapter 5 and Supplement 2 for requirements pertaining to central station supervising systems and central station service.

e. Municipal fire alarm systems

Public fire alarm reporting systems can be used to transmit fire alarm signals from a protected premises to the public fire service communications center. See the commentary under 1-3.1(3)a and Chapter 6 for requirements pertaining to public fire reporting systems.

1-3.2 A device or system having materials or forms that differ from those detailed in this code shall be permitted to be examined and tested according to the intent of the code and, if found equivalent, shall be approved.

A device or system that does not meet the specific requirements of the code may be submitted to a testing laboratory for listing by the laboratory to determine if the device or system meets the intent of the code. The authority having jurisdiction determines whether or not a product, method, or device is suitable.

1-3.3 The intent and meaning of the terms used in this code are, unless otherwise defined herein, the same as those of NFPA 70, *National Electrical Code®*.

1-4 Definitions

For the purposes of this code, the following terms are defined as follows.

All definitions that apply to subjects covered throughout the code are located in Section 1-4. Standard dictionary definitions apply to words not defined in Section 1-4.

    Acknowledge. To confirm that a message or signal has been received, such as by the pressing of a button or the selection of a software command.

    Active Multiplex System. A multiplexing system in which signaling devices such as transponders are employed to transmit status signals of each initiating device or initiating device circuit within a prescribed time interval so that the lack of receipt of such a signal can be interpreted as a trouble signal.

Active multiplex circuits use an interrogation (call and response) routine to determine the status of a device. Failure to receive a status signal from a device initiates a trouble signal. This interrogation routine serves to monitor the interconnecting path for integrity. See 1-5.8 for requirements pertaining to monitoring for integrity.

    Addressable Device. A fire alarm system component with discrete identification that can have its status individually identified or that is used to individually control other functions.

Addressable devices, as shown in Exhibits 1.1 and 1.2, may be either initiating devices or control appliances. An initiating device provides the fire alarm system control panel with its status and actual location. The control appliance receives operating commands from the fire alarm system control panel. Digital addresses for each device or appliance can be assigned by the system hardware or software.

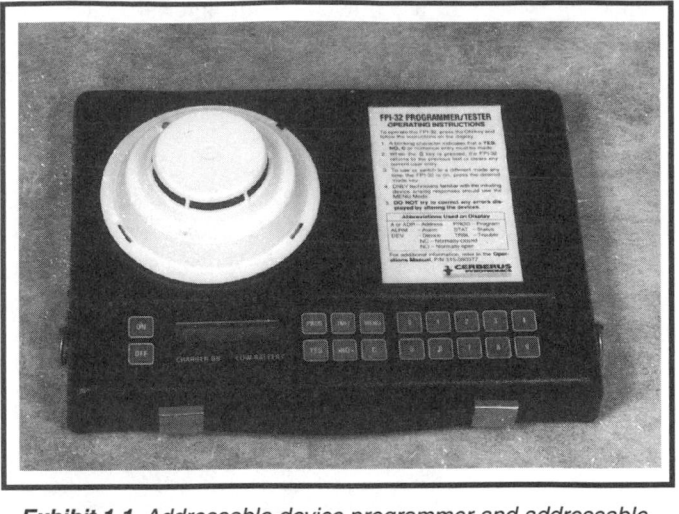

Exhibit 1.1 Addressable device programmer and addressable smoke detector. (Source: Seimens Cerberus Division, Cedar Knolls, NJ)

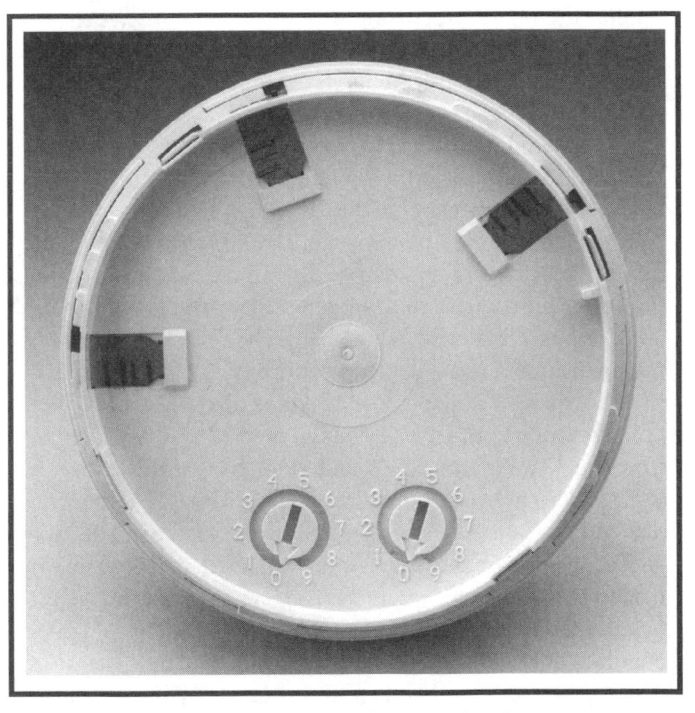

Exhibit 1.2 Addressable smoke detector showing programming switches. (Source: System Sensor Corp., St. Charles, IL)

    Adverse Condition. Any condition occurring in a communications or transmission channel that interferes with the proper transmission or interpretation, or both, of status change signals at the supervising station. *(Refer to Trouble Signal.)*

Examples of adverse conditions include circuits with open faults or ground faults, electrical or radio frequency interference on communications paths, and circuit wiring with short-circuit faults.

Air Sampling–Type Detector. A detector that consists of a piping or tubing distribution network that runs from the detector to the area(s) to be protected. An aspiration fan in the detector housing draws air from the protected area back to the detector through air sampling ports, piping, or tubing. At the detector, the air is analyzed for fire products.

There are passive and active air sampling–type detectors. Duct smoke detectors, as shown in Exhibit 1.3, are typically considered passive detection devices. Active sampling requires the creation of a negative pressure within a sampling tube to draw products of combustion from the protected area or protected space into the sampling network. Vacuum pumps or blower assemblies normally create this negative pressure. An active air-sampling smoke detector is illustrated in Exhibit 1.4.

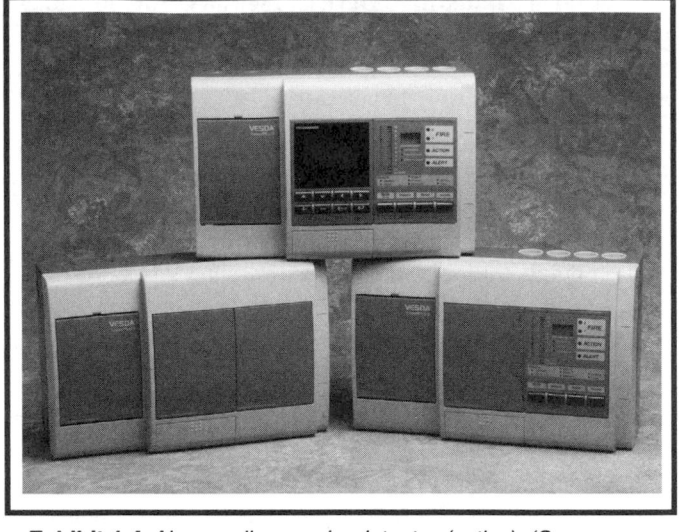

Exhibit 1.4 *Air-sampling smoke detector (active). (Source: Vision Systems, Inc., Hingham, MA)*

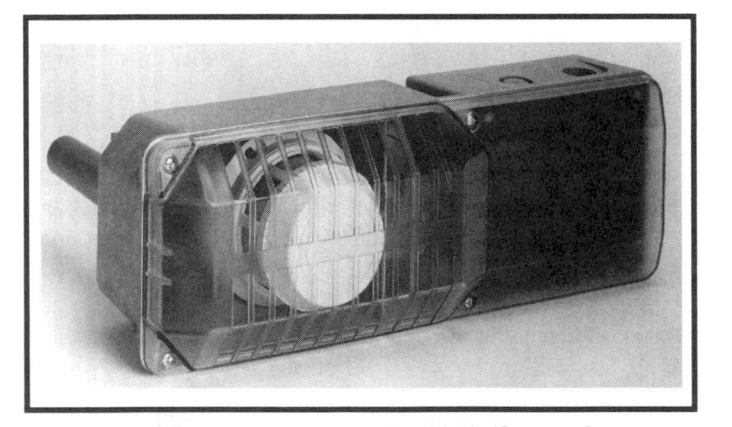

Exhibit 1.3 *Duct smoke detector (passive). (Source: System Sensor Corp., St. Charles, IL)*

Alarm. A warning of fire danger.

The word *alarm* indicates a fire alarm condition. The phrases *supervisory alarm* or *trouble alarm* are not appropriate replacements for the word *alarm*. See the definitions of Trouble Signal and Supervisory Signal in Section 1-4 of the code.

Alarm Service. The service required following the receipt of an alarm signal.

The action taken when an alarm signal is received is called the *alarm service*. The alarm service may include any or all of the following:

(1) Response by a private fire brigade or public fire department
(2) Dispatch of an alarm service provider's runner
(3) Notification to the building owner and occupant
(4) Notification to the authorities having jurisdiction

Alarm Signal. A signal indicating an emergency that requires immediate action, such as a signal indicative of fire.

Alarm Verification Feature. A feature of automatic fire detection and alarm systems to reduce unwanted alarms wherein smoke detectors report alarm conditions for a minimum period of time, or confirm alarm conditions within a given time period after being reset, in order to be accepted as a valid alarm initiation signal.

The alarm verification feature helps reduce false alarms where transient conditions actuate smoke detectors. The feature may reside within individual smoke detectors. Manufacturers of fire alarm systems may also include the feature in fire alarm system control units. Qualified testing laboratories define the timing sequence. Alarm verification may also be used where smoke detectors are installed in environments that are marginally acceptable for the application of smoke detectors.

Alert Tone. An attention-getting signal to alert occupants of the pending transmission of a voice message.

Voice/alarm communications systems precede a voice announcement with an alert tone in order to gain occupant attention and advise the occupants that a fire alarm voice announcement is about to be made. The alert tone is not considered an alarm signal. See 3-8.4.1.3.5.3.1 for applicable requirements.

Analog Initiating Device (Sensor). An initiating device that transmits a signal indicating varying degrees of condition as contrasted with a conventional initiating device, which can only indicate an on–off condition.

Analog devices measure and transmit a range of values of smoke density, temperature variation, water level, water pressure changes, and other variables to a fire alarm system control unit. Typically, the control unit software determines the set points for initiation of an alarm, supervisory, or trouble signal. By storing reported values over time, some smoke detector technology uses the analog feature to provide a warning signal to the owner when the detector is dirty or when the detector drifts outside of its listed sensitivity range. Some analog technology may be used as a substitute for smoke detector sensitivity testing per 7-3.2.1.

Annunciator. A unit containing one or more indicator lamps, alphanumeric displays, or other equivalent means in which each indication provides status information about a circuit, condition, or location.

Approved.* Acceptable to the authority having jurisdiction.

A-1-4 Approved. The National Fire Protection Association does not approve, inspect, or certify any installations, procedures, equipment, or materials; nor does it approve or evaluate testing laboratories. In determining the acceptability of installations, procedures, equipment, or materials, the authority having jurisdiction may base acceptance on compliance with NFPA or other appropriate standards. In the absence of such standards, said authority may require evidence of proper installation, procedure, or use. The authority having jurisdiction may also refer to the listings or labeling practices of an organization that is concerned with product evaluations and is thus in a position to determine compliance with appropriate standards for the current production of listed items.

The authority having jurisdiction may choose to grant approval on the basis of whether or not a product has received a listing and has been labeled by a qualified testing laboratory. However, the listing or labeling alone does not constitute approval. While, in 1-5.1.2, the code requires that all installed fire alarm equipment be listed, some jurisdictions place the ultimate decision making in the hands of the authority having jurisdiction. This may mean that in some jurisdictions, the authority having jurisdiction has the right to approve products or systems that are not labeled or listed.

Audible Notification Appliance. A notification appliance that alerts by the sense of hearing.

Authority Having Jurisdiction.* The organization, office, or individual responsible for approving equipment, materials, an installation, or a procedure.

A-1-4 Authority Having Jurisdiction. The phrase "authority having jurisdiction" is used in NFPA documents in a broad manner, since jurisdictions and approval agencies vary, as do their responsibilities. Where public safety is primary, the authority having jurisdiction may be a federal, state, local, or other regional department or individual such as a fire chief; fire marshal; chief of a fire prevention bureau, labor department, or health department; building official; electrical inspector; or others having statutory authority. For insurance purposes, an insurance inspection department, rating bureau, or other insurance company representative may be the authority having jurisdiction. In many circumstances, the property owner or his or her designated agent assumes the role of the authority having jurisdiction; at government installations, the commanding officer or departmental official may be the authority having jurisdiction.

Any given physical property may have multiple authorities having jurisdiction. These multiple authorities having jurisdiction may be concerned with life safety, property protection, mission continuity, heritage preservation, and environmental protection. Some authorities having jurisdiction may impose additional requirements beyond that of the code. If there are conflicting requirements for the installation of a specific fire alarm system, the installer must follow the most stringent requirements.

Automatic Extinguishing System Supervisory Device. A device that responds to abnormal conditions that could affect the proper operation of an automatic sprinkler system or other fire extinguishing system(s) or suppression system(s), including, but not limited to, control valves; pressure levels; liquid agent levels and temperatures; pump power and running; engine temperature and overspeed; and room temperature.

When an abnormal condition is detected, a supervisory signal is activated to warn the owner or attendant that the extinguishing system requires attention. Supervisory signals are distinct from alarm signals or trouble signals.

Automatic Fire Detector. A device designed to detect the presence of a fire signature and to initiate action. For the purpose of this code, automatic fire detectors are classified as follows: Automatic Fire Extinguishing or Suppression System Operation Detector, Fire-Gap Detector, Heat Detector, Other Fire Detectors, Radiant Energy-Sensing Fire Detector, Smoke Detector.

Automatic Fire Extinguishing or Suppression System Operation Detector. A device that automatically detects the operation of a fire extinguishing or suppression system by means appropriate to the system employed.

Examples of automatic fire extinguishing or suppression system operation alarm initiating devices are agent discharge flow switches and agent discharge pressure switches.

Auxiliary Box. A fire alarm box that can be operated from one or more remote actuating devices.

In an auxiliary fire alarm system, the auxiliary fire alarm box (commonly called a *master box*) is normally located on the outside of a building. The protected premises fire alarm system is connected to this box and will automatically actuate the box when an alarm occurs. The auxiliary box may also be actuated manually. Upon actuation, the box transmits a coded signal to the public fire service communications center. Also see the commentary following 1-3.1(3).

Auxiliary Fire Alarm System. A system connected to a municipal fire alarm system for transmitting an alarm of fire to the public fire service communications center. Fire alarms from an auxiliary fire alarm system are received at the public fire service communications center on the same equipment and by the same methods as alarms transmitted manually from municipal fire alarm boxes located on streets.

Auxiliary Fire Alarm System, Local Energy Type. An auxiliary system that employs a locally complete arrangement of parts, initiating devices, relays, power supply, and associated components to automatically trip a municipal transmitter or master box over electrical circuits that are electrically isolated from the municipal system circuits.

Auxiliary Fire Alarm System, Parallel Telephone Type. An auxiliary system connected by a municipally controlled individual circuit to the protected property to interconnect the initiating devices at the protected premises and the municipal fire alarm switchboard.

Auxiliary Fire Alarm System, Shunt Auxiliary Type. An auxiliary system electrically connected to an integral part of the municipal alarm system extending the municipal circuit into the protected premises to interconnect the initiating devices, which, when operated, open the municipal circuit shunted around the trip coil of the municipal transmitter or master box. The municipal transmitter or master box is thereupon energized to start transmission without any assistance from a local source of power.

Average Ambient Sound Level. The root mean square, A-weighted, sound pressure level measured over a 24-hour period.

Box Battery. The battery supplying power for an individual fire alarm box where radio signals are used for the transmission of box alarms.

Carrier. High-frequency energy that can be modulated by voice or signaling impulses.

Carrier System. A means of conveying a number of channels over a single path by modulating each channel on a different carrier frequency and demodulating at the receiving point to restore the signals to their original form.

Ceiling. The upper surface of a space, regardless of height. Areas with a suspended ceiling have two ceilings, one visible from the floor and one above the suspended ceiling.

Ceiling Height. The height from the continuous floor of a room to the continuous ceiling of a room or space.

Ceiling Surfaces, Beam Construction. Ceilings that have solid structural or solid nonstructural members projecting down from the ceiling surface more than 4 in. (100 mm) and spaced more than 3 ft (0.9 m), center to center.

Ceiling Surfaces, Girder. A support for beams or joists that runs at right angles to the beams or joists. If the top of the girder is within 4 in. (100 mm) of the ceiling, the girder is a factor in determining the number of detectors and is to be considered a beam. If the top of the girder is more than 4 in. (100 mm) from the ceiling, the girder is not a factor in detector location.

Ceiling Surfaces, Solid Joist Construction. Ceilings that have solid structural or solid nonstructural members projecting down from the ceiling surface for a distance of more than 4 in. (100 mm) and spaced at intervals of 3 ft (0.9 m) or less, center to center.

Solid joists, whether structural or nonstructural, impede the flow of products of combustion. Web or bar joists are not considered to be solid joists unless the top chord is over 4 in. (100 mm) deep. In that case, only the top chord is considered a ceiling obstruction.

Central Station. A supervising station that is listed for central station service.

The listed central station is the constantly attended location where signals from the central station fire alarm system are received. Operators take action on signals, including initiating a retransmission of the signals, and provide runner service. See Chapter 5 and Supplement 2 for requirements pertaining to central station systems and central station service.

Central Station Fire Alarm System. A system or group of systems in which the operations of circuits and devices are transmitted automatically to, recorded in, maintained by, and supervised from a listed central station that has competent and experienced servers and operators who, upon receipt of a signal, take such action as required by this code. Such service is to be controlled and operated by a person, firm, or corporation whose business is the furnishing, maintaining, or monitoring of supervised fire alarm systems.

A central station fire alarm system is shown in Exhibit 1.5.

Central Station Service. The use of a system or a group of systems in which the operations of circuits and devices at a protected property are signaled to, recorded in, and supervised from a listed central station that has competent and experienced operators who, upon receipt of a

Exhibit 1.5 *Central station fire alarm system. (Source: Simplex Time Recorder Company, Gardner, MA)*

signal, take such action as required by this code. Related activities at the protected property, such as equipment installation, inspection, testing, maintenance, and runner service, are the responsibility of the central station or a listed fire alarm service–local company. Central station service is controlled and operated by a person, firm, or corporation whose business is the furnishing of such contracted services or whose properties are the protected premises.

There are eight elements of central station service: installation; testing; maintenance; runner service at the protected premises; and management oversight, monitoring, retransmission, and record keeping at the central station. All eight elements of this service must be provided under contract to the subscriber by the prime contractor, either alone or in conjunction with their subcontractors. See Section 5-2 for requirements pertaining to central station service.

Certification. A systematic program that uses randomly selected follow-up inspections of the certified systems installed under the program that allows the listing organization to verify that a fire alarm system complies with all the requirements of this code. A system installed under such a program is identified by the issuance of a certificate and is designated as a certificated system.

Certification is a process whereby the prime contractor conspicuously indicates that the fire alarm system providing service at a protected premises complies with all the requirements of the code. In doing this, the contractor provides a means of third party verification in the form of a certificate. After the installation is complete, the prime contractor applies for a certificate from the organization that has listed the prime contractor. After the listing organization reviews the application and supporting documentation and finds this information acceptable, it issues a certificate. The prime contractor then posts the certificate in a conspicuous location near the fire alarm control unit. The listing organization may, at any time, conduct an inspection of any certificated system that the prime contractor has installed. This provides the means of verification by the third party. The code requires all central station fire alarm systems to be certificated or placarded. See 5-2.2.3 for third party verification requirements.

Certification of Personnel.* A formal program of related instruction and testing as provided by a recognized organization or the authority having jurisdiction.

A-1-4 Certification of Personnel. This definition of *certification of personnel* applies only to municipal fire alarm systems.

Channel. A path for voice or signal transmission that uses modulation of light or alternating current within a frequency band.

Circuit Interface. A circuit component that interfaces initiating devices or control circuits, or both; notification appliances or circuits, or both; system control outputs; and other signaling line circuits to a signaling line circuit.

Cloud Chamber Smoke Detection. The principle of using an air sample drawn from the protected area into a high-humidity chamber combined with a lowering of chamber pressure to create an environment in which the resultant moisture in the air condenses on any smoke particles present, forming a cloud. The cloud density is measured by a photoelectric principle. The density signal is processed and used to convey an alarm condition when it meets preset criteria.

Cloud chamber smoke detection is another form of an active air sampling–type smoke detector. Cloud chamber smoke detectors are extremely sensitive to low levels of combustion products and are frequently used to detect very small fires in vital equipment. Also see the definition of Air Sampling–Type Smoke Detector.

Code. A standard that is an extensive compilation of provisions covering broad subject matter or that is suitable for adoption into law independently of other codes and standards.

Coded. An audible or visible signal that conveys several discrete bits or units of information. Notification signal examples are numbered strokes of an impact-type appliance and numbered flashes of a visible appliance.

Table A-1-5.4.2.1 provides recommended assignments for simple zone coded signals.

In addition to the examples that are described in Table A-1-5.4.2.1, textual signals may use words that are familiar only to those concerned with response to the signal. This practice avoids general alarm notification disruption of the occupants. Hospitals often use this type of signal. To hospital occupants who do not know the code words, a typical message might sound like a normal paging announcement: "Paging Dr. Firestone, Dr. Firestone, Building 4 West Wing." In other words, there is a fire on the West Wing of Building 4.

Combination Detector. A device that either responds to more than one of the fire phenomenon or employs more than one operating principle to sense one of these phenomenon. Typical examples are a combination of a heat detector with a smoke detector or a combination rate-of-rise and fixed-temperature heat detector.

Exhibits 1.6 and 1.7 illustrate two common types of combination detectors.

Exhibit 1.7 *Typical combination smoke and heat detector. (Source: System Sensor; photo courtesy of Mammoth Fire Alarms, Inc., Lowell, MA)*

Combination System.* A fire alarm system in which components are used, in whole or in part, in common with a non-fire signaling system.

Combination systems are permitted by Chapter 3. Exhibit 1.8 illustrates a typical combination system.

Exhibit 1.6 *Combination rate-of-rise and fixed-temperature heat detector. (Source: Mammoth Fire Alarms, Inc., Lowell, MA)*

Combination Fire Alarm and Guard's Tour Box. A manually operated box for separately transmitting a fire alarm signal and a distinctive guard patrol tour supervisory signal.

Exhibit 1.8 *Combination burglary and fire alarm system control panel with keypad. (Source: Ademco, Syosett, NY)*

A-1-4 Combination System. Examples of non-fire systems are security, card access control, closed circuit television, sound reinforcement, background music, paging, sound masking, building automation, time, and attendance.

Communications Channel. A circuit or path connecting a subsidiary station(s) to a supervising station(s) over which signals are carried.

Compatibility Listed. A specific listing process that applies only to two-wire devices, such as smoke detectors, that are designed to operate with certain control equipment.

Two-wire, circuit-powered detectors must be listed for compatibility with a fire alarm control unit. This listing ensures that the required operating characteristics of the fire alarm system control unit's initiating device circuit matches the operating characteristics of the smoke detector. Detectors that are not listed for this purpose may appear to operate properly under normal operating conditions but may not work under adverse conditions such as a low voltage situation. Fire alarm system control units often have limits on the number of devices per circuit. In addition, addressable devices must be compatibility listed because most manufacturers use different software or communications protocols.

Compatible Equipment. Equipment that interfaces mechanically or electrically as manufactured without field modification.

Contiguous Property. A single-owner or single-user protected premises on a continuous plot of ground, including any buildings thereon, that is not separated by a public thoroughfare, transportation right-of-way, property owned or used by others, or body of water not under the same ownership.

Control Unit. A system component that monitors inputs and controls outputs through various types of circuits.

The definition of Control Unit includes fire alarm system control units that provide sprinkler supervisory service, multiplex interfaces (transponders), and fire alarm system control units with integral digital alarm communicator transmitters (DACTs).

Delinquency Signal. A signal indicating the need for action in connection with the supervision of guards or system attendants.

Delinquency signal applies to guard's tour systems. If the guard fails to initiate a signal from a tour station or fails to make a tour in a prescribed amount of time, the fire alarm system that provides guard's tour supervision generates a guard's tour supervisory off-normal signal.

Derived Channel. A signaling line circuit that uses the local leg of the public switched network as an active multiplex channel while simultaneously allowing that leg's use for normal telephone communications.

Detector. A device suitable for connection to a circuit that has a sensor that responds to a physical stimulus such as heat or smoke.

Digital Alarm Communicator Receiver (DACR). A system component that accepts and displays signals from digital alarm communicator transmitters (DACTs) sent over the public switched telephone network.

Exhibit 1.9 illustrates a typical DACR.

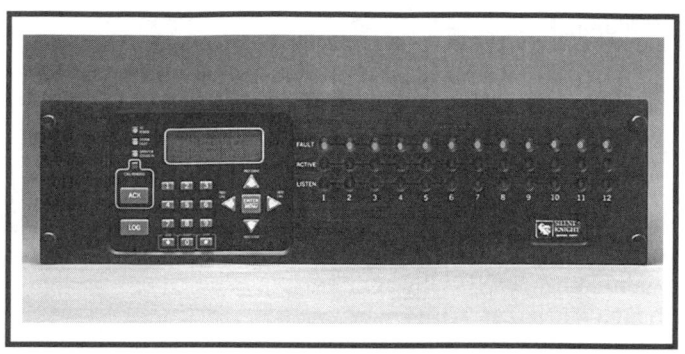

Exhibit 1.9 *Digital alarm communicator receiver (DACR). (Source: Silent-Knight, Maple Grove, MN)*

Digital Alarm Communicator System (DACS). A system in which signals are transmitted from a digital alarm communicator transmitter (DACT) located at the protected premises through the public switched telephone network to a digital alarm communicator receiver (DACR).

Digital Alarm Communicator Transmitter (DACT). A system component at the protected premises to which initiating devices or groups of devices are connected. The DACT seizes the connected telephone line, dials a preselected number to connect to a DACR, and transmits signals indicating a status change of the initiating device.

The communications portion of a DACT functions very similarly to a modem that allows a personal computer to connect to the Internet. When a fire alarm, supervisory, or trouble signal is initiated, the DACT dials one of two preprogrammed telephone numbers. Once a digital alarm communicator receiver (DACR) answers the incoming call, the DACT transmits digital information. The DACR interprets and displays the digital information as a fire alarm, supervisory, or trouble signal. The DACT, as shown in Exhibit 1.10, is the most popular type of transmission means for fire alarm signals to a supervising station.

Digital Alarm Radio Receiver (DARR). A system component composed of two subcomponents: one that receives and decodes radio signals, the other that annunciates the decoded data. These two subcomponents can be

Exhibit 1.10 *Digital alarm communicator transmitter (DACT).*
(Source: ESL Sentrol, Inc., Tualitin, OR)

Display. The visual representation of output data, other than printed copy.

Double Doorway.* A single opening that has no intervening wall space or door trim separating the two doors.

A-1-4 Double Doorway. Refer to Figure 2-10.6.5.3.1 for an illustration of detector location requirements for double doors.

Double Dwelling Unit. A building consisting solely of two dwelling units. *(See Dwelling Unit.)*

Dual Control. The use of two primary trunk facilities over separate routes or different methods to control one communications channel.

Dwelling Unit. One or more rooms for the permanent use of one or more persons as a space for eating, living, and sleeping, with permanent provisions for cooking and sanitation. For the purposes of this code, *dwelling unit* includes one- and two-family attached and detached dwellings, apartments, and condominiums but does not include hotel and motel rooms and guest suites, dormitories, or sleeping rooms in nursing homes.

Electrical Conductivity Heat Detector. A line-type or spot-type sensing element in which resistance varies as a function of temperature.

Ember.* A particle of solid material that emits radiant energy due either to its temperature or the process of combustion on its surface. *(Refer to Spark.)*

A-1-4 Ember. Class A and Class D combustibles burn as embers under conditions where the flame typically associated with fire does not necessarily exist. This glowing combustion yields radiant emissions in parts of the radiant energy spectrum that are radically different from those parts affected by flaming combustion. Specialized detectors that are specifically designed to detect those emissions should be used in applications where this type of combustion is expected. In general, flame detectors are not intended for the detection of embers.

coresident at the central station or separated by means of a data transmission channel.

Digital Alarm Radio System (DARS). A system in which signals are transmitted from a digital alarm radio transmitter (DART) located at a protected premises through a radio channel to a digital alarm radio receiver (DARR).

Exhibit 1.11 illustrates a typical digital alarm radio system arrangement.

Digital Alarm Radio Transmitter (DART). A system component that is connected to or an integral part of a digital alarm communicator transmitter (DACT) that is used to provide an alternate radio transmission channel.

Exhibit 1.12 illustrates a typical digital alarm radio transmitter.

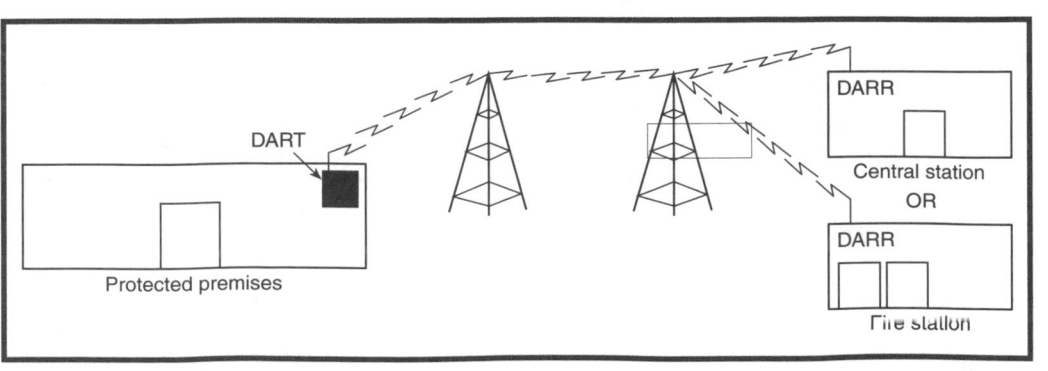

Exhibit 1.11 *Digital alarm radio system (DARS). (Source: Ademco AlarmNet; courtesy of AFA Protective Systems, Inc., North Brunswick, NJ)*

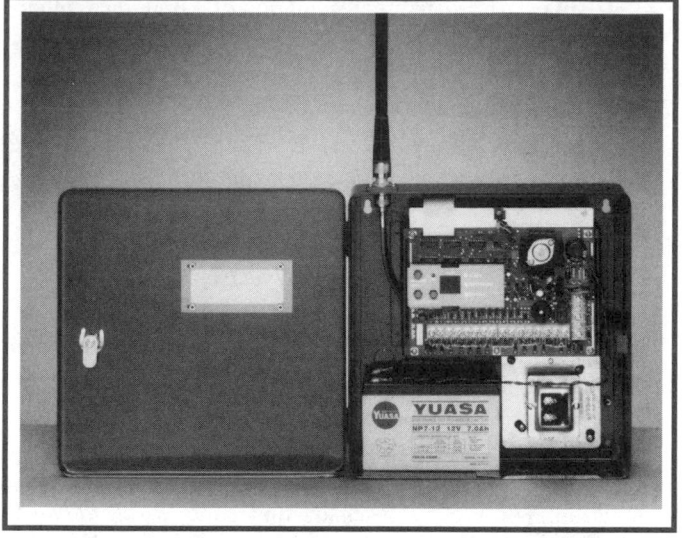

Exhibit 1.12 *Digital alarm radio transmitter (DART). (Source: Ademco AlarmNet; photo courtesy of AFA Protective Systems, Inc., Syosset, NY)*

Emergency Voice/Alarm Communications. Dedicated manual or automatic facilities for originating and distributing voice instructions, as well as alert and evacuation signals pertaining to a fire emergency, to the occupants of a building.

Authorities having jurisdiction often require an emergency voice/alarm communications system whenever the fire plan for a building calls for selective, partial evacuation of the building or relocation of the occupants to areas of refuge. Most often such buildings include high rise structures or buildings that cover very large areas.

Evacuation.* The withdrawal of occupants from a building.

A-1-4 Evacuation. Evacuation does not include the relocation of occupants within a building.

Evacuation Signal. Distinctive signal intended to be recognized by the occupants as requiring evacuation of the building.

The code requires fire alarm systems to use a three-pulse temporal pattern evacuation signal that meets the requirements of ANSI S3.41, *American National Standard Audible Emergency Evacuation Signal*, for commercial and industrial fire alarm systems and fire warning equipment in dwelling units.

Exit Plan. A plan for the emergency evacuation of the premises.

Field of View. The solid cone that extends out from the detector within which the effective sensitivity of the detector is at least 50 percent of its on-axis, listed, or approved sensitivity.

Field of view applies to radiant energy fire detectors. Designers who use these detectors need to understand that the field of view defines the line-of-sight area of coverage in which the detector can view a spark, ember, or flaming fire. It should also be noted that unintended sources within a field of view, such as welding arcs or sunlight, may cause nuisance alarms.

Fire Alarm Control Unit (Panel). A system component that receives inputs from automatic and manual fire alarm devices and might supply power to detection devices and to a transponder(s) or off-premises transmitter(s). The control unit might also provide transfer of power to the notification appliances and transfer of condition to relays or devices connected to the control unit. The fire alarm control unit can be a local fire alarm control unit or a master control unit.

Exhibit 1.13 illustrates a typical fire alarm control unit.

Exhibit 1.13 *Fire alarm control unit. (Source: Fire Control Instruments, Waltham, MA)*

Fire Alarm/Evacuation Signal Tone Generator. A device that produces a fire alarm/evacuation tone upon command.

Fire Alarm Signal. A signal initiated by a fire alarm-initiating device such as a manual fire alarm box, automatic fire detector, waterflow switch, or other device in which activation is indicative of the presence of a fire or fire signature.

Fire alarm signals are not permitted to indicate supervisory or trouble conditions. See 1-5.4.7 for requirements pertaining to distinctive signals.

Fire Alarm System. A system or portion of a combination system that consists of components and circuits arranged to monitor and annunciate the status of fire alarm or supervisory signal-initiating devices and to initiate the appropriate response to those signals.

The definition of Fire Alarm System includes fire alarm systems whose sole purpose is to provide sprinkler supervisory service.

Fire Command Center. The principal attended or unattended location where the status of the detection, alarm communications, and control systems is displayed and from which the system(s) can be manually controlled.

The term *fire command center* is specifically associated with fire alarm systems that meet the requirements of 3-8.4.1.3, Emergency Voice/Alarm Communications. Building codes or local ordinances may require that the building owner provide a fire command center, as shown in Exhibit 1.14, for certain types of configurations, such as high-rise buildings. The fire command center houses the fire alarm and other building systems, such as security, HVAC, and elevator and lighting system controls. During an emergency, the fire command center also serves as a command post for coordinating fire-fighting efforts.

Fire–Gas Detector. A device that detects gases produced by a fire.

Examples of fire gases include hydrogen chloride (HCl) and carbon monoxide (CO). Users should not confuse fire–gas detectors designed to detect CO with dwelling unit CO alarms that are designed to prevent CO poisoning by alerting occupants to the presence of CO gas in the home.

Fire Rating. The classification indicating in time (hours) the ability of a structure or component to withstand a standardized fire test. This classification does not necessarily reflect performance of rated components in an actual fire.

Fire Safety Function Control Device. The fire alarm system component that directly interfaces with the control system that controls the fire safety function.

Fire safety function control devices are used to control safety functions that increase the level of life safety and property protection. Fire safety functions may be operated manually or automatically by the fire alarm system control unit. These devices may control door locking and release systems, elevator recall, suppression system activation, and HVAC systems.

Fire Safety Functions. Building and fire control functions that are intended to increase the level of life safety for occupants or to control the spread of the harmful effects of fire.

Fire safety functions may include interconnection with the building fire alarm system to shut down air-handling systems; close HVAC dampers; and effect elevator recall, release of doors, and unlocking of doors. Also see Supplement 3, which provides further information on fire safety

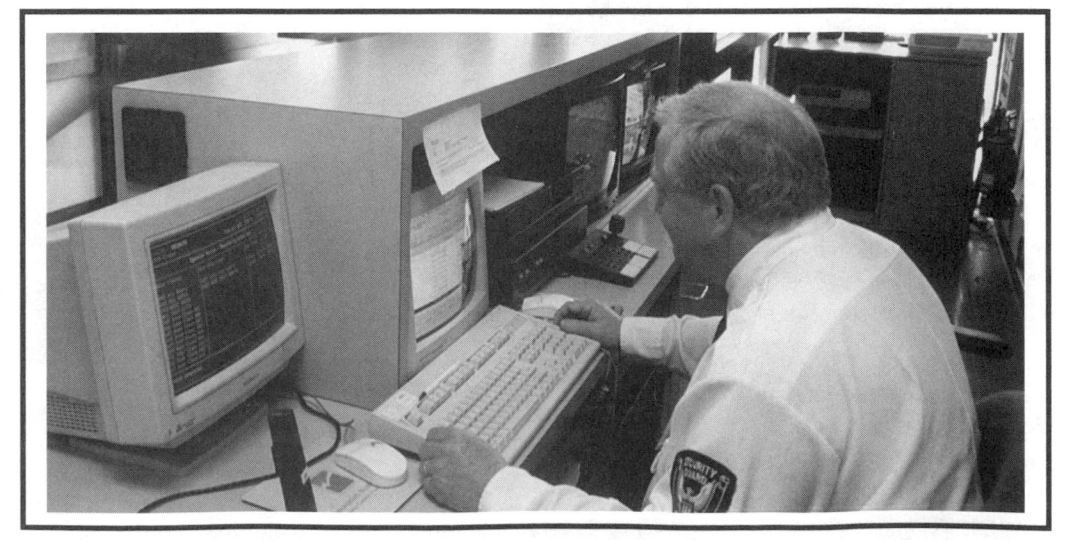

Exhibit 1.14 *A fire command and security center. (Source: Simplex Time Recorder Co., Gardner, MA)*

functions that are interconnected with building fire alarm systems.

Fire Warden. A building staff member or a tenant trained to perform assigned duties in the event of a fire emergency.

Fixed-Temperature Detector.* A device that responds when its operating element becomes heated to a predetermined level.

Exhibit 1.15 illustrates a typical fixed-temperature heat detector.

Exhibit 1.15 *Typical fixed-temperature (nonrestorable) heat detector. (Source: Mammoth Fire Alarms, Inc., Lowell, MA and Chemetronics Caribe, Ashland, MA)*

A-1-4 Fixed-Temperature Detector. The difference between the operating temperature of a fixed-temperature device and the surrounding air temperature is proportional to the rate at which the temperature is rising. The rate is commonly referred to as *thermal lag.* The air temperature is always higher than the operating temperature of the device.

Typical examples of fixed-temperature sensing elements are as follows.

(a) *Bimetallic.* A sensing element comprised of two metals that have different coefficients of thermal expansion arranged so that the effect is deflection in one direction when heated and in the opposite direction when cooled.

(b) *Electrical Conductivity.* A line-type or spot type sensing element in which resistance varies as a function of temperature.

(c) *Fusible Alloy.* A sensing element of a special composition metal (eutectic) that melts rapidly at the rated temperature.

(d) *Heat-Sensitive Cable.* A line-type device in which the sensing element comprises, in one type, two current-carrying wires separated by heat-sensitive insulation that softens at the rated temperature, thus allowing the wires to make electrical contact. In another type, a single wire is centered in a metallic tube, and the intervening space is filled with a substance that becomes conductive at a critical temperature, thus establishing electrical contact between the tube and the wire.

(e) *Liquid Expansion.* A sensing element comprising a liquid that is capable of marked expansion in volume in response to an increase in temperature.

Flame. A body or stream of gaseous material involved in the combustion process and emitting radiant energy at specific wavelength bands determined by the combustion chemistry of the fuel. In most cases, some portion of the emitted radiant energy is visible to the human eye.

Flame Detector.* A radiant energy-sensing fire detector that detects the radiant energy emitted by a flame. *(Refer to A-2-4.2.)*

Exhibit 1.16 illustrates a typical flame detector.

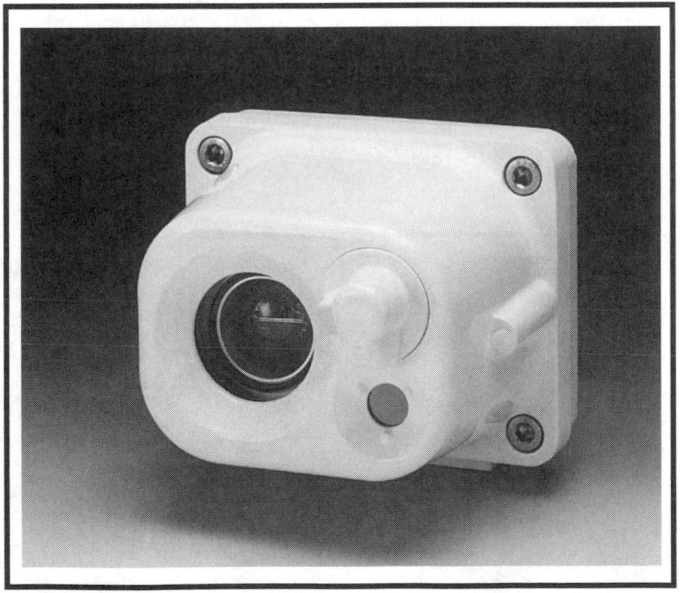

Exhibit 1.16 *Typical flame detector. (Source: Meggitt Avionics, Inc., Manchester, NH)*

A-1-4 Flame Detector. Flame detectors are categorized as ultraviolet, single wavelength infrared, ultraviolet infrared, or multiple wavelength infrared.

Flame Detector Sensitivity. The distance along the optical axis of the detector at which the detector can detect a fire of specified size and fuel within a given time frame.

Gateway. A device that is used in the transmission of serial data (digital or analog) from the fire alarm control unit to other building system control units, equipment, or networks and/or from other building system control units to the fire alarm control unit.

Guard's Tour Reporting Station. A device that is manually or automatically initiated to indicate the route being followed and the timing of a guard's tour.

Guard's Tour Supervisory Signal. A supervisory signal monitoring the performance of guard patrols.

Heat Alarm. A single or multiple station alarm responsive to heat.

Heat Detector. A fire detector that detects either abnormally high temperature or rate of temperature rise, or both.

There are many types of heat detectors. A typical spot-type heat detector is shown in Exhibit 1.17. For descriptions of other types of heat detectors, see definitions of Electrical Conductivity Heat Detector, Fixed-Temperature Detector (see Exhibit 1.15), Line-Type Detector, Nonrestorable Initiating Device, Rate Compensation Detector, and Rate-of-Rise Detector in Section 1-4.

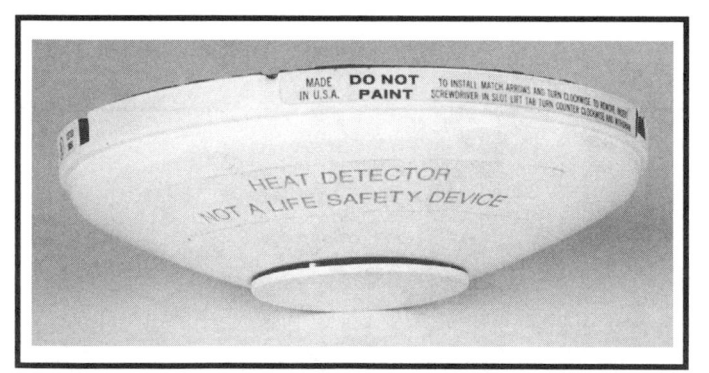

Exhibit 1.17 *Typical heat detector. (Source: EST, Cheshire, CT)*

Household Fire Alarm System. A system of devices that produces an alarm signal in the household for the purpose of notifying the occupants of the presence of a fire so that they will evacuate the premises.

Hunt Group. A group of associated telephone lines within which an incoming call is automatically routed to an idle (not busy) telephone line for completion.

Initiating Device. A system component that originates transmission of a change-of-state condition, such as in a smoke detector, manual fire alarm box, or supervisory switch.

Initiating Device Circuit. A circuit to which automatic or manual initiating devices are connected where the signal received does not identify the individual device operated.

Initiating device circuits are used with conventional fire alarm systems when individual devices are not identified at the control unit. Only the individual circuit or "zone" is identified. In contrast, each device on a signaling line circuit is usually identified by means of an address. See the definition of Signaling Line Circuit in Section 1-4 for more information.

Intermediate Fire Alarm or Fire Supervisory Control Unit. A control unit used to provide area fire alarm or area fire supervisory service that, where connected to the proprietary fire alarm system, becomes a part of that system.

Ionization Smoke Detection.* The principle of using a small amount of radioactive material to ionize the air between two differentially charged electrodes to sense the presence of smoke particles. Smoke particles entering the ionization volume decrease the conductance of the air by reducing ion mobility. The reduced conductance signal is processed and used to convey an alarm condition when it meets preset criteria.

A-1-4 Ionization Smoke Detection. Ionization smoke detection is more responsive to invisible particles (smaller than 1 micron in size) produced by most flaming fires. It is somewhat less responsive to the larger particles typical of most smoldering fires. Smoke detectors that use the ionization principle are usually of the spot type.

Though all listed smoke detectors must pass the same series of tests at a qualified testing laboratory, system designers typically use ionization-type smoke detectors where they expect a greater risk of a flaming rather than a smoldering fire scenario. Generally, fire scientists consider ionization detectors to be slightly more sensitive to the smaller particles of smoke produced by a flaming fire. In locations where smoldering fires are more likely to occur, photoelectric-type smoke detectors may offer better protection. Additionally, light scattering photoelectric-type smoke detectors respond best to light colored smoke rather than black particles because black particles absorb light. Exhibit 1.18 provides details of operation for ionization-type smoke detectors.

Labeled. Equipment or materials to which has been attached a label, symbol, or other identifying mark of an organization that is acceptable to the authority having jurisdiction and concerned with product evaluation, that maintains periodic inspection of production of labeled equipment or materials, and by whose labeling the manufacturer indicates compliance with appropriate standards or performance in a specified manner.

Leg Facility. The portion of a communications channel that connects not more than one protected premises to a

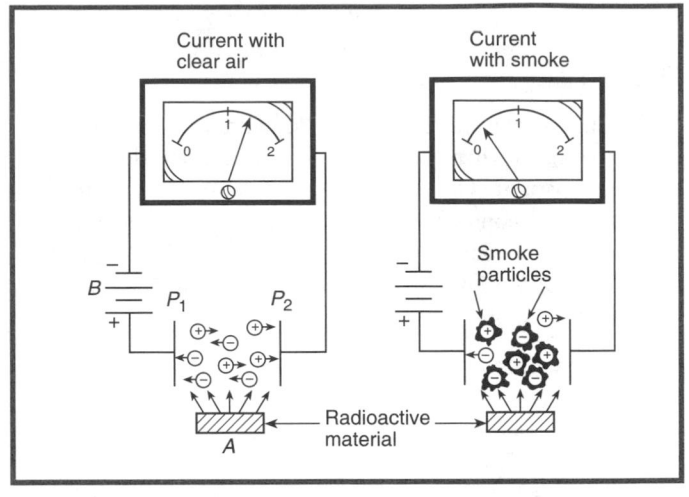

Exhibit 1.18 *Principle of operation for ionization smoke detector.*

primary or secondary trunk facility. The leg facility includes the portion of the signal transmission circuit from its point of connection with a trunk facility to the point where it is terminated within the protected premises at one or more transponders.

Level Ceilings. Ceilings that are level or have a slope of less than or equal to 1 in 8.

Life Safety Network. A type of combination system that transmits fire safety control data through gateways to other building system control units.

Line-Type Detector. A device in which detection is continuous along a path. Typical examples are rate-of-rise pneumatic tubing detectors, projected beam smoke detectors, and heat-sensitive cable.

Listed.* Equipment, materials, or services included in a list published by an organization that is acceptable to the authority having jurisdiction and concerned with evaluation of products or services, that maintains periodic inspection of production of listed equipment or materials or periodic evaluation of services, and whose listing states that either the equipment, material, or service meets appropriate designated standards or has been tested and found suitable for a specified purpose.

A-1-4 Listed. The means for identifying listed equipment may vary for each organization concerned with product evaluation; some organizations do not recognize equipment as listed unless it is also labeled. The authority having jurisdiction should utilize the system employed by the listing organization to identify a listed product.

Loading Capacity. The maximum number of discrete elements of fire alarm systems permitted to be used in a particular configuration.

Loading capacity applies to various transmission technologies used by supervising station fire alarm systems. The loading capacity of a system depends on the performance characteristic of the particular transmission technology employed. Chapter 5 provides the loading capacities for various types of supervising station transmission technologies.

Loss of Power. The reduction of available voltage at the load below the point at which equipment can function as designed.

Low-Power Radio Transmitter. Any device that communicates with associated control/receiving equipment by low-power radio signals.

Maintenance. Repair service, including periodic inspections and tests, required to keep the fire alarm system and its component parts in an operative condition at all times, and the replacement of the system or its components when they become undependable or inoperable for any reason.

Manual Fire Alarm Box. A manually operated device used to initiate an alarm signal.

Operation of a manual fire alarm box, shown in Exhibit 1.19, may require one action, such as pulling a lever, or two actions, such as lifting a cover and then pulling a lever. In some institutional occupancies, the building codes, the *Life Safety Code*, and local ordinances may permit the use of key-operated manual fire alarm boxes, such as the one shown in Exhibit 1.20.

Exhibit 1.19 *Typical single-action manual fire alarm box. (Source: Protectowire; photo courtesy of Mammoth Fire Alarms, Inc., Lowell, MA)*

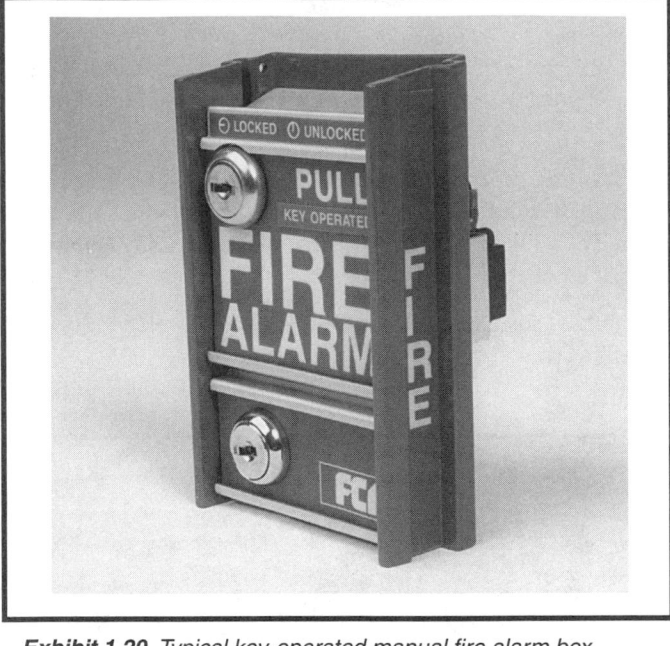

Exhibit 1.20 Typical key-operated manual fire alarm box. (Source: Fire Control Instruments; photo courtesy of Mammoth Fire Alarms, Inc., Lowell, MA)

Master Box. A municipal fire alarm box that can also be operated by remote means.

Public fire alarm reporting system master boxes, as shown in Exhibit 1.21, have an interface circuit that allows a protected premises fire alarm system control unit to actuate the master box whenever the system initiates a fire alarm signal.

Exhibit 1.21 Auxiliary (master) box. (Source: Gamewell; photo courtesy of Mammoth Fire Alarms, Inc., Lowell, MA)

Master Control Unit (Panel). A control unit that serves the protected premises or portion of the protected premises as a local control unit and accepts inputs from other fire alarm control units.

Where more than one fire alarm system control unit has been installed in a facility, one of the control units must act as the master control unit to monitor alarm, trouble, and supervisory conditions from the other satellite fire alarm system control units.

Multiple Dwelling Unit. A building containing three or more dwelling units. *(See Dwelling Unit.)*

Multiple-Station Alarm. A single station alarm capable of being interconnected to one or more additional alarms so that the actuation of one causes the appropriate alarm signal to operate in all interconnected alarms.

The definition of Multiple-Station Alarm was added to the 1996 edition of the code to help differentiate between automatic fire detectors connected to and powered by a fire alarm system control unit and single- and multiple-station smoke detectors powered by either ac power or by a battery. This definition corresponds with the terminology used internationally.

Multiple-Station Alarm Device. Two or more single-station alarm devices that can be interconnected so that actuation of one causes all integral or separate audible alarms to operate; or one single-station alarm device having connections to other detectors or to a manual fire alarm box.

Multiplexing. A signaling method characterized by simultaneous or sequential transmission, or both, and reception of multiple signals on a signaling line circuit, a transmission channel, or a communications channel, including means for positively identifying each signal.

Multiplexing includes two technologies: active and passive. An active multiplex system establishes two-way communication on a signaling line circuit. Generally, the multiplex fire alarm system control unit transmits a signal to the devices or appliances connected to the signaling line circuit. The devices or appliances then transmit a status signal to the fire alarm system control unit. This interrogation and response signaling provides a means to monitor the integrity of the signaling line circuit.

Devices connected to a passive multiplex system transmit multiple signals over the same signaling line circuit. However, the circuit must have some other means to monitor its integrity.

Municipal Fire Alarm Box (Street Box). An enclosure housing a manually operated transmitter used to send an alarm to the public fire service communications center.

Exhibit 1.22 displays a municipal fire alarm box.

Exhibit 1.22 Municipal fire alarm box (street box). (Source: Gamewell; photo courtesy of Mammoth Fire Alarms, Inc., Lowell, MA)

Municipal Fire Alarm System. A system of alarm-initiating devices, receiving equipment, and connecting circuits (other than a public telephone network) used to transmit alarms from street locations to the public fire service communications center.

Municipal Transmitter. A transmitter that can only be tripped remotely that is used to send an alarm to the public fire service communications center.

Noncoded Signal. An audible or visible signal conveying one discrete bit of information.

Noncontiguous Property. An owner- or user-protected premises where two or more protected premises, controlled by the same owner or user, are separated by a public thoroughfare, body of water, transportation right-of-way, or property owned or used by others.

Nonrequired System. A supplementary fire alarm system component or group of components that is installed at the option of the owner, and is not installed due to a building or fire code requirement.

Nonrestorable Initiating Device. A device in which the sensing element is designed to be destroyed in the process of operation.

One example of a nonrestorable initiating device is the fixed-temperature heat detector (see Exhibit 1.15) that uses a fusible element that melts when subjected to heat. When the element melts, the electrical contacts are shorted together and the alarm signal is activated.

Notification Appliance. A fire alarm system component such as a bell, horn, speaker, light, or text display that provides audible, tactile, or visible outputs, or any combination thereof.

Exhibits 1.23 and 1.24 illustrate two types of notification appliances.

Exhibit 1.23 Typical audible notification appliance. (Source: Mammoth Fire Alarms, Inc., Lowell, MA)

Notification Appliance Circuit. A circuit or path directly connected to a notification appliance(s).

Notification Zone. An area covered by notification appliances that are activated simultaneously.

Notification zones should coincide with the smoke and fire zones of a building. See the commentary for 1-5.7 for more information.

Nuisance Alarm. Any alarm caused by mechanical failure, malfunction, improper installation, or lack of proper maintenance, or any alarm activated by a cause that cannot be determined.

Fire officials often use the term *nuisance alarm* in place of the term *false alarm* to describe fire alarm signals initiated

Exhibit 1.24 *Typical visible notification appliance. (Source: Gentex, Corp., Zeeland, MI)*

when an otherwise properly functioning fire alarm system detects conditions that it interprets as a fire signature. Though nuisance alarms are, in fact, false alarms, the term *false alarm* is often reserved for a malicious act whereby some individual falsely initiates a fire alarm. Such acts are also called *malicious false alarms*.

Off-Hook. To make connection with the public-switched telephone network in preparation for dialing a telephone number.

When someone lifts a telephone handset from its normal resting position, the telephone instrument is considered to be off-hook. Digital alarm communicator transmitters use equipment to access the public switched network and automatically provide an off-hook condition prior to beginning a transmission.

On-Hook. To disconnect from the public-switched telephone network.

When someone returns a telephone handset to its normal resting position, the telephone instrument is considered to be on-hook. When a transmission is completed, digital alarm communicator transmitters and receivers accomplish the on-hook function automatically.

Open Area Detection (Protection). Protection of an area such as a room or space with detectors to provide early warning of fire.

Operating Mode, Private. Audible or visible signaling only to those persons directly concerned with the implementation and direction of emergency action initiation and procedure in the area protected by the fire alarm system.

At some locations, the fire alarm system uses the private operating mode to alert individuals who have responsibility to take prescribed action during a fire emergency. Such individuals may include operators in a supervising station, the telephone switchboard operator, the building receptionist, nurses at a nursing station, the building engineer, the plant manager, boiler room operators, emergency response team members, or other specially trained personnel. Some building codes, the *Life Safety Code*, and local ordinances may permit private operating mode notification to precede public operating mode notification of the general occupants.

Operating Mode, Public. Audible or visible signaling to occupants or inhabitants of the area protected by the fire alarm system.

The fire alarm system uses the public operating mode to notify the general occupants of a building to take specified action during a fire. This action may include complete evacuation of the building. Or, it may include selective, partial evacuation or relocation to areas of refuge within the building.

Operating System Software. The basic operating system software that can be altered only by the equipment manufacturer or its authorized representative. Operating system software is sometimes referred to as *firmware, BIOS,* or *executive program.*

Fire alarm control unit operating system software is similar to the main operating system software used in computers. This software is listed for use with the specific fire alarm system control unit. This software is generally not accessible to the end user. Any changes to this software must be thoroughly tested. This definition was added in the 1996 edition to provide a better understanding of the different software types. See 7-1.6.2.1 for maintenance and testing requirements following software alterations.

Other Fire Detectors. Devices that detect a phenomenon other than heat, smoke, flame, or gases produced by a fire.

Ownership. Any property or building or its contents under legal control by the occupant, by contract, or by holding of a title or deed.

Paging System. A system intended to page one or more persons by such means as voice over loudspeaker, coded audible signals or visible signals, or lamp annunciators.

Exhibits 1.25 and 1.26 illustrate examples of paging systems and their use.

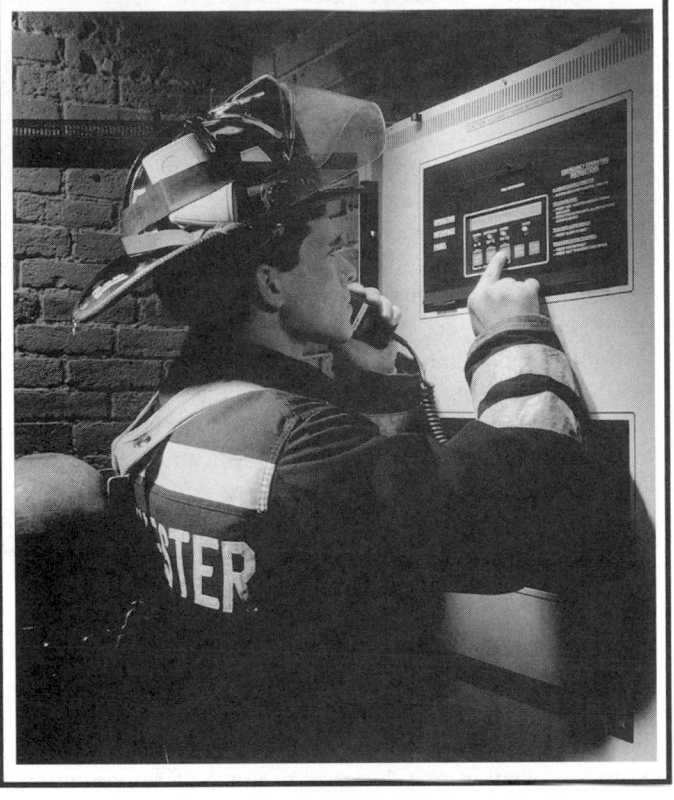

Exhibit 1.25 *Fire official using paging system. (Source: Simplex Time Recorder, Gardner, MA)*

Parallel Telephone System. A telephone system in which an individually wired circuit is used for each fire alarm box.

Path (Pathways). Any conductor, optic fiber, radio carrier, or other means for transmitting fire alarm system information between two or more locations.

Permanent Visual Record (Recording). An immediately readable, not easily alterable, print, slash, or punch record of all occurrences of status change.

Photoelectric Light Obscuration Smoke Detection.* The principle of using a light source and a photosensitive sensor onto which the principal portion of the source emissions is focused. When smoke particles enter the light path, some of the light is scattered and some is absorbed, thereby reducing the light reaching the receiving sensor. The light reduction signal is processed and used to convey an alarm condition when it meets preset criteria.

Exhibit 1.27 illustrates the principle of operation of a photoelectric-type light obscuration smoke detector.

A-1-4 Photoelectric Light Obscuration Smoke Detection. The response of photoelectric light obscuration smoke detectors is usually not affected by the color of smoke.

Smoke detectors that use the light obscuration principle are usually of the line type. These detectors are commonly referred to as "projected beam smoke detectors."

Though all listed smoke detectors must pass the same series of tests at a qualified testing laboratory, system designers typically use photoelectric-type smoke detectors where they expect a fire to produce larger smoke particles, such as with a smoldering fire or an aged smoke scenario. A photoelectric-type smoke detector is better at detecting the larger or lighter colored particles produced by smoldering fires or smoke particles that have agglomerated or "aged" as they move away from the thermal energy source at the fire. In locations where flaming fires are more likely to occur, ionization smoke detectors may offer better protection.

Photoelectric Light-Scattering Smoke Detection.* The principle of using a light source and a photosensitive sensor arranged so that the rays from the light source do not normally fall onto the photosensitive sensor. When smoke particles enter the light path, some of the light is scattered by reflection and refraction onto the sensor. The light signal is processed and used to convey an alarm condition when it meets preset criteria.

Exhibit 1.28 illustrates the principle of operation for a photoelectric light-scattering smoke detector.

A-1-4 Photoelectric Light-Scattering Smoke Detection. Photo-electric light-scattering smoke detection is more responsive to visible particles (larger than 1 micron in size) produced by most smoldering fires. It is somewhat less responsive to the smaller particles typical of most flaming fires. It is also less responsive to black smoke than to lighter colored smoke. Smoke detectors that use the light-scattering principle are usually of the spot type.

Placarded. A means to signify that the fire alarm system of a particular facility is receiving central station service in accordance with this code by a listed central station or listed fire alarm service–local company that is part of a systematic follow-up program under the control of an independent third party listing organization.

Plant. One or more buildings under the same ownership or control on a single property.

Pneumatic Rate-of-Rise Tubing Heat Detector. A line-type detector comprising small-diameter tubing, usually copper, that is installed on the ceiling or high on the walls throughout the protected area. The tubing is terminated in a detector unit containing diaphragms and associated contacts set to actuate at a predetermined pressure. The system is sealed except for calibrated vents that compensate for normal changes in temperature.

Positive Alarm Sequence. An automatic sequence that results in an alarm signal, even when manually delayed for investigation, unless the system is reset.

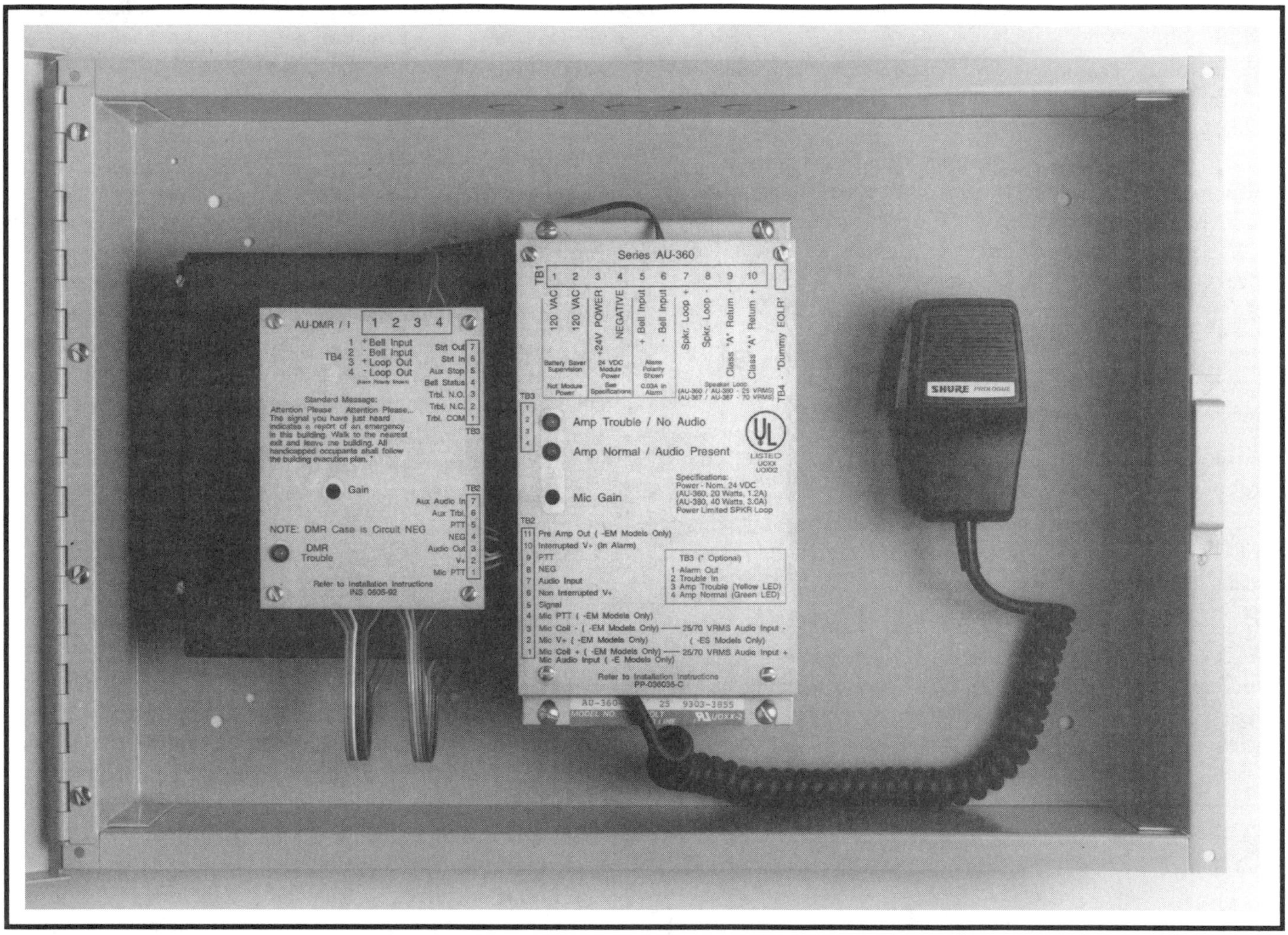

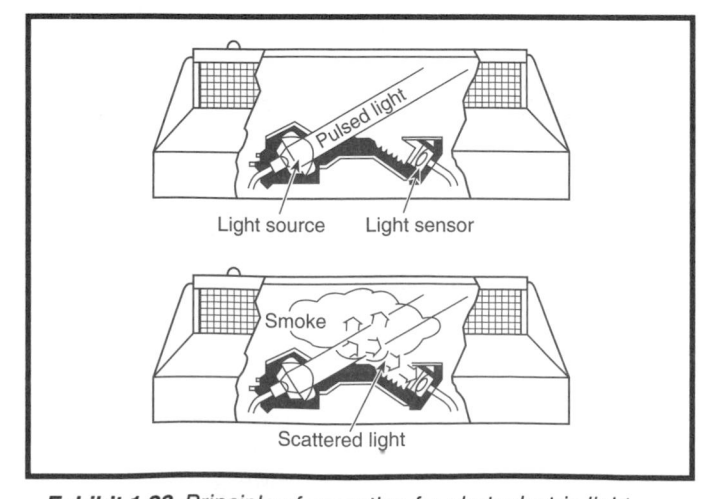

Exhibit 1.26 Typical single channel paging system. (Source: Audiosone Corp., Stratford, CT)

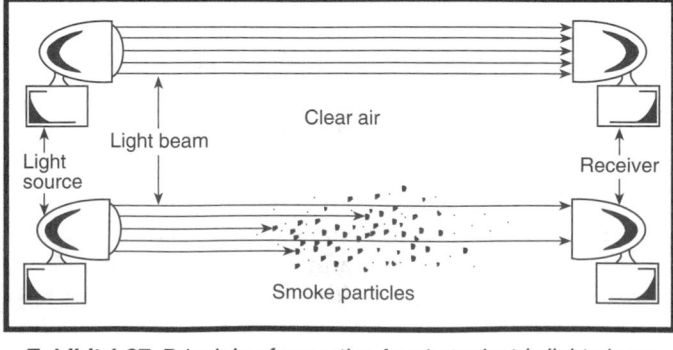

Exhibit 1.27 Principle of operation for photoelectric light obscuration smoke detector.

Power Supply. A source of electrical operating power, including the circuits and terminations connecting it to the dependent system components.

Exhibit 1.28 Principle of operation for photoelectric light-scattering smoke detector (top: clean air; bottom: with smoke).

Primary Battery (Dry Cell). A nonrechargeable battery requiring periodic replacement.

Primary Trunk Facility. That part of a transmission channel connecting all leg facilities to a supervising or subsidiary station.

Prime Contractor. The one company contractually responsible for providing central station services to a subscriber as required by this code. The prime contractor can be either a listed central station or a listed fire alarm service–local company.

The prime contractor is a person, firm, or corporation listed by a qualified testing laboratory to install, maintain, and test a central station fire alarm system. See Section 5-2 for further requirements on central station fire alarm systems.

Private Radio Signaling. A radio system under control of the proprietary supervising station.

Projected Beam–Type Detector. A type of photoelectric light obscuration smoke detector wherein the beam spans the protected area.

Projected beam–type detection is often used in large open areas such as atria, convention halls, auditoriums, gymnasiums, and where a building or portion of a building has a high ceiling. Exhibit 1.29 illustrates a typical projected beam–type smoke detector.

Exhibit 1.29 *Typical projected beam–type smoke detector. (Source: Detection Systems, Inc.; photo courtesy of Mammoth Fire Alarms, Inc., Lowell, MA)*

Proprietary Supervising Station. A location to which alarm or supervisory signaling devices on proprietary fire alarm systems are connected and where personnel are in attendance at all times to supervise operation and investigate signals.

Many large industrial plants, college campuses, large hospital complexes, and detention and correction facilities provide a proprietary supervising station, as shown in Exhibit 1.30, to monitor all portions of the contiguous or noncontiguous protected premises.

Proprietary Supervising Station Fire Alarm System. An installation of fire alarm systems that serves contiguous and noncontiguous properties, under one ownership, from a proprietary supervising station located at the protected property, at which trained, competent personnel are in constant attendance. This includes the proprietary supervising station; power supplies; signal-initiating devices; initiating device circuits; signal notification appliances; equipment for the automatic, permanent visual recording of signals; and equipment for initiating the operation of emergency building control services.

Protected Premises. The physical location protected by a fire alarm system.

Protected Premises (Local) Control Unit (Panel). A control unit that serves the protected premises or a portion of the protected premises and indicates the alarm via notification appliances inside the protected premises.

Protected Premises (Local) Fire Alarm System. A protected premises system that sounds an alarm at the protected premises as the result of the manual operation of a fire alarm box or the operation of protection equipment or systems, such as water flowing in a sprinkler system, the discharge of carbon dioxide, the detection of smoke, or the detection of heat.

Public Fire Alarm Reporting System. A system of fire alarm initiating devices, receiving equipment, and connecting circuits used to transmit alarms from street locations to the communications center.

Public Fire Alarm Reporting System, Type A. A system in which an alarm from a fire alarm box is received and is retransmitted to fire stations either manually or automatically.

Public Fire Alarm Reporting System, Type B. A system in which an alarm from a fire alarm box is automatically transmitted to fire stations and, if used, is transmitted to supplementary alerting devices.

Public Fire Service Communications Center. The building or portion of the building used to house the central operating part of the fire alarm system; usually the place where the necessary testing, switching, receiving, transmitting, and power supply devices are located.

In a large municipality, the public fire service communications center may be located at the main fire department station, at the public safety complex, or at a specially designed communications building. The communications center may or may not include the public service answering point (PSAP) for the community's 9-1-1 emergency telephone

Exhibit 1.30 *Proprietary supervising station. (Source: Simplex Time Recorder Co., Gardner, MA)*

system. NFPA 1221, *Standard for the Installation, Maintenance, and Use of Emergency Services Communications Systems*, provides requirements for the design, installation, and operation of the public fire service communications center. Public fire alarm reporting systems (see Chapter 6) terminate at the public fire service communications center.

Public Switched Telephone Network. An assembly of communications facilities and central office equipment operated jointly by authorized common carriers that provides the general public with the ability to establish communications channels via discrete dialing codes.

Radiant Energy-Sensing Fire Detector. A device that detects radiant energy (such as ultraviolet, visible, or infrared) that is emitted as a product of combustion reaction and obeys the laws of optics.

Radio Alarm Repeater Station Receiver (RARSR). A system component that receives radio signals and resides at a repeater station that is located at a remote receiving location.

Radio Alarm Supervising Station Receiver (RASSR). A system component that receives data and annunciates that data at the supervising station.

Radio Alarm System (RAS). A system in which signals are transmitted from a radio alarm transmitter (RAT) located at a protected premises through a radio channel to two or more radio alarm repeater station receivers (RARSR) and that are annunciated by a radio alarm supervising station receiver (RASSR) located at the central station.

Radio Alarm Transmitter (RAT). A system component at the protected premises to which initiating devices or groups of devices are connected that transmits signals indicating a status change of the initiating devices.

Radio Channel.* A band of frequencies of a width sufficient to allow its use for radio communications.

A-1-4 Radio Channel. The width of the channel depends on the type of transmissions and the tolerance for the frequency of emission. Channels normally are allocated for radio transmission in a specified type for service by a specified transmitter.

Rate Compensation Detector.* A device that responds when the temperature of the air surrounding the device reaches a predetermined level, regardless of the rate of temperature rise.

Exhibit 1.31 illustrates a typical rate-compensated heat detector.

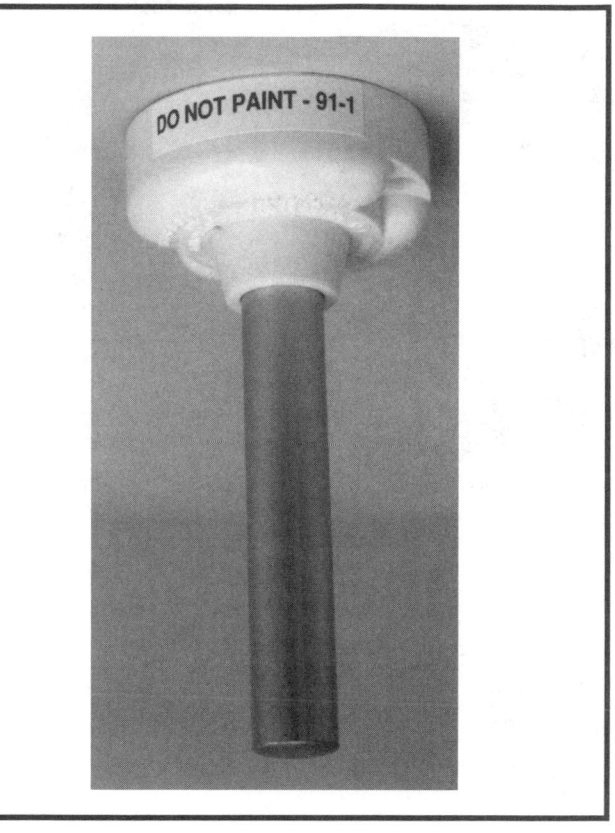

Exhibit 1.31 *Typical rate-compensated heat detector. (Source: Thermotech; photo courtesy of Mammoth Fire Alarms, Inc., Lowell, MA)*

A-1-4 Rate Compensation Detector. A typical example of a rate compensation detector is a spot-type detector with a tubular casing of a metal that tends to expand lengthwise as it is heated and an associated contact mechanism that closes at a certain point in the elongation. A second metallic element inside the tube exerts an opposing force on the contacts, tending to hold them open. The forces are balanced in such a way that, on a slow rate-of-temperature rise, there is more time for heat to penetrate to the inner element, which inhibits contact closure until the total device has been heated to its rated temperature level. However, on a fast rate-of-temperature rise, there is not as much time for heat to penetrate to the inner element, which exerts less of an inhibiting effect so that contact closure is achieved when the total device has been heated to a lower temperature. This, in effect, compensates for thermal lag.

Rate-of-Rise Detector.* A device that responds when the temperature rises at a rate exceeding a predetermined value.

A-1-4 Rate-of-Rise Detector. Typical examples of rate-of-rise detectors are as follows.

(a) *Pneumatic Rate-of-Rise Tubing.* A line-type detector comprising small-diameter tubing, usually copper, that is installed on the ceiling or high on the walls throughout the protected area. The tubing is terminated in a detector unit that contains diaphragms and associated contacts set to actuate at a predetermined pressure. The system is sealed except for calibrated vents that compensate for normal changes in temperature.

(b) *Spot-Type Pneumatic Rate-of-Rise Detector.* A device consisting of an air chamber, a diaphragm, contacts, and a compensating vent in a single enclosure. The principle of operation is the same as that described for pneumatic rate-of-rise tubing.

(c) *Electrical Conductivity–Type Rate-of-Rise Detector.* A line-type or spot-type sensing element in which resistance changes due to a change in temperature. The rate of change of resistance is monitored by associated control equipment, and an alarm is initiated when the rate of temperature increase exceeds a preset value.

Record Drawings. Drawings (as-built) that document the location of all devices, appliances, wiring sequences, wiring methods, and connections of the components of the fire alarm system as installed.

Record drawings (also called as-built drawings) provide information that is essential to those who test and maintain the fire alarm system. These drawings consist of original fire alarm system layout drawings that have been annotated during the installation of the system to show exactly where the fire alarm system components have been installed. These record drawings account for all field changes that were made during the installation. They also show details of how each conductor of each fire alarm system circuit was installed, the color codes used, the actual location of each device and appliance, terminal identifications, and dates of revisions. Any changes made throughout the life of the fire alarm system must be noted on the record drawings. The system owner is responsible for maintaining all record drawings. Record drawings must reflect the actual system installation.

Record of Completion. A document that acknowledges the features of installation, operation (performance), service, and equipment with representation by the property owner, system installer, system supplier, service organization, and the authority having jurisdiction.

Formerly called the certificate of completion, the record of completion, shown in Figure 1-6.2.1, documents the type of system; the names of installers; and the location of record drawings, owners' manuals, and test reports. It also provides

a confirming record of the acceptance test and gives details of the components and wiring of the system.

Relocation. The movement of occupants from a fire zone to a safe area within the same building.

In hospitals, high-rise buildings, or large facilities where it is often impossible to evacuate all occupants, occupants in the fire zone are usually relocated to a safe area in a separate wing or a more protected section of the building.

Remote Supervising Station Fire Alarm System. A system installed in accordance with this code to transmit alarm, supervisory, and trouble signals from one or more protected premises to a remote location where appropriate action is taken.

Remote supervising station fire alarm systems provide a supervising station connection for a protected premises fire alarm system when the building owner does not desire or is not required to provide a central station fire alarm system or a proprietary supervising station fire alarm system. Section 5-4 of the code presumes that fire alarm signals will be transmitted to the pubic fire service communications center. That center will serve as the remote supervising station. The code presumes that supervisory and trouble signals will be transmitted to some other constantly attended location. If a municipality is not willing to receive alarm signals, or is willing to permit the authority having jurisdiction to accept another suitable location, then fire alarm signals may also be transmitted to such a location. These locations may include telephone answering services, an alarm monitoring center, or any other constantly attended location acceptable to the authority having jurisdiction.

Repeater Station. The location of the equipment needed to relay signals between supervising stations, subsidiary stations, and protected premises.

Reset. A control function that attempts to return a system or device to its normal, non-alarm state.

Reset should not be confused with alarm signal deactivation, which only deactivates the alarm signal and does not return the fire alarm system to its normal standby quiescent condition.

Restorable Initiating Device. A device in which the sensing element is not ordinarily destroyed in the process of operation, whose restoration can be manual or automatic.

Runner. A person other than the required number of operators on duty at central, supervising, or runner stations (or otherwise in contact with these stations) available for prompt dispatching, when necessary, to the protected premises.

The code's intent is that the runner be qualified to perform the required duties at the protected premises, such as resetting equipment, investigating alarm signals, and taking cor-

rective action when necessary. This usually requires training runners so they will develop an in-depth knowledge of the systems and equipment within the protected premises. See 7-1.2.2 for requirements pertaining to qualifications of service personnel.

Runner Service. The service provided by a runner at the protected premises, including resetting and silencing of all equipment transmitting fire alarm or supervisory signals to an off-premises location.

Runner service is generally provided as part of central station service, or when a proprietary supervising station fire alarm system is installed. A runner is sent to the protected premises from which the signal was received and takes appropriate action as outlined in the commentary following the definition of Runner.

Satellite Trunk. A circuit or path connecting a satellite to its central or proprietary supervising station.

Scanner. Equipment located at the telephone company wire center that monitors each local leg and relays status changes to the alarm center. Processors and associated equipment might also be included.

Secondary Trunk Facility. That part of a transmission channel connecting two or more, but fewer than all, leg facilities to a primary trunk facility.

Separate Sleeping Area. An area of the family living unit in which the bedrooms (or sleeping rooms) are located. Bedrooms (or sleeping rooms) separated by other use areas, such as kitchens or living rooms (but not bathrooms), are considered as separate sleeping areas.

Shall. Indicates a mandatory requirement.

Shapes of Ceilings. The shapes of ceilings can be classified as sloping or smooth.

Should. Indicates a recommendation or that which is advised but not required.

Signal. A status indication communicated by electrical or other means.

Signal Transmission Sequence. A DACT that obtains dial tone, dials the number(s) of the DACR, obtains verification that the DACR is ready to receive signals, transmits the signals, and receives acknowledgment that the DACR has accepted that signal before disconnecting (going on-hook).

Signaling Line Circuit. A circuit or path between any combination of circuit interfaces, control units, or transmitters over which multiple system input signals or output signals, or both, are carried.

Signaling Line Circuit Interface. A system component that connects a signaling line circuit to any combination of initiating devices, initiating device circuits, notification appliances, notification appliance circuits, system control outputs, and other signaling line circuits.

A signaling line circuit interface (SLCI) may also be called a *transponder* or *data-gathering panel.* SLCIs, as shown in Exhibit 1.32, are frequently used to provide a single address for an initiating device circuit (IDC). This makes them useful in retrofits to existing nonaddressable devices such as waterflow alarm or tamper switches.

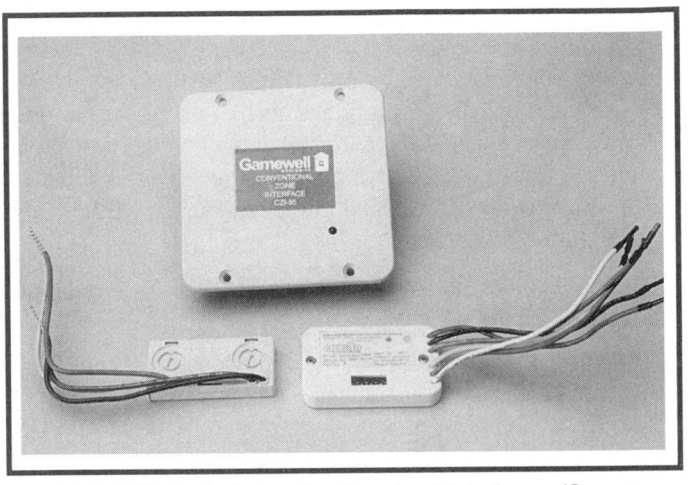

Exhibit 1.32 *Typical signaling line circuit interfaces. (Source: Mammoth Fire Alarms, Inc., Lowell, MA)*

Single Dwelling Unit. A building consisting solely of one dwelling unit. *(See Dwelling Unit.)*

Single-Station Alarm. A detector comprising an assembly that incorporates a sensor, control components, and an alarm notification appliance in one unit operated from a power source either located in the unit or obtained at the point of installation.

A single-station alarm may be powered by a battery, an ac power source, or both (ac with battery back-up). See the commentary following the definition of Multiple-Station Alarm in Section 1-4.

Single-Station Alarm Device. An assembly that incorporates the detector, the control equipment, and the alarm-sounding device in one unit operated from a power supply either in the unit or obtained at the point of installation.

Site-Specific Software. Software that defines the specific operation and configuration of a particular system. Typically, it defines the type and quantity of hardware modules, customized labels, and specific operating features of a system.

Sloping Ceiling. A ceiling that has a slope of more than 1 in 8.

Sloping Peaked-Type Ceiling.* A ceiling in which the ceiling slopes in two directions from the highest point. Curved or domed ceilings can be considered peaked with

the slope figured as the slope of the chord from highest to lowest point.

A-1-4 Sloping Peaked-Type Ceiling. Refer to Figure A-2-2.4.4.1 for an illustration of smoke or heat detector spacing on peaked-type sloped ceilings.

Sloping Shed-Type Ceiling.* A ceiling in which the high point is at one side with the slope extending toward the opposite side.

A-1-4 Sloping Shed-Type Ceiling. Refer to Figure A-2-2.4.4.2 for an illustration of smoke or heat detector spacing on shed-type sloped ceilings.

Smoke Alarm. A single or multiple station alarm responsive to smoke.

Exhibit 1.33 illustrates a typical single-station smoke alarm used in dwellings.

Exhibit 1.33 *Typical smoke alarm. (Source: Gentex, Inc., Zeeland, MI)*

Smoke Detector. A device that detects visible or invisible particles of combustion.

There are many types of smoke detectors. For examples, see the definitions of Ionization Smoke Detection, Photoelectric Smoke Detection, Air Sampling–Type Detector, and Spot-Type Smoke Detector (shown in Exhibit 1.34) in Section 1-4.

Smooth Ceiling.* A ceiling surface uninterrupted by continuous projections, such as solid joists, beams, or ducts, extending more than 4 in. (100 mm) below the ceiling surface.

Exhibit 1.34 *Typical spot-type smoke detector. (Source: System Sensor; photo courtesy of Mammoth Fire Alarms, Inc., Lowell, MA)*

A-1-4 Smooth Ceiling. Open truss constructions are not considered to impede the flow of fire products unless the upper member, in continuous contact with the ceiling, projects below the ceiling more than 4 in. (100 mm).

Spacing. A horizontally measured dimension related to the allowable coverage of fire detectors.

Spacing refers to the maximum linear horizontal distance, permitted by the code, between each automatic fire detection initiating device or each visible notification appliance. Spacing is based on the listing of the device for heat detectors, on the manufacturer's guidelines for smoke detectors, and on design characteristics for visible notification appliances.

Spacing of audible notification appliances is not an issue in the code, because audible notification requirements are performance based. Square foot coverage is not used to determine the spacing of any device or appliance except for smoke detectors installed in high air–movement areas. See Chapter 2 for spacing of initiating devices.

Spark.* A moving ember.

A-1-4 Spark. The overwhelming majority of applications involving the detection of Class A and Class D combustibles with radiant energy-sensing detectors involve the transport of particulate solid materials through pneumatic conveyor ducts or mechanical conveyors. It is common in the industries that include such hazards to refer to a moving piece of burning material as a *spark* and to systems for the detection of such fires as *spark detection systems*.

Spark/Ember Detector. A radiant energy-sensing fire detector that is designed to detect sparks or embers, or both. These devices are normally intended to operate in dark environments and in the infrared part of the spectrum.

Spark/Ember Detector Sensitivity. The number of watts (or the fraction of a watt) of radiant power from a point source radiator, applied as a unit step signal at the wavelength of maximum detector sensitivity, necessary to produce an alarm signal from the detector within the specified response time.

Spot-Type Detector. A device in which the detecting element is concentrated at a particular location. Typical examples are bimetallic detectors, fusible alloy detectors, certain pneumatic rate-of-rise detectors, certain smoke detectors, and thermoelectric detectors.

Story. The portion of a building included between the upper surface of a floor and the upper surface of the floor or roof next above.

Stratification. The phenomenon where the upward movement of smoke and gases ceases due to the loss of buoyancy.

As heat from a fire drives the products of combustion upward from the flame front, the air surrounding the fire begins to extract heat from the mass of material. Once the temperature of the plume reaches equilibrium with the temperature of the surrounding air, the plume loses its buoyancy, and upward movement ceases. The plume then will spread out horizontally, whether or not it has reached the ceiling of the space. Stratification can prevent smoke and fire gases from reaching the installed detection device(s). Care must be exercised when installing detection devices in areas subject to this phenomenon, such as an atrium, any other high ceiling space, or an area with unusually high upper level airflow. See 2-3.4 for more details on this phenomenon and detector placement.

Subscriber. The recipient of a contractual supervising station signal service(s). In case of multiple, noncontiguous properties having single ownership, the term refers to each protected premises or its local management.

Subsidiary Station. A subsidiary station is a normally unattended location that is remote from the supervising station and is linked by a communications channel(s) to the supervising station. Interconnection of signals on one or more transmission channels from protected premises with a communications channel(s) to the supervising station is performed at this location.

Supervising Station. A facility that receives signals and at which personnel are in attendance at all times to respond to these signals.

Supervisory Service. The service required to monitor performance of guard tours and the operative condition of fixed suppression systems or other systems for the protection of life and property.

Supervisory Signal. A signal indicating the need for action in connection with the supervision of guard tours, the fire suppression systems or equipment, or the maintenance features of related systems.

Supervisory Signal Initiating Device. An initiating device such as a valve supervisory switch, water level indicator, or low air pressure switch on a dry-pipe sprinkler system in which the change of state signals an off-normal condition and its restoration to normal of a fire protection or life safety system; or a need for action in connection with guard tours, fire suppression systems or equipment, or maintenance features of related systems.

Supplementary. As used in this code, supplementary refers to equipment or operations not required by this code and designated as such by the authority having jurisdiction.

For equipment to be designated "supplementary," it must meet two specific conditions. First, the equipment must not be required by the code. Secondly, the authority having jurisdiction must specifically declare in writing that the equipment is supplementary. This two-fold test helps limit the use of supplementary equipment. Use of supplementary equipment must be limited because such equipment enjoys somewhat relaxed requirements regarding the monitoring of the integrity of system interconnections and power supplies.

Switched Telephone Network. An assembly of communications facilities and central office equipment operated jointly by authorized service providers that provides the general public with the ability to establish transmission channels via discrete dialing.

System Unit. The active subassemblies at the central station used for signal receiving, processing, display, or recording of status change signals; a failure of one of these subassemblies causes the loss of a number of alarm signals by that unit.

Tactile Notification Appliance. A notification appliance that alerts by the sense of touch or vibration.

Tactile notification appliances include vibrating pagers and bed shakers used to notify persons with disabilities who are not able to respond to an audible or visual fire alarm notification appliance. These appliances must be listed for their intended purpose.

Textual Audible Notification Appliance. A notification appliance that conveys a stream of audible information. An example of a textual audible notification appliance is a speaker that reproduces a voice message.

Textual Visible Notification Appliance. A notification appliance that conveys a stream of visible information that displays an alphanumeric or pictorial message. Textual visible notification appliances provide temporary text, permanent text, or symbols. Textual visible notification appliances include, but are not limited to, annunciators, monitors, CRTs, displays, and printers.

Transmission Channel. A circuit or path connecting transmitters to supervising stations or subsidiary stations on which signals are carried.

Transmitter. A system component that provides an interface between signaling line circuits, initiating device circuits, or control units and the transmission channel.

Transponder. A multiplex alarm transmission system functional assembly located at the protected premises.

Trouble Signal. A signal initiated by the fire alarm system or device indicative of a fault in a monitored circuit or component.

Visible Notification Appliance. A notification appliance that alerts by the sense of sight.

Voice Intelligibility.* Audible voice information that is distinguishable and understandable.

A-1-4 Voice Intelligibility. As used in this code, intelligibility and intelligible are both applied to the description of voice communications systems intended to reproduce human speech. When a human being can clearly distinguish and understand human speech reproduced by such a system, the system is said to be intelligible. Satisfactory intelligibility requires adequate audibility and adequate clarity. Clarity is defined as freedom from distortion of all kinds (IEC 60849, *Sound Systems for Emergency Purposes*, Section 3.6). The following are three kinds of distortion responsible for the reduction of speech clarity in an electroacoustic system:

(1) Amplitude distortion, due to non-linearity in electronic equipment and transducers
(2) Frequency distortion, due to non-uniform frequency response of transducers and selective absorption of various frequencies in acoustic transmission
(3) Time domain distortion, due to reflections and reverberation in the acoustic domain

Of these three kinds of distortion, frequency distortion is partially, and time domain distortion is totally, a function of the environment in which the system is installed (size, shape, and surface characteristics of walls, floors, and ceilings) and the character and placement of the loudspeakers (transducers).

WATS (Wide Area Telephone Service). Telephone company service allowing reduced costs for certain telephone call arrangements. In-WATS or 800-number service calls can be placed from anywhere in the continental United States to the called party at no cost to the calling party. Out-WATS is a service whereby, for a flat-rate charge, dependent on the total duration of all such calls, a subscriber can make an unlimited number of calls within a prescribed area from a particular telephone terminal without the registration of individual call charges.

Wavelength.* The distance between the peaks of a sinusoidal wave. All radiant energy can be described as a wave having a wavelength. Wavelength serves as the unit of measure for distinguishing between different parts of the spectrum. Wavelengths are measured in microns (μM), nanometers (nM), or angstroms (Å).

A-1-4 Wavelength. The concept of wavelength is extremely important in selecting the proper detector for a particular application. There is a precise interrelation between the wavelength of light being emitted from a flame and the combustion chemistry producing the flame. Specific subatomic, atomic, and molecular events yield radiant energy of specific wavelengths. For example, ultraviolet photons are emitted as the result of the complete loss of electrons or very large changes in electron energy levels. During combustion, molecules are violently torn apart by the chemical reactivity of oxygen, and electrons are released in the process, recombining at drastically lower energy levels, thus giving rise to ultraviolet radiation. Visible radiation is generally the result of smaller changes in electron energy levels within the molecules of fuel, flame intermediates, and products of combustion. Infrared radiation comes from the vibration of molecules or parts of molecules when they are in the superheated state associated with combustion. Each chemical compound exhibits a group of wavelengths at which it is resonant. These wavelengths constitute the chemical's infrared spectrum, which is usually unique to that chemical.

This interrelationship between wavelength and combustion chemistry affects the relative performance of various types of detectors with respect to various fires.

Wireless Control Panel. A component that transmits/ receives and processes wireless signals.

Wireless Protection System. A system or a part of a system that can transmit and receive signals without the aid of wire. It can consist of either a wireless control panel or a wireless repeater.

Wireless Repeater. A component used to relay signals between wireless receivers or wireless control panels, or both.

Wireless control panel, wireless protection systems, and *wireless repeater* apply to systems covered by Section 3-10.

Zone. A defined area within the protected premises. A zone can define an area from which a signal can be received, an area to which a signal can be sent, or an area in which a form of control can be executed.

Zoning requirements are covered by 1-5.7 of the code. Additional zoning requirements are also found in building codes, NFPA *101, Life Safety Code,* and local ordinances.

1-5 Fundamentals

1-5.1 Common System Fundamentals.

The provisions of Chapter 1 shall apply to Chapters 2 through 7.

The basic requirements for all fire alarm systems, except fire warning systems in dwellings, are contained in Chapter 1. Chapter 8 is a stand-alone chapter with specific requirements for detection in residential family occupancies such as single-family homes, apartments, and condominiums.

1-5.1.1 General. The provisions of Chapter 1 shall cover the basic functions of a complete fire alarm system. These systems shall be primarily intended to provide notification of fire alarm, supervisory, and trouble conditions; to alert the occupants; to summon appropriate aid; and to control fire safety functions.

1-5.1.2 Equipment. Equipment constructed and installed in conformity with this code shall be listed for the purpose for which it is used.

Fire alarm products must be listed for the specific fire alarm system applications for which they are used. Because fire alarm systems are used for life safety, property protection, mission continuity, heritage preservation, and environmental protection, the listing requirements are often more stringent than for those products listed for electrical safety only.

Equipment listings generally contain information pertaining to the permitted use, required ambient conditions in the installed location, mounting orientation, voltage tolerances, compatibility, and so on. Equipment must be installed in conformance with the listing to meet the requirements of the code.

1-5.1.3* System Design. Fire alarm system plans and specifications shall be developed in accordance with this code by persons who are experienced in the proper design, application, installation, and testing of fire alarm systems. The system designer shall be identified on the system design documents. Evidence of qualifications shall be provided when requested by the authority having jurisdiction.

The code now requires that fire alarm system designers be qualified to perform this type of work through training, education, and experience. Requiring the system designer to be identified on the system design documents encourages the designer to feel a sense of ownership toward the design. This, in turn, provides an additional incentive for the designer to meet the requirements of the code.

A-1-5.1.3 Examples of qualified personnel include individuals who can demonstrate experience on similar systems and have the following qualifications:

(1) Factory trained and certified in fire alarm system design
(2) National Institute of Certification in Engineering Technologies (NICET) fire alarm certified—minimum level III
(3) Licensed or certified by a state or local authority

1-5.1.4 System Installation. Installation personnel shall be supervised by persons who are qualified and experienced in the installation, inspection, and testing of fire alarm systems. Examples of qualified personnel shall include, but not be limited to, the following:

(1) Factory trained and certified personnel
(2) National Institute of Certification in Engineering Technologies (NICET) fire alarm level II certified personnel
(3) Personnel licensed or certified by state or local authority

The examples of qualified personnel correlate with similar requirements in 7-1.2.2, which requires testing and maintenance personnel to be qualified.

1-5.2 Power Supplies.

1-5.2.1 Scope. The provisions of this section apply to power supplies used for fire alarm systems.

1-5.2.2 Code Conformance. All power supplies shall be installed in conformity with the requirements of NFPA 70, *National Electrical Code,* for such equipment and with the requirements indicated in this subsection.

1-5.2.3* Power Sources. Fire alarm systems shall be provided with at least two independent and reliable power supplies, one primary and one secondary (standby), each of which shall be of adequate capacity for the application.

Exception 1: Where the primary power is supplied by a dedicated branch circuit of an emergency system in accordance with NFPA 70, National Electrical Code, Article 700, or a legally required standby system in accordance with NFPA 70, National Electrical Code, Article 701, a secondary supply shall not be required.

Exception 2: Where the primary power is supplied by a dedicated branch circuit of an optional standby system in accordance with NFPA 70, National Electrical Code, Article 702, which also meets the performance requirements of Article 700 or Article 701, a secondary supply shall not be required.

Where dc voltages are employed, they shall be limited to no more than 350 volts above earth ground.

Prior to the 1993 edition, NFPA 72 required three sources of power: primary, secondary (standby), and trouble. The

requirement for a trouble signal power supply had an exception that permitted the secondary power supply to provide power to the trouble signal. Because the majority of fire alarm system control unit designs applied this exception, subsequent editions of the code deleted the requirement for a separate trouble power source.

The exceptions to 1-5.2.3 permit an elimination of the secondary power supply. However, few installations actually meet the exceptions.

Formal Interpretation 85-13 provides further clarification of power source requirements.

Formal Interpretation 85-13

Reference: 1-5.2.3, Exception No. 1

Background: In reading the 1999 code under NFPA 72, 1-5.2.3, Exception No. 1 stipulates that fire alarm system power supply can be fed from the emergency generator of an emergency system or a legally required standby system.

Question 1: When Exception No. 1 is complied with, are the requirements of 1-5.2.10.4 also required?

Answer: Yes, for generator option.

Question 2: When Exception No. 1 is complied with do the requirements for a 2-hour fuel supply stated in NFPA 70, Article 700-12(b)(2) apply?

Answer: No, 1-5.2.3 is more stringent.

Issue Edition: 1985 of NFPA 72A

Reference: 2-3.2.1, 2-3.4.1

Date: November 1986

A-1-5.2.3 The term *fire alarm systems* includes all equipment, regardless of location, that requires power but does not receive all power directly from the main control unit. Examples of fire alarm equipment that require power include, but are not limited to, the following multiple control units, remote power supplies, amplifiers, transponders, and required computer equipment.

1-5.2.4 Primary Supply. The primary supply shall have a high degree of reliability, shall have adequate capacity for the intended service, and shall consist of one of the following:

(1) Light and power service arranged in accordance with 1-5.2.5
(2) Where a person specifically trained in its operation is on duty at all times, an engine-driven generator or equivalent arranged in accordance with 1-5.2.10

An engine-driven generator is permitted as a primary power supply because light and power service may not be available at all locations.

1-5.2.5 Light and Power Service.

1-5.2.5.1 A light and power service employed to operate the system under normal conditions shall have a high degree of reliability and capacity for the intended service. This service shall consist of one of the following:

(a) *Two-Wire Supplies.* A two-wire supply circuit shall be permitted to be used for either the primary operating power supply or the trouble signal power supply of the signaling system.

(b) *Three-Wire Supplies.* A three-wire ac or dc supply circuit having a continuous unfused neutral conductor, or a polyphase ac supply circuit having a continuous unfused neutral conductor where interruption of one phase does not prevent operation of the other phase, shall be permitted to be used with one side or phase for the primary operating power supply and the other side or phase for the trouble signal power supply of the fire alarm system.

1-5.2.5.2 Connections to the light and power service shall be on a dedicated branch circuit(s). The circuit(s) and connections shall be mechanically protected. Circuit disconnecting means shall have a red marking, shall be accessible only to authorized personnel, and shall be identified as FIRE ALARM CIRCUIT CONTROL. The location of the circuit disconnecting means shall be permanently identified at the fire alarm control unit.

There are six separate requirements in 1-5.2.5.2. These requirements are intended to protect the power supply from tampering, to ensure reliability, to aid in troubleshooting, and to help ensure the safety of those who service the equipment.

1-5.2.5.3 Overcurrent Protection.

An overcurrent protective device of suitable current-carrying capacity and capable of interrupting the maximum short-circuit current to which it may be subject shall be provided in each ungrounded conductor. The overcurrent protective device shall be enclosed in a locked or sealed cabinet located immediately adjacent to the point of connection to the light and power conductors.

Connections ahead of the main disconnecting apparatus can be dangerous because, when improperly installed, these connections can allow high fault currents to be carried on the primary power supply conductors for the fire alarm system. NFPA 70, *National Electrical Code*®, provides requirements for proper installation. Proper connections must be made using service-rated equipment in accordance with 230-82(4) in the *NEC*. The service equipment used to interrupt the fault current must be listed for such use.

Circuit breaker locks are permitted, provided they are listed for use with the circuit breaker. Circuit breaker locks allow the breaker to trip, but do not allow tampering.

1-5.2.5.4 Circuit breakers or engine stops shall not be installed in such a manner as to cut off the power for lighting or for operating elevators.

1-5.2.6 Secondary Supply Capacity and Sources. The secondary supply shall automatically supply the energy to the system within 30 seconds, and without loss of signals, wherever the primary supply is incapable of providing the minimum voltage required for proper operation. The secondary (standby) power supply shall supply energy to the system in the event of total failure of the primary (main) power supply or when the primary voltage drops to a level insufficient to maintain functionality of the control equipment and system components. Under maximum quiescent load (system functioning in a non-alarm condition), the secondary supply shall have sufficient capacity to operate a protected premises, central station, or proprietary system for 24 hours, or an auxiliary or remote station system for 60 hours; and, at the end of that period, shall be capable of operating all alarm notification appliances used for evacuation or to direct aid to the location of an emergency for 5 minutes. The secondary power supply for emergency voice/alarm communications service shall be capable of operating the system under maximum quiescent load for 24 hours and then shall be capable of operating the system during a fire or other emergency condition for a period of 2 hours. Fifteen minutes of evacuation alarm operation at maximum connected load shall be considered the equivalent of 2 hours of emergency operation.

The proper amount of battery standby capacity can be calculated. These calculations should include the normal standby supervisory quiescent load for a specified period of time as well as the load during the specified period of alarm. If combination systems are used, the secondary supply must be able to power the entire system for the required 24-hour period. Other loads, such as security or building management systems, must be figured into the secondary power calculations.

The emergency voice/alarm communications system must be capable of operating for 2 hours during the emergency condition because communication to occupants must continue throughout the duration of the fire.

The difference in normal standby time for auxiliary fire alarm systems and remote supervising station fire alarm systems stems from previous editions of NFPA fire alarm standards when the standards did not require trouble signals for these systems to be transmitted off premises. The idea behind the 60-hour requirement was that the longest time period when a building was likely to be unoccupied would begin at 6 p.m. on Friday night and end at 6 a.m. on Monday morning. If an auxiliary or remote station fire alarm system

experienced a trouble during this time — or during any similar period of time when the building was not occupied — the extra standby power capacity would allow the trouble signal to persist until the building was once again occupied. In contrast, trouble signals for central station fire alarm systems and proprietary supervising station fire alarm systems were required to transmit trouble signals to the supervising station. Today, the 60-hour standby power supply capacity for remote supervising station and auxiliary fire alarm systems has become a distinction that sets these systems apart from other fire alarm systems.

For a combination system, the secondary supply capacity required above shall include the load of any non-fire related equipment, functions, or features which are not automatically disconnected upon transfer of operating power to the secondary supply.

The secondary supply shall consist of one of the following:

(a) A storage battery arranged in accordance with 1-5.2.9.

(b) An automatic starting, engine-driven generator arranged in accordance with 1-5.2.10 and storage batteries with 4 hours of capacity under maximum normal load followed by 5 minutes of alarm/emergency capacity arranged in accordance with 1-5.2.9.

The engine-driven generator must be capable of supplying power to the fire alarm system even with all other loads energized. The four hours of batteries are necessary to power the fire alarm system in case the engine-driven generator fails to start.

(c) Multiple engine-driven generators, one of which is arranged for automatic starting, arranged in accordance with 1-5.2.10, and capable of supplying the energy required herein, with the largest generator out of service. The second generator shall be permitted to be started by pushbutton.

The code permits pushbutton starting means, but key starting generators are not permitted because keys can be lost.

Operation on secondary power shall not affect the required performance of a fire alarm system. The system shall produce the same alarm, supervisory, and trouble signals and indications (excluding the ac power indicator) when operating from the standby power source as are produced when the unit is operating from the primary power source.

Manufacturers have supplied systems in the past that, in order to save battery power, eliminated annunciation of additional trouble conditions and eliminated some supplementary functions when in the standby power mode. The code prevents this practice by requiring the system to operate with all of the same features as when it is powered by the primary power source.

1-5.2.7 Continuity of Power Supplies.

(a) Where signals could be lost on transfer of power between the primary and secondary sources, rechargeable batteries of sufficient capacity to operate the system under maximum normal load for 15 minutes shall assume the load in such a manner that no signals are lost where either of the following conditions exists:

(1) Secondary power is supplied in accordance with 1-5.2.6(a) or 1-5.2.6(b), and the transfer is made manually
(2) Secondary power is supplied in accordance with 1-5.2.6(c)

(b) Where signals will not be lost due to transfer of power between the primary and secondary sources, one of the following arrangements shall be made:

(1) The transfer shall be automatic.
(2) Special provisions shall be made to allow manual transfer within 30 seconds of loss of power.
(3) The transfer shall be arranged in accordance with 1-5.2.6(a).

(c)* Where a computer system of any kind or size is used to receive or process signals, an uninterruptible power supply (UPS) with sufficient capacity to operate the system for at least 15 minutes, or until the secondary supply is capable of supplying the UPS input power requirements, shall be required where either of the following conditions apply:

(1) The status of signals previously received will be lost upon loss of power.
(2) The computer system cannot be restored to full operation within 30 seconds of loss of power.

The Institute of Electronics and Electrical Engineers (IEEE) *Standard Dictionary of Electrical and Electronics Terms* (1996) defines a computer as, "A functional programmable unit that consists of one or more associated processing units and peripheral equipment, that is controlled by internally stored programs, and that can perform substantial computation, including numerous arithmetic operations or logic operations, without human intervention." A computer system is defined as, "A system containing one or more computers and associated software." For the purposes of the code, a computer system employs a microprocessor or a system that uses software.

A-1-5.2.7(c) An engine-driven generator without standby battery supplement should not be assumed to be capable of a reliable power transfer within 30 seconds of a primary power loss.

(d)* A positive means for disconnecting the input and output of the UPS system while maintaining continuity of power supply to the load shall be provided.

The requirement for disconnection of the UPS is to provide power to the fire alarm system during maintenance and testing of the UPS.

A-1-5.2.7(d) UPS equipment often contains an internal bypass arrangement to supply the load directly from the line. These internal bypass arrangements are a potential source of failure. UPS equipment also requires periodic maintenance. It is, therefore, necessary to provide a means of promptly and safely bypassing and isolating the UPS equipment from all power sources while maintaining continuity of power supply to the equipment normally supplied by the UPS.

1-5.2.8 Power Supply for Remotely Located Control Equipment.

1-5.2.8.1 Additional power supplies, where provided for control units, circuit interfaces, or other equipment essential to system operation, located remote from the main control unit, shall be comprised of a primary and secondary power supply that shall meet the same requirements as those of 1-5.2.1 through 1-5.2.8 and 1-5.8.6.

1-5.2.8.2 Power supervisory devices shall be arranged so as not to impair the receipt of fire alarm or supervisory signals.

1-5.2.9* Storage Batteries.

A-1-5.2.9 Rechargeable-Type Batteries, or Storage Batteries. The following newer types of rechargeable batteries are normally used in protected premises applications.

(a) *Vented Lead-Acid, Gelled, or Starved Electrolyte Battery.* This rechargeable-type battery is generally used in place of primary batteries in applications that have a relatively high current drain or that require the extended standby capability of much lower currents. The nominal voltage of a single cell is 2 volts, and the battery is available in multiples of 2 volts (e.g., 2, 4, 6, 12). Batteries should be stored according to the manufacturer's recommendations.

(b) *Nickel-Cadmium Battery.* The sealed-type nickel-cadmium battery generally used in applications where the battery current drain during a power outage is low to moderate (typically up to a few hundred milliamperes) and is fairly constant. Nickel-cadmium batteries are also available in much larger capacities for other applications. The nominal voltage per cell is 1.42 volts, with batteries available in multiples of 1.42 (e.g., 12.78, 25.56). Batteries in storage can be stored in any state of charge for indefinite periods. However, a battery in storage will lose capacity (will self-discharge), depending on storage time and temperature. Typically, batteries stored for more than 1 month require an 8-hour to 14-hour charge period to restore capacity. In service, the battery should receive a continuous, constant-charging current that is sufficient to keep it fully charged. (Typically, the charge rate equals $1/10$ to $1/20$ of the ampere-hour rating of the battery.) Because batteries are made up of individual cells connected in series, the possibility exists that, during deep discharge, one or more cells that are low in capacity will reach complete discharge prior to other cells. The cells with remaining life tend to charge the depleted cells, causing a polarity reversal resulting in permanent battery damage. This condition can be determined by measuring the open cell voltage of a fully charged battery (voltage should be a minimum of 1.28 volts per cell multiplied by the number of cells). Voltage depression effect is a minor change in discharge voltage level caused by constant current charging below the system discharge rate.

In some applications of nickel-cadmium batteries, for example, battery-powered shavers, a memory characteristic also exists. Specifically, if the battery is discharged daily for 1 minute, followed by a recharge, operation for 5 minutes will not result in the rated ampere-hour output because the battery has developed a 1-minute discharge memory.

(c) *Sealed Lead-Acid Battery.* In a sealed lead-acid battery, the electrolyte is totally absorbed by the separators, and no venting normally occurs. Gas evolved during recharge is internally recombined, resulting in minimal loss of capacity life. A high-pressure vent, however, is provided to avoid damage under abnormal conditions.

1-5.2.9.1 Location. Storage batteries shall be located so that the fire alarm equipment, including overcurrent devices, are not adversely affected by battery gases and shall conform to the requirements of NFPA 70, *National Electrical Code*, Article 480. Cells shall be suitably insulated against grounds and crosses and shall be mounted securely in such a manner so as not to be subject to mechanical injury. Racks shall be suitably protected against deterioration. If not located in or adjacent to the fire alarm control panel, the batteries and their charger location shall be permanently identified at the fire alarm control unit.

The requirement for identification of the location of remotely located batteries or chargers, or both, is intended to make system inspections and tests easier. Long runs of conductors to remote batteries may create unacceptable voltage drops that may affect system performance. Voltage drop calculations must be conducted to ensure that the system has adequate voltage under full load.

Battery gases can cause severe corrosion of terminals and contacts in equipment enclosures. Sealed lead-acid batteries are generally permitted inside control units; however, vented lead-acid batteries are not permitted inside control units. See Exhibit 1.35 for examples of sealed lead-acid batteries. If large batteries are necessary, a separate battery cabinet, as shown in Exhibit 1.36, may be required to adequately house the batteries.

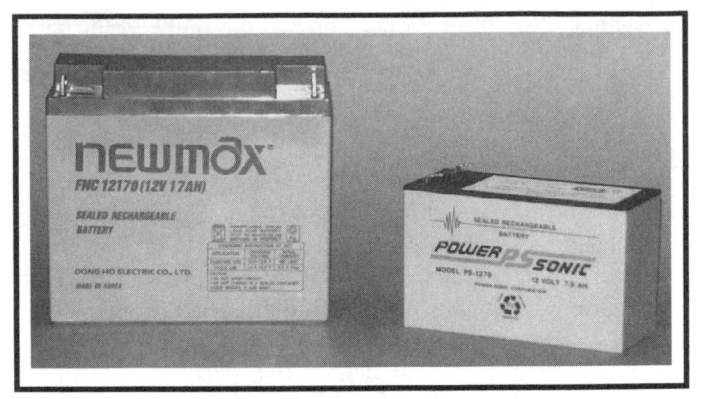

Exhibit 1.35 *Typical sealed lead-acid batteries. (Source: Mammoth Fire Alarms, Inc., Lowell, MA)*

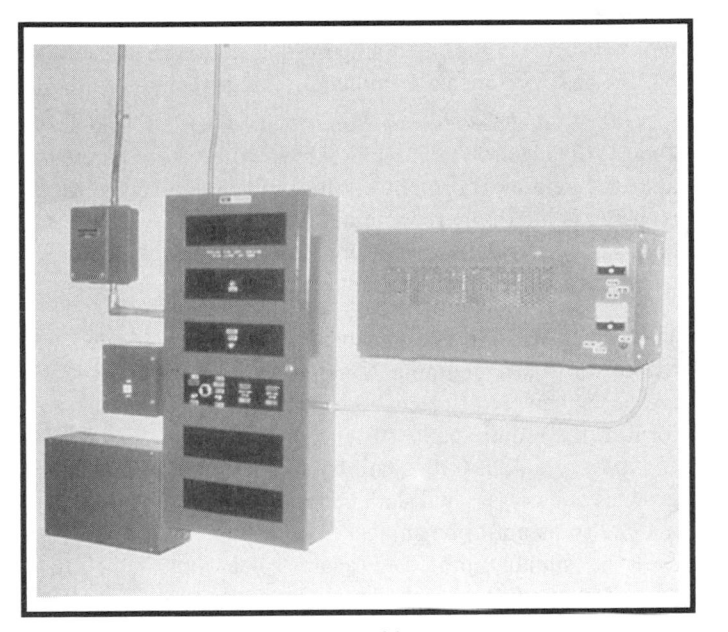

Exhibit 1.36 *Separate battery cabinet.*

1-5.2.9.2 Battery Charging.

1-5.2.9.2.1 Adequate facilities shall be provided to automatically maintain the battery fully charged under all conditions of normal operation and, in addition, to recharge batteries within 48 hours after fully charged batteries have been subject to a single discharge cycle as specified in 1-5.2.5.3. Upon attaining a fully charged condition, the charge rate shall not be so excessive as to result in battery damage.

Unless the capacity of the battery charger has been carefully calculated, systems with large batteries may have difficulty meeting this requirement. The manufacturer's data sheets should provide maximum charging capabilities.

1-5.2.9.2.2 Supervising stations shall maintain spare parts or units available, which shall be used to restore failed charging capacity prior to the consumption of one-half of the capacity of the batteries for the supervising station equipment.

1-5.2.9.2.3* Batteries shall be either trickle- or float-charged.

A-1-5.2.9.2.3 Batteries are trickle-charged if they are off-line and waiting to be put under load in the event of a loss of power.

Float-charged batteries are fully charged and connected across the output of the rectifiers to smooth the output and to serve as a standby source of power in the event of a loss of line power.

1-5.2.9.2.4 A rectifier employed as a battery charging supply source shall be of adequate capacity. A rectifier employed as a charging means shall be energized by an isolating transformer.

1-5.2.9.3 Overcurrent Protection. The batteries shall be protected against excessive load current by overcurrent devices having a rating not less than 150 percent and not more than 250 percent of the maximum operating load in the alarm condition. The batteries shall be protected from excessive charging current by overcurrent devices or by automatic current-limiting design of the charging source.

1-5.2.9.4 Metering. The charging equipment shall provide either integral meters or readily accessible terminal facilities for the connection of portable meters by which the battery voltage and charging current can be determined.

1-5.2.9.5 Charger Supervision. Supervision means appropriate for the batteries and charger employed shall be provided to detect a failure of battery charging and initiate a trouble signal in accordance with 1-5.4.6.

The requirement in 1-5.2.9.5 was part of the metering requirement of the 1989 edition of NFPA 71, *Standard for the Installation, Maintenance, and Use of Signaling Systems for Central Station Service.* The battery charging circuits of all systems are now required to be monitored and to produce a trouble signal upon failure. See 1-5.8.7 for monitoring integrity of power supplies.

1-5.2.10 Engine-Driven Generator.

1-5.2.10.1 Installation. The installation of engine-driven generators shall conform to the provisions of NFPA 110, *Standard for Emergency and Standby Power Systems.*

Exception: Where restricted by the provisions of 1-5.2.

1-5.2.10.2 Capacity. The unit shall be of a capacity that is sufficient to operate the system under the maximum normal

load conditions in addition to all other demands placed upon the unit, such as those of emergency lighting.

1-5.2.10.3 Fuel.

1-5.2.10.3.1 Fuel shall be stored in outside underground tanks wherever possible, and gravity feed shall not be used. If gasoline-driven generators are used, fuel shall be supplied from a frequently replenished "working" tank, or other means provided, to ensure that the gasoline is always fresh because gasoline deteriorates with age.

Gravity feed is not permitted because of the potential for fuel leaks, which may pose a fire hazard.

1-5.2.10.3.2 Sufficient fuel shall be available in storage for 6 months of testing plus the capacity specified in 1-5.2.5. For public fire alarm reporting systems, the requirements of Chapter 6 shall apply.

Exception 1: If a reliable source of supply is available at any time on a 2-hour notice, sufficient fuel shall be in storage for 12 hours of operation at full load.

Exception 2: Fuel systems using natural or manufactured gas supplied through reliable utility mains shall not be required to have fuel storage tanks unless located in seismic risk zone 3 or greater as defined in ANSI A-58.1, Building Code Requirements for Minimum Design Loads in Buildings and Other Structures.

NFPA 110, *Standard for Emergency Power Systems*, provides requirements for testing of engine-driven generators. The manufacturer's equipment data sheets should provide fuel consumption rates for the engine-driven generator.

1-5.2.10.4 Battery and Charger. A separate storage battery and separate automatic charger shall be provided for starting the engine-driven generator and shall not be used for any other purpose.

1-5.3 Compatibility.

All fire detection devices that receive their power from the initiating device circuit or signaling line circuit of a fire alarm control unit shall be listed for use with the control unit.

A two-wire smoke detector obtains its power from the control unit initiating device circuit. Analog addressable devices on signaling line circuits communicate with the control unit using manufacturer specific protocols. Therefore, it is mandatory that these smoke detector devices be listed for use with the control unit and its associated initiating device or signaling line circuit. The listing organizations have developed specific requirements for this listing process and should be consulted if there is any doubt as to the detector's compatibility with a specific control unit.

1-5.4 System Functions.

1-5.4.1 Protected Premises Fire Safety Functions.

1-5.4.1.1 Fire safety functions shall be permitted to be performed automatically. The performance of automatic fire safety functions shall not interfere with power for lighting or for operating elevators. The performance of automatic fire safety functions shall not preclude the combination of fire alarm services with other services requiring monitoring of operations.

1-5.4.1.2* The time delay between the activation of an initiating device and the automatic activation of a local fire safety function shall not exceed 20 seconds.

Effective on January 1, 2002, the time delay between the activation of an initiating device and the automatic activation of a local fire safety function shall not exceed 10 seconds.

The requirement in 1-5.4.1.2 was added to the 1996 edition of the code because these systems are used for life safety. The time delay requirement was changed from 90 seconds to 20 seconds for the 1999 code because critical functions, such as smoke control, elevator recall, and suppression system actuation, must occur quickly. Changing the requirement again in 2002 will allow manufacturers the time to develop the technology to meet the new 10-second requirement.

A-1-5.4.1.2 It is not the intent of this paragraph to dictate the time frame for the local fire safety devices to complete their function, such as fan wind-down time, door closure time, or elevator travel time.

1-5.4.2 Alarm Signals.

1-5.4.2.1* Coded Alarm Signals. A coded alarm signal shall consist of not less than three complete rounds of the number transmitted. Each round shall consist of not less than three impulses.

A-1-5.4.2.1 Coded Alarm Signal Designations. The recommended coded signal designations for buildings that have four floors and multiple basements are provided in Table A-1-5.4.2.1.

Table A-1-5.4.2.1 Recommended Coded Signal Designations

Location	Coded Signal
Fourth floor	2–4
Third floor	2–3
Second floor	2–2
First floor	2–1
Basement	3–1
Sub-basement	3–2

1-5.4.2.2* Actuation of alarm notification appliances or emergency voice communications and annunciation at the protected premises shall occur within 20 seconds after the activation of an initiating device.

Effective on January 1, 2002, actuation of alarm notification appliances or emergency voice communications and annunciation at the protected premises shall occur within 10 seconds after the activation of an initiating device.

The requirement in 1-5.4.2.2 was revised for the 1999 code to clearly indicate that actuation of emergency voice/alarm communications or actuation of notification appliances must occur within 20 seconds. In 2002, the requirement will require the time to reduce to 10 seconds. The more distant effective date will allow manufacturers the time to develop the technology needed to meet the new 10-second requirement.

A-1-5.4.2.2 Actuation of an initiating device is usually the instant at which a complete digital signal is achieved at the device, such as a contact closure. For smoke detectors or other automatic initiating devices, which may involve signal processing and analysis of the signature of fire phenomena, actuation means the instant when the signal analysis requirements are completed by the device or control unit software.

A separate control unit contemplates a network of control units forming a single large system as defined in Section 3-8.

For some analog initiating devices, actuation is the moment that the fire alarm control unit interprets that the signal from an initiating device has exceeded the alarm threshold programmed into the control unit.

For smoke detectors working on a system with alarm verification, where the verification function is performed in the fire alarm control unit, the moment of actuation of smoke detectors is sometimes determined by the fire alarm control unit.

1-5.4.3 Supervisory Signals.

1-5.4.3.1 Coded Supervisory Signal. A coded supervisory signal shall be permitted to consist of two rounds of the number transmitted to indicate a supervisory off-normal condition, and one round of the number transmitted to indicate the restoration of the supervisory condition to normal.

1-5.4.3.2 Combined Coded Alarm and Supervisory Signal Circuits.

1-5.4.3.2.1 Where both coded sprinkler supervisory signals and coded fire or waterflow alarm signals are transmitted over the same signaling line circuit, provision shall be made either to obtain alarm signal precedence or sufficient repetition of the alarm signal to prevent the loss of an alarm signal.

1-5.4.3.2.2 Visible and audible supervisory signals and visible indication of their restoration to normal shall be indicated within 90 seconds at the following locations:

(1) Control unit (central equipment) for local fire alarm systems
(2) Building fire command center for emergency voice/alarm communications systems
(3) Supervising station location for systems installed in compliance with Chapter 5

The requirement in 1-5.4.3.2.2(3) provides reporting requirements for supervisory signals. The 90-second requirement is considered adequate because supervisory signals do not represent immediate life threatening conditions.

1-5.4.4 Distinctive Signals. Fire alarms, supervisory signals, and trouble signals shall be distinctively and descriptively annunciated.

1-5.4.5 Fire Safety Function Status Indicators.

1-5.4.5.1 All controls provided specifically for the purpose of manually overriding any automatic fire safety function shall provide visible indication of the status of the associated control circuits.

This requirement in 1-5.4.5.1 was relocated from Chapter 3 for the 1999 code to consolidate the requirements for manual controls. The required visible status indication can be by a labeled annunciator (or equivalent means) or by the labeled position of a toggle or rotary switch.

1-5.4.5.2* Where status indicators are provided for emergency equipment or fire safety functions, they shall be arranged to reflect the actual status of the associated equipment or function.

A-1-5.4.5.2 The operability of controlled mechanical equipment (e.g., smoke and fire dampers, elevator recall arrangements, and door holders) should be verified by periodic testing. Failure to test and properly maintain controlled mechanical equipment can result in operational failure during an emergency, with potential consequences up to and including loss of life.

1-5.4.6 Trouble Signals. Trouble signals and their restoration to normal shall be indicated within 200 seconds at the locations identified in 1-5.4.6.1 or 1-5.4.6.2. Trouble signals required to indicate at the protected premises shall be indicated by distinctive audible signals. These audible trouble signals shall be distinctive from alarm signals. If an intermittent signal is used, it shall sound at least once every 10 seconds, with a minimum duration of $^{1}/_{2}$ second. An audible trouble signal shall be permitted to be common to several supervised circuits. The trouble signal(s) shall be located in an area where it is likely to be heard.

1-5.4.6.1 Visible and audible trouble signals and visible indication of their restoration to normal shall be indicated at the following locations:

(1) Control unit (central equipment) for protected premises fire alarm systems

(2) Building fire command center for emergency voice/alarm communications service

(3) Central station or remote station location for systems installed in compliance with Chapter 5

1-5.4.6.2 Trouble signals and their restoration to normal shall be visibly and audibly indicated at the proprietary supervising station for systems installed in compliance with Chapter 5.

1-5.4.6.3 Audible Trouble Signal Silencing Means.

1-5.4.6.3.1 A means for silencing the trouble notification appliance(s) shall be permitted only if it is key-operated, located within a locked enclosure, or arranged to provide equivalent protection against unauthorized use. Such a means shall be permitted only if it transfers the trouble indication to a suitably identified lamp or other acceptable visible indicator. The visible indication shall persist until the trouble condition has been corrected. The audible trouble signal shall sound when the silencing means is in its silence position and no trouble exists.

The word *means* recognizes that it is possible to perform this function with alpha-numeric keypads, switches, or touch screens.

1-5.4.6.3.2 If an audible trouble notification appliance is also used to indicate a supervisory condition, as permitted in 1-5.4.7(b), a trouble signal silencing switch shall not prevent subsequent sounding of supervisory signals.

1-5.4.6.3.3* An audible trouble signal that has been silenced at the protected premises shall automatically re-sound every 24 hours or less until fault conditions are restored to normal. The audible trouble signal shall sound until it is manually silenced or acknowledged. The re-sounded trouble signal shall also be automatically retransmitted to any supervising station to which the original trouble signal was transmitted.

Trouble signals indicate a fault that may impair system operation. The code requires that a silenced trouble signal re-sound at least once every 24 hours until the source of the trouble signal has been identified and corrected by the system operator. This requirement helps ensure that trouble signals are not ignored. Additionally, if a supervising station fire alarm system is provided, the re-sound of the trouble signal must also be transmitted *daily* to the supervising station.

A-1-5.4.6.3.3 The purposes for automatic trouble re-sound is to remind owners, or those responsible for the system, that the system remains in a fault condition. A secondary benefit is to possibly alert occupants of the building that the fire alarm system is in a fault condition.

1-5.4.6.3.4* If permitted by the authority having jurisdiction, the requirement for a 24-hour re-sound of an audible trouble signal shall be permitted to occur only at a supervising station that meets the requirements of Chapter 5 and not at the protected premises.

A-1-5.4.6.3.4 In large, campus-style arrangements with proprietary supervising stations monitoring protected premises systems, and in other situations where off-premises monitoring achieves the desired result, the authority having jurisdiction is permitted to allow the re-sound to occur only at the supervising station. Approval by the authority having jurisdiction is required so it can consider all fire safety issues and make a determination that there are procedures in place to ensure that the intent is met, in other words, someone is available to take action to correct the problem.

1-5.4.7 Distinctive Signals. Audible alarm notification appliances for a fire alarm system shall produce signals that are distinctive from other similar appliances used for other purposes in the same area. The distinction among signals shall be as follows:

(a) Fire alarm signals shall be distinctive in sound from other signals. Their sound shall not be used for any other purpose. The requirements of 3-8.4.1.2.1 shall apply.

(b)* Supervisory signals shall be distinctive in sound from other signals. Their sound shall not be used for any other purpose.

A-1-5.4.7(b) A valve supervisory, low-pressure switch or other device intended to cause a supervisory signal when actuated should not be connected in series with the end-of-line supervisory device of initiating device circuits, unless a distinctive signal, different from a trouble signal, is indicated.

The code permits the use of different alarm signals throughout a protected premises. However, unless there are very specific reasons for doing otherwise, every effort should be made to use the same type of signal throughout a protected premises to avoid confusion among the occupants. Some facilities, however, are better suited for the use of different signals in different areas. For example, hospitals may use coded signals in patient care areas because of concerns for patient safety, and use non-coded signals in other non-patient areas to notify all occupants of an alarm condition.

Supervisory initiating devices, such as valve tamper switches, that produce an alarm signal violate the requirement in 1-5.4.7(b) of the code. Therefore, supervisory initiating devices must be on separate initiating device circuits, unless used with a technology that permits distinction between an alarm signal and a supervisory signal.

Formal Interpretations 87-3, 85-9, 85-3 and the June 1974 FI referencing 1-5.4.7 provide further clarification of distinctive signal requirements.

Formal Interpretation

Reference: 1-5.4.7

Statement: Is it proper, within the meaning of the code, to interconnect the gate valve signal with the trouble signal of the fire alarm system?

Question: Is it proper, within the meaning of the code, to utilize a common audible device to indicate a closed gate valve on a sprinkler system supervisory circuit, as well as to indicate trouble on a separate waterflow alarm circuit, with the understanding that there will be visual means for identifying the specific circuit involved?

Answer: Yes.

Question: Is it proper, within the meaning of the code, to interconnect the gate valve switch(es) on the waterflow alarm circuit so that a closed gate valve will be indicated as a trouble on the waterflow alarm circuit?

Answer: No.

Issue Edition: 1972 of NFPA 72A

Reference: 3610

Date: June 1974

Formal Interpretation 87-3

Reference: 1-5.4.7

Background: An initiating device circuit has a water flow device and a valve supervisory device connected to it and by using current limiting techniques provides the distinctive signals (i.e., separate alarm, trouble and supervisory signals) required by NFPA 72, 1-5.4.7.

Question 1: Does this meet the intent of NFPA 72, 1-5.4.7?

Answer: Yes.

Issue Edition: 1987 of NFPA 72A

Reference: 2-8.5

Issue Date: April 30, 1990

Effective Date: May 20, 1990

Formal Interpretation 85-9

Reference: 1-5.4.7, 3-8.3.3.1.3, 3-8.3.3.3.2

Question 1: Is it the intent of the Committee to prohibit the use of a dedicated closed loop circuit to which only normally closed supervisory switches (one or more) are connected where a break in the line or an off normal supervisory switch produce the same signal at the control unit?

Answer: Yes.

Question 2: Is it the intent of the Committee to permit the use of the same audible signal for both a supervisory signal and a trouble signal?

Answer: Yes.

Question 3: Is it the intent of the Committee to permit silencing an audible supervisory signal?

Answer: Yes.

Question 4: Is it the intent of the Committee to permit an arrangement where silencing an audible trouble signal would prevent the receipt of the first (in case there are several supervisory circuits and the answer to question 3 is "yes") audible supervisory signal?

Answer: No.

Question 5: Is it the intent of the Committee to prohibit the use of a common trouble signal silencing switch to silence both trouble and supervisory audible signals when operation of the switch to silence the audible signal caused by a trouble condition will prevent the receipt of an audible signal associated with a supervisory signal?

Answer: Yes.

Issue Edition: 1985 of NFPA 72A

Reference: 2-5.5, 3-5.4.2 et al.

Date: October 1985

Formal Interpretation 85-3

Background: Previous interpretations by this Committee and actions by the membership at the fall of 1984 meeting still leave unclear the Committee's intent on the questions of permitted means to connect supervisory devices to fire alarm control units.

Question 1: If a control unit is arranged to sound the same audible signal for trouble indication as it does for a supervisory signal, is it the intent of the Committee that a supervisory device be permitted to be connected in such a manner that it is not possible to differentiate between an actuated supervisory device or an open circuit trouble condition on the same circuit?

Answer: No.

Question 2: If the answer to Question 1 is "no," would the answer be "yes" if the circuit involved was individually annunciated in some manner?

Answer: No

Question 3: In a control unit arranged as in Question 1, is it the intent of the Committee to permit an audible trouble signal silencing switch to prevent subsequent sounding of audible supervisory signals?

Answer: No.

Question 4: In a control unit arranged as in Question 1, is it the intent of the Committee to permit an audible supervisory signal silencing switch to prevent subsequent sounding of audible trouble signals or supervisory signals?

Answer: Yes.

Issue Edition: 1985 of NFPA 72A

Reference: 2-5.5, 3-5.4.2, et al.

Date: June 1985

Exception: A supervisory signal sound shall be permitted to be used to indicate a trouble condition. If the same sound is used for both supervisory signals and trouble signals, the distinction between signals shall be by other appropriate means such as visible annunciation.

(c) Fire alarm, supervisory, and trouble signals shall take precedence, in that respective order of priority, over all other signals.

Exception: Signals from hold-up alarms or other life-threatening signals shall be permitted to take precedence over supervisory and trouble signals if acceptable to the authority having jurisdiction.

1-5.4.8 Alarm Signal Deactivation. A means for turning off activated alarm notification appliances shall be permitted only where it is key-operated, located within a locked cabinet, or arranged to provide equivalent protection against unauthorized use. Such means shall be permitted only if a visible zone alarm indication or the equivalent has been pro-

vided as specified in 1-5.7.1, and subsequent actuation of initiating devices on other initiating device circuits or subsequent actuation of addressable initiating devices on signaling line circuits cause the notification appliances to reactivate. A means that is left in the "off" position when there is no alarm shall operate an audible trouble signal until the means is restored to normal. If automatically turning off the alarm notification appliances is permitted by the authority having jurisdiction, the alarm shall not be turned off in less than 5 minutes.

Exception 1: If otherwise permitted by the authority having jurisdiction, the 5-minute requirement shall not apply.

Exception 2: If permitted by the authority having jurisdiction, subsequent actuation of another addressable initiating device of the same type in the same room or space shall not be required to cause the notification appliance(s) to reactivate.

The intent of 1-5.4.8 is to allow the audible appliances to be silenced to facilitate communication by emergency responders. Exception No. 1 clearly permits automatic deactivation of all notification appliances after the time limit for deactivation determined by the authority having jurisdiction. However, it must be understood that once the notification appliances are deactivated, such action presumes that trained personnel have taken over the control of the premises. As a part of this control, entrance to the facility must be secured so that persons will not inadvertently enter a building during the emergency. For this reason, the audible and visible notification appliances should not be deactivated until the emergency responders have completely secured the building. This Section permits notification appliances to be deactivated. Therefore, it is permitted, but not required, to silence audible appliances while visible appliances remain energized.

This section has been revised to require reactivation of the alarm notification appliances when a subsequent alarm signal from another initiating device circuit or a subsequent alarm signal from an addressable initiating device on any signaling line circuit is received. A new Exception No. 2 allows the system to be arranged—if the authority having jurisdiction permits—so that subsequent alarms from other addressable initiating devices in the same area as the device that initiated the first alarm need not reactivate the alarm notification appliances.

1-5.4.9 Supervisory Signal Silencing. A means for silencing a supervisory signal notification appliance(s) shall be permitted only if it is key-operated, located within a locked enclosure, or arranged to provide equivalent protection against unauthorized use. Such a means shall be permitted only if it transfers the supervisory indication to a lamp or other visible indicator and subsequent supervisory signals in

other zones cause the supervisory notification appliance(s) to re-sound. A means that is left in the "silence" position where there is no supervisory off-normal signal shall operate a visible signal silence indicator and cause the trouble signal to sound until the silencing means is restored to normal position.

The requirements for supervisory signal silencing are very similar to those for alarm signal deactivation. See 1-5.4.8 and Exhibit 1.37, which illustrates this function.

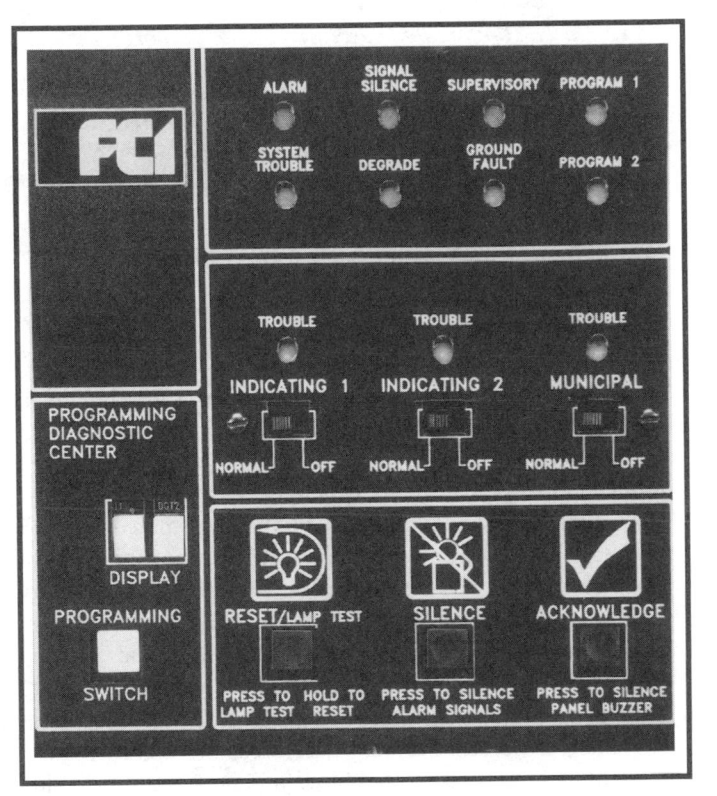

Exhibit 1.37 *Interior of a fire alarm control unit showing alarm, supervisory, and trouble signal indications. (Source: Fire Control Instruments, Inc., Waltham, MA)*

1-5.4.10* Presignal Feature. If permitted by the authority having jurisdiction, systems shall be permitted to have a feature that allows initial fire alarm signals to sound only in department offices, control rooms, fire brigade stations, or other constantly attended central locations and for which human action is subsequently required to activate a general alarm, or a feature that allows the control equipment to delay the general alarm by more than 1 minute after the start of the alarm processing. If there is a connection to a remote location, the transmission of the alarm signal to the supervising station shall activate upon the initial alarm signal.

The remote location referred to in 1-5.4.10 is a supervising station or other location where signals are transmitted. Pre-

signal systems rely on human action, which can be unreliable. NFPA *101, Life Safety Code,* provides additional guidance for using this feature. Because this feature delays the general alarm more than 1 minute, specific permission of the authority having jurisdiction is required. Caution is recommended when delaying alarm signals.

A-1-5.4.10 A system provided with an alarm verification feature as permitted by 3-8.3.2.3.1 is not considered a presignal system, since the delay in the signal produced is 60 seconds or less and requires no human intervention.

1-5.4.11 Positive Alarm Sequence.

1-5.4.11.1 Systems that have positive alarm features complying with 1-5.4.11 shall be permitted if approved by the authority having jurisdiction.

A positive alarm sequence provides a timed delay of a general alarm signal in a building and at a supervising station. This gives a trained responder up to 3 minutes to investigate the cause of an alarm signal. This feature is usually used only in special occupancies where fire does not necessarily pose an immediate threat to the occupants. A positive alarm sequence feature can only be used with the specific approval of the authority having jurisdiction. Exhibit 1.38 illustrates a positive alarm sequence flow chart, which provides for a safe use of this feature.

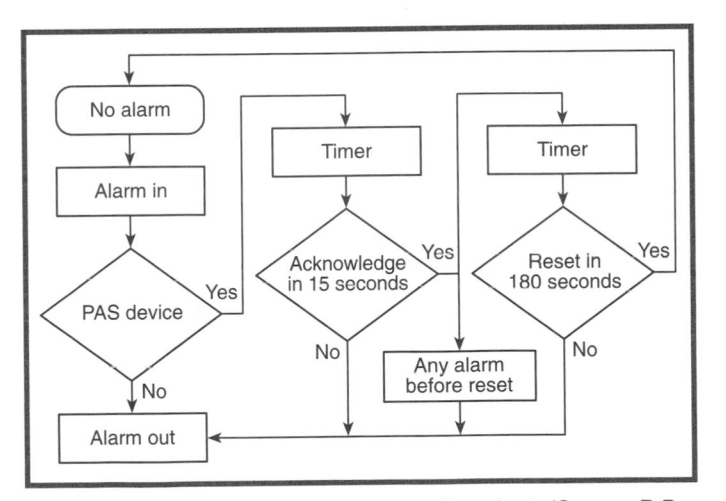

Exhibit 1.38 *Positive alarm sequence flow chart. (Source: R.P. Schifiliti Associates, Inc., Reading, MA)*

1-5.4.11.1.1 The signal from an automatic fire detection device selected for positive alarm sequence operation shall be acknowledged at the control unit by trained personnel within 15 seconds of annunciation in order to initiate the alarm investigation phase. If the signal is not acknowledged within 15 seconds, all building and remote signals shall be activated immediately and automatically.

1-5.4.11.1.2 Trained personnel shall have up to 180 seconds during the alarm investigation phase to evaluate the fire condition and reset the system. If the system is not reset during this investigation phase, all building and remote signals shall be activated immediately and automatically.

1-5.4.11.2 If a second automatic fire detector selected for positive alarm sequence is actuated during the alarm investigation phase, all normal building and remote signals shall be activated immediately and automatically.

1-5.4.11.3 If any other initiating device is actuated, all building and remote signals shall be activated immediately and automatically.

The requirements of 1-5.4.11.1 through 1-5.4.11.3 help to eliminate the human unreliability factor from the use of the positive alarm sequence feature by requiring time limitations and automatic activation when additional devices initiate an alarm.

1-5.4.11.4* The system shall provide means for bypassing the positive alarm sequence.

A-1-5.4.11.4 The bypass means is intended to enable automatic or manual day and night and weekend operation.

1-5.5 Performance and Limitations.

1-5.5.1 Voltage, Temperature, and Humidity Variation. Equipment shall be designed so that it is capable of performing its intended functions under the following conditions:

(1)* At 85 percent and at 110 percent of the nameplate primary (main) and secondary (standby) input voltage(s)

A-1-5.5.1(1) The requirement of 1-5.5.1(1) does not preclude transfer to secondary supply at less than 85 percent of nominal primary voltage, provided the requirements of 1-5.2.5 are met.

(2) At ambient temperatures of 32°F (0°C) and 120°F (49°C)

(3) At a relative humidity of 85 percent and an ambient temperature of 86°F (30°C)

Equipment not listed for use outside these limits must be relocated or the space must be conditioned to meet these parameters. If the space must be artificially conditioned, standby power to operate that artificial conditioning should be provided to ensure that the artificial conditioning continues during a power outage for at least as long as the standby power required for the fire alarm system.

1-5.5.2 Installation and Design.

1-5.5.2.1* All systems shall be installed in accordance with the specifications and standards approved by the authority having jurisdiction.

A-1-5.5.2.1 Fire alarm specifications can include some or all of the following:

(1) Address of the protected premises
(2) Owner of the protected premises
(3) Authority having jurisdiction
(4) Applicable codes, standards, and other design criteria to which the system is required to comply
(5) Type of building construction and occupancy
(6) Fire department response point(s) and annunciator location(s)
(7) Type of fire alarm system to be provided
(8) Calculations, for example, secondary supply and voltage drop calculations
(9) Type(s) of fire alarm-initiating devices, supervisory alarm-initiating devices, and evacuation notification appliances to be provided
(10) Intended area(s) of coverage
(11) Complete list of detection, evacuation signaling, and annunciator zones
(12) Complete list of fire safety control functions
(13) Complete sequence of operations detailing all inputs and outputs

1-5.5.2.2 Devices and appliances shall be located and mounted so that accidental operation or failure is not caused by vibration or jarring.

1-5.5.2.3 All apparatus requiring rewinding or resetting to maintain normal operation shall be restored to normal as promptly as possible after each alarm and kept in normal condition for operation.

1-5.5.2.4 Equipment shall be installed in locations where conditions do not exceed the voltage, temperature, and humidity limits specified in 1-5.5.1.

Exception: Equipment specifically listed for use in locations where conditions can exceed the upper and lower limits specified in 1-5.5.1.

1-5.5.3 To reduce the possibility of damage by induced transients, circuits and equipment shall be properly protected in accordance with the requirements of NFPA 70, *National Electrical Code*, Article 800.

Qualified testing laboratories subject fire alarm equipment to a 15,000-volt static discharge test. This test helps ensure that a basic level of transient protection has been built into the equipment. Section 760-7 of NFPA 70, *National Electrical Code*, requires circuits extending beyond one building to meet the requirements of *NEC* Article 225 or Article 800. Article 800 pertains to aerial circuits and requires that circuits be equipped with surge and lightning protection. Article 225 pertains to underground feeder circuits.

1-5.5.4* Wiring. The installation of all wiring, cable, and equipment shall be in accordance with NFPA 70, *National Electrical Code*, and specifically with Articles 760, 770, and 800, where applicable. Optical fiber cables shall be protected against mechanical injury in accordance with Article 760.

A-1-5.5.4 The installation of all fire alarm system wiring should take into account the fire alarm system manufacturer's published installation instructions and the limitations of the applicable product listings or approvals.

1-5.5.5 Grounding. All systems shall test free of grounds.

Exception: Parts of circuits or equipment that are intentionally and permanently grounded to provide ground-fault detection, noise suppression, emergency ground signaling, and circuit protection grounding.

1-5.5.6 Initiating Devices.

1-5.5.6.1 Initiating devices of the manual or automatic type shall be selected and installed so as to minimize nuisance alarms.

1-5.5.6.2 Fire alarm boxes of the manually operated type shall comply with 3-8.3.2.1.

1-5.6* Protection of Fire Alarm Control Unit(s).

In areas that are not continuously occupied, automatic smoke detection shall be provided at the location of each fire alarm control unit(s) to provide notification of fire at that location.

Exception: Where ambient conditions prohibit installation of automatic smoke detection, automatic heat detection shall be permitted.

This requirement applies even in areas protected by automatic sprinklers. The exception allows the use of heat detectors where conditions are not suitable for smoke detectors. However, areas that are not suitable for smoke detectors are most often not suitable for a fire alarm system control unit. The listing of the control equipment should always be checked to determine suitable locations. Additionally, the term *continuously occupied* means that there is *always* a person at the location (7 days, 24 hours). See the commentary following the definition of Control Equipment in Section 1-4 of the code for more information.

A-1-5.6 The intent of 1-5.6 is to have the fire alarm system respond before it is incapacitated by fire. There have been several fatal fires where the origin and path of the fire resulted in destruction of the control unit before a detector responded.

Caution: The exception to 1-5.6 permits use of a heat detector if ambient conditions are not suitable for smoke detection. It is important to also evaluate whether the area is suitable for the control unit.

The code intends that only one smoke detector is required at the control unit even when the area of the room would require more than one detector if installed according to the spacing rules in Chapter 2.

1-5.7 Zoning and Annunciation.

1-5.7.1 Visible Zone Alarm Indication. If required, the location of an operated initiating device shall be visibly indicated by building, floor, fire zone, or other approved subdivision by annunciation, printout, or other approved means. The visible indication shall not be canceled by the operation of an audible alarm silencing means.

Building codes; NFPA *101, Life Safety Code*; and local ordinances often require each floor of a building to be zoned separately for smoke detectors, waterflow switches, manual fire alarm boxes, and other initiating devices. Addressable systems typically provide individual device status on the fire alarm system control unit. Addressable devices may satisfy those requirements. Exhibits 1.39 and 1.40 illustrate typical annunciators used to provide zone information.

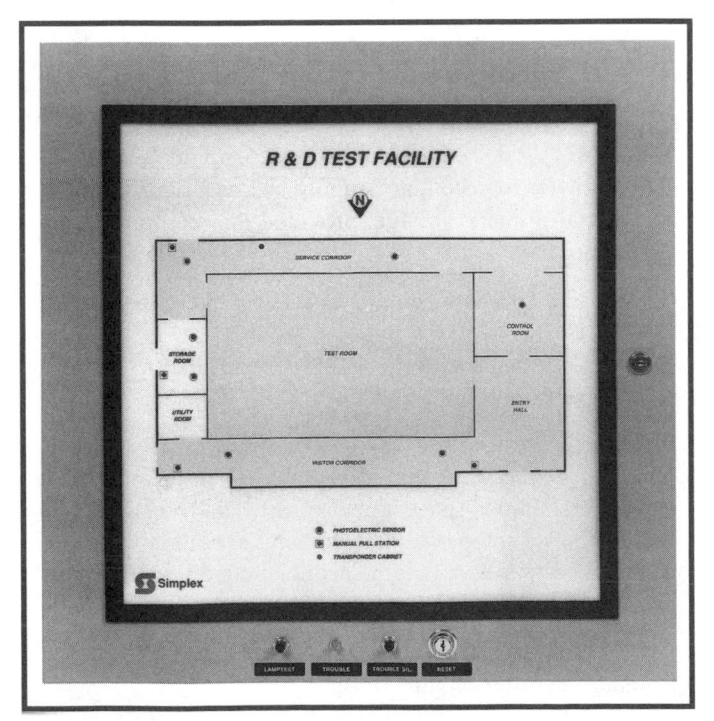

Exhibit 1.39 *Typical remote fire alarm annunciator. (Source: Simplex Time Recorder Co., Gardner, MA)*

Exhibit 1.40 *Typical back-lit labeled annunciators. (Source: ESL Sentrol, Inc., Tualitin, OR)*

1-5.7.1.1 The primary purpose of fire alarm system annunciation is to enable responding personnel to identify the location of a fire quickly and accurately and to indicate the status of emergency equipment or fire safety functions that might affect the safety of occupants in a fire situation. All required annunciation means shall be readily accessible to responding personnel and shall be located as required by the authority having jurisdiction to facilitate an efficient response to the fire situation.

The authority having jurisdiction determines the type and location of any required annunciation. Common locations for annunciation are lobbies, guard's desks, and fire command centers.

1-5.7.1.2* Zone of Origin. Fire alarm systems serving two or more zones shall identify the zone of origin of the alarm initiation by annunciation or coded signal.

A-1-5.7.1.2 Fire alarm system annunciation should, as a minimum, be sufficiently specific to identify the origin of a fire alarm signal in accordance with the following.

(a) If a floor exceeds 20,000 ft² (1860 m²) in area, the floor should be subdivided into detection zones of 20,000 ft² (1860 m²) or less, consistent with the existing smoke and fire barriers on the floor.

(b) If a floor exceeds 20,000 ft² (1860 m²) in area and is undivided by smoke or fire barriers, detection zoning should be determined on a case-by-case basis in consultation with the authority having jurisdiction.

(c) Waterflow switches on sprinkler systems that serve multiple floors, areas exceeding 20,000 ft² (1860 m²), or areas inconsistent with the established detection system zoning should be annunciated individually.

(d) In-duct smoke detectors on air-handling systems that serve multiple floors, areas exceeding 20,000 ft²

(1860 m²), or areas inconsistent with the established detection system zoning should be annunciated individually.

(e) If a floor area exceeds 20,000 ft² (1860 m²), additional zoning should be provided. The length of any zone should not exceed 300 ft (91 m) in any direction. If the building is provided with automatic sprinklers throughout, the area of the alarm zone should be permitted to coincide with the allowable area of the sprinkler zone.

1-5.7.1.3 Visual annunciators shall be capable of displaying all zones in alarm. If all zones in alarm are not displayed simultaneously, there shall be visual indication that other zones are in alarm.

The requirement in 1-5.7.1.3 ensures that when systems require scrolling to view all of the zones in alarm, the system will provide an indication that there are more alarms to view than are currently displayed. The intent is to aid emergency responders in quickly obtaining complete information from the system.

1-5.7.2 Alarm annunciation at the fire command center shall be by means of audible and visible indicators.

1-5.7.3 For the purpose of alarm annunciation, each floor of the building shall be considered as a separate zone. If a floor is subdivided by fire or smoke barriers and the fire plan for the protected premises allows relocation of occupants from the zone of origin to another zone on the same floor, each zone on the floor shall be annunciated separately for purposes of alarm location.

Fire alarm system notification zones should correlate with building smoke and fire zones. This is especially important if an emergency voice/alarm communications system is used to selectively, partially evacuate occupants or to relocate occupants to areas of refuge during a fire.

1-5.7.4 If the system serves more than one building, each building shall be indicated separately.

1-5.8 Monitoring Integrity of Installation Conductors and Other Signaling Channels.

1-5.8.1* All means of interconnecting equipment, devices, and appliances and wiring connections shall be monitored for the integrity of the interconnecting conductors or equivalent path so that the occurrence of a single open or a single ground-fault condition in the installation conductors or other signaling channels and their restoration to normal shall be automatically indicated within 200 seconds.

Connections to devices and appliances must be made so that the opening of any installer's connection to the device or appliance causes a trouble signal. Many installers loop the conductor around the terminal without cutting the conductor and making the necessary two connections. If the wire is

disconnected from the terminal, there may be no indication of trouble. This practice is in violation of the code. If a listed device installed on an initiating device circuit is furnished with pigtail connections, the installer must use separate in/out wires for each circuit passing into or through the device in order to prevent T-tapping of the device connections.

However, addressable devices on signaling line circuits typically use an interrogation/response routine to monitor for integrity. Some types of signaling line circuits may be wired without duplicate terminals; they are often T-tapped. The control unit interrogates each device on a regular basis and "knows" when a device has become disconnected. Therefore, T-tapping is an acceptable practice for Class B signaling line circuits, when the designer allows it. Exhibits 1.41 and 1.42 illustrate typical field-wired equipment with duplicate terminals.

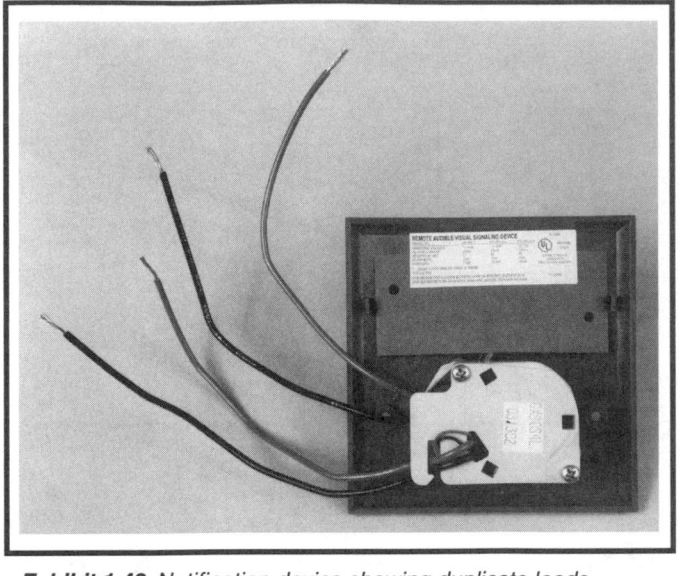

Exhibit 1.42 *Notification device showing duplicate leads. (Source: Gentex; photo courtesy of R.P. Schifilitti Associates Inc., Reading, MA.)*

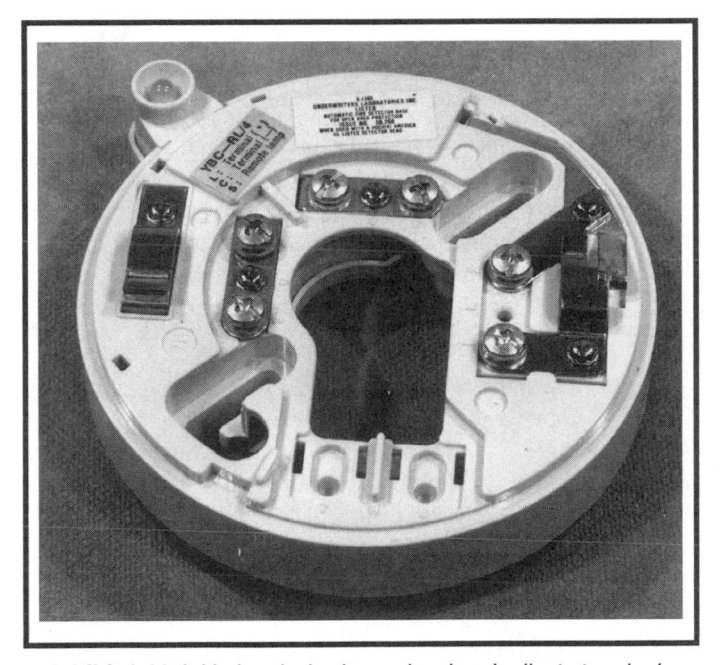

Exhibit 1.41 *Initiating device base showing duplicate terminals. (Source: Radionics, Salinas, CA)*

A-1-5.8.1 The provision of a double loop or other multiple path conductor or circuit to avoid electrical monitoring is not acceptable.

Exception 1: Styles of initiating device circuits, signaling line circuits, and notification appliance circuits tabulated in Table 3-5, Table 3-6, and Table 3-7 that do not have an "X" under "Trouble" for the abnormal condition indicated.

Exception 2: Shorts between conductors, other than as required by 1-5.8.5, 1-5.8.6, and 1-5.8.7.1 and Table 3-5, Table 3-6, and Table 3-7, shall not be subject to this requirement.

Exception 3: A noninterfering shunt circuit, provided that a fault circuit condition on the shunt circuit wiring results only in the loss of the noninterfering feature of operation.

Exception 4: Connections to and between supplementary system components, provided that single open, ground, or short-circuit conditions of the supplementary equipment or interconnecting means, or both, do not affect the required operation of the fire alarm system.

See the commentary under the definition of Supplementary for further explanation of the term.

Exception 5: The circuit of an alarm notification appliance installed in the same room with the central control equipment, provided that the notification appliance circuit conductors are installed in conduit or are equivalently protected against mechanical injury.

Exception 6: A trouble signal circuit.

Exception 7:* Interconnection between listed equipment within a common enclosure.

The requirement for monitoring applies only to installation conductors. The wiring within equipment, devices, or appliances is not required to be monitored for integrity.

A-1-5.8.1 Exception No. 7. This code does not have jurisdiction over the monitoring integrity of conductors within equipment, devices, or appliances.

Exception 8: Interconnection between enclosures containing control equipment located within 20 ft (6 m) of each other where the conductors are installed in conduit or equivalently protected against mechanical injury.

Exception 9: Conductors for ground detection where a single ground does not prevent the required normal operation of the system.

Exception 10: Central station circuits serving notification appliances within a central station.

Exception 11: Pneumatic rate-of-rise systems of the continuous line type in which the wiring terminals of such devices are connected in multiple across electrically supervised circuits.

Exception 12: Interconnecting wiring of a stationary computer and the computer's keyboard, video monitor, mouse-type device, or touch screen, so long as the interconnecting wiring does not exceed 8 ft (2.4 m) in length; is a listed computer/data processing cable as permitted by NFPA 70, National Electrical Code; and failure of cable does not cause the failure of the required system functions not initiated from the keyboard, mouse, or touch screen.

This new exception recognizes that the interconnection wiring of certain specific listed equipment does not have to be monitored for integrity if a stated length of a particular type of cable is used and if a cable failure does not prevent the fire alarm system from performing a required system function.

Formal Interpretation 75-5 clarifies the intent of 1-5.8.1.

Formal Interpretation 75-5

Reference: 1-5.8.1

Question: Is it the intent of 1-5.8.1, with reference to initiating device circuits, that all wires installed by the installer be supervised?

Answer: Yes.

a) In Figure 1 the two field installed wires to the screw terminals of the normally open device are not supervised and therefore unacceptable.

b) In Figure 2 all four of the installed wires connected to the normally open device are supervised and therefore acceptable.

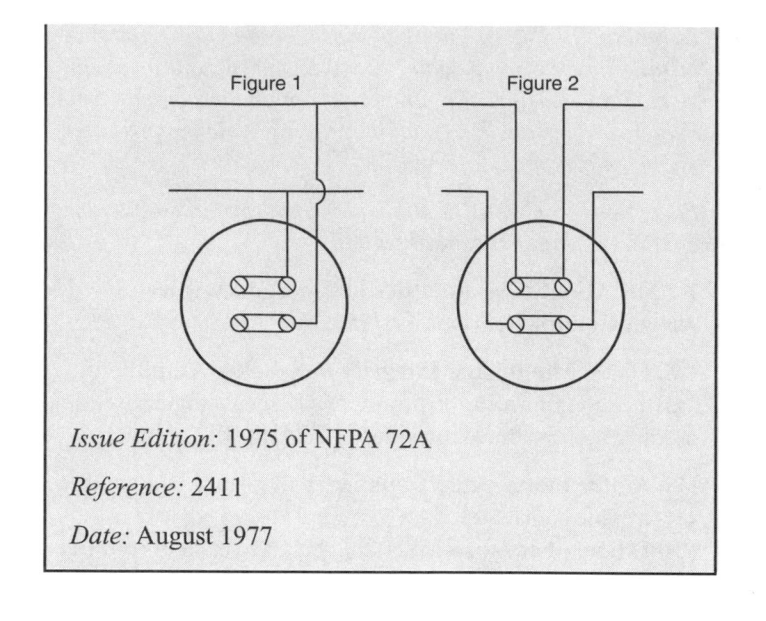

Issue Edition: 1975 of NFPA 72A

Reference: 2411

Date: August 1977

Exception 13: Communications and transmission channels extending from a supervising station to a subsidiary station(s) or protected premises, or both, which are compliant with the requirements of Chapter 5 and electrically isolated from the fire alarm system (or circuits) by a transmitter(s), are not required to monitor the integrity of installation conductors for a ground-fault condition, provided that a single ground condition does not affect the required operation of the fire alarm system.

1-5.8.2 Interconnection means shall be arranged so that a single break or single ground fault does not cause an alarm signal.

1-5.8.3 Unacknowledged alarm signals shall not be interrupted if a fault on an initiating device circuit or a signaling line circuit occurs while there is an alarm condition on that circuit.

Exception: Circuits used to interconnect fire alarm control panels.

1-5.8.4 An open, ground, or short-circuit fault on the installation conductors of one alarm notification appliance circuit shall not affect the operation of any other alarm notification circuit.

1-5.8.5 The occurrence of a wire-to-wire short-circuit fault on any alarm notification appliance circuit shall result in a trouble signal at the protected premises.

Exception 1: A circuit employed to produce a supplementary local alarm signal, provided that the occurrence of a short circuit on the circuit in no way affects the required operation of the fire alarm system.

Exception 2: The circuit of an alarm notification appliance installed in the same room with the central control equipment, provided that the notification appliance circuit conductors are installed in conduit or are equivalently protected against mechanical injury.

Exception 3: Central station circuits serving notification appliances within a central station.

1-5.8.6 Monitoring Integrity of Emergency Voice/ Alarm Communications Systems.

1-5.8.6.1* Monitoring Integrity of Speaker Amplifier and Tone-Generating Equipment. If speakers are used to produce audible fire alarm signals, the following shall apply:

(1) Failure of any audio amplifier shall result in an audible trouble signal.
(2) Failure of any tone-generating equipment shall result in an audible trouble signal.

Exception: Tone-generating and amplifying equipment enclosed as integral parts and serving only a single, listed loudspeaker shall not be required to be monitored.

A-1-5.8.6.1 Backup amplifying and evacuation signal-generating equipment is recommended with automatic transfer upon primary equipment failure to ensure prompt restoration of service in the event of equipment failure.

1-5.8.6.2 Where a two-way telephone communications circuit is provided, its installation wires shall be monitored for a short-circuit fault that would cause the telephone communications circuit to become inoperative.

1-5.8.7 Monitoring Integrity of Power Supplies.

1-5.8.7.1 All primary and secondary power supplies shall be monitored for the presence of voltage at the point of connection to the system. Failure of either supply shall result in a trouble signal in accordance with 1-5.4.6. The trouble signal also shall be visually and audibly indicated at the protected premises. Where the DACT is powered from a protected premises fire alarm system control unit, power failure indication shall be in accordance with this paragraph.

This requirement means that the failure of either the primary or secondary power supply initiates a trouble signal.

Exception 1: A power supply for supplementary equipment.

Because supplementary equipment is not required to meet code requirements, monitoring for integrity is not required.

Exception 2: The neutral of a three-, four-, or five-wire ac or dc supply source.

Exception 3: In a central station, the main power supply, provided the fault condition is otherwise indicated so as to be obvious to the operator on duty.

Exception 4: The output of an engine-driven generator that is part of the secondary power supply, provided the generator is tested weekly in accordance with Chapter 7.

When an engine-driven generator is not running, no voltage will be present on the output terminals. Therefore, monitoring for integrity is impossible.

1-5.8.7.2* Power supply sources and electrical supervision for digital alarm communications systems shall be in accordance with 1-5.2 and 1-5.8.1.

A-1-5.8.7.2 Because digital alarm communicator systems establish communications channels between the protected premises and the central station via the public switched telephone network, the requirement to supervise circuits between the protected premises and the central station *(refer to 1-5.8.1)* is considered to be met if the communications channel is periodically tested in accordance with 5-5.3.2.1.10.

1-5.8.7.3 The primary power failure trouble signal for the DACT shall not be transmitted until the actual battery capacity is depleted by at least 25 percent, but not by more than 50 percent.

Because different batteries discharge at different rates, the DACTs will transmit their associated trouble signals at different times. This requirement prevents jamming of telephone lines at the supervising station during the first moments of a widespread power outage.

1-6 Documentation

1-6.1 Approval and Acceptance.

1-6.1.1 The authority having jurisdiction shall be notified prior to installation or alteration of equipment or wiring. At its request, complete information regarding the system or system alterations, including specifications, wiring diagrams, battery calculation, and floor plans shall be submitted for approval.

Many AHJs require a permit when installing or modifying a fire alarm system. It is strongly recommended that the authority having jurisdiction be invited to participate in the fire alarm system design process as early as possible. This practice often saves a building owner a great deal of money. It allows the AHJ to provide guidance throughout the design process rather than simply providing a list of additional requirements after reviewing the plans.

1-6.1.2 Before requesting final approval of the installation, if required by the authority having jurisdiction, the installing contractor shall furnish a written statement stating that the system has been installed in accordance with approved plans

and tested in accordance with the manufacturer's specifications and the appropriate NFPA requirements.

1-6.1.3* The record of completion form, Figure 1-6.2.1, shall be permitted to be a part of the written statement required in 1-6.1.2. When more than one contractor has been responsible for the installation, each contractor shall complete the portions of the form for which that contractor had responsibility.

A-1-6.1.3 Protected premises fire alarm systems are often installed under construction or remodeling contracts and subsequently connected to a supervising station fire alarm system under a separate contract. All contractors should complete the portions of the record of completion form for the portions of the connected systems for which they are responsible. Several partially completed forms might be accepted by the authority having jurisdiction provided that all portions of the connected systems are covered in the set of forms.

1-6.1.4 The Record of Completion Form, Figure 1-6.2.1, shall be permitted to be a part of the documents that support the requirements of 1-6.2.3.

1-6.2 Completion Documents.

1-6.2.1* A record of completion *(Figure 1-6.2.1)* shall be prepared for each system. Parts 1, 2, and 4 through 10 shall be completed after the system is installed and the installation wiring has been checked. Part 3 shall be completed after the operational acceptance tests have been completed. A preliminary copy of the record of completion shall be given to the system owner and, if requested, to other authorities having jurisdiction after completion of the installation wiring tests. A final copy shall be provided after completion of the operational acceptance tests.

Formerly called the certificate of completion, the record of completion documents the type of system; the names of installers; and the location of record drawings, owners' manuals, and test reports. It also provides a confirming record of the acceptance test and gives details of the components and wiring of the system. A record of completion is required for all installed fire alarm systems.

A-1-6.2.1 The requirements of Chapter 7 should be used to perform the installation wiring and operational acceptance tests required when completing the record of completion.

The record of completion form shall be permitted to be used to record decisions reached prior to installation regarding intended system type(s), circuit designations, device types, notification appliance type, power sources, and the means of transmission to the supervising station.

1-6.2.1.1 All fire alarm systems that are modified after the intial installation shall have the original record of completion revised to show all changes from the original information and shall include a revision date.

1-6.2.2 Every system shall include the following documentation, which shall be delivered to the owner or the owner's representative upon final acceptance of the system:

(1)* An owner's manual and installation instructions covering all system equipment
(2) Record drawings

A-1-6.2.2(1) The owner's manual should include the following:

(a) A detailed narrative description of the system inputs, evacuation signaling, ancillary functions, annunciation, intended sequence of operations, expansion capability, application considerations, and limitations

(b) Operator instructions for basic system operations, including alarm acknowledgment, system reset, interpretation of system output (LEDs, CRT display, and printout), operation of manual evacuation signaling and ancillary function controls, and change of printer paper

(c) A detailed description of routine maintenance and testing as required and recommended and as would be provided under a maintenance contract, including testing and maintenance instructions for each type of device installed. This information should include the following:

(1) Listing of the individual system components that require periodic testing and maintenance
(2) Step-by-step instructions detailing the requisite testing and maintenance procedures, and the intervals at which these procedures shall be performed, for each type of device installed
(3) A schedule that correlates the testing and maintenance procedures recommended by A-1-6.2.2(c)(2) with the listing recommended by A-1-6.2.2(c)(1)

(d) Detailed troubleshooting instructions for each trouble condition generated from the monitored field wiring, including opens, grounds, and loop failures [These instructions should include a list of all trouble signals annunciated by the system, a description of the condition(s) that causes such trouble signals, and step-by-step instructions describing how to isolate such problems and correct them (or how to call for service, as appropriate).]

(e) A service directory, including a list of names and telephone numbers of those who provide service for the system

1-6.2.3 Central Station Fire Alarm Systems. It shall be conspicuously indicated by the prime contractor *(see Chapter 4)* that the fire alarm system providing service at a protected premises complies with all applicable requirements of this code by providing a means of verification as specified in either 1-6.2.3.1 or 1-6.2.3.2.

1-6.2.3.1 The installation shall be certificated.

FIRE ALARM SYSTEM
RECORD OF COMPLETION

Name of protected property: _____

Address: _____

Representative of protected property (name/phone): _____

Authority having jurisdiction: _____

Address/telephone number: _____

1. Type(s) of System or Service

_____ NFPA 72, Chapter 3 — Local

    If alarm is transmitted to location(s) off premises, list where received: _____

    _____

_____ NFPA 72, Chapter 3 — Emergency Voice/Alarm Service

    Quantity of voice/alarm channels: _____ Single: _____ Multiple: _____

    Quantity of speakers installed: _____ Quantity of speaker zones: _____

    Quantity of telephones or telephone jacks included in system: _____

_____ NFPA 72, Chapter 6 — Auxillary

    Indicate type of connection:

    _____ Local energy _____ Shunt _____ Parallel telephone

    Location of telephone number for receipt of signals: _____ _____

    _____

_____ NFPA 72, Chapter 5 — Remote Station

    Alarm: _____

    Supervisory: _____

_____ NFPA 72, Chapter 5 — Proprietary

    If alarms are retransmitted to public fire service communications centers or others, indicate location and telephone numbers of the organization receiving alarm: _____

    _____

    Indicate how alarm is retransmitted: _____

_____ NFPA 72, Chapter 5 — Central Station

    Prime contractor: _____

    Central station location: _____

    Means of transmission of signals from the protected premises to the central station:

    _____ McCulloh _____ Multiplex _____ One-way radio

    _____ Digital alarm communicator _____ Two-way radio _____ Others

    Means of transmission of alarms to the public fire service communications center:

    (a) _____

    (b) _____

    System location: _____

(NFPA Record of Completion 1 of 4)

Figure 1-6.2.1 *Record of completion.*

	Organization name/phone	Representative name/phone
Installer		
Supplier		
Service organization		

Location of record (as-built) drawings: _____

Location of owners manuals: _____

Location of test reports: _____

A contract, dated _____ , for test and inspection in accordance with NFPA standard(s)

No(s). _____ , dated _____ , is in effect.

2. Record of System Installation

(Fill out after installation is complete and wiring checked for opens, shorts, ground faults, and improper branching, but prior to conducting operational acceptance tests.)

This system has been installed in accordance with the NFPA standards as shown below, was inspected

by _____ on _____ , includes the devices shown below, and has been in service

since _____ .

___ NFPA 72, Chapters 1 2 3 4 5 6 7 (circle all that apply)

___ NFPA 70, *National Electrical Code*, Article 760

___ Manufacturer's instructions

___ Other (specify): _____

Signed: _____ Date: _____

Organization: _____

3. Record of System Operation

All operational features and functions of this system were tested by _____ on _____ ,
and found to be operating properly in accordance with the requirements of:

___ NFPA 72, Chapters 1 2 3 4 5 6 7 (circle all that apply)

___ NFPA 70, *National Electrical Code*, Article 760

___ Manufacturer's instructions

___ Other (specify): _____

Signed: _____ Date: _____

Organization: _____

4. Alarm-Initiating Devices and Circuits

Quantity and class of initiating device circuits (*see NFPA 72, Table 3-5*) Quantity: ____ Style: _____ Class: _____

MANUAL

(a) _____ Manual stations _____ Noncoded, activating _____ Transmitters _____ Coded

(b) _____ Combination manual fire alarm and guard's tour coded stations

AUTOMATIC

Coverage: Complete: _____ Partial: _____

(a) _____ Smoke detectors _____ Ion _____ Photo

(b) _____ Duct detectors _____ Ion _____ Photo

(c) _____ Heat detectors _____ FT _____ RR _____ FT/RR _____ RC

(NFPA Record of Completion 2 of 4)

Figure 1-6.2.1 Continued.

(continues)

(d) _____ Sprinkler waterflow switches: _____ Transmitters _____ Noncoded, activating _____ Coded

(e) _____ Other (list): _____

5. **Supervisory Signal-Initiating Devices and Circuits** (use blanks to indicate quantity of devices)

GUARD'S TOUR

(a) _____ Coded stations

(b) _____ Noncoded stations, activating _____ transmitters

(c) _____ Compulsory guard tour system comprised of _____ transmitter stations and _____ intermediate stations

Note: Combination devices are recorded under 4(b) and 5(a).

SPRINKLER SYSTEM

(a) _____ Coded valve supervisory signaling attachments

Value supervisory switches, activating _____ transmitters

(b) _____ Building temperature points

(c) _____ Site water temperature points

(d) _____ Site water supply level points

Electric fire pump:

(e) _____ Fire pump power

(f) _____ Fire pump running

(g) _____ Phase reversal

Engine-driven fire pump:

(h) _____ Selector in auto position

(i) _____ Engine or control panel trouble

(j) _____ Fire pump running

Engine-driven generator:

(k) _____ Selector in auto position

(l) _____ Control panel trouble

(m) _____ Transfer switches

(n) _____ Engine running

Other supervisory function(s) (specify): _____

_____

_____

_____

6. **Alarm Notification Appliances and Circuits**

Quantity and class *(see NFPA 72, Table 3-7)* of notification appliance circuits connected to the system:

Types and quantities of notification appliances installed: Quantity: _____ Style: _____ Class: _____

(a) _____ Bells _____ Inch

(b) _____ Speakers

(c) _____ Horns

(d) _____ Chimes

(e) _____ Other: _____

(NFPA Record of Completion 3 of 4)

Figure 1-6.2.1 Continued.

(f) _____ Visual signals   Type: _____

_____ with audible   _____ w/o audible

(g) _____ Local annunciator

7. Signaling Line Circuits

Quantity and class *(see NFPA 72, Table 3-6)* of signaling line circuits connected to system:

Quantity: _____   Style: _____   Class: _____

8. System Power Supplies

(a) Primary (main): _____   Nominal voltage: _____   Current rating: _____

Overcurrent protection: Type: _____   Current rating: _____

Location: _____

(b) Secondary (standby):

_____ Storage battery: Amp-hour rating: _____

_____ Calculated capacity to drive system, in hours: _____ 24 _____ 60

_____ Engine-driven generator dedicated to fire alarm system:

Location of fuel storage: _____

(c) Emergency or standby system used as backup to primary power supply, instead of using a secondary power supply:

_____ Emergency system described in NFPA 70, Article 700

_____ Legally required standby system described in NFPA 70, Article 701

_____ Optional standby system described in NFPA 70, Article 702, which also meets the performance requirements of Article 700 or 701

9. System Software

(a) Operating system software revision level(s): _____

(b) Application software revision level(s): _____

(c) Revision completed by: _____
(name)   (firm)

10. Comments:

_____
(signed) for central station or alarm service company or installation contractor/supplier   (title)   (date)

Frequency of routine tests and inspections, if other than in accordance with the referenced NFPA standard(s):

_____
_____

System deviations from the referenced NFPA standard(s) are: _____
_____

_____
(signed) for central station or alarm service company or installation contractor/supplier   (title)   (date)

Upon completion of the system(s) satisfactory test(s) witnessed (if required by the authority having jurisdiction):

_____
(signed) representative of the authority having jurisdiction   (title)   (date)

(NFPA Record of Completion 4 of 4)

Figure 1-6.2.1 Continued.

1-6.2.3.1.1 Central station fire alarm systems providing service that complies with all requirements of this code shall be certificated by the organization that has listed the prime contractor, and a document attesting to this certification shall be located on or near the fire alarm system control unit or, where no control unit exists, on or near a fire alarm system component.

1-6.2.3.1.2 A central repository of issued certification documents, accessible to the authority having jurisdiction, shall be maintained by the organization that has listed the central station.

1-6.2.3.2 The installation shall be placarded.

1-6.2.3.2.1 Central station fire alarm systems providing service that complies with all requirements of this code shall be conspicuously marked by the prime contractor to indicate compliance. The marking shall be by means of one or more securely affixed placards.

1-6.2.3.2.2 The placard(s) shall be 20 in.^{2} (130 cm^2) or larger, shall be located on or near the fire alarm system control unit or, where no control unit exists, on or near a fire alarm system component, and shall identify the central station and, where applicable, the prime contractor by name and telephone number.

Formal Interpretations 89-1 and 89-3 provide further information about central station fire alarm system certification.

Formal Interpretation 89-1

Reference: 1-6.2.3.2, 5-2.2.3.2

Question: Is placarding (which is not defined as is certification) intended to be an independent method of verification by the central station with no third party agency being involved as assurance?

Answer: No.

Issue Edition: 1989 of NFPA 71

Reference: 1-2.3.1

Issue Date: June 22, 1992

Effective Date: July 13, 1992

Formal Interpretation 89-3

Reference: 1-6.2.3.1.1

Question: Is it the intent of the Committee that this section provide a means for the authority having jurisdiction to require all central station signaling systems to be certified to verify compliance with this standard, where the central station is listed and provides a certification service in accordance with its listing?

Answer: Yes.

Issue Edition: 1989 of NFPA 71

Reference: 1-2.3.1

Issue Date: March 15, 1993

Effective Date: April 3, 1993

1-6.3 Records.

A complete, unalterable record of the tests and operations of each system shall be kept until the next test and for 1 year thereafter. The record shall be available for examination and, if required, reported to the authority having jurisdiction. Archiving of records by any means shall be permitted if hard copies of the records can be provided promptly when requested.

Exception: If off-premises monitoring is provided, records of all signals, tests, and operations recorded at the supervising station shall be maintained for not less than 1 year.

References Cited in Commentary

ANSI S3.41, *American National Standard Audible Emergency Evacuation Signal,* 1990 edition.
IEEE Standard Dictionary of Electrical and Electronics Terms, 1993 edition.
NFPA 71, *Standard for the Installation, Maintenance, and Use of Signaling Systems for Central Station Service,* 1989 edition.
NFPA 70, *National Electrical Code®,* 1999 edition.
NFPA *101®, Life Safety Code®,* 1997 edition.
NFPA 110, *Standard for Emergency and Standby Power Systems,* 1999 edition.
NFPA 1221, *Standard for the Installation, Maintenance, and Use of Emergency Services Communications Systems,* 1994 edition.

CHAPTER 2

Initiating Devices

Chapter 2 deals with the sensors that provide input to the fire alarm system control unit. Fire detectors initiate the fire alarm system response to the fire. The term *initiating device* applies to all types of sensors, ranging from manually operated fire alarm boxes to switches that detect the operation of a fire extinguishing or fire suppression system. This chapter covers initiating devices, that is, any device that provides an incoming signal to the fire alarm system control unit.

Initiating devices can be ranked by speed of response. Because small fires are easier to extinguish and produce less damage than large fires, the sooner the fire is detected, the better. However, with increased sensitivity and speedier response comes an increased probability of unwarranted operation and reduced stability. The criteria established here for each general type of detector reflect an effort to balance the speed and stability trade-off.

Every requirement in Chapter 2 of the *National Fire Alarm Code*® stems from the need for speed and surety of response to a fire with minimal probability that an alarm signal will result from a non-fire stimulus. System designers can only achieve this objective if they select the proper type of detector for each application. This selection process requires a thorough understanding of how each type of detector operates. The mission-effectiveness of the fire alarm system depends greatly on this choice.

Fire detection devices do not actually respond to the fire itself, but to some change in the ambient conditions in the immediate vicinity of the detector as a result of the fire. A heat detector responds to an increase in the ambient temperature in its immediate vicinity. A smoke detector responds to the presence of smoke in the air in its immediate vicinity. A flame detector responds to the influx of radiant energy that has traveled from the fire to the detector. In each case, either heat, smoke, or light travels from the fire to the detector before the device initiates an alarm signal.

The placement and spacing of both smoke detectors and heat detectors depend on the transfer of combustion products (e.g., heat, smoke, etc.) from the location of the fire to the vicinity of the detector. A set of physical principles generally called *fire plume dynamics* describes this transfer of smoke, aerosol, or heated combustion product gases and air. The combustion reactions of the fire heat the air immediately above it as hot combustion product gases and radiant energy are released. The hot air and combustion product gas mixture rises in an expanding column from the fire to the ceiling. As the fire continues to produce more hot air and combustion product gases and these gases flow upward, the plume turns and forms a ceiling jet. This jet consists of a layer of hot air and combustion gases that expand radially away from the fire plume centerline as shown in Exhibit 2.1.

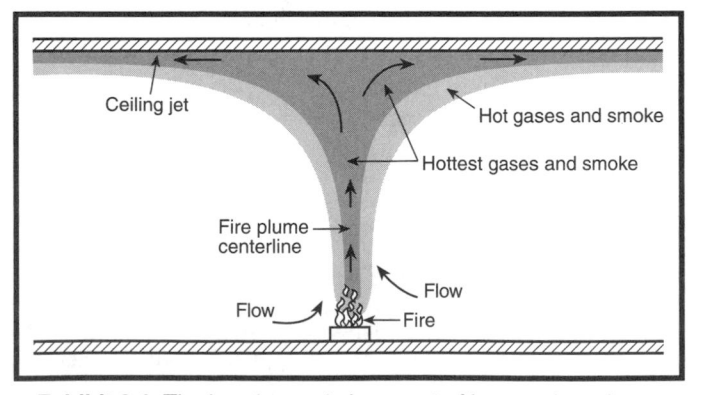

Exhibit 2.1 *The location and placement of heat and smoke detectors are determined by fire plume.*

The ceiling jet carries the heat and combustion product gases (smoke) to the heat detector or smoke detector mounted on the ceiling. The location and spacing criteria in Chapter 2 of the code derive from an understanding of how the ceiling jet forms and how it behaves.

Radiant energy-sensing detectors respond to the radiation from the fire. The fire emits radiation in all directions. All materials in the environment, including the air through which the radiation must travel, reflects, diffracts, absorbs, and transmits radiant energy. As the distance increases between the fire and the detector, the intensity of the radiant energy available to the detector diminishes.

2-1 Introduction

2-1.1 Scope.

Chapter 2 shall cover minimum requirements for performance, selection, use, and location of automatic fire detection devices, sprinkler waterflow detectors, manually activated fire alarm stations, and supervisory signal initiating devices, including guard tour reporting used to ensure timely warning for the purposes of life safety and the protection of a building, space, structure, area, or object.

Detector requirements in dwelling units shall be determined in accordance with Chapter 8.

An understanding of the scope limitations of Chapter 2 is crucial. When the code is referenced in laws or ordinances, the requirements of the code assume the effect of law. The scope statement determines whether the provisions of this chapter apply to a device in question.

As used here, the term *initiating device* has been broadened to cover not only traditional fire detection devices, but also other devices monitoring conditions related to fire safety. These devices include sprinkler system waterflow switches, pressure switches, valve tamper switches, manual fire alarm boxes, municipal fire alarm boxes, and any signaling switches used to monitor special extinguishing systems. The requirements in Section 2-1 apply to all monitoring devices covered in 2-1.1 that provide information, either in the form of a digital or analog transmission, to a fire alarm control unit.

2-1.2 Purpose.

2-1.2.1 The material in Chapter 2 shall be intended for use by persons knowledgeable in the application of fire detection and fire alarm systems and services.

The phrase "persons knowledgeable in the application of fire detection and fire alarm systems and services" refers to someone of a higher skill and knowledge level than that which is necessary to simply read the code. The user is expected to understand the role that the fire alarm system plays in the overall fire safety strategy for the site. Knowing the limitations of the types of detectors is as important as knowing which detector is the right choice for the application. Understanding which type of system will meet the

owner's goals is as important as understanding how the system will operate.

2-1.2.2 Automatic and manual initiating devices shall contribute to life safety, fire protection, and property conservation only if used in conjunction with other equipment. The interconnection of these devices with control equipment configurations and power supplies, or with output systems responding to external actuation, shall be detailed elsewhere in this code or in other NFPA codes and standards.

Chapter 2 of the *National Fire Alarm Code* covers the requirements relevant to the installation of fire alarm and supervisory initiating devices. However, these initiating devices are connected to fire alarm control units whose requirements are covered in Chapter 3 of the code. Consequently, the designer must refer to Chapter 3 for requirements relating to the means of connection.

Furthermore, Chapter 2 does not require that a building owner install detectors of any particular type. If some other code or standard, such as NFPA *101®*, *Life Safety Code®*, requires a building owner to install fire detectors, then Chapter 2 establishes the installation requirements for them. The requirements also apply to any detection devices installed as part of a new or existing non-required fire alarm system.

Chapter 2 establishes the selection and placement criteria that determine the necessary number and type of detectors, but it does not address which types of facilities need initiating devices. The requirement for detection or some form of initiating device is established in the codes and standards that cover a specific class of occupancy or, in some cases, a specific class of fire protection system. The property owner, property insurance carrier, or other authority having jurisdiction may also establish requirements for detection. Once this requirement has been established for detection devices to be installed in a property, the designer would refer to the code for the specifics of selection, installation, and placement.

For example, NFPA 664, *Standard for the Prevention of Fires and Explosions in Wood Processing and Woodworking Facilities*, requires the use of spark/ember detectors in certain specific instances. The designer using NFPA 664 must then refer to Chapter 2 of the *National Fire Alarm Code* for the relevant installation requirements for spark/ember detectors.

Another example, Chapter 28 of NFPA *101*, which covers industrial facilities, requires a fire alarm system where there are more than 100 people on site or more than 25 people on a floor other than the ground floor. The fire protection designer of such an industrial site must then refer to NFPA 72 for the relevant requirements for that fire alarm system. The designer must refer to Chapter 2 for the determination of the type, quantity, and placement of the fire detection devices. The designer must also refer to Chapter 2 for the

installation requirements for waterflow switches, pressure switches, and other initiating devices that may be required by NFPA 13, *Standard for the Installation of Sprinkler Systems*, NFPA 12, *Standard on Carbon Dioxide Extinguishing Systems*, or other such standards. When a designer is placing detection in a specific area or in a manner to protect from a specific hazard, the detection devices to be installed must follow the requirements outlined in the code.

2-1.3 Installation and Required Location of Initiating Devices.

This portion of Chapter 2 begins with basic requirements that apply to all initiating devices, regardless of type. These requirements have come from years of experience relating to the installation of heat, smoke, or radiant energy-sensing detectors. However, Chapter 2 also covers general requirements for the use of supervisory switches, manual fire alarm boxes, and other types of initiating devices.

2-1.3.1 Where subject to mechanical damage, an initiating device shall be protected. A mechanical guard used to protect a smoke or heat detector shall be listed for use with the detector being used.

A prudent designer and installer would apply this requirement to every component of the fire alarm system. The cause of many unwanted alarms as well as system failures has been found to be the result of damage to a detector or other initiating device. See Exhibits 2.2 and 2.3 for examples of protected detectors.

Mechanical damage is not necessarily limited to catastrophic destruction. Mechanical damage can occur over an extended period of time from vibration, extremes in temperature, corrosive atmospheres, other chemical reactions, or excessive humidity. The designer and installer must be sure that the initiating device will be appropriate for the environment in which it is to be installed.

Paragraph 2-1.3.1 requires that mechanical guards used to protect smoke detectors and heat detectors be listed for that purpose. Since both smoke detectors and heat detectors rely on the ceiling jet to convey smoke and hot, combustion product gases from the fire plume to the detector, any object that impedes that flow retards the response of the detector. The only means to be certain the mechanical guard is not a material impediment to detector response is to require that a qualified testing laboratory test and list the guard for the specific make and model detector. The listing will indicate the reduction in spacing or sensitivity that will result from use of the guard.

2-1.3.2 In all cases, initiating devices shall be supported independently of their attachment to the circuit conductors.

Paragraph 2-1.3.2 applies to all types of initiating devices. The copper used in the wiring conductors is not formulated

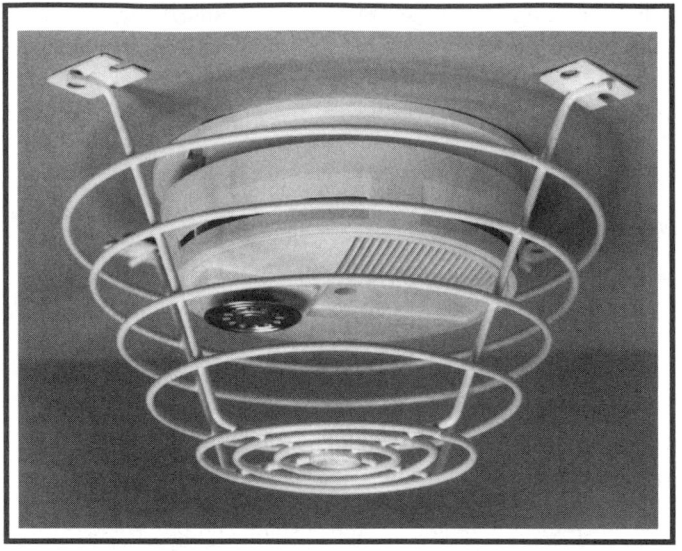

Exhibit 2.2 *Combination smoke and heat detector with protective mechanical guard. (Source: Mammoth Fire Alarms, Inc., Lowell, MA)*

Exhibit 2.3 *Rate-of-rise heat detector with protective mechanical guard. (Source: Mammoth Fire Alarms, Inc., Lowell, MA)*

to serve as a mechanical support. Copper fatigues over time if placed under a mechanical stress, resulting in increasing brittleness and increasing electrical resistance. Ultimately, the fatigued conductor either breaks or its resistance becomes too high to allow the initiating device circuit to function properly. In either case, the operation of the circuit is impaired, and a loss of life or property could conceivably result because of fire alarm system failure.

Initiating devices should always be mounted as shown in the manufacturer's instructions. The requirements for listing include a method for mounting that adequately supports

the initiating device so that no mechanical stresses are applied to the circuit conductors. Furthermore, listing also requires that no electrical shock hazard exists when the device is mounted according to the instructions. When the instructions show use of an electrical backbox then installation of the device with a backbox is a requirement of the listing, and the specific type of backbox shown must be used. If not shown, the use of an electrical backbox is determined by field conditions and the requirements of NFPA 70, *National Electrical Code®*.

2-1.3.3 Initiating devices shall be installed in all areas where required by other NFPA codes and standards or the authority having jurisdiction. Each installed initiating device shall be accessible for periodic maintenance and testing.

The first requirement of 2-1.3.3 provides correlation between the *National Fire Alarm Code* and other codes and standards. Initiating devices must be used wherever required by another code or standard. Chapter 2 answers the questions of how many devices and how they should be installed. It should be noted that the authority having jurisdiction may require initiating devices in areas where they are not necessarily required by other codes or standards. If the authority having jurisdiction makes such a requirement, those initiating devices must also be installed in a manner consistent with Chapter 2 of the code.

The second requirement addresses the issue of accessibility of initiating devices. Accessibility is defined in NFPA 70, *National Electrical Code,* as "admitting close approach: not guarded by locked doors, elevation, and other effective means." The prudent designer or installer should apply this to all system components. The term *accessible* is often subject to debate. For example, if initiating devices such as smoke detectors are mounted on the ceiling of an auditorium, one individual can assert that they are accessible with a scaffold whereas another can disagree, asserting that erecting a scaffold precludes the use of the facility for its intended purpose and is, therefore, not a viable alternative. In addition, NFPA 70 (*NEC*) defines *readily accessible* and by definition, smoke detectors that must be reached using scaffolding do not comply with the definition. The accessibility of a detector or other initiating device will ultimately be reflected in the ability of service personnel to perform maintenance at the required frequency, as outlined in Chapter 7.

2-1.3.4* Connection to the Fire Alarm System. Duplicate terminals, leads, or connectors that provide for the connection of installation wiring, shall be provided on each initiating device for the express purpose of connecting into the fire alarm system to monitor the integrity of the signaling and power wiring.

Exception: Initiating devices connected to a system that provides the required monitoring.

Traditionally, fire alarm system control units have used a small monitoring current to recognize a break in a conductor or the removal of a detector from the circuit. Under normal conditions, the monitoring current flows through the circuit. When a detector is removed or a conductor is broken, the current path is interrupted and the flow of current stops. The control unit translates this into a trouble signal.

Common practice in the electrical trade when installing initiating devices has been to remove a short section of insulation from the conductor and to loop the wire beneath the screw terminal without ever cutting the conductor. This is an unacceptable method of installation. If this method is used, the connection to the initiating device (detector) could loosen over time, and the control unit would not be able to recognize this as a break in the circuit. Paragraph 2-1.3.4 was incorporated into the code to preclude this practice.

Recently, systems using "smart" detectors and even "smarter" control units have been introduced. A microcomputer in the control unit maintains a list of the names of all of the initiating devices in the system. It sequentially addresses each device by name (location) and verifies the response from that device. Thus the control unit recognizes when an initiating device fails to respond and indicates either a device failure or a break in the wiring. This method does not depend on the continuous flow of current. Therefore, these systems are exempt from the duplicate terminal requirement.

See Exhibits 2.4 and 2.5 for examples of duplicate terminals.

Exhibit 2.4 *Smoke detector with base showing incoming and outgoing terminals. (Source: Mammoth Fire Alarms Inc., Lowell, MA)*

A-2-1.3.4 The monitoring of circuit integrity relies on the interruption of the wiring continuity when the connection to the initiating device is lost. Terminals and leads, as illustrated in Figures A-2-1.3.4(a) and (b) monitor the presence of the device on the initiating device circuit.

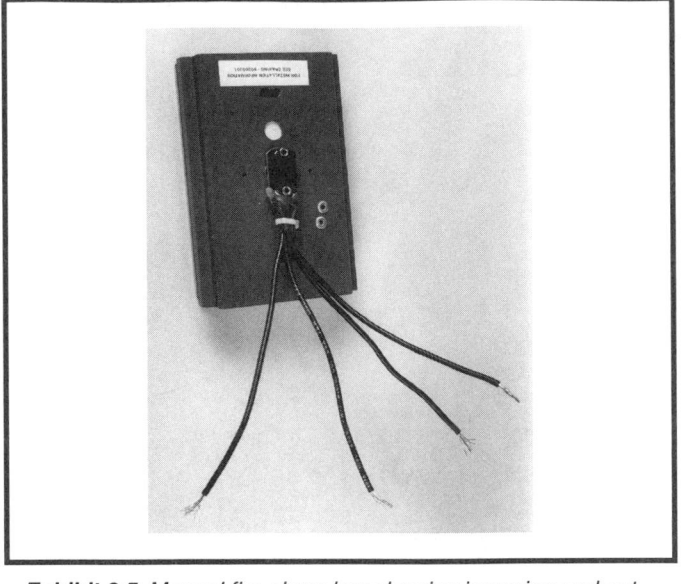

Exhibit 2.5 *Manual fire alarm box showing incoming and outgoing leads. (Source: Mammoth Fire Alarms Inc., Lowell, MA)*

nals, terminals for a normally closed trouble contact, and terminals for a normally open alarm contact. Exhibit 2.6 shows the equivalent schematic of the detector (initiating device). The numbering of the terminals shown in Exhibit 2.6 is strictly illustrative and will not necessarily be consistent with the numbering of commercially available detectors.

Using the designations in Exhibit 2.6, the operating potential (voltage) for the detector is supplied to terminals 7 and 8. Within the detector there is a connection from terminal 7 to terminal 4, and from terminal 8 to terminal 3. Terminals 4 and 3 are wired to terminals 7 and 8, respectively, of the next detector on the circuit, providing operating potential (voltage) to the subsequent detectors. The application of operating potential (voltage) in the proper polarity closes a normally closed trouble contact (n.c.) between terminals 5 and 6. Within the detector there is a jumper between terminals 1 and 2. Thus, under normal operational conditions, terminals 1, 2, 5, and 6 provide a circuit path for the monitoring current. Between terminals 1 and 6, there is a normally open alarm contact (n.o.) that closes when the detector senses the by-products of fire.

The normally closed contacts allow a monitoring current to flow from the control unit into terminal 6, out terminal 5, on through each detector, through the end-of-line device, and

A review of the equivalent circuit inside the detector is helpful when considering Figures A-2-1.3.4(a) and (b). The generic four-wire detector has power supply termi-

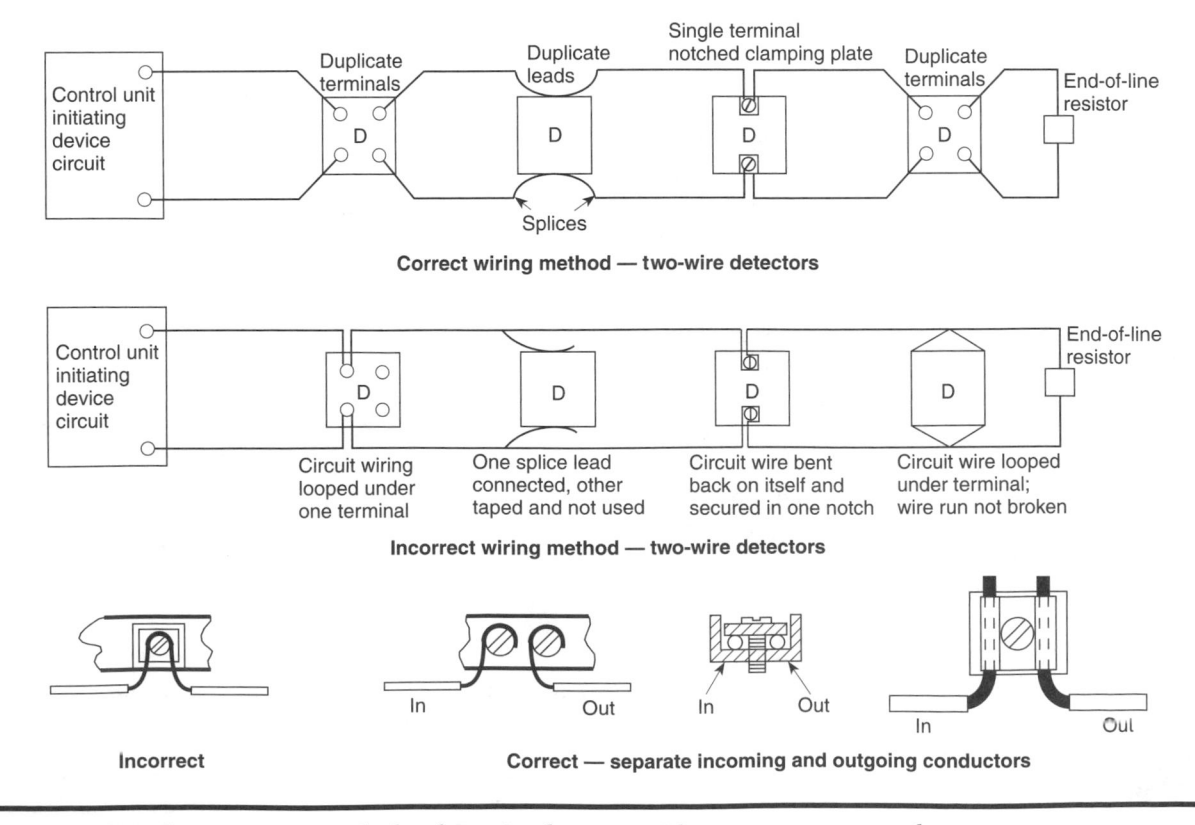

Figure A-2-1.3.4(a) *Correct wiring methods—four-wire detectors with separate power supply.*

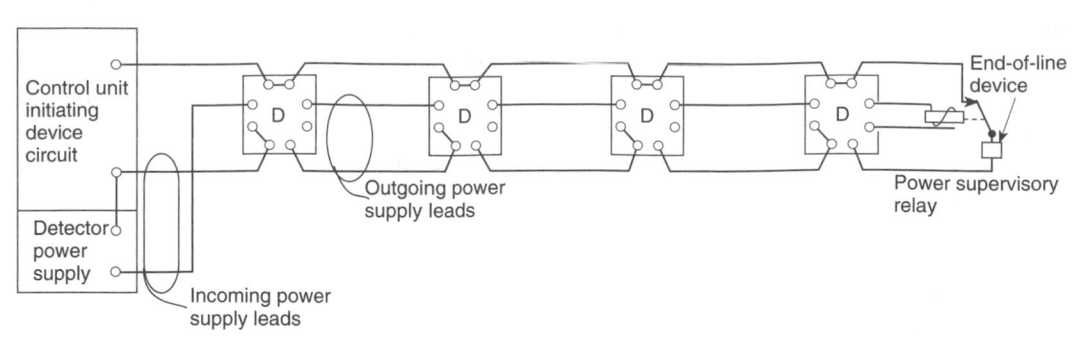

Illustrates four-wire smoke detector employing a three-wire connecting arrangement. One side of power supply is connected to one side of initiating device circuit. Wire run broken at each connection to smoke detector to provide supervision.

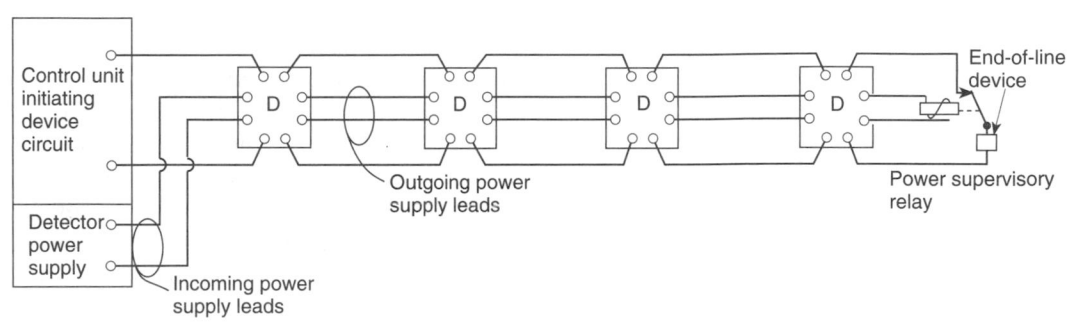

Illustrates four-wire smoke detector employing a four-wire connecting arrangement. Incoming and outgoing leads or terminals for both initiating device and power supply connections. Wire run broken at each connection to provide supervision.

D = Detector

Figure A-2-1.3.4(b) Wiring arrangements for three- and four-wire detectors.

back through terminals 1 and 2 of each detector to the control unit. If an initiating device (detector) loses its source of operating potential (voltage), the trouble contact between terminals 5 and 6 opens, interrupting the current flow. If an initiating device senses a fire, the alarm contact between terminals 1 and 6 closes, bypassing the end-of-line device, which increases the current flowing through the initiating device circuit. The control unit interprets the larger flow of current as a fire alarm.

The ability to use the three-wire format or the four-wire format is determined by the initiating device input circuit of the fire alarm control unit, not the detector. Some control units use one side of the power supply as part of the initiating device circuit, others do not. The system must be wired according to the instructions provided by the manufacturer of the fire alarm system control unit. In addition, the only circuit-powered detectors that are permitted to be connected to a fire alarm control unit are those that have been listed as being compatible with the specific make and model control unit.

2-1.4 Requirements for Smoke and Heat Detectors.

2-1.4.1 Detectors shall not be recessed into the mounting surface in any manner.

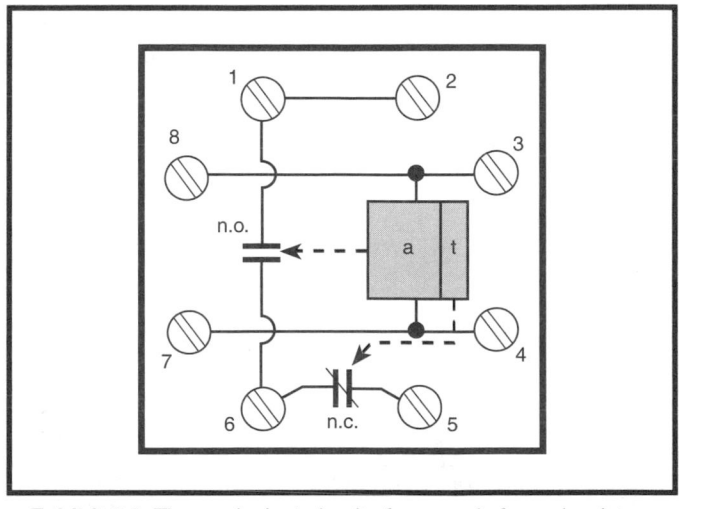

Exhibit 2.6 *The equivalent circuit of a generic four-wire detector. The circuitry (shaded area) is divided into two parts, a sensing part, a, which operates the alarm operated contact and a trouble indicating part, t, which operates the trouble operated contact. (Source: J. M. Cholin Consultants, Inc., Oakland, NJ)*

Exception: If tested and listed for such recessed mounting.

Recessing fire detectors has an adverse effect on the ability of the detector to perform as intended. A heat detector must absorb heat from the hot gases of the ceiling jet, as shown in Exhibit 2.1, before it can respond. Approximately 92 percent to 98 percent of the heat a detector receives is carried to the detector in the hot air and combustion product gases of the ceiling jet, that are created by the fire. This process is called convection or convective heat transfer. When a detector is recessed, it is removed from the flow of air, and, consequently, the quantity of heat it receives per unit of time is reduced. Its response slows, which allows the fire to grow larger before it is detected. A heat detector also receives a small percentage of radiated heat. If the detector is recessed, this radiated heat energy cannot strike the detector. Consequently, the results of recessing a heat detector, contrary to the code, are a very slow response and a fire that could grow very large before it is detected.

Other modes of fire detection are also less sensitive if they are recessed. Smoke detectors depend on air movement to convey smoke from the fire to the detector. Usually this air movement is the ceiling jet from the fire. Smoke detectors are typically mounted on the ceiling to take advantage of the fire plume dynamics and the ceiling jet. However, because there is frictional energy loss between the ceiling jet and the ceiling surface, there is a very thin layer of air immediately below the ceiling surface that is less involved in the ceiling jet flow created by the fire plume and, consequently, contains less smoke. If a smoke detector is recessed, this more slowly moving, relatively clean air immediately below the ceiling surface could impede the flow of smoke into the detector sensing chamber, retarding response to the fire.

2-1.4.2* Detector Coverage.

A-2-1.4.2 The requirement of 2-1.4.2 recognizes that there are several different types of detector coverage.

For years locally adopted codes have required smoke detection in specific parts of the building but not in all compartments, as sound fire protection engineering usually dictates. Paragraph 2-1.4.2 of the code recognizes this reality and endeavors to characterize the different detection coverage concepts, allowing the designer alternatives for the application under consideration.

The designer should consider that whenever the fire is ignited in a building compartment that is not equipped with detection, there will be a substantial and often critical delay in the detection of the fire. This allows the fire the time to grow much larger before detection than would have been the case if detection were installed in the compartment of fire origin.

2-1.4.2.1 Total (Complete) Coverage. If required, total coverage shall include all rooms, halls, storage areas, base-

ments, attics, lofts, spaces above suspended ceilings, and other subdivisions and accessible spaces; and the inside of all closets, elevator shafts, enclosed stairways, dumbwaiter shafts, and chutes. Inaccessible areas shall not be required to be protected by detectors.

Exception 1: If inaccessible areas contain combustible material, they shall be made accessible and shall be protected by a detector(s).

Exception 2: Detectors shall not be required in combustible blind spaces if any of the following conditions exist:

(a) Where the ceiling is attached directly to the underside of the supporting beams of a combustible roof or floor deck

(b) Where the concealed space is entirely filled with a noncombustible insulation (In solid joist construction, the insulation shall be required to fill only the space from the ceiling to the bottom edge of the joist of the roof or floor deck.)

(c) Where there are small, concealed spaces over rooms, provided any space in question does not exceed 50 ft² (4.6 m²) in area

(d) In spaces formed by sets of facing studs or solid joists in walls, floors, or ceilings where the distance between the facing studs or solid joists is less than 6 in. (150 mm)

Exception 3: Detectors shall not be required below open grid ceilings if all of the following conditions exist:

(a) Openings of the grid are $^1/_4$ in. (6.4 mm) or larger in the least dimension.

(b) Thickness of the material does not exceed the least dimension.

(c) Openings constitute at least 70 percent of the area of the ceiling material.

Exception 4: Concealed, accessible spaces above suspended ceilings that are used as a return air plenum meeting the requirements of NFPA 90A, Standard for the Installation of Air Conditioning and Ventilating Systems, where equipped with smoke detection at each connection from the plenum to the central air-handling system.

Exception 5: Detectors shall not be required underneath open loading docks or platforms and their covers and for accessible underfloor spaces if all of the following conditions exist:

(a) Space is not accessible for storage purposes or entrance of unauthorized persons and is protected against the accumulation of windborne debris.

(b) Space contains no equipment such as steam pipes, electric wiring, shafting, or conveyors.

(c) Floor over the space is tight.

(d) No flammable liquids are processed, handled, or stored on the floor above.

Formal Interpretations 78-3 and 78-2 provide further clarification of the requirements of total coverage.

Formal Interpretation 78-3

Reference: 2-1.4.2.1

Question 1: Does Exception No. 1 apply only to the last sentence of 2-1.4.2.1, referring to "inaccessible areas which contain combustible material"?

Answer: Yes.

Question 2: If the answer to Question 1 is "yes," does this indicate that inaccessible areas which do not contain combustible material need not have detectors?

Answer: Yes.

Question 3: If the answer to Question 2 is "yes," does the installation of small access doors for servicing of smoke or fire dampers make the space above an otherwise nonaccessible ceiling accessible within the intent of the code?

Answer: Yes.

Issue Edition: 1978 of NFPA 72E

Reference: 2-6.5

Date: September 1980

Formal Interpretation 78-2

Reference: 2-1.4.2.1

Paragraph 2-1.4.2.1. it states "If required, total coverage shall include all rooms . . ., spaces above suspended ceilings, . . . and chutes." I am interested in the "spaces above suspended ceilings."

There are many buildings which have suspended ceilings—acoustic tile exposed T-bar ceilings.

Question 1: Does the above mean that detectors are required on the underside of the suspended ceiling for area protection of the room and also detectors required above the ceiling to protect space between ceiling and roof or space between ceiling and floor above?

Answer: Yes.

Question 2: Is there any criteria to consider that would not require detectors in the spaces above suspended ceilings?

Answer: Yes. If the space contained no combustible material as defined by NFPA 220 and the ceiling tiles were secured to their T-bar by clips or other methods of fixing such as in an approved fire resistant ceiling-roof assembly or, if the authority having jurisdiction does not require total coverage.

Issue Edition: 1978 of NFPA 72E

Reference: 2-6.5

Date: September 1980

The concept of total or complete detector coverage is addressed in 2-1.4.2.1. The code defines exactly what total coverage entails in 2-1.4.2.1.

Total or complete coverage as established in 2-1.4.2.1 means installing detectors in all accessible compartments or spaces. The underlying premise is that if an enclosed compartment is accessible, it might be used to store combustible materials. There is a parallel between the requirement set forth in 2-1.4.2.1 and that of 1-6.1 of NFPA 13, *Standard for the Installation of Sprinkler Systems.*

Inaccessible, noncombustible compartments do not need detectors. These compartments include a number of blind, boxed-in spaces that are common in stud-wall, curtain-wall, and frame construction, which, if not excepted from this requirement, would result in detectors being placed within walls, etc. The basis for the exceptions to 2-1.4.2.1 is that these spaces contain limited combustibles, and the probability of an ignition originating in these spaces is remote.

Exception No. 1 requires that inaccessible, combustible spaces must be made accessible and be provided with detection. Applying the criteria in Exception No. 1 requires engineering judgment.

Exception No. 2 covers boxed-in spaces that often occur in modern construction and renovation where a void space results. Experience has shown that such spaces are not significant as a fire ignition location and hence do not warrant detectors.

Exception No. 3 pertains to open grid ceilings where all of the stated criteria are met. Where true open grid ceilings exist, the ceiling does not represent a significant barrier to the movement of smoke and fire gases. In most facilities where there are suspended ceilings, the above-ceiling space contains combustibles and does not comply with Exception No. 3. These spaces must, therefore, be equipped with detection when total coverage is required. If the open grid

ceiling does not meet all of the criteria in Exception No. 3 in a site where complete coverage is required, detectors must be placed in the above-ceiling space.

Exception No. 4 addresses above-ceiling spaces used as what NFPA 70 defines as "other space used for environmental air." When above-ceiling air spaces meet the requirements of NFPA 90A, *Air Conditioning and Ventilating Equipment*, and are equipped with smoke detection at each of the exhaust duct connections, again per NFPA 90A, then Exception No. 4 waives the requirement for detectors throughout the above-ceiling space. This exception applies both when detectors have been installed to comply with NFPA 90A and when they have been installed at each of the exhaust duct connections as an alternative to regularly spaced area detectors in the above-ceiling space. The detectors in the exhaust duct will not be effective under no-air-flow conditions, and area detection at the ceiling plane is still required. The relevant sections of NFPA 90A limit the types and quantities of combustible materials that may be included within the return air space.

Exception No. 5 deletes detectors in areas underneath open loading docks or platforms and their covers and for accessible underfloor spaces when the stated conditions are met.

2-1.4.2.2* Partial Coverage. If required, partial detection systems shall be provided in all common areas and work spaces, such as corridors, lobbies, storage rooms, equipment rooms, and other tenantless spaces in those environments suitable for proper detector operation in accordance with this code.

This is the first time that partial detection has been defined by the code. Subparagraph 2-1.4.2.2 provides a name for the level of detection often required by locally adopted codes, outlining where detection must be installed in order to meet the requirements of partial detection.

A-2-1.4.2.2 If there are no detectors in the room or area of fire origin, the fire could exceed the design objectives before being detected by remotely located detectors.

A building owner should very carefully consider the logic of placing detection only in part of the building. A fire in a building compartment that does not have detectors will grow undetected until it becomes large. As the fire grows, plume buoyancy and ceiling jet momentum may eventually force combustion products (smoke and heat) into an adjoining compartment that is equipped with detectors. Doors, ceiling irregularities, ventilation supplies and returns, and distance all retard the flow of smoke and heat toward a building compartment equipped with detection. The resulting delays in response have been critical in some fires.

2-1.4.2.3* Selective Coverage. Where codes, standards, laws, or authorities having jurisdiction require the protection

of selected areas only, the specified areas shall be protected in accordance with this code.

Subparagraph 2-1.4.2.3 is new to the code. It introduces the concept of selective coverage. Selective coverage allows for the protection of selected compartments (areas) without requiring additional detection in other compartments (areas) of the building. However, when a selected compartment (area) has detectors installed, detectors must be installed throughout the entire compartment. These detectors must be installed in the locations and in the quantities required by the relevant sections of Chapter 2.

Paragraph 2-1.4.2.3 establishes that whenever and wherever an initiating device is installed, regardless of its ultimate purpose or function, it must be installed in accordance with the code. Even if detection is installed in only a single room within a building, the detectors in that room must be installed in accordance with the spacing requirements of the code.

A-2-1.4.2.3 If there are no detectors in the room or area of fire origin, the fire could exceed the design objectives before being detected by remotely located detectors. The intent of selective coverage is to address a specific hazard only.

The cautions expressed in the commentary for A-2.1.4.2.2 regarding potential delayed response to a fire are again relevant. If ignition occurs in a building compartment that has not been equipped with detection, the fire emergency will not be recognized until fire products reach a fire detection initiating device in another compartment.

2-1.4.2.4* Supplementary (Nonrequired) Coverage.
Where installed, detection that is not required by an applicable law, code, or standard, whether total (complete), partial, or selective coverage, shall conform to the requirements of this code.

Exception: Spacing requirements of Chapter 2.

Supplementary (nonrequired) coverage is a new concept introduced in this edition of the code. This requirement addresses the possible placement of a very valuable asset in one portion of a much larger compartment. When the performance objective is to attain early warning of a fire involving that valuable asset, but not necessarily some other portion of the compartment, it might be sufficient only to provide detection for the portion of the compartment where the asset is actually located, rather than throughout the entire compartment. Subparagraph 2-1.4.2.4 states that whether or not some applicable law, code, or standard requires detectors, 2-1.4.2.4 requires that the detectors be installed in accordance with the requirements of this chapter. Because this type of detection coverage is specifically intended to permit intensive detection in a portion of the compartment, the spacing requirements of this chapter are waived for this detection strategy only.

A-2-1.4.2.4 The requirement of 2-1.4.2.4 recognizes there will be instances where, for example, a facility owner would want to apply detection to meet certain performance goals and to address a particular hazard or need, but that detection is not required. Once installed, of course, acceptance testing, annual testing, and ongoing maintenance in accordance with this code is anticipated.

Where any detection coverage other than total or complete detection is used in a system design, the designer should be aware that the interconnecting wiring between the devices, appliances, and the fire alarm control unit will often pass through unprotected areas of the building. The designer must expect that such wiring may be subject to the thermal impact of fire possibly before detection. The designer may wish to protect those areas through which fire alarm system wiring passes, require the fire alarm system wiring to be installed in a 2-hour rated enclosure, or require the use of 2-hour rated circuit integrity cables. See also 2-1.4.4.

2-1.4.3 Where nonrequired detection devices are installed for a specific hazard, additional nonrequired detection devices shall not be required to be installed throughout an entire room or building.

Paragraph 2-1.4.3 clarifies and reinforces the concept of supplementary coverage introduced in 2-1.4.2.4 of the code. Exhibits 2.7 and 2.8 represent special application detection devices.

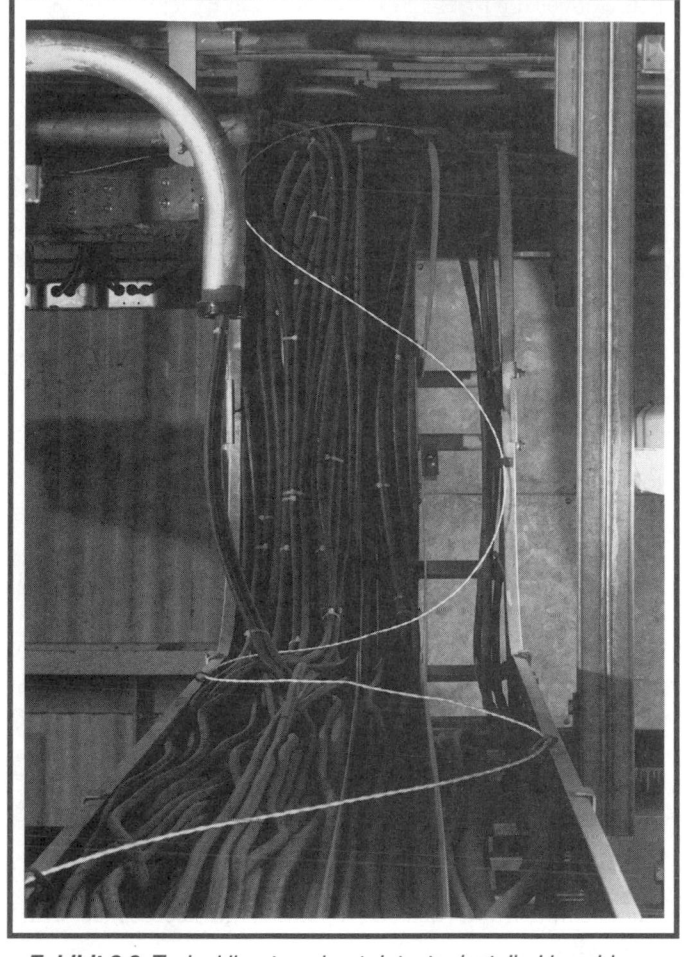

Exhibit 2.8 *Typical line-type heat detector installed in cable tray applications. (Source: Protectowire Company, Hanover, MA)*

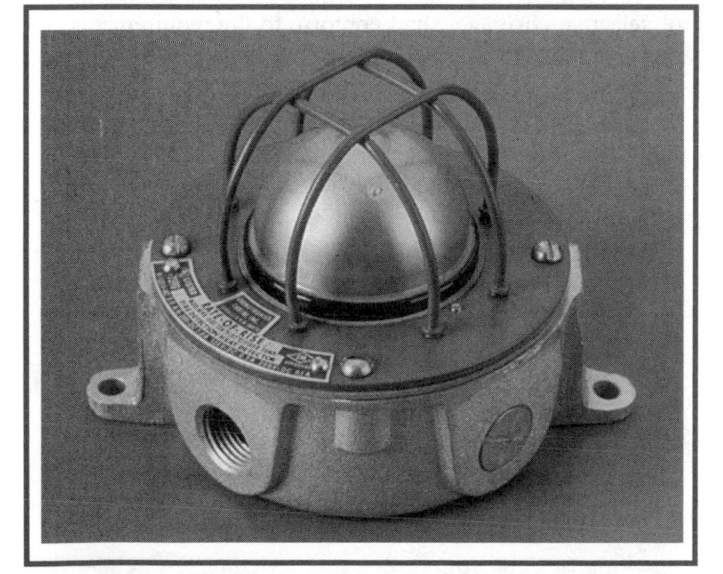

Exhibit 2.7 *Explosionproof spot-type heat detector. (Source: Chemtronics Caribe, Ashland, MA)*

2-2 Heat-Sensing Fire Detectors

Heat-sensing fire detectors shall be installed in all areas where required by other NFPA codes and standards or by the authority having jurisdiction.

The relationship between heat and temperature must be understood if heat-sensing detectors are to be applied properly. Heat is energy and is quantified in terms of an amount, usually British thermal units (Btu) or Joules (J). Temperature is a measure of the quantity of heat in a given mass of material and is measured as an intensity, quantified in terms of extent, usually in degrees Farenheit or degrees Celsius. The majority of the heat flowing into a heat detector is from the hot gases of the ceiling jet. This is called convective heat transfer. A much smaller portion of the heat absorbed by a heat detector is transferred by radiation. This is called radi-

ant heat transfer. Heat detectors operate on one or more of three different principles. These operating principles are categorized as fixed temperature, rate compensation, and rate-of-rise. Each principle has its performance advantages and can be used in either a spot-type device or a line-type device. Heat detectors are devices that change in some way when the temperature at the detector achieves a particular level, or a set point, as in a fixed-temperature type. Other detectors respond to the rate of temperature change, as in a rate-of-rise heat detector. The increase in temperature of the sensing element of a heat detector is due to the absorption of heat from a fire.

Heat detectors are available in two general types: spot-type, which are devices that occupy a specific spot or point, and line-type, which are linear devices that extend over a distance, sensing temperature along their entire length.

See Exhibits 2.9 through 2.16 for examples of typical heat detectors.

A number of different technologies can be used to detect the heat from a fire, including the following:

(1) Expanding bimetallic components
(2) Eutectic solders
(3) Eutectic salts
(4) Melting insulators
(5) Thermistors
(6) Temperature-sensitive semiconductors
(7) Expanding air volume
(8) Expanding liquid volume

(9) Temperature-sensitive resistors
(10) Thermopiles

The code has been written to allow the development and use of new technologies. The designer must be careful not to confuse the terms *type* and *principle* with *technology*, which is the method used to achieve heat detection.

A precise definition and explanation of the mode of operation for each type of heat detector can be found in Section 1-4.

The code does not require detectors to be installed in any building; that requirement comes from other codes, laws, or standards.

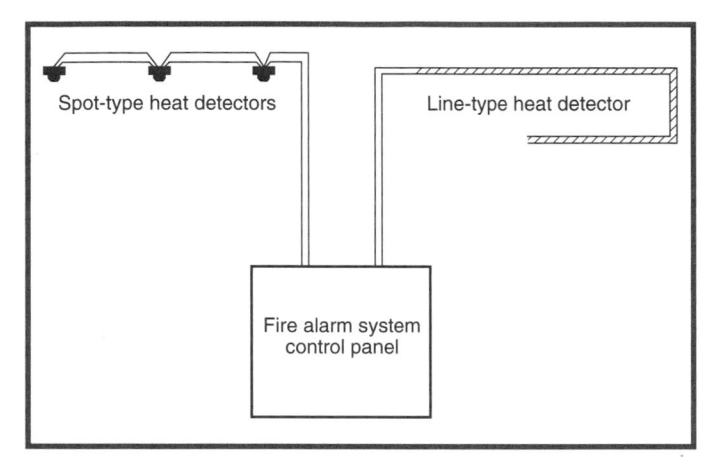

Exhibit 2.10 *The two types of heat detectors, spot-type and line-type. (Source: J. M. Cholin Consultants, Inc., Oakland, NJ)*

Exhibit 2.9 *Electronic spot-type heat detector. (Source: Mammoth Fire Alarms, Inc., Lowell, MA)*

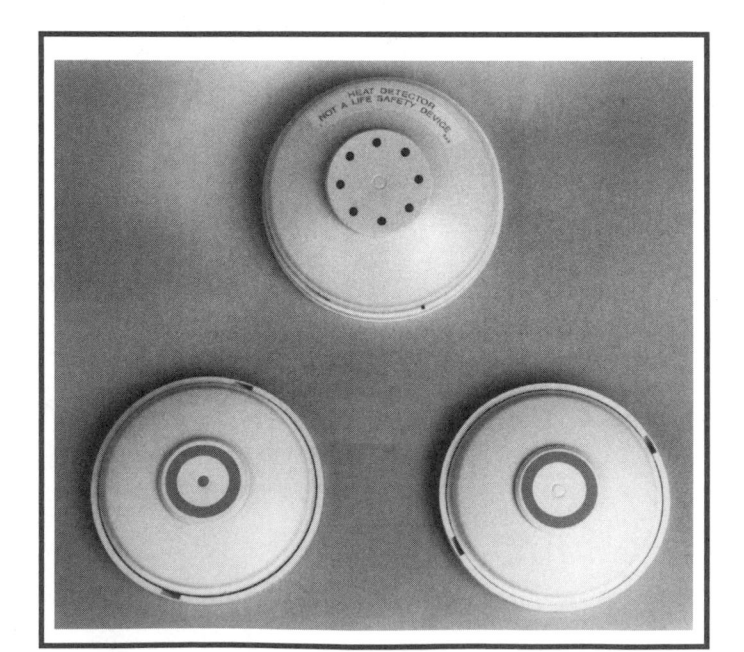

Exhibit 2.11 *Typical "low-profile" rate-of-rise and fixed-temperature heat detectors. (Source: Mammoth Fire Alarms, Inc., Lowell, MA)*

Exhibit 2.12 *Typical spot-type fixed-temperature heat detector. (Source: Mammoth Fire Alarms, Inc., Lowell, MA)*

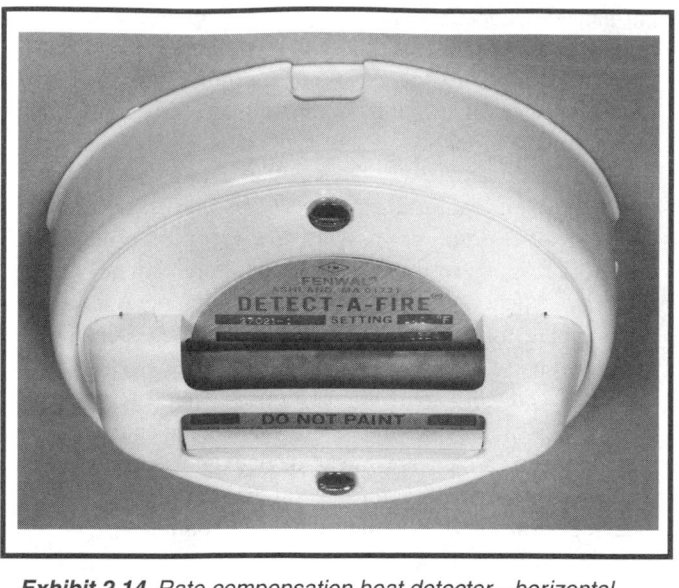

Exhibit 2.14 *Rate compensation heat detector—horizontal mounting. (Source: Kidde-Fenwal Protection Systems, Ashland, MA)*

Exhibit 2.13 *Line-type heat detectors. (Source: Protectowire Company, Hanover, MA)*

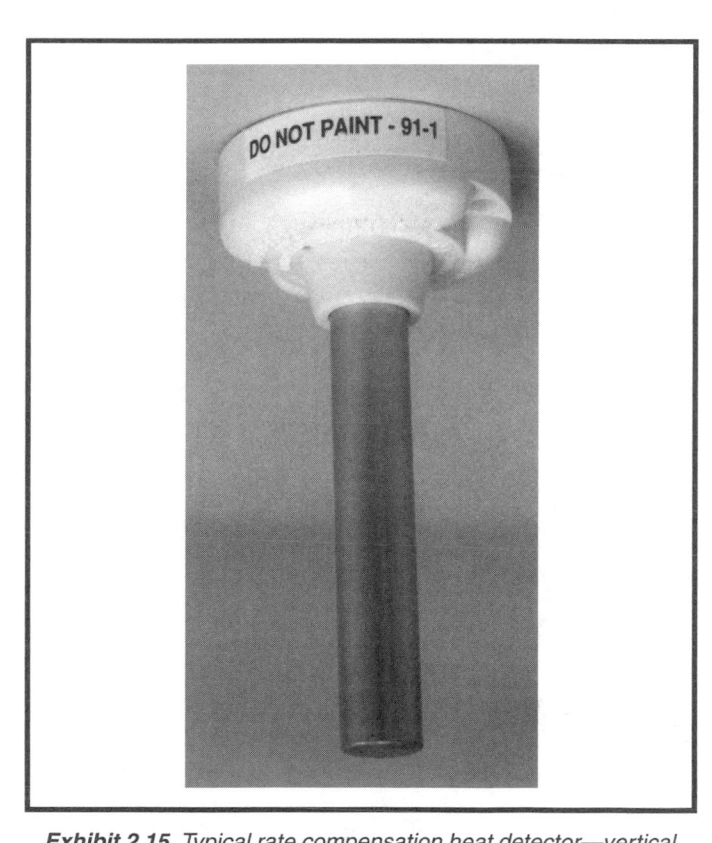

Exhibit 2.15 *Typical rate compensation heat detector—vertical mounting. (Source: Mammoth Fire Alarms, Inc., Lowell, MA)*

Exhibit 2.16 *Typical spot-type combination rate-of-rise fixed-temperature heat detector. (Source: Mammoth Fire Alarms, Inc., Lowell, MA)*

2-2.1 Temperature Classification.

The performance of a heat detector depends on two parameters: its temperature classification and its time-dependent thermal response characteristics. Traditionally, the temperature classification has been the principal parameter used in selecting the proper detector for a given site. An additional objective is to select a detector that will be stable in the environment in which it will be installed.

2-2.1.1 Color Coding.

2-2.1.1.1 Heat-sensing fire detectors of the fixed-temperature or rate-compensated, spot-type shall be classified as to the temperature of operation and marked with a color code in accordance with Table 2-2.1.1.1.

Spot-type heat detectors are the most popular type of heat detector in general use. Table 2-2.1.1.1 presents specific criteria for the nominal temperature classification versus the maximum expected normal temperature for the location of the detector. It is necessary to provide at least a 20°F (11.1°C) difference between the temperature classification of the detector and the maximum expected normal temperature. This requirement technically applies only to low-temperature detectors, not to detectors with other temperature classifications. Nevertheless, it is still good practice to provide at least a 20°F (11.1°C) difference between maximum ambient and detector temperature classification when using heat detectors with the higher temperature classifications. Also, it is crucial not to select a detector temperature classification any higher than necessary. The higher the temperature classification, the longer it will take for the detector to initiate an alarm because a larger fire is needed to produce the higher temperatures at the detector location.

The unified color coding of heat detectors facilitates inspections, making it possible to identify the temperature rating of a ceiling-mounted heat detector while standing on the floor. The color code for heat detectors is very similar to that used for sprinkler heads, as described in 3-2.5.1 of NFPA 13. Manufacturers also provide this information in their data sheets for each type of heat detector.

Exception: Heat-sensing fire detectors where the alarm threshold is field adjustable and that are marked with the temperature range.

Table 2-2.1.1.1 Temperature Classification for Heat-Sensing Fire Detectors

Temperature Classification	Temperature Rating Range		Maximum Ceiling Temperature		Color Code
	°F	°C	°F	°C	
Low*	100 – 134	39 – 57	20 below	11 below	Uncolored
Ordinary	135 – 174	58 – 79	100	38	Uncolored
Intermediate	175 – 249	80 – 121	150	66	White
High	250 – 324	122 – 162	225	107	Blue
Extra high	325 – 399	163 – 204	300	149	Red
Very extra high	400 – 499	205 – 259	375	191	Green
Ultra high	500 – 575	260 – 302	475	246	Orange

*Intended only for installation in controlled ambient areas. Units shall be marked to indicate maximum ambient installation temperature.

The commercial availability of solid state thermal sensors and analog/addressable fire alarm system control units has made possible the development of analog/addressable heat detectors. These detectors permit the designer to adjust the alarm threshold temperature at the fire alarm control unit and select a unique threshold temperature based on an analysis of the compartment and the fire hazard. A color code would be meaningless for this technology and consequently is waived. Clearly, when analog/addressable technology is used, an alternative means for facilitating inspection should be in place.

2-2.1.1.2 If the overall color of a heat-sensing fire detector is the same as the color code marking required for that detector, one of the following arrangements, applied in a contrasting color and visible after installation, shall be employed:

(1) Ring on the surface of the detector
(2) Temperature rating in numerals at least $^3/_8$ in. (9.5 mm) high

See Exhibits 2.17 and 2.18 for examples of heat detector marking.

Exhibit 2.18 *Heat detector with temperature marked with numerals. (Source: Mammoth Fire Alarms, Inc., Lowell, MA)*

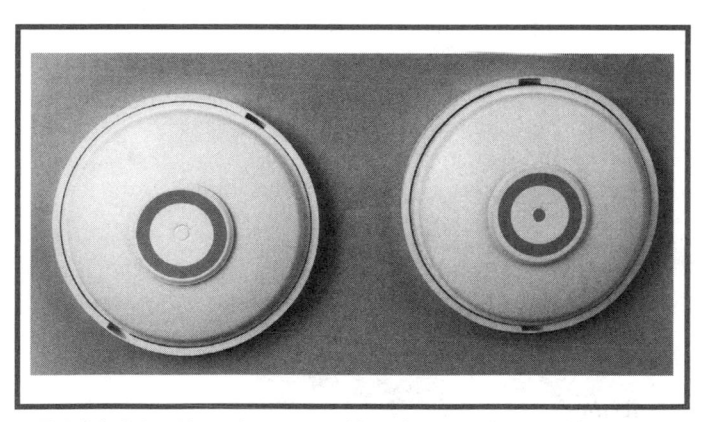

Exhibit 2.17 *Heat detectors with color-coded rings. (Source: Mammoth Fire Alarms, Inc., Lowell, MA)*

2-2.1.2* A heat-sensing fire detector integrally mounted on a smoke detector shall be listed or approved for not less than 50-ft (15-m) spacing.

Many common smoke detectors are equipped with an integral heat sensor. Because the recommended spacing of smoke detectors is 30 ft (9.1 m), and because smoke detectors are primarily used for early warning, the heat detector, if deemed necessary, should be more sensitive. In order for the heat sensor portion of the detector to comply with the code, it must have a 50-ft (15-m) spacing factor. See Exhibit 2.19 for an example of a combination heat and smoke detector.

Exhibit 2.19 *Smoke detector with 50-ft (15-m) listed heat detection. (Source: Mammoth Fire Alarms, Inc., Lowell, MA)*

A-2-2.1.2 The linear space rating is the maximum allowable distance between heat detectors. The linear space rating is also a measure of the heat detector response time to a standard test fire where tested at the same distance. The higher the rating, the faster the response time. This code rec-

ognizes only those heat detectors with ratings of 50 ft (15 m) or more.

2-2.1.3* Heat-sensing fire detectors shall be marked with their operating temperature and thermal response coefficient as determined by the organization listing the device. The requirement for the marking of the thermal response coefficient shall have an effective date of July 1, 2002.

The new requirement in 2-2.1.3 in this edition of the code has been precipitated by the growing acceptance of performance-based design for fire protection systems. The response of a heat detector is determined by two factors: its temperature rating and the speed with which it can absorb heat from the surrounding air. The second factor is called the thermal response coefficient. Without a verified value for the thermal response coefficient of a heat detector, predicting when the heat detector will operate in relation to the development of a fire is impossible. Consequently, the lack of a thermal response coefficient prohibits the use of such heat detectors in performance based designs.

A-2-2.1.3 In order to predict the response of a heat detector using current fire modeling programs and currently published equations describing plume dynamics, two parameters must be known: operating temperature and thermal response coefficient. The thermal response coefficient is the quantification of the rate of heat transfer from the ceiling jet to the detector sensing element per unit of time, expressed as a function of ceiling jet temperature, ceiling jet velocity, and time. Response time index is a commonly used thermal response coefficient for sprinklers.

Some researchers have used the response time index (RTI) as a measure of the thermal response of a heat detector [Alpert, 1972; Evans and Stroup, 1986]. RTI is determined by means of a plunge test that was originally designed to test the response characteristics of sprinkler heads. The plunge test uses a simulated ceiling jet velocity of 1.5 m/sec (5.0 ft/sec). Other researchers have raised questions regarding this measurement, because heat flow from the ceiling jet to a detector is proportional to ceiling jet velocity [Brozovski, 1989; Heskestad and Delichatsios, 1989]. Because the spacing between heat detectors is often much greater than the spacing between sprinkler heads, the jet velocities normally encountered at the heat detectors are an order of magnitude smaller than those encountered at the sprinklers. Therefore, using RTI as a measure of heat detector sensitivity presumes a much larger fire and could introduce important inaccuracies. Consequently, a thermal response coefficient for heat detectors derived from a test specifically designed for the purpose is necessary.

The adoption of the concept of a thermal response coefficient presumes the development of a test method to determine the coefficient. That test would become part of the

listing evaluation for heat detectors. Because the test is still under development, a delayed effective date was adopted. This will provide sufficient time for the test method to be refined and for detectors to be tested.

2-2.2 Location.

Subsection 2-2.2 of the code prescribes the proper location of heat detectors for general purpose, open area detection. The location stipulated takes maximum benefit of the ceiling jet produced by a fire. Because the occurrence of a ceiling jet causes the hot combustion product gases to flow outward and away from the fire plume centerline, a ceiling location provides for the maximum flow across the detector and the maximum speed of response to a growing fire.

2-2.2.1* Spot-type heat-sensing fire detectors shall be located on the ceiling not less than 4 in. (100 mm) from the sidewall or on the sidewalls between 4 in. and 12 in. (100 mm and 300 mm) from the ceiling.

A-2-2.2.1 Figure A-2-2.2.1 illustrates the proper mounting placement for detectors.

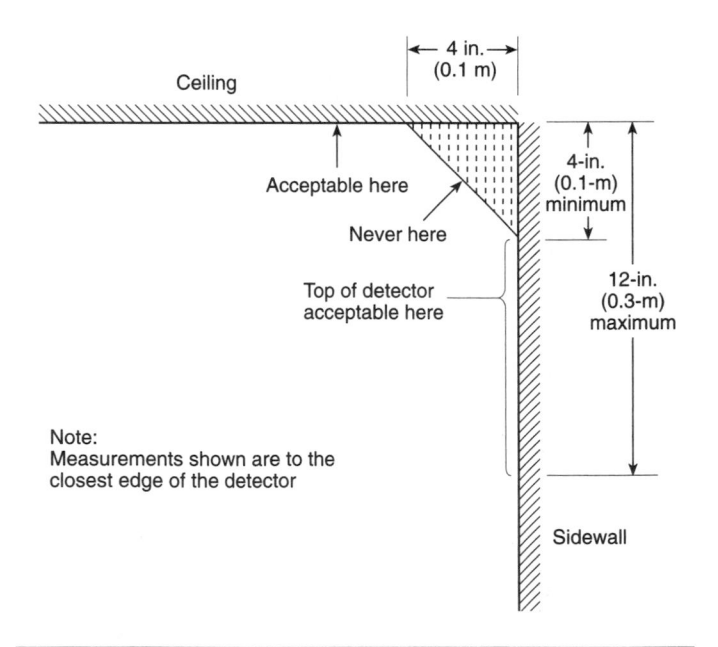

Figure A-2-2.2.1 Example of proper mounting for detectors

Exception 1: In the case of solid joist construction, detectors shall be mounted at the bottom of the joists.

Exception 2: In the case of beam construction where beams are less than 12 in. (300 mm) in depth and less than 8 ft (2.4 m) on center, detectors shall be permitted to be installed on the bottom of beams.

Paragraph 2-2.2.1 applies only to spot-type heat detectors. Section 1-4 defines the terms *ceiling surfaces, solid joist construction*, and *beam construction* for use in this context.

Section 1-4 defines a ceiling surface as the upper surface of a space, regardless of height. Paragraph 2-1.4.2 identifies those spaces where detectors must be placed when total coverage is required.

In compartments equipped with heat detection, spot-type heat detectors must be located on the ceiling at a distance 4 in. (100 mm) or more from a vertical side wall, or on the side wall between 4 in. (100 mm) and 12 in. (300 mm) from the ceiling, measured to the top of the detector.

The ceiling location derives the maximum benefit from the upward flow of the fire plume and the flow of the ceiling jet beneath the ceiling plane. The best currently available research data support the existence of a dead air space where the walls meet the ceiling in a typical room [National Fire Protection Research Foundation, 1993]. Figure A-2-2.2.1 shows this dead air space extending 4 in. (100 mm) in from the wall and 4 in. (100 mm) down from the ceiling. Consequently, the code excludes detectors from those areas. The prudent designer will keep detectors further from the wall than the 4 in. (100 mm) minimum distance criterion established by the code. As the ceiling jet approaches the wall, its velocity declines. Lower ceiling jet velocities result in slower heat transfer to the detector and, therefore, a retarded response.

The type of ceiling configuration in the protected space determines the applicability of the exceptions. But the definitions of the terms *joist* and *beam* must be inferred from the definition of Solid Joist Construction and Beam Construction in Section 1-4. Joists are solid projections, whether structural or not, extending downward from the ceiling, which are more than 4.0 in. (100 mm) in depth and are spaced on 3.0 ft (0.9 m) centers or less. The commonly encountered 2 in. × 10 in. (50 mm × 250 mm) installed on 16-in. (400-mm) centers supporting a roof deck is typical of solid joist construction.

The structural component commonly called a bar-joist is actually an open web beam. If the upper web member of an open web beam is less than 4 in.(100 mm) deep, the beam is ignored. If it is more than 4 in. (100 mm) deep, it is called either a joist or a beam depending on the center-to-center spacing.

The narrow spacing between joists [usually 16 in. (400 mm)] creates air pockets. These air pockets have two effects on the flow of the ceiling jet. First, the air pockets tend to slow the ceiling jet; second, the air pockets force the ceiling jet to flow across the bottoms of the joists, as depicted in Exhibit 2.20. Exception No. 1 requires that heat detectors be placed on the bottoms of joists rather than up in the pockets between them. This placement puts the detectors in the region of maximum ceiling jet flow.

Beams are defined as solid projections, whether structural or not, extending downward from the ceiling, which are more than 4.0 in. (100 mm) in depth and are spaced on centers of more than 3.0 ft (0.9 m). In the context of the code, the principal distinction between a joist and a beam is the center-to-center spacing. Exception No. 2 allows the detectors to be placed on the beam bottoms only when the beams are less than 12 in. (300 mm) deep, and only when the beams are on centers of less than 8 ft (2.4 m). If the beams are more than 12 in. (300 mm) deep or if they are spaced more than 8 ft (2.4 m) apart, the detectors must be placed on the ceiling surface between the beams.

Finally, the only permitted location for spot-type heat detectors is at or in close proximity to the ceiling plane, consistent with the stipulations in 2-2.2.1. There is no research to provide guidance for detector placement in areas without ceilings. By inference, if there is no ceiling in the hazard area on which to locate heat detectors, then heat detection cannot be installed in compliance with the prescriptive requirements of the code.

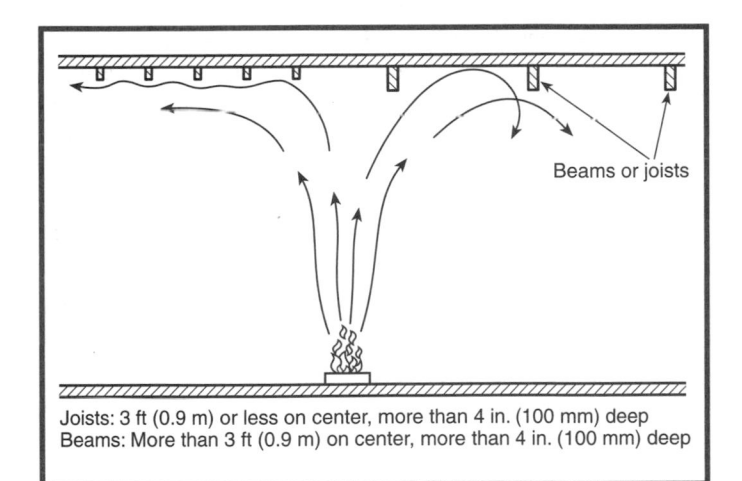

Joists: 3 ft (0.9 m) or less on center, more than 4 in. (100 mm) deep
Beams: More than 3 ft (0.9 m) on center, more than 4 in. (100 mm) deep

Exhibit 2.20 *The effect of joists and beams on the ceiling jet. (Source: J.M. Cholin Consultants, Inc., Oakland, NJ)*

2-2.2.2 Line-type heat detectors shall be located on the ceiling or on the sidewalls not more than 20 in. (500 mm) from the ceiling.

Exception 1: In the case of solid joist construction, detectors shall be mounted at the bottom of the joists.

Exception 2: In the case of beam construction where beams are less than 12 in. (300 mm) in depth and less than 8 ft (2.4 m) on center, detectors shall be permitted to be installed on the bottom of beams.

Exception 3: If a line-type detector is used in an application other than open area protection, the manufacturer's installation instructions shall be followed.

Paragraph 2-2.2.2 only applies to linear heat detectors. It does not prohibit the installation of a line-type detector in the portion of the ceiling within 4 in. (100 mm) of a vertical wall or on the vertical wall within 4 in. (100 mm) of the ceiling (the area often referred to as the dead air space). Although installation in this area is not prohibited, the prudent designer avoids the ceiling/wall corner, even when using a line-type heat detector.

Also it is important to note that Exception No. 3 recognizes the installation of linear heat detection for uses other than open area protection. Linear heat detection can be used for special application purposes where the installation is consistent with the manufacturer's instructions. Where linear heat detection is used in this way, there is no ceiling jet moving the hot combustion product gases horizontally to the detector location. Consequently, the spacing criteria provided in the code does not apply in these special applications.

2-2.3* Temperature.

Detectors having fixed-temperature or rate-compensated elements shall be selected in accordance with Table 2-2.1.1.1 for the maximum expected ambient ceiling temperature that can be expected.

The temperature rating of the detector shall be at least 20°F (11°C) above the maximum expected temperature at the ceiling.

A-2-2.3 Detectors should be selected to minimize this temperature difference in order to minimize response time. However, a heat detector with a temperature rating that is somewhat in excess of the highest normally expected ambient temperature is specified in order to avoid the possibility of premature operation of the heat detector to non-fire conditions.

2-2.4* Spacing.

A-2-2.4 In addition to the special requirements for heat detectors that are installed on ceilings with exposed joists, reduced spacing also might be required due to other structural characteristics of the protected area, such as possible drafts or other conditions that could affect detector operation.

2-2.4.1* Smooth Ceiling Spacing.

A-2-2.4.1 Maximum linear spacings on smooth ceilings for spot-type heat detectors are determined by full-scale fire tests. *[See Figure A-2-2.4.1(c).]* These tests assume that the detectors are to be installed in a pattern of one or more squares, each side of which equals the maximum

spacing as determined in the test, as illustrated in Figure A-2-2.4.1(a). The detector to be tested is placed at a corner of the square so that it is positioned at the farthest possible distance from the fire while remaining within the square. Thus, the distance from the detector to the fire is always the test spacing multiplied by 0.7 and can be calculated as shown in Table A-2-2.4.1.

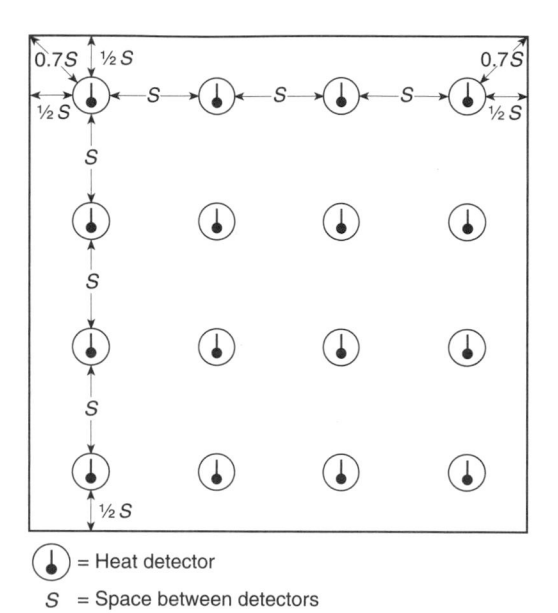

Figure A-2-2.4.1(a) *Spot-type heat detectors.*

Table A-2-2.4.1 Test Spacing for Spot-Type Heat Detectors

Test Spacing		Maximum Test Distance from Fire to Detector (0.7 × D)	
ft	m	ft	m
50 × 50	15.24 × 15.24	35.0	10.67
40 × 40	12.19 × 12.19	28.0	8.53
30 × 30	9.10 × 9.10	21.0	6.40
25 × 25	7.62 × 7.62	17.5	5.33
20 × 20	6.10 × 6.10	14.0	4.27
15 × 15	4.57 × 4.57	10.5	3.20

Once the correct maximum test distance has been determined, it is valid to interchange the positions of the fire and the detector. The detector is now in the middle of the square, and the listing specifies that the detector is adequate to detect a fire that occurs anywhere within that square—even out to the farthest corner.

In laying out detector installations, designers work in terms of rectangles, as building areas are generally rectangular in shape. The pattern of heat spread from a fire source, however, is not rectangular in shape. On a smooth ceiling, heat spreads out in all directions in an ever-expanding circle. Thus, the coverage of a detector is not, in fact, a square, but rather a circle whose radius is the linear spacing multiplied by 0.7. [See also Figure A-2-2.4.1(b).]

This is graphically illustrated in Figure A-2-2.4.1(d). With the detector at the center, by rotating the square, an infinite number of squares can be laid out, the corners of which create the plot of a circle whose radius is 0.7 times the listed spacing. The detector will cover any of these squares and, consequently, any point within the confines of the circle.

So far this explanation has considered squares and circles. In practical applications, very few areas turn out to be exactly square, and circular areas are extremely rare. Designers deal generally with rectangles of odd dimensions and corners of rooms or areas formed by wall intercepts, where spacing to one wall is less than one-half the listed spacing. To simplify the rest of this explanation, the use of a detector with a listed spacing of 30 ft × 30 ft (9.1 m × 9.1 m) should be considered. The principles derived are equally applicable to other types.

Figure A-2-2.4.1(f) illustrates the derivation of this concept. In Figure A-2-2.4.1(f), a detector is placed in the center of a circle with a radius of 21 ft (0.7 × 30 ft) [6.4 m (0.7 × 9.1 m)]. A series of rectangles with one dimension less than the permitted maximum of 30 ft (9.1 m) is constructed within the circle. The following conclusions can be drawn.

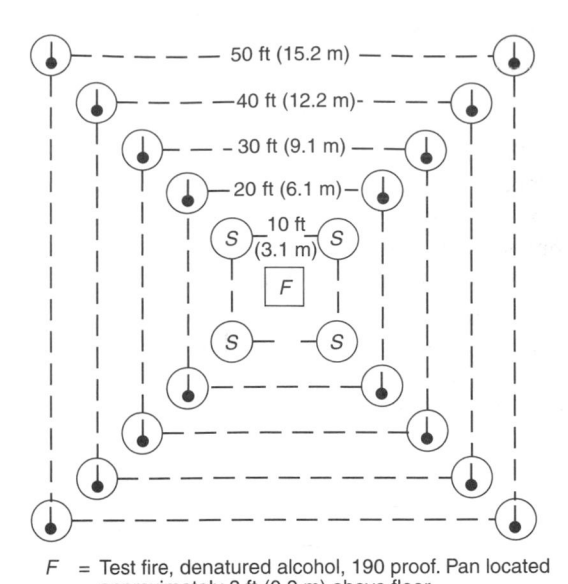

F = Test fire, denatured alcohol, 190 proof. Pan located approximately 3 ft (0.9 m) above floor.

S = Indicates normal sprinkler spacings on 10-ft (3.1-m) schedules.

= Indicates normal heat detector spacing on various spacing schedules.

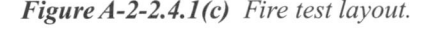

Figure A-2-2.4.1(c) *Fire test layout.*

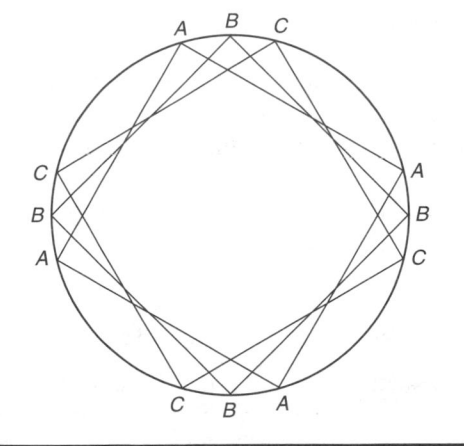

Figure A-2-2.4.1(d) *Detector covering any square laid out in the confines of a circle in which the radius is 0.7 times the listed spacing.*

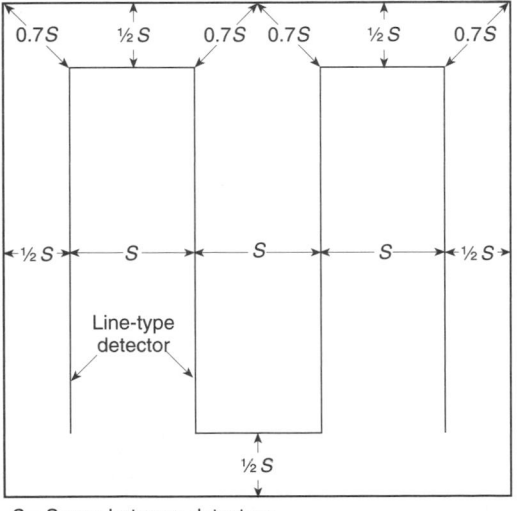

S = Space between detectors

Figure A-2-2.4.1(b) *Line-type detectors—spacing layouts, smooth ceiling.*

(a) As the smaller dimension decreases, the longer dimension can be increased beyond the linear maximum spacing of the detector with no loss in detection efficiency.

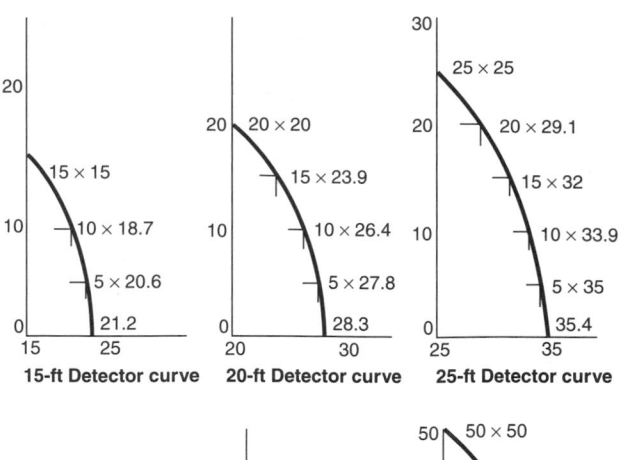

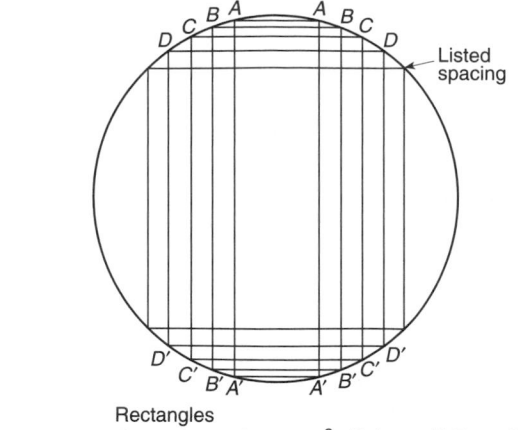

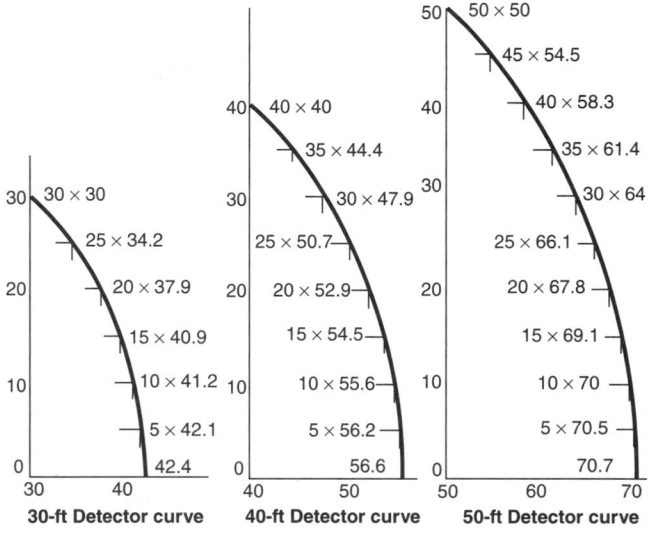

Figure A-2-2.4.1(e) *Typical rectangles for detector curves of 15 ft to 50 ft (4.57 m to 15.24 m).*

Rectangles
A = 10 ft × 41 ft = 410 ft² (3.1 m × 12.5 m = 38.1 m²)
B = 15 ft × 39 ft = 585 ft² (4.6 m × 11.9 m = 54.3 m²)
C = 20 ft × 37 ft = 740 ft² (6.1 m × 11.3 m = 68.8 m²)
D = 25 ft × 34 ft = 850 ft² (7.6 m × 10.4 m = 78.9 m²)
Listed spacing = 30 ft × 30 ft = 900 ft² (9.1 m × 9.1 m = 83.6 m²)

Figure A-2-2.4.1(f) *Detector spacing, rectangular areas.*

dimensions that fit most appropriately. For example, refer to Figure A-2-2.4.1.2. A corridor 10 ft (3 m) wide and up to 82 ft (25 m) long can be covered with two 30-ft (9.1-m) spot-type detectors. An area 40 ft (12.2 m) wide and up to 74 ft (22.6 m) long can be covered with four spot-type detectors. Irregular areas need more careful planning to make certain that no spot on the ceiling is more than 21 ft (6.4 m) away from a detector. These points can be determined by striking arcs from the remote corner. Where any part of the area lies beyond the circle with a radius of 0.7 times the listed spacings, additional detectors are required.

The spacing criteria established by 2-2.4.1 determine how many detectors of a given type are necessary to provide heat detection for a compartment of a given area. The design must reduce the listed spacing to compensate for the impact that variations in the specific compartment can have on the temperature and velocity of the ceiling jet. These spacing reductions will compensate for environmental impacts on the detector performance and provide response roughly equivalent to that attainable from the same detectors when installed on smooth level ceilings using the listed spacing.

The number of detectors required is a function of the spacing factor, S, of the chosen detector. The spacing is established through a series of fire tests conducted in the listing evaluation by the qualified testing laboratory listing the detector. The spacing indicates the relative sensitivity of the detector.

The spacing derived from the fire tests relates heat detectors to the response of a specially chosen 160°F

(b) A single detector covers any area that fits within the circle. For a rectangle, a single, properly located detector may be permitted, provided the diagonal of the rectangle does not exceed the diameter of the circle.

(c) Relative detector efficiency actually is increased, because the area coverage in square feet is always less than the 900 ft² (83.6 m²) permitted if the full 30 ft × 30 ft (9.1 m × 9.1 m) square were to be utilized. The principle illustrated here allows equal linear spacing between the detector and the fire, with no recognition for the effect of reflection from walls or partitions, which in narrow rooms or corridors is of additional benefit. For detectors that are not centered, the longer dimension should always be used in laying out the radius of coverage.

Areas so large that they exceed the rectangular dimensions given in Figure A-2-2.4.1(f) require additional detectors. Often proper placement of detectors can be facilitated by breaking down the area into multiple rectangles of the

(71.1°C) automatic sprinkler head. The fire test room has a ceiling height of 15 ft, 9 in. (4.8 mm) above the floor and has no airflow. The test fire is situated at the center of a square array of the test sprinkler heads, installed on 10 ft × 10 ft (3 m × 3 m) centers. This places the centerline of the test fire 7.07 ft (2.2 m) from the test sprinklers.

Heat detectors are mounted in a square array that is centered about the test fire with increased spacings. The fire is located approximately 3.0 ft (0.9 m) above the floor and consists of a number of pans of an ethanol/methanol mixture yielding an output of approximately 1200 kW. The height of the test fire and the fire area are adjusted to produce a time versus temperature curve at the test sprinklers that falls within the envelope established for the test and causes the activation of the test sprinkler at 2 minutes ± 10 seconds. The greatest detector spacing that produces an alarm signal before a test sprinkler actuates is the listed spacing for the heat detector.

Heat detector performance is defined relative to the distance at which it could detect the same fire that fused the test sprinkler head in 2 minutes ± 10 seconds. For example, a heat detector installed on a 50 ft by 50 ft (15.2 m by 15.2 m) array receives a 50 ft (15.2 m) listed spacing if it responds to the test fire just before the test sprinkler head operates.

It is important to keep in mind that the listed spacing for a heat detector is a lumped parameter, a number of variables, including fire size, fire growth rate, ambient temperature, ceiling height, and thermal response coefficient are lumped into a single parameter called *listed spacing*. The listed spacing is sufficiently accurate to compare two heat detectors to each other, but it cannot be used to predict when a given detector will respond, except in the context of the fire test. Outside the context of the listing test, the listed spacing is only a relative indication of the detector thermal response.

When a quantitative prediction of detector performance is needed for either analysis or the basis of a design, the alternative design method in Appendix B should be used. (See 2-2.4.6.) Because Appendix B is not part of the body of the code, it is generally considered an optional alternative method. The designer should obtain the approval or acceptance of the authority having jurisdiction prior to using Appendix B for the design of a required system.

The number of detectors necessary for a given application also depends on the ceiling height, the type of ceiling (whether it has exposed joists or beams), and other features that may affect the flow of air or the accumulation of heat from a fire. All of these factors enter into the spacing design rules that are provided in Section 2-2.4.5.

2-2.4.1.1 One of the following requirements shall apply:

(1) The distance between detectors shall not exceed their listed spacing, and there shall be detectors within a distance of one-half the listed spacing, measured at a right angle, from all walls or partitions extending to within 18 in. (460 mm) of the ceiling.

(2) All points on the ceiling shall have a detector within a distance equal to 0.7 times the listed spacing (0.7S). This is useful in calculating locations in corridors or irregular areas.

2-2.4.1.2* Irregular Areas. For irregularly shaped areas, the spacing between detectors shall be permitted to be greater than the listed spacing, provided the maximum spacing from a detector to the farthest point of a sidewall or corner within its zone of protection is not greater than 0.7 times the listed spacing.

A-2-2.4.1.2 Figure A-2-2.4.1.2 illustrates smoke or heat detector spacing layouts in irregular areas.

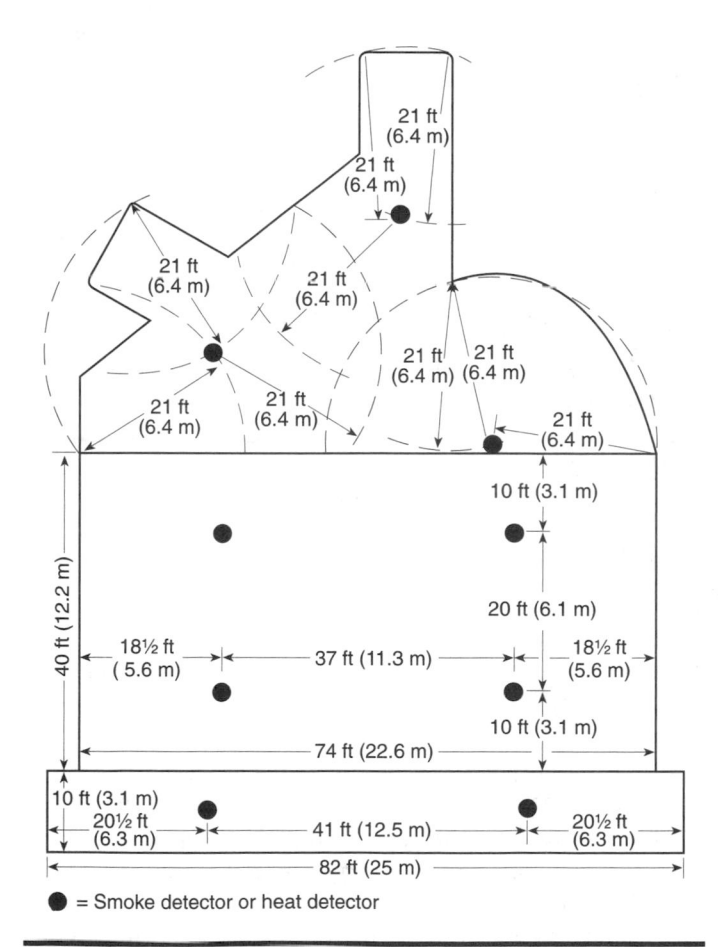

● = Smoke detector or heat detector

Figure A-2-2.4.1.2 Smoke or heat detector spacing layout in irregular areas.

The Formal Interpretation that follows provides further clarification of the requirements of irregular areas.

2-2.4.2 Solid Joist Construction. The spacing of heat detectors, where measured at right angles to the solid joists, shall not exceed 50 percent of the smooth ceiling spacing permitted under 2-2.4.1.1 and 2-2.4.1.2.

Subsection 2-2.2 establishes the requirement to locate heat detectors on the bottom of joists. In paragraph 2-2.4.2, the effect of joists on detector spacing is established. The hot combustion product gases and smoke from a fire rise vertically in a plume until the plume impinges on the ceiling. There, the hot combustion product gases and entrained air of the fire plume change direction and move horizontally across the ceiling, becoming a ceiling jet. Where the joists are running parallel to the direction of travel of the ceiling jet, they have little effect on the speed with which the hot gases of the ceiling jet move across the ceiling. However, where the joists are perpendicular to the direction of gas flow from the fire to the detector, they produce turbulence and thus reduce the ceiling jet velocity. Thus, a closer spacing for heat detectors in the direction perpendicular to the joists is necessary to attain uniform performance.

It is important to remember that joists are solid members extending more than 4 in. (100 mm) down from the ceiling and are installed on centers of less than 3 ft (0.9 m). If the solid members extending down from the ceiling are on 3 ft (0.9 m) centers or larger, they are beams. Also, bar-joists have no effect on spacing unless the top cord is greater than 4 in. (100 mm).

2-2.4.3* Beam Construction. A ceiling shall be treated as a smooth ceiling if the beams project no more than 4 in.

(100 mm) below the ceiling. If the beams project more than 4 in. (100 mm) below the ceiling, the spacing of spot-type heat detectors at right angles to the direction of beam travel shall be not more than two-thirds of the smooth ceiling spacing permitted under 2-2.4.1.1 and 2-2.4.1.2. If the beams project more than 18 in. (460 mm) below the ceiling and are more than 8 ft (2.4 m) on center, each bay formed by the beams shall be treated as a separate area.

Beams create barriers to the horizontal flow of the ceiling jet; they project more than 4 in. (10.2 cm) from the ceiling and are on center-to-center spacing greater than 3 ft (0.9 m). The bay created by the beams and the walls at either end, or the cross beams extending from beam to beam, fill up with smoke and hot combustion product gases before spilling into the next bay. This fill and spill progression of the ceiling jet is slower than the velocity attained on a smooth flat ceiling.

Because the rate of heat transfer from the ceiling jet gases to the detector is proportional to the velocity of the ceiling jet flow, slower flow results in slower detector response. Consequently, for the design to attain consistent performance, the detector spacing in the direction perpendicular to the beams must be reduced to compensate for this reduced ceiling jet velocity and reduced speed of response.

Open web beams and trusses have little effect on the passage of air currents that are caused by fire. Generally, they are not considered in determining the proper spacing of detectors unless the solid part of the top cord extends more than 4 in. (100 mm) down from the ceiling. See Exhibit 2.21 for an example of beam effects on detection.

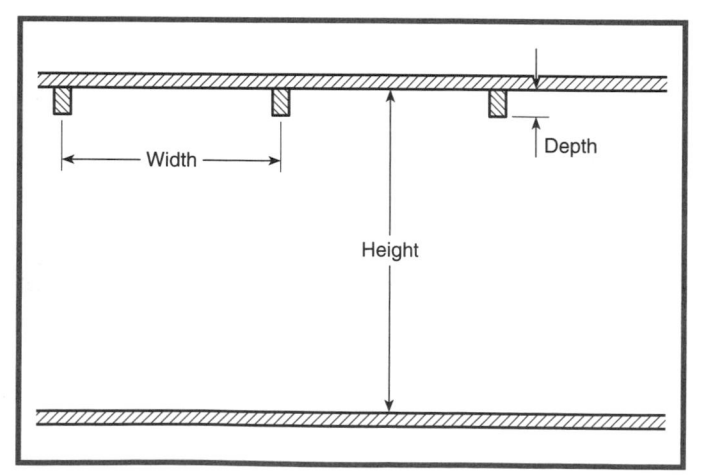

Exhibit 2.21 *The depth and center-to-center spacing of beams relative to the ceiling height affects the preferred location for heat detectors per A-2-2.4.3. (Source: J.M. Cholin Consultants, Inc., Oakland, NJ)*

A-2-2.4.3 The location and spacing of heat detectors should consider beam depth, ceiling height, beam spacing, and fire size.

If the ratio of beam depth *(D)* to ceiling height *(H)*, *(D/H)*, is greater than 0.10 and the ratio of beam spacing *(W)* to ceiling height *(H)*, *(W/H)*, is greater than 0.40, heat detectors should be located in each beam pocket.

If either the ratio of beam depth to ceiling height *(D/H)* is less than 0.10 or the ratio of beam spacing to ceiling height *(W/H)* is less than 0.40, heat detectors should be installed on the bottom of the beams.

The criteria included in A-2-2.4.3 make some tacit assumptions regarding the thickness of the ceiling jet under varied conditions. In general [Alpert, 1972; Heskestad and Delichatsios, 1989], research has shown that, to a first order approximation, the ceiling jet can be thought of as occupying the upper 10 percent of the compartment volume. If the downward extension of the beams is less than 10 percent of the ceiling height, then the impact of the beams on the flow of the hot combustion product gases in the ceiling jet will be lessened, since a significant portion of the ceiling jet will pass beneath the beams.

Research [Heskestad, 1975; Morton, Taylor, and Turner, 1956; Schifiliti, 1986] also shows that as the plume rises from the fire, it expands. Generally, a first order approximation of the plume diameter at the ceiling is 0.4 times the ceiling height. Therefore, when relatively narrow center-to-center beam spacing is encountered with a width-to-height ratio of less than 0.4, the plume will be wider than the bay formed by the beams and purlins in at least one direction. Consequently, more than one bay will be filling from the plume and these bays will fill very rapidly, making the fill part of the fill and spill propagation of the ceiling jet, a relatively short delay in time. Where the beam depths are relatively large, or the bay volumes that are proportional to beam center-to-center spacing are large, the fill delay is significant and detectors must be located in each bay. See Exhibit 2.21 for an example of beam effects on detection.

2-2.4.4 Sloping Ceilings.

When the fire plume impinges on a sloped ceiling, the development of the ceiling jet is affected by the slope of the ceiling. The plume is buoyant and is flowing upward due to the force of cooler and denser ambient air. Because it takes less energy to turn the flow of combustion product gases and entrained air less than 90 degrees, the ceiling jet moves more rapidly up a sloped ceiling and more slowly across the slope than it would across a level ceiling.

When the ceiling jet reaches the peak of the roof, its flow stops. This affects the placement and spacing of heat detectors. At the peak, the ceiling jet collides with a mass of the hottest air that normally exists beneath the ceiling. The ceiling jet will not penetrate this air mass unless and until the plume is hotter than the air at the peak of the roof. Furthermore, because the jet cannot continue flowing with-

out moving down the opposite side of the roof, the ceiling jet velocity decreases almost to zero. Because the rate of heat flow into the heat detector sensing element is a function of ceiling jet velocity, where the jet velocity decreases so does the flow of heat into the detector. This creates thermal lag. When the ceiling jet stops, the thermal lag increases significantly. The spacing adjustment rules provided in Section 2-2.4 derive from past experience and the study of ceiling jet flows under these ceilings.

2-2.4.4.1* Peaked. A row of detectors shall first be spaced and located at or within 3 ft (0.9 m) of the peak of the ceiling, measured horizontally. The number and spacing of additional detectors, if any, shall be based on the horizontal projection of the ceiling in accordance with the type of ceiling construction.

A-2-2.4.4.1 Figure A-2-2.4.4.1 illustrates smoke or heat detector spacing for peaked-type sloped ceilings.

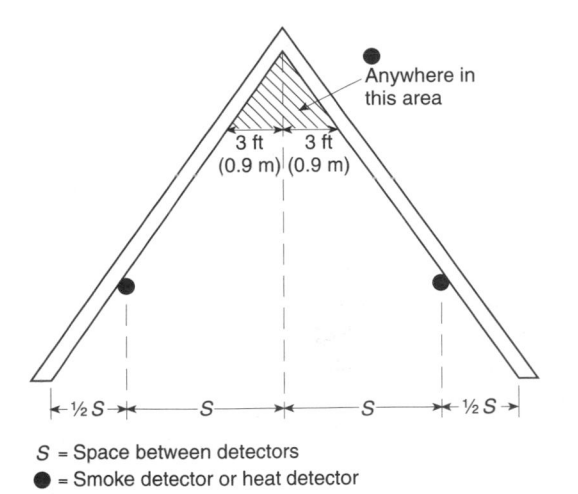

S = Space between detectors
● = Smoke detector or heat detector

Figure A-2-2.4.4.1 *Smoke or heat detector spacing layout, sloped ceilings (peaked type).*

2-2.4.4.2* Shed. Sloping ceilings having a rise greater than 1 ft in 8 ft (1 m in 8 m) shall have a row of detectors located on the ceiling within 3 ft (0.9 m) of the high side of the ceiling measured horizontally, spaced in accordance with the type of ceiling construction. The remaining detectors, if any, shall be located in the remaining area on the basis of the horizontal projection of the ceiling.

A-2-2.4.4.2 Figure A-2-2.4.4.2 illustrates smoke or heat detector spacing for shed-type sloped ceilings.

2-2.4.4.3 For a roof slope of less than 30 degrees, all detectors shall be spaced using the height at the peak. For a roof

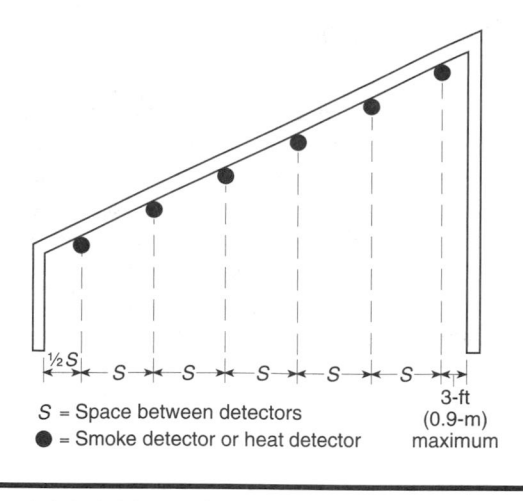

Figure A-2-2.4.4.2 Smoke or heat detector spacing layout, sloped ceilings (shed type).

Table 2-2.4.5.1 Heat Detector Spacing Reduction Based on Ceiling Height

Ceiling Height Above		Up to and Including		Multiply Listed Spacing by
ft	m	ft	m	
0	0	10	3.05	1.00
10	3.05	12	3.66	0.91
12	3.66	14	4.27	0.84
14	4.27	16	4.88	0.77
16	4.88	18	5.49	0.71
18	5.49	20	6.10	0.64
20	6.10	22	6.71	0.58
22	6.71	24	7.32	0.52
24	7.32	26	7.93	0.46
26	7.93	28	8.54	0.40
28	8.54	30	9.14	0.34

slope of greater than 30 degrees, the average slope height shall be used for all detectors other than those located in the peak.

2-2.4.5 High Ceilings.

The computational method of detection system design presented in Appendix B is based on first principles of physics. It serves as an alternative design method to the prescriptive criteria in 2-2.4.5 of the code. Designers who elect to use Appendix B should involve the authorities having jurisdiction. The design method in Appendix B can lead to detector spacings that exceed the listed spacing of the detector. This is to be expected when a designer uses the methods of Appendix B; the performance that characterizes a particular fire will likely be quite different than the characteristics of the test fire used to evaluate the detector for listing.

2-2.4.5.1* On ceilings 10 ft to 30 ft (3 m to 9.1 m) high, heat detector linear spacing shall be reduced in accordance with Table 2-2.4.5.1, prior to any additional reductions for beams, joists, or slope, where applicable.

Exception 1: Table 2-2.4.5.1 shall not apply to the following detectors, which rely on the integration effect:

   (a) Line-type electrical conductivity detectors (see A-1-4)

   (b) Pneumatic rate-of-rise tubing (see A-1-4)

In these cases, the manufacturer's recommendations shall be followed for appropriate alarm point and spacing.

A-2-2.4.5.1 Both 2-2.4.5.1 and Table 2-2.4.5.1 are constructed to provide detector performance on higher ceil-

ings [to 30 ft (9.1 m) high] that is essentially equivalent to that which would exist with detectors on a 10-ft (3-m) ceiling.

The Fire Detection Institute Fire Test Report *(refer to Appendix C)* is used as a basis for Table 2-2.4.5.1. The report does not include data on integration-type detectors. Pending development of such data, the manufacturer's recommendations will provide guidance. *(Refer to Figure A-2-2.4.5.1.)*

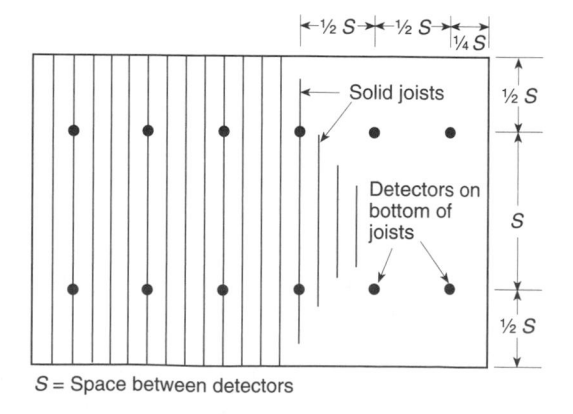

Figure A-2-2.4.5.1 Detector spacing layout, solid joist construction.

Table 2-2.4.5.1 provides for spacing modification to take into account different ceiling heights for generalized fire conditions. Information regarding a design method that

allows the designer to take into account ceiling height, fire size, and ambient temperatures is provided in Appendix B.

The spacing factor for a given detector is a rough measure of how far the ceiling jet can travel from the test fire (used in the listing evaluation) before the jet has cooled and slowed down too much to provide reliable detection in the required time period. As hot combustion product gases in the fire plume rise from the fire, the gases expand, giving off energy and cooling down. Furthermore, cool air becomes entrained into the plume cooling it more. This leaves less energy available to continue accelerating the plume toward the ceiling. Once the ceiling plane has been reached, the remaining plume momentum is the only force available to accelerate the ceiling jet horizontally across the ceiling.

Increasing the ceiling height has a very significant effect on the ceiling jet temperature and velocity. The reduction of spacing with increased ceiling height places detectors closer to the fire plume centerline, thus allowing the hot combustion product gas, air, and radiated heat to travel a shorter distance before encountering a detector.

The inverse square law predicts that when the distance between the fire and the detector is doubled, the amount of radiated heat reaching the detector will be reduced by a factor of 4. To the extent that there may be some contribution from radiant heat transfer, this also increases the need to reduce the spacing as the ceiling height is increased.

It is important to note that Table 2-2.4.5.1 covers ceiling heights up to 30 ft (9.1 m). This is the highest ceiling for which the Technical Committee on Initiating Devices for Fire Alarm Systems had test data. (See references in Appendix B.) Where ceilings are higher than 30 ft (9.1 m), the designer must act with the knowledge that those conditions are beyond the limits of the testing that provided the basis for the requirements of the code. The temptation is to extrapolate for higher ceiling heights. However, a theoretical basis for doing so has yet to be reviewed by the Technical Committee on Initiating Devices for Fire Alarm Systems. This is currently an area of considerable research that eventually may yield important new insights.

The code neither prohibits nor permits the use of heat detectors on ceilings higher than 30 ft (9.1 m). Computer models such as *FPETool* and *Hazard 1* have been used to predict detector performance at higher ceiling heights. Some studies have confirmed the predictions derived from these models. However, it should be understood that in the context of an exponentially growing fire, a much larger fire will be necessary to actuate the detectors on higher ceilings. The fire allowed by the delayed detection may be considerably larger than that normally assumed. In some cases, this will mean that the detection system will not meet the protection goals of the owner or the intent of the code. The final decision as to whether or not a design is acceptable rests with the authority having jurisdiction.

2-2.4.5.2* The minimum spacing of heat detectors shall not be required to be less than 0.4 times the height of the ceiling.

Research also shows that as the plume rises from the fire it expands. Generally, a first order approximation of the plume diameter at the ceiling is 0.4 times the ceiling height. Consequently, no performance advantage is accrued when heat detectors are installed with spacings smaller than 0.4 times the ceiling height. In some cases, a design will not have an equivalent level of performance; for example, when a design uses an installed spacing smaller than 0.4 times the ceiling height, derived by modifying the listed spacing using Table 2-2.4.5.1. When the installed spacing for detectors is based on 2-2.4.5.2, equivalent performance to normal ceiling heights might not be achieved.

A-2-2.4.5.2 The width of uniform temperature of the plume when it impinges on the ceiling is approximately 0.4 times the height above the fire, so reducing spacing below this level will not increase response time. For example, a detector with a listed spacing of 15 ft (4.6 m) or 225 ft^2 (21 m^2) need not be spaced closer than 12 ft (3.7 m) on a 30-ft (9.1-m) ceiling, even though Table 2-2.4.5.1 states that the spacing should be 0.34 × 15, which equals 5.1 ft (1.6 m).

2-2.4.6 Appendix B shall be permitted to be used as one alternative design method for determining detector spacing.

2-3 Smoke-Sensing Fire Detectors

2-3.1 General.

Definitions and descriptions of the mode of operation of each type of smoke detector can be found in Section 1-4.

2-3.1.1* The purpose of Section 2-3 shall be to provide information to assist in design and installation of reliable early warning smoke detection systems for protection of life and property.

A-2-3.1.1 The addition of a heat detector to a smoke detector does not enhance its performance as an early warning device.

2-3.1.2 Section 2-3 shall cover general area application of smoke detectors in ordinary indoor locations.

Subsection 2-3.1.2 limits the applicability of the requirements and recommendations of Section 2-3 to "general area application. . . in ordinary indoor locations." The authority having jurisdiction must decide whether or not a hazard area falls into this category. The code assumes that users understand what is meant by "ordinary indoor locations."

Some authorities having jurisdiction establish additional requirements for specific types of occupancies. These requirements go above and beyond the requirements of Sec-

tion 2-3. To protect extremely valuable assets, the designer may also choose closer spacing in certain areas, such as in a data center or where the owner's fire protection goals demand closer spacing of smoke detectors.

Finally, the common interpretation of 2-3.1.2 usually does not include special compartments, such as switchgear enclosures or aircraft lavatories. While a particular design may use detectors in these and similar locations, the designer should use engineering judgment.

2-3.1.3 For information on use of smoke detectors for control of smoke spread, the requirements of Section 2-10 shall apply.

Early in the second half of this century, there were several fires in high-rise buildings that demonstrated the futility of trying to evacuate all of the occupants. Designers developed a new strategy of protecting in place or using areas of refuge. Concurrently, it became well known that smoke inhalation was the principal cause of death associated with fires. To protect occupants in place, the heating, ventilating, and air conditioning (HVAC) system or an engineered smoke control system automatically controls the flow of smoke. Designers typically employ smoke detectors to actuate such systems. Section 2-10 covers the use of smoke detectors for that purpose.

2-3.1.4 For additional guidance in the application of smoke detectors for flaming fires of various sizes and growth rates in areas of various ceiling heights, refer to Appendix B.

Traditionally, designers use smoke detectors to provide early warning. Appendix B gives two methods for predicting the actuation of smoke detectors for fires that produce a buoyant plume, such as flaming fires. This limitation on the applicability of Appendix B for smoke detector design comes from the fact that both methods use conservation of momentum and energy relationships to infer temperature at a given location relative to the fire centerline. The methodology then uses correlations to infer the probable optical density or mass density at the detector locations. Since both methods assume a buoyant plume and a ceiling jet as the mechanism of smoke transfer, the validity of these methods is limited to a flaming fire.

Popular computer models such as *FPETool* and *FastLite* model the smoke detector as a very sensitive (RTI = 1) heat detector. The models deem that a smoke detector will actuate when a temperature rise of 13°C (20°F) occurs at the detector. This simplified assumption of temperature to smoke density correlation, introduced in early research [Schifiliti, 1986] has persisted since the 1980s. Today, many researchers consider this an extremely conservative estimate.

In some applications, the extension of the normal 30 ft (9.1 m) spacing for smoke detectors for flaming fire detection permitted by Appendix B seems appropriate. However,

designers should not confuse this spacing allowance with the 30 ft (9.1 m) spacing normally used in life safety or early warning applications.

Testing performed under the auspices of the Fire Detection Institute [Heskestad and Delichatsios, 1986 and 1995; Heskestad and Delichatsios, 1989] have provided the basis for the new method for predicting detector response. This testing gave rise to the computational procedure outlined in Appendix B for smoke detectors applied to detect flaming fires. This procedure provides a more analytical and precise method of determining detector spacing.

2-3.2* Smoke detectors shall be installed in all areas where required by applicable laws, codes, or standards.

Subsection 2-3.2 has been included to correlate with national and state adopted codes that reference the code. NFPA 72 does not stipulate where, or in which occupancies, detection shall be installed. Rather, it establishes how detection shall be designed and installed once an applicable law, code, or standard has established the requirement for detection in the occupancy in question.

A-2-3.2 The person designing an installation should keep in mind that in order for a smoke detector to respond the smoke has to travel from the point of origin to the detector. In evaluating any particular building or location, likely fire locations should be determined first. From each of these points of origin, paths of smoke travel should be determined. Wherever practicable, actual field tests should be conducted. The most desired locations for smoke detectors are the common points of intersection of smoke travel from fire locations throughout the building.

> Note: This is one of the reasons that specific spacing is not assigned to smoke detectors by the testing laboratories.

The designer must understand the behavior of the fire plume and ceiling jet in order to understand how the building structure can affect the flow of smoke through the compartment and from one compartment to others. The site evaluation includes an audit of all combustibles within the compartment, as well as all ignition sources, including transient ones [Babrauskas, Lawson, Walton, and Twilley, 1982; Heskestad and Delichatsios, 1986 and 1995; *CFR* 47, 1934]. The designer models the fires to obtain an estimate of the rate of fire growth for each combustible and ignition source scenario. The designer then compares fire scenarios to the performance objectives for the compartment, this procedure leads to a basis for design.

Without sound performance metrics and validated modeling methods for smoke detectors, the selection of detector locations often becomes more of an art than a science. Most manufacturers recommend a spacing of 30 ft (9.1 m) on center. This recommendation derives from the size of the

room in which the testing laboratories perform full-scale fire tests. These tests supply listing criteria, but do not provide a truly predictive performance model.

2-3.3* Sensitivity.

A-2-3.3 Throughout this code, smoke detector sensitivity is referred to in terms of the percent obscuration required to alarm or produce a signal. Smoke detectors are tested using various smoke sources that have different characteristics (for example, color, particle size, number of particles, particle shape). Unless otherwise specified, this code, the manufacturers, and the listing agencies report and use the percent obscuration produced using a specific type of gray smoke. Actual detector response will vary when the characteristics of the smoke reaching the detector is different from the smoke used in testing and reporting detector sensitivity.

Qualified testing laboratories base the listings of smoke detectors on repeatable laboratory tests. These tests do not correlate to actual applications and do not provide a listed spacing per se. Rather, each manufacturer will recommend a spacing under ordinary conditions.

The tests conducted in the course of a listing investigation include a sensitivity measurement in a "smoke box." The manufacturer marks the detector with its sensitivity based on the listing investigation. However, the sensitivity measurements obtained from the smoke relate only to the smoke box. They are not intended to predict performance in any other context. Consequently, a marking of a nominal smoke obscuration of 1 percent to 4 percent obscuration per foot does not necessarily mean that an installed detector will respond at that level. Therefore, a designer should not base a design on this marked sensitivity.

Full-scale room fire tests are also conducted using smoke detectors. The detectors must render alarms when subjected to fires that ultimately produce smoke obscurations of 37 percent per foot for the paper fire, 17 percent per foot for the wood fire, 21 percent per foot for the heptane/toluene fire, and 10 percent per foot for a smoldering wood fire. These pass/fail tests do not provide a meaningful basis for predicting smoke detector performance either. Consequently, the designer can only rely on the manufacturers recommended spacing and the qualitative requirements provided in Section 2-3 of the code.

In most fires, smoke detectors respond much sooner than either automatic sprinklers or heat detectors. In flaming fire tests, smoke detectors actuate long before typical heat detectors. The difference in the speed of response becomes even more dramatic with low-energy smoldering fires. Because of this profound difference in the speed of response, adding a heat detector to a smoke detector adds little to overall fire detection performance; particularly when the design criteria call for early warning.

2-3.3.1 Smoke detectors shall be marked with their nominal production sensitivity (percent per foot obscuration), as required by the listing. The production tolerance around the nominal sensitivity also shall be indicated.

Because the mission of most smoke detection systems is the protection of human life, the response of a smoke detector is usually defined in human terms. The percent per foot obscuration method of measuring sensitivity relates to a person's ability to see well enough to escape from a fire. Smoke is composed of both visible and invisible particulate matter. While the portion of the smoke that is invisible has little immediate impact on an individual's ability to escape, it can constitute the majority of the smoke mass under some fire conditions.

The tests conducted during a listing investigation include a sensitivity measurement in a smoke box using a cotton lamp wick that produces a gray smoke. However, the sensitivity measurements obtained from the smoke box and marked on the detector only relate to the smoke box. A marking of a nominal smoke obscuration of 0.6 percent to 4.0 percent per ft does not necessarily mean that the detector will respond at that level because other fuels may not produce the same type or quantity of smoke.

Qualified testing laboratories also conduct full-scale room fire tests. The detectors must render alarms when subjected to fires that ultimately produce smoke obscurations of 37 percent per ft for the paper fire, 17 percent per ft for the wood fire, 21 percent per ft for the heptane/toluene fire, and 10 percent per ft for a smoldering wood fire. These pass/fail tests do not provide a meaningful basis for predicting smoke detector performance either. Consequently, the designer can only rely on the manufacturer's recommended spacing and the qualitative requirements provided in this section of the code.

2-3.3.2 Smoke detectors that have provision for field adjustment of sensitivity shall have an adjustment range of not less than 0.6 percent per foot obscuration. If the means of adjustment is on the detector, a method shall be provided to restore the detector to its factory calibration. Detectors that have provision for program-controlled adjustment of sensitivity shall be permitted to be marked with their programmable sensitivity range only.

The adjustment of detector sensitivity over a range of less than 0.6 percent per ft has little, if any, practical benefit. Even when smoke detectors are used for property protection (as in data centers), the difference in response represented by an adjustment range of less than 0.6 percent per ft is minor.

Some smoke detectors have a feature allowing the adjustment of detector sensitivity to accommodate the immediate ambient conditions in the area of the detector. Other smoke detectors send a voltage or current value back

to the control unit that is proportional to the smoke-sensing signal in the detector. In such a case, the trip point of the detector is a voltage or current level stored in the control unit memory. In either case, there may be occasion to adjust the detector sensitivity, either at the detector or at the control unit. The system should provide a means to restore the detector to its factory sensitivity. The manufacturer must mark the detector to show the sensitivity range. In some cases, maintenance personnel may use the adjustment feature between cleaning intervals to maintain stability. Once they clean the detector, they should restore it to its original design sensitivity. Chapter 7 covers the maintenance of smoke detectors.

2-3.4 Location and Spacing.

2-3.4.1* General.

A-2-3.4.1 For operation, all types of smoke detectors depend on smoke entering the sensing chamber or light beam. If sufficient concentration is present, operation is obtained. Since the detectors are usually mounted on the ceiling, response time depends on the nature of the fire. A hot fire rapidly drives the smoke up to the ceiling. A smoldering fire, such as in a sofa, produces little heat; therefore, the time for smoke to reach the detector is increased.

2-3.4.1.1 The location and spacing of smoke detectors shall result from an evaluation based on the guidelines detailed in this code and on engineering judgment. Some of the conditions that shall be included in the evaluation are the following:

(1) Ceiling shape and surface
(2) Ceiling height
(3) Configuration of contents in the area to be protected
(4) Burning characteristics of the combustible materials present
(5) Ventilation
(6) Ambient environment

These general criteria are far less specific than those established for heat detectors. The reason for this can be understood by reviewing the importance of fire plume dynamics in the location and spacing of heat detectors versus smoke detectors.

Heat detectors depend on the fire plume and ceiling jet to carry hot combustion product gases and entrained air to the detector where heat can flow from the ceiling jet into the detector, resulting in an alarm. The fire liberates significant quantities of energy. This serves as the engine that creates its own air currents. The energy from the fire propels the hot air and smoke mixture across the ceiling. Under these circumstances, it is possible to model the flow of the fire plume and ceiling jet with computer programs that apply the rules of fluid flow physics and thermodynamics.

The Fire Detection Institute has served as an important guiding force in the development of these models. The models predict that smoke detectors provide response significantly before heat detectors. This prediction has been verified experimentally.

However, under smoldering, low-energy-output fire conditions, the fire does not achieve an energy output (heat release rate) sufficient to serve as the primary source of propulsion for the smoke. Existing air currents through the hazard area dominate the flow of smoke, with little, if any, contribution from the fire. This makes the prediction of flow far more dependent on site-specific airflow variables, meaning that prediction of detection becomes much more difficult.

Consequently, the location and spacing of smoke detectors must be determined subject to the judgment of the designer on how the site-specific environment will affect the flow of smoke from these early-stage, low-energy-output fires.

2-3.4.1.2 If the intent is to protect against a specific hazard, the detector(s) shall be permitted to be installed closer to the hazard in a position where the detector can intercept the smoke.

The code specifically allows the designer to add detectors where he or she expects the pre-existing, normal air currents to convey the smoke from an early-stage fire. Usually, the design process begins by locating detectors so that they will provide general area protection. Then, additional detectors are added, or positions adjusted, to take into account known or anticipated ignition sources and known air currents. Subparagraph 2-3.4.1.2 of the code is generally cited when additional detectors are placed above switchgear enclosures, power supplies, and similar assets with known histories of ignition, as well as high dollar value assets.

2-3.4.2 Air Sampling–Type Smoke Detector. Each sampling port of an air sampling–type smoke detector shall be treated as a spot-type detector for the purpose of location and spacing. Maximum air sample transport time from the farthest sampling point shall not exceed 120 seconds.

Section 1-4 defines an air sampling–type smoke detector. These detectors use a sampling tube and draw a sample of air from the hazard area to the detector where the presence of visible smoke or invisible combustion products is determined. The air transport time criterion places an effective limit on the design of the fan and the maximum distance from the detector to the farthest sampling port, as well as the size and layout of the sampling tubes. The manufacturer's listing and instructions provide the detail on how the particular product must be used in order to comply with this limitation. Some air sampling–type smoke detectors have a means to detect changes in airflow, which provides some measure of monitoring the integrity of the tubing or piping network.

See Exhibit 2.22 for an example of an "active" air sampling or "aspirating" type smoke detector.

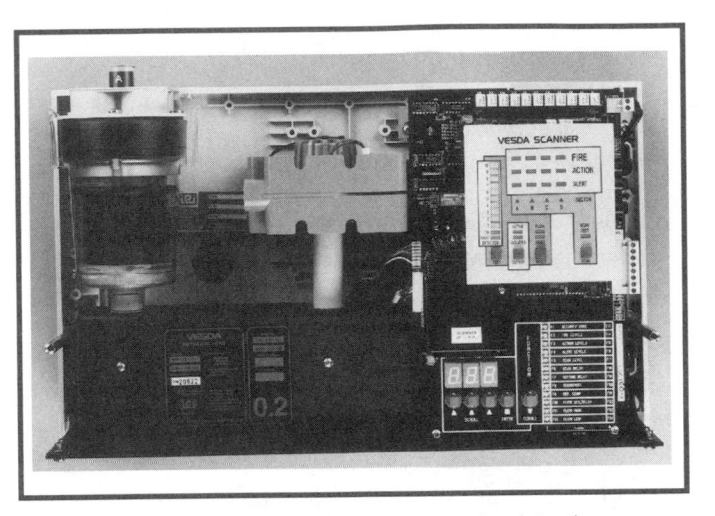

Exhibit 2.22 *Typical air sampling–type smoke detection apparatus. (Source: Vision Systems, Inc., Hingham, MA)*

2-3.4.3* Spot-Type Smoke Detectors.

A-2-3.4.3 In high-ceiling areas, such as atriums, where spot-type smoke detectors are not accessible for periodic maintenance and testing, projected beam–type or air sampling–type detectors should be considered where access can be provided.

The issue of accessibility and the maintenance of a smoke detection system cannot be over-emphasized. The designer must exercise judgment and discretion to provide a system that can be maintained pursuant to the criteria established in Chapter 7. Subsection A-2-3.4.3 clarifies subsection A-2-1.4.2.4, which requires that all initiating devices, including smoke detectors, be installed in such a manner that they can be effectively maintained.

Atria and other areas with exceptionally high ceilings (such as auditoriums, gymnasiums, exhibit halls, storage facilities, and some manufacturing facilities) represent very difficult situations for the use of spot-type smoke detection. Stratification, maintenance concerns, accessibility for testing, and smoke dissipation may warrant the use of other types of detection. Subsection A-2-3.4.3 advises the designer to consider either air sampling or linear projected beam–type photoelectric light obscuration smoke detection as alternatives. However, it is important to note that the air-sampling ports of an air sampling–type detector are treated as individual spot-type detectors. Air sampling–type detectors rely on the plume and ceiling jet to carry smoke to the sampling ports. Consequently, where stratification is a concern, this type of detection (air sampling detectors) might not represent an advantage over traditional spot-type detectors.

2-3.4.3.1 Spot-type smoke detectors shall be located on the ceiling not less than 4 in. (100 mm) from a sidewall to the near edge or, if on a sidewall, between 4 in. and 12 in. (100 mm and 300 mm) down from the ceiling to the top of the detector. *(Refer to Figure A-2-2.2.1.)*

Fluid dynamics predict that there will be dead air space where the wall joins the ceiling. Therefore, this area does not provide a location that is conducive to proper smoke detector performance. This location requirement is valid for both the low-energy incipient fire as well as a high-energy-output fire—one that is immediately life threatening. Either the normally existing air currents or the fire plume and ceiling jet from the larger fire convey smoke to ceiling-mounted detectors. For the detectors to be able to respond they must be installed in the working air volume of the compartment. This logic is identical to that used in locating heat detectors. See 2-2.2.1.

2-3.4.3.2* To minimize dust contamination, smoke detectors, where installed under raised floors, shall be mounted only in an orientation for which they have been listed.

A-2-3.4.3.2 Figure A-2-3.4.3.2 illustrates under-floor mounting installations.

The fast moving air in a data center underfloor space has sufficient energy to suspend dust. As that air enters the detector, it slows down and the suspended dust settles in the detector. The accumulation of dust within a smoke detector has a similar effect to that of smoke. In an ionization smoke detector, the dust impedes the flow of current within the chamber. In a spot-type photoelectric detector, the dust increases the reflectance within the chamber. Thus, dust causes each type of detector to become more sensitive, increasing the likelihood of false alarms. The permitted orientations shown in Figure A-2-3.4.3.2 (top) minimize the possibility of dust falling into the detector from the floor and also minimize the effect of air-conveyed dust on the detector.

There are other concerns that reinforce the benefits of positioning detectors as shown in Figure A-2-3.4.3.2 (top).

The detector is placed in the upper half of the sub-floor volume. Because the purpose of the sub-floor space is to allow for the routing of cables between machines, the floor is usually covered with cable. This cable has the same effect on the flow of air in the underfloor volume that joists have on airflow in a room. The cables create turbulence and force the flow to be concentrated in the upper half of the under-floor volume. Placing the detector in the upper half of the underfloor improves the system's ability to respond to an early-stage fire.

Another reason for positioning detectors as shown in Figure A-2-3.4.3.2 (top) is that detectors mounted in the

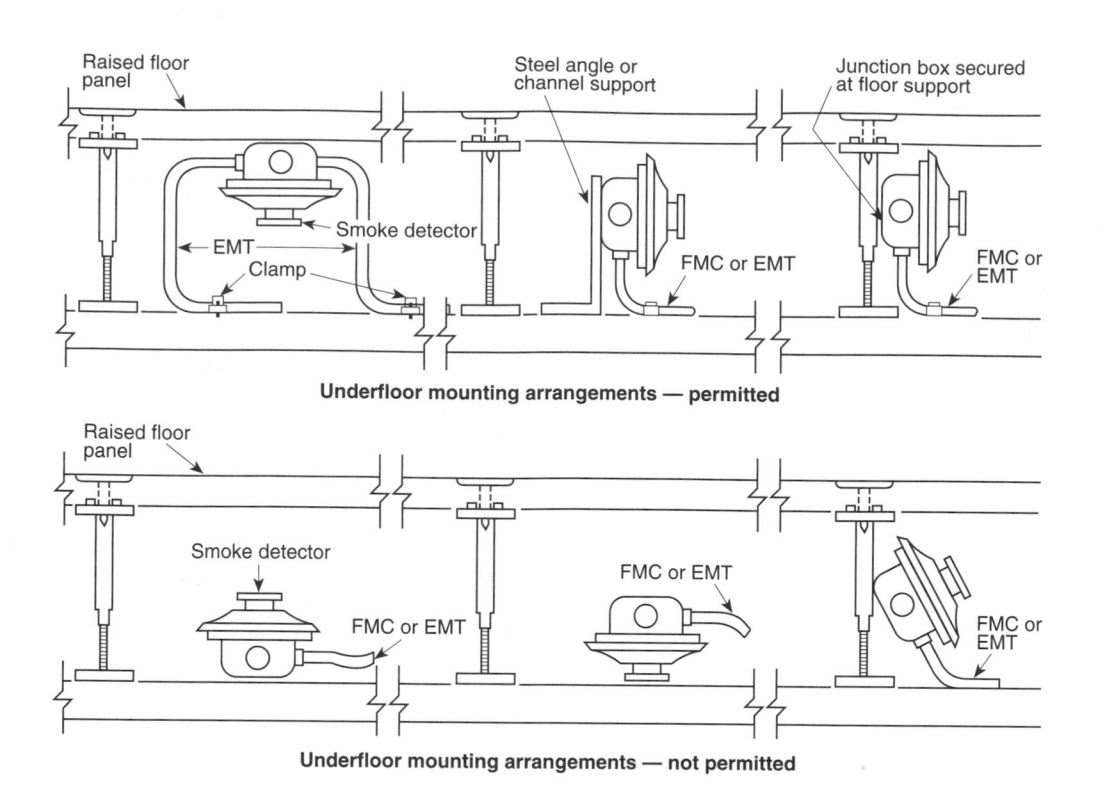

Figure A-2-3.4.3.2 *Mounting installations permitted (top) and not permitted (bottom).*

upper half of the underfloor volume are far less likely to be damaged as new cables are installed or old cables are rerouted through the underfloor space. If water-cooled computers are in use, the detectors are less likely to become wet if the computer cooling system leaks. Also, when there is no airflow, the detectors will be in the best orientation for detection. Finally, Figure A-2-3.4.3.2 shows the detectors in the orientation for which they have been tested and listed.

2-3.4.4 Projected Beam–Type Smoke Detectors. Projected beam–type smoke detectors shall be located with their projected beams parallel to the ceiling and in accordance with the manufacturer's documented instructions. The effects of stratification shall be evaluated when locating the detectors.

Exception: Beams shall be permitted to be installed vertically or at any angle needed to afford protection of the hazard involved (for example, vertical beams through the open shaft area of a stairwell where there is a clear vertical space inside the handrails).

2-3.4.4.1 The beam length shall not exceed the maximum permitted by the equipment listing.

Linear projected-beam smoke detectors have limitations on both the minimum and maximum beam length over which they will operate properly. The minimum beam length limi-

tation is established by the lowest smoke concentration that can be detected at that minimum beam length. The maximum beam length is determined by the maximum distance at which the detector can maintain its design stability even when some normal light obscuration is present. The linear projected-beam smoke detector must be able to identify a low concentration of smoke distributed along a substantial portion of the beam and a high concentration of smoke localized in a short segment of the beam. Each manufacturer obtains a listing from a qualified testing laboratory that sets the upper and lower limits on the beam length. Failure to observe these limits could result in an unstable detector or the failure to detect a fire consistent with the performance objectives. Mirrors used with linear projected-beam smoke detectors must also be listed for use with the detector.

2-3.4.4.2 If mirrors are used with projected beams, the mirrors shall be installed in accordance with the manufacturer's documented instructions.

2-3.4.5 Smooth Ceiling Spacing.

2-3.4.5.1 Spot-Type Detectors.

2-3.4.5.1.1 On smooth ceilings, spacing of 30 ft (9.1 m) shall be permitted to be used as a guide. In all cases, the manufacturer's documented instructions shall be followed.

Other spacing shall be permitted to be used depending on ceiling height, different conditions, or response requirements. For the detection of flaming fires, the guidelines in Appendix B shall be permitted to be used.

This spacing requirement is based on experience and anecdotal information regarding smoke detector performance in fires and laboratory fire tests.

Two analytical methods are provided in Appendix B of the code. These methods rely on plume and ceiling jet dynamics, and the use of these methods is limited to flaming fires that produce a buoyant plume. These methods are very useful in determining detector spacing and placement for flaming fire scenarios and have become very important tools for the fire detection systems designer.

Furthermore, several available computer models (including *FPETool*, *FastLite*, and *Hazard 1*) predict smoke detector activation. However, it must be noted that these computer models use a temperature rise model, not optical density or mass density, to predict the activation of smoke detectors.

In this regard, the Fire Detection Institute has sponsored a research paper entitled, *Fire Detection Modeling, State of the Art*, by Robert P. Schifliti and William E. Pucci. This paper analyzes the various ways the computer fire models predict smoke detector operation and points out the advantages and disadvantages of each method (Schifiliti and Pucci, 1998).

2-3.4.5.1.2* For smooth ceilings, all points on the ceiling shall have a detector within a distance equal to 0.7 times the selected spacing.

A-2-3.4.5.1.2 This is useful in calculating locations in corridors or irregular areas *(refer to A-2-2.4.1 and Figure A-2-2.4.1.2)*. For irregularly shaped areas, the spacing between detectors can be greater than the selected spacing, provided the maximum spacing from a detector to the farthest point of a sidewall or corner within its zone of protection is not greater than 0.7 times the selected spacing (0.7*S*).

The concepts behind the spacing of smoke detectors follow directly from the concepts developed for heat detectors. Subsection A-2-2.4.1 develops the concepts that enable a designer to determine the area that will be covered by a detector. That area can vary in shape as long as the distance from the detector to the farthest point to be covered by the detector does not exceed 0.7 times the selected spacing. See A-2-2.4.1(a), (b), (d), (e) and (f) and Figure A-2-2.4.1.2 for a graphical representation of these mathematical concepts.

2-3.4.5.2* Projected Beam–Type Detectors. For location and spacing of projected beam–type detectors, the manufacturer's documented installation instructions shall be followed.

A-2-3.4.5.2 On smooth ceilings, a spacing of not more than 60 ft (18.3 m) between projected beams and not more than one-half that spacing between a projected beam and a sidewall (wall parallel to the beam travel) should be used as a guide. Other spacing should be determined based on ceiling height, airflow characteristics, and response requirements.

In some cases, the light beam projector is mounted on one end wall, with the light beam receiver mounted on the opposite wall. However, it is also permitted to suspend the projector and receiver from the ceiling at a distance from the end walls not exceeding one-quarter the selected spacing (S). *(Refer to Figure A-2-3.4.5.2.)*

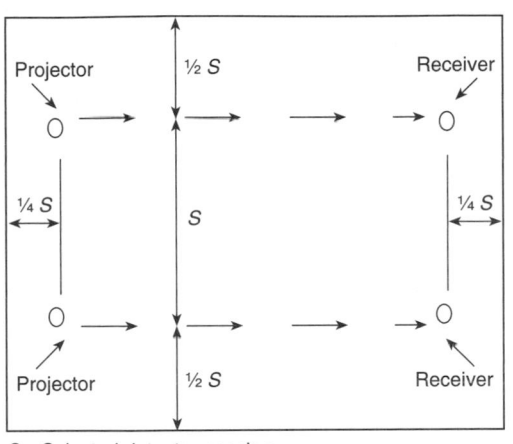

S = Selected detector spacing

Figure A-2-3.4.5.2 *Maximum distance at which ceiling-suspended light projector and receiver can be positioned from end wall is one-quarter selected spacing (S).*

The similarity between the installation and spacing concept developed for line-type heat detectors and line-type (linear projected-beam) smoke detectors should be noted. The logic behind the design rules remains consistent. When developing spacing strategies, just as a line-type heat detector can be thought of as a row of spot-type heat detectors, it is often helpful to think of a linear projected-beam detector as equivalent to a row of spot-type smoke detectors. The distance between the linear projected beams is analogous to the distance between rows of spot-type smoke detectors. Obviously, in high ceiling areas where stratification is probable and is a serious concern, linear projected beams can be positioned at several levels.

2-3.4.6* Solid Joist and Beam Construction. Solid joists over 1 ft (0.3 m) in depth shall be treated as beams for smoke detector spacing purposes.

As a result of computer modeling research conducted by the National Institute of Standards and Technology (NIST) under the auspices of the National Fire Protection Research Foundation and the Fire Detection Institute, the rules for smoke detector placement when joists or beams are present have changed. The research determined that when solid joists are greater than 1 ft (0.3 m) in depth, smoke detection efficiency is reduced in the same manner as when beams are present. Therefore, when solid joists are greater than 1 ft (0.3 m) in depth they should be treated as beams.

A-2-3.4.6 Detectors are placed at reduced spacings at right angles to joists or beams in an attempt to ensure that detection time is equivalent to that which would be experienced on a flat ceiling. It takes longer for the combustion products (smoke or heat) to travel at right angles to beams or joists because of the phenomenon wherein a plume from a relatively hot fire with significant thermal lift tends to fill the pocket between each beam or joist before moving to the next beam or joist.

Though it is true that this phenomenon might not be significant in a small smoldering fire where there is only enough thermal lift to cause stratification at the bottom of the joists, reduced spacing is still recommended to ensure that detection time is equivalent to that which would exist on a flat ceiling, even in the case of a hotter type of fire.

2-3.4.6.1* Flat Ceilings. For ceiling heights of 12 ft (3.66 m) or lower, and beam or solid joist depths of 1 ft (0.3 m) or less, smooth ceiling spaces running in the direction parallel to the run of the beams or solid joists shall be used and one-half the smooth ceiling spacing shall be in the direction perpendicular to the run of the beams or solid joists. For beams over 1 ft (0.3 m) in depth, spot-type detectors shall be permitted to be located either on the ceiling or on the bottom of the beams.

For beam depths exceeding 1 ft (0.3 m) or for ceiling heights exceeding 12 ft (3.66 m), spot-type detectors shall be located on the ceiling in every beam pocket.

For solid joists, the detectors shall be located on the bottom of the joists.

A-2-3.4.6.1 The spacing guidelines in 2-3.4.6.1 are based on a detection design fire of 100 kW. For detection at a larger 1-MW fire and ceiling heights of 28 ft (8.53 m) or less, smooth ceiling spacings should be used and the detectors may be located on the ceiling or the bottom of the beams.

The concept behind these recommendations is analogous to that regarding heat detectors even though the dimension criteria regarding the depth of the beams and joists is different for smoke detectors.

Where the beams or joists are less than 12 in. (300 mm) deep and the ceiling is less than 12 ft (3.7 m) high, the spacing perpendicular to the beams must be reduced by 50 percent. The spacing running parallel to the beams is the same as smooth ceilings. Because the ceiling jet is usually taken to be approximately one-tenth the thickness of the floor to ceiling height, it is expected to be approximately 12 in. (300 mm) in depth. This permits detectors to be installed either on the beam bottom or on the ceiling. The air currents that convey smoke and heat, whether a ceiling jet or ambient air movement, ensure that the concentration of smoke at a detector mounted on the bottom of a beam less than 12 in. (300 mm) deep will be sufficient to achieve an alarm. As the smoke plume from a fire impinges on the ceiling and begins expanding horizontally, a beam becomes a dam of sorts. Time and the expenditure of smoke energy force the smoke downward far enough to spill over the dam (actually flow under the dam), encountering the smoke detector in the process. Energetic fires that produce large thermal outputs quickly force smoke across beams due to the large amounts of thermal energy available. Low-energy-output fires provide less energy, and the movement across beams is slower. Where beams are less than 12 in. (300 mm) deep, detectors are permitted (not required) to be mounted on the beam bottoms.

Where the beam depth exceeds 12 in. (300 mm) or when the ceiling height exceeds 12 ft (3.7 m) smoke detectors must be placed in each beam pocket and mounted on the ceiling surface between the beams. By definition the space between joists is small and, therefore, the smoke detectors are mounted on the bottom of the joists.

2-3.4.6.2* Sloped Ceilings. For beamed ceilings with beams running parallel to (up) the slope, the spacing for flat beamed ceilings shall be used. The ceiling height shall be taken as the average height over the slope. For slopes greater than 10 degrees, the detectors located at one-half the spacing from the low end shall not be required. Spacings shall be measured along a horizontal projection of the ceilings.

For beamed ceilings with beams running perpendicular to (across) the slope, the spacing for flat beamed ceilings shall be used. The ceiling height shall be taken as the average height over the slope.

For solid joists, the detectors shall be located on the bottom of the joists.

First, beams that are parallel to the slope are perpendicular to the ridge beam of the roof. Beams that are perpendicular to the slope are parallel to the ridge beam.

As with heat detector placement, when a buoyant plume from a flaming fire impinges on a sloped ceiling it will progress rapidly upward toward the ridge beam. This rapid upward flow reduces the lateral flow parallel to the ridge beam. The spacing criteria in 2-3.4.6.2 represent the consensus opinion of the Technical Committee on Initiating Devices for Fire Alarm Systems as to how to attain response

equivalent to that for a flat ceiling of the same height. It should be noted that when the slope exceeds 10 degrees, the row of detectors nearest the lower wall can be deleted. This is because there is virtually no downward flow from the centerline of the fire toward detectors mounted at lower heights than the plume impingement point.

A-2-3.4.6.2 The spacing guidelines in 2-3.4.6.2 are based on a detection design fire of 100 kW. For detection at a larger 1-MW fire, the following spacings should be used.

(a) For beamed ceilings with beams running parallel to (up) the slope, with slopes 10 degrees or less, spacing for flat-beamed ceilings should be used. For ceilings with slopes greater that 10 degrees, twice the smooth ceiling spacing should be used in the direction parallel to (up) the slopes, and one-half the spacing should be used in the direction perpendicular to (across) the slope. For slopes greater than 10 degrees, the detectors located at a distance of one-half the spacing from the low end are not required. Spacing should be measured along the horizontal projection of the ceiling.

It is important to remember that the design fire for a smoke detection system is usually orders of magnitude smaller than that for heat detectors. Fires of 1.0 MW in heat release rate are usually used as a design criterion for heat detection.

These design criteria are the result of research conducted under auspices of the National Fire Protection Research Foundation and the Fire Detection Institute. Copies of the reports of the International Fire Detection Research Project (vol. 1–4) are available from the National Fire Protection Research Foundation.

(b) For beamed ceilings with beams running perpendicular to (across) the slope, for any slope, smooth ceiling spacing should be used in the direction parallel to the beams (across the slope), and one-half the smooth ceiling spacing should be used in the direction perpendicular to the beams (up the slope).

2-3.4.6.3 A projected beam–type smoke detector shall be equivalent to a row of spot-type smoke detectors for flat and sloped ceiling applications.

2-3.4.7* Peaked. Detectors shall first be spaced and located within 3 ft (0.9 m) of the peak, measured horizontally. The number and spacing of additional detectors, if any, shall be based on the horizontal projection of the ceiling.

A-2-3.4.7 Refer to Figure A-2-2.4.4.1.

2-3.4.8* Shed. Detectors shall first be spaced and located within 3 ft (0.9 m) of the high side of the ceiling, measured horizontally. The number and spacing of additional detectors, if any, shall be based on the horizontal projection of the ceiling.

A-2-3.4.8 Refer to Figure A-2-2.4.4.2.

2-3.4.9 Raised Floors and Suspended Ceilings. Spaces beneath raised floors and above suspended ceilings shall be treated as separate rooms for smoke detector spacing purposes. Detectors installed beneath raised floors or above suspended ceilings, or both, including raised floors and suspended ceilings used for environmental air, shall not be used in lieu of providing detection within the room.

When total coverage is required by the authority having jurisdiction or other codes, 2-1.4.2.1 requires detection in all accessible spaces (combustible or noncombustible and in inaccessible combustible spaces). The spaces beneath raised floors and above suspended ceilings usually fall into that category and, hence, require detection using the same location and spacing concepts as required for the occupied portion of a building.

2-3.4.9.1 Raised Floors. Detectors installed beneath raised floors shall be spaced in accordance with 2-3.4.1, 2-3.4.1.2, and 2-3.4.3.2. If the area beneath the raised floor is also used for environmental air, detector spacing shall also conform to 2-3.5.1 and 2-3.5.2.

2-3.4.9.2 Suspended Ceilings. Detector spacing above suspended ceilings shall conform to the requirements of 2-3.4 for the ceiling configuration. If detectors are installed in ceilings used for environmental air, detector spacing shall also conform to 2-3.5.1 and 2-3.5.2.

2-3.4.10 Partitions. Where partitions extend upward to within 18 in. (460 mm) of the ceiling, they shall not influence the spacing. Where the partition extends to within less than 18 in. (460 mm) of the ceiling, the effect of smoke travel shall be evaluated in the reduction of spacing.

Research on fire plumes and ceiling jets indicates that the thickness of the ceiling jet under most conditions is approximately 10 percent of the distance from the floor to the ceiling in the fire compartment. However, the ceiling jet does not have an abrupt boundary. The dimension used for its thickness depends on the velocity criterion used for the jet boundary.

When considering partitions, in the case of a fire with an established plume, the important factor is whether the partition impedes the flow of the ceiling jet across the ceiling. Or, in the case of a small, low-energy fire, whether the partition impedes the flow of smoke entrained in the normal air currents. In spaces where the ceiling is approximately 10 ft (3 m) high, allowing a 50 percent margin of error on the impact on the ceiling jet and environmental air, a designer should see that a partition extending to within 18 in. (460 mm) of the ceiling will very likely affect the ceiling jet. The partition will restrict the horizontal flow of smoke across the ceiling.

The treatment of partitions in the code is very different from the treatment of partitions in NFPA 13, *Installation of Sprinkler Systems*, where the principal concern is the discharge pattern of the sprinkler head and the impact of the

partition on that discharge pattern and, thus, on the control of the fire.

2-3.5 Heating, Ventilating, and Air-Conditioning (HVAC).

2-3.5.1* In spaces served by air-handling systems, detectors shall not be located where airflow prevents operation of the detectors.

A-2-3.5.1 Detectors should not be located in a direct airflow nor closer than 3 ft (1 m) from an air supply diffuser or return air opening. Supply or return sources larger than those commonly found in residential and small commercial establishments can require greater clearance to smoke detectors. Similarly, smoke detectors should be located farther away from high velocity air supplies.

Paragraph A-2-3.5.1 recommends a separation of at least 3 ft (1 m) between an air supply diffuser and the detector as well as between an air return and a detector. The application of the 3-ft-rule to air returns is new in this revision of the code. There may be situations where even a 3 ft (1 m) separation is not adequate, depending on the air velocity, and the throw characteristics of the diffuser and diffuser size.

The computer modeling research conducted by NIST as part of the International Fire Detection Research Project identified situations where areas of non-actuation extended almost 11 ft (3.4 m) from some diffusers. The designer should consider evaluating the HVAC system effects with a velometer, comparing ambient ceiling velocities to the expected ceiling jet velocities from the design fire.

2-3.5.2 Plenums.

2-3.5.2.1 In under-floor spaces and above-ceiling spaces that are used as HVAC plenums, detectors shall be listed for the anticipated environment as required by 2-3.6.1.1. Detector spacings and locations shall be selected based on anticipated airflow patterns and fire type.

In order to cool a room to 70°F (21°C) it may be necessary to introduce extremely frigid air into the room. Heating a room sometimes requires introducing superheated air into a room. Consequently, HVAC plenums usually have ambient conditions that are far more extreme than the spaces they support.

Smoke detectors are electronic sensors. Ambient temperature, the relative humidity, and, especially in the case of spot-type ionization detectors, the velocity of the air around the detector, all affect detector operation. Not all smoke detectors are listed for the range of conditions found in HVAC plenums. It is the designer's responsibility to ensure the detector has the correct operating characteristics for the environment.

2-3.5.2.2* Detectors placed in environmental air ducts or plenums shall not be used as a substitute for open area

detectors. If detectors are used for the control of smoke spread, the requirements of Section 2-10 shall apply. If open area protection is required, 2-3.4 shall apply.

A-2-3.5.2.2 Smoke might not be drawn into the duct or plenums when the ventilating system is shut down. Furthermore, when the ventilating system is operating, the detector(s) can be less responsive to a fire condition in the room of fire origin due to dilution by clean air.

2-3.6 Special Considerations.

It is important for the designer to recognize that in presenting minimum requirements, the code may not cover special considerations unique to a specific application. Nor will the code address a particular product that may allow the system to fulfill its design objective yet are not listed as minimum compliance criteria in the code. While the code makes every effort to establish minimum compliance criteria to address problems that have a documented history of affecting smoke detection systems, it cannot be assumed that this list is exhaustive and covers every conceivable contingency. The designer should be aware of any factor in the protected area that could contribute to unwanted alarms or could prevent the successful conveyance of smoke to the detector.

2-3.6.1 The selection and placement of smoke detectors shall take into account both the performance characteristics of the detector and the areas into which the detectors are to be installed to prevent nuisance alarms or improper operation after installation. Paragraphs 2-3.6.1.1 through 2-3.6.1.3 shall apply.

2-3.6.1.1* Smoke detectors shall not be installed if any of the following ambient conditions exist:

(1) Temperature below 32°F (0°C)
(2) Temperature above 100°F (38°C)
(3) Relative humidity above 93 percent
(4) Air velocity greater than 300 ft/min (1.5 m/sec)

Exception: Detectors specifically designed for use in ambient conditions beyond the limits of 2-3.6.1.1(1) through (4) and listed for the temperature, humidity, and air velocity conditions expected.

A-2-3.6.1.1 Product-listing standards include tests for temporary excursions beyond normal limits. In addition to temperature, humidity, and velocity variations, smoke detectors should operate reliably under such common environmental conditions as mechanical vibration, electrical interference, and other environmental influences. Tests for these conditions are also conducted by the testing laboratories in their listing program. In those cases in which environmental conditions approach the limits shown in Table A-2-3.6.1.1, the detector manufacturer should be consulted for additional information and recommendations.

Table A-2-3.6.1.1 Environmental Conditions That Influence Smoke Detector Response

Detection Protection	Air Velocity >300 ft (>91.44 m)/min	Altitude >3000 ft (>914.4 m)	Humidity >93% RH	Temp. <32°F >100°F (<0°C >37.8°C)	Color of Smoke
Ion	X	X	X	X	O
Photo	O	O	X	X	X
Beam	O	O	X	X	O
Air Sampling	O	O	X	X	O

X = Can affect detector response.
O = Generally does not affect detector response.

Different detection technologies are affected differently by these environmental extremes. Different makes and models within each group may be affected more or less than others. Apart from the generalities presented here, it is beyond the scope of this handbook to identify these effects. However, the designer must recognize that some detector designs are inherently more forgiving than others.

The tests performed in the process of listing ascertain that a detector meets minimum performance criteria. Design features in specific devices that allow them to be effectively used in extreme environments can be beyond those considered in the listing evaluation. The manufacturer should be consulted when such an application is contemplated.

These environmental limits may require the designer to consider alternative detection methods. Although smoke detection may be preferable from an early warning standpoint, heat or radiant energy detection may be a better choice where the hazard area is one that undergoes too broad a range of environmental conditions to allow the use of smoke detection.

2-3.6.1.2* The location of smoke detectors shall be based on an evaluation of potential ambient sources of smoke, moisture, dust, or fumes, and electrical or mechanical influences to minimize nuisance alarms.

A-2-3.6.1.2 Smoke detectors can be affected by electrical and mechanical influences and by aerosols and particulate matter found in protected spaces. The location of detectors should be such that the influences of aerosols and particulate matter from sources such as those in Table A-2-3.6.1.2(a) are minimized. Similarly, the influences of electrical and mechanical factors shown in Table A-2-3.6.1.2(b) should be minimized. While it might not be possible to isolate environmental factors totally, an awareness of these factors during system layout and design favorably affects detector performance.

In applications where the factors outlined in Tables A-2-3.6.1.2(a) and A-2-3.6.1.2(b) cannot be sufficiently limited to allow reasonable stability and response times, alternate modes of fire detection should be considered.

Table A-2-3.6.1.2(a) Common Sources of Aerosols and Particulate Matter Moisture

Moisture
Humid outside air
Humidifiers
Live steam
Showers
Slop sink
Steam tables
Water spray

Combustion Products and Fumes
Chemical fumes
Cleaning fluids
Cooking equipment
Curing
Cutting, welding, and brazing
Dryers
Exhaust hoods
Fireplaces
Machining
Ovens
Paint spray

Atmospheric Contaminants
Corrosive atmospheres
Dust or lint
Excessive tobacco smoke
Heat treating
Linen and bedding handling
Pneumatic transport
Sawing, drilling, and grinding
Textile and agricultural processing

Engine Exhaust
Diesel trucks and locomotives
Engines not vented to the outside
Gasoline forklift trucks

Heating Element with Abnormal Conditions
Dust accumulations
Improper exhaust
Incomplete combustion

Table A-2-3.6.1.2(b) Sources of Electrical and Mechanical Influences on Smoke Detectors

Electrical Noise and Transients	Airflow
Vibration or shock	Gusts
Radiation	Excessive velocity
Radio frequency	
Intense light	
Lightning	
Electrostatic discharge	
Power supply	

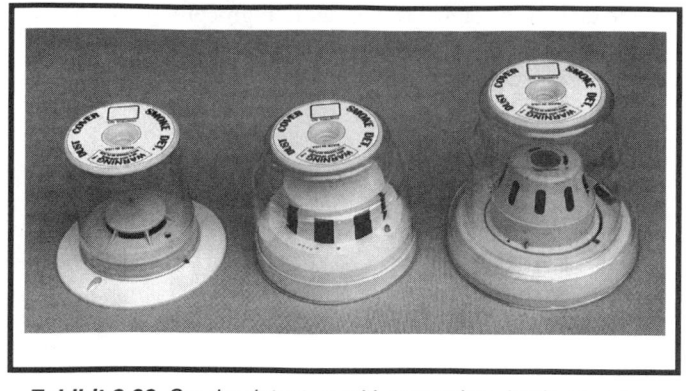

Exhibit 2.23 *Smoke detectors with protective plastic covers installed. (Source: PDH System Co., Braintree, MA)*

2-3.6.1.3 Detectors shall not be installed until after the construction cleanup of all trades is complete and final.

Exception: Where required by the authority having jurisdiction for protection during construction. Detectors that have been installed during construction and found to have a sensitivity outside the listed and marked sensitivity range shall be cleaned or replaced in accordance with Chapter 7 at completion of construction.

Many needless alarms have been caused by the early installation of smoke detectors. Construction activities produce airborne dust that inevitably finds its way into detectors, contaminating them and making them prone to false alarms. Subparagraph 2-3.6.1.3 forbids that practice unless the authority having jurisdiction requires it.

In the latter case, detectors installed prior to the completion of final finish work must be measured for their normal operating sensitivity. Those detectors found outside their design sensitivity range must be either replaced or cleaned.

The authority having jurisdiction may allow the installation of smoke detectors with protective covers. These covers cannot be relied on to keep the detector entirely free of contaminants. Therefore, sensitivity measurement and cleaning of the detectors after all construction trades have finished their work may still be necessary. If these covers are used, the contractor must ensure that they are all removed when the construction trades have completed their work. If the authority having jurisdiction requires the covers to be removed at the end of each day, it is good practice to number the covers to ensure all have been removed and then replaced the next morning. Again, if the covers are removed during the construction process, it will be necessary to inspect the detectors closely, cleaning them when necessary. See Exhibits 2.23 and 2.24 for examples of smoke detector protective covers.

2-3.6.1.4* Stratification. The effect of stratification below the ceiling shall be taken into account. The guidelines in Appendix B shall be permitted to be used.

Exhibit 2.24 *Smoke detectors with protective plastic cover installed. (Source: Mammoth Fire Alarms, Inc., Lowell, MA)*

A-2-3.6.1.4 Stratification of air in a room can hinder air containing smoke particles or gaseous combustion products from reaching ceiling-mounted smoke detectors or fire–gas detectors.

Stratification occurs when air containing smoke particles or gaseous combustion products is heated by smoldering or burning material and, becoming less dense than the surrounding cooler air, rises until it reaches a level at which there is no longer a difference in temperature between it and the surrounding air.

Stratification also can occur when evaporative coolers are used, because moisture introduced by these devices can condense on smoke, causing it to fall toward the floor. Therefore, to ensure rapid response, it might be necessary to install smoke detectors on sidewalls or at locations below the ceiling.

In installations where detection of smoldering or small fires is desired and where the possibility of stratification exists, consideration should be given to mounting a portion

of the detectors below the ceiling. In high-ceiling areas, projected beam–type or air sampling–type detectors at different levels also should be considered. *(Refer to Figure A-2-3.6.1.4.)*

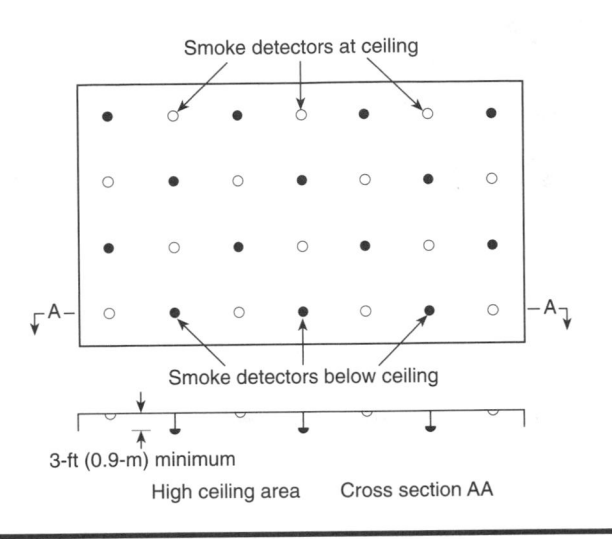

Figure A-2-3.6.1.4 Smoke detector layout accounting for stratification.

When the combustion product gases (smoke) form in a fire, they are hot and, consequently, begin expanding. These expanded gases are less dense than the surrounding air and are buoyed upward. Because the gases are still hot, they continue to expand and a V-shaped fire plume results that is small at the bottom and grows larger the higher it rises. According to the ideal gas law, as a gas expands, it loses heat. In addition, the rising plume gases mix with the surrounding air, entraining the air as it flows upward. The entrainment of ambient air also contributes to the cooling of the plume. Eventually, the combustion product gases in the fire plume decrease in temperature until they are no longer hotter than the surrounding air. At this height, there is no longer an upward push on the plume, and it spreads out in a layer.

If this height is reached before the fire plume impinges on the ceiling, then there is no force causing the fire plume to turn and form a ceiling jet. All of the spacing criteria for smoke and heat detectors are based on the existence of a ceiling jet that moves the smoke and heat horizontally across the ceiling.

The height at which the stratification occurs depends on both the size of the fire and the ambient temperature of the space. The relationship is not constant. Where the stratification layer begins to form in a space depends entirely on the temperature gradient in the space and the heat release rate of the fire.

Stratification impacts the performance of the detection system. It is most likely to occur when the fires are small and the ceilings are high. However, HVAC systems designed to form a layer of cool air at some given distance above the floor can create exactly the same conditions as naturally occurring stratification. This phenomenon can have the same profound effects on the performance of a detection system.

Where stratification can be expected, the location and spacing of smoke detectors must be adjusted. The design of a smoke detection system must address both the spectrum of ambient conditions, as well as the range of fire scenarios for the space. In areas of high ceilings, this often necessitates layers of detectors or combining detectors to address all possible fire scenarios.

The objective of detecting the fire before it has achieved a high-energy output requires additional insight into the placement of detectors. The high-energy-output flaming fire produces a fire plume that propels smoke and hot air upward. The larger the fire, the higher the plume extends and the greater the air velocity within the plume.

In the low-energy-output smoldering fire often encountered in residential (e.g., homes, hotels, apartments), institutional (e.g., hospitals, nursing homes, schools), and commercial (e.g., offices, stores) occupancies, significant quantities of smoke may be produced before an energetic fire plume develops. This smoke may lack the energy to rise up to ceiling-mounted smoke detectors where the ceilings are higher than normally encountered. This situation must be addressed in any fire alarm system designed for residential, institutional, or commercial occupancies. The addition of smoke detectors at some distance below the ceiling might not eliminate the requirement for ceiling-mounted detectors.

2-3.6.2 Spot-Type Detectors.

2-3.6.2.1 Smoke detectors that have a fixed temperature element as part of the unit shall be selected in accordance with Table 2-2.1.1.1 for the maximum ceiling temperature expected in service.

Subparagraph 2-3.6.2.1 reminds the code user to consider temperature (in addition to the ambient conditions that affect smoke detection,) when selecting a heat detector for combination smoke and heat detection in an environment with a higher than normal ambient temperature.

2-3.6.2.2* Holes in the back of a detector shall be covered by a gasket, sealant, or equivalent means, and the detector shall be mounted so that airflow from inside or around the housing does not prevent the entry of smoke during a fire or test condition.

A-2-3.6.2.2 Airflow through holes in the rear of a smoke detector can interfere with smoke entry to the sensing chamber. Similarly, air from the conduit system can flow around the outside edges of the detector and interfere with smoke

reaching the sensing chamber. Additionally, holes in the rear of a detector provide a means for entry of dust, dirt, and insects, each of which can adversely affect the detector's performance.

The conditions stated in A-2-3.6.2.2 have been encountered frequently enough to warrant inclusion of the requirements in 2-3.6.2.2 into the code. However, the list of installation related problems in A-2-3.6.2.2 cannot be assumed to be exhaustive. Once again, the designer should be aware of any factor in the protected area that could contribute to unwanted alarms or could prevent the successful conveyance of smoke to the detector.

2-3.6.3 Projected Beam–Type Detectors.

2-3.6.3.1 Projected beam–type detectors and mirrors shall be mounted on stable surfaces to prevent false or erratic operation due to movement. The beam shall be designed so that small angular movements of the light source or receiver do not prevent operation due to smoke and do not cause nuisance alarms.

Contrary to popular belief, buildings move. Portions of buildings vibrate due to passing traffic on nearby streets. They sway due to wind or uneven thermal expansion; even the ebb and flow of the tides can cause oceanfront buildings to flex. Modern curtain-wall/steel-frame buildings are designed to flex. This, however, places a demand on fire alarm systems, especially fire alarm systems using linear projected-beam smoke detection. The detectors must be able to accommodate the natural or designed movement of the building. The manufacturers of linear projected-beam detectors provide installation instructions that address the potential for this type of difficulty. Some manufacturers do not allow the use of mirrors due to the physical instability of mounting surfaces and building movement.

2-3.6.3.2* The light path of projected beam–type detectors shall be kept clear of opaque obstacles at all times.

A-2-3.6.3.2 Where the light path of a projected beam–type detector is abruptly interrupted or obscured, the unit should not initiate an alarm. It should give a trouble signal after verification of blockage.

Modern linear projected-beam detectors use obscuration algorithms in their software that can distinguish the progressive obscuration that occurs during a fire with the step-wise obscuration that usually indicates interference in the path of the beam by an opaque object. However, Christmas decorations, party balloons, and hanging plants have been known to cause problems in spite of the most sophisticated software. Obstructions that can gradually grow and block a beam detector, such as trees in an atrium, should also be considered a potential problem.

2-3.6.4 Air Sampling–Type Detectors.

In addition to the cloud chamber–type of smoke detector, several varieties of aspirating-type air-sampling smoke detectors exist. These detectors are essentially photoelectric smoke detectors with aspirating fans that draw smoke in through a network of small diameter pipes and deliver the smoke to a smoke detector. These detectors are used in a variety of applications where the designer is concerned with the effects of high airflow on smoke detection. Because of their sensitivity ranges, air-sampling detectors are also used in areas that house very valuable equipment. See Exhibits 2.25, 2.26, and 2.27(a) and (b) for examples of air sampling–type smoke detectors.

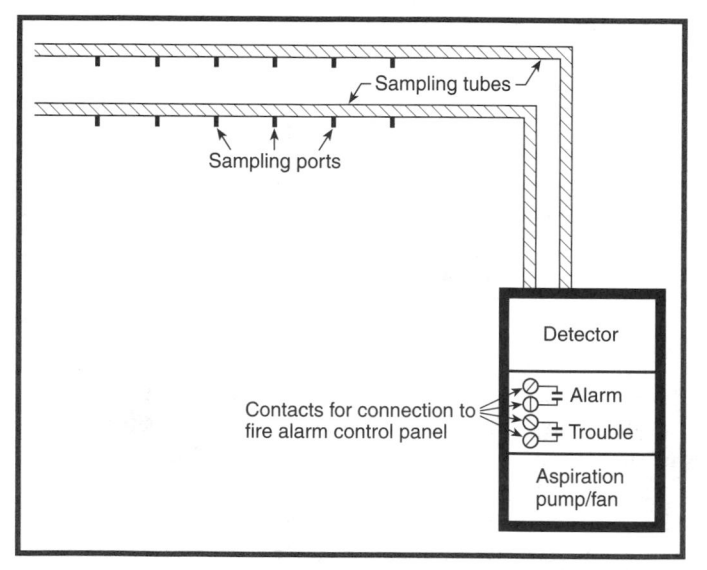

Exhibit 2.25 *The air-sampling detector uses sampling tubes to convey smoke-laden air to the central detection unit. (Source: J.M. Cholin Consultants, Inc., Oakland, NJ)*

2-3.6.4.1* Sampling pipe networks shall be designed on the basis of and shall be supported by sound fluid dynamic principles to ensure required performance. Network design details shall include calculations showing the flow characteristics of the pipe network and each sample port.

A-2-3.6.4.1 A single-pipe network has a shorter transport time than a multiple-pipe network of similar length pipe; however, a multiple-pipe system provides a faster smoke transport time than a single-pipe system of the same total length. As the number of sampling holes in a pipe increases, the smoke transport time increases. Where practicable, pipe run lengths in a multiple-pipe system should be nearly equal, or the system should be otherwise pneumatically balanced.

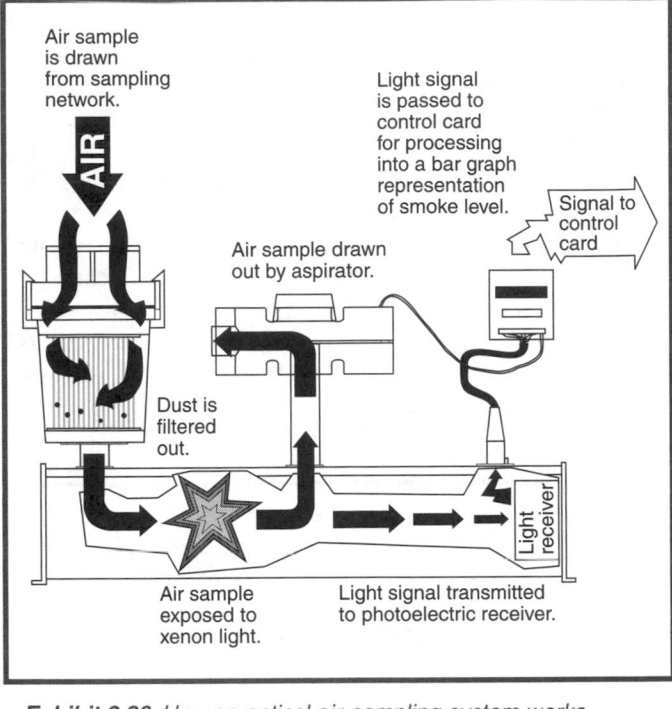

Exhibit 2.26 *How an optical air-sampling system works. (Source: Vision Systems Inc., Hingham, MA)*

The manufacturers of this type of smoke detector provide engineering guidelines in their installation manuals that ensure that the products meet the criteria of 2-3.6.4.1. These guidelines are evaluated by testing laboratories as part of the listing evaluation procedure. The factors in A-2-3.6.4.1 are generalizations the designer can use as guidance in deciding the type of piping network that best serves the application under consideration.

2-3.6.4.2* Air-sampling detectors shall give a trouble signal if the airflow is outside the manufacturer's specified range. The sampling ports and in-line filter, if used, shall be kept clear in accordance with the manufacturer's documented instructions.

A-2-3.6.4.2 The air sampling–type detector system should be able to withstand dusty environments by either air filtering or electronic discrimination of particle size. The detector should be capable of providing optimal time delays of alarm outputs to eliminate nuisance alarms due to transient smoke conditions. The detector should also provide facilities for the connection of monitoring equipment for the recording of background smoke level information necessary in setting alert and alarm levels and delays.

2-3.6.4.3 Air-sampling network piping and fittings shall be airtight and permanently fixed. Sampling system piping shall be conspicuously identified as SMOKE DETECTOR SAMPLING TUBE. DO NOT DISTURB, as follows:

Exhibit 2.27(a) *Typical air sampling–type smoke detectors with covers in place. (Source: Top—Kidde-Fenwal Inc., Ashland, MA; Bottom—Vision Systems Inc., Hingham, MA)*

(1) At changes in direction or branches of piping
(2) At each side of penetrations of walls, floors, or other barriers
(3) At intervals on piping that provide visibility within the space, but no greater than 20 ft (6 m)

2-3.6.5*   **High Rack Storage.** Where smoke detectors are installed to actuate a suppression system, NFPA 13, *Standard for the Installation of Sprinkler Systems,* shall apply.

Fire protection for high rack storage warehouses is a particularly difficult problem. The fuel load per unit of floor area is extremely high; the accessibility to the fuel is relatively low. Also, the combustibility of the materials in any given rack can vary from nominally noncombustible to flammable.

The orientation of the fuel also creates vertical flues between the combustibles that produce ideal conditions for the propagation of the fire and the worst possible conditions for extinguishment. Likewise, the presence of solid shelving

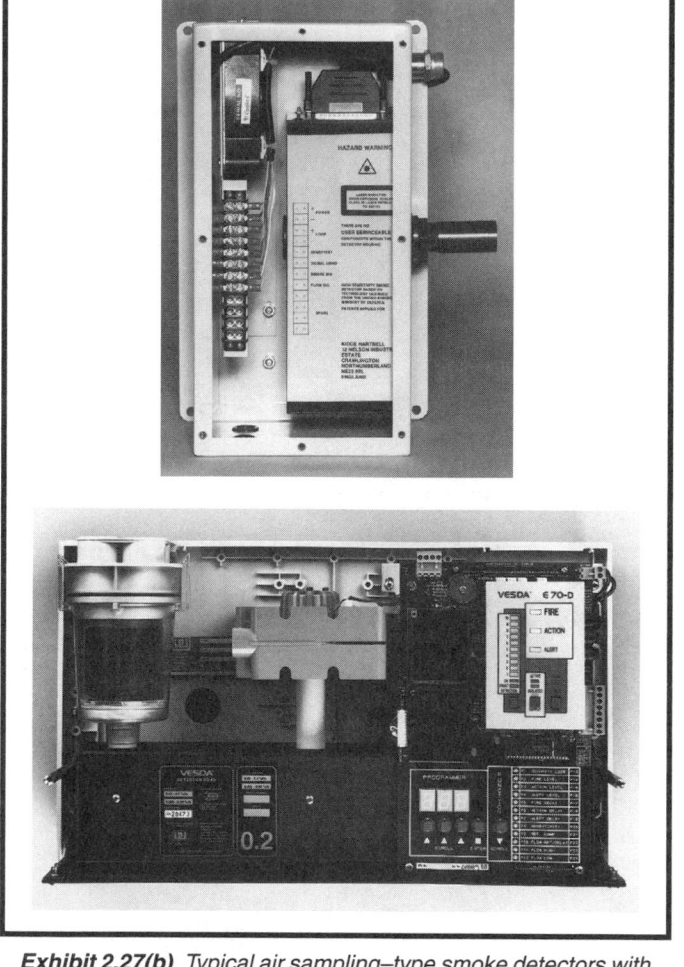

Exhibit 2.27(b) *Typical air sampling–type smoke detectors with covers removed. (Source: Top—Kidde-Fenwal Inc., Ashland, MA; Bottom—Vision Systems Inc., Hingham, MA)*

can create horizontal flues that materially aid in horizontal fire spread. The shelves can also shield the fire from water discharged to extinguish the fire. This makes early detection and rapid extinguishment of the fire in the incipient stages critical. Once the fire becomes well established, it is virtually impossible to extinguish. A number of catastrophic total losses have occurred in high rack storage facilities in the past decade.

The guidance provided for locating detectors in rack storage arrays strives to assure that any flue spaces created by the stored commodities are covered with a detector at some level. Care must also be used in installing detectors in these applications. The detectors are vulnerable to damage as commodities are moved in and out of the storage racks.

Although it may seem impossible to maintain accessibility for service and maintenance while locating detectors for both maximum speed of response and minimum expo-

sure to damage from operations, it is not. System designs exist that have satisfied all three of these apparently conflicting requirements.

Air sampling–type smoke detectors, with the piping network extended throughout each rack, have been used successfully in this application.

A-2-3.6.5 For the most effective detection of fire in high-rack storage areas, detectors should be located on the ceiling above each aisle and at intermediate levels in the racks. This is necessary to detect smoke that is trapped in the racks at an early stage of fire development when insufficient thermal energy is released to carry the smoke to the ceiling. Earliest detection of smoke is achieved by locating the intermediate level detectors adjacent to alternate pallet sections as shown in Figures A-2-3.6.5(a) and (b). The detector manufacturer's recommendations and engineering judgment should be followed for specific installations.

A projected beam–type detector can be permitted to be used in lieu of a single row of individual spot-type smoke detectors.

Sampling ports of an air sampling–type detector can be permitted to be located above each aisle to provide coverage that is equivalent to the location of spot-type detectors. The manufacturer's recommendations and engineering judgment should be followed for the specific installation.

2-3.6.6 High Air-Movement Areas.

2-3.6.6.1 General. The purpose and scope of 2-3.6.6 shall be to provide location and spacing guidance for smoke detectors intended for early warning of fire in high air-movement areas.

Exception: Detectors provided for the control of smoke spread are covered by the requirements of Section 2-10.

The most regularly encountered example of a high air-movement area is the data center (computer room), specifically its underfloor and above-ceiling spaces that are used for environmental air. This is by no means the only area that falls into this category. In general, areas where the air velocity across the detector exceeds 300 ft per minute (1.5 m per second) are considered high air movement ambients.

Table 2-3.6.6.3 and Figure 2-3.6.6.3 provide the detector spacing for high air-movement ambients in areas other than above-ceiling and underfloor spaces. Where high air-flow is encountered in spaces other than above the ceiling or under the floor, consideration should be given to reducing the spacing of spot-type detectors or using detectors that are not as affected by high airflow. In the majority of very high airflow areas, spot-type detectors are not the best detectors for the application. Air-sampling detectors have been used for such spaces quite successfully. It is important to remember that detectors listed for high airflow environments are tested to ensure they do not false alarm in high-airflow

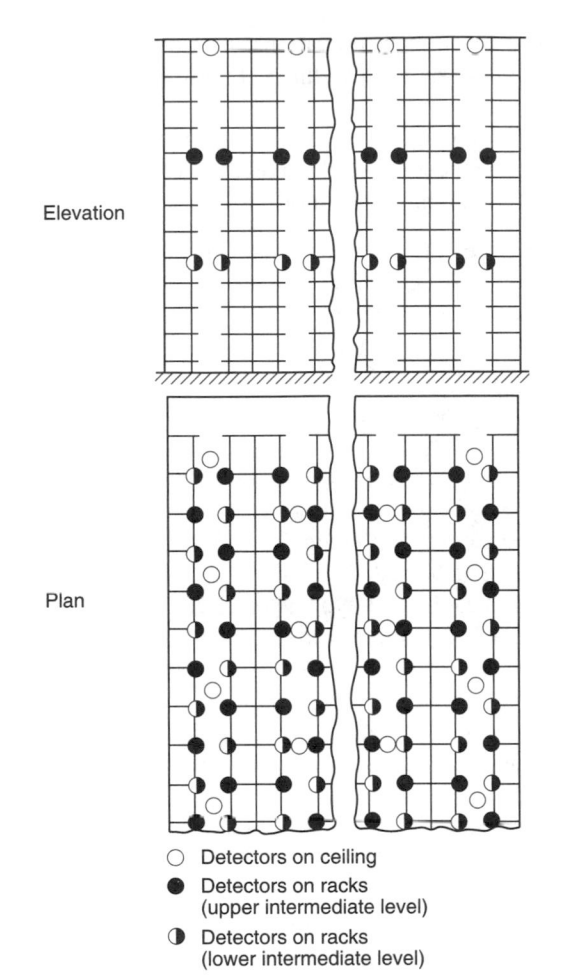

Elevation

Plan

○ Detectors on ceiling
● Detectors on racks
    (upper intermediate level)
◑ Detectors on racks
    (lower intermediate level)

Figure A-2-3.6.5(a) Detector location for solid storage (closed rack) in which transverse and longitudinal flue spaces are irregular or nonexistent, as for slatted or solid shelved storage.

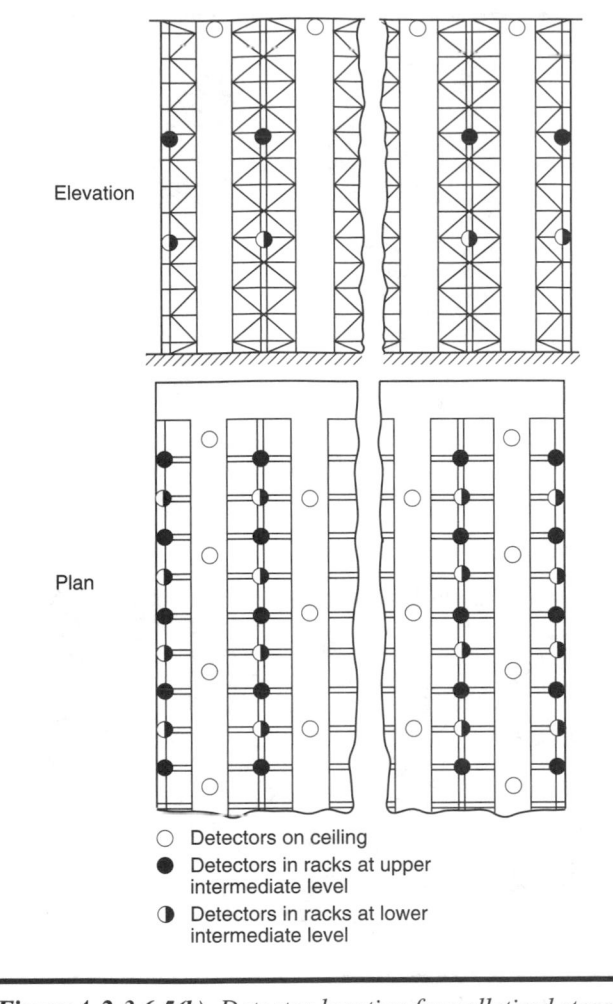

Elevation

Plan

○ Detectors on ceiling
● Detectors in racks at upper
    intermediate level
◑ Detectors in racks at lower
    intermediate level

Figure A-2-3.6.5(b) Detector location for palletized storage (open rack) or no shelved storage in which regular transverse and longitudinal flue spaces are maintained.

conditions, not to ensure that they will detect fires as quickly as they would in a non-airflow condition.

2-3.6.6.2 Location. Smoke detectors shall not be located directly in the airstream of supply registers.

2-3.6.6.3* Spacing. Smoke detector spacing shall be in accordance with Table 2-3.6.6.3 and Figure 2-3.6.6.3.

Exception: Air-sampling or projected beam smoke detectors installed in accordance with the manufacturer's documented instructions.

Because of the very high value of a data center, reducing the spacing of spot-type smoke detectors is common. This spacing may be derived from Table 2-3.6.6.3 and Figure 2-3.6.6.3. It should be still noted that the table and figure are not to be used for spaces under the floor or above the ceiling.

Some authorities having jurisdiction will compute the rate of air change based on the whole air volume, including the room, underfloor plenum and above-ceiling plenum. In other circumstances, the above-ceiling space is not part of the working air volume of the hazard area and only the volume of the room and the underfloor are used to compute air changes per hour. Before the design process is begun the HVAC system must be well understood and the designer and the authority having jurisdiction must agree on what air volume the calculations are to be based.

The reduced spacing for spot-type detectors is based on the concerns about the dilution of smoke. Instead of forming a localized plume of relatively high smoke concentration, the smoke is uniformly mixed throughout the entire air volume by the HVAC system, retarding the development of a detectable concentration until after the fire has increased in intensity.

Table 2-3.6.6.3 Smoke Detector Spacing Based on Air Movement

Minimum per Air Change	Air Changes per Hour	Spacing per Detector ft²	Spacing per Detector m²
1	60	125	11.61
2	30	250	23.23
3	20	375	34.84
4	15	500	46.45
5	12	625	58.06
6	10	750	69.68
7	8.6	875	81.29
8	7.5	900	83.61
9	6.7	900	83.61
10	6	900	83.61

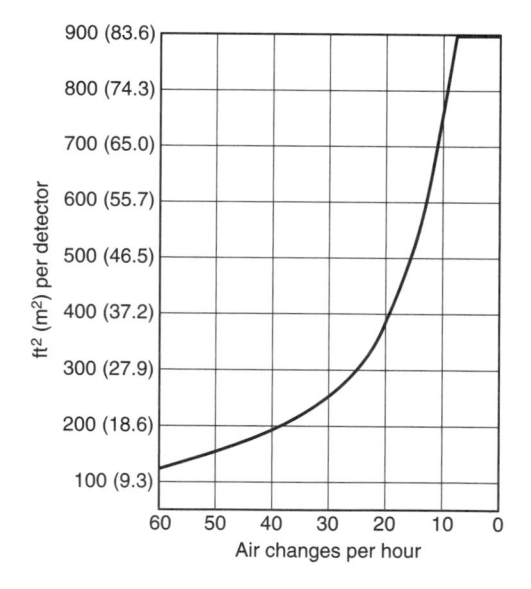

Figure 2-3.6.6.3 High air-movement areas (not to be used for under-floor or above-ceiling spaces).

Data has not yet been presented to the Technical Committee on Initiating Devices for Fire Alarm Systems that would enable the committee to determine the effect of high air movement on linear projected-beam or air-sampling detectors. The designer must depend on the data provided by the manufacturer.

A-2-3.6.6.3 Smoke detector spacing depends on the movement of air within the room.

2-4 Radiant Energy–Sensing Fire Detectors

Radiant energy-sensing fire detectors is the term used to encompass both flame detectors and spark/ember detectors.

2-4.1* General. The purpose and scope of Section 2-4 shall be to provide standards for the selection, location, and spacing of fire detectors that sense the radiant energy produced by burning substances. These detectors are categorized as flame detectors and spark/ember detectors.

The radiant emissions from an ember and a flame are very different. Furthermore, flame detectors and spark/ember detectors are used in very different contexts. Although they share similar physical principles, the way they are applied differs. Section 1-4 provides definitions for these types of detectors.

A-2-4.1 For the purpose of this code, radiant energy includes the electromagnetic radiation emitted as a by-product of the combustion reaction, which obeys the laws of optics. This includes radiation in the ultraviolet, visible, and infrared portions of the spectrum emitted by flames or glowing embers. These portions of the spectrum are distinguished by wavelengths as shown in Table A-2-4.1.

Table A-2-4.1 Spectrum Wavelength Ranges

Radiant Energy	μm
Ultraviolet	0.1–0.35
Visible	0.36–0.75
Infrared	0.76–220

Conversion Factors: 1.0 μm = 1000; nM = 10,000 Å.

Subsection A-2-4.1 clarifies the distinction drawn in the code between heat (which is commonly detected with heat detectors using convective heat transfer), and radiant energy (which is detected with either flame or spark/ember detectors using electro-optical methods to sense sparks, embers, and flames). See Section 1-4 for the definition of Wavelength and the associated appendix material in A-1-4, Wavelength.

2-4.2* Fire Characteristics and Detector Selection.

When using radiant energy-sensing detectors, the designer must match the detector to the signature of the fire or spark/ember fire to be detected. The designer must do so with a degree of precision and attention to detail that is not generally required with other types of detectors.

In an effort to reduce unwanted alarms from non-fire radiant emission sources, detector designers have developed

detectors that look for very specific radiant emission wavelengths that are uniquely associated with the combustion process of particular fuels. This has resulted in detectors that will detect one type of radiant emissions from one class of fuels but virtually blind to fires involving other combustibles. A thorough understanding of how these detectors operate is necessary if they are to be properly applied. See Exhibits 2.28 through 2.33 for examples of spectral response characteristics to various fuels.

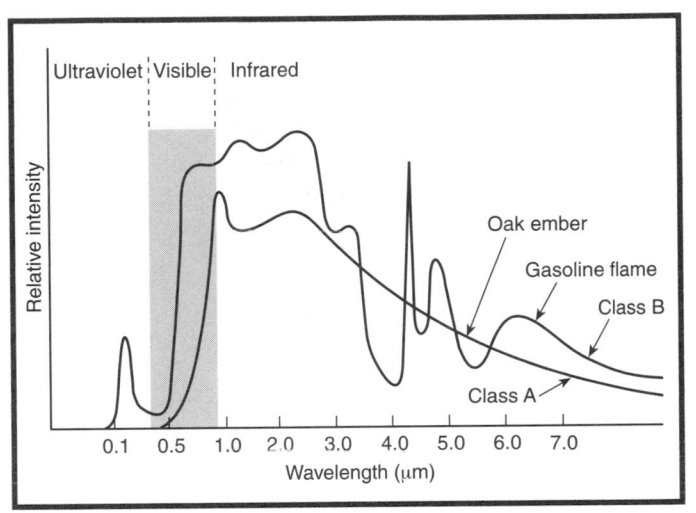

Exhibit 2.28 *Emission spectral of Class A and Class B combustibles. (Source: J.M. Cholin Consultants, Inc., Oakland, NJ)*

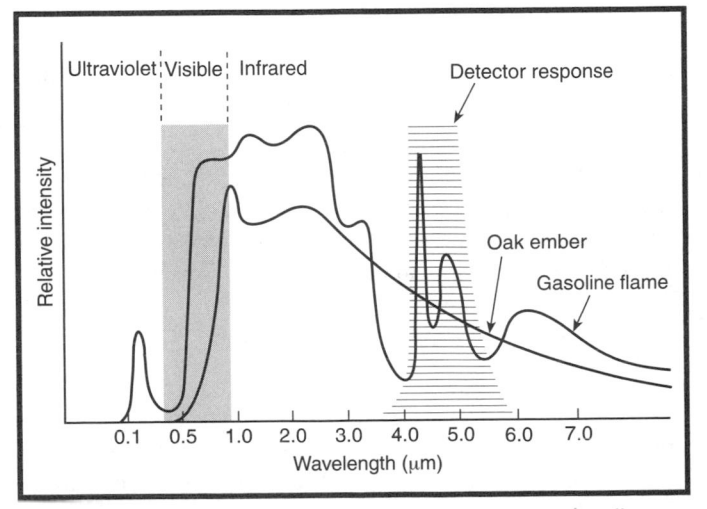

Exhibit 2.29 *The spectral response of a single wavelength infrared flame detector superimposed on the spectrum of typical radiators. (Source: J.M. Cholin Consultants, Inc., Oakland, NJ)*

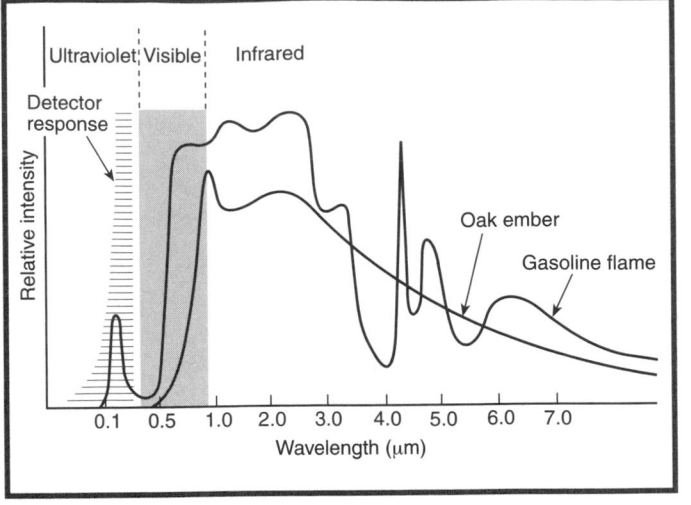

Exhibit 2.30 *The spectral response of a UV flame detector superimposed on the spectrum of typical radiators. (Source: J.M. Cholin Consultants, Inc., Oakland, NJ)*

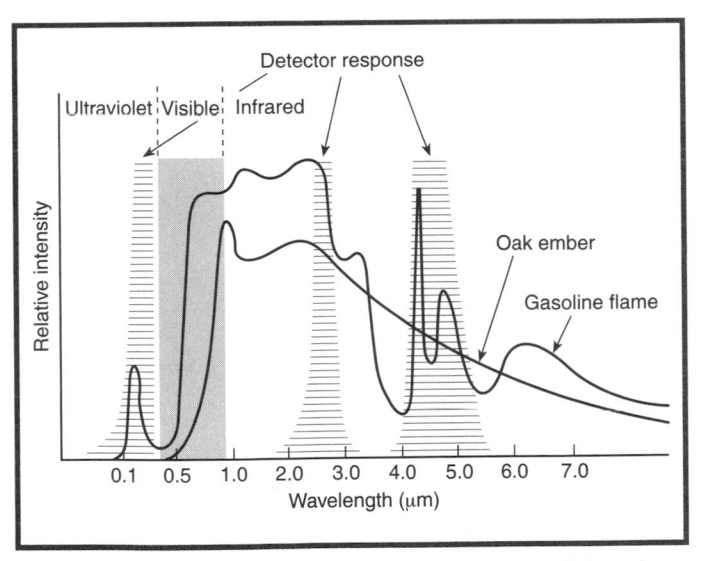

Exhibit 2.31 *The spectral response of an ultraviolet/infrared (UV/IR) flame detector superimposed on the spectrum of typical radiators. (Source: J.M. Cholin Consultants, Inc., Oakland, NJ)*

A-2-4.2 Following are operating principles for two types of detectors.

(a) *Flame Detectors.* Ultraviolet flame detectors typically use a vacuum photodiode Geiger-Muller tube to detect the ultraviolet radiation that is produced by a flame. The photodiode allows a burst of current to flow for each ultraviolet photon that hits the active area of the tube. When the

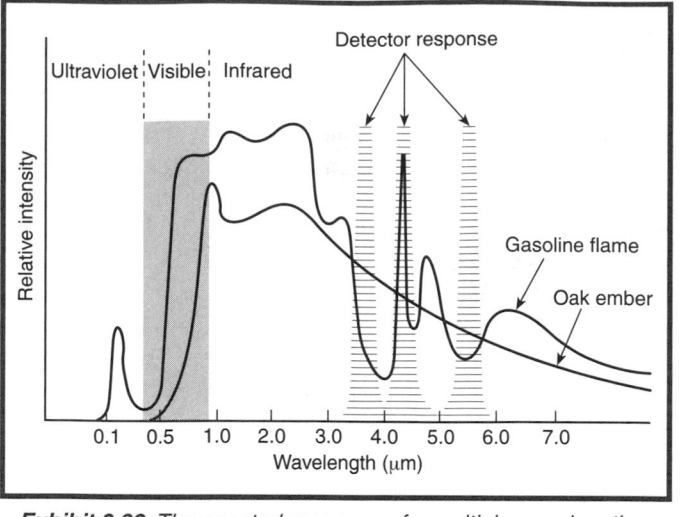

Exhibit 2.32 *The spectral response of a multiple wavelength infrared (IR/IR) flame detector superimposed on the spectrum of typical radiators. (Source: J.M. Cholin Consultants, Inc., Oakland, NJ)*

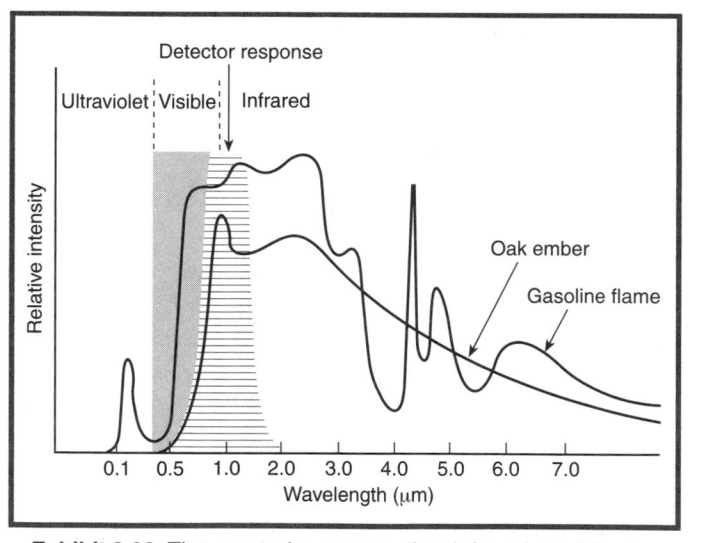

Exhibit 2.33 *The spectral response of an infrared spark/ember detector superimposed on the spectrum of typical radiators. (Source: J.M. Cholin Consultants, Inc., Oakland, NJ)*

number of current bursts per unit time reaches a predetermined level, the detector initiates an alarm.

Some UV/IR flame detectors require radiant emissions at 0.2 μm (UV) and 2.5 μm (IR). Other UV/IR flame detectors require radiant emissions at 0.2 μm (UV) and nominal 4.7 μm (IR).

Some IR/IR flame detectors compare radiant emissions at 4.3 μm (IR) to a reference at nominal 3.8 μm (IR). Other IR/IR flame detectors use a nominal 5.6 μm (IR) reference.

A single wavelength infrared flame detector uses one of several different photocell types to detect the infrared emissions in a single wavelength band that are produced by a flame. These detectors generally include provisions to minimize alarms from commonly occurring infrared sources such as incandescent lighting or sunlight.

An ultraviolet/infrared (UV/IR) flame detector senses ultraviolet radiation with a vacuum photodiode tube and a selected wavelength of infrared radiation with a photocell and uses the combined signal to indicate a fire. These detectors need exposure to both types of radiation before an alarm signal can be initiated.

A multiple wavelength infrared (IR/IR) flame detector senses radiation at two or more narrow bands of wavelengths in the infrared spectrum. These detectors electronically compare the emissions between the bands and initiate a signal where the relationship between the two bands indicates a fire.

(b) *Spark/Ember Detectors.* A spark/ember-sensing detector usually uses a solid state photodiode or phototransistor to sense the radiant energy emitted by embers, typically between 0.5 microns and 2.0 microns in normally dark environments. These detectors can be made extremely sensitive (microwatts), and their response times can be made very short (microseconds).

2-4.2.1* The type and quantity of radiant energy-sensing fire detectors shall be determined based on the performance characteristics of the detector and an analysis of the hazard, including the burning characteristics of the fuel, the fire growth rate, the environment, the ambient conditions, and the capabilities of the extinguishing media and equipment.

A-2-4.2.1 The radiant energy from a flame or spark/ember is comprised of emissions in various bands of the ultraviolet, visible, and infrared portions of the spectrum. The relative quantities of radiation emitted in each part of the spectrum are determined by the fuel chemistry, the temperature, and the rate of combustion. The detector should be matched to the characteristics of the fire.

Almost all materials that participate in flaming combustion emit ultraviolet radiation to some degree during flaming combustion, whereas only carbon-containing fuels emit significant radiation at the 4.35-micron (carbon dioxide) band used by many detector types to detect a flame. *(Refer to Figure A-2-4.2.1.)*

The radiant energy emitted from an ember is determined primarily by the fuel temperature (Planck's Law Emissions) and the emissivity of the fuel. Radiant energy from an ember is primarily infrared and, to a lesser degree, visible in wavelength. In general, embers do not emit ultraviolet energy in significant quantities (0.1 percent of total emissions) until the ember achieves temperatures of 2000 K (1727°C or 3240°F). In most cases, the emissions are

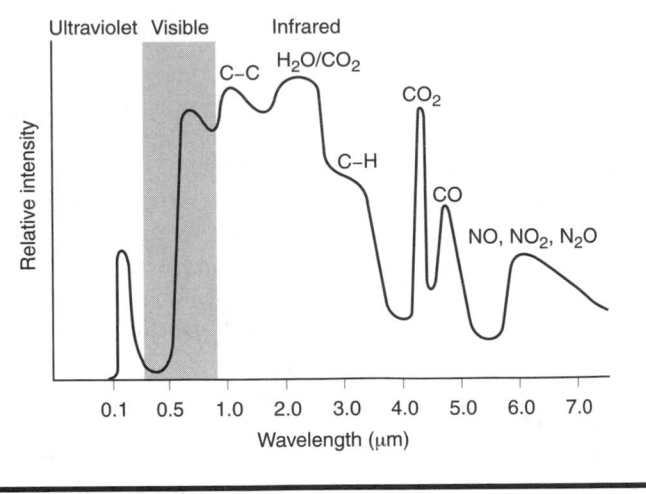

Figure A-2-4.2.1 *Spectrum of a typical flame (free-burning gasoline).*

included in the band of 0.8 microns to 2.0 microns, corresponding to temperatures of approximately 750°F to 1830°F (398°C to 1000°C).

Most radiant energy detectors have some form of qualification circuitry within them that uses time to help distinguish between spurious, transient signals and legitimate fire alarms. These circuits become very important where the anticipated fire scenario and the ability of the detector to respond to that anticipated fire are considered. For example, a detector that uses an integration circuit or a timing circuit to respond to the flickering light from a fire might not respond well to a deflagration resulting from the ignition of accumulated combustible vapors and gases, or where the fire is a spark that is traveling up to 328 ft/sec (100 m/sec) past the detector. Under these circumstances, a detector that has a high-speed response capability is most appropriate. On the other hand, in applications where the development of the fire is slower, a detector that uses time for the confirmation of repetitive signals is appropriate. Consequently, the fire growth rate should be considered in selecting the detector. The detector performance should be selected to respond to the anticipated fire.

The radiant emissions are not the only criteria to be considered. The medium between the anticipated fire and the detector is also very important. Different wavelengths of radiant energy are absorbed with varying degrees of efficiency by materials that are suspended in the air or that accumulate on the optical surfaces of the detector. Generally, aerosols and surface deposits reduce the sensitivity of the detector. The detection technology used should take into account those normally occurring aerosols and surface deposits to minimize the reduction of system response between maintenance intervals. It should be noted that the

smoke evolved from the combustion of middle and heavy fraction petroleum distillates is highly absorptive in the ultraviolet end of the spectrum. If using this type of detection, the system should be designed to minimize the effect of smoke interference on the response of the detection system.

The environment and ambient conditions anticipated in the area to be protected impact the choice of detector. All detectors have limitations on the range of ambient temperatures over which they will respond, consistent with their tested or approved sensitivities. The designer should make certain that the detector is compatible with the range of ambient temperatures anticipated in the area in which it is installed. In addition, rain, snow, and ice attenuate both ultraviolet and infrared radiation to varying degrees. Where anticipated, provisions should be made to protect the detector from accumulations of these materials on its optical surfaces.

2-4.2.2* The selection of the radiant energy-sensing detectors shall be based on the following:

(1) Matching of the spectral response of the detector to the spectral emissions of the fire or fires to be detected
(2) Minimizing the possibility of spurious nuisance alarms from non-fire sources inherent to the hazard area

Once the type of combustion has been determined and the decision regarding type of detector to be used has been made, the decision regarding the most appropriate model or technology must be selected. This is the second stage of the decision tree.

The expected emission spectrum from the fuel is matched to the wavelength bands of the candidate detector to assure response to the fire, using the criteria stated in the detector manufacturer's engineering manual. The performance capabilities of the detector must be matched with the known radiant emissions of the fuel. To ascertain that the detector is appropriate for the fuels to be detected, the designer may use the performance attributes that were verified by a qualified testing laboratory during the listing evaluation.

Then the candidate detector must be evaluated for its unwanted alarm immunity with respect to the ambient or false alarm sources anticipated in the hazard area. The information provided in A-2-4.2.1 also relates to 2-4.2.2(2) and should be used for this stage of the decision process.

Finally the designer must consider the impact of the full range of expected ambient conditions on both the detection capability, as well as on the stability of the candidate detector. Both flame detectors and spark/ember detectors are routinely installed outdoors where they are exposed to the weather and fluctuations in temperature.

Special attention must be given to the temperature range limits and other limiting weather-related conditions that are provided by the manufacturer. Such attention will

help ensure that the detector has been qualified for the anticipated extremes. The prudent designer will document his or her decision-making process in writing for future reference.

A-2-4.2.2 Normal radiant emissions that are not from a fire can be present in the hazard area. When selecting a detector for an area other potential sources of radiant emissions should be evaluated. Refer to A-2-4.2.1 for additional information.

2-4.3 Spacing Considerations.

The spacing considerations for radiant energy-sensing fire detectors are derived from the physics of light transmission. This contrasts with the fire plume dynamics and the fluid flow physics that govern the spacing of heat and smoke detectors. Consequently, when using radiant energy-sensing fire detectors, the designer must determine the spacing of the detectors by the location and aiming of the devices.

The location and aiming of the detectors is in turn determined by two critical factors: the field of view of the detector (see definition in Section 1-4) and the sensitivity of the detector (see definition in Section 1-4).

2-4.3.1 General Rules.

2-4.3.1.1* Radiant energy-sensing fire detectors shall be employed consistent with the listing or approval and the inverse square law, which defines the fire size versus distance curve for the detector.

The inverse square law relates the size of the fire, the detector sensitivity, and the distance between the fire and the detector. It is applicable to all radiant energy-sensing detectors. However, there are some tacit assumptions made when the inverse square law is used for modeling the performance of flame detectors.

The first assumption is that the fire is small and far away from the detector. This permits modeling the fire as a point source. When the fire is modeled as a point source all of the radiant power is thought of as emanating from a single point. The alternative would be to model the fire as a portion of the field of view. This alternative approach requires the use of advanced calculus and is far more difficult for the average designer.

The second assumption is that the flame is assumed to be optically dense, meaning that radiation from the back side of the flame does not pass through the flame. Generally, because flame intermediates absorb radiation at the same wavelengths at which they emit radiation, this assumption holds true.

Using the inverse square law enables the design engineer to compute with considerable precision how large the fire must get before there is enough radiant energy hitting the detector to cause an alarm. These calculations are critical because they help determine the number of detectors of

given sensitivity, location, and aiming that are necessary to detect a fire of given size.

It is important to remember that fire size is quantified in units of power output, either Btu/second or kilowatts, regardless of the type of radiant energy-sensing fire detector under consideration. Also the normally assumed 35 percent radiative fraction used in other fire calculations is not used when quantifying the power output of a fire in this context. Radiant energy-sensing detectors' sensitivity derives from experiments conducted by a qualified testing laboratory that performs the listing investigation. These experiments use the whole fire output as the metric.

A-2-4.3.1.1 All optical detectors respond according to the following theoretical equation:

$$S = \frac{kpe\zeta d}{d^2}$$

where:

k = proportionality constant for the detector

p = radiant power emitted by the fire

e = Naperian logarithm base (2.7183)

ζ = extinction coefficient of air

d = distance between the fire and the detector

S = radiant power reaching the detector

The sensitivity *(S)* typically is measured in nanowatts. This equation yields a family of curves similar to the one shown in Figure A-2-4.3.1.1.

The curve defines the maximum distance at which the detector consistently detects a fire of defined size and fuel. Detectors should be employed only in the shaded area above the curve.

Under the best of conditions, with no atmospheric absorption, the radiant power reaching the detector is reduced by a factor of 4 if the distance between the detector and the fire is doubled. For the consumption of the atmospheric extinction, the exponential term *zeta* (ζ) is added to the equation. Zeta is a measure of the clarity of the air at the wavelength under consideration. Zeta is affected by humidity, dust, and any other contaminants in the air that are absorbent at the wavelength in question. Zeta generally has values between −0.001 and −0.1 for normal ambient air.

2-4.3.1.2 Detector quantity shall be based on the detectors being positioned so that no point requiring detection in the hazard area is obstructed or outside the field of view of at least one detector.

A flame detector or spark/ember detector cannot detect what it cannot "see." The definition of the term *field of view* in Section 1-4 has a sensitivity criterion attached to it. It is the angle off the optical axis of the detector where the effective sensitivity is 50 percent of the on-axis sensitivity.

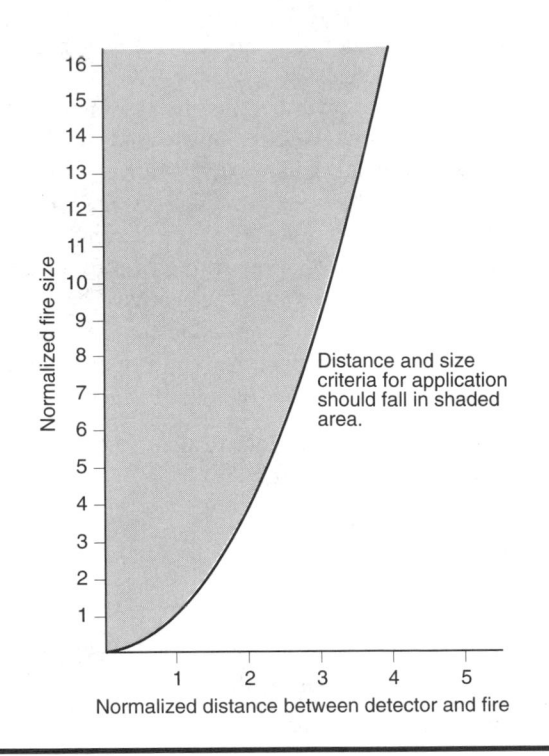

Figure A-2-4.3.1.1 Normalized fire size vs. distance.

All points where a fire can exist in the hazard area must be within the field of view of at least one detector. This requirement also effectively demands that the manufacturer provide sensitivity versus angle of incidence data in its engineering manual.

When flame detectors are used to release extinguishing agents, such as aqueous film-forming foam (AFFF), alarm signals from two or more detectors are usually required before the agent is released. Under those circumstances, the designer should apply 2-4.3.2.5 in a manner that requires all points in the hazard area where a fire can exist to be within the fields of view of the number of detectors required to discharge the extinguishing agent. Otherwise, the fire could occur in a portion of the hazard area that is within the field of view of only one detector. The release of the extinguishing agent would be delayed until the fire grows to a size sufficient to alarm the additional confirmation detector(s). This would result in far more fire damage and a greater threat of loss of life.

The requirements in 2-4.3.1.1 and 2-4.3.1.2 pertain to both flame and spark/ember detectors. There are other design considerations that are more specific to one type of detector or the other. These are addressed in 2-4.3.2 for flame detectors and 2-4.3.3 for spark/ember detectors, respectively.

2-4.3.2 Spacing Considerations for Flame Detectors.

2-4.3.2.1* The location and spacing of detectors shall be the result of an engineering evaluation that includes the following:

(1) Size of the fire that is to be detected
(2) Fuel involved
(3) Sensitivity of the detector
(4) Field of view of the detector
(5) Distance between the fire and the detector
(6) Radiant energy absorption of the atmosphere
(7) Presence of extraneous sources of radiant emissions
(8) Purpose of the detection system
(9) Response time required

In the context of 2-4.3.2.1, the term *spacing* includes the number, location, and aiming of the detectors selected for the hazard area. In every system design using flame detectors, the location of each unit in the system must address the criteria listed in 2-4.3.2.1.

Product development in the field of radiant energy-sensing fire detection has been vigorous. New design concepts are being introduced frequently. Consequently, at the current rate of change, it is impossible for the code or the code handbook to provide an exhaustive list of the available technologies.

Recently, microcomputer based multispectrum flame detectors have become available that use a microcomputer to evaluate emissions from four, five, and possibly six different bands in the UV, visible, and IR regions. However, the wavelength bands and operational architecture of these multispectrum devices have not yet been disclosed in sufficient detail to provide the Technical Committee on Automatic Fire Detectors with the requisite information for inclusion in this edition of the code.

The requirements of 2-4.3 effectively direct the system designer to work through a decision tree to arrive at the most appropriate detector for the fire hazard under consideration. The first decision is whether flame detection or spark/ember detection is the most appropriate type of radiant energy-sensing fire detector. The type of detector is often determined by the physical state of the material involved in the fire.

Combustion occurs in the gas phase and in the solid phase. Flammable gases, flammable liquids, combustible liquids, and many combustible solids will form a flame (see definition in Section 1-4). The combustion takes place in the gas phase, regardless of the physical state of the unburned fuel. The heat from the combustion gasifies the fuel allowing it to mix with air, supporting the flame. Because gas molecules are free to vibrate in free space, the flame spectra show typical emission spikes that indicate flame intermediates and products.

Many solids also burn in the solid phase as embers. In solid phase combustion, the molecules on the surface of the fuel particle are oxidized off the surface of the particle without the development of a layer of gasified fuel, which could produce a true flame. Therefore, combustion intermediates (partially oxidized molecules), and, often combustion products, are locked-up on the surface of the fuel particle and are not free to assume the diverse vibrational states of a gas phase molecule. Consequently, the radiant emissions are profoundly different in solid phase combustion than they are in gas phase combustion. This difference in combustion radiant emissions necessitates different types of radiant energy-sensing detectors for the different physical combustion states.

Subsection 2-4.3.2.1 states the criteria that must be considered during the decision-making process. The material of 2-4.3.3.2 provides additional insight into how the decision-making process is driven by the detector performance criteria, and the anticipated fire and hazard environment.

A-2-4.3.2.1 The following are types of application for which flame detectors are suitable:

(1) High-ceiling, open-spaced buildings such as warehouses and aircraft hangers
(2) Outdoor or semioutdoor areas where winds or draughts can prevent smoke from reaching a heat or smoke detector
(3) Areas where rapidly developing flaming fires can occur, such as aircraft hangars, petrochemical production areas, storage and transfer areas, natural gas installations, paint shops, or solvent areas
(4) Areas needing high fire risk machinery or installations, often coupled with an automatic gas extinguishing system
(5) Environments that are unsuitable for other types of detectors

Some extraneous sources of radiant emissions that have been identified as interfering with the stability of flame detectors include the following:

(1) Sunlight
(2) Lightning
(3) X-rays
(4) Gamma rays
(5) Cosmic rays
(6) Ultraviolet radiation from arc welding
(7) Electromagnetic interference (EMI, RFI)
(8) Hot objects
(9) Artificial lighting

No detector type or model is susceptible to all or even a majority of these unwanted alarm sources. Different types and models of flame detectors exhibit different degrees of susceptibility to some of these sources. Despite the best

intentions and ardent efforts of flame detector manufacturers, the completely "nuisance alarm–proof" radiant-energy sensing detector has not yet been invented.

2-4.3.2.2 The system design shall specify the size of the flaming fire of given fuel that is to be detected.

This is a performance based code requirement. Because of the complexities inherent in the design of flame detection systems, a performance criterion must drive the design. The performance criterion is the detection of a fire of specified size and fuel.

Fire size is usually measured in kilowatts (kW) or British thermal units per second (Btu/second). But more information is necessary in this context because flames are optically dense radiators. This means that the radiation from the back side of the flame does not travel through the flame toward the detector. Instead, it is reabsorbed by the flame. Consequently, the flame detector only "sees" the profile of the fire—that is, its width and height.

The flame height is proportional to the heat release rate (kW or Btu/sec). Consequently, both fire width and heat release rate are necessary to quantify the size of a fire. Many designers have not yet made the conversion from simply stipulating a fire size criterion in terms of a pool fire of given fuel and area.

Appendix B outlines a detailed design method for flame detection systems. The design fire is specified. The fire flame height is calculated. The radiating area of the fire is then calculated. The radiant output of the fire is then correlated to the sensitivity tests performed by a testing laboratory in the course of the listing evaluation. The correlated radiant density per unit of flame area is then assigned to the design fire and the radiant output is calculated based on the radiant output per unit area times the radiating area of the fire. The design fire is then modeled as a point source radiator having the calculated radiant output.

2-4.3.2.3* In applications where the fire to be detected could occur in an area not on the optical axis of the detector, the distance shall be reduced or detectors added to compensate for the angular displacement of the fire in accordance with the manufacturer's documented instructions.

A-2-4.3.2.3 The greater the angular displacement of the fire from the optical axis of the detector, the larger the fire must become before it is detected. This phenomenon establishes the field of view of the detector. Figure A-2-4.3.2.3 shows an example of the effective sensitivity versus angular displacement of a flame detector.

2-4.3.2.4* In applications in which the fire to be detected is of a fuel that differs from the test fuel used in the process of listing or approval, the distance between the detector and the fire shall be adjusted consistent with the fuel specificity of the detector as established by the manufacturer.

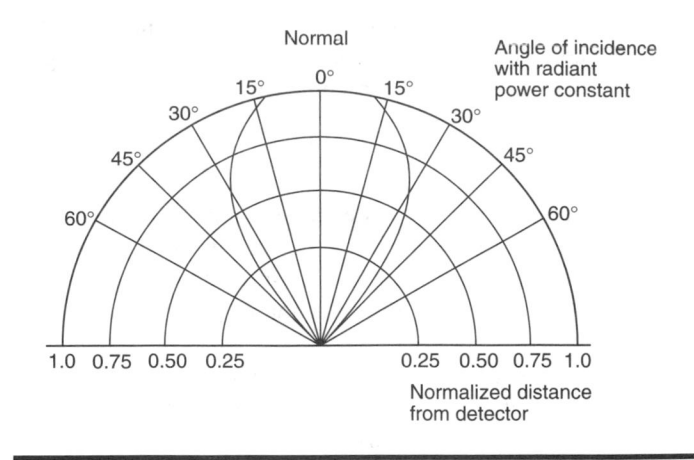

Figure A-2-4.3.2.3 Normalized sensitivity vs. angular displacement.

In an effort to make flame detectors more sensitive yet more immune to unwanted alarms, manufacturers began designing detectors that concentrated on very specific features of the flame spectrum. These include the emissions of the flame across the range of wavelengths from ultraviolet to infrared. In concept, such flame detectors infer that a flame exists if an emission of a specific wavelength or set of wavelengths is detected. However, one fuel emits a different radiant intensity at a given wavelength than another fuel. This gives rise to detectors that are fuel specific. There are cases where a flame detector may be several times more sensitive to one fuel than another.

The language of 2-4.3.2.4 effectively requires the designer to obtain flame spectra of potential fuels in the hazard area and response curves from the detector manufacturer to make certain the detector will respond to the fuel(s) involved. Furthermore, if the detector chosen for the system is less sensitive to one of the fuels in the hazard area, the spacing (including quantity, location, and aiming) of the detectors must be adjusted accordingly.

A-2-4.3.2.4 Virtually all radiant energy-sensing detectors exhibit some kind of fuel specificity. If burned at uniform rates [J/sec (W)], different fuels emit different levels of radiant power in the ultraviolet, visible, and infrared portions of the spectrum. Under free-burn conditions, a fire of given surface area but of different fuels burns at different rates [J/sec (W)] and emits varying levels of radiation in each of the major portions of the spectrum. Most radiant energy detectors designed to detect flame are qualified based on a defined fire under specific conditions. If employing these detectors for fuels other than the defined fire, the designer should make certain that the appropriate adjustments to the maximum distance between the detector and the fire are made consistent with the fuel specificity of the detector.

2-4.3.2.5 Because flame detectors are line-of-sight devices, their ability to respond to the required area of fire in the zone that is to be protected shall not be compromised by the presence of intervening structural members or other opaque objects or materials.

Some atmospheric contaminants including vapors and gases may be opaque at the wavelengths used by some flame detectors. This can have a significant effect on the performance of the system. See A-2-4.3.1.1 for the relationship of fire size and distance from a detector. Also, a window material that is clear in the visible portion of the spectrum might be opaque in either the ultraviolet (UV) or infrared (IR) portions of the spectrum. Common glass is opaque in both the UV and IR. Consequently, 2-4.3.2.5 must be applied to any window material that is not specifically listed for use with the detector in question.

2-4.3.2.6* Provisions shall be made to sustain detector window clarity in applications where airborne particulates and aerosols coat the detector window between maintenance intervals and affect sensitivity.

A-2-4.3.2.6 This requirement has been satisfied by the following means:

(1) Lens clarity monitoring and cleaning where a contaminated lens signal is rendered
(2) Lens air purge

The need to clean detector windows can be reduced by the provision of air purge devices. These devices are not foolproof, however, and are not a replacement for regular inspection and testing. Radiant energy-sensing detectors should not be placed in protective housings (for example, behind glass) to keep them clean, unless such housings are listed for the purpose. Some optical materials are absorptive at the wavelengths used by the detector.

2-4.3.3 Spacing Considerations for Spark/Ember Detectors.

2-4.3.3.1* The location and spacing of detectors shall be the result of an engineering evaluation that includes the following:

(1) Size of the spark or ember that is to be detected
(2) Fuel involved
(3) Sensitivity of the detector
(4) Field of view of the detector
(5) Distance between the fire and the detector
(6) Radiant energy absorption of the atmosphere
(7) Presence of extraneous sources of radiant emissions
(8) Purpose of the detection systems
(9) Response time required

A-2-4.3.3.1 Spark/ember detectors are installed primarily to detect sparks and embers that could, if allowed to con-

tinue to burn, precipitate a much larger fire or explosion. Spark/ember detectors are typically mounted on some form of duct or conveyor, monitoring the fuel as it passes by. Usually, it is necessary to enclose the portion of the conveyor where the detectors are located, as these devices generally require a dark environment. Extraneous sources of radiant emissions that have been identified as interfering with the stability of spark/ember detectors include the following:

(1) Ambient light
(2) Electromagnetic interference (EMI, RFI)
(3) Electrostatic discharge in the fuel stream

Exhibit 2.34 shows typical applications where spark/ember detectors are used. It should be noted that the detectors are located at a point along the duct or conveyor, monitoring the cross section of the duct or conveyor at that one point by essentially "looking across" the duct. Commercially available, listed spark/ember detectors are designed to monitor a fuel stream as it moves past the detector. They are not designed to "look down the duct." The capacitive nature of the circuitry of this type of detector generally makes them incapable of detecting a slowly growing radiator; the radiator must move past the detector rapidly if it is to be detected.

Appendix B provides a more detailed design guide for spark detection system design.

2-4.3.3.2* The system design shall specify the size of the spark or ember of the given fuel that the detection system is to detect.

The size of an ember is measured in terms of watts or milliwatts. The radiant energy from an ember and hence its size, cannot be accurately inferred from a description that states diameter and temperature only. See the definition for Spark/Ember Detector Sensitivity in Section 1-4. Furthermore, the equation for the inverse square law in A-2-4.3.1.1 cannot be used to calculate the ability of the detector to detect the ember in question unless both the detector sensitivity and the ember size are specified in the same terms of radiant power: watts, milliwatts, or microwatts.

As with 2-4.3.2.2 regarding flame detectors, 2-4.3.3.2 is a performance based design criterion that drives the entire system design.

In general terms, this material provides the applicable criteria for radiant energy-sensing fire detector selection. The hazard analysis must, therefore, begin with the determination as to whether the combustible will burn in the solid phase as an ember or in the gas phase as a flame. That determination then points the designer toward the spark/ember detector (for solid phase combustion) or the flame detector (for gas phase combustion). The engineering manuals provided by the manufacturers of the various detectors under consideration should be used to determine the usefulness of a particular device for the hazard under consideration.

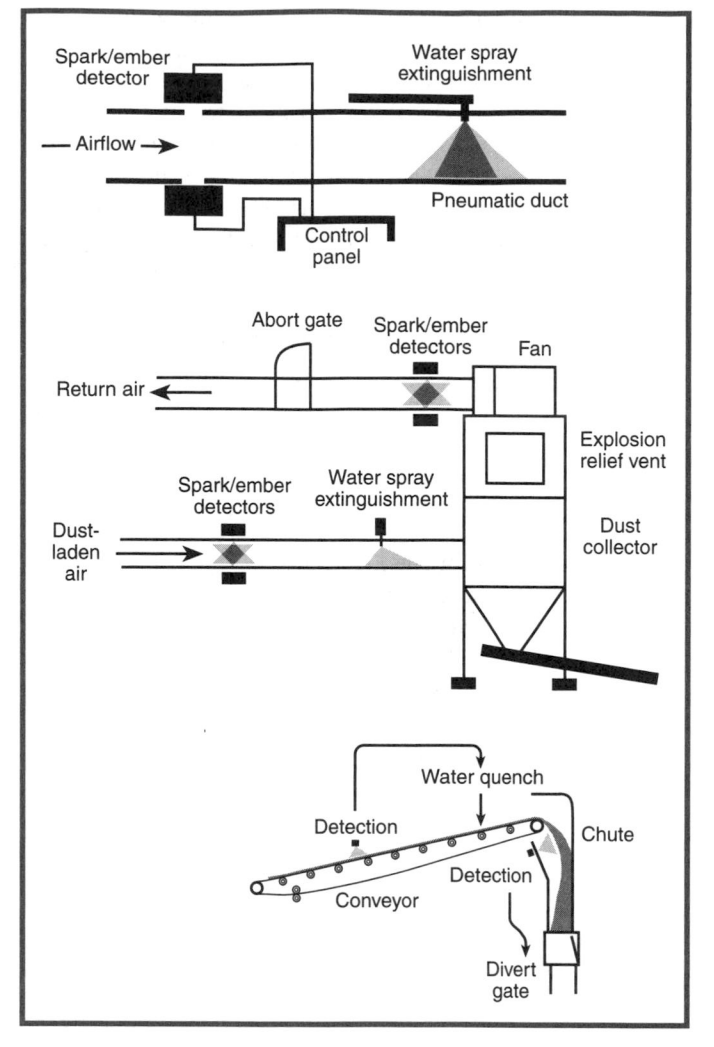

Exhibit 2.34 Spark/ember detectors are usually used on conveyance ducts and conveyors to detect embers in particulate solids as they are transported. The top drawing shows the general concept of spark/ember detectors. The middle drawing illustrates the application of spark/ember detectors to protect a dust collector. The bottom drawing illustrates the protection of a conveyor. (Source: J.M. Cholin Consultants, Inc., Oakland, NJ)

A-2-4.3.3.2 There is a minimum ignition power (watts) for all combustible dusts. If the spark or ember is incapable of delivering that quantity of power to the adjacent combustible material (dust), an expanding dust fire cannot occur. The minimum ignition power is determined by the fuel chemistry, fuel particle size, fuel concentration in air, and ambient conditions such as temperature and humidity.

2-4.3.3.3 Spark detectors shall be positioned so that all points within the cross section of the conveyance duct, conveyor, or chute where the detectors are located are within the field of view (as defined in Section 1-4) of at least one detector.

Most makes of spark detection require a minimum of two detectors at each location on a pneumatic conveyance duct requiring detection. The need for this quantity of detectors is determined by the field of view of the detector. Unless the field of view is 180 degrees, two devices are needed to cover the inside of a duct. As the duct diameter increases, it becomes necessary to use that portion of the field of view where the detector is most sensitive to offset the absorption of the radiant emission from the spark by the nonburning material. Consequently, most spark detection systems require additional detectors as duct size increases.

2-4.3.3.4* The location and spacing of the detectors shall be adjusted using the inverse square law, modified for the atmospheric absorption and the absorption of nonburning fuel suspended in the air in accordance with the manufacturer's documented instructions.

The equation used for spark detection design is the same as that used for flame detection design. However, the atmospheric extinction coefficient, ζ (zeta), is determined by the optical absorbance of the nonburning material in the band of wavelengths used by the detector and by the concentration of the nonburning material per unit of air volume. The conservative design approach is to assume an emissivity (absorbance) of 1.0. This means that the material is 100 percent absorbant and does not reflect any radiation that strikes a nonburning fuel particle.

A-2-4.3.3.4 As the distance between the fire and the detector increases, the radiant power reaching the detector decreases. Refer to A-2-4.3.1.1 for additional information.

2-4.3.3.5* In applications where the sparks to be detected could occur in an area not on the optical axis of the detector, the distance shall be reduced or detectors added to compensate for the angular displacement of the fire in accordance with the manufacturer's documented instructions.

A-2-4.3.3.5 The greater the angular displacement of the fire from the optical axis of the detector, the larger the fire must become before it is detected. This phenomenon establishes the field of view of the detector. Figure A-2-4.3.2.3 shows an example of the effective sensitivity versus angular displacement of a flame detector.

2-4.3.3.6* Provisions shall be made to sustain the detector window clarity in applications where airborne particulates and aerosols coat the detector window and affect sensitivity.

A-2-4.3.3.6 This requirement has been satisfied by the following means:

(1) Lens clarity monitoring and cleaning where a contaminated lens signal is rendered
(2) Lens air purge

2-4.4 Other Considerations.

2-4.4.1 Radiant energy-sensing detectors shall be protected either by way of design or installation to ensure that optical performance is not compromised.

Since these types of detectors are usually installed where they must endure the rigors of difficult industrial environments, the designer is cautioned to consider the long-term impact of the environment on the optical performance of the detectors.

Atmospheric contaminants are often opaque at detector wavelengths. Structures are often modified after detector placement and aiming. This may affect the clear view of the hazard area or impede the required routine maintenance of detectors. Finally, unless a detector has been specifically listed for use with a particular window material, the installation of a detector behind a protective window violates the listing of the detector.

2-4.4.2 If necessary, radiant energy-sensing detectors shall be shielded or otherwise arranged to prevent action from unwanted radiant energy.

In some cases, shielding a detector from radiant emissions coming from a portion of its field of view—where the sole source of radiant emissions is a spurious source—can be an effective way of dealing with the source of unwanted alarms. Many detectors are available with scoops or baffles to limit the field of view to a small portion of the total viewing area. This provides the ability to operate in spite of the presence of a spurious alarm source. When considering such methods, the designer should consult the manufacturer. It is important to remember that reflected radiant emissions can also cause alarms.

All surfaces are not uniformly reflective at all wavelengths. Unwanted alarms are often traced to reflections from radiant sources that are outside the actual field of view of the detector.

2-4.4.3 Where used in outdoor applications, radiant energy-sensing detectors shall be shielded or otherwise arranged in a fashion to prevent diminishing sensitivity by conditions such as rain or snow and yet allow a clear field of vision of the hazard area.

Both water and snow are highly absorptive in both the ultraviolet and infrared portions of the spectrum. Where detectors are exposed to interference from streaming water or snow, their ability to respond to the design fire can be seriously compromised. It is also possible for water to initiate false alarms by causing the modulation of background radiant emissions, simulating the modulated emissions of a flame.

2-4.4.4 A radiant energy-sensing fire detector shall not be installed in a location where the ambient conditions are

known to exceed the extremes for which the detector has been listed.

2-5 Other Fire Detectors

It is the intent of the code to provide for the development of new technologies and to allow the use of such technologies consistent with sound principles of fire protection engineering. The requirements in Section 2-5 provide for methods not explicitly described in other sections of Chapter 2.

2-5.1 Detectors that operate on principles different from those covered by Sections 2-2, 2-3, and 2-4 shall be classified as *other fire detectors*. Such detectors shall be installed in all areas where they are required either by other NFPA codes and standards or by the authority having jurisdiction.

Chapter 7 outlines the required maintenance procedures and schedules for all components of a fire alarm system, including the initiating devices. Initiating devices covered by 2-5 must be maintained pursuant to Chapter 7 and the manufacturer's recommendations.

2-5.2 Other fire detectors" shall operate where subjected to the abnormal concentration of combustion effects that occur during a fire, such as water vapor, ionized molecules, or other phenomena for which they are designed. Detection depends on the size and intensity of fire to provide the necessary quantity of required products and related thermal lift, circulation, or diffusion for operation.

2-5.3* Room sizes and contours, airflow patterns, obstructions, and other characteristics of the protected hazard shall be taken into account.

A-2-5.3 The performance characteristics of the detector and the area into which it is to be installed should be evaluated to minimize nuisance alarms or conditions that would interfere with operation.

2-5.4 Location and Spacing.

2-5.4.1 The location and spacing of detectors shall be based on the principle of operation and an engineering survey of the conditions anticipated in service. The manufacturer's technical bulletin shall be consulted for recommended detector uses and locations.

2-5.4.2 Detectors shall not be spaced beyond their listed or approved maximums. Closer spacing shall be used where the structural or other characteristics of the protected hazard warrant.

2-5.4.3 The location and sensitivity of the detectors shall be the result of an engineering evaluation that includes the following:

(1) Structural features, size, and shape of the rooms and bays
(2) Occupancy and uses of the area
(3) Ceiling height
(4) Ceiling shape, surface, and obstructions
(5) Ventilation
(6) Ambient environment
(7) Burning characteristics of the combustible materials present
(8) Configuration of the contents in the area to be protected

2-6 Sprinkler Waterflow Alarm-Initiating Devices

2-6.1 The provisions of Section 2-6 shall apply to devices that initiate an alarm indicating a flow of water in a sprinkler system.

2-6.2* Initiation of the alarm signal shall occur within 90 seconds of waterflow at the alarm-initiating device when flow occurs that is equal to or greater than that from a single sprinkler of the smallest orifice size installed in the system. Movement of water due to waste, surges, or variable pressure shall not be indicated.

The water in a wet pipe automatic sprinkler system riser is not static, but moves up and down depending on different pressures in the municipal main and the sprinkler system. Air trapped in the sprinkler piping provides a compressible cushion, enhancing the tendency for flow to occur as a result of variance of pressures in the water supply system. The alarm check valve in the sprinkler system tends to reduce, but not eliminate this flow. Fortunately, dry pipe, preaction, and deluge sprinkler systems do not suffer from this phenomenon because of their design.

Some water flow alarm initiations are obtained from pressure switches installed on the wet pipe sprinkler system alarm trim, rather than from paddle or vane-type switches located on the riser itself. If the pressure switch is mounted on top of the retard chamber, care must be taken to meet the requirements of 2-6.3 and 3-8.3.3.3.2. False alarms can arise if the retard chamber drain on the sprinkler system trim becomes clogged.

A-2-6.2 The waterflow device should be field adjusted so that an alarm is initiated no more than 90 seconds after a sustained flow of at least 10 gpm (40 L/min).

Features that should be investigated to minimize alarm response time include the following:

(1) Elimination of trapped air in the sprinkler system piping
(2) Use of an excess pressure pump,
(3) Use of pressure drop alarm-initiating devices
(4) A combination thereof

Care should be used when choosing waterflow alarm-initiating devices for hydraulically calculated looped systems and those systems using small orifice sprinklers. Such systems might incorporate a single point flow of significantly less than 10 gpm (40 L/min). In such cases, additional waterflow alarm-initiating devices or the use of pressure drop-type waterflow alarm-initiating devices might be necessary.

Care should be used when choosing waterflow alarm-initiating devices for sprinkler systems that use on–off sprinklers to ensure that an alarm is initiated in the event of a waterflow condition. On-off sprinklers open at a predetermined temperature and close when the temperature reaches a predetermined lower temperature. With certain types of fires, waterflow might occur in a series of short bursts of a duration of 10 seconds to 30 seconds each. An alarm-initiating device with retard might not detect waterflow under these conditions. An excess pressure system or a system that operates on pressure drop should be considered to facilitate waterflow detection on sprinkler systems that use on–off sprinklers.

Excess pressure systems can be used with or without alarm valves. The following is a description of one type of excess pressure system with an alarm valve.

An excess pressure system with an alarm valve consists of an excess pressure pump with pressure switches to control the operation of the pump. The inlet of the pump is connected to the supply side of the alarm valve, and the outlet is connected to the sprinkler system. The pump control pressure switch is of the differential type, maintaining the sprinkler system pressure above the main pressure by a constant amount. Another switch monitors low sprinkler system pressure to initiate a supervisory signal in the event of a failure of the pump or other malfunction. An additional pressure switch can be used to stop pump operation in the event of a deficiency in water supply. Another pressure switch is connected to the alarm outlet of the alarm valve to initiate a waterflow alarm signal when waterflow exists. This type of system also inherently prevents false alarms due to water surges. The sprinkler retard chamber should be eliminated to enhance the detection capability of the system for short duration flows.

In many facilities, the sprinkler system is used as both a suppression system and a detection system. The flow of water initiates an alarm. In a large wet pipe system with large sprinkler risers, depending on the amount of trapped air, the flow from a single head has proven to be hard to detect. The air acts as a gas cushion, allowing pulsating variations in water pressure within the riser when a single head discharges. This can prevent the vane of a vane-type waterflow switch from lifting, or the clapper of an alarm check valve from opening, long enough to overcome the pneumatic, electronic, or mechanical retard mechanism. Rising and falling water supply pressure over the course of any 24-hour period causes the pressure entering the wet pipe sprinkler system to change. This pressure change can cause unwanted alarms. Consequently, most flow switches are equipped with a retard feature that delays the transmission of a signal until after stable waterflow has been achieved.

Meeting the 90-second criterion can be challenging. If the sprinkler system uses on/off sprinkler heads, a waterflow alarm-initiating device must be used that can sense a possible flow of shorter duration.

Meeting the 90-second criterion also can be a challenge with large systems. Where the waterflow alarm-initiating device is installed on a simple fire alarm control unit, the retard adjustment is simple and straightforward. However, where addressable fire alarm systems are used to monitor waterflow alarm-initiating devices, the worst-case polling delay of the addressable initiating device circuit must be added to the retard adjustment of the waterflow alarm-initiating device. This adjustment is made to ensure that the total accumulated delay, from the moment that waterflow is begun to the time that the fire alarm notification begins, is not longer than 90 seconds.

Care must be exercised when adjusting waterflow alarm-initiating device retard delays on large sprinkler systems that are monitored by large fire alarm systems. The retard delay needed to prevent false waterflow alarm indications, due to the extent of the sprinkler system added to the worst-case polling delay of the fire alarm system, may result in an excessive delay in the activation of alarm notification.

Finally, the designer must be familiar with NFPA 13, *Standard for the Installation of Sprinkler Systems*.

2-6.3 Piping between the sprinkler system and a pressure actuated alarm-initiating device shall be galvanized or of nonferrous metal or other approved corrosion-resistant material of not less than $^3/_8$ in. (9.5 mm) nominal pipe size.

These requirements stem from the experiences associated with piping corrosion and with the mechanical strength necessary to endure the environment of the sprinkler system.

2-7* Detection of the Operation of Other Automatic Extinguishing Systems

The operation of fire extinguishing systems or suppression systems shall initiate an alarm signal by alarm-initiating devices installed in accordance with their individual listings.

A-2-7 Alarm initiation can be accomplished by devices that detect the following:

(1) Flow of water in foam systems
(2) Pump activation
(3) Differential pressure
(4) Pressure (for example, clean agent systems, carbon dioxide systems, and wet/dry chemical systems)
(5) Mechanical operation of a release mechanism

Many extinguishing systems include emergency mechanical manual release capability. This provides for the release of the extinguishing agent, bypassing the operation of the fire alarm system. However, the fire alarm system usually oversees all of the active fire protection systems for the protected hazard area.

The fire alarm system houses the electrical switching that is used to sound warnings, terminate fuel flow, release automatic door closers, actuate emergency power interrupts, energize smoke management systems and similar functions that secure the hazard area. There must be a means to ensure that these critical functions are achieved if the extinguishing system discharge is caused by the mechanical release.

Discharge pressure switches or actuation mechanism microswitches are the usual means of providing the extinguishing system operation signal to the fire detection control unit. Due to the critical function these switches perform, they must be listed for use with the specific make and model of extinguishing system. The connection of a listed extinguishing agent release initiating device to an appropriate initiating device circuit on the fire alarm control unit, consistent with the listings of both the unit and the initiating device, is critical for the successful operation of most special extinguishing systems. See Exhibit 2.35 for an example of a manual agent release device.

The designer should review the following:

NFPA 12, *Standard on Carbon Dioxide Extinguishing Systems*

NFPA 12A, *Standard on Halon 1301 Fire Extinguishing Systems*

NFPA 16, *Standard for the Installation of Foam-Water Sprinkler and Foam-Water Spray Systems*

NFPA 17, *Standard for Dry Chemical Extinguishing Systems*

NFPA 2001, *Standard on Clean Agent Fire Extinguishing Systems*

See Exhibit 2.36 for an example of a pressure-actuated switch.

Exhibit 2.35 *Emergency manual cable release for extinguishing systems. (Source: Kidde-Fenwal Protection Systems, Ashland, MA)*

2-8 Manually Actuated Alarm-Initiating Devices

Manual fire alarm boxes shall be used only for fire alarm-initiating purposes. However, combination manual fire alarm boxes and guard's signaling stations shall be permitted.

This requirement stems from two concerns: credibility and reliability. If the manual fire alarm box is incorporated into some other non-fire-related assembly (with the single exception of guard's tour supervisory stations), the probability of unwarranted operation is increased. This leads to false alarms and erodes the occupants' confidence in the system. Also, when manual fire alarms are combined with non-fire-related functions, there is an increased probability that a

Exhibit 2.36 *Typical extinguishing system pressure-actuated discharge switch. (Source: Kidde-Fenwal Protection Systems, Ashland, MA)*

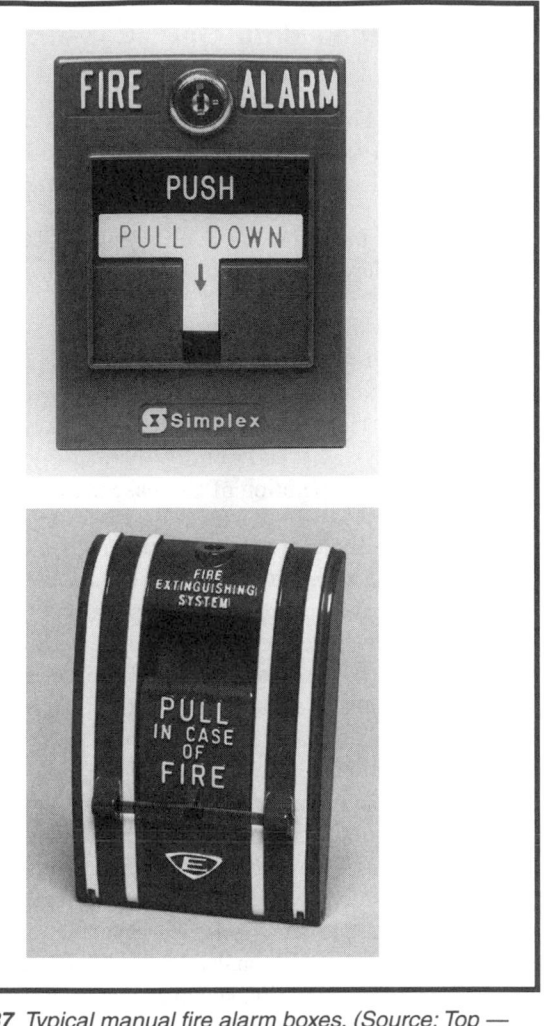

Exhibit 2.37 *Typical manual fire alarm boxes. (Source: Top — Simplex Time Recorder Co., Gardner, MA; Bottom — Edwards Systems Technology/EST, Cheshire, CT)*

failure in the non-fire function will compromise the fire alarm system.

2-8.1 Mounting.

Each manual fire alarm box shall be securely mounted. The operable part of each manual fire alarm box shall be not less than $3^1/_2$ ft (1.1 m) and not more than $4^1/_2$ ft (1.37 m) above floor level.

The *National Fire Alarm Code* has addressed the fire alarm box "side reach" accessibility requirements that resulted from the adoption of the Americans with Disabilities Act (ADA). The "front reach" ADA requirement permits a maximum mounting height of 48 in. (1.3 m).

See Exhibit 2.37 for examples of manual fire alarm boxes.

Manual fire alarm boxes mounted in damp or wet locations must be listed for such use.

2-8.2 Location and Spacing.

2-8.2.1 Manual fire alarm boxes shall be located throughout the protected area so that they are unobstructed and accessible.

Plastic covers are permitted to protect manual fire alarm boxes and provide relief from false alarms. These covers

must be listed for such use. A local alarm signal sounds at the device when lifted. Care must be taken to educate users that the fire alarm box must still be actuated. See Exhibit 2.38 for an example of a manual fire alarm box protective cover.

2-8.2.2 Manual fire alarm boxes shall be located within 5 ft (1.5 m) of the exit doorway opening at each exit on each floor.

2-8.2.3 Manual fire alarm boxes shall be mounted on both sides of group openings over 40 ft (12.2 m) in width. Manual fire alarm boxes shall be mounted within 5 ft (1.5 m) of each side of the opening.

The Technical Committee on Initiating Devices for Fire Alarm Systems has clarified this requirement to ensure the number and location of the manual fire alarm boxes needed will be consistent where there are groups of exit doors.

Exhibit 2.38 *Protective cover for manual fire alarm box. (Source: STI, Inc., Waterford, MI)*

2-8.2.4* Additional manual fire alarm boxes shall be provided so that the travel distance to the nearest fire alarm box will not be in excess of 200 ft (61 m) measured horizontally on the same floor.

This criterion is derived from the requirements established in NFPA *101*, *Life Safety Code*.

A-2-8.2.4 It is not the intent of 2-8.2.4 to require manual fire alarm boxes to be attached to moveable partitions or to equipment, nor to require the installation of permanent structures for mounting purposes only.

2-8.3* A coded manual fire alarm box shall produce at least three repetitions of the coded signal, with each repetition to consist of at least three impulses.

A-2-8.3 Recommended coded signal designations for buildings that have four floors and multiple basements are provided in Table A-2-8.3.

Table A-2-8.3 Recommended Coded Signal Designations

Location	Coded Signal
Fourth floor	2–4
Third floor	2–3
Second floor	2–2
First floor	2–1
Basement	3–1
Sub-basement	3–2

This is one of many possible coded signal schemes. Other signal schemes have been used for large buildings that divide the building up into a north and south end, for example, and use the first digit of the coded signal to specify the floor (1, 2, 3, 4, etc.) and the second digit to specify either north (1) or south (2) wing. Coding of the signals allows the parties responsible for emergency response to proceed directly to the area where the alarm was actuated. There are both benefits and disadvantages to coding manual fire alarm boxes. The decision to use coded signals must be part of the overall fire prevention and protection plan for the site.

2-9 Supervisory Signal-Initiating Devices

2-9.1 Control Valve Supervisory Signal-Initiating Device.

Control valve supervisory signal–initiating devices have traditionally been switches specifically designed and listed for service as valve-monitoring devices. See Exhibit 2.39 for an example of a gate valve supervising switch. The requirement for two distinct signals does not necessarily mean two switches. A switch that transfers when the valve begins to close and stays transferred while the valve remains closed, then returns to normal when the valve is reopened, satisfies the requirement. The initial transfer is the first signal. The return to normal is the second signal.

For example, assume the switch on the valve is a normally open contact. As the operator begins to turn the valve, the switch closes, indicating an off-normal condition. The switch stays in the closed, off-normal position as the operator continues to close the valve. When the operator reopens the valve, the closed contacts transfer back to the open state as the valve is completely open. The opening of the contacts provides the second, distinct signal.

2-9.1.1 Two separate and distinct signals shall be initiated: one indicating movement of the valve from its normal position and the other indicating restoration of the valve to its normal position. The off-normal signal shall be initiated during the first two revolutions of the hand wheel or during one-fifth of the travel distance of the valve control apparatus from its normal position. The off-normal signal shall not be restored at any valve position except normal.

2-9.1.2 An initiating device for supervising the position of a control valve shall not interfere with the operation of the valve, obstruct the view of its indicator, or prevent access for valve maintenance.

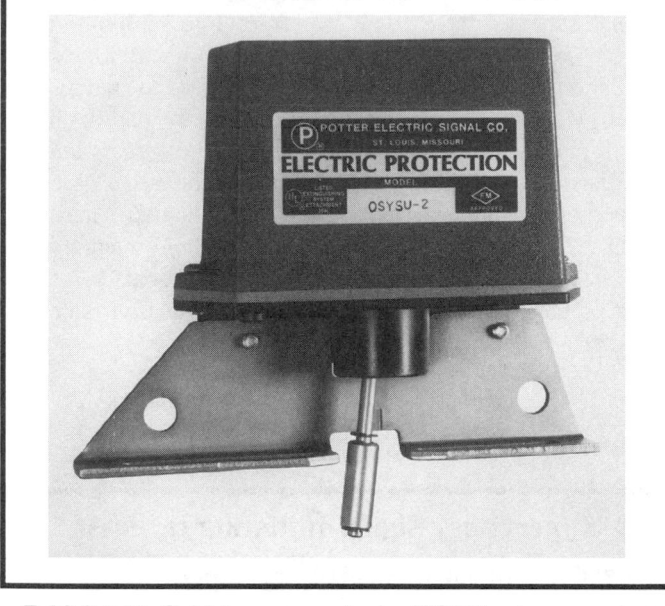

Exhibit 2.39 Outside screw and yoke (OS&Y) valve supervisory switch. (Source: Potter Electric Signal Co., St. Louis, MO)

2-9.2 Pressure Supervisory Signal-Initiating Device.

Two separate and distinct signals shall be initiated: one indicating that the required pressure has increased or decreased and the other indicating restoration of the pressure to its normal value. The following requirements shall apply to pressure supervisory signal-initiating devices:

(a) A pressure tank supervisory signal-initiating device for a pressurized limited water supply, such as a pressure tank, shall indicate both high- and low-pressure conditions. A signal shall be initiated when the required pressure increases or decreases by 10 psi (70 kPa).

(b) A pressure supervisory signal-initiating device for a dry-pipe sprinkler system shall indicate both high- and low-pressure conditions. A signal shall be initiated when the pressure increases or decreases by 10 psi (70 kPa).

(c) A steam pressure supervisory signal-initiating device shall indicate a low-pressure condition. A signal shall be initiated prior to the pressure falling below 110 percent of the minimum operating pressure of the steam-operated equipment supplied.

(d) An initiating device for supervising the pressure of sources other than those specified in 2-9.2(a) through (c) shall be provided as required by the authority having jurisdiction.

As with supervisory initiating devices for valve operation, water pressure supervisory initiating devices may consist of a single switch.

See Exhibit 2.40 for an example of a pressure supervisory switch.

Exhibit 2.40 Pressure supervisory switch. (Source: Potter Electric Signal Co., St. Louis, MO)

2-9.3 Water Level Supervisory Signal-Initiating Device.

Two separate and distinct signals shall be initiated: one indicating that the required water level has been lowered or raised and the other indicating restoration.

As with supervisory initiating devices for valve operation and water pressure, water level supervisory initiating devices may also consist of a single switch. See Exhibit 2.41 for an example of a water level supervisory switch.

2-9.3.1 A pressure tank signal-initiating device shall indicate both high- and low-water level conditions. A signal shall be initiated when the water level falls 3 in. (76 mm) or rises 3 in. (76 mm).

2-9.3.2 A supervisory signal-initiating device for other than pressure tanks shall initiate a low–water level signal when the water level falls 12 in. (300 mm).

2-9.4 Water Temperature Supervisory Signal-Initiating Device.

A temperature supervisory device for a water storage container exposed to freezing conditions shall initiate two separate and distinctive signals. One signal shall indicate a

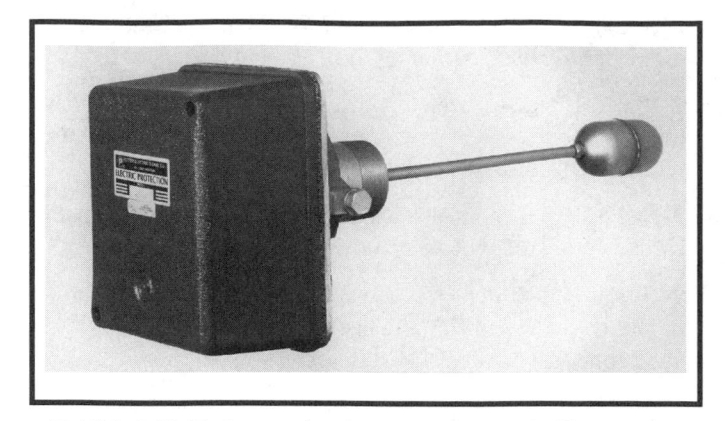

Exhibit 2.41 *Tank water level supervisory switch. (Source: Potter Electric Signal Co., St. Louis, MO)*

Exhibit 2.43 *Room temperature supervisory switch. (Source: Potter Electric Signal Co., St. Louis, MO)*

decrease in water temperature to 40°F (4.4°C) and the other shall indicate its restoration to above 40°F (4.4°C).

Water temperature supervisory initiating devices may consist of a single switch. See Exhibit 2.42 for a water temperature supervisory switch.

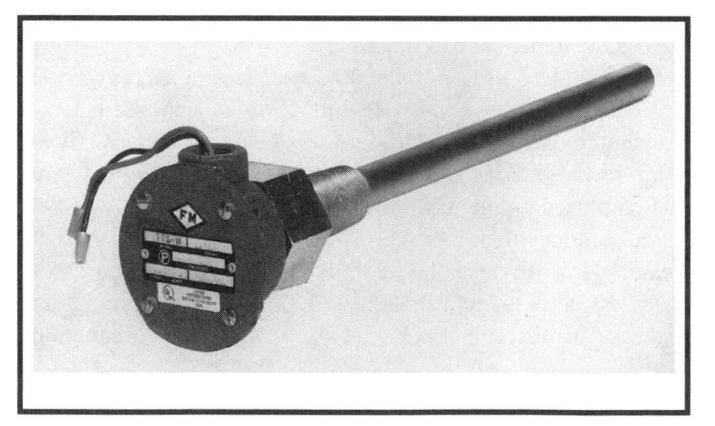

Exhibit 2.42 *Tank water temperature supervisory switch. (Source: Potter Electric Signal Co., St. Louis, MO)*

2-9.5 Room Temperature Supervisory Signal-Initiating Device.

A room temperature supervisory device shall indicate a decrease in room temperature to 40°F (4.4°C) and its restoration to above 40°F (4.4°C).

As with other supervisory initiating devices mentioned in Section 2-9, room temperature supervisory initiating devices may also consist of a single switch. See Exhibit 2.43 for a typical room temperature supervisory switch.

2-10* Smoke Detectors for Control of Smoke Spread

A-2-10 Refer to NFPA *101*®, *Life Safety Code*®, for the definition of *smoke compartment;* NFPA 90A, *Standard for the Installation of Air Conditioning and Ventilating Systems,* for the definition of *duct systems;* and NFPA 92A, *Recommended Practice for Smoke-Control Systems,* for the definition of *smoke zone.*

Between 1960 and late 1970 there were several fires in high-rise buildings that demonstrated the futility of trying to evacuate an entire building when a fire occurs. Not only did occupants incur injuries during the evacuation, but also the means of egress often became untenable due to heavy smoke concentrations within them.

As the improved building codes resulted in structures that could maintain their integrity in spite of the complete combustion of the interior fire load through passive fire-resistive construction and compartmentation, the option of defending occupants in place became viable. Strategies of establishing smoke compartments and areas of refuge, and managing the flow of smoke by directing it away from the occupants were developed. Experiences with high-rise fires indicate that the proactive control of smoke with either automatic smoke detectors and HVAC systems or engineered smoke control systems is a viable strategy for occupant protection in high-rise buildings.

2-10.1* Smoke detectors installed and used to prevent smoke spread by initiating control of fans, dampers, doors, and other equipment shall be classified in the following manner:

(1) Area detectors that are installed in the related smoke compartments

(2) Detectors that are installed in the air duct systems

Section 2-10 does not require the installation of smoke detectors for smoke control. The purpose of Section 2-10 is to describe the performance and installation requirements for smoke detectors when they are used for smoke control, as required by some other code or standard.

A-2-10.1 Smoke detectors located in an open area(s) should be used rather than duct-type detectors because of the dilution effect in air ducts. Active smoke management systems installed in accordance with NFPA 92A, *Recommended Practice for Smoke-Control Systems,* or NFPA 92B, *Guide for Smoke Management Systems in Malls, Atria, and Large Areas,* should be controlled by total coverage open area detection.

Paragraph 2-1.4.2 identifies all of the spaces that must have smoke detectors if total coverage is to be achieved.

2-10.2* Detectors that are installed in the air duct system per 2-10.1(2) shall not be used as a substitute for open area protection. If open area protection is required, 2-3.4 shall apply.

All too often, uninformed designers will attempt to use air duct–type smoke detectors to provide open area protection. Such a strategy does not address the potential for a fire during those times when the HVAC system is not running, nor does it address the delay in detection due to smoke dilution. Subsection 2-10.2 specifically prohibits the use of duct smoke detection in lieu of area detection installed pursuant to Section 2-3 of the code.

A-2-10.2 Dilution of smoke-laden air by clean air from other parts of the building or dilution by outside air intakes can allow high densities of smoke in a single room with no appreciable smoke in the air duct at the detector location. Smoke might not be drawn from open areas if air-conditioning systems or ventilating systems are shut down.

2-10.3* Purposes.

A-2-10.3 Smoke detectors can be applied in order to initiate control of smoke spread for the following purposes:

(1) Prevention of the recirculation of dangerous quantities of smoke within a building
(2) Selective operation of equipment to exhaust smoke from a building
(3) Selective operation of equipment to pressurize smoke compartments
(4) Operation of doors and dampers to close the openings in smoke compartments

2-10.3.1 To prevent the recirculation of dangerous quantities of smoke, a detector approved for air duct use shall be installed on the supply side of air-handling systems as

required by NFPA 90A, *Standard for the Installation of Air Conditioning and Ventilating Systems,* and 2-10.4.2.1.

2-10.3.2 If smoke detectors are used to initiate selectively the operation of equipment to control smoke spread, the requirements of 2-10.4.2.2 shall apply.

2-10.3.3 If detectors are used to initiate the operation of smoke doors, the requirements of 2-10.6 shall apply.

2-10.3.4 If duct detectors are used to initiate the operation of smoke dampers within ducts, the requirements of 2-10.5 shall apply.

2-10.4 Application.

2-10.4.1 Area Detectors within Smoke Compartments. Area smoke detectors within smoke compartments shall be permitted to be used to control the spread of smoke by initiating appropriate operation of doors, dampers, and other equipment.

Paragraph 2-10.4.1 allows specific area detectors to control the spread of smoke. From an engineering standpoint, smoke detectors are needed where they are intended to identify the presence of smoke at a particular location or the movement of smoke past a particular location. The necessary locations for area smoke detectors are a function of building geometry, anticipated fire locations, and intended goals of smoke control functions.

Complete area smoke detection is not necessary to provide for such control features, which can be accomplished by many possible detector locations for any given fire scenario. An example is the smoke detectors that are often placed at the perimeter of an atrium opening to detect smoke movement through the floor opening. Another example is the releasing of smoke doors only as their associated smoke detector is actuated, thus avoiding premature release of all other doors, which may impede evacuation during an emergency.

Paragraph 2-10.4.1 also allows complete area coverage to be used if the designer wishes to do so. In this case, when a compartment detector actuates in the smoke compartment, it signals the fire alarm control unit, which, in turn, signals the HVAC control system or smoke door release system. The HVAC controller operates or controls fans and dampers to prevent the introduction of smoke into other smoke compartments and to vent the smoke from the fire compartment, facilitating occupant egress. The smoke door release system either closes all doors in the building or all doors in the smoke zone.

2-10.4.2* Smoke Detection for the Air Duct System.

A-2-10.4.2 Smoke detectors are designed to sense the presence of particles of combustion, but depending on the sens-

ing technology and other design factors, different detectors respond to different types of particles. Detectors based on ionization detection technology are most responsive to smaller, invisible sub-micron sized particles. Detectors based on photoelectric technology, by contrast, are most responsive to larger visible particles.

It is generally accepted that particle size distribution varies from sub-micron diameter particles predominant in the proximity of the flame of a flaming fire to particles one or more orders of magnitude larger, which are characteristic of smoke from a smoldering fire. The actual particle size distribution depends on a host of other variables including the fuel and its physical make-up, the availability of oxygen including air supply and fire gas discharge, and other ambient conditions, especially humidity. Moreover, the particle size distribution is not constant, but as the fire gases cool, the sub-micron particles agglomerate and the very large ones precipitate. In other words, as smoke travels away from the fire source, the particle size distribution shows a relative decrease in smaller particles. Water vapor, which is abundantly present in most fires, when cooled sufficiently will condense to form fog particles—an effect frequently seen above tall chimneys. Because water condensation is basically clear in color, when it is mixed with other smoke particles, it can be expected to lighten the color of the mixture.

In almost every fire scenario in an air-handling system, the point of detection will be some distance from the fire source, therefore, the smoke will be cooler and more visible because of the growth of sub-micron particles into larger particles due to agglomeration and recombination. For these reasons, photoelectric detection technology has advantages over ionization detection technology in air duct system applications.

2-10.4.2.1 Supply Air System. Where the detection of smoke in the supply air system is required by other NFPA standards, a detector(s) listed for the air velocity present and that is located in the supply air duct downstream of both the fan and the filters shall be installed.

Exception: Additional smoke detectors shall not be required to be installed in ducts where the air duct system passes through other smoke compartments not served by the duct.

The NFPA standards relevant to 2-10.4.2.1 are NFPA 90A, *Standard for the Installation of Air-Conditioning and Ventilating Systems*; NFPA 92A, *Recommended Practice for Smoke Control Systems*; and NFPA *101, Life Safety Code*. The purpose of supply-side smoke detection is to sense smoke that may be contaminating the area served by the duct but not as a result of a fire in that area. The smoke might be coming from the area via return air ducts, from outside via fresh air mixing ducts, or from a fire within the duct (such as in a filter or fan belt). If the source of the

smoke is from outside or from within the duct, a fire alarm response for area detection within the space would not normally be expected to produce the most appropriate set of responses to the fire within the duct.

Different airflow management programs are required for supply-side smoke in-flow as opposed to smoke generated within the compartment. Furthermore, one could not rely on compartment area detection to respond to a supply duct smoke in-flow, because of the expected dilution of smoke-laden air with fresh air as it enters the smoke compartment where the area detection is installed. This expected condition necessitates the use of detectors downstream from the fan and filters in the supply air duct.

The exception is based on the fire resistance of HVAC ducts and the unlikelihood of smoke escaping from the HVAC duct into a compartment not served by the duct. If this exception were not provided, an additional detector would be necessary every time the duct passed from one smoke compartment into another.

Figure A-2-10.4.2.2(c) shows an air duct passing through a smoke compartment without serving the compartment. The middle air supply duct serves the center smoke compartment. The top air supply duct serves only the left compartment and passes through the center and right compartments without serving them.

2-10.4.2.2* Return Air System. If the detection of smoke in the return air system is required by other NFPA standards, a detector(s) listed for the air velocity present shall be located where the air leaves each smoke compartment, or in the duct system before the air enters the return air system common to more than one smoke compartment.

Exception 1: Where total coverage smoke detection is installed in all areas of the smoke compartment served by the return air system, installation of air duct detectors in the return air system shall not be required, provided their function is accomplished by the design of the area detection system.

Exception 2: Additional smoke detectors shall not be required to be installed in ducts where the air duct system passes through other smoke compartments not served by the duct.

A-2-10.4.2.2 Detectors listed for the air velocity present can be permitted to be installed at the opening where the return air enters the common return air system. The detectors should be installed up to 12 in. (0.3 m) in front of or behind the opening and spaced according to the following opening dimensions *[refer to Figures A-2-10.4.2.2(a), (b), and (c)]:*

(a) *Width.*

(1) Up to 36 in. (914 mm)—One detector centered in opening

(2) Up to 72 in. (1829 mm)—Two detectors located at the one-quarter points of the opening

(3) Over 72 in. (1829 mm)—One additional detector for each full 24 in. (610 mm) of opening

(b) *Depth.* The number and spacing of the detector(s) in the depth (vertical) of the opening should be the same as those given for the width (horizontal) above.

(c) *Orientation.* Detectors should be oriented in the most favorable position for smoke entry with respect to the direction of airflow. The path of a projected beam–type detector across the return air openings should be considered equivalent in coverage to a row of individual detectors.

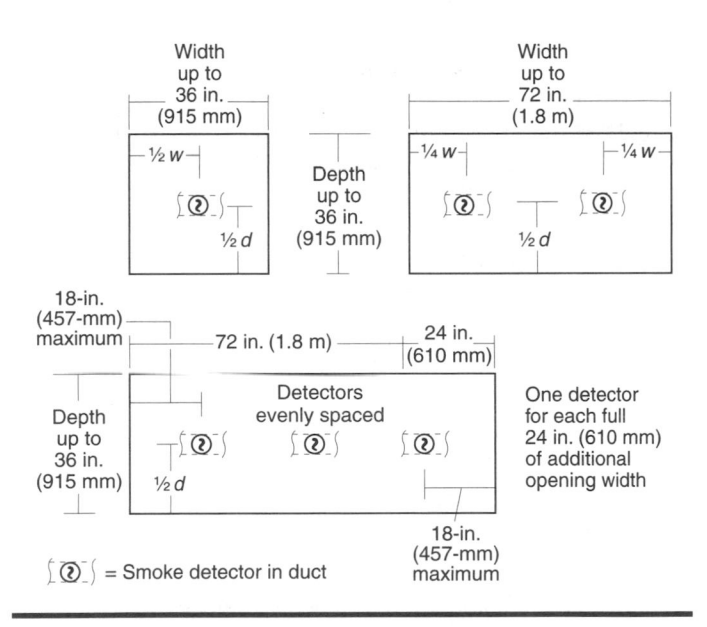

Figure A-2-10.4.2.2(a) *Location of a smoke detector(s) in return air system openings for selective operation of equipment.*

When duct detection is used for control of smoke spread, detectors must be installed only where the return air duct leaves the smoke compartment or before the duct joins a return air plenum serving more than one smoke compartment. This is intended to minimize the effects of smoke dilution.

The key phrase in Exception No. 1 is "provided their function is accomplished…" When an engineering analysis shows that the area smoke detection addresses all of the smoke ingress paths from the compartment into the return air duct, this exception is operative.

Exception No. 2 is based on the same reasoning used in the Exception to 2-10.4.2.1. Based on this exception and referring to Figure A-2-10.4.2.2(c), the top duct does not need additional detectors and/or dampers where it passes through either the center compartment or the right compartment.

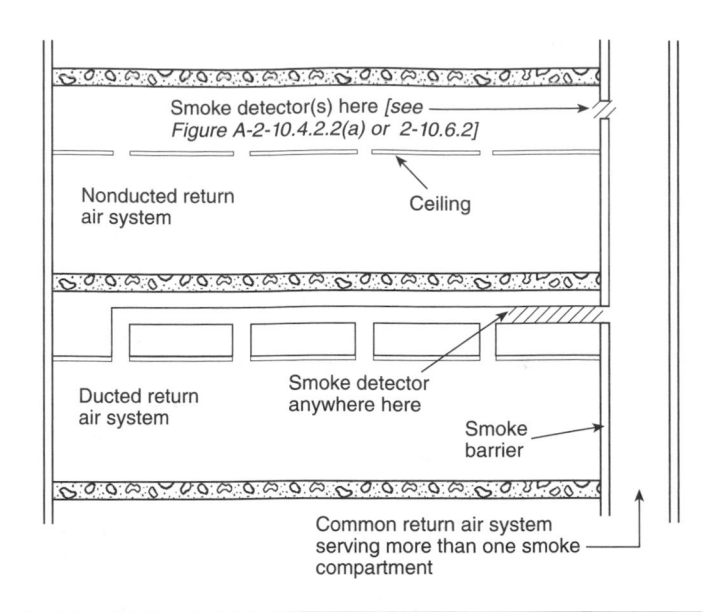

Figure A-2-10.4.2.2(b) *Location of a smoke detector(s) in return air systems for selective operation of equipment.*

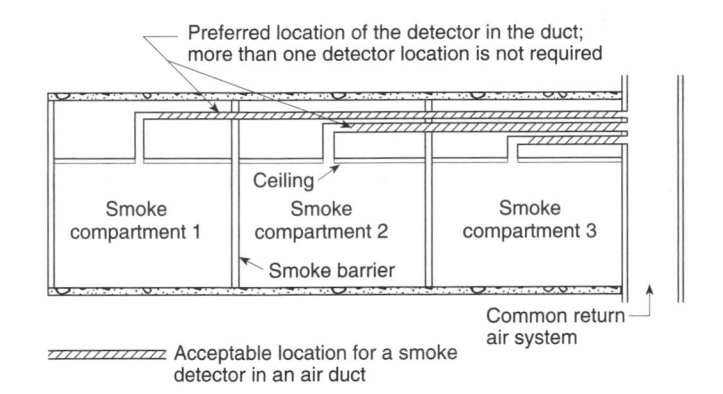

Figure A-2-10.4.2.2(c) *Detector location in a duct that passes through smoke compartments not served by the duct.*

2-10.5 Location and Installation of Detectors in Air Duct Systems.

2-10.5.1 Detectors shall be listed for the purpose for which they are being used.

The listing of the detector stipulates the range of air velocities over which it can operate, as well as the temperature and relative humidity range. These last two criteria are particularly important where a general purpose detector is being

installed in a duct detector housing. Often HVAC system fans and ducts are located in penthouses and mechanical rooms where comfort heating and cooling are not provided. Consequently, it is possible to inadvertently install a smoke detector where the ambient conditions exceed its design range. The location of the duct detector must be maintained within the operating range of the detector used.

2-10.5.2* Air duct detectors shall be installed in such a way as to obtain a representative sample of the airstream. This installation shall be permitted to be achieved by any of the following methods:

(1) Rigid mounting within the duct
(2) Rigid mounting to the wall of the duct with the sensing element protruding into the duct
(3) Installation outside the duct with rigidly mounted sampling tubes protruding into the duct
(4) Installation through the duct with projected light beam

See Exhibits 2.44 and 2.45 for examples of typical duct type smoke detectors.

The flow of air through a duct is not necessarily uniform. Bends in the duct produce regions of reduced flow velocity and, hence, reduced flow volume. The flow in a duct can also become laminar, resulting in smoke being concentrated in a portion of the duct cross section, not uniformly dispersed across the duct area. Consequently, options (1) and (2) are most appropriate for smaller ducts or where an engineering analysis shows that smoke concentrations will be even across the duct cross section and that laminar flow is not going to produce a non-uniform smoke concentration.

A-2-10.5.2 If duct detectors are used to initiate the operation of smoke dampers, they should be located so that the detector is between the last inlet or outlet upstream of the damper and the first inlet or outlet downstream of the damper.

In order to obtain a representative sample, stratification and dead air space should be avoided. Such conditions could be caused by return duct openings, sharp turns, or connections, as well as by long, uninterrupted straight runs. For this reason, duct smoke detectors should be located in the zone between 6 and 10 duct-equivalent diameters of straight, uninterrupted run. In return air systems, the requirements of 2-10.4.2.2 take precedence over these considerations. *[Refer to Figures A-2-10.5.2(a), (b), and (c).]*

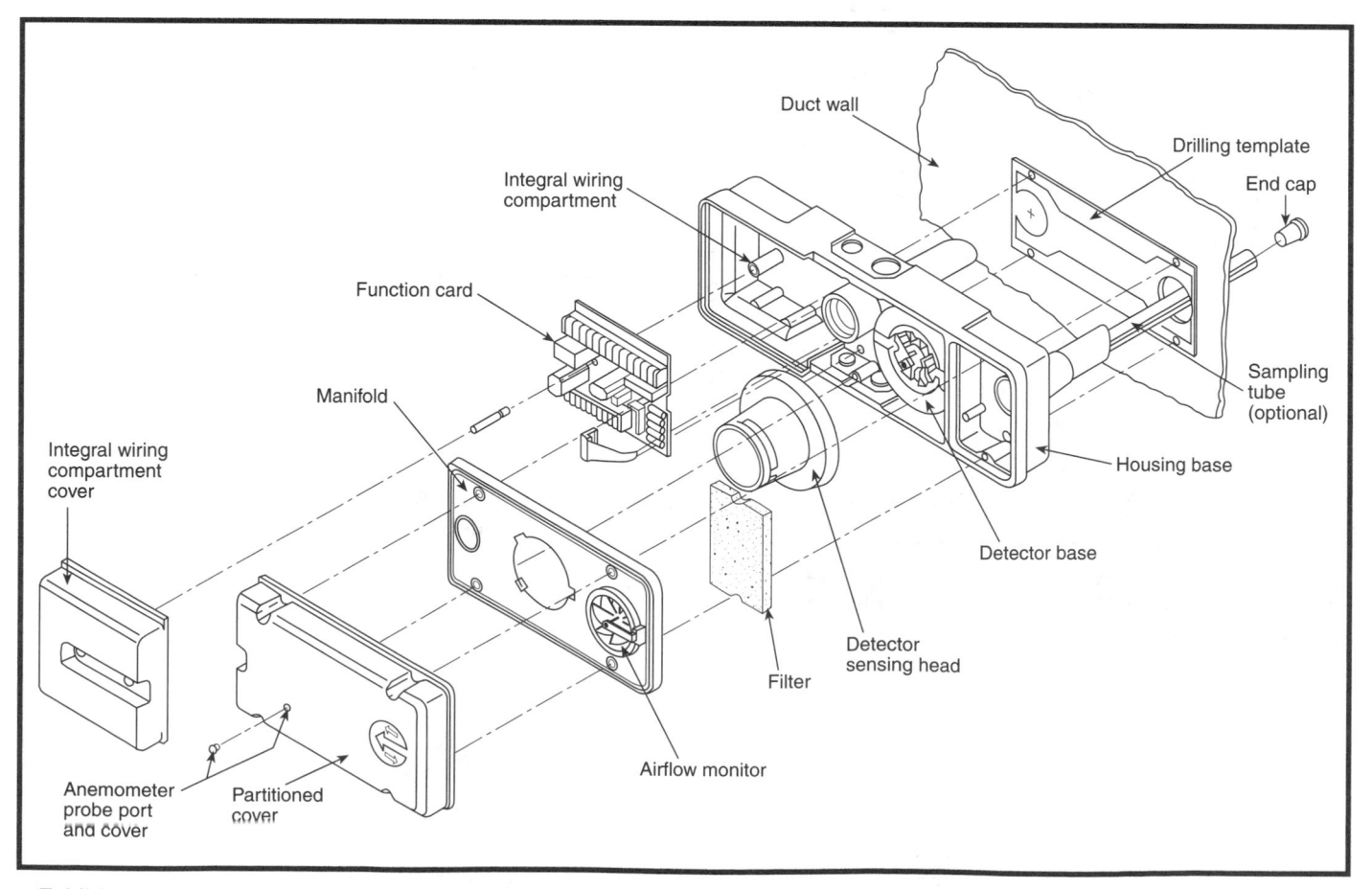

Exhibit 2.44 *Exploded view of a duct smoke detector. (Source: ESL Sentrol, Inc., Portland, OR)*

Exhibit 2.45 *Duct-type smoke detectors. (Source: Top—Kidde-Fenwal, Inc., Ashland, MA; Bottom—Simplex Time Recorder Company, Gardner, MA)*

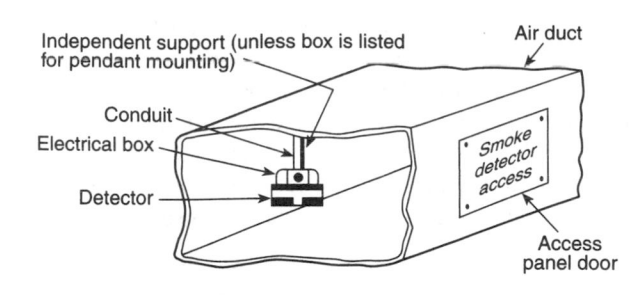

Figure A-2-10.5.2(a) *Pendant-mounted air duct installation.*

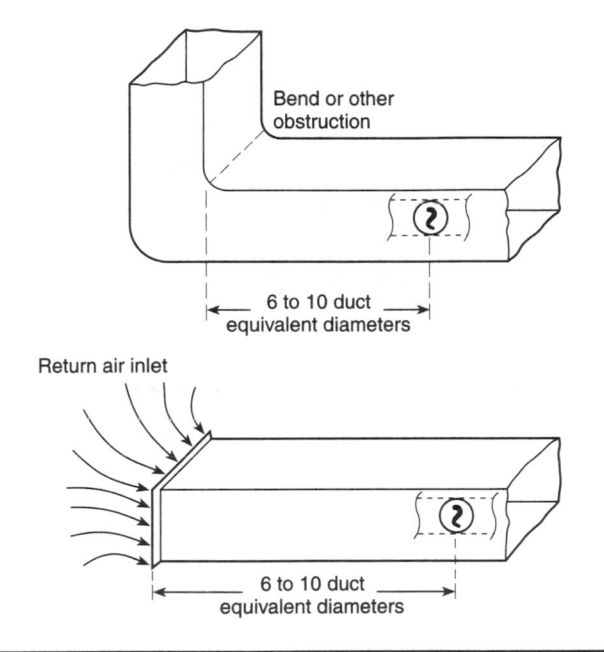

Figure A-2-10.5.2(b) *Typical duct detector placement.*

It is important to note that the text in the appendix is not a requirement of the code. It is generally considered good practice to observe the distance limitations to reduce the potential for non-uniform smoke distribution in the duct. However, it is also recognized that the design of the air-handling ductwork may make it impossible to achieve a distance of 6 to 10 duct widths from a bend or opening.

Support of the detector by the conduit or raceway containing its wires is not permitted by NFPA 70, *National Electrical Code*, unless the box is specifically listed for the purpose and installed in accordance with the listing.

The requirements in 2-10.5.2 and the guidance in A-2-10.5.2 are provided to ensure that the detectors in the air duct are suitably located to obtain an adequate sampling of air. To be certain that the detectors are placed where there is the highest probability that smoke will be evenly distributed throughout the duct cross section, these location guidelines should be followed.

2-10.5.3 Detectors shall be accessible for cleaning and shall be mounted in accordance with the manufacturer's instruc-

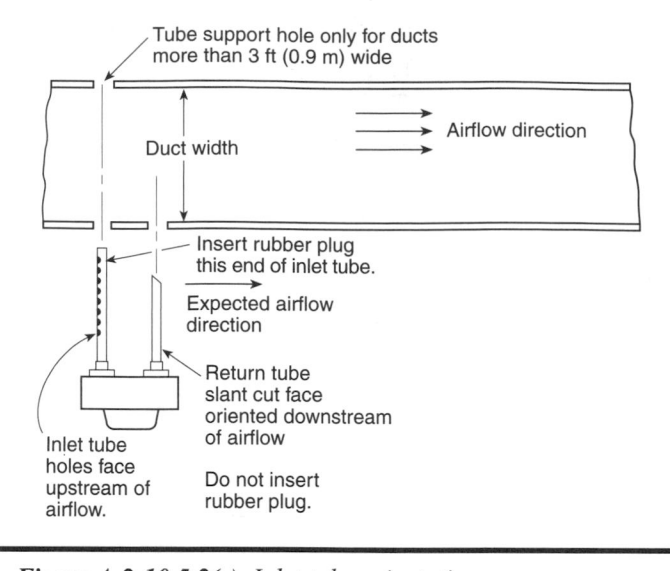

Figure A-2-10.5.2(c) Inlet tube orientation.

tions. Access doors or panels shall be provided in accordance with NFPA 90A, *Standard for the Installation of Air Conditioning and Ventilating Systems.*

Chapter 7 provides recommended maintenance schedules for each type of detector. It is critical that the detectors be accessible in order to facilitate cleaning, which is critical for reliable operation of the detector.

2-10.5.4 The location of all detectors in air duct systems shall be permanently and clearly identified and recorded.

It is advisable to place permanent placards outside the first point of access, indicating that a detector is accessible from that point. For example, the placard may be mounted on the wall beneath the ceiling tile that must be removed to access the duct. HVAC and fire alarm drawings should clearly show the actual as-built locations of the detectors. In most cases it is useful to generate one drawing that shows only the smoke detector locations. The location could also be in the display descriptor of addressable systems.

2-10.5.5 Detectors mounted outside of a duct that employs sampling tubes for transporting smoke from inside the duct to the detector shall be designed and installed to allow verification of airflow from the duct to the detector.

Sampling tubes can provide a flow of air through the detector enclosure due to a pressure differential that results from the flow of air across the tubes. Small errors in the orientation of the sampling tubes can reduce the pressure differential, rendering them ineffective in drawing air into the detector enclosure.

For sampling tubes to take a representative sample of the air passing through the duct, they must be fabricated and installed in a manner consistent with their listing. Not all detectors are listed for use in a sampling tube enclosure. If the flow of air through the sampling tube and detector enclosure assembly cannot be verified there is no basis to presume that the air within the duct is being sampled by the detector.

2-10.5.6 Detectors shall be listed for operation over the complete range of air velocities, temperature, and humidity expected at the detector when the air-handling system is operating.

Often HVAC system fans and ducts are located in penthouses and mechanical rooms where comfort heating and cooling are not provided. Consequently, it is possible that the environment of the detector might exceed the limits observed in the listing investigation. In this case, ambient conditions can exceed the design range of the detector. The location of the duct detector must be maintained within its operating range.

2-10.5.7 All penetrations of a return air duct in the vicinity of detectors installed on or in an air duct shall be sealed to prevent entrance of outside air and possible dilution or redirection of smoke within the duct.

2-10.5.8 Where in-duct smoke detectors are installed in concealed locations more than 10 ft (3 m) above the finished floor or in arrangements where the detector's alarm indicator is not visible to responding personnel, the detectors shall be provided with remote alarm indicators. Remote alarm indicators shall be installed in an accessible location and shall be clearly labeled to indicate both their function and the air-handling unit(s) associated with each detector (for example, In-Duct Smoke Detector Alarm).

Exception: Where the specific detector in alarm is indicated at the control unit.

Identification of which in-duct smoke detector is in alarm is a chronic problem, due to the difficulties typically involved in getting to the detectors. Because the fire causing the alarm may be located far from the in-duct smoke detector location, rapid identification of the individual detector in alarm and the associated air handling unit is critical.

2-10.6 Smoke Detectors for Door Release Service.

There are two general methods of controlling doors with smoke detectors. The first is to use area smoke detectors to control the doors for that area. Either smoke detectors served by a selected circuit of a fire alarm control unit or specific addressable detectors are programmed to operate

magnetic door release devices via the fire alarm system control unit. See Exhibit 3.26. When one of the area smoke detectors renders an alarm, the control unit transfers to the alarm state and energizes the output circuit that controls the door holders. The requirements for such a system are addressed in Chapter 3.

The second method is to control the door holder mechanism directly with a dedicated smoke detector or smoke detectors. The requirements in subsection 2-10.6 apply equally to both design concepts. When the open area protection system is used, 2-10.6.1 allows the spacing in the corridors as normally required to be considered acceptable for smoke door release service, and the requirements of 2-10.6.2 do not apply.

2-10.6.1 Smoke detectors that are part of an open area protection system covering the room, corridor, or enclosed space on each side of the smoke door and that are located and spaced as required by 2-3.4 shall be permitted to accomplish smoke door release service.

Area detection installed in accordance with 2-3.4 is permitted to be used as long as area detection is provided on both sides of the doors to be closed. Discrete and dedicated smoke detectors separate from the area protection are not required to be used. The requirements of 2-10-6.5 do not apply where both sides of the door are protected by open area detectors.

2-10.6.2 Smoke detectors that are used exclusively for smoke door release service shall be located and spaced as required by 2-10.6.

Where area detection per Section 2-3 is not provided and where automatic closure of doors upon the presence of smoke is required, then they must be installed according to the requirements in 2-10.6.3 through 2-10.6.6.2.

2-10.6.3 When smoke door release is accomplished directly from the smoke detector(s), the detector(s) shall be listed for releasing service.

2-10.6.4 Smoke detectors shall be of the photoelectric, ionization, or other approved type.

2-10.6.5 Number of Detectors Required.

The placement requirements outlined in 2-10.6.5 have been derived from the ceiling jet dynamics that have served as the physical principles from which the rules for location and placement area smoke detection were derived. As research continues, there may be additional insight developed for this application.

Although it is not explicitly required, it is recommended that smoke detectors installed only for door release be connected to a fire alarm control unit to actuate notification appliances when smoke is detected.

2-10.6.5.1 If doors are to be closed in response to smoke flowing in either direction, the requirements of 2-10.6.5.1.1 through 2-10.6.5.1.3 shall apply.

2-10.6.5.1.1 If the depth of wall section above the door is 24 in. (610 mm) or less, one ceiling-mounted detector shall be required on one side of the doorway only. Figure 2-10.6.5.1.1, parts B and D, shall apply.

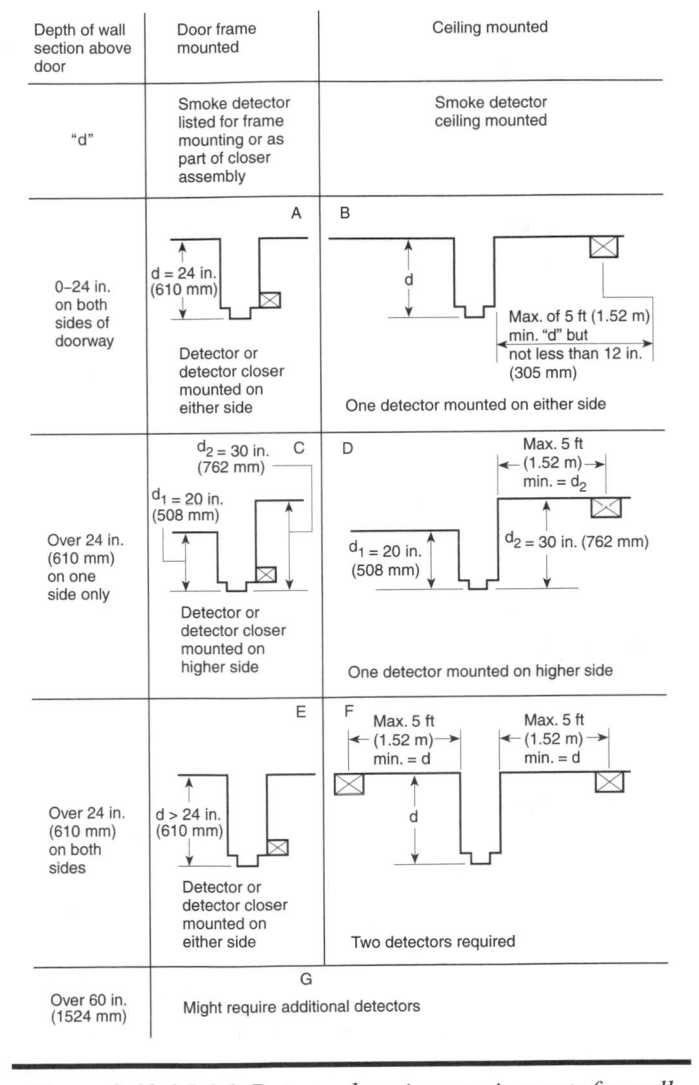

Figure 2-10.6.5.1.1 *Detector location requirements for wall sections.*

This requirement is similar to the requirements regarding smoke detectors and ceilings with deep beams (see 2-3.4.6). The difference in the depth requirements comes from the fact that doors for smoke control are typically located in

corridors that are narrower than the bays encountered in deep beam ceilings.

2-10.6.5.1.2* If the depth of wall section above the door is greater than 24 in. (610 mm), two ceiling-mounted detectors shall be required, one on each side of the doorway. Figure 2-10.6.5.1.1, part F, shall apply.

A-2-10.6.5.1.2 If the depth of wall section above the door is 60 in. (1520 mm) or greater, additional detectors might be required as indicated by an engineering evaluation.

Since the average door height is a nominal 84 in. to 96 in. (2.1 m to 2.4 m), the addition of 60 in. (1.5 m) above the door results in a ceiling height greater than 148 in. (3.8 m). This is above the height for which reduced spacing is required for heat detectors. The data in Appendix B indicate that when the ceiling height exceeds 10 ft (3 m), reduced spacing is required. An engineering evaluation should be performed to determine if reduced smoke detector spacing is appropriate for the specific application under consideration.

2-10.6.5.1.3 If a detector is specifically listed for door frame mounting or if a listed combination or integral detector–door closer assembly is used, only one detector shall be required if installed in the manner recommended by the manufacturer.

Formal Interpretation 78-1 provides further clarification of the requirements of 2-10.6.5.1.3.

Formal Interpretation 78-1

Reference: 2-10.6.5.1.3

Question: Does a door frame mounted combination automatic door closer incorporating a smoke detector listed by a testing laboratory for limited open area protection meet the requirements of 2-10.6.5.1.3?

Answer: Yes.

Issue Edition: 1978 of NFPA 72E

Reference: 9-2.2

Date: February 1979

2-10.6.5.2 If door release is intended to prevent smoke transmission from one space to another in one direction only, one detector located in the space to which smoke is to be confined shall be required, regardless of the depth of wall section above the door. Alternatively, a smoke detector conforming with 2-10.6.5.1.3 shall be permitted to be used.

2-10.6.5.3 If there are multiple doorways, additional ceiling-mounted detectors shall be required as specified in 2-10.6.5.3.1 through 2-10.6.5.3.3.

2-10.6.5.3.1 If the separation between doorways exceeds 24 in. (610 mm), each doorway shall be treated separately. Figure 2-10.6.5.3.1, part E, shall apply.

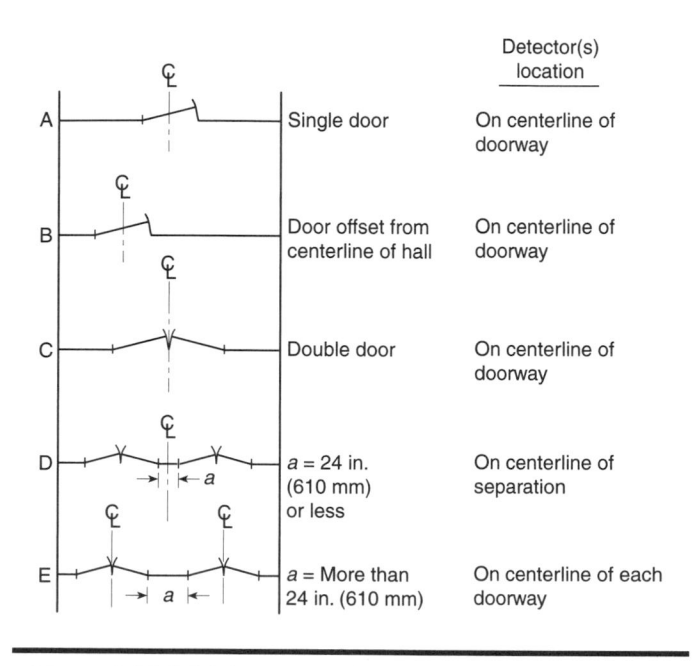

Figure 2-10.6.5.3.1 *Detector location requirements for single and double doors.*

2-10.6.5.3.2 Each group of three doorway openings shall be treated separately. Figure 2-10.6.5.3.2 shall apply.

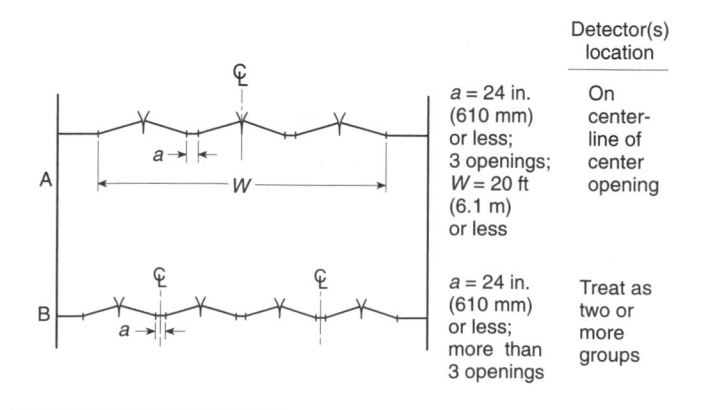

Figure 2-10.6.5.3.2 *Detector location requirements for group doorways.*

2-10.6.5.3.3 Each group of doorway openings that exceeds 20 ft (6.1 m) in width measured at its overall extremes shall be treated separately. Figure 2-10.6.5.3.3 shall apply.

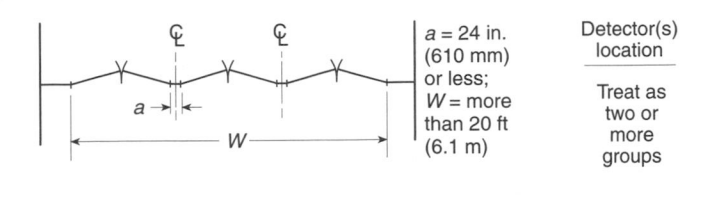

Figure 2-10.6.5.3.3 Detector location requirements for group doorways over 20 ft (6.1 m) in width.

2-10.6.5.4 If there are multiple doorways and listed door frame-mounted detectors or if listed combination or integral detector–door closer assemblies are used, there shall be one detector for each single or double doorway.

2-10.6.6 Location.

2-10.6.6.1 If ceiling-mounted smoke detectors are to be installed on a smooth ceiling for a single or double doorway, they shall be located as follows *(Figure 2-10.6.5.3.1 shall apply.)*:

(1) On the centerline of the doorway
(2) No more than 5 ft (1.5 m) measured along the ceiling and perpendicular to the doorway *(Figure 2-10.6.5.1.1 shall apply.)*
(3) No closer than shown in Figure 2-10.6.5.1.1, parts B, D, and F

2-10.6.6.2 If ceiling-mounted detectors are to be installed in conditions other than those outlined in 2-10.6.6.1, an engineering evaluation shall be made.

References Cited in Commentary

Alpert, R., 1972, "Ceiling Jets," *Fire Technology* (August).
Babrauskas, V.; Lawson, J. R.; Walton, W. D.; and Twilley, W. H., 1982, "Upholstered Furniture Heat Release Rates Measured with a Furniture Calorimeter" [NBSIR 82-2604 (December)], National Institute of Standards and Technology (formerly National Bureau of Standards), Center for Fire Research, Gaithersburg, MD 20889.
Brozovski, E., 1989, "A Preliminary Approach to Siting Smoke Detectors Based on Design Fire Size and Detector Aerosol Entry Lag Time," Master's Thesis, Worcester Polytechnic Institute, Worcester, MA.

Evans, D. D. and Stroup, D. W., 1986, "Methods to Calculate Response Time of Heat and Smoke Detectors Installed Below Large Unobstructed Ceilings" [NBSIR 85-3167 (February 1985, issued July 1986]. National Institute of Standards and Technology (formerly National Bureau of Standards), Center for Fire Research, Gaithersburg, MD 20889.
Heskestad, G., 1975, "Characterization of Smoke Entry and Response for Products-of-Combustion Detectors." Proceedings, 7th International Conference on Problems of Automatic Fire Detection, Rheinish-Westfalischen Technischen Hochschule Aachen (March).
Heskestad, G., and Delichatsios, M. A., "Environments of Fire Detectors - Phase 1: Effect of Fire Size, Ceiling Height and Material," Measurements vol. I (NBS-GCR-77-86), Analysis vol. II (NBS-GCR-77-95), National Technical Information Service (NTIS), Springfield, VA 22151.
Heskestad, G., and Delichatsios, M. A., 1985, "Update: The Initial Convective Flow in Fire," *Fire Safety Journal*, vol. 15, No. 5.
Morton, B. R.; Taylor, Sir Geoffrey; and Turner, J. S., 1956, "Turbulent Gravitational Convection from Maintained and Instantaneous Sources," Proc. Royal Society A, 234:1–23.
National Fire Protection Research Foundation, 1993, International Fire Detection Research Project, vol. 3 (April).
NFPA 12, *Standard on Carbon Dioxide Extinguishing Systems*, National Fire Protection Association, Quincy, MA.
NFPA 12A, *Standard on Halon 1301 Fire Extinguishing Systems*, National Fire Protection Association, Quincy, MA.
NFPA 13, *Standard for the Installation of Sprinkler Systems*, National Fire Protection Association, Quincy, MA.
NFPA 16, *Standard for the Installation of Foam-Water Sprinkler and Foam-Water Spray Systems*, National Fire Protection Association, Quincy, MA.
NFPA 17, *Standard for Dry Chemical Extinguishing Systems*, National Fire Protection Association, Quincy, MA.
NFPA 70, *National Electrical Code*®, National Fire Protection Association, Quincy, MA.
NFPA 90A, *Standard for the Installation of Air-Conditioning and Ventilating Systems*, National Fire Protection Association, Quincy, MA.
NFPA 92A, *Recommended Practice for Smoke-Control Systems*, National Fire Protection Association, Quincy, MA.
NFPA *101*®, *Life Safety Code*®, National Fire Protection Association, Quincy, MA.
NFPA 664, *Standard for the Prevention of Fires and Explosions in Wood Processing and Woodworking Facilities*, National Fire Protection Association, Quincy, MA.

NFPA 2001, *Standard on Clean Agent Fire Extinguishing Systems*, National Fire Protection Association, Quincy, MA.

Schifiliti, R., 1986, "Use of Fire Plume Theory in the Design and Analysis of Fire Detector and Sprinkler Response." Master's Thesis, Worcester Polytechnic Institute, Center for Firesafety Studies, Worcester, MA.

Schifiliti, Robert P., and William E. Pucci, 1998, *Fire Detection Modeling, State of the Art.* Fire Detection Institute, Windsor, CT.

Title 47, *Code of Federal Regulations*, Communications Act of 1934, Amended.

CHAPTER 3

Protected Premises Fire Alarm Systems

Chapter 3 has been totally revised and restructured in the 1999 code to promote user friendliness. Several key areas, such as general system requirements, system inputs, and system outputs have been grouped together for a more logical arrangement.

Chapter 3 is completely reorganized as well as incorporating changes to the requirements found in the 1996 edition of the code. Some requirements previously found in the 1996 edition of Chapter 3 have been moved to other chapters. Other requirements have been merged together. Table 3.1 is designed to assist the reader in finding the new location of previous requirements in the chapter. It does not include new requirements or previous requirements that were not carried forward in the 1999 version of Chapter 3.

Table 3.1 Chapter 3 Cross References to 1996 Code

1996 Code Section	1999 Code Section	1996 Code Section	1999 Code Section	1996 Code Section	1999 Code Section
3-1	3-1	3-8	3-8	3-8.6.8	3-8.3.3.1.2 & 2-9.4
3-2	3-2	3-8.1	3-8.3.2	3-8.7	3-2.2
3-2.1	3-2	3-8.1.1	3-8.3.2.1 & 3-8.3.1.1	3-8.8	3-8.3.2.5
3-2.2	3-2.1	3-8.1.2	3-8.3.1.2	3-8.8.1	3-8.3.2.5.1
3-2.3	3-8.4.1	3-8.1.3	3-8.3.2.1	3-8.8.2	3-8.3.3.3.1
3-2.4	3-8.4.1.1	3-8.2	3-8.3.2.2	3-8.8.3	3-8.3.2.5.2
3-2.5	3-2.3	3-8.2.1	3-8.3.1.1	3-8.9	3-8.3.3.2
3-2.5.1	3-2.3.1	3-8.2.2	3-8.3.2.2	3-8.9.1	3-8.3.3.2.1
3-2.5.2	3-2.3.2	3-8.2.3	3-8.3.2.3.1	3-8.9.2	3-8.3.3.2.2
3-2.5.3	3-2.3.3	3-8.2.4	–	3-8.10	3-8.3.4
3-2.5.4	3-2.3.4	3-8.2.5	3-8.3.2.3.3	3-8.10.1	3-8.3.4.1
3-3	3-3	3-8.3	3-8.4.2.2	3-8.10.2	3-8.3.3.3.2
3-4	3-4	3-8.4	3-8.3.2.3.2	3-8.11	3-8.5
3-4.1	3-4.1	3-8.5	3-8.3.2.4	3-8.11.1	3-8.5.1
3-4.2	3-8.4.1	3-8.5.1	1-5.1.2 & 2-9	3-8.11.2	3-8.5.2
3-4.3	3-4.2	3-8.5.2	3-8.3.2.4.1	3-8.11.3	3-8.5.3
3-4.3.1	3-4.2.1	3-8.5.3	3-8.3.2.4.2	3-8.12	3-8.6
3-4.3.2	3-4.2.2	3-8.6	3-8.3.3	3-8.12.1	3-8.6.1
3-4.4	3-4.2.2.2	3-8.6.1	3-8.3.3.1	3-8.12.2	3-8.6.2
3-4.5	3-4.3	3-8.6.1.1	1-5.1.2 & 2-9	3-8.12.3	3-8.6.3
3-4.5.1	3-4.3.1	3-8.6.1.2	3-8.3.3.1.1	3-8.12.4	3-8.6.4
3-4.5.2	3-4.3.2	3-8.6.2	3-8.3.3.1.2	3-8.12.5	3-8.6.5
3-5	3-5	3-8.6.3	3-8.3.3.1.3	3-8.12.6	3-8.6.6
3-6	3-6	3-8.6.4	3-8.3.3.1.3	3-8.13	3-8.2
3-7	3-7	3-8.6.5	3-8.3.3.1.2 & 2-9	3-8.13.1	3-8.2.1
3-7.1	3-7	3-8.6.6	Lost	3-8.13.2	3-8.2.2
3-7.2	3-8.4.1.2	3-8.6.7	3-8.3.3.1.2 & 2-9	3-8.13.3	3-8.2.3

(continues)

Table 3.1 Continued

1996 Code Section	1999 Code Section	1996 Code Section	1999 Code Section	1996 Code Section	1999 Code Section
3-8.13.4	3-8.2.4	3-10.3	3-8.4.3.3	3-12.7.2	3-8.4.1.3.6.2
3-8.13.5	3-8.2.5	3-10.4	3-8.4.3.4	3-12.8	3-8.4.1.3.7
3-8.13.6	3-8.2.6	3-10.5	3-8.4.3.5	3-12.8.1	3-8.4.1.3.7.1
3-8.14	3-9.3	3-10.6	3-8.4.3.6	3-12.8.2	3-8.4.1.3.7.2
3-8.14.1	3-9.3.1	3-10.7	3-8.4.3.7	3-12.8.3	3-8.4.1.3.7.3
3-8.14.2	3-9.3.2	3-11	3-8.1	3-12.8.4	3-8.4.1.3.7.4
3-8.14.3	3-9.3.4	3-11.1	3-8.1.1	3-12.8.5	3-8.4.1.3.7.5
3-8.14.4	3-9.3.5	3-11.2	3-8.1.2	3-12.8.6	3-8.4.1.3.7.6
3-8.14.5	3-9.3.6	3-11.2.1	3-8.1.2.1	3-12.8.7	3-8.4.1.3.7.7
3-8.14.6	3-9.3.7	3-11.2.2	3-8.1.2.2	3-12.8.8	3-8.4.1.3.7.8
3-8.15	3-9.4	3-11.3	3-8.1.3	3-12.8.9	3-8.4.1.3.7.9
3-8.15.1	3-9.4.1	3-12	3-8.4.1.3	3-13	3-10
3-8.15.2	3-9.4.2	3-12.1	3-8.4.1.3	3-13.1	3-10.1
3-8.15.3	3-9.4.3	3-12.2	3-8.4.1.3.1	3-13.2	3-10.2
3-8.16	3-8.4.4.2	3-12.3	3-8.4.1.3.2	3-13.3	3-10.3
3-8.16.1	3-8.4.4.2.1	3-12.4	3-8.4.1.3.3	3-13.3.1	3-10.3.1
3-8.16.2	3-8.4.4.2.2	3-12.4.1	3-8.4.1.3.3.1	3-13.3.2	3-10.3.2
3-9	3-9	3-12.4.2	3-8.4.1.3.3.2	3-13.3.3	3-10.3.3
3-9.1	3-9.1	3-12.4.3	3-8.4.1.3.3.3	3-13.3.4	3-10.3.4
3-9.2	3-9.2	3-12.5	3-8.4.1.3.4	3-13.3.5	3-10.3.5
3-9.2.1	3-9.2.1	3-12.5.1	3-8.4.1.3.4.1	3-13.4	3-10.4
3-9.2.2	3-9.2.2	3-12.5.2	3-8.4.1.3.4.2	3-13.4.1	3-10.4.1
3-9.2.3	3-9.2.3	3-12.6	3-8.4.1.3.5	3-13.4.2	3-10.4.2
3-9.2.4	3-9.2.4	3-12.6.1	3-8.4.1.3.5.1	3-13.4.3	3-10.4.3
3-9.2.5	3-9.2.6	3-12.6.2	3-8.4.1.3.5.2	3-13.4.4	3-10.4.4
3-9.2.6	1-5.4.5.1	3-12.6.3	3-8.4.1.3.5.3	3-13.4.5	3-10.4.5
3-9.3.1	3-9.5.1	3-12.6.3.1	3-8.4.1.3.5.3.1	3-13.4.6	3-10.4.6
3-9.3.2	3-9.5.2	3-12.6.3.2	3-8.4.1.3.5.3.2		
3-9.3.3	3-9.5.3	3-12.6.3.3	3-8.4.1.3.5.3.3		
3-9.3.4	3-9.5.4	3-12.6.3.4	3-8.4.1.3.5.3.4		
3-9.4	3-9.6	3-12.6.4	3-8.4.1.3.5.4		
3-9.4.1	3-9.6.1	3-12.6.5	3-8.4.1.3.5.5		
3-9.4.2	3-9.6.2	3-12.6.5.1	3-8.4.1.3.5.5.1		
3-9.4.3	3-9.6.3	3-12.6.5.2	3-8.4.1.3.5.5.2		
3-9.4.4	3-9.6.4	3-12.6.5.3	3-8.4.1.3.5.5.3		
3-9.5	3-9.7	3-12.6.6	3-8.4.1.3.5.6		
3-9.5.1	3-9.7.1	3-12.6.6.1	3-8.4.1.3.5.6.1		
3-9.5.2	3-9.7.2	3-12.6.6.2	3-8.4.1.3.5.6.2		
3-9.5.3	3-9.7.3	3-12.6.6.3	3-8.4.1.3.5.6.3		
3-10	3-8.4.3	3-12.6.6.4	3-8.4.1.3.5.6.4		
3-10.1	3-8.4.3.1	3-12.7	3-8.4.1.3.6		
3-10.2	3-8.4.3.2	3-12.7.1	3-8.4.1.3.6.1		

3-1 Scope

Chapter 3 shall provide requirements for the application, installation, and performance of fire alarm systems, including fire alarm and supervisory signals, within protected premises.

3-2 General

The systems covered in Chapter 3 shall be used for the protection of life by automatically indicating the necessity for evacuation of the building or fire area, and for the protection of property through the automatic notification of responsible

persons and for the automatic activation of fire safety functions. The requirements of Chapters 1, 2, 4, and 5 shall also apply, unless they are in conflict with this chapter.

Section 3-2 clearly states that the primary purpose of Chapter 3 is "for the protection of life and property." Therefore, Section 3-2 gives the protection of both life and property equal and full consideration.

Section 3-2 also ensures that all fire alarm systems installed within the protected premises must first comply with Chapter 3 and then comply with the requirements of other chapters. The other chapters may add to the requirements for protected premises system installations, but must not replace, or conflict with, the requirements of Chapter 3.

Fire alarm systems as discussed in Chapter 3 can also be used in single living units or as dwelling fire warning systems, provided that the requirements of Chapter 8 are satisfied. However, the fire alarm systems covered under Chapter 8 are usually more economical and easier to operate for household applications.

Chapter 5 applies to signals transmitted to off-premises locations, using central supervising stations, proprietary supervising stations, and remote supervising station systems. Chapter 5 requirements apply to the transmitter located at the protected premises, the transmission channel between the protected premises, and the remotely located supervising station. Auxiliary systems are covered by Chapter 6.

Essentially, two choices can be made where a multiple-building, contiguous property has its proprietary supervising station in one of the on-site buildings. Each building can have its own protected premises system and be connected to its on-site supervisory station through a transmitter and transmission channel that meet the requirements of Section 5-5. Or, the alternative, the individual building systems can be directly connected to a master fire alarm control unit that is co-located within the supervising station. This arrangement uses signaling line circuits (SLCs) for the interconnections. These systems must comply with the requirements of 3-8.1 for interconnected fire alarm control units.

For a single-building property, the initiating devices and notification appliances are either directly connected or connected through zone or floor fire alarm system control units to the supervising station using initiating device circuits or signaling line circuits. Where other interconnected control units are used, the requirements of 3-8.1 apply. Regardless of the method of connection, for both single- and multiple-building properties, the proprietary supervising station facilities must comply with the requirements of Section 5-3.

3-2.1 Testing.

All protected premises fire alarm systems shall be maintained and tested in accordance with Chapter 7.

Testing of fire alarm systems is of paramount importance to ensure system reliability or mission effectiveness. Subsection 3-2.1 applies to all installed fire alarm systems. Because the testing of fire alarm systems is critical, many jurisdictions feel the need to develop and enforce their own fire alarm system testing requirements. The purpose of 3-2.1 is to provide the authorities having jurisdiction with an enforceable mandatory requirement to test all fire alarm systems in accordance with the code.

The requirements of the code do not apply retroactively, except for Chapter 7. See 7-1.1.4.

3-2.2 Signal Annunciation.

Protected premises fire alarm systems shall be arranged to annunciate alarm, supervisory, and trouble signals in accordance with 1-5.7.

The requirement for three separate and distinct signals has been included in previous editions of the code. All protected premises systems must distinctively indicate fire alarm, supervisory, and trouble signals. Where a protected premises has fire alarm, supervisory, and trouble signals, all three signals must indicate in a distinctive manner.

To comply with this requirement, the system trouble Light-Emitting Diode (LED) visible notification appliance cannot be used to indicate a supervisory condition (see 1-5.4.7). However, the code does permit a trouble signal to share the audible supervisory notification appliance. But, in that case, both supervisory signals and trouble signals must have distinctive visible notification to comply with 3-2.2 and 1-5.4.6.3.2.

3-2.3 Software and Firmware Control.

3-2.3.1 All software and firmware provided with a fire alarm system shall be listed for use with the fire alarm control unit.

3-2.3.2 A record of installed software and firmware version numbers shall be maintained at the location of the fire alarm control unit.

3-2.3.3* All software and firmware shall be protected from unauthorized changes.

A-3-2.3.3 A commonly used method of protecting against unauthorized changes can be described as follows (in ascending levels of access):

(a) *Access level 1:* Access by persons who have a general responsibility for safety supervision, who might be expected to investigate and initially respond to a fire alarm or trouble signal

(b) *Access level 2:* Access by persons who have a specific responsibility for safety, and who are trained to operate the control unit

(c) *Access level 3:* Access by persons who are trained and authorized to do the following:

(1) Reconfigure the site specific data held within the control unit, or controlled by it
(2) Maintain the control unit in accordance with the manufacturer's published instructions and data

(d) *Access level 4:* Access by persons who are trained and authorized either to repair the control unit or to alter its site specific data or operating system program, thereby changing its basic mode of operation

3-2.3.4 All changes shall be tested in accordance with 7-1.6.2.

Subsection 3-2.3 provides special requirements that must be followed when using computerized or microprocessor-based fire alarm systems. The term *firmware* is covered by the definition of Operating System Software and is synonymous with it. The term *software* is not specifically defined in the code, but it means programming that is specific to the fire alarm system within the protected premises. Site-specific software is the software that contains such items as the operations matrix and device addresses. Site-specific software is programmed by the installer. Firmware is similar to the operating system of a personal computer; it is generally not easily modified in the field and can only be modified by the manufacturer.

The term *firmware* is defined by the *IEEE Standard Dictionary of Electrical and Electronics Terms* as "The combination of a hardware device and computer instructions and data that reside as read-only software on that device." Firmware usually resides on a Read Only Memory (ROM) integrated circuit that is programmed at the factory and is not modified by the installer or user. The *IEEE Dictionary* defines software as "Computer programs, procedures, and possibly associated documentation and data pertaining to the operation of a computer system."

Because a single programming change could affect the entire operation of the fire alarm system, the software and firmware must be protected from unauthorized use, and the revision (REV) number of the installed software and firmware must be recorded. Generally, the revision number is recorded on the permanently attached diagram in the Fire Alarm Control Unit (FACU) and on the as-built drawings. If, upon subsequent inspections, the authority having jurisdiction finds a REV number of software or firmware currently installed different from that which was installed at the time of the acceptance test, then the additional tests performed in accordance with 7-1.6.2 should be reviewed.

3-2.4 Nonrequired Systems.

Subsection 3-2.4 was added for the 1999 edition of the code. Nonrequired systems are those that are installed to meet specific performance criteria desired by the owner. These performance criteria may not be mandated by a building code, fire code, or other NFPA standard. However, there is a need to document the intended performance so that the authority having jurisdiction can approve the final installation. Systems that do not meet requirements of the code are likely to fail when needed. These systems only create a false sense of security among occupants who think they are protected by a fire alarm system that is compliant with the code. Therefore, systems not deemed as supplementary by the authority having jurisdiction must meet code requirements.

3-2.4.1 Nonrequired protected premises systems shall meet the requirements of this code.

3-2.4.2 Nonrequired systems shall meet performance standards approved by the authority having jurisdiction.

3-3 Applications

Protected premises fire alarm systems shall include one or more of the following features:

(1) Manual alarm signal initiation
(2) Automatic alarm signal initiation
(3) Monitoring of abnormal conditions in fire suppression systems
(4) Activation of fire suppression systems
(5) Activation of fire safety functions
(6) Activation of alarm notification appliances
(7) Emergency voice/alarm communications
(8) Guard's tour supervisory service
(9) Process monitoring supervisory systems
(10) Activation of off-premises signals
(11) Combination systems
(12) Integrated systems

3-4 System Performance and Integrity

3-4.1 Purpose.

Section 3-4 shall provide information to be used in the design and installation of protected premises fire alarm systems for the protection of life and property.

3-4.2* Circuit Designations.

Initiating device, notification appliance, and signaling line circuits shall be designated by class or style, or both, depending on the circuit's capability to continue to operate during specified fault conditions.

Subsection 3-4.2 requires the class or style of circuits to be designated based on the designer's (or owner's) requirements.

Therefore, unless another code, the authority having jurisdiction, or the project's specifications has designated a class or style of circuit to be used, it is the designer's responsibility to designate the circuit classifications. Although 3-4.2 states that "circuits shall be designated by class or style, or both," if a circuit performance is specified as a certain style, it would be redundant to classify the circuit using both class and style. Because style provides a more complete performance description, many designers prefer to designate circuits by style.

A-3-4.2 Class A circuits are considered to be more reliable than Class B circuits because they remain fully operational during the occurrence of a single open or a single ground fault, while Class B circuits remain operational only up to the location of an open fault. However, neither Class A nor Class B circuits remain operational during a wire-to-wire short.

For both Class A and Class B initiating device circuits, a wire-to-wire short is permitted to cause an alarm on the system based on the rationale that a wire-to-wire short is the result of a double fault (for example, both circuit conductors have become grounded), while the code only considers the consequences of single faults. For many applications, an alarm caused by a wire-to-wire short is not permitted, and limitation to a simple Class A designation is not adequate. Introducing the style designation has made it possible to specify the exact performance required during a variety of possible fault conditions.

Limitation to Class A and Class B circuits only poses a more serious problem for signaling line circuits. Though a Class A signaling line circuit remains fully operational during the occurrence of a single open or single ground fault, a wire-to-wire short disables the entire circuit. The risk of such a catastrophic failure is unacceptable to many system designers, users, and authorities having jurisdiction. Once again, using the style designation makes it possible to specify either full system operation during a wire-to-wire short (Style 7) or a level of performance in between that of a Style 7 and a minimum function Class A circuit (Style 2).

A specifier can specify a circuit as either Class A or Class B where system performance during wire-to-wire shorts is of no concern, or it can specify, by the appropriate style designation, where the system performance during a wire-to-wire short and other multiple fault conditions *is* of concern.

The type of circuit to be used by the designer often depends on the number of devices connected to the circuit, the amount of detection that would be lost during a fault condition, and the impact that the loss of detection would have on life safety or property protection. Furthermore, the class and style selections can be and often are based on the number and condition of occupants, the length of the circuit, and other mitigating factors. See 3-4.3.2 and Supplement 1.

The code does not require a specific class or style of circuit to be used, but simply defines the operation of each and leaves the decision to the designer or owner. Reliable operation under the environmental and physical conditions present in the specific protected premises is one of the most important factors to be considered for circuit class or style selection.

3-4.2.1 Class. Initiating device, notification appliance, and signaling line circuits shall be permitted to be designated as either Class A or Class B, depending on the capability of the circuit to transmit alarm and trouble signals during nonsimultaneous single circuit fault conditions as specified by the following:

The performance of either class of circuit is dependent on two factors: how the circuit is physically wired and how the fire alarm control unit operates during the specified fault conditions. The class of a circuit is based on its ability to operate under a specified single fault.

(1) Circuits capable of transmitting an alarm signal during a single open or a nonsimultaneous single ground fault on a circuit conductor shall be designated as Class A.

(2) Circuits incapable of transmitting an alarm beyond the location of the fault conditions specified in 3-4.2.1(a) shall be designated as Class B.

Faults on both Class A and Class B circuits shall result in a trouble condition on the system in accordance with the requirements of 1-5.8.

3-4.2.2 Style.

3-4.2.2.1 Initiating device, notification appliance, and signaling line circuits shall be permitted to be designated by style so as to describe requirements in addition to the requirements shown for Class A or Class B according to the following:

(1) An initiating device circuit shall be permitted to be designated as either Style A, B, C, D, or E, depending on its ability to meet the alarm and trouble performance requirements shown in Table 3-5, during a single open, single ground, wire-to-wire short, and loss-of-carrier fault condition.

(2) A notification appliance circuit shall be permitted to be designated as either Style W, X, Y, or Z, depending on its ability to meet the alarm and trouble performance requirements shown in Table 3-7, during a single open, single ground, and wire-to-wire short fault condition.

(3) A signaling line circuit shall be permitted to be designated as either Style 0.5, 1, 2, 3, 3.5, 4, 4.5, 5, 6, or 7, depending on its ability to meet the alarm and trouble performance requirements shown in Table 3-6, during a single open, single ground, wire-to-wire short, simultaneous wire-to-wire short and open, simultaneous wire-

to-wire short and ground, simultaneous open and ground, and loss-of-carrier fault conditions.

Again, the style of circuit chosen by the designer depends on the number of devices on the circuit, the amount of detection that would be lost during the specified fault conditions, and the impact that the loss of detection would have on life safety or property protection. The code does not require a specific style of circuit to be used, but simply defines the operation of each circuit style and leaves the choice to the designer or owner. Reliable operation under the environmental and physical conditions present in the specific protected premises is one of the most important factors to be considered for circuit style selection.

The performance of any of the circuit styles is dependent on two factors: how the circuit is physically wired and how the fire alarm control unit operates during the specified fault conditions. The style of a circuit is based on its ability to operate under specified multiple faults.

3-4.2.2.2* All styles of Class A circuits using physical conductors (for example, metallic, optical fiber) shall be installed such that the outgoing and return conductors, exiting from and returning to the control unit, respectively, are routed separately. The outgoing and return (redundant) circuit conductors shall not be run in the same cable assembly (that is, multiconductor cable), enclosure, or raceway.

This requirement minimizes the chances that the Class A circuit's operation could be defeated by the cutting of the cable or the failure of a section of cable due to fire damage in a single location. The use of plenum, riser, or circuit integrity cable also provides better resistance to fire damage and subsequent failure of the circuit. See A-3-4.2.2.2.

Exception 1: The outgoing and return (redundant) circuit conductors shall be permitted to be run in the same cable assembly, enclosure, or raceway under any of the following conditions:

(a) For a distance not to exceed 10 ft (3 m) where the outgoing and return conductors enter or exit the initiating device, notification appliance, or control unit enclosures.

(b) Where the vertically run conductors are contained in a 2-hour rated cable assembly or enclosed (installed) in a 2-hour rated enclosure.

It is acceptable to use cable assemblies that have been listed with a minimum fire rating of 2 hours. Article 760 of NFPA 70, *National Electrical Code*® contains wiring requirements for fire alarm systems. Only mineral insulated (MI) cable and circuit integrity (CI) cables have the necessary 2-hour rating. The exception for 2-hour rated stairwells has been removed from the 1999 edition of the code because NFPA 101®, *Life Safety Code*®, now permits the fire alarm cables to be installed in the stairwell provided certain provisions

are met. Exception No. 2 of 5-1.3.2.1(e) of NFPA *101* permits penetrations in stairwells for fire alarm circuits, provided that the conductors are installed in metallic conduit and are protected where the penetrations are made.

(c) Where looped conduit/raceway systems are provided, single conduit/raceway drops to individual devices or appliances shall be permitted.

A drop to a single device or appliance limits exposure of the conductors. Even if all four conductors to the device or appliance were cut or damaged by fire, only the single device or appliance would be lost. It should be noted that this exception only applies to conduit/raceway systems, and does not permit the use of cables.

(d) Where looped conduit/raceway systems are provided, single conduit/raceway drops to multiple devices or appliances installed within a single room not exceeding 1000 ft² (92.9 m²) in area shall be permitted.

A drop to a room of 1000 ft² (92.9 m²) or less in size limits exposure of the conductors. Even if all four conductors to the room were cut or damaged by fire, it is likely that only a a small number of devices or appliances would be lost. It should be noted that this exception only permits the use of conduit/raceway systems and does not permit the use of cables.

A-3-4.2.2.2 A goal of 3-4.2.2.2 is to provide adequate separation between the outgoing and return cables. This separation is required to help ensure protection of the cables from physical damage. The recommended minimum separation to prevent physical damage is 1 ft (0.305 m) where the cable is installed vertically and 4 ft (1.22 m) where the cable is installed horizontally.

The amount of separation distance recommendation is in Appendix A because each installation presents different physical conditions, and the amount of separation distance depends on those conditions. The information in the appendix is provided for guidance only. Separation distances may be less or more than the recommendation depending on the protection afforded by the building construction or wiring installation. Where Class A circuits are required by the designer, reliability of operation under the specified fault conditions is obviously a design consideration. The separation of the outgoing and return conductors is an additional requirement to help ensure reliable operation.

3-4.3 Signaling Paths.

3-4.3.1 The class or style of signaling paths (circuits) shall be determined from an evaluation based on the path performance detailed in this code and on engineering judgment.

It is important to evaluate the type of circuit (Initiating Device Circuit or Signaling Line Circuit) that is to be used

based on the planned installation technique. If T-Tapping is used, the performance of the type of signaling line circuit originally chosen may not be valid with the T-taps in the circuit. T-Tapping is only allowed on Class B Signaling Line Circuits and is not allowed with initiating device circuits or Notification Appliance Circuits. Engineering judgment *must* be used in conjunction with the requirements of 3-4.3.1 to determine the overall reliability desired from the class or style circuit needed to meet the fire protection goals of the owner. Exhibit 3.1 illustrates T-Tapping.

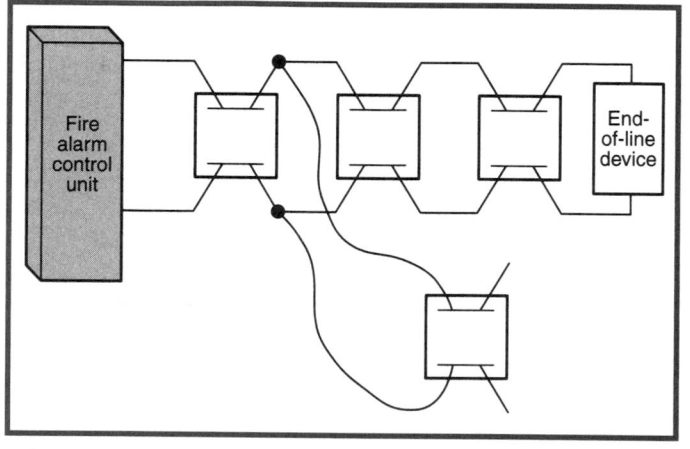

Exhibit 3.1 *Illustration of T-Tapping.*

3-4.3.2 When determining the integrity and reliability of the interconnecting signaling paths (circuits) installed within the protected premises, the following influences shall be considered:

(1) Transmission media used
(2) Length of the circuit conductors
(3) Total building area covered by and the quantity of initiating devices and notification appliances connected to a single circuit
(4) Effect of a fault in the fire alarm system on the objectives stated in Section 3-2
(5) Nature of the hazard present within the protected premises
(6) Functional requirements of the system necessary to provide the level of protection required for the system
(7) Size and nature of the population of the protected premises

Paragraph 3-4.3.1 requires an evaluation of the hazards involved in the protected premises when selecting circuit class and style. Designers should avoid "putting all the eggs in one basket" (for example, installing the entire detection system on one signaling line circuit) when designing a fire alarm system. Paragraph 3-4.3.3 is new, and it requires the

system designer to record this evaluation in the system documentation required by Section 1-6. Also see Supplement 1.

3-4.3.3 Results of the evaluation required by 3-4.3.1 shall be included with the documentation required by Section 1-6.

3-5* Performance of Initiating Device Circuits (IDC)

The assignment of class designations or style designations, or both, to initiating device circuits shall be based on their performance capabilities under abnormal (fault) conditions in accordance with the requirements of Table 3-5.

A-3-5 Table 3-5 and Table 3-6 should be used as follows.

(a) It should be determined if the initiating devices are directly connected as follows:

(1) To the initiating device circuit
(2) To a signaling line circuit interface on a signaling line circuit
(3) To an initiating device circuit, which in turn is connected to a signaling line circuit interface on a signaling line circuit

(b) Tables 3-5 and 3-6 of signaling performance required should be determined. The columns marked A through Eα in Table 3-5, and 0.5 through 7α in Table 3-6 are arranged in ascending order of performance.

(c) The prime purpose of the tables is to enable identification of minimum performance for styles of initiating device circuits and signaling line circuits. It is not the intention that the styles be construed as grades. That is, a Style 3 system is not superior to a Style 2 system, or vice versa. In fact, a particular style might better provide adequate and reliable signaling for an installation than a more complex style.

(d) Tables 3-5 and 3-6 allow users, designers, manufacturers, and the authority having jurisdiction to identify minimum performance of present and future systems by determining the trouble and alarm signals received at the control unit for the specified abnormal conditions.

(e) The number of automatic fire detectors connected to an initiating device circuit is limited by good engineering practice and the listing of the detectors. If a large number of detectors are connected to an initiating device circuit, locating the detector in alarm or locating a faulty detector becomes difficult and time consuming.

On certain types of detectors, a trouble signal results from faults in the detector. When this occurs where there are large numbers of detectors on an initiating device circuit, locating the faulty detector also becomes difficult and time consuming.

Table 3-5 Performance of Initiating Device Circuits (IDC)

Class	B			B			B			A			A		
Style	A			B			C			D			Eα		
Abnormal condition	Alarm	Trouble	Alarm receipt capability during abnormal condition	Alarm	Trouble	Alarm receipt capability during abnormal condition	Alarm	Trouble	Alarm receipt capability during abnormal condition	Alarm	Trouble	Alarm receipt capability during abnormal condition	Alarm	Trouble	Alarm receipt capability during abnormal condition
	1	2	3	4	5	6	7	8	9	10	11	12	13	14	15
Single open	—	X	—	—	X	—	—	X	—	—	X	X	—	X	X
Single ground	—	X	—	—	X	R	—	X	R	—	X	R	—	X	R
Wire-to-wire short	X	—	—	X	—	—	—	X	—	X	—	—	—	X	—
Loss of carrier (if used)/ channel interface	—	—	—	—	—	—	—	X	—	—	—	—	—	X	—

R = Required capacity
X = Indication required at protected premises and as required by Chapter 5
α = Style exceeds minimum requirements of Class A

3-6* Performance of Signaling Line Circuits (SLC)

The assignment of class designations or style designations, or both, to signaling line circuits shall be based on their performance capabilities under abnormal (fault) conditions in accordance with the requirements of Table 3-6.

Formal Interpretations 79-8, 87-1, and 86-1 provide additional information relating to the use of Tables 3-5 and 3-6.

Formal Interpretation 79-8

Reference: 3-5, 3-6

Background: This is a request for a formal interpretation in regards to the use of addressable initiating devices. Since the control unit or the central supervising station is in two way communication with these devices, then:

Question 1: Is it the intent to categorize the description of performance of the circuit these devices are on as a "Signaling Line Circuit," rather than an "Initiating Device Circuit"?

Answer: Yes.

Question 2: If the style of this addressable communication circuit has a different performance (style number) than the remaining portions of the multiplex pathways, would these different circuit performance levels have to be individually specified in order to adequately describe the system?

Answer: Yes.

Issue Edition: 1979 of NFPA 72D

Reference: 3-9, 3-10

Date: June 1985

Formal Interpretation 87-1

Reference: Tables 3-5, 3-6

Question 1: Is it the intent of the Committee to categorize the description of performance of the circuit these devices are on as a "Signaling Line Circuit" illustrated in Table 3-6 rather than an Initiating Device Circuit as described in Table 3-5?

Answer. Yes.

Question 2: If the style of this communication circuit has a different performance (style number as illustrated in Table

Table 3-6 Performance of Signaling Line Circuits (SLC)

Class	B			B			A			B			B			B			B			A			A			A		
Style	0.5			1			2α			3			3.5			4			4.5			5α			6α			7α		
	Alarm	Trouble	Alarm receipt capability during abnormal conditions	Alarm	Trouble	Alarm receipt capability during abnormal conditions	Alarm	Trouble	Alarm receipt capability during abnormal conditions	Alarm	Trouble	Alarm receipt capability during abnormal conditions	Alarm	Trouble	Alarm receipt capability during abnormal conditions	Alarm	Trouble	Alarm receipt capability during abnormal conditions	Alarm	Trouble	Alarm receipt capability during abnormal conditions	Alarm	Trouble	Alarm receipt capability during abnormal conditions	Alarm	Trouble	Alarm receipt capability during abnormal conditions	Alarm	Trouble	Alarm receipt capability during abnormal conditions
Abnormal condition	1	2	3	4	5	6	7	8	9	10	11	12	13	14	15	16	17	18	19	20	21	22	23	24	25	26	27	28	29	30
Single open	—	X	—	—	X	—	—	X	R	—	X	—	—	X	—	—	X	—	—	X	R	—	X	R	—	X	R	—	X	R
Single ground	—	X	—	—	X	R	—	X	R	—	X	R	—	X	—	—	X	R	—	X	—	—	X	R	—	X	R	—	X	R
Wire-to-wire short	—	—	—	—	—	—	—	—	M	—	X	—	—	X	—	—	X	—	—	X	—	—	X	—	—	X	—	—	X	R
Wire-to-wire short & open	—	—	—	—	—	—	—	—	M	—	X	—	—	X	—	—	X	—	—	X	—	—	X	—	—	X	—	—	X	—
Wire-to-wire short & ground	—	—	—	—	—	—	—	X	M	—	X	—	—	X	—	—	X	—	—	X	—	—	X	—	—	X	—	—	X	—
Open and ground	—	—	—	—	—	—	—	X	R	—	X	—	—	X	—	—	X	—	—	X	—	—	X	—	—	X	X	—	X	R
Loss of carrier (if used)/ channel interface	—	—	—	—	—	—	—	—	—	—	—	—	—	X	—	—	X	—	—	X	—	—	X	—	—	X	—	—	X	—

M = May be capable of alarm with wire-to-wire short
R = Required capability
X = Indication required at protected premises and as required by Chapter 5
α = Style exceeds minimum requirements for Class A

3-6) than the remaining portions of the multiplex pathways, would these different circuit performance levels have to be individually specified in order to adequately describe the system?

Answer: Yes.

Issue Edition: 1987 of NFPA 72D

Reference: Tables 2-12.1, 2-13.1

Date: June 1987

Reprinted to correct error: January 1989

Formal Interpretation 86-1

Reference: Table 3-6

Question: For systems where the main (primary) and standby (secondary) power are transmitted over the same circuit separate from the signaling circuit, is it the intent of 72 to require redundancy (i.e. two such circuits) of the power circuit, when a Style 7 signaling line circuit is required?

Answer: Yes.

Issue Edition: 1986 of NFPA 72D

Reference: Entire Standard

Date: August 1986

Replaces F.I. 86-1 issued February 1986

A-3-6 Refer to A-3-5.

The capacity limitations of initiating device and signaling line circuits have not been defined. Many manufacturers publish the capacity limitations of their systems, but if there is a question about the number of devices allowed, common sense should prevail. The capacities allowed on a single circuit should be determined based on the amount of detection that would be lost during a fault condition and the resulting impact on life safety or property protection. Reliable operation under the environmental and physical conditions present in the specific protected premises is one of the most important factors to be considered when determining the number of devices to be installed on a single circuit. The original Table 3-5 and Table 3-6 that appeared in the 1993 edition of NFPA 72, and in previous editions of NFPA 72, included device capacity limitations based on the class or style of the circuit used. However, the capacity sections only applied to circuits used as part of a proprietary supervising station fire alarm system. When so used, the circuit loading capacity section defined an equivalent performance for each circuit when the circuit was fully loaded. This feature was an attempt to make the circuits equal.

3-7 Performance of Notification Appliance Circuits (NAC)

The assignment of class designations or style designations, or both, to notification appliance circuits shall be based on their performance capabilities under abnormal (fault) conditions in accordance with the requirements of Table 3-7.

3-8 System Requirements

3-8.1* Fire Alarm Control Units.

Fire alarm systems shall be permitted to be either integrated systems combining all detection, notification, and auxiliary functions in a single system or a combination of component subsystems. Fire alarm system components shall be permitted to share control equipment or shall be able to operate as stand alone subsystems, but, in any case, they shall be arranged to function as a single system. All component subsystems shall be capable of simultaneous, full load operation without degradation of the required, overall system performance.

Table 3-7 Notification Appliance Circuits (NAC)

Class	B	B	B	A
Style	W	X	Y	Z
	Trouble indication at protected premises	Alarm capability during abnormal conditions	Trouble indication at protected premises	Alarm capability during abnormal conditions	Trouble indication at protected premises	Alarm capability during abnormal conditions	Trouble indication at protected premises	Alarm capability during abnormal conditions
Abnormal condition	1	2	3	4	5	6	7	8
Single open	X	—	X	X	X	—	X	X
Single ground	X	—	X	—	X	X	X	X
Wire-to-wire short	X	—	X	—	X	—	X	—

X = Indication required at protected premises

Subsection 3-8.1 applies both where the interconnected control units are the products of a single manufacturer and where they are the products of two or more manufacturers, regardless of when installed. The subsection covers the requirements for the interconnection, monitoring, and compatibility of the control units.

A-3-8.1 This code addresses field installations that interconnect two or more listed control units, possibly from different manufacturers, that together fulfill the requirements of this code.

Such an arrangement should preserve the reliability, adequacy, and integrity of all alarm, supervisory, and trouble signals and interconnecting circuits intended to be in accordance with the provisions of this code.

Where interconnected control units are in separate buildings, consideration should be given to protecting the interconnecting wiring from electrical and radio frequency interference.

There are many reasons for interconnecting control units. In some cases, an existing building may need additional power supplies and notification appliances to conform to a new law, such as the Americans with Disabilities Act (ADA), and the original fire alarm system may not be able to accommodate the necessary changes. It also may not be feasible to modify or expand the existing fire alarm control unit due to lack of parts, economics, or the system configuration. Some

newer systems consist of two or more subsystems (control units) connected to a single- or multiple-master control unit(s). In other cases, a new wing is added to a building, and the contractor plans to use a new, separate, and different manufacturer's system for the new wing. However, the new wing's fire alarm system must operate as if it were part of the original system installed many years before.

Until subsection A-3-8.1 appeared in the code, there had been no guidance regarding the correct procedures to follow when interconnecting the control panels or fire alarm systems as described in these examples. The goal of 3-8.1 and A-3-8.1 is to advise the fire alarm system designer, the installer, and the authority having jurisdiction to consider the appropriate and most reliable method of interconnection.

It is not the intent of subsection 3-8.1 to allow two small fire alarm control units to be interconnected to avoid the installation of a single larger fire alarm control unit in a newly constructed building. It is the intent of the code that the system designer should use the appropriately sized equipment that is designed to meet the fire alarm system needs of the owner of a newly constructed or renovated building.

3-8.1.1 The method of interconnection of control units shall meet the monitoring requirements of 1-5.8 and NFPA 70, *National Electrical Code*®, Article 760, and shall be achieved by the following recognized means:

(1) Electrical contacts listed for the connected load
(2) Listed digital data interfaces (such as serial communications ports and gateways)
(3) Other listed methods

Increasingly, digital and analog data are being sent to HVAC controls, elevator controls, security system controls and other building system controls in a serial format. Paragraph 3-8.1.1 was revised for the 1999 code to include the term *gateways*, defined in Chapter 1, which implies the use of Building Automation Control networks such as BACnet, LONWorks, or other communications protocols. In cases where gateways are used, the communications protocol or interface must be specifically listed for such use.

3-8.1.2 If approved by the authority having jurisdiction, interconnected control units providing localized detection, evacuation signaling, and auxiliary functions shall be permitted to be monitored by a fire alarm system as initiating devices.

Paragraph 3-8.1.2 covers such installations where, for example, multiple buildings on a contiguous property, under single ownership, and with individual protected premises fire alarm systems, may be monitored by a single fire alarm control unit. This control unit may also serve the building in which it is located as part of the protected premises fire alarm system. The purpose of this arrangement may be to have only one off-premises connection. In any case, the

building with the master control unit that is providing the monitoring of the other systems should be identified at the building, at the remote annunciator, and at the supervising station receiving the fire alarm signal.

3-8.1.2.1 Each interconnected control unit shall be separately monitored for alarm, trouble, and supervisory conditions.

Each interconnected fire alarm control unit must be monitored for alarm signals first. Then each interconnected fire alarm control unit must be monitored for supervisory conditions. This requirement means that if the satellite fire alarm control unit (FACU) interconnected to the master FACU experiences a trouble condition for any reason, that trouble condition reports to the master FACU as a supervisory condition, indicating the interconnected fire alarm control unit is off-normal. The interconnection between the fire alarm control units is monitored for integrity and if that circuit experiences a fault condition, a trouble condition for that circuit (zone or point) is indicated at the master fire alarm control unit. See Exhibit 3.2 for an example of a typical master fire alarm control unit.

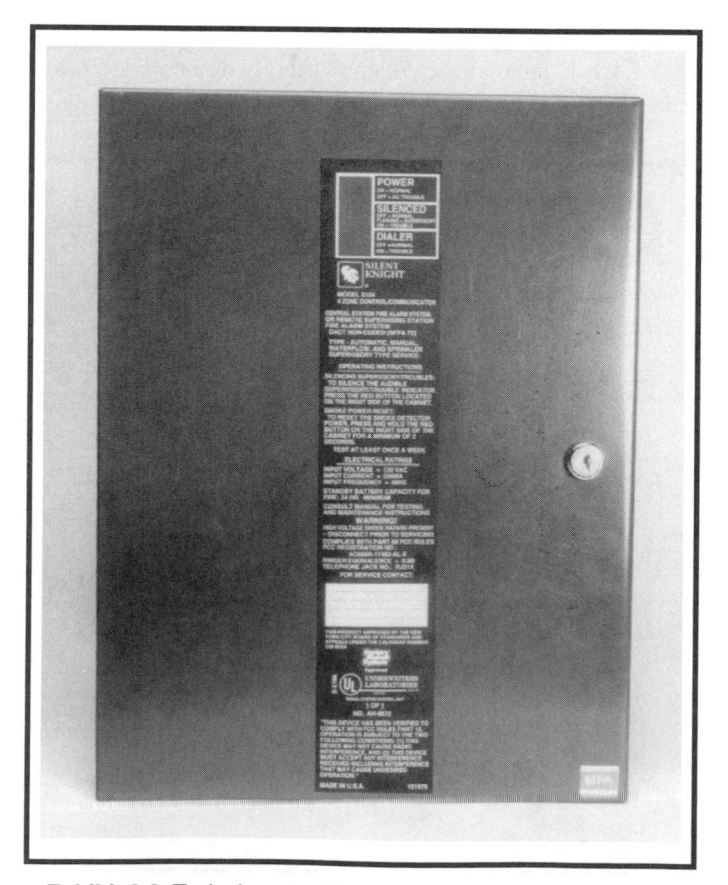

Exhibit 3.2 *Typical master fire alarm control unit. (Source: Silent Knight Corp., Maple Grove, MN)*

3-8.1.2.2 Interconnected control unit alarm signals shall be permitted to be monitored by zone or by combined common signals.

3-8.1.3 Protected premises fire alarm control units shall be capable of being reset or silenced only from the control unit at the protected premises.

Exception: Where otherwise specifically permitted by the authority having jurisdiction.

Remotely resetting fire alarm control equipment without first investigating the premises is a very dangerous practice. Paragraph 3-8.1.3 requires the on-site restoration to normal of fire alarm systems. If a fire alarm system is in alarm or trouble, a technician should be required to investigate the causes of that alarm or trouble, and then reset the fire alarm control unit. See Exhibit 3.3 for an example of a case where the control unit was remotely reset, which resulted in a large loss.

3-8.1.4 Protected Premises Fire Alarm Systems Interconnected with Dwelling Unit Fire Warning Equipment.

3-8.1.4.1 A protected premises fire alarm system shall be permitted to be interconnected to dwelling unit fire warning equipment only for the purposes of transmitting an alarm from the protected premises fire alarm system to the dwelling unit fire warning system.

3-8.1.4.2 If interconnected, an alarm condition at the protected premises fire alarm system shall cause the alarm notification appliance(s) within the family living unit of the dwelling unit fire warning system to become energized. The notification appliances shall remain energized until the protected premises fire alarm system is silenced or reset.

3-8.1.4.3 The interconnection circuit or path from the protected premises fire alarm system to the dwelling unit fire warning system shall be monitored for integrity by the protected premises fire alarm system in accordance with 1-5.8.

3-8.1.4.4 An alarm condition occurring at the dwelling unit fire warning system or the operation of any test switches provided as part of the dwelling unit fire warning equipment shall not cause an alarm condition at the protected premises fire alarm system.

3-8.2 Combination Systems.

Combination systems can be a combination of fire alarm and burglary, HVAC control, access control, paging, lighting, and more. New communications technologies such as BACnet and LONWorks have further blurred the lines between fire alarm systems and other non-fire systems. Of all of the systems that can be interconnected, only the fire alarm system performance requirements are typically regulated by codes and standards. Wiring of other systems is covered by Article 725 of NFPA 70, *National Electrical Code*. Special care must be taken to ensure that the nonregulated systems do not interfere with the operation of the fire

Exhibit 3.3 *A large loss resulted at this facility when the central unit was remotely reset. (Photo courtesy of Jim Regan, copyright © Jim Regan, Hinsdale, IL)*

alarm system. For more information on integrated systems, see Supplement 3.

3-8.2.1* Fire alarm systems shall be permitted to share components, equipment, circuitry, and installation wiring with non-fire alarm systems.

A-3-8.2.1 The provisions of 3-8.2.1 apply to the types of equipment used in common for fire alarm systems, such as fire alarm, sprinkler supervisory, or guard's tour service, and for other systems, such as burglar alarm or coded paging systems, and to methods of circuit wiring common to both types of systems.

3-8.2.2 If common wiring is used for combination systems, the equipment for non-fire alarm systems shall be permitted to be connected to the common wiring of the system. Short circuits, open circuits, or grounds in this equipment or between this equipment and the fire alarm system wiring shall not interfere with the monitoring for integrity of the fire alarm system or prevent alarm, supervisory, or fire safety control signal transmissions.

Common wiring between fire alarm and non-fire alarm systems means receiving signal inputs from or providing outputs to either system on the same wires. This common wiring could include circuits supplying device power, initiating device circuits, signaling line circuits, or notification appliance circuits. In any case, a short, a ground, or an open circuit in the common wiring caused by the non-fire equipment must not prevent the receipt of an alarm, a trouble signal, or a supervisory signal or prevent the fire alarm system's notification appliances from operating.

3-8.2.3 To maintain the integrity of fire alarm system functions, the provision for removal, replacement, failure, or maintenance procedure on any supplementary hardware, software, or circuit(s) shall not impair the required operation of the fire alarm system.

Exception: Where the hardware, software, or circuit(s) is listed for fire alarm use.

Paragraph 3-8.2.3 addresses the problem of field maintenance or equipment failure of non-fire alarm system components. Equipment that is not required for the operation of the fire alarm system that is modified, removed, or is malfunctioning in any way must not impair the operation of the fire alarm system. The exception to 3-8.2.3 implies that fire alarm systems may share components with other systems, and that these shared components might not be listed for fire alarm use. Such a practice is dangerous unless the system design carefully protects the integrity of the fire alarm system. Most applications of this permitted practice involve interconnection of the listed fire alarm system with paging systems, burglar alarm systems, HVAC control systems, and process monitoring systems.

Users have also connected fire alarm systems to supplementary equipment (such as business computers and monitors) not listed for fire alarm use to display system conditions to operators in more detail than may be available from the fire alarm systems alone.

However, authorities having jurisdiction have found that some installations do not satisfy the intent of 3-8.2 of the code covering combination systems. In some installations, software or firmware changes or other repairs to the non-fire alarm equipment have delayed fire alarm signals or prevented their display altogether. In general, incorrectly applied and interconnected systems may prevent one or more of the fire alarm system functions from operating as intended. To guard against such failures, reacceptance testing the fire alarm system in accordance with the requirements of 7-1.6.2 is usually necessary after all changes and repairs have been made to the non-fire alarm components of the system.

Where a non-fire alarm system component is listed for fire alarm use, the listing agency has investigated the compatibility of its integration with a fire alarm system, as well as temperature characteristics and other extensive fire alarm safety factors, and the listing agency has found the product suitable for the purpose. Consequently, the exception to 3-8.2.3 exempts this specifically listed equipment from the requirements of 3-8.2.3.

3-8.2.4 Speakers used as alarm notification appliances on fire alarm systems shall not be used for nonemergency purposes.

Exception 1:* If the fire command center is constantly attended by a trained operator, selective paging shall be permitted as approved by the authority having jurisdiction.

A-3-8.2.4 Exception No. 1 If the building paging system can be controlled by personnel at the fire command center, and if permitted by the authority having jurisdiction, the building paging system can be used as a supplementary notification system to provide selective and all-call fire alarm evacuation voice messages and messages for occupants to relocate to safe areas in a building.

Exception 2:* If all of the following conditions are met:

(a) The speakers and associated audio equipment are installed or located with safeguards to prevent tampering or misadjustment of those components essential to intended operation for fire.

(b) The monitoring integrity requirements of 1-5.8 and 3-8.4.1.3.2 shall continue to be met while the system is used for nonemergency purposes.

(c) It is permitted by the authority having jurisdiction.

The fire alarm notification appliances cannot be used for general paging functions unless the exceptions apply. Conversely,

standard background music or paging speakers cannot be used as fire alarm notification appliances. The speakers must be listed for fire alarm use. The term *constantly attended* means that personnel are in the area where the fire alarm system control unit is located 24 hours a day, seven days a week, not only during the operation of the facility. In both exceptions, the approval of the authority having jurisdiction is required.

A-3-8.2.4 Exception No. 2 An emergency voice/alarm communications system can be used for nonemergency purposes provided the performance and supervision requirements of an emergency voice/alarm communications system are still complied with. The building operator, system designer, and authority having jurisdiction should be aware that in some situations such a system could be subject to deliberate tampering. Tampering is usually attempted to reduce the output of a sound system that is in constant use as a music or paging system and is a source of annoyance to employees. The likelihood of tampering can be reduced through proper consideration of loudspeaker accessibility and system operation. Access can be reduced through the use of hidden or nonadjustable transformer taps (which can reduce playback levels), use of vandal resistant listed loudspeakers, and placement in areas that are difficult to access, such as high ceilings (any ceiling higher than could be reached by standing on a desk or chair). Nonemergency operation of the system should always consider that an audio system that annoys an employee potentially reduces employee productivity and can also annoy the public in a commercial environment. Most motivations for tampering can be eliminated through appropriate use of the system and employee discipline. Access to amplification equipment and controls should be limited to those in authority to make adjustments to such equipment. It is common practice to install such equipment in a manner that allows adjustment of nonemergency audio signal levels while defaulting to a fixed, preset level of playback when operating in emergency mode. Under extreme circumstances, certain zones of a protected area might require a dedicated emergency voice/alarm communications zone.

3-8.2.5 In combination systems, fire alarm signals shall be distinctive, clearly recognizable, and take precedence over any other signal even when a non-fire alarm signal is initiated first.

The requirement of 3-8.2.5 does not mean that two separate notification appliances must be used. A single appliance may be used if it can supply two different, distinctive signals; the fire alarm signal always takes precedence. However, the notification appliance circuit must comply with 3-8.2.2, which requres non-interference by faults on non-fire alarm equipment and monitoring for integrity. It

should be noted that 3-8.4.1.2.2 requires the national standard (temporal three-pulse pattern) evacuation signal where evacuation is planned (see 3-8.4.1.2).

3-8.2.6 If the authority having jurisdiction determines that the information being displayed or annunciated on a combination system is excessive and is causing confusion and delayed response to a fire emergency, the authority having jurisdiction shall be permitted to require that the display or annunciation of information for the fire alarm system be separate from and have priority over information for the non-fire alarm systems.

Confusion or delays in response to a fire alarm signal while scrolling through many non-fire alarm system events, such as burglary signals, is unacceptable. If fire alarm signals cannot be displayed on a priority basis, the authority having jurisdiction may require a separate display for the fire alarm signals.

3-8.3 Fire Alarm System Inputs.

3-8.3.1 General.

3-8.3.1.1 Fire alarm boxes, automatic alarm-initiating devices, and waterflow initiating devices shall be listed for the intended application, installed in accordance with Chapter 2, and tested in accordance with Chapter 7.

The phrases "listed for the purpose" and "listed for the intended application" have similar meanings. "Listed for the intended application" means, for example, that if a device is to be used in a low-temperature environment or a wet location, the device should be listed for that environment. Also, if a fire alarm system control unit device is to be used to release a fire extinguishing or suppression system, it should be listed for that application as well. "Listed for the purpose," as used in this code, means listed for the specific fire alarm system application.

3-8.3.1.2 For fire alarm systems employing automatic fire detectors or waterflow detection devices, at least one fire alarm box shall be provided to initiate a fire alarm signal. This fire alarm box shall be located where required by the authority having jurisdiction.

Exception: Fire alarm systems dedicated to elevator recall control and supervisory service as permitted in 3-9.3.1.

One reason that at least one fire alarm box is required is to allow an alarm to be transmitted if the automatic fire detectors or sprinkler system are out of service during repairs or during a test. This requirement presumes there is a contingency plan to address a fire emergency during the out-of-service time or during a test. It also presumes that personnel within the facility and at any supervising station to which the premises is connected are aware of the plan and will acknowledge receipt of the alarm. Typical locations for

the manual fire alarm box could be in the immediate vicinity of the automatic sprinkler system control valves, at the building engineer's office, or at some location that will likely be contacted in the event of a fire, such as at the telephone operator's station. However, this manual fire alarm box must be located where required by the authority having jurisdiction.

Another reason the fire alarm box is required is to permit a building occupant to initiate an alarm signal prior to the actuation of an automatic initiating device, providing earlier warning of a fire emergency. Manual fire alarm boxes should be located electrically ahead of all other initiating devices. This location ensures that the alarm signal will be initiated even if there is an open circuit downstream from the manual fire alarm box.

The exception to 3-8.3.1.2 exempts only fire alarm control units dedicated to elevator recall and supervisory service as covered by 3-9.3.1.

3-8.3.1.3 Supervisory devices shall be installed in accordance with Chapter 2.

3-8.3.2 Alarm Signal Initiation.

3-8.3.2.1 Manual Fire Alarm Signal Initiation. Manual fire alarm signal initiation shall comply with the requirements of Section 2-8. If signals from fire alarm boxes and other fire alarm initiating devices within a building are transmitted over the same signaling line circuit, there shall be no interference with fire alarm box signals when both types of initiating devices are operated at the same time. Provision of the shunt noninterfering method of operation shall be permitted for this performance.

The requirement in 3-8.3.2.1 applies only to systems that have devices reporting to the control panel using a signaling line circuit arrangement. The requirement does not apply to systems using initiating device circuits, since these circuits do not distinguish which device initiated the alarm. The requirement was first introduced to apply to spring-wound coded devices that could only transmit a fixed number of rounds of code. If the first device was interfered with by the simultaneous alarm transmission of another device on the same circuit, the first transmission could be lost or garbled. Section 3-8.3.2.1 does not preclude the installation of manual fire alarm boxes on initiating device circuits with other initiating devices such as smoke detectors or heat detectors.

3-8.3.2.2 Automatic Fire Alarm Signal Initiation. Automatic fire alarm-signal initiation devices that have integral trouble signal contacts shall be connected to the initiating device circuit so that a trouble condition within a device does not impair alarm transmission from any other initiating device.

Exception: Where the trouble condition is caused by electrical disconnection of the device or by removing the initiating device from its plug-in base.

It is possible to disable part or all of an initiating device circuit beyond the initiating device in trouble if the integral trouble contacts are connected improperly. Most currently manufactured automatic alarm-initiating devices do not have integral trouble contacts; therefore, this requirement does not apply.

At one time, photoelectric smoke detectors used a tungsten filament lamp as a light source. The code required the detector to monitor the integrity of the filament. An open filament would cause a relay within the detector to open a normally-closed trouble contact. This contact was wired in series with the initiating device circuit.

To comply with 3-8.3.2.2, the initiating device circuit must first connect to all of the alarm contacts of the initiating devices. Then, after the alarm contacts of the last initiating device, the circuit must route back through the trouble contacts. This will connect the trouble contacts in series with one side of the initiating device circuit and terminate at the end-of-line resistor. See Exhibits 3.4 and 3.5 for examples of this type of arrangement.

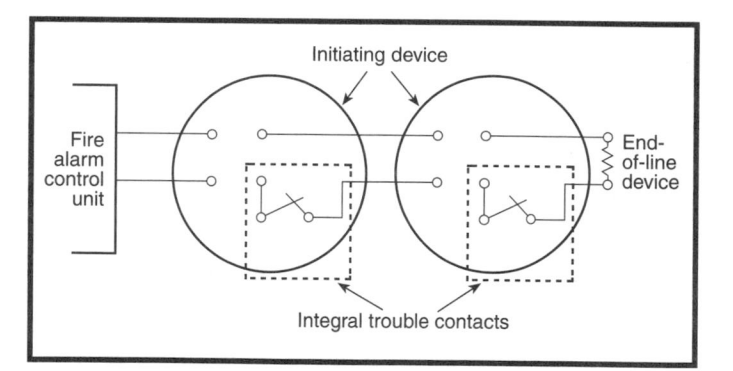

Exhibit 3.4 *Incorrect method of connection of integral trouble contacts. (Source: Hughes Associates, Inc., Warwick, RI)*

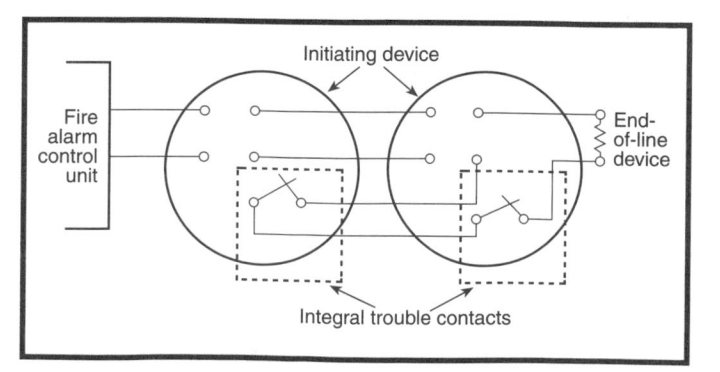

Exhibit 3.5 *Correct method of connection of integral trouble contacts. (Source: Hughes Associates, Inc., Warwick, RI)*

The exception following 3-8.3.2.2 exempts initiating devices from the requirement when they have been electrically disconnected or unplugged from a plug-in base. However, it is important to remember that disconnection or the removal of a plug-in initiating device from its base will interrupt the initiating device circuit (IDC) and could affect the alarm receipt capability at the fire alarm control unit of devices located downstream from the removed device. See commentary following 3-8.3.1.2, Exception. Formal Interpretation 85-2 provides further clarification of 3-8.3.2.2.

3-8.3.2.3 Detection Devices.

3-8.3.2.3.1* Systems equipped with alarm verification features shall be permitted, under the following conditions:

(a) The alarm verification feature is not initially enabled unless conditions or occupant activities that are expected to cause nuisance alarms are anticipated in the area that is protected by the smoke detectors. Enabling of the alarm verification feature shall be protected by password or limited access.

(b) A smoke detector that is continuously subjected to a smoke concentration above alarm threshold does not delay the system functions of 1-5.4 by more than 1 minute.

(c) Actuation of an alarm initiating device other than a smoke detector causes the system functions of 1-5.4 without additional delay.

(d) When the alarm verification feature is enabled, disabled, or changed, the comments section of the Record of Completion (Figure 1-6.2.1, item 10) shall be used to record the status or change to system operation.

Subparagraph 3-8.3.2.3.1 was revised for the 1999 edition of the code to reflect that smoke detectors manufactured today are far more stable and less prone to nuisance alarms. Therefore, alarm verification should be used only where absolutely necessary. In some cases the feature will have been automatically programmed into the fire alarm system control unit. As 3-8.2.3.1(d) indicates, they may be changed to defeat the alarm verification feature. However, the end user must not have access to the programming with this feature.

When the alarm verification feature is enabled, disabled, or changed, a Record of Completion (see Figure 1-6.2.1) must be completed prior to acceptance or reacceptance testing. After a successful retest, a Record of Inspection, Maintenance, and Testing (see Figure 7-5.2.2) must also be completed.

Formal Interpretation 85-2

Reference: 3-8.3.2.2

Background: The figure below illustrates four wire smoke detectors being supervised for the absence of operating power by an end-of-line power supervision relay.

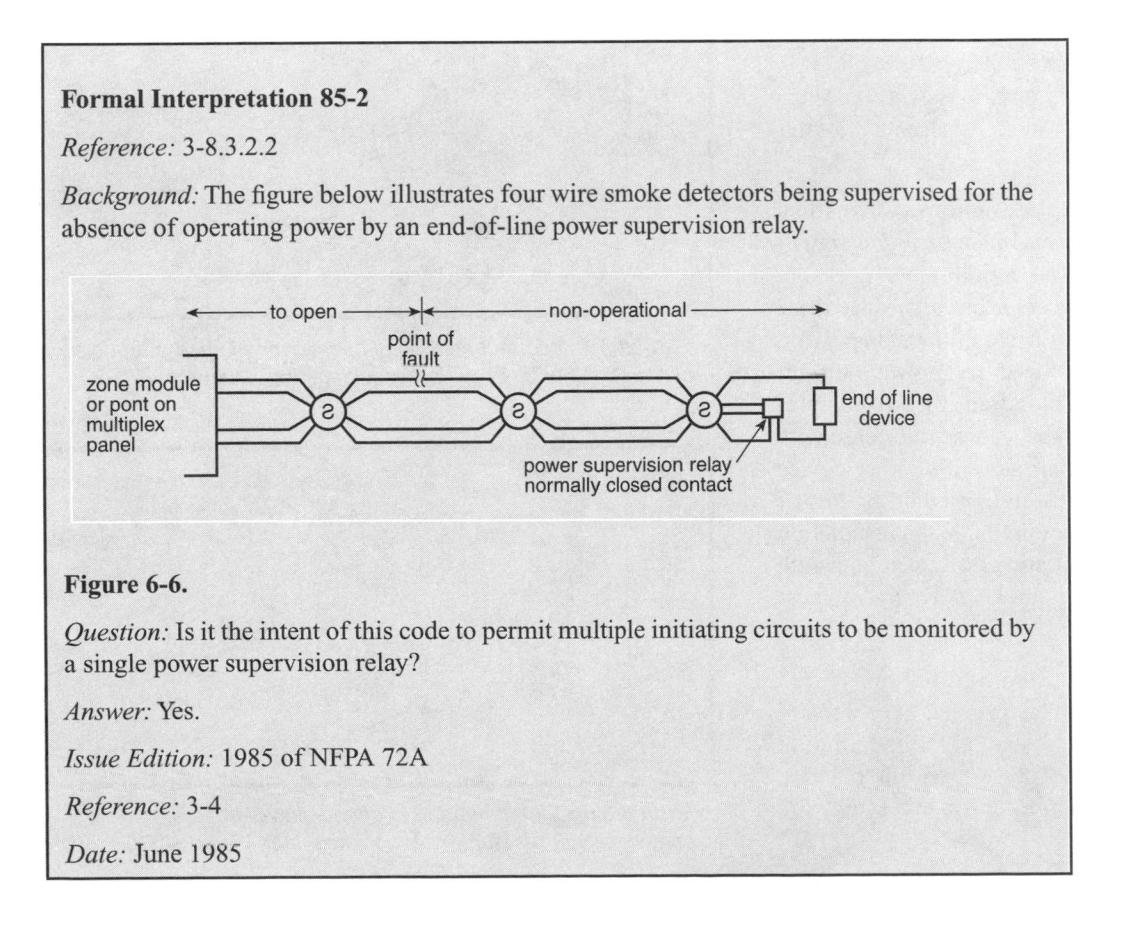

Figure 6-6.

Question: Is it the intent of this code to permit multiple initiating circuits to be monitored by a single power supervision relay?

Answer: Yes.

Issue Edition: 1985 of NFPA 72A

Reference: 3-4

Date: June 1985

A-3-8.3.2.3.1 The alarm verification feature should not be used as a substitute for proper detector location/applications or regular system maintenance. Alarm verification features are intended to reduce the frequency of false alarms caused by transient conditions. They are not intended to compensate for design errors or lack of maintenance.

Alarm verification can be very useful in reducing smoke detector false alarms caused by transient conditions. Verification does not reduce false alarms from conditions that remain relatively constant, such as an environment with high humidity, an environment subject to insect infestation, or where people maliciously and persistently initiate false alarms. The alarm verification feature can reduce both malicious and accidental false alarms caused by such acts as the casual spraying of aerosols into a smoke detector or a gust of wind blowing dust or contaminants into the detector. The feature should not be installed/programmed in a system until a thorough investigation of the causes of false alarms has been accomplished. Alarm verification refers to specific timing sequences of smoke detector/system operation. As with any component of a fire alarm system, alarm verification must be listed as part of the control unit, device, or circuit card. See Exhibit 3.6 for an illustration of an alarm verification timing diagram.

3-8.3.2.3.2 Automatic Drift Compensation. If automatic drift compensation of sensitivity for a fire detector is pro-

vided, the control unit shall identify the affected detector when the limit of compensation is reached.

Automatic drift compensation of sensitivity is used to help eliminate false alarms from smoke detectors that experience dust or dirt build-up within the detection chamber or that respond to minor changes in the environment. The feature allows the detector to maintain its original sensitivity by compensating for effects caused by outside sources. If the compensated value places the detector's sensitivity outside its listed window of sensitivity, the control unit indicates that maintenance is needed.

3-8.3.2.3.3 Systems that require the operation of two automatic detectors to initiate the alarm response shall be permitted as follows:

(1) The systems shall not be prohibited by the authority having jurisdiction.
(2) There shall be at least two automatic detectors in each protected space.
(3) The alarm verification feature shall not be used.

The common names used for the type of configuration described in 3-8.3.2.3.3 are cross zoning and priority matrix zoning. The most common use of this configuration is with special hazard extinguishing systems. In that configuration, false alarms are minimized because more than one detector in alarm is required to activate the special hazard system.

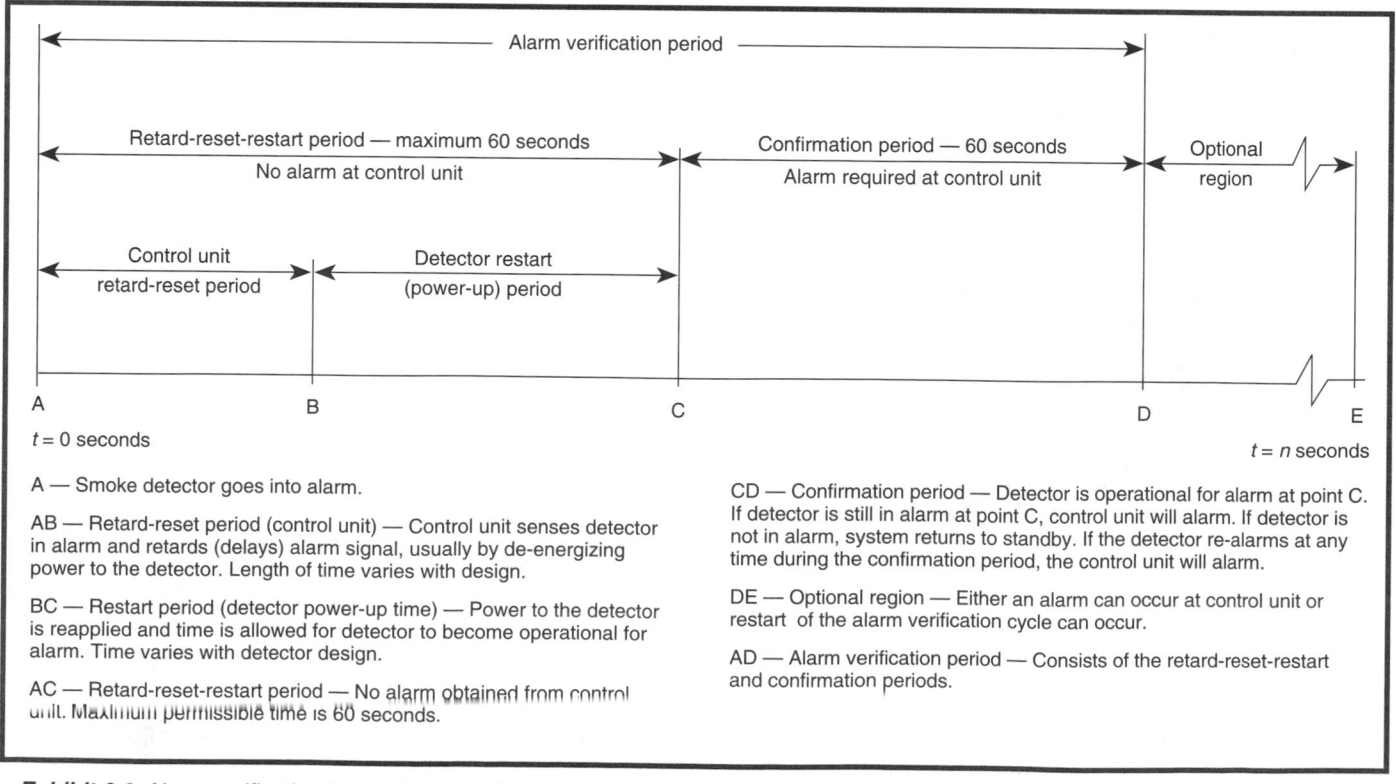

Exhibit 3.6 *Alarm verification timing diagram. (Source: Underwriters Laboratories, Inc., Northbrook, IL)*

Another similar design option requires any second detector in alarm to be on the same zone before sounding the general alarm.

3-8.3.2.3.3.1 For systems that require the operation of two automatic detectors to initiate fire safety functions or to activate fire extinguishing or suppression systems, the detectors shall be installed at the spacing determined in accordance with Chapter 2.

3-8.3.2.3.3.2 For systems that require the operation of two automatic detectors to activate public mode notification, the detectors shall be installed at a linear spacing not more than 0.7 times the linear spacing determined in accordance with Chapter 2.

3-8.3.2.4 Waterflow Alarm Signal Initiation.

Waterflow devices are also required by NFPA 13, *Standard for the Installation of Sprinkler Systems*, and additional information regarding their placement and use can be found in NFPA 13. See Exhibits 3.7 and 3.8 for examples of typical waterflow switches.

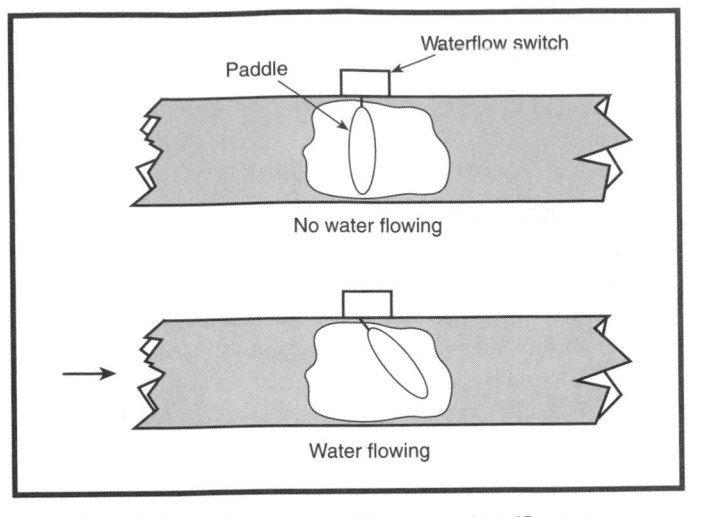

Exhibit 3.8 *Typical waterflow switch operation. (Source: FIREPRO Incorporated, Andover, MA)*

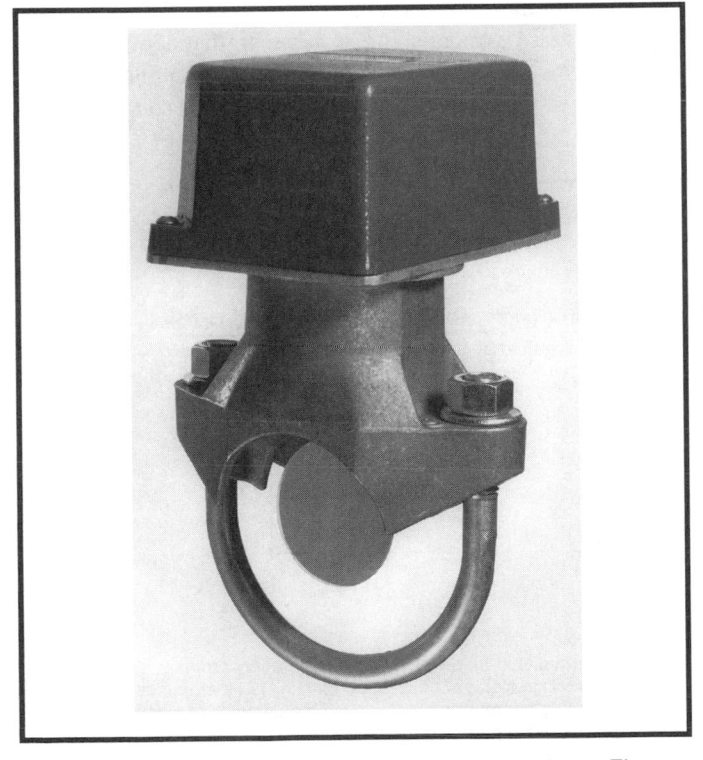

Exhibit 3.7 *Vane-type waterflow switch. (Source: Potter Electric Signal Company, St. Louis, MO)*

3-8.3.2.4.1 A dry-pipe or preaction sprinkler system that is supplied with water by a connection beyond the alarm-initiating device of a wet-pipe system shall be equipped with a separate waterflow alarm-initiating pressure switch or other approved means to initiate a waterflow alarm.

3-8.3.2.4.2 The number of waterflow switches permitted to be connected to a single initiating device circuit shall not exceed five.

The code allows a maximum of five waterflow switches on a single IDC. In contrast, there is no limit on the number of waterflow switches on a signaling line circuit, except for manufacturer's limitations. However, design considerations for a given installation or requirements by the authority having jurisdiction might further limit the number. The reason is to avoid the potential loss of all waterflow signals in a large sprinklered facility and to avoid confusion as to the location of the activated sprinkler area. In a large protected area, consideration should be given to separately zoning each main waterflow switch.

The maximum floor area that can be protected by an automatic sprinkler system supplied by any one sprinkler riser ranges from 40,000 ft^2 to 52,000 ft^2 (3721 m^2 to 4831 m^2). See Section 5-2 of NFPA 13. The size of the building may warrant a smaller area breakdown by zone to assist the fire department in locating the source of the fire quickly.

3-8.3.2.5 Signal Initiation from Automatic Fire Suppression System Other than Waterflow.

3-8.3.2.5.1 The operation of an automatic fire suppression system installed within the protected premises shall cause an alarm signal at the protected premises fire alarm control unit.

Paragraph 3-8.3.2.5.1 requires that actuation of an automatic fire suppression system causes an alarm initiation at the protected premises fire alarm control unit. The automatic fire suppression system should be connected to the protected

premises control unit as a separate zone or discrete point. When an actuation occurs on an automatic fire suppression system, the protected premises fire alarm system operates as intended and actuates all the notification appliances and other system features. This requirement applies to any automatic fire suppression system, including kitchen range hood extinguishing systems.

As required by 3-8.3.2.5, the actuation of a fire extinguishing or suppression system must initiate an alarm on the protected premises fire alarm system. In addition, a condition that places the fire extinguishing or suppression system in an off-normal condition, or one that restores the system to normal, must also initiate a supervisory signal on the protected premises fire alarm system. This supervisory signal must be able to be differentiated from a trouble condition such as a broken wire in the interconnection wiring between the protected premises fire alarm system and the automatic suppression system. See 3-8.3.4.2.

3-8.3.2.5.2 The integrity of each fire suppression system actuating device and its circuit shall be supervised in accordance with 1-5.8.1 and with other applicable NFPA standards.

The term *supervised* as used here means the same as *monitored for integrity* as used elsewhere in the code.

Because fire suppression and alarm systems seldom actuate, the protected premises fire alarm system must monitor the integrity of the actuating means.

As with all interconnected fire alarm system control units (see 3-8.1.1), trouble signals from a suppression system control unit must appear as supervisory signals at the master control unit. A fault on a circuit at the suppression system control unit might affect the operational integrity of the suppression system.

3-8.3.3 Supervisory Signal Initiation.

3-8.3.3.1 General. The provisions of 3-8.3.3 shall apply to the monitoring of sprinkler systems, other fire suppression systems, and other systems for the protection of life and property for the initiation of a supervisory signal indicating an off-normal condition that could adversely affect the performance of the system.

3-8.3.3.1.1 The number of supervisory devices permitted to be connected to a single initiating device circuit shall not exceed 20.

The code allows up to 20 supervisory devices on a single supervisory initiating device circuit. The number is unlimited for signaling line circuits, except as limited by the manufacturer's design. However, design considerations for a given installation might limit the number even further. Locating a supervisory device that is in an off-normal position may prove to be difficult and time-consuming where 20 devices are connected to one circuit. Because 3-8.3.3.1.3 requires distinct indication of the particular supervisory fea-

ture that is off normal, all 20 devices installed on a supervisory initiating device circuit must be of the same type. That is, all must be valve supervisory switches, or all must be low-temperature supervisory switches. Care should be taken when determining the number and type of supervisory devices installed on an initiating device circuit.

3-8.3.3.1.2* Provisions shall be made for supervising the conditions that are essential for the operation of sprinkler and other fire suppression systems.

Exception: Those conditions related to water mains, tanks, cisterns, reservoirs, and other water supplies controlled by a municipality or a public utility.

A-3-8.3.3.1.2 Supervisory systems are not intended to provide indication of design, installation, or functional defects in the supervised systems or system components and are not a substitute for regular testing of those systems in accordance with the applicable standard.

Supervised conditions should include, but should not be limited to the following:

(1) Size of control valves [$1^{1}/_{2}$ in. (38.1 mm) or larger]
(2) Pressure, including dry-pipe system air, pressure tank air, pre-action system supervisory air, steam for flooding systems, and public water
(3) Water tanks, including water level and temperature
(4) Building temperature, including areas such as valve closet and fire pump house
(5) Electric fire pumps, including running (alarm or supervisory), power failure, and phase reversal
(6) Engine-driven fire pumps, including running (alarm or supervisory), failure to start, controller off "automatic," and trouble (for example, low oil, high temperature, overspeed)
(7) Steam turbine fire pumps, including running (alarm or supervisory), steam pressure, and steam control valves
(8) Fire suppression systems appropriate to the system employed

Included in the intent of 3-8.3.3.1.2 is the monitoring of the equipment or conditions that affect the operation of suppression systems. Conditions that might affect the suppression system will result in the receipt of a supervisory signal prior to loss of favorable conditions for the operation of the suppression system.

There are two methods of supervision allowed by NFPA 13, *Standard for the Installation of Sprinkler Systems*. Subparagraph 5-14.1.1.3 of the 1999 edition of NFPA 13 requires sprinkler control valves to be either locked or electrically supervised. However, 3-8.3.3.1.2 of the code requires electrical supervision by the fire alarm system, even if the valve is locked. See Exhibits 3.9 and 3.10.

The requirements described in 3-8.3.3.1.2 are satisfied by the requirements of 5-14.1.1.3(1) and (2) of NFPA 13.

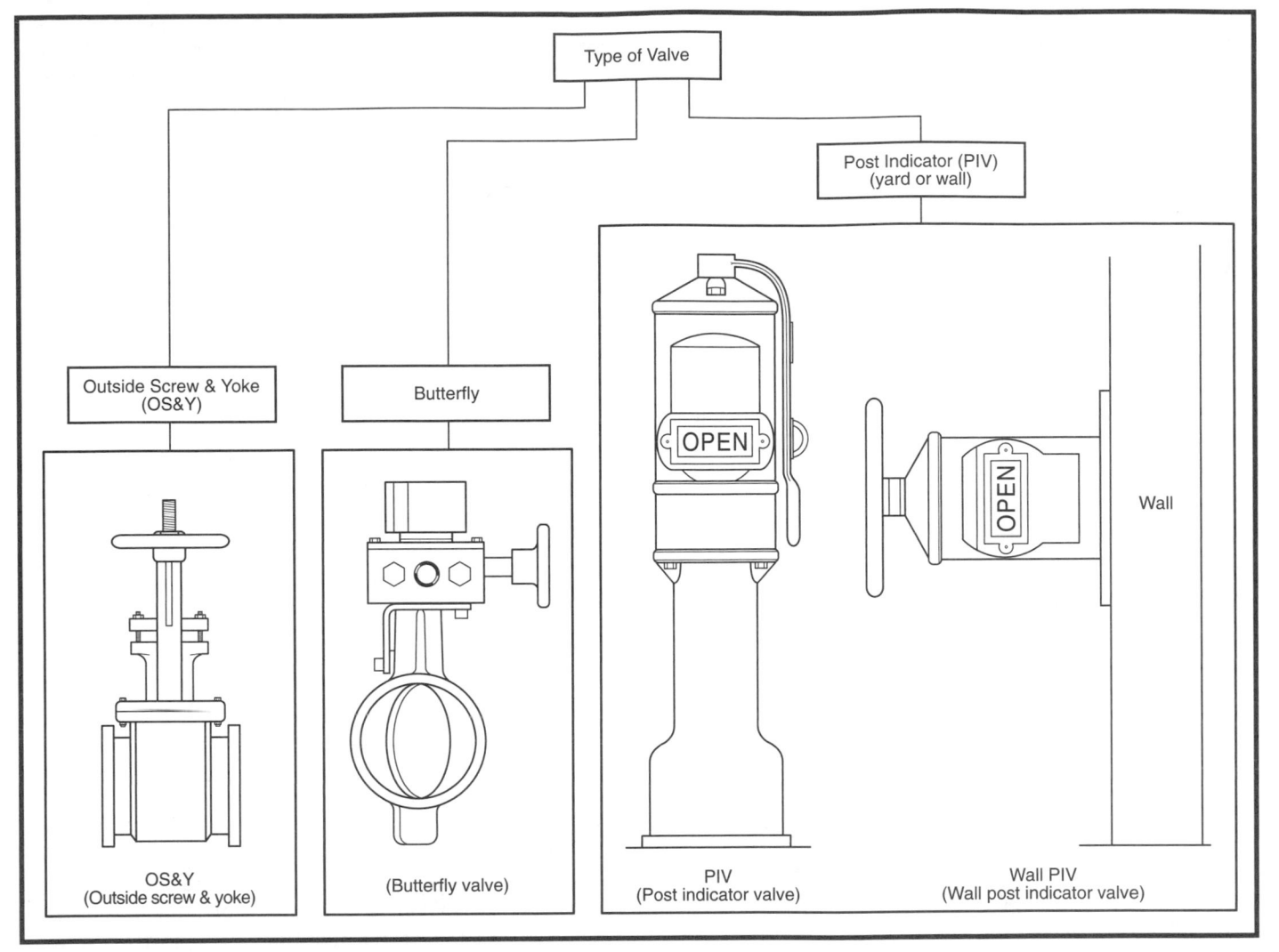

Exhibit 3.9 *Types of control valves that require a supervisory switch.*
(Source: Potter Electric Signal Company, St. Louis, MO)

5-14.1.1.3* Valves on connections to water supplies, sectional control and isolation valves, and other valves in supply pipes to sprinklers and other fixed water-based fire suppression systems shall be supervised by one of the following methods:

(1) Central station, proprietary, or remote station signaling service

(2) Local signaling service that will cause the sounding of an audible signal at a constantly attended point

(3) Valves locked in the correct position

(4) Valves located within fenced enclosures under the control of the owner, sealed in the open position, and inspected weekly as part of an approved procedure

Floor control valves in high-rise buildings and valves controlling flow to sprinklers in circulating closed loop systems shall comply with 5-14.1.1.3(1) or (2).

Exception: Supervision of underground gate valves with roadway boxes shall not be required.

Exhibit 3.10 *Excerpt from NFPA 13, Standard for the Installation of Sprinkler Systems, 1999 edition.*

Some property insurers may require that the valves on connections to water supplies be locked in the open position using either frangible shackle locks or hardened shackle locks. However, electrical supervision does provide more information to management regarding the position of these valves. As is the case with all of these requirements, the primary goal is to increase the reliability of water availability to the fire suppression system. See Exhibits 3.9 and 3.11 for examples of control valves and supervisory switches. See Exhibit 3.12 for an example of a pressure switch for supervising suppression system pressure.

(See Appendix A of NFPA 13, *Standard for the Installation of Sprinkler Systems*, for further information.)

3-8.3.3.1.3* Signals shall distinctively indicate the particular function (e.g., valve position, temperature, or pressure) of the system that is off-normal and also indicate its restoration to normal.

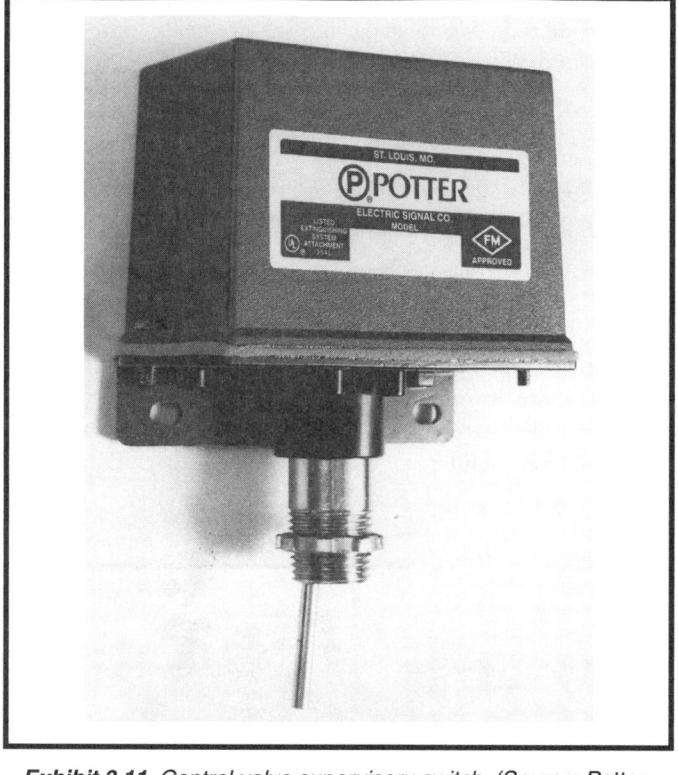

Exhibit 3.11 Control valve supervisory switch. (Source: Potter Electric Signal Company, St. Louis, MO)

Exhibit 3.12 High/low pressure supervisory switch. (Source: Potter Electric Signal Company, St. Louis, MO)

A-3-8.3.3.1.3 Cancellation of the off-normal signal can be permitted as a restoration signal, unless separate recording of all changes of state is a specific requirement. *(Refer to Chapter 5.)*

3-8.3.3.2 Pump Supervision. Automatic fire pumps and special service pumps shall be supervised in accordance with NFPA 20, *Standard for the Installation of Centrifugal Fire Pumps,* and the authority having jurisdiction.

3-8.3.3.2.1 Supervision of electric power supplying the pump shall be made on the line side of the motor starter. All phases and phase reversal shall be supervised.

The term *supervision* used in 3-8.3.3.2 and 3-8.3.3.2.1 means the same as the phrase *monitoring for integrity.* (See the commentary following 3-8.3.2.5.2.)

3-8.3.3.2.2 If both sprinkler supervisory signals and pump running signals are transmitted over the same path, the pump running signal shall have priority.

Exception: Where the path is so arranged such that simultaneous signals are not lost.

The code gives a pump running signal priority because such a signal may indicate that the fire pump has started due to a fire or due to a break in the fire protection water supply piping.

3-8.3.3.3 Automatic Fire Suppression System Panel Supervision.

3-8.3.3.3.1 A supervisory signal shall indicate the off-normal condition and its restoration to normal as required by the system employed.

3-8.3.3.3.2* If a valve is installed in the connection between an alarm-initiating device intended to signal activation of a fire suppression system and the fire suppression system, the valve shall be supervised in accordance with the requirements of Chapter 2.

Supervision of the valve in this case means monitoring the status of the valve and initiating a supervisory signal when the valve operates in an off-normal position. When the valve returns to its normal position, the supervisory device initiates a restoration to normal signal.

A-3-8.3.3.3.2 Sealing or locking such a valve in the open position, or removing the handle from the valve, does not meet the intent of the supervision requirement.

3-8.3.4 Trouble Signal Initiation.

3-8.3.4.1 Automatic fire suppression system alarm-initiating devices and supervisory signal-initiating devices and their circuits shall be designed and installed so that they cannot be subject to tampering, opening, or removal without initiating a signal. This provision shall include junction boxes installed outside of buildings to facilitate access to the initiating device circuit.

Exception 1: Covers of junction boxes inside of buildings.

Exception 2: Tamperproof screws or other equivalent mechanical means shall be permitted for preventing access to junction boxes installed outside buildings.

Junction boxes installed outside of buildings must either be equipped with tamperproof screws or some mechanical means that prevents access to the junction box, or they must have a device to initiate a trouble signal when the box is opened.

3-8.3.4.2 The integrity of each fire suppression system actuating device and its circuit shall be supervised in accordance with 1-5.8.1 and with other applicable NFPA standards.

3-8.4 Fire Alarm System Outputs.

3-8.4.1 Occupant Notification. Fire alarm systems provided for evacuation or relocation of occupants shall have one or more notification appliances listed for the purpose on each floor of the building and so located such that they have the characteristics described in Chapter 4 for public mode or private mode, as required.

Notification zones shall be consistent with the emergency response or evacuation plan for the protected premises. The boundaries of notification zones shall be coincident with building outer walls, building fire or smoke compartment boundaries, floor separations, or other fire safety subdivisions.

As stated in 3-8.4.1, *notification zone* has been defined. The boundaries described in 3-8.4.1 further define the term *evacuation zone* as used in 3-8.4.1. It is possible to have multiple notification appliance circuits in an evacuation zone. In such cases, the notification appliances must all sound within that evacuation zone. Also, if the evacuation plan is for selective evacuation, then the requirements of 3-8.4.1.1 must be followed regarding survivability.

When the protected premises fire alarm system's purpose is to notify and evacuate the occupants of the protected premises, Chapter 6 requirements for public mode signaling must be followed.

The phrase *listed for the purpose* appears in 1-5.1.2. Many devices and appliances are listed, but they may be listed for other non-fire alarm applications. For example, there are listed notification appliances that do not meet all the requirements for fire alarm use and are intended for background music or other applications. Only those notification appliances specifically listed for fire alarm use meet the listed-for-the-purpose requirement of 3-8.4.1.

3-8.4.1.1 Survivability.

Subparagraph 3-8.4.1.1 applies to fire alarm systems that provide relocation or selective evacuation of a protected premises. Survivability refers to the ability of the circuit to operate during a fire. Occupants may be evacuated floor by floor such as in a high-rise or may be relocated, such as in a hospital. Therefore, it is imperative that the occupant notification system continue to operate during the fire so the responders can continue to instruct the occupants. Formerly

3-2.4, 3-8.4.1.1 has been completely revised for the 1999 edition of the code to make the requirements more easily understood.

3-8.4.1.1.1 Paragraph 3-8.4.1.1 applies only to systems used for partial evacuation or relocation of occupants.

3-8.4.1.1.2 A single notification appliance circuit shall not serve more than one notification zone.

An evacuation zone can be an area of a floor, an entire floor, or several floors that are always intended to be evacuated simultaneously. One notification appliance circuit cannot serve more than one notification zone because of the potential of a fire in one zone eliminating communications to another zone. See 3-8.4.1.3. Exhibit 3.13 illustrates a typical multiple-zone splitter.

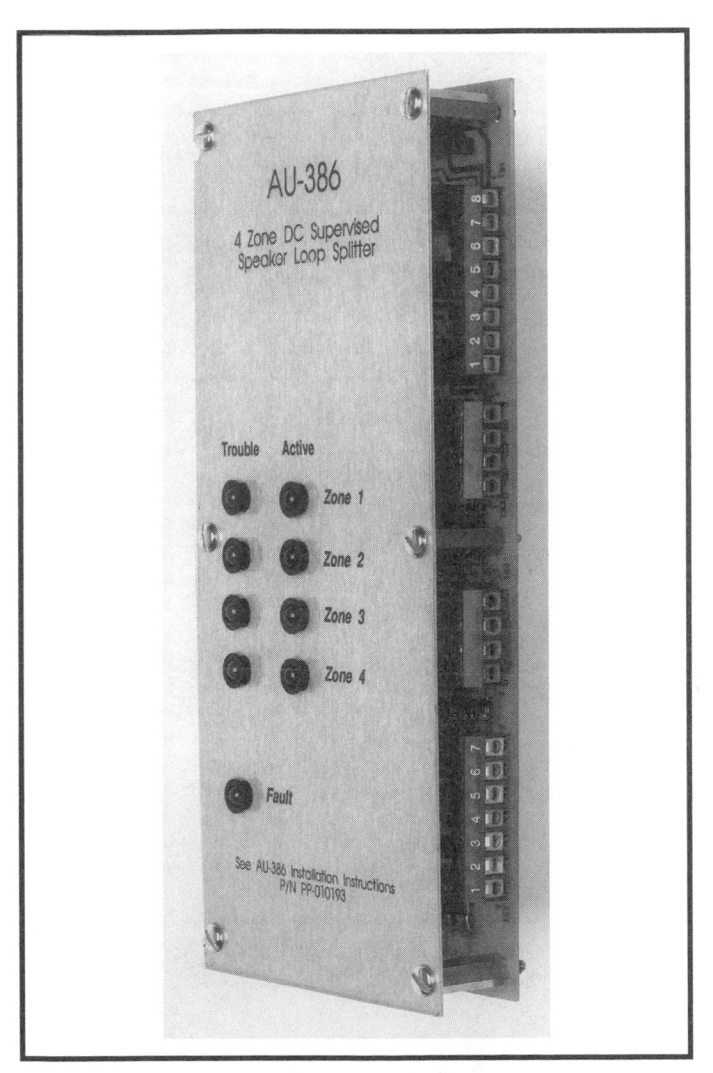

Exhibit 3.13 *Supervised speaker loop splitters can be used to separate speaker circuits to evacuation zones. (Source: Audiosone Corp., Stratford, CT)*

3-8.4.1.1.3* The system shall be designed so that failure of equipment or a fault on one or more installation wiring conductors of one notification appliance circuit shall not result in functional loss of any other notification appliance circuit.

Subparagraph 3-8.4.1.1.3 essentially requires the use of separate communications paths, circuits, or some other arrangement in each evacuation zone to avoid the total loss of communications when the fire attacks the installation wiring conductors of one communications path sending signals to an evacuation zone. For example, if a fire attacked a communications path riser that connected all of the notification appliances on multiple floors, it is possible that there would be a total loss of communications to all the floors connected to that riser. However, if the notification appliances on each floor were served by a separate riser, then fire damage to one communications path riser would not result in a total loss of communications to all the connected floors.

A-3-8.4.1.1.3 Paragraph 3-8.4.1.1.3 requires that the equipment used operate in a certain manner during fault conditions. For example, it is necessary that a fault such as a short circuit does not open a fuse or damage components common to other circuits.

3-8.4.1.1.4* Notification appliance circuits and any other circuits necessary for the operation of the notification appliance circuits shall be protected from the point at which they exit the control unit until the point that they enter the notification zone that they serve using one or more of the following methods:

(1) A 2-hour rated cable assembly

Article 760 of NFPA 70, *National Electrical Code,* contains wiring requirements for fire alarm systems. Only Mineral Insulated (MI) cable and Circuit Integrity (CI) cables have the necessary 2-hour rating. See Exhibits 3.14(a) and 3.14(b) for examples of MI and CI cables.

(2) A 2-hour rated shaft or enclosure

(3) A 2-hour rated stairwell in a building fully sprinklered in accordance with NFPA 13, *Standard for the Installation of Sprinkler Systems*

The survivability requirements of 3-8.4.1.1 apply to all types of systems installed in buildings where the fire response plan permits either selective evacuation or relocation of the building occupants to a safe area during a fire emergency. The intent is to have survivability requirements apply to non-voice systems used to evacuate the building occupants by floor or zone. If occupants are allowed or required to remain in the building, it is essential that the fire alarm system remain operational so that additional floors or zones can be evacuated as needed.

A-3-8.4.1.1.4 Paragraph 3-8.4.1.1.4 requires the protection of circuits as they pass through fire areas other than the one served. This is to delay possible damage to the circuits from fires in areas other than those served by the circuits. This is done to increase the likelihood that circuits serving areas remote from the original fire will have the opportunity to be activated and serve their purpose. Note that the protection requirement would also apply to a signaling line circuit that extends from a master fire alarm control unit to another remote fire alarm control unit where notification appliance circuits might originate.

3-8.4.1.2* Distinctive Evacuation Signal.

A-3-8.4.1.2 Paragraph 1-5.4.7 requires that fire alarm signals be distinctive in sound from other signals and that this sound not be used for any other purpose. The use of the distinctive three-pulse temporal pattern fire alarm evacuation signal required by 3-8.4.1.2.2 became effective July 1, 1996, for new systems installed after that date. It had been previously recommended for this purpose by this code since 1979. It has since been adopted as both an American National Standard (ANSI S3.41, *Audible Emergency Evacuation Signal*) and an International Standard (ISO 8201, *Audible Emergency Evacuation Signal*).

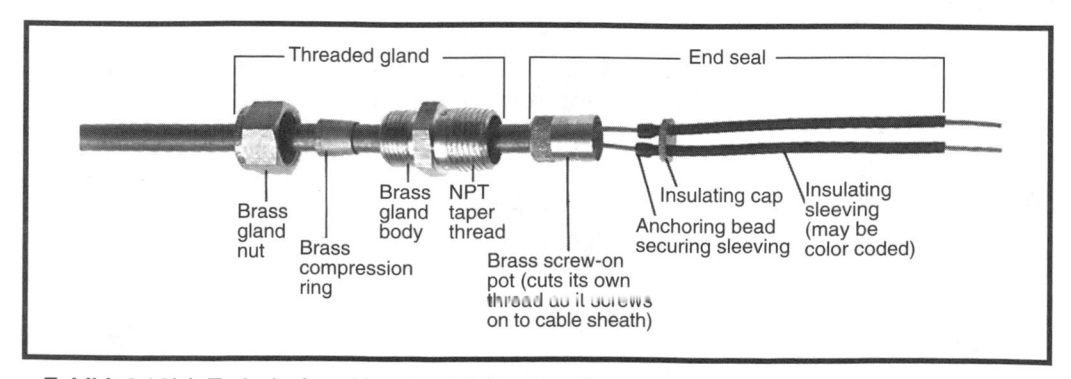

Exhibit 3.14(a) *Typical mineral insulated (MI) cable. (Source: Pyrotenax USA, Inc.)*

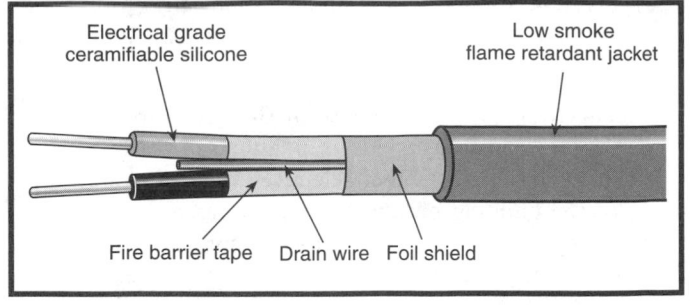

Exhibit 3.14(b) *Typical circuit integrity (CI) cable. (Source: Rockbestos-Suprenant Cable Corp., Clinton, MA)*

Copies of both of these standards are available from the Standards Secretariat, Acoustical Society of America, 335 East 45th Street, New York, NY 10017-3483.

The standard fire alarm evacuation signal is a three-pulse temporal pattern using any appropriate sound. The pattern consists of the following in this order:

(a) An on phase lasting 0.5 seconds ± 10 percent

(b) An off phase lasting 0.5 seconds ± 10 percent, for three successive on periods

(c) An off phase lasting 1.5 seconds ± 10 percent *[refer to Figures A-3-8.4.1.2(a) and A-3-8.4.1.2(b)]*. The signal should be repeated for a period that is appropriate for the purposes of evacuation of the building, but for not less than 180 seconds. A single-stroke bell or chime sounded at "on" intervals lasting 1 second ± 10 percent, with a 2-second ± 10 percent "off" interval after each third "on" stroke, is permitted *[refer to Figure A-3-8.4.1.2(c)]*.

The minimum repetition time is permitted to be manually interrupted.

It is important to note that the American National Standard Audible Emergency Evacuation Signal is required only where immediate evacuation of a building or a zone of a building is desired. The goal is to have anyone hearing the signal in any building in the United States (or in other countries that adopt ISO standards requiring the same evacuation signal) immediately recognize the signal as a fire alarm evacuation signal.

3-8.4.1.2.1 To meet the requirements of 1-5.4.7, the fire alarm signal used to notify building occupants of the need to evacuate (leave the building) shall be in accordance with ANSI S3.41, *Audible Emergency Evacuation Signal.*

3-8.4.1.2.2 The use of the American national standard evacuation signal shall be restricted to situations where it is desired that all occupants hearing the signal evacuate the building immediately. It shall not be used where, with the approval of the authority having jurisdiction, the planned

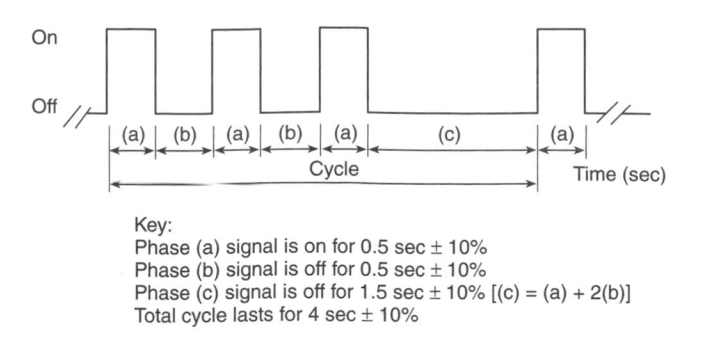

Key:
Phase (a) signal is on for 0.5 sec ± 10%
Phase (b) signal is off for 0.5 sec ± 10%
Phase (c) signal is off for 1.5 sec ± 10% [(c) = (a) + 2(b)]
Total cycle lasts for 4 sec ± 10%

Figure A-3-8.4.1.2(a) Temporal pattern parameters.

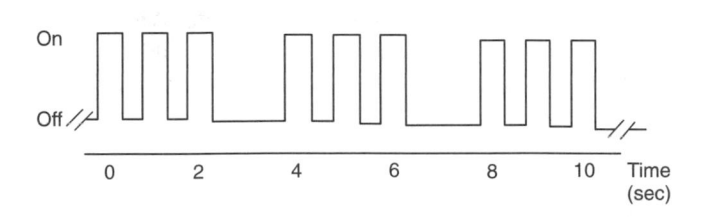

Figure A-3-8.4.1.2(b) Temporal pattern imposed on audible notification appliances that otherwise emit a continuous signal while energized

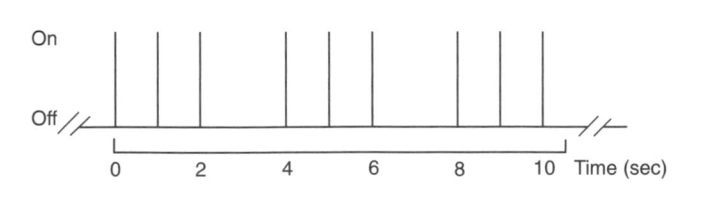

Figure A-3-8.4.1.2(c) Temporal pattern imposed on a single-stroke bell or chime.

action during a fire emergency is not evacuation, but rather is the relocation of occupants or their protection in place as directed by the building fire protection plan or as directed by fire fighting personnel.

3-8.4.1.2.3* The American national standard evacuation signal shall be synchronized within a notification zone.

A-3-8.4.1.2.3 Coordination or synchronization of the audible signal within a notification zone is needed to preserve the temporal pattern. It is unlikely that the audible signal in one evacuation/notification zone will be heard in another at a level that will destroy the temporal pattern. Thus, it would not normally be necessary to provide coordination/synchronization for an entire system. Caution should be used in spaces such as atriums where the sounds produced in one

notification zone can be sufficient to cause confusion regarding the temporal pattern.

3-8.4.1.3 Emergency Voice/Alarm Communications. Emergency voice/alarm communications service shall be provided by a system with automatic or manual voice capability that is installed to provide voice instructions to the building occupants where it is intended that there be only partial or selective evacuation or directed relocation of building occupants in the event of a fire.

Exception: If emergency voice/alarm communications are used to automatically and simultaneously notify all occupants to evacuate the protected premises during a fire emergency, manual or selective paging shall not be required, but, if provided, shall meet the requirements of Section 3-8.4.1.3.

Many designers now routinely specify fire alarm systems equipped with notification appliances that reproduce the sound of a human voice. These systems serve buildings that are neither high-rise, nor large in area. Upon initiation of a fire alarm signal, these systems notify the occupants throughout all areas of the building to evacuate. As such, these systems need not meet the rather stringent requirements for systems serving the fire alarm needs of those structures where the fire plan for the building requires relocation or selective and partial evacuation of the occupants. However, where features, such as a fire command center, are provided, they must meet the appropriate requirements of 3-8.4.1.3.3 and 3-8.4.1.3.5.5.

3-8.4.1.3.1 Application. Subparagraph 3-8.4.1.3 describes the requirements for emergency voice/alarm communications. The primary purpose is to provide dedicated manual and automatic facilities for the origination, control, and transmission of information and instructions pertaining to a fire alarm emergency to the occupants (including fire department personnel) of the building. It shall be the intent of 3-8.4.1.3 to establish the minimum requirements for emergency voice/alarm communications.

3-8.4.1.3.2 Monitoring of the integrity of speaker amplifiers, tone-generating equipment, and two-way telephone communications circuits shall be in accordance with 1-5.8.6.

3-8.4.1.3.3 Fire Command Center Survivability.

3-8.4.1.3.3.1 A fire command center shall be provided in accordance with 3-8.4.1.3.3.

Exception: If emergency voice/alarm communications are used to automatically and simultaneously notify all occupants to evacuate the protected premises during a fire emergency, a fire command center shall not be required, but, if provided, shall meet the requirements of 3-8.4.1.3.

3-8.4.1.3.3.2 The fire command center and the central control unit shall be located within a minimum 1-hour rated fire-resistive area and shall have a minimum 3-ft (1-m) clearance from the front of the fire command center control equipment.

Exception: If approved by the authority having jurisdiction, the fire command center control equipment shall be permitted to be located in a lobby or other approved space.

The minimum fire rating helps ensure that the fire command center will remain operational. This will enable personnel in the fire command center to inform and instruct the occupants throughout the duration of a fire even if the fire lasts longer than one hour. The 3-ft (1-m) clearance will allow for the fire ground commander to have adequate room to access the fire alarm control unit operating controls.

3-8.4.1.3.3.3 If the fire command center control equipment is remote from the central control equipment, the following requirements shall apply:

(1) The interconnecting wiring shall be provided with mechanical protection by installing the wiring in metal conduit or metal raceway.
(2) The interconnecting wiring shall be provided with resistance to attack from a fire by routing the wiring through areas whose characteristics are at least equal to the limited-combustible characteristics defined in NFPA 90A, *Standard for the Installation of Air Conditioning and Ventilating Systems.*
(3) If the interconnecting wiring exceeds 100 ft (30 m), additional resistance to attack from a fire shall be provided by doing either of the following:
   a. Installing the wiring in metal conduit or metal raceway in a 2-hour fire-rated enclosure
   b. Enclosing the wiring in a 2-hour fire-rated cable assembly and installing the cable in metal conduit or metal raceway

Emergency voice/alarm communications service is generally used where relocation or selective and partial evacuation (by floor or zone) of the building occupants is part of the fire response plan. If occupants are allowed to remain in the building, system survivability must be considered. It is essential that the fire alarm system remain operational during the fire emergency so that additional floors or zones can be evacuated as needed. See 3-8.4.1.1 for additional system survivability requirements.

3-8.4.1.3.4 Power Supplies.

3-8.4.1.3.4.1 The wiring between the central control equipment and the primary power supply shall be routed through areas whose characteristics are at least equal to the limited-combustible characteristics as defined in NFPA

90A, *Standard for the Installation of Air Conditioning and Ventilating Systems.*

3-8.4.1.3.4.2 The secondary (standby) power supply shall be provided in accordance with 1-5.2.5.

Exception: Where emergency voice/alarm communications are used to notify all occupants automatically and simultaneously to evacuate the protected premises during a fire emergency in meeting the requirements of 1-5.2.5, the secondary supply shall be required to be capable of operating the system during a fire or other emergency condition for a period of 5 minutes rather than 2 hours.

As in the exception to 3-8.4.1.3, the fire plan for a building may use the prerecorded announcement from an emergency voice/alarm communications system to notify the occupants in all areas of the building to evacuate. In this case, the battery standby requirements must match the fire alarm system type used in the building (24 hours or 60 hours) and the alarm must sound for not less than 5 minutes.

3-8.4.1.3.5 Voice/Alarm Signaling Service.

3-8.4.1.3.5.1* General. The purpose of the voice/alarm signaling service shall be to provide an automatic response to the receipt of a signal indicative of a fire emergency. Subsequent manual control capability of the transmission and audible reproduction of evacuation tone signals, alert tone signals, and intelligible voice directions on a selective and all-call basis, as determined by the authority having jurisdiction, shall also be required from the fire command center.

Exception 1: If the fire command center or remote monitoring location is constantly attended by trained operators, and operator acknowledgment of receipt of a fire alarm signal is received within 30 seconds, automatic response shall not be required.

Exception 2: If emergency voice/alarm communications are used to notify all occupants automatically and simultaneously to evacuate the protected premises during a fire emergency, the ability to give voice directions on a selective basis shall not be required but, if provided, shall meet the requirements of 3-8.4.1.3.

A-3-8.4.1.3.5.1 It is not the intention that emergency voice/alarm communications service be limited to English-speaking populations. Emergency messages should be provided in the language of the predominant building population. If there is a possibility of isolated groups that do not speak the predominant language, multilingual messages should be provided. It is expected that small groups of transients unfamiliar with the predominant language will be picked up in the traffic flow in the event of an emergency and are not likely to be in an isolated situation.

3-8.4.1.3.5.2 Multichannel Capability. If required by the authority having jurisdiction, the system shall allow the application of an evacuation signal to one or more zones and, at the same time, shall allow voice paging to the other zones selectively or in any combination.

3-8.4.1.3.5.3 Functional Sequence.

3-8.4.1.3.5.3.1 In response to an initiating signal indicative of a fire emergency, the system shall automatically transmit the following either immediately or after a delay acceptable to the authority having jurisdiction.

(a) If the emergency voice/alarm communications service is used to transmit a voice evacuation message, the voice message shall be preceded and followed by a minimum of two cycles of the audible emergency evacuation signal specified in 3-8.4.1.2.

(b) If the emergency voice/alarm communications service is used to transmit relocation instructions or other non-evacuation messages, a continuous alert tone of 3-second to 10-second duration followed by a message (or messages where multichannel capability is provided) shall be repeated at least three times to direct the occupants of the alarm signal initiation zone and other zones in accordance with the building's fire evacuation plan.

(c) An evacuation signal shall be transmitted to the alarm signal initiation zone and other zones in accordance with the building's fire evacuation plan.

Exception: If emergency voice/alarm communications are used to notify all occupants automatically and simultaneously to evacuate the protected premises during a fire emergency, and the functional sequence described in 3-8.4.1.3.5.3.1(a) is provided, the capability to notify portions of the protected premises selectively shall not be required, but, if provided, shall meet the requirements of 3-8.4.1.3.

3-8.4.1.3.5.3.2 Failure of the message described by 3-8.4.1.3.5.3.1(a), if used, shall sound the evacuation signal automatically. Provisions for manual initiation of voice instructions or evacuation signal generation shall be provided.

Exception 1: Other functional sequences shall be permitted if approved by the authority having jurisdiction.

Exception 2: If emergency voice/alarm communications are used to notify all occupants automatically and simultaneously to evacuate the protected premises during a fire emergency, provision for manual initiation of voice instructions shall not be required, but, if provided, shall meet the requirements of 3-8.4.1.3.

3-8.4.1.3.5.3.3 Live voice instructions shall override all previously initiated signals on that channel and shall have priority over any subsequent automatically initiated signals on that channel. If multichannel application is required, subsequent alarms shall be activated in accordance with 3-8.4.1.3.5.2.

Exception: If emergency voice/alarm communications are used to notify all occupants automatically and simultaneously to evacuate the protected premises during a fire emergency, the ability to give live voice instructions shall not be required, but, if provided, shall meet the requirements of 3-8.4.1.3.

3-8.4.1.3.5.3.4 If provided, manual controls for emergency voice/alarm communications shall be arranged to provide visible indication of the on–off status for their associated evacuation zones.

3-8.4.1.3.5.4 Voice and Tone Devices. The alert tone preceding any message shall be permitted to be a part of the voice message or to be transmitted automatically from a separate tone generator.

3-8.4.1.3.5.5 Fire Command Center.

3-8.4.1.3.5.5.1* A fire command center shall be provided at a building entrance or other location approved by the authority having jurisdiction. The fire command center shall provide a communications center for the arriving fire department and shall provide for control and display of the status of detection, alarm, and communications systems. The fire command center shall be permitted to be physically combined with other building operations and security centers as permitted by the authority having jurisdiction. Operating controls for use by the fire department shall be clearly marked.

A-3-8.4.1.3.5.5.1 The choice of the location(s) for the fire command center should also take into consideration the ability of the fire alarm system to operate and function during any probable single event.

3-8.4.1.3.5.5.2 The fire command center shall control the emergency voice/alarm communications signaling service and, if provided, the two-way telephone communications service. All controls for manual initiation of voice instructions and evacuation signals shall be located or secured to restrict access to trained and authorized personnel.

All of the requirements in 3-8.4.1.3.5.5 are in addition to the obvious coordination necessary with the responding fire department. Voice/alarm communication systems can be very effective both by calming occupants in areas remote from the fire and by directing others toward safety. All of the requirements in 3-8.4.1.3.5.5.2 must be enhanced with adequate training of the fire service personnel who will be responsible for using the equipment. Also see Supplement 4 for a tutorial on human behavior and response to alarm signals.

3-8.4.1.3.5.5.3* If there are multiple fire command centers, the center in control shall be identified by a visible indication at that center.

A-3-8.4.1.3.5.5.3 The operation of a fire command center in systems with multiple fire command centers should also consider visible indications at all locations to assist opera-

tors in understanding that manual system operation has been established by the fire command center in use.

3-8.4.1.3.5.6 Loudspeakers.

3-8.4.1.3.5.6.1 Loudspeakers and their enclosures shall be installed in accordance with Chapter 4.

Loudspeakers used for background music, and so forth, are not acceptable unless they have been specifically listed for fire alarm system use. Also see commentary following 3-8.2.4.

3-8.4.1.3.5.6.2* There shall be at least two loudspeakers located in each paging zone of the building. Each of the loudspeakers shall meet the requirements of Chapter 4.

Paging zone is not defined in the code. A common definition of *paging zone* is a zone where all speakers are selected as a group simultaneously. (See the definition of Notification Zone.) It can be inferred from 3-8.4.1.3.5.6.2 that a paging zone can be different from an evacuation zone. However, 3-8.4.1.3.6.1 requires that undivided fire or smoke areas must not be divided into multiple evacuation zones. This means that once the Emergency Voice/Alarm Communications Service (EVACS) must initiate evacuation of a floor or other fire division within a building, the system must notify the entire evacuation zone, even when the evacuation zone has been divided into several paging zones. The two-loudspeaker requirement is part of the requirements for survivability as discussed in 3-8.4.1.1. See Exhibit 3.15 for an illustration of a typical paging zone.

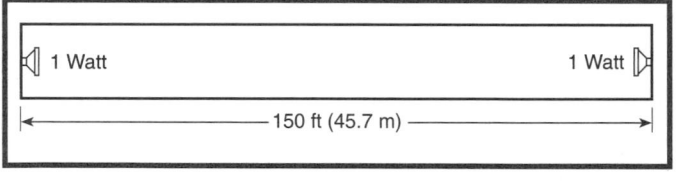

Exhibit 3.15 *Paging zone showing typical speaker location in a corridor. (Source: Simplex Time Recorder Company, Gardner, MA)*

A-3-8.4.1.3.5.6.2 The design and layout of the loudspeaker audible notification appliances should be arranged such that they do not interfere with the operations of the emergency response personnel. Speakers located in the vicinity of the fire command center should be arranged so they do not cause audio feedback when the system microphone is used. Speakers installed in the area of two-way telephone stations should be arranged so that the sound pressure level emitted does not preclude the effective use of the two-way telephone system. Circuits for paging zones and telephone zones should be separated, shielded, or otherwise arranged to prevent audio cross-talk between circuits.

The problem discussed in A-3-8.4.1.3.5.6.2 is often overlooked. Loudspeakers located too near a microphone at a

command center or a fire fighters' telephone on a floor or in a stairwell can interfere with critical communications during a fire emergency.

3-8.4.1.3.5.6.3 Each elevator car shall be equipped with a single loudspeaker connected to the paging zone that serves the elevator group in which the elevator is located.

Exception: If permitted by the authority having jurisdiction, or if in existing elevator cars where two-way communications by approved means are provided between each elevator car and the fire command center.

3-8.4.1.3.5.6.4 Each enclosed stairway exceeding two stories in height shall be equipped with loudspeakers connected to a separate paging zone.

3-8.4.1.3.6 Evacuation Signal Zoning.

3-8.4.1.3.6.1* Undivided fire or smoke areas shall not be divided into multiple evacuation signaling zones.

A-3-8.4.1.3.6.1 Paragraph 3-8.4.1.3.6.1 does not prohibit the provision of multiple notification appliance circuits within an evacuation zone.

Evacuation signaling zones are not defined in this code. (See the definition of Notification Zone.) As stated in 3-8.4.1.1, an evacuation zone can be an area of a floor, an entire floor, or several floors that are always intended to be evacuated simultaneously. Subparagraph 3-8.4.1.3.6.1 requires evacuation signaling zones (also termed *evacuation zones* elsewhere in the code) be consistent with the fire or smoke barriers within the protected premises. See Exhibit 3.16 for an illustration of a typical evacuation zone.

3-8.4.1.3.6.2 If multiple notification appliance circuits are provided within a single evacuation signaling zone, all of the notification appliances within the zone shall be arranged to activate simultaneously, either automatically or by actuation of a common, manual control.

Exception: Where the different notification appliance circuits within an evacuation signaling zone perform separate functions, for example, presignal and general alarm signals, predischarge and discharge signals.

3-8.4.1.3.7 Two-Way Telephone Communications Service.

3-8.4.1.3.7.1 Two-way telephone communications equipment shall be listed for two-way telephone communications service and installed in accordance with 3-8.4.1.3.7.

3-8.4.1.3.7.2 Two-way telephone communications service, if provided, shall be for use by the fire service. Additional uses, if specifically permitted by the authority having juris-

Exhibit 3.16 *Typical evacuation signal zone—one floor of office occupancy with ceiling-mounted speakers. (Source: EST, Inc., Cheshire, CT)*

diction, shall be permitted to include signaling and communications for a building fire warden organization, signaling and communications for reporting a fire and other emergencies (for example, voice call box service, signaling, and communications for guard's tour service), and other uses. Variation of equipment and system operation provided to facilitate additional use of the two-way telephone communications service shall not adversely affect performance when used by the fire service.

Two-way telephone communications service is normally provided because fire department handheld radios may be ineffective in buildings with a great deal of structural steel, or when there is a large amount of radio traffic. The authority having jurisdiction can waive this requirement if the handheld radios used by the fire department personnel in the jurisdiction work effectively in the specific building in question. See Exhibits 3.17(a) and 3.17(b) for an example of a two-way telephone communications system.

Exhibit 3.17(a) *Typical fire emergency phone/cabinet assembly. (Source: Simplex Time Recorder Company, Gardner, MA)*

3-8.4.1.3.7.3* Two-way telephone communications service shall be capable of permitting the simultaneous operation of any five telephone stations in a common talk mode.

Exhibit 3.17(b) *Two-way telephone communications service in use.*

A-3-8.4.1.3.7.3 Consideration should be given to the type of telephone handset that fire fighters use in areas where high ambient noise levels exist or areas where high noise levels could exist during a fire condition. Push-to-talk handsets, handsets that contain directional microphones, or handsets that contain other suitable noise-canceling features can be used.

3-8.4.1.3.7.4 A notification signal at the fire command center, distinctive from any other alarm or trouble signal, shall indicate the off-hook condition of a calling telephone circuit. If a selective talk telephone communications service is supplied, a distinctive visible indicator shall be furnished for each selectable circuit so that all circuits with telephones off-hook are continuously and visibly indicated.

Exception: If emergency voice/alarm communications are used to notify all occupants automatically and simultaneously to evacuate the protected premises during a fire emergency, signals from the two-way telephone system shall be required to indicate only at a location approved by the authority having jurisdiction.

3-8.4.1.3.7.5 A means for silencing the audible call-in signal sounding appliance shall be permitted, provided it is key-operated, in a locked cabinet, or provided with protection to prevent use by unauthorized persons. The means shall operate a visible indicator and sound a trouble signal whenever the means is in the silence position and there are no telephone circuits in an off-hook condition. If a selective talk system is used, such a switch shall be permitted, provided subsequent telephone circuits going off-hook operate the distinctive off-hook audible signal sounding appliance.

The term *means for silencing* includes switches, touch pads, and touch screens.

3-8.4.1.3.7.6 Minimum Systems. As a minimum (for fire service use only), two-way telephone systems shall be common talk (that is, a conference or party line circuit), providing at least one telephone station or jack per floor and at least one telephone station or jack per exit stairway. In buildings equipped with a fire pump(s), a telephone station or jack shall be provided in each fire pump room.

3-8.4.1.3.7.7 Fire Warden Use. If the two-way telephone system is intended to be used by fire wardens in addition to the fire service, the minimum requirement shall be a selective talk system (where phones are selected from the fire command center). Systems intended for fire warden use shall provide telephone stations or jacks as required for fire service use and additional telephone stations or jacks as necessary to provide at least one telephone station or jack in each voice paging zone. Telephone circuits shall be selectable from the fire command center either individually or, if approved by the authority having jurisdiction, by floor or stairwell.

The paging zone is normally the same as the evacuation zone. See commentary for 3-8.4.1.3.5.6.2. In most designs, a fire fighters' telephone station or jack is located in each stairwell and elevator lobby.

3-8.4.1.3.7.8 If the control equipment provided does not indicate the location of the caller (common talk systems), each telephone station or telephone jack shall be clearly and permanently labeled to allow the caller to identify his or her location to the fire command center by voice.

3-8.4.1.3.7.9 If telephone jacks are provided, two or more portable handsets, as determined by the authority having jurisdiction, shall be stored at the fire command center for use by emergency responders.

The required quantity of portable handsets is often a percentage of the number of telephone jacks in the building. The authority having jurisdiction should be consulted for the acceptable quantity of portable handsets.

3-8.4.2* Signal Annunciation.

A-3-8.4.2 Embossed plastic tape, pencil, ink, or crayon should not be considered to be a permanently attached placard.

3-8.4.2.1 Protected premises fire alarm systems shall be arranged to annunciate alarm, supervisory, and trouble signals in accordance with 1-5.7.

3-8.4.2.2 Concealed Detectors. If a remote alarm indicator is provided for an automatic fire detector in a concealed location, the location of the detector and the area protected by the detector shall be prominently indicated at the remote alarm indicator by a permanently attached placard or by other approved means.

In a conventional fire alarm system (non-addressable), a remote alarm indicator, usually a red Light-Emitting Diode (LED) mounted on a single gang plate, is the most common method of indicating an alarm from a concealed detector. It is important to locate and mark the remote alarm indicator so that the detector in alarm can be found easily. Engraved phenolic plates permanently attached to the remote alarm indicator are generally considered the most appropriate way of complying with this requirement. See Exhibits 3.18 and 3.19 for examples of remote indicators and concealed detectors.

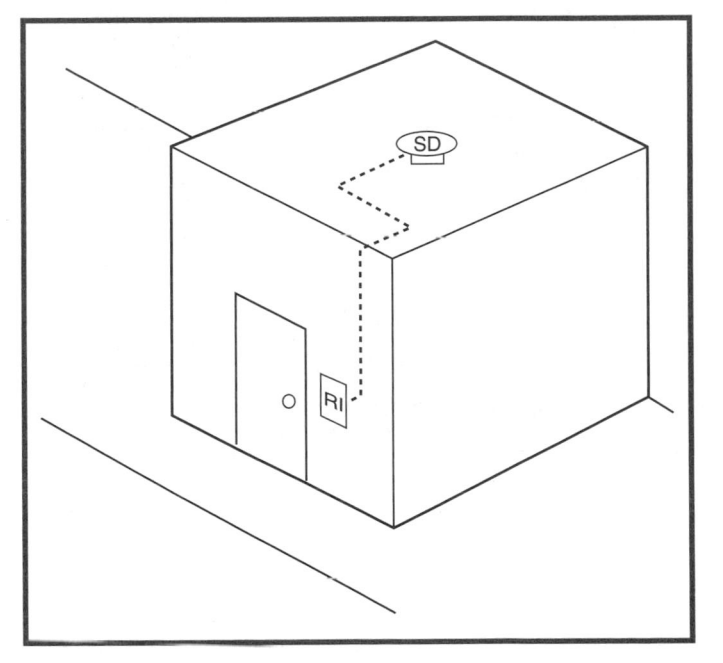

Exhibit 3.18 Concealed smoke detector (SD) in locked room with remote indicator (RI). (Source: FIREPRO Incorporated, Andover, MA)

Exhibit 3.19 Remote indicator used for concealed detectors. (Source: Mammoth Fire Alarms, Inc., Lowell, MA)

In an addressable fire alarm system, the Liquid Crystal Display (LCD) or video screen, located at the fire alarm control unit or remote annunciator, provides detailed detector location information and could be an acceptable alternative for meeting the intent of 3-8.4.2.2.

3-8.4.3 Suppression System Actuation.

3-8.4.3.1 If automatic or manual activation of a fire suppression system is to be performed through a fire alarm control unit, the control unit shall be listed for releasing service.

3-8.4.3.2 Each releasing device (for example, solenoid, relay) shall be monitored for integrity (supervised) in accordance with applicable NFPA standards.

See commentary following 3-8.3.2.5.2. and Exhibit 3.20 for interconnection of control units, monitoring for integrity, and a typical suppression system.

3-8.4.3.3 The integrity of the installation wiring shall be monitored in accordance with the requirements of Chapter 1.

3-8.4.3.4 Fire alarm systems used for fire suppression releasing service shall be provided with a disconnect switch to allow the system to be tested without activating the fire suppression systems. Operation of the disconnect switch shall cause a trouble signal at the fire alarm control unit.

This feature is extremely important. Very often the contractor who will be testing the fire alarm system is not an expert in the operation of fire suppression systems. The supervised

Exhibit 3.20 Wet chemical kitchen hood and duct cylinder with control head. (Source: Kidde-Fenwal Protection Systems, Ashland, MA)

disconnect switch allows the fire alarm system contractor to perform maintenance or tests on the fire alarm system without inadvertently activating the suppression system.

3-8.4.3.5 Sequence of operation shall be consistent with the applicable suppression system standards.

3-8.4.3.6* Each space protected by an automatic fire suppression system actuated by the fire alarm system shall contain one or more automatic fire detectors installed in accordance with Chapter 2.

A-3-8.4.3.6 Automatic fire suppression systems referred to in 3-8.4.3.6 include, but are not limited to, preaction and deluge sprinkler systems, carbon dioxide systems, halon systems, and dry chemical systems.

3-8.4.3.7 Suppression systems or groups of systems shall be controlled by a single control unit that monitors the associated initiating device(s), actuates the associated releasing device(s), and controls the associated agent release notification appliances. If the releasing panel is located in a protected premises having a separate fire alarm system, it shall be monitored for alarm, supervisory, and trouble signals by, but shall not be dependent on or affected by, the operation or failure of the protected premises fire alarm system.

There have been several instances where multitier releasing arrangements have resulted in inadvertent system discharges. These inadvertent discharges have occurred during normal system testing and maintenance, and occasionally because of system wiring faults unrelated to the required operation of the releasing system.

Another issue addressed by 3-8.4.3.7 deals with the interconnection of system control units. One control unit is listed for releasing service and is connected to actuate the fire extinguishing or suppression system. The other control unit is not listed for releasing service and has connected fire alarm-initiating devices that serve the area protected by the fire extinguishing or suppression system. The initiating devices actuate the control unit not listed for releasing service. That control unit, in turn, actuates the control unit that is listed for releasing device service. The latter control unit actuates the fire extinguishing or suppression system. These arrangements are particularly prone to inadvertent suppression system release and border on misapplication of equipment.

Section 3-8.4.3 was revised for the 1999 edition of the code to require all three signals (alarm, supervisory and trouble) to be sent to the master control unit. Since these subsystems control suppression systems that are remotely located from the master control unit, it is imperative that these signals be sent. See commentary following 3-8.3.2.5.2.

Exception: If the configuration of multiple control units is listed for releasing device service, and if a trouble condition or manual disconnect on either control unit causes a trouble or supervisory signal, the initiating devices on one control unit shall be permitted to actuate releasing devices on another control unit.

3-8.4.3.8 Fire alarm systems performing suppression system releasing functions shall be installed in such a manner that they are effectively protected from damage caused by activation of the suppression system(s) they control.

3-8.4.4 Off-Premises Signals.

3-8.4.4.1 Systems requiring transmission of signals to continuously attended locations providing supervising station service (for example, central station, proprietary, supervising station, remote supervising station) shall also comply with the applicable requirements of Chapter 5.

3-8.4.4.2 Trouble Signals to Supervising Station.

There have been fire alarm systems that have used normally de-energized relays to transmit this information off-premises; it was found that certain trouble conditions involving loss of operating power prevented transmission of a trouble signal. For protected premises that are normally not occupied, this could result in a system being out of service for an unacceptable amount of time. In addition, where addressable relays or modules are used to provide this information off-

premises, faults such as short circuit faults on the signaling line circuit could prevent activation of the addressable relay.

3-8.4.4.2.1 Relays or modules providing transmission of trouble signals to a supervising station shall be arranged to provide fail-safe operation.

3-8.4.4.2.2 Means provided to transmit trouble signals to supervising stations shall be arranged so as to transmit a trouble signal to the supervising station for any trouble condition received at the protected premises control unit, including loss of primary or secondary power.

3-8.5 Guard's Tour Supervisory Service.

Guard's tour supervisory service is used to provide fire protection surveillance during the hours when occupants are not in a building; to facilitate and control the movement of persons into, out of, and within a building; and to carry out procedures for the orderly conduct of specific operations in a building or on the surrounding property.

Guard's tour supervisory services designed to continually report the performance of a guard are often found in connection with protected premises fire alarm systems using off-premises reporting through central or proprietary supervising stations. Cases have been reported where a guard has failed to complete prescribed rounds due to illness or other impairment. Runners sent to investigate the failure to complete rounds have actually saved the guard's life. See Exhibit 3.21 for an illustration of a guard's tour supervisory system.

3-8.5.1 Guard's tour reporting stations shall be listed for the application.

3-8.5.2 The number of guard's tour reporting stations, their locations, and the route to be followed by the guard for operating the stations shall be approved for the particular installation in accordance with NFPA 601, *Standard for Security Services in Fire Loss Prevention*.

3-8.5.3 A permanent record indicating every time each signal-transmitting station is operated shall be made at the main control unit. Where intermediate stations that do not transmit a signal are employed in conjunction with signal-transmitting stations, distinctive signals shall be transmitted at the beginning and end of each tour of a guard, and a signal-transmitting station shall be provided at intervals not exceeding 10 stations. Intermediate stations that do not transmit a signal shall be capable of operation only in a fixed sequence.

3-8.6 Suppressed (Exception Reporting) Signal System.

This tour arrangement is somewhat more flexible than supervised tours but has the advantages of the absence of interconnected wires between the preliminary stations and

Exhibit 3.21 *Guard's tour; supervisory services ensure the safe completion of rounds. (Source: Simplex Time Recorder Company, Gardner, MA)*

the reduction of signal traffic. The usual arrangement is to have the guard transmit only start and finish signals that must be received at the central point at programmed reception times.

3-8.6.1 The suppressed signal system shall comply with the provisions of 3-8.5.2.

3-8.6.2 The system shall transmit a start signal to the signal-receiving location and shall be initiated by the guard at the start of continuous tour rounds.

3-8.6.3 The system shall automatically transmit a delinquency signal within 15 minutes after the predetermined actuation time if the guard fails to actuate a tour station as scheduled.

3-8.6.4 A finish signal shall be transmitted within a predetermined interval after the guard's completion of each tour of the premises.

3-8.6.5 For periods of over 24 hours during which tours are continuously conducted, a start signal shall be transmitted at least every 24 hours.

3-8.6.6 The start, delinquency, and finish signals shall be recorded at the signal-receiving location.

3-9 Protected Premises Fire Safety Functions

The control of preprogrammed protected premises fire safety functions is automatically initiated by the fire alarm system in response to fire alarm signals. Depending on the system configuration, either all of the fire safety functions are initiated by any alarm signal, or selected fire safety functions may be initiated in response to a specific initiating device or zone in alarm (e.g., doors released or fans shut down on the floor of fire origin only). In some cases, as in the exception to 3-9.3.6, a supervisory signal from the fire alarm control unit may be permitted to initiate the fire safety control function.

3-9.1 Scope.

The provisions of Section 3-9 shall cover the minimum requirements for the interconnection of protected premises fire safety functions (for example, fan control, door control) to the fire alarm system in accordance with 1-5.4.1.

3-9.2 General.

3-9.2.1 A listed relay or other listed appliance connected to the fire alarm system used to initiate control of protected premises fire safety functions shall be located within 3 ft (1 m) of the controlled circuit or appliance. The relay or other appliance shall function within the voltage and current limitations of the fire alarm control unit. The installation wiring between the fire alarm control unit and the relay or other appliance shall be monitored for integrity.

For example, the fan control for a fan located on the fourth floor may be located in the basement. Paragraph 3-9.2.1 requires the monitoring for integrity of the wiring from the fire alarm system control unit to the fire alarm system fan control relay, which actuates the fan control in the basement, not the fan located on the fourth floor. The distance between the fire alarm system fan control relay and the fan control should not exceed 3 ft (1 m). The code permits the relay to be located within 3 ft (1 m) of the controlled circuit or device. Therefore, it is not necessary to have the relay within 3 ft (1 m) of the fan. However, increasing the distance increases the potential for common mode failure.

Paragraph 3-9.2.1 and its exception are based on Chapter 7 (7-6.5.5) of NFPA *101, Life Safety Code.* The addition of paragraph 3-9.2.1 has been made to ensure that where auxiliary relays are used to cause the operation of smoke dampers, fire dampers, fan controls, smoke doors, and fire doors, and where they are connected to the building fire alarm system, the requirements of monitoring the interconnecting wiring for integrity apply. The exception covers devices that are installed in a fail-safe manner such as magnetic door hold-open devices. If a wire connection providing power to one of these devices is broken, the door closes.

This is the expected operation during a fire condition, therefore the device fails "safe," and the wiring does not need to be monitored for integrity.

Exception: Relays or appliances that operate on loss of power shall be considered self-monitoring for integrity.

3-9.2.2 Fire safety functions shall not interfere with other operations of the fire alarm system.

This requirement is similar to the requirements for combination systems covered by 3-8.2. One way to ensure that the fire safety functions do not interfere with other operations of the fire alarm system is to use auxiliary relays, listed for use with the fire alarm control unit, to isolate the fire safety function from the control unit. Fire safety functions connected directly to a notification appliance circuit, for instance, may disable the notification appliance circuit if a fault condition occurs in the fire safety function control equipment.

3-9.2.3 The method(s) of interconnection between the fire alarm system and controlled electrical and mechanical systems shall be monitored for integrity in accordance with 1-5.8; shall comply with the applicable provisions of NFPA 70, *National Electrical Code,* Article 760; and shall be achieved by one of the following recognized means:

(1) Electrical contacts listed for the connected load
(2) Listed digital data interfaces, such as serial communications ports and gateways
(3) Other listed methods

3-9.2.4 Fire safety function control devices and gateways shall be listed as compatible with the fire alarm control unit so as to prevent interference with control unit operation caused by controlled devices and to ensure transmission of data to operate the controlled devices.

Generally, the fire safety control function devices are auxiliary relays. These relays must be listed specifically to operate with the fire alarm control unit and not be off-the-shelf items from an electronics supply store.

Increasingly, digital and analog data are being sent to HVAC controls, elevator controls, security system controls and other building system controls in a serial format. Paragraph 3-9.2.4 was revised for the 1999 edition of the code to include the term *gateways,* which implies the use of BACnet, LONWorks, or other communications protocols. In cases where gateways are used, the communications protocol or interface must be listed for such use.

3-9.2.5 If a fire alarm system is a component of a life safety network, and it communicates data to other systems providing life safety functions or it receives data from such systems, the following shall apply:

(a) The path used for communicating data shall be monitored for integrity. This shall include monitoring the physical communication media and the ability to maintain intelligible communications.

(b) Data received from the network shall not affect the operation of the fire alarm system in any way other than to display the status of life safety network components.

(c) Where non-fire alarm systems are interconnected to the fire alarm system using a network or other digital communication technique, a signal (for example, heartbeat, poll, ping, query) shall be generated between the fire alarm system and the non-fire alarm system. Failure of proper receipt by the fire alarm system of confirmation of the transmission shall indicate a trouble signal within 200 seconds.

Paragraph 3-9.2.5 is new; it addresses the issue of interconnecting fire alarm systems with building automation, HVAC, elevator control, security, access control, public address, lighting control, and other similar non-fire alarm systems that may be performing fire safety functions based on input from the fire alarm system. Paragraph 3-9.2.5 requires a level of monitoring for integrity that is consistent with that for other fire alarm circuits.

3-9.2.6 The operation of all fire safety functions shall be verified by an operational test at the time of system acceptance.

The testing of the interfaced systems is extremely important. The proper interface and subsequent proper operation of the fire safety function is often dependent on more than one contractor and design engineer.

To ensure that the individuals involved communicate and work together to interface the fire safety function properly, all parties involved should be present when the authority having jurisdiction witnesses the acceptance test. See Exhibit 3.22 for an illustration of a typical interfaced system.

3-9.3 Elevator Recall for Fire Fighters' Service.

3-9.3.1* System-type smoke detectors or other automatic fire detection as permitted by 3-9.3.5 located in elevator lobbies, elevator hoistways, and elevator machine rooms used to initiate fire fighters' service recall shall be connected to the building fire alarm system. In facilities without a building fire alarm system, these smoke detectors or other automatic fire detection as permitted by 3-9.3.5 shall be connected to a dedicated fire alarm system control unit that shall be designated as "elevator recall control and supervisory panel," permanently identified on the control unit and on the record drawings. Unless otherwise required by the authority having jurisdiction, only the elevator lobby, elevator hoistway, and the elevator machine room smoke detectors or other automatic fire detection as permitted by 3-9.3.5 shall be used to recall elevators for fire fighters' service.

Elevator lobby, elevator hoistway, and elevator machine room smoke detectors are the only smoke detectors required

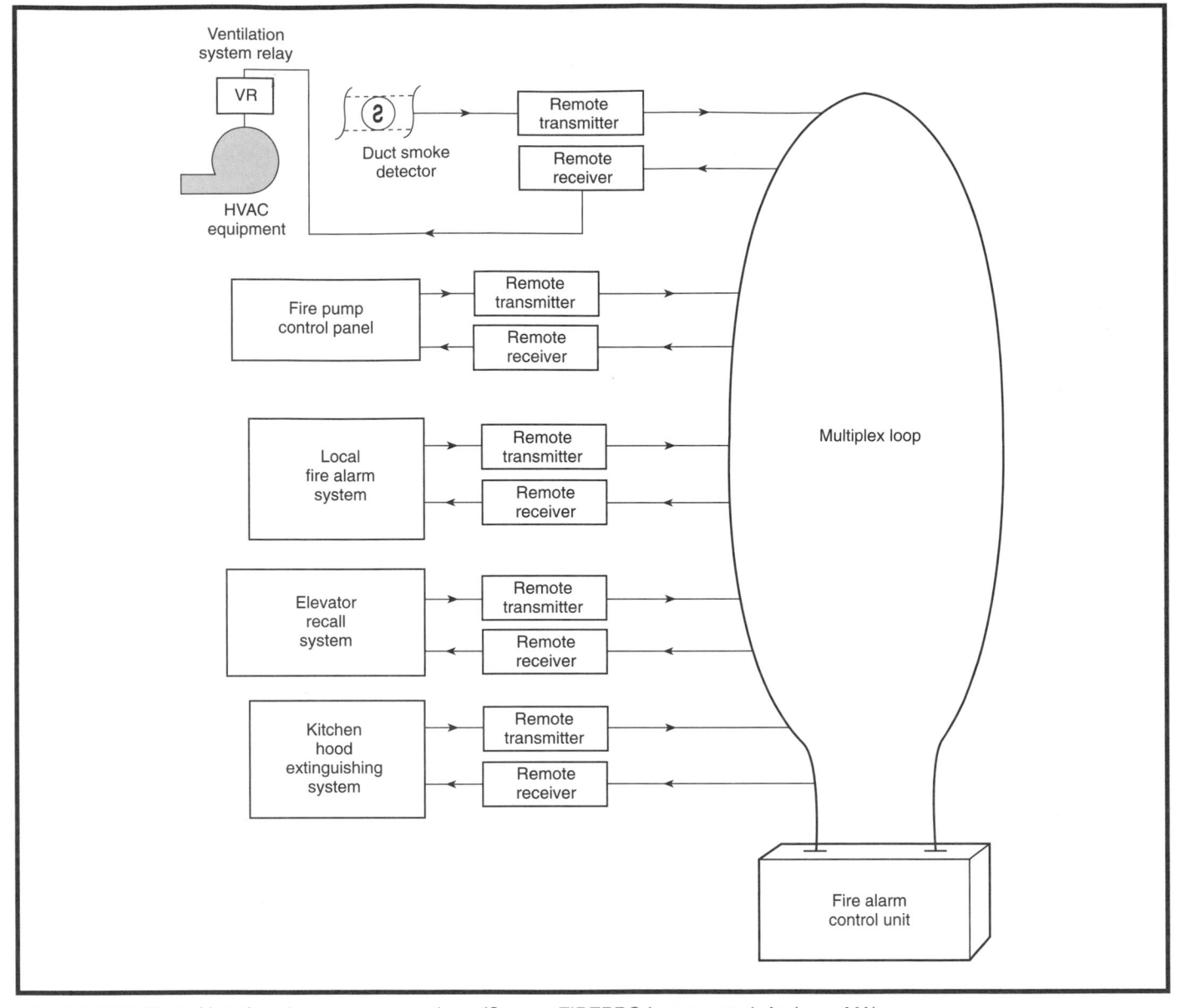

Exhibit 3.22 *Typical interfaced systems connections. (Source: FIREPRO Incorporated, Andover, MA)*

to initiate elevator recall in accordance with ANSI/ASME A17.1-1995, *Safety Code for Elevators and Escalators*, (See Exhibit 3.23, excerpt from A17.1-1997, Section 211.3b), which requires recall of elevators to the designated or alternate recall level when these detectors are actuated.

Buildings that do not have and are not required to have a fire alarm system, use a type of control unit designated and permanently labeled as the "Elevator Recall Control and Supervisory Panel," that serves the smoke detectors that initiate elevator recall.

A-3-9.3.1 In facilities without a building alarm system, dedicated fire alarm system control units are required by

3-9.3.1 for elevator recall in order that the elevator recall systems be monitored for integrity and have primary and secondary power meeting the requirements of this code.

The control unit used for this purpose should be located in an area that is normally occupied and should have audible and visible indicators to annunciate supervisory (elevator recall) and trouble conditions; however, no form of general occupant notification or evacuation signal is required or intended by 3-9.3.1.

The elevator recall control and supervisory unit should be placed in an area that is constantly attended for monitoring, especially when it is installed as a stand-alone control. See

211.3b Phase I Fire Alarm Activation. Fire alarm initiating devices shall be installed at each elevator floor, associated elevator machine room, and, where required, elevator hoistway in compliance with the requirements in ANSI/NFPA 72.

(1) The activation of a Phase I Emergency Recall fire alarm initiating device at any floor, other than at the designated level, shall cause all cars that serve that floor to return nonstop to the designated level. The activation of a Phase I Emergency Recall fire alarm initiating device in any elevator machine room shall cause all elevators having any equipment located in that machine room, and any associated elevators of a group automatic operation, to return nonstop to the designated level. The activation of a Phase I Emergency Recall fire alarm initiating device in any elevator hoistway shall cause all elevators having any equipment located in the hoistway, and any associated elevators of a group automatic operation, to return nonstop to the designated level, except that Phase I Emergency Recall fire alarm initiating device in hoistways installed at or below the lowest landing of recall, when activated, shall cause the car to be sent to the upper level of recall. The operation shall conform to the requirements of Rule 211.3a.

(2) When the Phase I Emergency Recall fire alarm initiating device at the designated level is activated, the operation shall conform to the requirements of Rule 211.3a, except that the cars shall return to an alternate level approved by the authority having jurisdiction, unless the designated-level three-position Phase I switch (Rule 211.3a) is in the "ON" position.

(3) Elevators shall only react to the first Phase I Emergency Recall fire alarm initiating device zone which is activated for that group.

(4) Phase I operation, when initiated by a fire alarm initiating device, shall be maintained until canceled by moving the Phase I switch to the "BYPASS" position [see also Rule 211.3a(10)].

(5) When activated, a fire alarm initiating device in the machine room shall cause the visual signal (Fig. 211.3a) to illuminate intermittently only in a car(s) with equipment in that machine room. When activated, a fire alarm initiating device in the hoistway shall cause the visual signal (Fig. 211.3a) to illuminate intermittently only in a car(s) with equipment in that hoistway.

Exhibit 3.23 Excerpt from ANSI/ASME A17.1a-1997. Courtesy of the American Society of Mechanical Engineers. Reprinted with permission.

Exhibit 3.24 for an example of an elevator recall control and supervisory system.

3-9.3.2 Each elevator lobby, elevator hoistway, and elevator machine room smoke detector or other automatic fire detection as permitted by 3-9.3.5 shall be capable of initiating elevator recall when all other devices on the same initiating device circuit have been manually or automatically placed in the alarm condition.

Generally, unless the required smoke detectors are installed on individual fire alarm-initiating device circuits without any other fire alarm devices installed on those circuits, the smoke detectors should be powered separately from the initiating device circuit. Smoke detectors installed on signaling line circuits will not be affected.

3-9.3.3 A lobby smoke detector shall be located on the ceiling within 21 ft (6.4 m) of the centerline of each elevator door within the elevator bank under control of the detector.

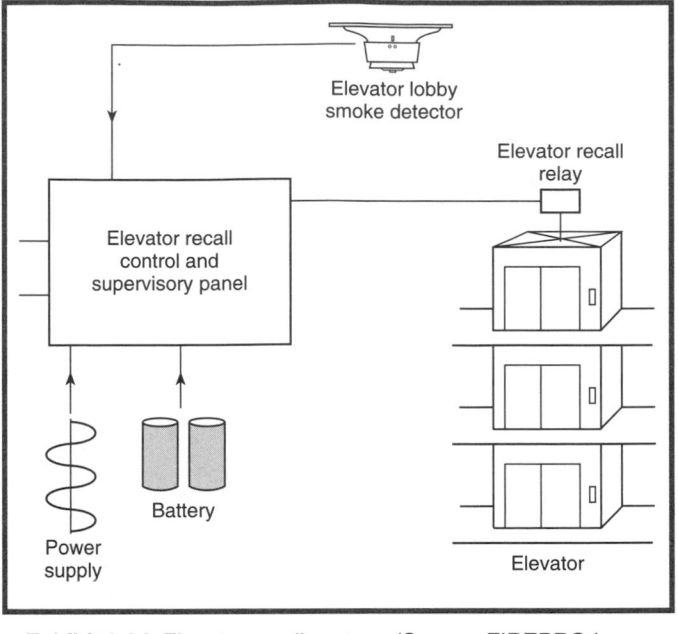

Exhibit 3.24 Elevator recall system. (Source: FIREPRO Incorporated, Andover, MA)

Exception: For lobby ceiling configurations exceeding 15 ft (4.6 m) in height or that are other than flat and smooth, detector locations shall be determined in accordance with Chapter 2.

Paragraph 3-9.3.3 covers a new requirement for the 1999 edition of the code. Chapter 3 requires a smoke detector to be within 0.7 times the selected spacing of the detector. On smooth ceilings under 15 ft (4.6 m), the selected spacing is typically permitted to be 30 ft (9.1 m). This ensures that a smoke detector will be within 21 ft (6.4 m) of the elevator door. High or nonsmooth ceilings may require a different spacing. There is no requirement that the smoke detector be located immediately adjacent to the elevator doors.

3-9.3.4 Smoke detectors shall not be installed in elevator hoistways.

Exception 1: Where the top of the elevator hoistway is protected by automatic sprinklers.

Exception 2: Where a smoke detector is installed to activate the elevator hoistway smoke relief equipment.

Even though 2-1.4.2 mentions elevator hoistways as one of the areas where detection must be installed where total coverage is specified, smoke detectors installed in elevator hoistways require continuous maintenance and are a source of numerous false or nuisance alarms. For this reason, the code has specifically excepted them from the elevator hoistway unless the top of the hoistway is protected by an automatic sprinkler system. If sprinklers are installed at the top

of the hoistway, then the smoke detector is needed to provide the recall feature before the heat detector or waterflow switch on the hoistway sprinkler system actuates. (See Exhibit 3.23, Rule 211.3b of ANSI/ASME A-17.1a-1997.) Section 5-13.6.1 of NFPA 13, *Standard for the Installation of Automatic Sprinkler Systems*, determines if the hoistway is required to be sprinklered. See in Exhibit 3.25 the excerpt from NFPA 13.

5-13.6.1* Sidewall spray sprinklers shall be installed at the bottom of each elevator hoistway not more than 2 ft (0.61 m) above the floor of the pit.

Exception: For enclosed, noncombustible elevator shafts that do not contain combustible hydraulic fluids, the sprinklers at the bottom of the shaft are not required.

A-5-13.6.1 The sprinklers in the pit are intended to protect against fires cause by debris, which can accumulate over time. Ideally, the sprinklers should be located near the side of the pit below the elevator doors, where most debris accumulates. However, care should be taken that the sprinkler location does not interfere with the elevator toe guard, which extends below the face of the door opening.

ASME A17.1, *Safety Code for Elevators and Escalators,* allows the sprinklers within 2 ft (0.65 m) of the bottom of the pit to be exempted from the special arrangements of inhibiting waterflow until elevator recall has occurred.

5-13.6.2* Automatic sprinklers in elevator machine rooms or at the tops of hoistways shall be of ordinary- or intermediate-temperature rating.

A-5-13.6.2 ASME A17.1, *Safety Code for Elevators and Escalators,* requires the shutdown of power to the elevator upon or prior to the application of water in elevator machine rooms or hoistways. This shutdown can be accomplished by a detection system with sufficient sensitivity that operates prior to the activation of the sprinklers (see also NFPA 72, *National Fire Alarm Code®*). As an alternative, the system can be arranged using devices or sprinklers capable of effecting power shutdown immediately upon sprinkler activation, such as a waterflow switch without a time delay. This alternative arrangement is intended to interrupt power before significant sprinkler discharge.

5-13.6.3* Upright or pendent spray sprinklers shall be installed at the top of elevator hoistways.

A-5-13.6.3 Passenger elevator cars that have been constructed in accordance with ASME A17.1, *Safety Code for Elevators and Escalators,* Rule 204.2a (under A17.1a-1985 and later editions of the code) have limited combustibility. Materials exposed to the interior of the car and the hoistway, in their end-use composition, are limited to a flame spread rating of 0 to 75 and a smoke development rating of 0 to 450.

Exhibit 3.25 *Excerpt from NFPA 13, Standard for the Installation of Sprinkler Systems, 1999 edition.*

Exception No. 2 was added to the 1999 edition of the code to allow for smoke relief equipment, such as a smoke hatch.

3-9.3.5 If ambient conditions prohibit installation of automatic smoke detection, other automatic fire detection shall be permitted.

The intent of paragraph 3-9.3.5 is to prevent nuisance alarms from smoke detectors installed in areas that are inappropriate for their use such as unheated areas. Where the designer, the authority having jurisdiction or another code requires a detector in areas where the ambient conditions are unsuitable for a smoke detector, 3-9.3.5 allows the use of any other type of detector that would be stable and still provide necessary detection.

3-9.3.6 When actuated, each elevator lobby, elevator hoistway, and elevator machine room smoke detector or other automatic fire detection as permitted by 3-9.3.5 shall initiate an alarm condition on the building fire alarm system and shall visibly indicate, at the control unit and required remote annunciators, the alarm initiation circuit or zone from which the alarm originated. Actuation from elevator hoistway and elevator machine room smoke detectors or other automatic fire detection as permitted by 3-9.3.5 shall cause separate and distinct visible annunciation at the control unit and required annunciators to alert fire fighters and other emergency personnel that the elevators are no longer safe to use. Actuation of these detectors shall not be required to actuate the system notification appliances where the alarm signal is indicated at a constantly attended location.

The intent of paragraph 3-9.3.6 is to ensure that the area, or zone of alarm (floor, room, etc.,) be indicated on the fire alarm control unit and the remote annunciator. The code requires that the elevator hoistway smoke detector (if one is present) and the elevator machine room smoke detector(s) be connected to the fire alarm control unit and remote annunciator as a separate zone or point of alarm indication. The detectors located in the elevator hoistway and elevator machine room actuate Phase I elevator recall, but are not required to sound the building evacuation alarm. Their annunciation, however, must conform to the requirements as stated.

Exception: If approved by the authority having jurisdiction, the elevator hoistway and machine room smoke detectors shall be permitted to initiate a supervisory signal.

The exception to 3-9.3.6 is provided to minimize the nuisance alarms from smoke detectors in these areas. The elevator recall system would still operate, but the fire alarm signal would not sound. The option should be used only where trained personnel are constantly in attendance and can immediately respond to the supervisory signal and investigate the cause of the signal. Means should be provided for initiating the fire alarm signal if the investigation of the cause of the supervisory signal indicates that building evacuation is necessary.

In addition to having trained personnel constantly in attendance, it is recommended that if the supervisory signal is not acknowledged within a given period of time (3 minutes to 10 minutes), the fire alarm system will automatically and immediately initiate an alarm.

3-9.3.7* For each group of elevators within a building, three separate elevator control circuits shall be terminated at the designated elevator controller within the group's elevator machine room(s). The operation of the elevators shall be in accordance with Rules 211.3 through 211.8 of ANSI/ASME A17.1, *Safety Code for Elevators and Escalators*. The smoke detectors or other automatic fire detection as permitted by 3-9.3.5 shall actuate the three elevator control circuits as follows:

(a) The smoke detector or other automatic fire detection as permitted by 3-9.3.5 located in the designated elevator recall lobby shall actuate the first elevator control circuit. In addition, if the elevator is equipped with front and rear doors, the smoke detectors in both lobbies at the designated level shall actuate the first elevator control circuit.

(b) The smoke detectors or other automatic fire detection as permitted by 3-9.3.5 in the remaining elevator lobbies shall actuate the second elevator control circuit.

(c) The smoke detectors or other automatic fire detection as permitted by 3-9.3.5 in elevator hoistways and the elevator machine room(s) shall actuate the third elevator control circuit. In addition, if the elevator machine room is located at the designated level, its smoke detector or other automatic fire detection as permitted by 3-9.3.5 shall also actuate the first elevator control circuit.

Three elevator control circuits are needed for proper operation of the recall sequence and for safe use of the elevators by the fire department. The first circuit is needed to prevent recalling the elevators and discharging passengers to the designated floor when the designated floor is the fire location, and to provide for an alternate recall location (determined by the authority having jurisdiction) when the designated floor is reporting a fire condition.

The second circuit configuration [part (b)] provides for standard recall to the designated floor when any other elevator lobby smoke detector is in alarm.

The operation of the third circuit, [part (c)], although not defined in the code, is intended for the safety of fire fighters who may be using the elevators to bring equipment to staging areas in a high-rise building. The third circuit's feature is intended to recall the cab(s) to the designated level during Phase I operation and to warn fire fighters of a fire in the hoistway or machine room during Phase II operation. The third circuit will sound a warning in the elevator cab to notify the fire department personnel using the elevators on Phase II Operation during the fire to immediately move to a safe floor and exit the elevator. See Exhibit 3.26 for an illustration of a connection of the fire alarm system to the elevator controller.

A-3-9.3.7 It is recommended that the installation be in accordance with Figures A-3-9.3.7(a) and (b). Figure A-3-9.3.7(a) should be used where the elevator is installed at the same time as the building fire alarm system. Figure A-3-9.3.7(b) should be used where the elevator is installed after the building fire alarm system.

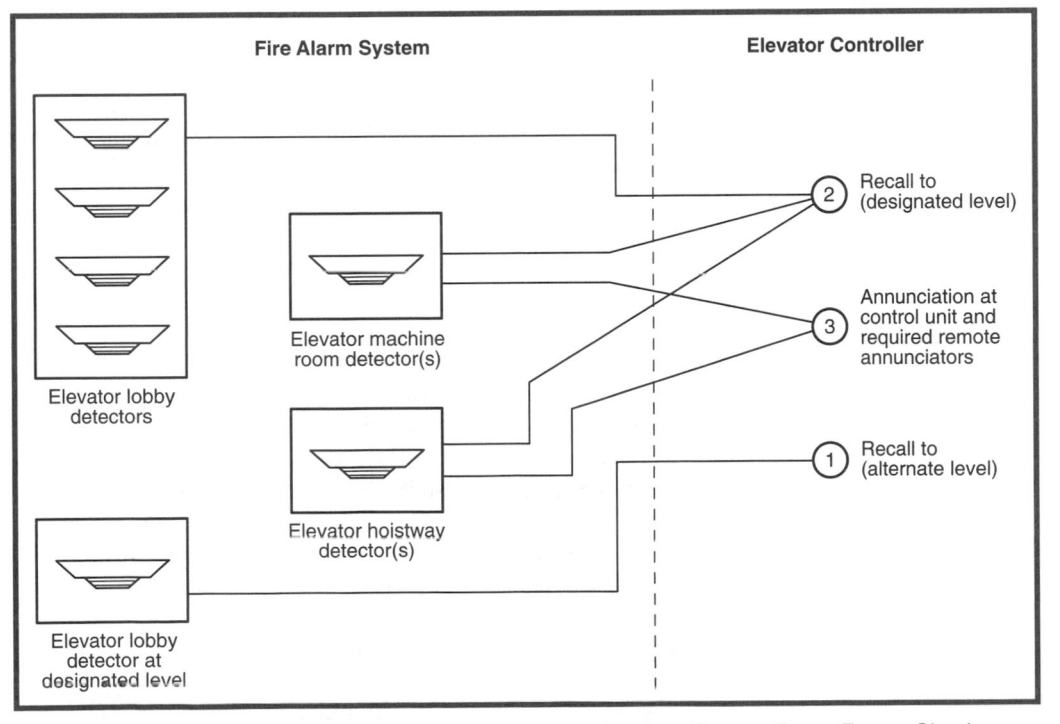

Exhibit 3.26 *Connection to third circuit for fire fighter notification. (Source: Bruce Fraser, Simplex Time Recorder Company, Gardner, MA)*

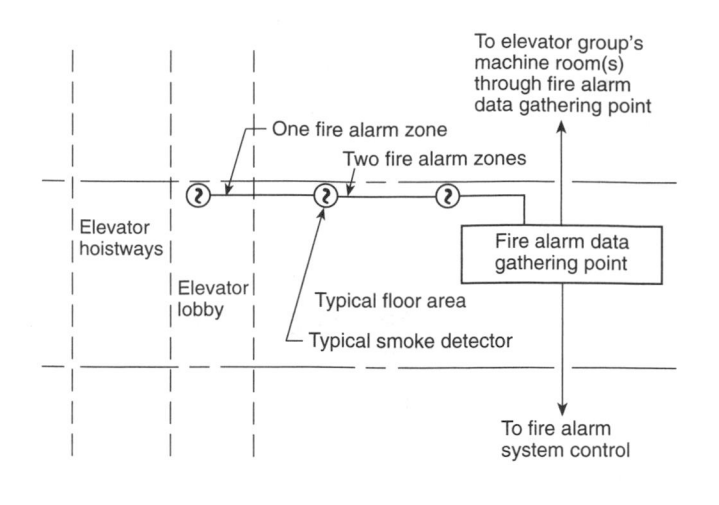

Figure A-3-9.3.7(a) Elevator zone—elevator and fire alarm system installed at same time.

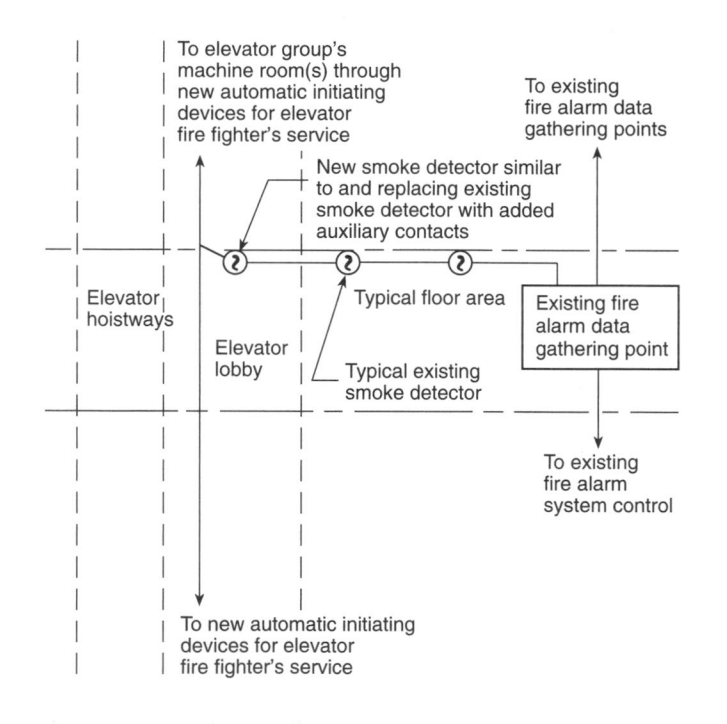

Figure A-3-9.3.7(b) Elevator zone—elevator installed after fire alarm system.

3-9.4 Elevator Shutdown.

Subsection 3-9.4 is the result of additional sprinkler protection requirements of other codes for elevator machine rooms and hoistways. See the excerpt from ANSI/ASME A17.1-1997, Rule 102.2, (c) (3), in Exhibit 3.27.

Rule 102.2
(c)(3) Means shall be provided to automatically disconnect the main line power supply to the affected elevator upon or prior to the application of water shall not disconnect the main line power supply.

Exhibit 3.27 *Excerpt from ANSI/ASME A17.1-1997. Courtesy of the American Society of Mechanical Engineers. Reprinted with permission.*

The purpose of elevator shutdown prior to sprinkler operation is to avoid the hazards of a wet elevator braking system and electrical shock. If the elevator brakes are wet, there is the danger of the elevator rising uncontrollably to the top of the hoistway or to the bottom of the hoistway, depending on the load in the cab.

3-9.4.1* Where heat detectors are used to shut down elevator power prior to sprinkler operation, the detector shall have both a lower temperature rating and a higher sensitivity as compared to the sprinkler.

Paragraph 3-9.4.1 is extremely important. Often, it is not understood that a 135°F (57.2°C) heat detector may not respond prior to a 165°F (73°C) sprinkler head despite the obvious differences in temperature sensitivity. The response time is based on the RTI of both devices and must be known prior to design and installation of heat detectors for elevator shutdown. Because the RTI for heat detectors is not readily available, one must use a sensitive fixed temperature heat detector with a listed spacing equal to or greater than 40 ft (12.2 m) on center (or use a rate-of-rise-type heat detector) in the locations where heat detectors are required.

The 1998 edition of A17.1 no longer requires elevator shutdown for waterflow from less than 24 in. (610 mm) from the floor of the hoistway pit. Therefore, heat detectors are not required in those cases.

A-3-9.4.1 A lower response time index is intended to provide detector response prior to the sprinkler response, because a lower temperature rating alone might not provide earlier response. The listed spacing rating of the heat detector should be 25 ft (7.6 m) or greater.

3-9.4.2 If heat detectors are used to shut down elevator power prior to sprinkler operation, they shall be placed within 2 ft (610 mm) of each sprinkler head and be installed in accordance with the requirements of Chapter 2. Alternatively, engineering methods, such as specified in Appendix B, shall be permitted to be used to select and place heat detectors to ensure response prior to any sprinkler head operation under a variety of fire growth rate scenarios.

3-9.4.3* If pressure or waterflow switches are used to shut down elevator power immediately upon or prior to the discharge of water from sprinklers, the use of devices with

time delay switches or time delay capability shall not be permitted.

The intent is to shut down elevator power as soon as water flows in the automatic sprinkler system that is protecting the elevator hoistway and machine room. These waterflow devices should be installed in the cross main or branch line serving the automatic sprinkler system in those areas. The requirement is for a device with no retard or time delay mechanism, not a device with its time delay or retard feature set to zero.

A-3-9.4.3 Care should be taken to ensure that elevator power cannot be interrupted due to water pressure surges in the sprinkler system. The intent of the code is to ensure that the switch and the system as a whole do not have the capability of introducing a time delay into the sequence. The use of a switch with a time delay mechanism set to zero does not meet the intent of the code, because it is possible to introduce a time delay after the system has been accepted. This might occur in response to unwanted alarms caused by surges or water movement, rather than addressing the underlying cause of the surges or water movement (often due to air in the piping). Permanently disabling the delay in accordance with the manufacturer's printed instructions should be considered acceptable. Systems that have software that can introduce a delay in the sequence should be programmed to require a security password to make such a change.

3-9.4.4* Control circuits to shut down elevator power shall be monitored for presence of operating voltage. Loss of voltage to the control circuit for the disconnecting means shall cause a supervisory signal to be indicated at the control unit and required remote annunciators.

There have been cases where the operating power for elevator shunt trip circuits has been de-energized. This is a dangerous condition because the elevator will not be shut down in the event of waterflow in the machine room or hoistway. The new requirement in 3-9.4.4 was added to the 1999 edition of the code to monitor the integrity of the operating power for the shunt trip control circuit. Monitoring the integrity of the control power is similar to monitoring the integrity of the power for an electric motor-driven fire pump.

A-3-9.4.4 Figure A-3-9.4.4 illustrates one method of monitoring elevator shunt trip control power for integrity.

3-9.5 Heating, Ventilation, and Air-Conditioning (HVAC) Systems.

3-9.5.1 The provisions of 3-9.5 shall apply to the basic method by which a fire alarm system interfaces with the HVAC systems.

3-9.5.2 If connected to the fire alarm system serving the protected premises, all detection devices used to cause the

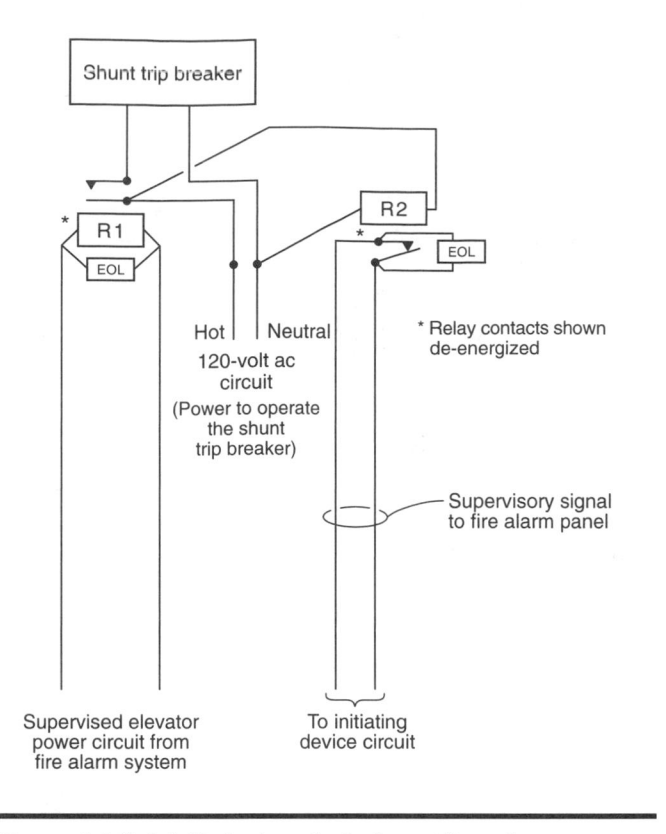

Figure A-3-9.4.4 *Typical method of providing elevator power shunt trip supervisory signal.*

operation of HVAC systems smoke dampers, fire dampers, fan control, smoke doors, and fire doors shall be monitored for integrity in accordance with 1-5.8.

Where the devices are connected to the fire alarm system, the wiring to these devices is required by 1-5.8 to be monitored for integrity. In many cases, this will be accomplished by smoke detectors connected to the fire alarm system. Stand alone detectors that are not connected to the fire alarm system cannot be monitored for integrity. This includes single-station smoke detectors that are 120-Vac-powered and are used to control the HVAC equipment.

3-9.5.3 Connections between fire alarm systems and the HVAC system for the purpose of monitoring and control shall operate and be monitored in accordance with applicable NFPA standards. Smoke detectors mounted in the air ducts of HVAC systems shall initiate either an alarm signal at the protected premises or a supervisory signal at a constantly attended location or supervising station.

The second sentence of 3-9.5.3 was added to the 1999 edition of the code to correlate with NFPA 90A, *Standard for the Installation of Air-Conditioning and Ventilating Systems.* This option is provided to minimize the nuisance alarms from smoke detectors in these areas. The HVAC system would still shut down as required by NFPA 90A, but the fire

alarm signal would not sound. The option should be used only where trained personnel are constantly in attendance or the signal is transmitted to a supervising station. See Exhibit 3.28 for an excerpt from NFPA 90A, 1999 edition.

4-4.4.2 In addition to the requirements of 4-4.3, where an approved fire alarm system is installed in a building, the smoke detectors required by the provisions of Section 4-4 shall be connected to the fire alarm system in accordance with the requirements of NFPA 72, *National Fire Alarm Code.* Smoke detectors used solely for closing dampers or for heating, ventilating, and air-conditioning system shutdown shall not be required to activate the building evacuation alarm.

Exhibit 3.28 *Excerpt from NFPA 90A, Standard for the Installation of Air-Conditioning and Ventilating Systems, 1999 edition.*

3-9.5.4 If the fire alarm control unit activates the HVAC system for the purpose of smoke control, the automatic alarm-initiating zones shall be coordinated with the smoke-control zones they actuate.

Paragraph 3-9.5.4 requires coordination between the fire alarm system designer and installer and the smoke control system designer and installer. Because the smoke zones are often smaller and more numerous than fire zones, this requirement most often leads to additional zone requirements for the fire alarm system.

3-9.6 Door Release Service.

3-9.6.1 The provisions of 3-9.6 shall apply to the methods of connection of door hold-open release devices and to integral door hold-open release, closer, and smoke detection devices.

3-9.6.2 All detection devices used for door hold-open release service shall be monitored for integrity in accordance with 1-5.8.

Exception: Smoke detectors used only for door release and not for open area protection.

Detectors that are integral to the door assembly or stand alone detectors not connected to the fire alarm system are not required to be monitored for integrity.

3-9.6.3 All door hold-open release and integral door release and closure devices used for release service shall be monitored for integrity in accordance with 3-9.2.

Generally, magnetic door release appliances are installed in a fail-safe manner so that they release on loss of power. If these types of appliances are connected in such a fashion to the fire alarm system, the wiring to the control relay or cir-

cuit does not need to be monitored for integrity. (See the exception to 3-9.2.1.)

3-9.6.4 Magnetic door holders that allow doors to close upon loss of operating power shall not be required to have a secondary power source.

The purpose of magnetic door release appliances is to keep doors open under normal conditions and allow the doors to close in smoke and fire conditions. If the designer or the authority having jurisdiction wishes for the doors to remain open even under a primary power failure, then the magnetic door holders must be placed on a circuit with secondary power. The code recognizes that this is optional and, therefore, does not require secondary power for these devices. See Exhibit 3.29 for an example of typical magnetic door hold release devices.

Exhibit 3.29 *Typical magnetic door hold release devices. (Source: ESL/Sentrol, Tualatin, OR)*

3-9.7 Door Unlocking Devices.

3-9.7.1 Any device or system intended to actuate the locking or unlocking of exits shall be connected to the fire alarm system serving the protected premises.

3-9.7.2 All exits connected in accordance with 3-9.7.1 shall unlock upon receipt of any fire alarm signal by means of the fire alarm system serving the protected premises.

Exception: Where otherwise required or permitted by the authority having jurisdiction or other codes.

3-9.7.3* All exits connected in accordance with 3-9.7.1 shall unlock upon loss of the primary power to the fire alarm system serving the protected premises. The secondary power supply shall not be utilized to maintain these doors in the locked condition.

Paragraph 3-9.7.3 requires the doors to unlock upon loss of power to the fire alarm system. It also prohibits using the fire alarm system secondary power supply to keep the doors locked. Some may attempt to accomplish these requirements by using the primary power circuit feeding the fire alarm system to also feed power to the locks. This action would violate the requirements of 1-5.2.5.2.

A-3-9.7.3 A problem could exist when batteries are used as a secondary power source if a control unit having 24 hours of standby operating power were to lose primary power and be operated for more than 24 hours from the secondary power source (batteries). It is possible that sufficient voltage would be available to keep the doors locked but not enough voltage available to operate the fire alarm system to release the locks. For systems requiring primary power that meets the requirements of 1-5.2.3, such as a hospital system, door locking would not be an issue even with batteries provided in the fire alarm control unit, because the primary power (emergency generator) would operate the fire alarm control unit and secondary power would not be required.

3-9.7.4 If exit doors are unlocked by the fire alarm system, the unlocking function shall occur prior to or concurrent with activation of any public-mode notification appliances in the area(s) served by the normally locked exits.

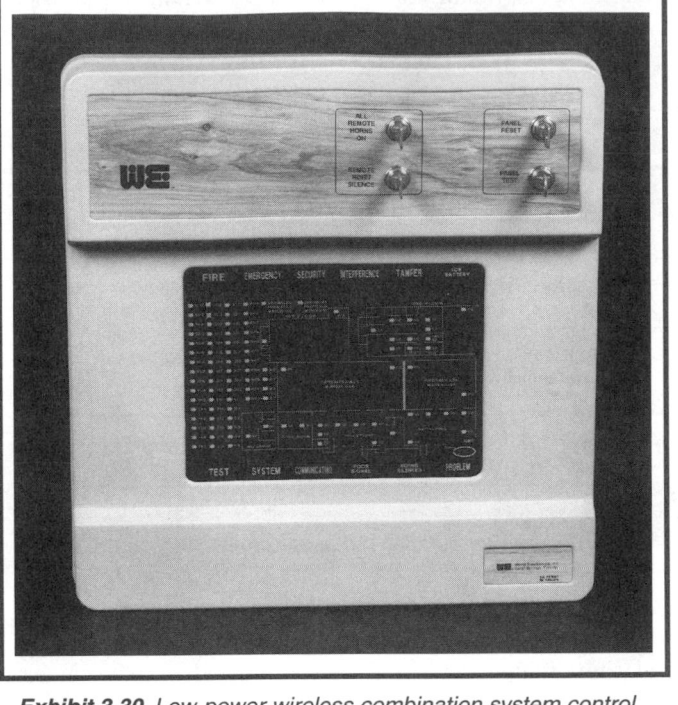

Exhibit 3.30 *Low-power wireless combination system control unit. (Source: World Electronics, Coral Springs, FL)*

3-10* Special Requirements for Low-Power Radio (Wireless) Systems

A-3-10 The term *wireless* has been replaced with the term *low-power radio* to eliminate potential confusion with other transmission media such as optical fiber cables.

Low-power radio devices are required to comply with the applicable low-power requirements of Title 47, *Code of Federal Regulations*, Part 15.

Listed low-power wireless fire alarm systems have numerous applications. Historical buildings where wire or cable installation will damage the building or impact the historical significance of the property have used wireless fire alarm systems successfully. Industrial buildings that use corrosive materials that can affect the integrity of the wiring used to interconnect a fire alarm system can often benefit from low-power wireless systems. Likewise, any buildings that are remote from the main facility can also be well served by a low-power wireless fire alarm system application. See Exhibits 3.30 and 3.31 for examples of low-power wireless equipment.

3-10.1 Compliance with Section 3-10 shall require the use of low-power radio equipment specifically listed for the purpose.

A-3-10.1 Equipment listed solely for dwelling unit use would not comply with this requirement.

3-10.2 Power Supplies.

A primary battery (dry cell) shall be permitted to be used as the sole power source of a low-power radio transmitter where all of the following conditions are met:

(a) Each transmitter shall serve only one device and shall be individually identified at the receiver/control unit.

(b) The battery shall be capable of operating the low-power radio transmitter for not less than 1 year before the battery depletion threshold is reached.

(c) A battery depletion signal shall be transmitted before the battery has depleted to a level insufficient to support alarm transmission after 7 additional days of non-alarm operation. This signal shall be distinctive from alarm, supervisory, tamper, and trouble signals; shall visibly identify the affected low-power radio transmitter; and, when silenced, shall automatically re-sound at least once every 4 hours.

(d) Catastrophic (open or short) battery failure shall cause a trouble signal identifying the affected low-power radio transmitter at its receiver/control unit. When silenced, the trouble signal shall automatically re-sound at least once every 4 hours.

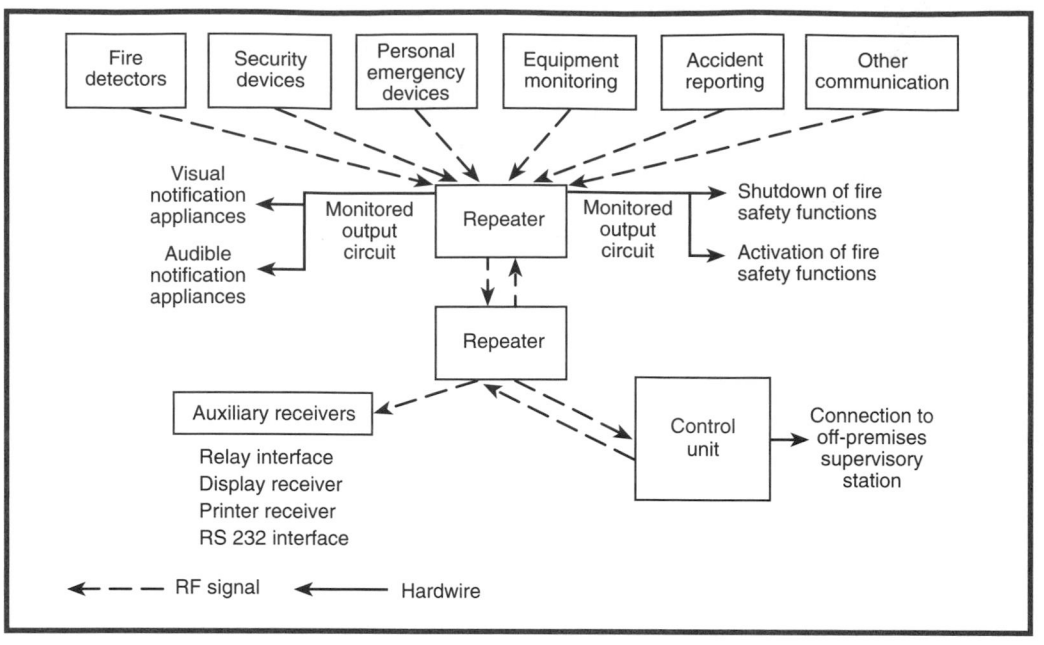

Exhibit 3.31 *Combination system including a low-power radio (wireless) fire alarm system with other non-fire equipment. (Source: World Electronics, Inc., Coral Springs, FL)*

(e) Any mode of failure of a primary battery in a low-power radio transmitter shall not affect any other low-power radio transmitter.

3-10.3 Alarm Signals.

3-10.3.1* When actuated, each low-power radio transmitter shall automatically transmit an alarm signal.

A-3-10.3.1 This requirement is not intended to preclude verification and local test intervals prior to alarm transmission.

3-10.3.2 Each low-power radio transmitter shall automatically repeat alarm transmission at intervals not exceeding 60 seconds until the initiating device is returned to its non-alarm condition.

3-10.3.3 Fire alarm signals shall have priority over all other signals.

3-10.3.4 Low-power wireless systems shall comply with the requirements of 1-5.4.1.2 and 1-5.4.2.2.

3-10.3.5 An alarm signal from a low-power radio transmitter shall latch at its receiver/control unit until manually reset and shall identify the particular initiating device in alarm.

3-10.4 Monitoring for Integrity.

3 10.4.1 The low-power radio transmitter shall be specifically listed as using a transmission method that is highly resistant to misinterpretation of simultaneous transmissions

and to interference (for example, impulse noise and adjacent channel interference).

3-10.4.2 The occurrence of any single fault that disables transmission between any low-power radio transmitter and the receiver/control unit shall cause a latching trouble signal within 200 seconds.

Exception: Where Federal Communications Commission (FCC) regulations prevent meeting the 200-second requirement, the time period for a low-power radio transmitter with only a single, connected alarm-initiating device shall be permitted to be increased to four times the minimum time interval permitted for a 1-second transmission up to the following:

(a) Four hours maximum for a transmitter serving a single initiating device

(b) Four hours maximum for a retransmission device (repeater) where disabling of the repeater or its transmission does not prevent the receipt of signals at the receiver/control unit from any initiating device transmitter

3-10.4.3 A single fault on the signaling channel shall not cause an alarm signal.

3-10.4.4 The periodic transmission required to comply with 3-10.4.2 from a low-power radio transmitter shall ensure successful alarm transmission capability.

3-10.4.5 Removal of a low-power radio transmitter from its installed location shall cause immediate transmission of a

distinctive supervisory signal that indicates its removal and individually identifies the affected device.

Exception: This requirement shall not apply to dwelling unit fire warning systems.

3-10.4.6 Reception of any unwanted (interfering) transmission by a retransmission device (repeater) or by the main receiver/control unit, for a continuous period of 20 seconds or more, shall cause an audible and visible trouble indication at the main receiver/control unit. This indication shall identify the specific trouble condition as an interfering signal.

3-10.5 Output Signals from Receiver/Control.

When the receiver/control is used to activate remote appliances, such as notification appliances and relays, by wireless means, the remote appliances shall meet the following requirements:

(1) Power supplies shall comply with Chapter 1 or the requirements of 3-10.2.
(2) All supervision requirements of Chapter 1, Chapter 3, or 3-10.4 shall apply.
(3) The maximum allowable response delay from activation of an initiating device to activation of required alarm functions shall be 90 seconds.

(4) Each receiver/control shall automatically repeat alarm transmission at intervals not exceeding 60 seconds or until confirmation that the output appliance has received the alarm signal.
(5) The appliances shall continue to operate (latch-in) until manually reset at the receiver/control.

References Cited in Commentary

IEEE *Standard Dictionary of Electrical and Electronics Terms*, 1993.

ANSI/ASME A17.1, *Safety Code for Elevators and Escalators*, 1997.

NFPA 13, *Standard for the Installation of Sprinkler Systems*, National Fire Protection Association, Quincy, MA, 1999.

NFPA 70, *National Electrical Code®*, National Fire Protection Association, Quincy, MA, 1999.

NFPA 90A, *Standard for the Installation of Air-Conditioning and Ventilating Systems*, National Fire Protection Association, Quincy, MA, 1999.

NFPA *101®*, *Life Safety Code®*, National Fire Protection Association, Quincy, MA, 1997.

CHAPTER 4

Notification Appliances for Fire Alarm Systems

4-1* Scope

Chapter 4 applies to notification appliances for fire alarm systems, which includes audible and visible appliances. This material was previously contained in Chapter 6 of the 1996 edition of the code and was relocated as part of a restructuring of the code for user friendliness.

A-4-1 Notification appliances should be sufficient in quantity, audibility, intelligibility, and visibility so as to reliably convey the intended information to the intended building occupants in a fire emergency.

Notification appliances in conventional commercial and industrial applications should be installed in accordance with the specific requirements of Sections 4-3 and 4-4.

The code recognizes that it is not possible to identify specific criteria sufficient to ensure effective occupant notification in every conceivable application. If the specific criteria of Sections 4-3 and 4-4 are determined to be inadequate or inappropriate to provide the performance recommended above, approved alternative approaches or methods or are permitted to be used.

4-1.1 Requirements.

Chapter 4 shall cover requirements for the performance, location, and mounting of fire alarm system notification appliances used to initiate evacuation or relocation of the occupants, or for providing information to occupants or staff. Chapter 4 shall also cover the requirements for the performance, location, and mounting of annunciators, displays, and printers used to display or record information for use by protected premises occupants or staff, responding emergency personnel, or supervising station personnel.

Subsection 4-1.1 recognizes that a building's fire emergency plan may require evacuation of the building or relocation of occupants within the protected premises.

In the 1999 edition of the code, the scope of Chapter 4 was supplemented to include other types of notification appliances that provide information to occupants, staff, responding emergency personnel, or supervising station personnel.

With the exception of 4-3.1.1, the requirements for having notification appliances are found in other codes, such as NFPA *101*®, *Life Safety Code*®. NFPA 72 covers installation and performance requirements, as stated in 4-1.1.

4-1.2 Use.

These requirements shall be used with other NFPA standards that deal specifically with fire alarm, extinguishment, or control systems. Notification appliances for fire alarm systems shall add to fire protection by providing stimuli for initiating emergency action.

The audible or visible notification appliances referred to in other chapters of the code are described in detail in Chapter 4.

For example, protected premises fire alarm systems refer to Chapter 4 for requirements on the use of notification appliances to alert occupants of the need for evacuation or relocation. Similarly, notification appliances required by Chapter 5 to alert supervising station personnel must meet Chapter 4's requirements.

4-1.3 All notification appliances installed in conformity with Chapter 4 shall be listed for the purpose for which they are used.

The requirement in 4-1.3 states that the listing of notification appliances be use-specific. This means that the listing of an appliance must relate to the exact manner in which it will be used. See Exhibit 4.1 for examples of typical notification appliances listed for fire alarm use.

For example, strobe lights listed for wall mounting are not permitted to be installed on a ceiling because they are

Exhibit 4.1 *Typical listed audible notification appliances (bells). (Source: EST, Inc., Cheshire, CT)*

designed, tested, and listed to cover a specific area. Mounting them on ceilings results in a signal that covers only half of the room or area.

4-1.4 The requirements of Chapter 4 shall be intended to address the reception of a notification signal and not the signal's information content.

The requirements in Chapter 4 do not address the content of a textual message or a message contained in a coded audible or visible signal. Rather, the requirements address the ability of notification appliances to deliver a message. Chapter 3 requires the use of the ANSI S3.41-1990 (R 1996), *American National Standard Audible Emergency Evacuation Signal*, where evacuation of an area is intended. Chapter 4 requires that signal to meet certain audibility and installation requirements.

4-1.5 Interconnection of Appliances.

The interconnection of appliances, the control configurations, the power supplies, and the use of the information provided by notification appliances for fire alarm systems shall be described in Chapter 1 and Chapter 3.

4-2 General

4-2.1 Nameplates.

4-2.1.1 Notification appliances shall include on their nameplates reference to electrical requirements and rated audible or visible performance, or both, as defined by the listing authority.

4-2.1.2 The audible appliances shall include on their nameplates reference to their parameters or reference to installation documents (supplied with the appliance) that include the parameters in accordance with 4-3.2. The visible appliances shall include on their nameplates reference to their parameters or reference to installation documents (supplied with the appliance) that include the parameters in accordance with 4-4.2.1.

To guide designers and installers of fire alarm systems so that the system will deliver audible and visible information with appropriate intensity, the nameplate must state the capabilities of the appliance, as determined through tests conducted by the listing organization. The nameplate information also assists inspectors in verifying compliance with approved documents.

Notification appliance circuits require special treatment to ensure that all the connected appliances will operate under adverse (low) voltage conditions. Low voltage, which is outside the operating range of the appliance, can cause the appliances to produce lower visible signal intensities or sound pressure levels (SPLs). Voltage at any appliance on the notification appliance circuit (NAC) should not drop below the limits of the appliance design to ensure correct intensity and audibility. The designer of the notification appliance circuit should consider these interrelated questions: How many appliances can be connected to the NAC? What is the size of the field-wiring conductors? What is the total length of the NAC? It is therefore apparent that voltage drop calculations must be made.

The following examples consider 24-V fire alarm systems. The same methodology of calculating voltage drop can be applied to 12-V fire alarm systems using the appropriate corresponding values of 12-V control units and appliances.

When the control unit's primary power supply has failed and the battery capacity is at its lowest point, the voltage on the NAC must be sufficient to operate all of the notification appliances so that they deliver the proper signal intensity. UL 864, *Standard for Control Units for Fire Signaling Systems*, indicates a minimum value of 20.4 V (end of useful battery life). This value then becomes the starting point for the voltage drop calculations.

Note: Should unique power supplies be used that maintain control unit voltage higher than 20.4 V for the conditions mentioned in the previous paragraph, then consult manufacturer's instructions for the allowable voltage to be used as the starting voltage.

The particular edition of the UL Standards to which the appliance has been tested and listed (UL 1971, *Signaling Devices for the Hearing Impaired*, and UL 464, *Audible Signal Appliances*) determines the voltage range that can be used in voltage drop calculations.

For manufactured appliances that don't have to comply with the published test Standard until May 1, 2004: Notification appliances are marked with a rated (nameplate) range (e.g., 22-29 V dc). For instance, in the case of visible appliances (strobes), testing laboratories test notification appliances at 80 percent and 110 percent of their rated (nameplate) voltage, to ensure proper signal intensity and flash rate. This testing provides a reasonable level of assurance that the appliances will operate at lower voltages, which may occur when incoming ac power nears brownout conditions or when the system has been operating on battery near the end of a required 24-hour or 60-hour time period. For example, if an appliance is rated for operation between 22 V and 29 V (nameplate voltage range), testing laboratories would test the output of the notification appliance at 17.6 V and 31.9 V. This range is called the *operating range* and is different (wider range) from the nameplate range. In this example, this particular NAC must be designed and installed to provide no less than 17.6 V at any appliance in order to deliver the required light output (intensity) and flash rate. In this example, the maximum voltage drop between the NAC terminals and the last appliance must be 2.8 V or less. (20.4 starting voltage at the control unit less 17.6 V required at appliance = 2.8 V maximum voltage drop).

For appliances manufactured to the May 1, 2004, UL test requirements: Newer versions of UL standards relating to notification appliances will eliminate the 80 percent–110 percent testing that established the operating range and will instead require a standard operating voltage range for notification appliances. In the case of 24-V appliances, when calculating voltage drop, both the listed and the nameplate operating voltage range will be 16 V–33 V. Therefore, 16 V should be considered the minimum voltage that must be delivered to any appliance. (Appliances for 12-V systems will have a standard operating range of 8–17.5 V.)

It should be noted that, even though not required until May 1, 2004, some manufacturers might obtain product listing to those requirements prior to that date. It is therefore essential for care to be taken to ensure compatibility when adding new appliances to older NACs or when replacing the fire alarm control unit. The appliance manufacturer should be contacted if there is any question relating to the electrical specifications or listing of the product.

Calculation method: There are several methods of calculating the voltage drop between the control unit and the last notification appliance on the NAC. Two methods include *center load* calculations and *point-to-point* calculations. However, these methods require actual appliance current draws at the minimum operating voltage and require fairly accurate measurements of conductor length between the appliances as well as total conductor length. Because these data are generally not reliable during the design phase, the *lump sum* method is recommended and should be used because of the margin of safety it provides for unknowns.

The simplest method is to lump sum all appliance loads at the end of the circuit. In the case of the earlier UL standards, most manufacturers do not usually provide the current draw of an appliance at the minimum listed operating voltage. Therefore, the current draw at the minimum nameplate rated voltage should be used for these calculations. Ohm's law is used to calculate the voltage drop for the circuit. The relationship is as follows:

$$V_{load} = V_{terminals} - (I_{load})(R_{conductors})$$

where:

V_{load} = minimum operating voltage of the appliance

$V_{terminals}$ = 20.4 (unless otherwise specified by the manufacturer)

I_{load} = total current draw of the connected appliances

$R_{conductors}$ = total conductor resistance

Solving for $R_{conductors}$ and using Table 8 of Chapter 9 in NFPA 70, *National Electrical Code*®, the required conductor size can be determined. Table 5 provides resistance per 1000 feet for stranded conductors at 75°C (167°F). The conductor resistance at other temperatures can be calculated or obtained from the wire manufacturer's data sheet. However, the temperature used should be representative of the ambient temperature where the conductors are located. Note that the calculated resistance is the total circuit resistance of the conductors and the total length of the circuit will be half of the conductor length used for the calculation.

4-2.2 Physical Construction.

Appliances intended for use in special environments, such as, outdoors versus indoors, high or low temperatures, high humidity, dusty conditions, and hazardous locations, or where subject to tampering shall be listed for the intended application.

It is essential to maintain the operational integrity of audible and visible notification appliances, despite their possible location in relatively hostile environments. Use of appliances not listed for use in the same type of environment in which the appliance is to be placed is a violation of 1-5.1.2 of the code.

4-2.3* Mechanical Protection.

If subject to obvious mechanical damage, appliances shall be suitably protected. If guards or covers are employed, they shall be listed for use with the appliance. Their effect on the appliance's field performance shall be in accordance with the listing requirements.

The protection described in 4-2.3 is usually provided by an enclosure that protects the actual audible or visible mechanism. In the case of speakers, a mechanical baffle protects the cone from being punctured by a sharp object. See Exhibit 4.2 for examples of typical notification appliances with mechanical baffles.

Exhibit 4.3 *Notification appliance showing external mechanical protection. (Source: Safety Technology International, Inc., Waterford, MI)*

Exhibit 4.2 *Notification appliances showing mechanical baffles. (Source: Wheelock, Inc., Long Branch, NJ)*

Any guards placed over an audible or visible appliance may degrade the level of the audible signal or the light intensity of the appliance. For this reason, the guard or protective device must be tested with the specific appliance, and its effect must be measured and reported. Then, system designers, installers, and inspectors can de-rate the appliance performance and make corrections when using the appliance with the guard in a design. See Exhibit 4.3 for an example of typical notification appliance with a protective guard.

A-4-2.3 Situations exist where supplemental enclosures are necessary to protect the physical integrity of a notification appliance. Protective enclosures should not interfere with the performance characteristics of the appliance. If the enclosure degrades the performance, methods should be detailed in the installation instructions of the enclosure that clearly identify the degradation. For example, where the appliance signal is attenuated, it might be necessary to adjust the appliance spacings or appliance output.

4-2.4 Mounting.

In all cases, appliances shall be supported independently of their attachments to the circuit conductors and shall be mounted in accordance with the manufacturer's instruction.

It is not permitted to physically support the appliance by means of the conductors that connect the appliance to the

notification appliance circuit of the fire alarm system. Constant strain on terminal connections may cause conductors to pull free or to break. See Exhibit 4.4 for an example of independent support for a notification appliance.

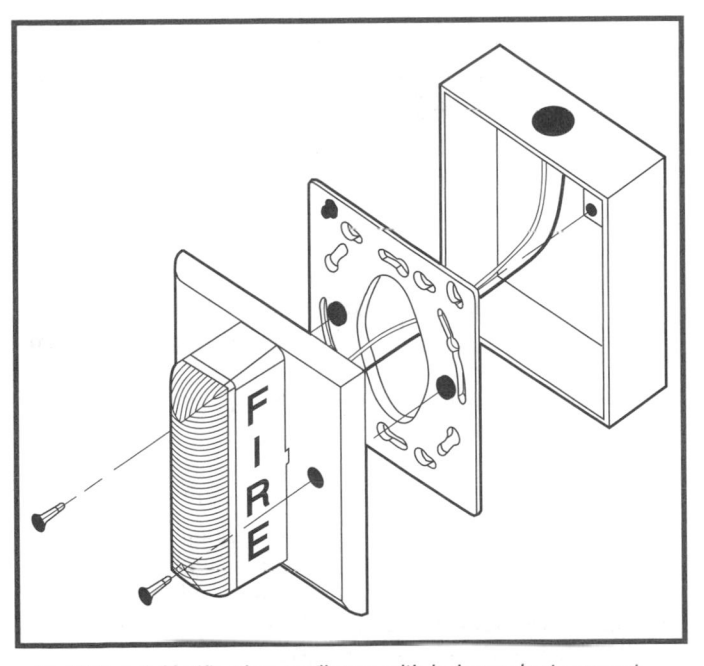

Exhibit 4.4 *Notification appliance with independent support. (Source: Gentex Corp., Zeeland, MI)*

4-2.5* Connection to the Fire Alarm System.

Terminals or leads, or their equivalent, shall be provided on each notification appliance for the express purpose of connecting into the fire alarm system to monitor the integrity of the connections.

To provide system reliability and availability, notification appliance circuits (NACs) are monitored for integrity in accordance with the requirements of Chapter 1. The appliances themselves are neither monitored nor supervised. To comply with the requirements of Chapter 1, the appliance must have the correct number and type of screw terminals or pigtail leads to permit proper connection to the circuit. See Figure A-2-1.3.4 (a). Although Figure A-2-1.3.4 (a) shows initiating devices, it is equally applicable to notification appliances. The correct type of terminals or leads, combined with correct installation practice, results in the circuit opening if a connection to an appliance is broken. This open circuit results in a trouble signal at the fire alarm control unit.

A-4-2.5 For hardwired appliances, terminals or leads, as described in 4-2.5, are necessary to ensure that the wire run is broken and that the individual connections are made to the leads or other terminals for signaling and power.

A common terminal can be used for connection of incoming and outgoing wires. However, the design and construction of the terminal should not permit an uninsulated section of a single conductor to be looped around the terminal and to serve as two separate connections. For example, a notched clamping plate under a single securing screw is acceptable only if separate conductors of a notification circuit are intended to be inserted in each notch.

4-3 Audible Characteristics

4-3.1 General Requirements.

4-3.1.1* An average ambient sound level greater than 105 dBA shall require the use of a visible signal appliance(s) in accordance with Section 4-4.

In some occupancies, the ambient sound level is so high that it would be impractical to rely solely on audible notification appliances. A drop forge shop, a large casino, a rock music dance hall, or a newspaper press room are all candidates for the addition of visible signal appliances to help ensure that the signals will be perceived by the occupants. See Exhibits 4.5 and 4.6 for examples of typical visible notification appliances.

Visible signaling is required by this code when the ambient sound pressure levels exceed 105 dBA because it is difficult, and possibly harmful, to try to overcome that level with audible fire alarm signals. In some occupancies, such

Exhibit 4.5 *Visible notification appliance. (See 4-3.1.1.) (Source: Gentex Corp., Zeeland, MI)*

Exhibit 4.6 *Visible notification appliance. (See 4-3.1.4.) (Source: Simplex Time Recorder Company, Gardner, MA)*

as theaters, it may be possible to turn off the ambient noise when the fire alarm system is activated. The value of 105 dBA was chosen because an audible signal 15 dBA above the limit set here would produce 120 dBA. The limit of 120 dBA has been set as the upper allowable limit by the Occupational, Health and Safety Act (OSHA) and the Americans with Disabilities Act (ADA).

A-4-3.1.1 The code does not require that all audible notification appliances within a building be of the same type. However, a mixture of different types of audible notification

appliances within a space is not the desired method. Audible notification appliances that convey a similar audible signal are preferred. For example, a space that uses mechanical horns and bells might not be desirable. A space that is provided with mechanical horns and electronic horns with similar audible signal output is preferred.

However, the cost of replacing all existing appliances to match new appliances can impose substantial economic impact where other methods can be used to avoid occupant confusion of signals and signal content. Examples of other methods used to avoid confusion include, but are not limited to, training of occupants, signage, consistent use of temporal code signal pattern, and fire drills.

4-3.1.2 The total sound pressure level produced by combining the ambient sound pressure level with all audible signaling appliances operating shall not exceed 120 dBA anywhere in the occupied area.

4-3.1.3 Sound within the occupied area, from a temporary or abnormal source, shall not be required to be included in measuring maximum ambient sound level.

Temporary sound sources, such as construction noise, are not required to be considered when measuring the maximum ambient sound pressure level. However, whether or not a sound source is "not normally found continuously" in a space must be carefully determined. In a college dormitory room, for example, a student's stereo may be portable, yet because it is usually part of the permanent equipment in the room, or a part of the normal occupancy, many authorities having jurisdiction do not believe it should be considered a temporary sound source.

4-3.1.4* Mechanical Equipment Rooms. If audible appliances are installed in mechanical equipment rooms, the average ambient sound level used for design guidance shall be at least 85 dBA for all occupancies.

The required value of 85 dBA is only a minimum value. Some mechanical equipment rooms may have an average ambient sound level that exceeds this value. Based on the average ambient sound level of 85 dBA, the audible notification appliances would need to deliver 100 dBA throughout the room. See Exhibit 4.7 for an example of a typical audible notification appliance for high ambient noise areas.

A-4-3.1.4 In determining maximum ambient sound levels, it is not necessary to include temporary or abnormal sources. For example, in a typical office environment, sound sources that should be considered include air-handling equipment, office cleaning equipment (vacuum cleaners), and background music. Examples of temporary or abnormal sound sources that can be excluded would be sound from internal or external construction activities, that is, office rearrangements and construction equipment.

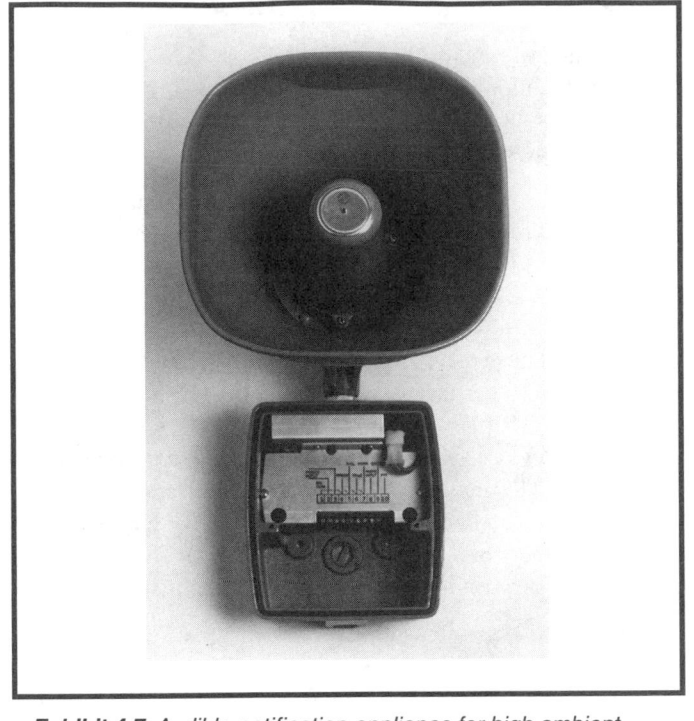

Exhibit 4.7 *Audible notification appliance for high ambient noise areas. (See 4-3.1.4.) (Source: Audiosone Corp., Stratford, CT)*

4-3.1.5* Emergency voice/alarm communications systems shall be capable of the reproduction of prerecorded, synthesized, or live (for example, microphone, telephone handset, and radio) messages with voice intelligibility.

Although additional requirements for textual audible appliances are covered in Section 4-7, paragraph 4-3.1.5 can be used by an authority having jurisdiction to enforce the intelligibility of textual signals. If a textual audible notification appliance produces a signal of adequate sound level, but the message is not intelligible, then such a signal is not adequate.

Although the code text requires voice reproduction to be intelligible, the technical definition of intelligibility was placed in the appendix. An intelligible system is one that has a measured value equivalent to a Common Intelligibility Scale (CIS) value of 0.70 or greater. Similarly, Chapter 7 includes reference to the intelligibility measurement methods, but does not include any requirement for the measurement to be done as a part of the required initial or periodic testing. The omission of a measurement requirement permits owners, designers, and authority having jurisdiction to decide if measurement of intelligibility is desired or required for a particular system. Many occupancies do not require intelligibility testing. However, some occupancies such as atria, gymnasiums, ice rinks, and auditoriums usually require intelligibility testing.

A-4-3.1.5 Voice intelligibility should be measured in accordance with the guidelines in Annex A of IEC 60849, Second Edition: 1998, *Sound Systems for Emergency Purposes*. When tested in accordance with Annex B, Clause B1, of IEC 60849, the system should exceed the equivalent of a common intelligibility scale (CIS) score of 0.70. Intelligibility is achieved when the quantity I_{av-s}, as specified in B3 of IEC 60849, exceeds this value. I_s is the arithmetical average of the measured intelligibility values on the CIS and σ (sigma) is the standard deviation of the results.

Objective means of determining intelligibility are found in IEC 60268, Part 16, Second Edition: 1998, *The Objective Rating of Speech Intelligibility by Speech Transmission Index*. Subject-based techniques for measuring intelligibility are defined by ANSI S3.2-1989, *Method for Measuring the Intelligibility of Speech Over Communications Systems*. ANSI S3.2-1989 should be considered an acceptable alternative to ISO TR 4870, where referenced in IEC 60268, Part 16, Second Edition: 1998, *The Objective Rating of Speech Intelligibility by Speech Transmission Index*.

4-3.2* Public Mode Audible Requirements.

A-4-3.2 The typical average ambient sound level for the occupancies specified in Table A-4-3.2 are intended only for design guidance purposes.

The typical average ambient sound levels specified should not be used in lieu of actual sound level measurements.

Table A-4-3.2 Average Ambient Sound Level According to Location

Location	Average Ambient Sound Level (dBA)
Business occupancies	55
Educational occupancies	45
Industrial occupancies	80
Institutional occupancies	50
Mercantile occupancies	
Piers and water-surrounded structures	40
Places of assembly	55
Residential occupancies	35
Storage occupancies	30
Thoroughfares, high density urban	70
Thoroughfares, medium density urban	55
Thoroughfares, rural and suburban	40
Tower occupancies	35
Underground structures and windowless buildings	40
Vehicles and vessels	50

4-3.2.1 Audible notification appliances intended for operation in the public mode shall have a sound level of not less than 75 dBA at 10 ft (3 m) or more than 120 dBA at the minimum hearing distance from the audible appliance.

Paragraph 4-3.2.1 is an equipment rating requirement, rather than an installation requirement. Audible appliance ratings, as measured by the manufacturer and the qualified testing laboratories, are provided a decibel rating at a predetermined distance, usually 10 ft (3 m). The rule of thumb is that the output of an audible notification appliance is reduced by 6 dBA if the distance between the appliance and the listener is doubled. The accuracy of this rule of thumb depends on many intervening variables, particularly the acoustic properties of the materials in the listening space, such as ceiling materials and floor and wall coverings.

The use of the appliance's rating along with this rule allows system designers to estimate audible levels in occupied spaces before a system is installed. See Exhibit 4.8 for an example of how this rule of thumb is applied.

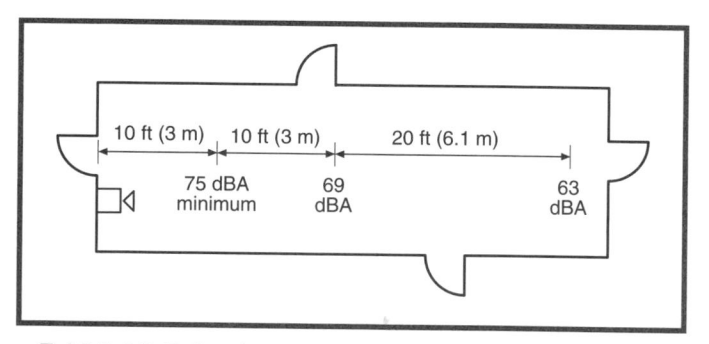

Exhibit 4.8 *Estimating audible levels using the 6 dBA rule of thumb method. (Source: R.P. Schifiliti Associates, Inc., Reading, MA)*

More complex situations require calculating sound attenuation through doors and walls. See the SFPE *Handbook of Fire Protection Engineering* for appropriate calculation methods.

4-3.2.2* To ensure that audible public mode signals are clearly heard, they shall have a sound level at least 15 dBA above the average ambient sound level or 5 dBA above the maximum sound level having a duration of at least 60 seconds, whichever is greater, measured 5 ft (1.5 m) above the floor in the occupiable area.

Exception 1: Audible alarm notification appliances installed in elevator cars shall be permitted to use the audibility criteria for private mode appliances detailed in 4-3.3.2.

Exception 2: If approved by the authority having jurisdiction, audible alarm notification appliances installed in

restrooms shall be permitted to use the audibility criteria for private mode appliances detailed in 4-3.3.2.

Exception 3: If permitted by the authority having jurisdiction, a fire alarm system arranged to stop or reduce ambient noise shall be permitted to produce a sound level at least 15 dBA above the reduced average ambient sound level or 5 dBA above the maximum sound level having a duration of at least 60 seconds after reduction of the ambient noise level, whichever is greater, measured 5 ft (1.5 m) above the floor in the occupiable area. Visible notification appliances shall be installed in the affected areas in accordance with Sections 4-4 or 4-5. Relays, circuits, or interfaces necessary to stop or reduce ambient noise shall meet the requirements of Chapters 1 and 3.

Most authorities having jurisdiction will measure the sound level of audible notification appliances to ensure that it is at least 5 dBA above the maximum sound level. It is more difficult to determine the average ambient sound level than it is to determine the maximum sound level that lasts at least 60 seconds. Care must be exercised in selecting the source of the maximum sound level for each occupancy.

The measurement is made at 5 ft (1.5 m) above the floor to reduce the effects of walls and surfaces on the signal level.

Additional appliances may be required so that the signal will be clearly heard throughout the occupiable area. These measurements are very important, especially when planning to use the building fire alarm signal to warn the occupants.

Where acceptable to the authority having jurisdiction, reducing the background noise is a viable alternative to providing a fire alarm system with a high level of audio output. However, care must be exercised to ensure that the shutdown mechanism is reliable, it will not damage the equipment being shut down, and it is safe.

A-4-3.2.2 The constantly changing nature of pressure waves, which are detected by ear, can be measured by electronic sound meters, and the resulting electronic waveforms can be processed and presented in a number of meaningful ways.

Most simple sound level meters quickly average a sound signal and present a root mean square (RMS) level to the meter movement or display. However, this quick average of impressed sound results in fast movements of the meter's output that are best sent when talking into the microphone; the meter quickly rises and falls with speech. However, when surveying the ambient sound levels to establish the increased level at which a notification appliance will properly function, the sound source needs to be averaged over a longer period of time. Moderately priced sound level meters have such a function, usually called Leq or "equivalent sound level." For example, an Leq of speech in a quiet room would cause the meter movement to rise gradually to a peak reading and slowly fall well after the speech is over.

Leq readings can be misapplied in situations where the background ambient noises vary greatly during a 24-hour period. Leq measurements should be taken over the period of occupancy.

In areas where the background noise is generated by machinery and is fairly constant, a frequency analysis can be warranted. It might be found that the high sound levels are predominately in one or two frequency bandwidths—often lower frequencies. Fire alarm notification appliances producing sound in one or two other frequency bandwidths can adequately penetrate the background noise and provide notification. The system would still be designed to produce or have a sound level at the particular frequency or frequency bandwidth of at least 15 dB above the average ambient sound level or 5 dB above the maximum sound level having a duration of at least 60 seconds, whichever is greater.

In very high noise areas, such as theaters, dance halls, nightclubs, and machine shops, sound levels during occupied times can be 100 dBA and higher. Peak sounds might be 110 dBA or greater. At other occupied times, the sound level might be below 50 dBA. A system designed to have a sound level of at least 15 dBA above the average ambient sound level or 5 dBA above the maximum sound level having a duration of at least 60 seconds might result in a required fire alarm level in excess of the maximum of 115 dBA. A viable option is to reduce or eliminate the background noise. Professional theaters or other entertainment venues can have road show connection panels *(refer to NFPA 70, National Electrical Code, Section 520-50)* for troupes to connect their light and sound systems to. These power sources can be controlled by the fire alarm system. In less formal applications, such as many nightclubs, designated power circuits could be controlled. Diligence needs to be exercised to ensure that the controlled circuits are used.

Also, in occupancies such as machine shops or other production facilities, care must be exercised in the design to ensure that the removal of power to the noise source does not create some other hazard. As with other fire safety functions, control circuits and relays would be monitored for integrity in accordance with Chapters 1 and 3.

Appropriate audible signaling in high ambient noise areas is often difficult. Areas such as automotive assembly areas, machining areas, paint spray areas, and so on, where the ambient noise is caused by the manufacturing process itself require special consideration. Adding additional audible notification appliances that merely contribute to the already noisy environment might not be appropriate. Other alerting techniques such as visible notification appliances, for example, could be more effectively used.

These sound levels, previously known as *equivalent sound levels,* are defined as the root mean square A-weighted sound pressure level measured over a 24-hour period. This value can be experimentally determined for a particular occupancy by using a recording sound pressure level meter. The technician taking the measurement obtains the mean square result for the area under the curve by using integral

calculus. The levels listed in A-4-3.2 were derived from acoustic reference literature and were originally based on this kind of calculation. See also 4-3.1.3 and 4-3.1.4.

It should be noted that measurements taken by a major manufacturer of fire alarm equipment in a large sampling of hotel rooms with through-the-wall air-conditioning units determined that the average ambient sound level in those rooms, with the air conditioners operating, was 55 dBA.

4-3.3 Private Mode Audible Requirements.

4-3.3.1 Private Mode. Audible notification appliances intended for operation in the private mode shall have a sound level of not less than 45 dBA at 10 ft (3 m) or more than 120 dBA at the minimum hearing distance from the audible appliance.

4-3.3.2 To ensure that audible private mode signals are clearly heard, they shall have a sound level at least 10 dBA above the average ambient sound level or 5 dBA above the maximum sound level having a duration of at least 60 seconds, whichever is greater, measured 5 ft (1.5 m) above the floor in the occupiable area.

A hospital patient care area is one example of where a code or authority having jurisdiction may permit private mode signaling. The public occupants include patients who may not be able to respond to a fire alarm signal. In some cases, it may even be dangerous to alert them directly with audible (and possibly visible) signals. For this reason, the system is designed to alert trained staff.

Areas that use private mode signaling (such as in a hospital) often have a less intense average ambient sound level and a lower maximum sound level, making the reduced level cited in 4-3.3 appropriate. In delivering private mode signals it is important that the sound level of the audible notification appliance be adequate, but not so loud as to startle the occupants.

Lower audible levels are permitted because part of the staff's job is to listen for, and respond appropriately to the fire alarm signals. In addition, they must communicate among themselves to be able to implement their emergency procedures; a louder alarm might interfere with this communication. See also Supplement 4.

In a few cases, such as operating rooms or critical care patient areas, other codes and authorities having jurisdiction may permit elimination of audible signaling altogether.

4-3.4 Sleeping Areas.

Where audible appliances are installed to provide signals for sleeping areas, they shall have a sound level of at least 15 dBA above the average ambient sound level or 5 dBA above the maximum sound level having a duration of at least 60 seconds or a sound level of at least 70 dBA, whichever is

greater, measured at the pillow level in the occupiable area. If any barrier, such as a door, curtain, or retractable partition, is located between the notification appliance and the pillow, the sound pressure level shall be measured with the barrier placed between the appliance and the pillow.

Subsection 4-3.4 requires that, in rooms where people sleep, the sound level delivered by the audible notification appliance must be 15 dBA above the average ambient sound level, or 5 dBA above any peak sound level lasting 60 seconds or more, or at least 70 dBA. If the average ambient sound level in the sleeping area is 40 dBA, then the audible notification appliances must deliver at least 70 dBA (40 + 15 < 70). If the average ambient sound level in the sleeping area is 60 dBA, then the audible notification appliances must deliver at least 75 dBA (60 + 15 = 75 > 70).

It should be noted that 70 dBA is a minimum requirement of the code. Some studies (Schifiliti 1988; Butler 1981; Myles 1979) suggest a minimum of 75 dBA. These levels are for people without any hearing impairments and without any incapacitation due to drugs, alcohol, or exhaustion. Also, a certain sound pressure level does not instantly awaken all test subjects. There is a distribution of time to alert some or all of the occupants. Obviously, as the sound level increases, the time to alert the majority of people decreases. The time it takes to awaken someone and the time it takes for them to act must be considered by designers with respect to the development of hazardous conditions.

The requirement to measure the SPL where barriers exist is new to the 1999 code.

4-3.5 Location of Audible Notification Appliances.

4-3.5.1 If ceiling heights allow, wall-mounted appliances shall have their tops above the finished floors at heights of not less than 90 in. (2.30 m) and below the finished ceilings at heights of not less than 6 in. (152 mm). This requirement shall not preclude ceiling-mounted or recessed appliances.

Exception: Different mounting heights shall be permitted by the authority having jurisdiction provided the sound pressure level requirements of 4-3.2 and 4-3.3 are met.

The purpose of mounting height requirements for audible appliances is to prevent common furnishings from blocking appliances. However, the required sound pressure levels (4-3.2, 4-3.3, and 4-3.4) are performance requirements. Thus, the exception permits designers to place audible appliances at other mounting heights as long as their designs ultimately provide the required sound pressure level. Ultimately, the design must pass the testing requirements of Chapter 7. It is important to remember that the appliances must also be accessible for repair and maintenance. This exception applies only to audible appliances, not to visible or combination appliances.

4-3.5.2 If combination audible/visible appliances are installed, the location of the installed appliance shall be determined by the requirements of 4-4.4.

Exception: Where the combination audible/visible appliance serves as an integral part of a smoke detector, the mounting location shall be in accordance with Chapter 8.

Paragraph 4-3.5.2 requires that the location of a combination audible/visible notification appliance complies with the requirements of 4-4.4. The entire lens of the visible appliance must be no less than 6 ft 8 in. (2.03 m) above the floor, or greater than 8 ft (2.44 m) above the floor. This height limitation is intended to keep visible notification appliances in the same configuration in which they were tested at the laboratory. For additional information relating to the exception, see 4-4.4.3.1, which specifies that smoke detectors in sleeping areas must be installed in accordance with Chapter 2 and Chapter 8.

4-4* Visible Characteristics, Public Mode

Following passage of the Americans with Disabilities Act (ADA), there was a great deal of debate about visible signaling requirements. In the past, the *National Fire Alarm Code* has differed from the ADA and from other accessibility standards such as ANSI 117.1-1998, *Accessible and Usable Buildings and Facilities*. The fire alarm industry has worked with the various code and advocacy groups to develop reasonable, safe, and effective visible notification requirements.

The requirements contained in the 1999 edition of the *National Fire Alarm Code* have been accepted as "equivalent facilitation" (and in some cases superior) to the original ADA requirements. The ADA *Accessibility Guidelines* (ADAAG) and ANSI 117.1 are being revised, and it is expected that they will adopt visible signaling performance and location requirements as specified in Chapter 4.

A-4-4 The mounting height of the appliances affects the distribution pattern and level of illumination produced by an appliance on adjacent surfaces. It is this pattern, or effect, that provides occupant notification by visible appliances. If mounted too high, the pattern is larger, but at a lower level of illumination (measured in lumens per square foot or footcandles). If mounted too low, the illumination is greater (brighter), but the pattern is smaller and might not overlap correctly with that of adjacent appliances.

A qualified designer could choose to present calculations to an authority having jurisdiction showing that it is possible to use a mounting height greater than 96 in. or less than 80 in. provided an equivalent level of illumination is achieved on the adjacent surfaces. This can be accomplished by using listed higher intensity appliances or closer spacing, or both.

Engineering calculations should be prepared by qualified persons and should be submitted to the authority having jurisdiction showing how the proposed variation achieves the same or greater level of illumination provided by the prescriptive requirements of Section 4-4.

The calculations require knowledge of calculation methods for high intensity strobes. In addition, the calculations require knowledge of the test standards used to evaluate and list the appliance.

4-4.1* There are two methods of visible signaling. These are methods in which the message of notification of an emergency condition is conveyed by direct viewing of the illuminating appliance or by means of illumination of the surrounding area. Public mode visible signaling shall meet the requirements of Section 4-4 using visible notification appliances.

A-4-4.1 One method of determining compliance with Section 4-4 is that the product be listed in accordance with UL 1971, *Standard for Safety Signaling Devices for the Hearing Impaired*.

4-4.2 Light Pulse Characteristics.

The flash rate shall not exceed two flashes per second (2 Hz) nor be less than one flash every second (1 Hz) throughout the listed voltage range of the appliance.

4-4.2.1 A maximum pulse duration shall be 0.2 seconds with a maximum duty cycle of 40 percent. The pulse duration shall be defined as the time interval between initial and final points of 10 percent of maximum signal.

The light intensity of a pulsed source may be graphed as a bell-shaped curve. The duration of the pulse is measured beginning at the point where the upward side of the curve exceeds 10 percent of the maximum intensity to the point where the downward side of the curve drops below 10 percent of the maximum intensity. See Exhibit 4.9 for a graph showing these phenomena.

Exhibit 4.9 *Peak versus effective intensity. (Source: R.P. Schifiliti Associates, Inc., Reading, MA)*

4-4.2.2* The light source color shall be clear or nominal white and shall not exceed 1000 cd (effective intensity).

Source intensity is a measure of the light output of the appliance. The unit of measure is the candela (cd). (This unit was formerly called candlepower. There is a one-to-one relationship between candela and candlepower.) As you move away from any light source, its illumination decreases. Illumination is measured in units of lumens (lm) per square meter (also called lux), or lumens per square foot. (Formerly, the unit used to describe illumination was the foot-candle. One foot-candle equals one lumen per square foot. One lumen per square meter equals 0.926 foot-candles.) See Exhibit 4.10 for graphic definitions of these terms and a mathematical relationship showing their use.

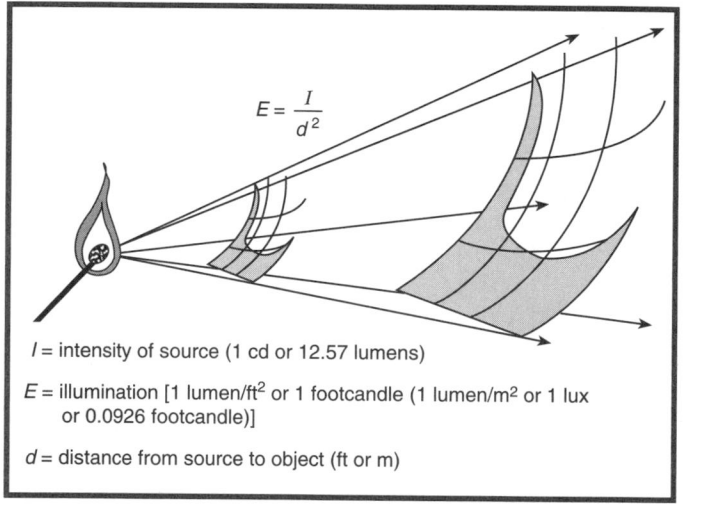

$$E = \frac{I}{d^2}$$

I = intensity of source (1 cd or 12.57 lumens)

E = illumination [1 lumen/ft^2 or 1 footcandle (1 lumen/m^2 or 1 lux or 0.0926 footcandle)]

d = distance from source to object (ft or m)

Exhibit 4.10 *Definitions of light source, intensity, and illumination. (Source: R.P. Schifiliti Associates, Inc., Reading, MA)*

A-4-4.2.2 Effective intensity is the conventional method of equating the brightness of a flashing light to that of a steady-burning light as seen by a human observer. The units of effective intensity are expressed in candelas (or candlepower, which is equivalent to candelas). For example, a flashing light that has an effective intensity of 15 cd has the same apparent brightness to an observer as a 15-cd steady-burning light source.

Measurement of effective intensity is usually done in a laboratory using specialized photometric equipment. Accurate field measurement of effective intensity is not practical. Other units of measure for the intensity of flashing lights, such as peak candela or flash energy, do not correlate directly to effective intensity and are not used in this standard.

Because strobe lights flash very briefly, the perceived brightness can vary depending on the actual peak source strength and duration of the flash. One appliance might reach a peak intensity of 1000 cd in 0.1 seconds, whereas another might

reach 750 cd in 0.2 seconds. Nevertheless, the human eye might perceive both as being equally bright. A mathematical relationship is used to relate the perceived brightness of a strobe light to that of a constantly burning light. The result is called the effective intensity (candela, effective, or cd, eff.). Exhibit 4.9 shows this relationship.

4-4.3* Appliance Photometrics.

A-4-4.3 The prescriptive requirements of Section 4-4 assume the use of appliances having very specific characteristics of light color, intensity, distribution, and so on. The appliance and application requirements are based on extensive research. However, the research was limited to typical residential and commercial applications such as school classrooms, offices, hallways, and hotel rooms. While these specific appliances and applications will likely work in other spaces, their use might not be the most effective solution and might not be as reliable as other visible notification methods.

For example, in large warehouse spaces and large distribution spaces such as super stores, it is possible to provide visible signaling using the appliances and applications of this chapter. However, mounting strobe lights at a height of 80 in. to 96 in. along aisles with rack storage subjects the lights to frequent mechanical damage by fork lift trucks and stock. Also, the number of appliances required would be very high. It might be possible to use other appliances and applications not specifically addressed by this chapter at this time. Alternative applications must be carefully engineered for reliability and function and would require permission of the authority having jurisdiction.

Visible notification using the methods contained in 4-4.4.1 is achieved by indirect signaling. This means the viewer need not actually see the appliance, just the effect of the appliance. This can be achieved by producing minimum illumination on surfaces near the appliance such as the floor, walls, and desks. There must be a sufficient change in illumination to be noticeable. The tables and charts in Section 4-4 specify a certain candela effective light intensity for certain size spaces. The data were based on extensive research and testing. Appliances do not typically produce the same light intensity when measured off-axis. To ensure that the appliance produces the desired illumination (effect), it must have some distribution of light intensity to the areas surrounding the appliance. UL 1971, *Standard for Safety Signaling Devices for the Hearing Impaired,* specifies the distribution of light shown to provide effective notification by indirect visible signaling.

4-4.3.1 Visible notification appliances used in the public mode shall be located and shall be of a type, size, intensity, and number so that the operating effect of the appliance is seen by the intended viewers regardless of the viewer's orientation.

In the same manner that signals produced by audible notification appliances must be clearly heard, the signals produced by visible notification appliances must be clearly seen without regard to the viewer's position within the protected area. This requirement does not mean that an appliance must be seen from any location in a space, but rather that the operating effect must be seen. An example that would meet this requirement is a single strobe in an L-shaped area. If properly located and sized, the visible appliance may not be seen in all parts of the room, but the flash from the appliance will be seen.

4-4.3.2 The light output shall comply with the polar dispersion requirements of UL 1971, *Standard for Safety Signaling Devices for the Hearing Impaired,* or equivalent.

The polar distribution characteristics of the appliance are very important for compliance with Chapter 4, because the effectiveness of visible signaling is based on tests where the viewers responded to the illumination of their surroundings. Thus, it is important that the appliance produces a pattern of light on adjacent surfaces such as walls, floors, desks, and so forth. Appliances listed to standards other than UL and not having specified polar distribution requirements can produce most of their light on axis and very little down or off to the side. Thus, they may not produce a noticeable pattern sufficient to alert occupants.

4-4.4 Appliance Location.

Wall-mounted appliances shall be mounted such that the entire lens is not less than 80 in. (2.03 m) and not greater than 96 in. (2.43 m) above the finished floor.

In rooms with sufficient ceiling height, the entire lens of the visible appliance must be at least 6 ft 8 in. (2.03 m) above the floor, but not more than 8 ft (2.43 m) above the floor.

The minimum mounting height is intended to locate appliances so they are not blocked by common furnishings or equipment. The maximum mounting height is important because the illumination from a visible appliance reduces drastically with distance and angle from a horizontal plane through the appliance. Proof of this can be determined by the mathematical relationship in Exhibit 4.10. For this reason, wall-mounted appliances are limited to 96 in. (2.43 m) above the floor. The maximum mounting height is also important because the appliances are tested at this height by qualified testing laboratories. Ceiling mounting is permitted; however, the appliances must be specifically listed for ceiling mounting. See also A-4-4.

4-4.4.1* Spacing in Rooms.

A-4-4.4.1 Areas large enough to exceed the rectangular dimensions given in Figures A-4-4.4.1(a), (b), and (c) require additional appliances. Often, proper placement of appliances can be facilitated by breaking down the area into multiple squares and dimensions that fit most appropriately

[refer to Figures A-4-4.4.1(a), (b), (c), and (d)]. An area that is 40 ft (12.2 m) wide and 74 ft (22.6 m) long can be covered with two 60-cd appliances. Irregular areas and areas with dividers or partitions need more careful planning to make certain that at least one 15-cd appliance is installed for each 20-ft × 20-ft (6.09-m × 6.09-m) area and that light from the appliance is not blocked.

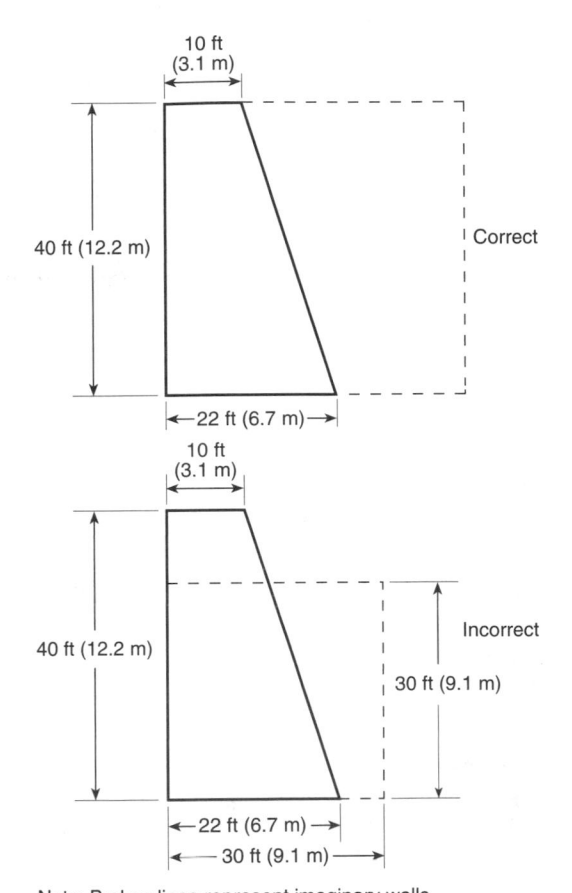

Note: Broken lines represent imaginary walls.

Figure A-4-4.4.1(a) *Irregular area spacing.*

These figures were added to avoid misinterpretation of the text. Figure A-4-4.4.1(a) demonstrates how a non-square or nonrectangular room can be fitted into the spacing allocation of Tables 4-4.4.1(a) and (b). Figure A-4-4.4.1(b) demonstrates how to divide a room or area into smaller areas to enable the use of lower intensity lights. Figures A-4-4.4.1(c) and (d) show the correct and incorrect placement of multiple visible notification appliances in a room.

4-4.4.1.1 Spacing shall be in accordance with Tables 4-4.4.1.1(a) and (b) and Figure 4-4.4.1.1.

Visible notification appliances shall be installed in accordance with Table 4-4.4.1.1(a), using one of the following:

(1)*A single visible notification appliance.

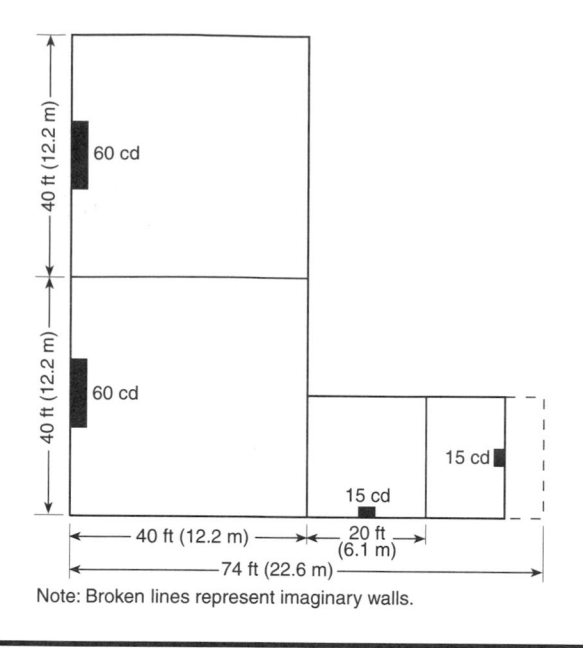

Figure A-4-4.4.1(b) *Spacing of wall-mounted visible appliances in rooms.*

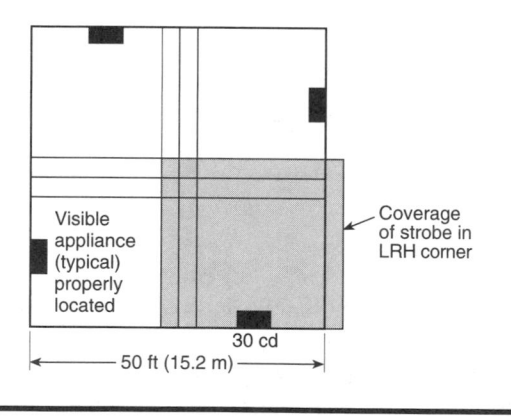

Figure A-4-4.4.1(c) *Room spacing allocation—correct.*

A-4-4.4.1.1(1) A design that delivers a minimum illumination of 0.0375 lumens/ft^2 (footcandles) or 0.4037 lumens/m^2 (lux) to all occupiable spaces where visible notification is required is considered to meet the minimum light intensity requirements of 4-4.4.1.1(1). This level of illumination has been shown to alert people by indirect viewing (reflected light) in a large variety of rooms with a wide range of ambient lighting conditions.

The illumination from a visible notification appliance at a particular distance is equal to the effective intensity of the appliance divided by the distance squared (the inverse square law). Tables 4-4.4.1.1(a) and (b) are based on applying the inverse square law to provide an illumination of at least 0.0375 lumens/ft^2 throughout each room size. For

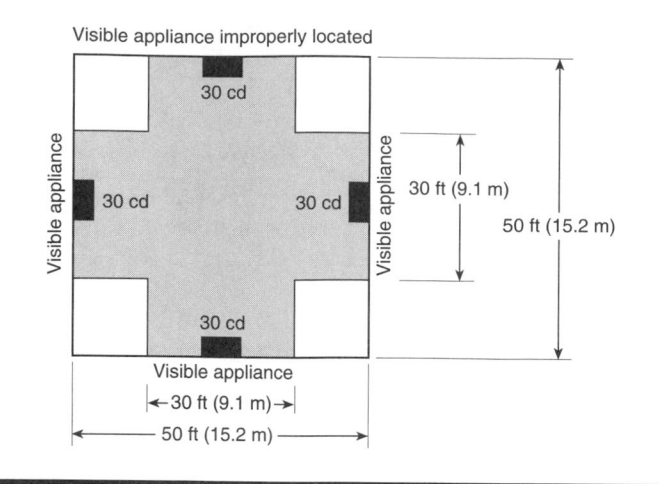

Figure A-4-4.4.1(d) *Room spacing allocation—incorrect.*

example, a 60-cd effective intensity appliance in a 40-ft × 40-ft room produces 0.0375 lumens/ft^2 on the opposite wall 40 ft away (60 ÷ 40 ft^2). This same 60-cd effective intensity appliance produces 0.0375 lumens/ft^2 on the adjacent wall 20 ft away (60 × 25% ÷ 20 ft^2) where the minimum light output of the appliance at 90 degrees off-axis is 25 percent of rated output per ANSI/UL 1971, *Standard for Safety Signaling Devices for the Hearing Impaired.* Similarly, a 110-cd strobe will produce at least 0.0375 lumens/ft^2 in a 54-ft × 54-ft room. Calculated intensities in Tables 4-4.4.1.1(a) and (b) have been adjusted to standardize the intensity options of presently available products and take into account additional reflections in room corners and higher direct viewing probability when there is more than one appliance in a room.

The application of visible notification appliances in outdoor areas has not been tested and is not addressed in this standard. Visible appliances that are mounted outdoors should be listed for outdoor use (under ANSI/UL 1638, *Visual Signaling Appliances—Private Mode Emergency and General Utility Signaling,* for example) and should be located for direct viewing because reflected light will usually be greatly reduced.

(2) Two visible notification appliances located on opposite walls.

(3)*More than two appliances in any field of view, spaced a minimum of 55 ft (16.76 m) from each other in rooms 80 ft × 80 ft (24.4 m × 24.4 m) or greater.

A-4-4.4.1.1(3) The field of view is based on the focusing capability of the human eye specified as 120 degrees in the Illuminating Engineering Society (IES) *Lighting Handbook Reference and Application.* The apex of this angle is the viewer's eye. In order to ensure compliance with the requirements of 4-4.4.1.1, this angle should be increased to approximately 135 degrees.

Table 4-4.4.1.1(a) Room Spacing for Wall-Mounted Visible Appliances

Maximum Room Size		Minimum Required Light Output (Effective Intensity) (cd)		
ft	m	One Light per Room	Two Lights per Room (Located on Opposite Wall)	Four Lights per Room; One Light per Wall
20 × 20	6.1 × 6.1	15	NA	NA
30 × 30	9.14 × 9.14	30	15	NA
40 × 40	12.2 × 12.2	60	30	15
50 × 50	15.2 × 15.2	95	60	30
60 × 60	18.3 × 18.3	135	95	30
70 × 70	21.3 × 21.3	185	95	60
80 × 80	24.4 × 24.4	240	135	60
90 × 90	27.4 × 27.4	305	185	95
100 × 100	30.5 × 30.5	375	240	95
110 × 110	33.5 × 33.5	455	240	135
120 × 120	36.6 × 36.6	540	305	135
130 × 130	39.6 × 39.6	635	375	185

NA: Not allowable.

Table 4-4.4.1.1(b) Room Spacing for Ceiling-Mounted Visible Appliances

Maximum Room Size		Maximum Ceiling Height		Minimum Required Light Output (Effective Intensity); One Light (cd)
ft	m	ft	m	
20 × 20	6.1 × 6.1	10	3.05	15
30 × 30	9.14 × 9.14	10	3.05	30
40 × 40	12.2 × 12.2	10	3.05	60
50 × 50	15.2 × 15.2	10	3.05	95
20 × 20	6.1 × 6.1	20	6.1	30
30 × 30	9.14 × 9.14	20	6.1	45
40 × 40	12.2 × 12.2	20	6.1	80
50 × 50	15.2 × 15.2	20	6.1	115
20 × 20	6.1 × 6.1	30	9.14	55
30 × 30	9.14 × 9.14	30	9.14	75
40 × 40	12.2 × 12.2	30	9.14	115
50 × 50	15.2 × 15.2	30	9.14	150

Testing has shown that high flash rates of high intensity strobe lights can pose a potential risk of seizure to people with photosensitive epilepsy. To reduce this risk, more than two visible appliances are not permitted in any field of view unless they are separated by at least 55 ft (16.8 m) or unless their flashes are synchronized.

(4) More than two visible notification appliances in the same room or adjacent space within the field of view that flash in synchronization. This requirement shall not preclude synchronization of appliances that are not within the same field of view.

In 1996, the code was modified to reduce the chances that strobe lights would induce seizures in persons with photosensitive epilepsy. The flash rate has been adjusted so that one or even two appliances not flashing in unison cannot produce a flash rate that is considered dangerous. If more than two appliances can be viewed at the same time, they must either be synchronized or located far enough apart so that their intensity at the viewer's location is low enough to be considered safe.

Prior to the 1999 edition, Table 4-4.4.1.1(a) did not permit more than two strobes in a room until the room was at least 80 ft × 80 ft (25.5 m × 25.5 m), even if they were synchronized. The 1999 edition permits the use of more than two strobes in smaller spaces, provided they are synchronized.

Visible signaling is a very complex topic. For this reason, the code presents prescriptive requirements rather than the performance requirements, such as those for audible signaling. In essence, the code provides preset designs that can be used for a variety of actual field conditions requiring these devices. The prescriptive requirements contained in the code are based, in part, on extensive tests performed by Underwriters Laboratories Inc. in developing UL 1971.

4-4.4.1.2 Room spacing for wall-mounted appliances shall be based on locating the visible notification appliance at the halfway distance of the longest wall. In square rooms with appliances not centered or nonsquare rooms, the effective intensity (cd) from one visible notification appliance shall be determined by maximum room size dimensions obtained either by measuring the distance to the farthest wall or by doubling the distance to the farthest adjacent wall, whichever is greater, as required by Table 4-4.4.1.1(a) and Figure 4-4.4.1.1.

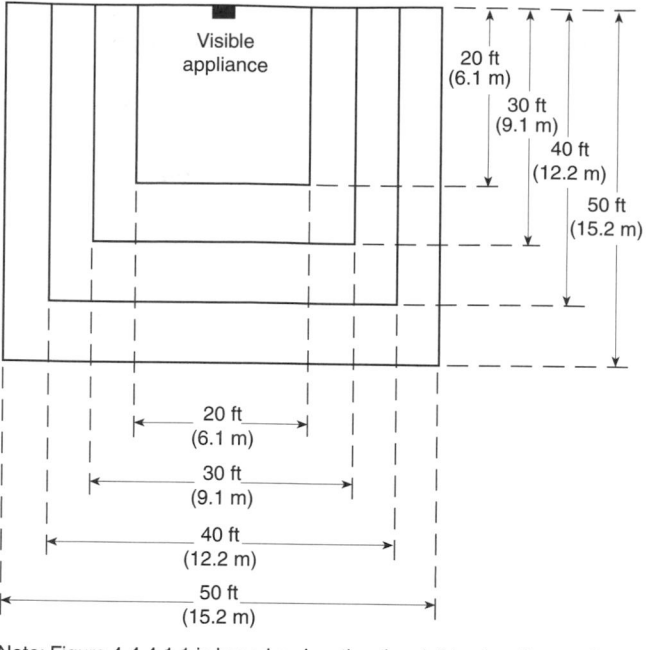

Note: Figure 4-4.4.1.1 is based on locating the visible signaling appliance at the halfway distance of the longest wall. In square rooms with appliances not centered or nonsquare rooms, the effective intensity (cd) from one visible signaling appliance shall be determined by maximum room size dimensions obtained either by measuring the distance to the farthest wall or by doubling the distance to the farthest adjacent wall, whichever is greater, as shown in Table 4-4.4.1.1(a).

Figure 4-4.4.1.1 Room spacing for wall-mounted visible appliances.

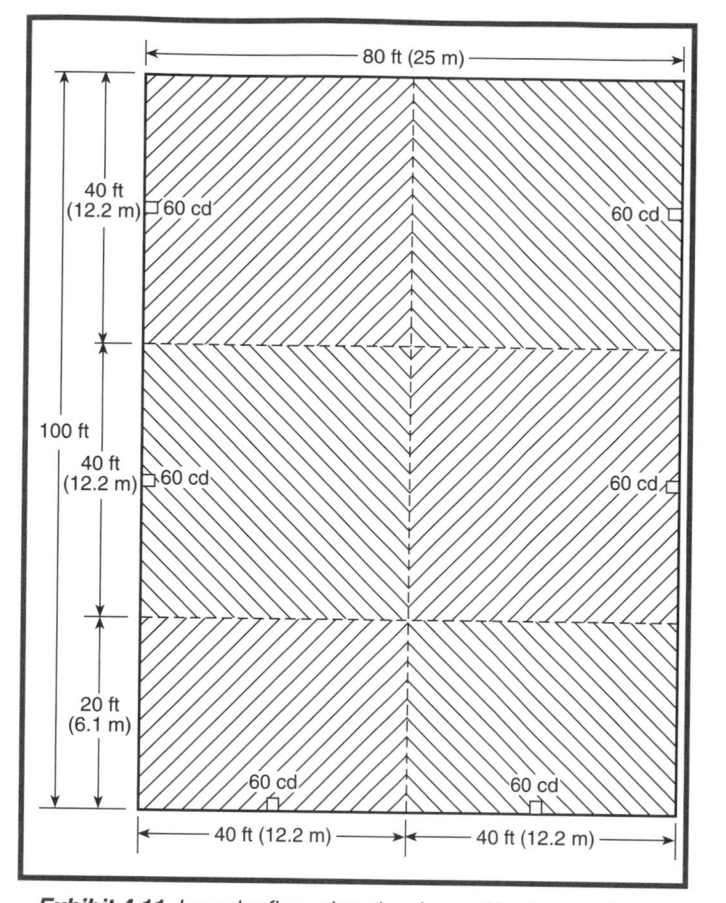

Exhibit 4.11 Irregular floor plan showing notification appliances for required locations.

Figure 4-4.4.1.1 and Tables 4-4.4.1.1(a) and (b) help to ensure that a sufficient number of properly sized visible notification appliances are installed in each protected space to provide complete coverage. The key to proper coverage in irregular spaces is to divide the space into a series of squares and provide proper coverage for each square as if it were an independent space. Exhibit 4.11 illustrates this concept. Synchronization may be required per 4-4.4.1.1.

4-4.4.1.3 If a room configuration is not square, the square room size that allows the entire room to be encompassed or allows the room to be subdivided into multiple squares shall be used.

4-4.4.1.4 If ceiling heights exceed 30 ft (9.14 m), visible notification appliances shall be suspended at or below 30 ft (9.14 m) or wall-mounted in accordance with Table 4-4.4.1.1(a).

The code does not presently have guidance or requirements for spaces with high ceilings. In some high ceiling spaces, such as a gymnasium or large atrium, it may not be feasible to suspend or wall mount appliances. Alternative methods for notification may need to be considered. Sample alternatives include high intensity revolving beacons, high inten-

sity indirect viewing appliances, or even the flashing of some or all of the building lights. Most of these methods are not presently recognized by the code because the committees have not seen test data to support their use or because of other potential issues of reliability. Nevertheless, careful engineering may show them to be effective and more reliable than suspending standard appliances from the ceiling or wall mounting them in very large congested spaces, such as warehouse stores or convention halls.

4-4.4.1.5 Table 4-4.4.1.1(b) shall be used if the visible notification appliance is at the center of the room. If the visible notification appliance is not located at the center of the room, the effective intensity (cd) shall be determined by doubling the distance from the appliance to the farthest wall to obtain the maximum room size.

4-4.4.2* Spacing in Corridors.

A-4-4.4.2 See Figure A-4-4.4.2 for corridor and elevator spacing allocation.

4-4.4.2.1 Table 4-4.4.2.1 shall apply to corridors not exceeding 20 ft (6.1 m) in width. For corridors greater than

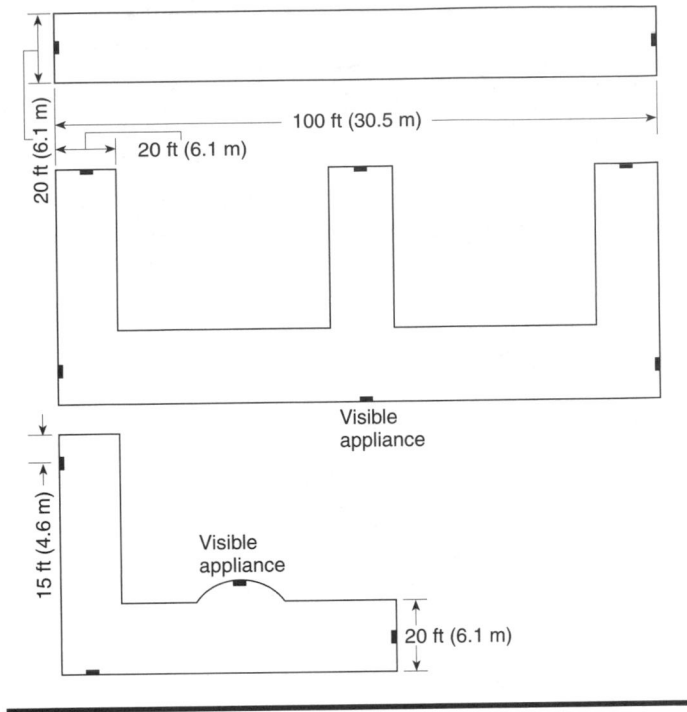

Figure A-4-4.4.2 *Corridor and elevator area spacing allocation.*

20 ft (6.1 m) wide, Figure 4-4.4.1.1 and Tables 4-4.4.1.1(a) and (b) shall apply. In a corridor application, visible appliances shall be rated not less than 15 cd.

Table 4-4.4.2.1 ***Corridor Spacing for Visible Appliances***

Corridor Length		Minimum Number of 15-cd Visible Appliances Required
ft	m	
0 – 30	0 – 9.14	1
31 – 130	9.45 – 39.6	2
131 – 230	39.93 – 70	3
231 – 330	70.4 – 100.6	4
331 – 430	100.9 – 131.1	5
431 – 530	131.4 – 161.5	6

The intensity and spacing requirements for visible appliances located in corridors [less than 20 ft (6.1 m) wide] are less stringent than for those in rooms. A person in a corridor is usually moving and alert. Because the occupants are usually alert, fewer appliances are required, which results in greater spacing in long corridors. Corridors that are more than 20 ft (6.1 m) wide are treated the same as rooms. Only 15 cd appliances are required in corridors less than 20 ft (6.1 m) wide.

4-4.4.2.2* Visible notification appliances shall be located not more than 15 ft (4.57 m) from the end of the corridor with a separation not greater than 100 ft (30.4 m) between appliances. If there is an interruption of the concentrated viewing path, such as a fire door, an elevation change, or any other obstruction, the area shall be treated as a separate corridor.

It may be possible to share notification appliances under some conditions where corridors change direction or intersect. This will result in the proper coverage with fewer appliances.

A-4-4.4.2.2 Visible appliances in corridors are permitted to be mounted on walls or on ceilings in accordance with 4-4.4.2.2. When there are more than two appliances in a field of view, they need to be at least 55 ft (16.8 m) apart or they need to be synchronized.

4-4.4.2.3 In corridors where there are more than two visible notification appliances in any field of view, they shall be spaced a minimum of 55 ft (16.76 m) from each other or they shall flash in synchronization.

The requirements for synchronization of strobes have been extended to apply to corridor strobes in the 1999 edition of the code. In most cases where room strobes are required to be synchronized and where corridor strobes require synchronization, the corridor strobes will also have to be synchronized with the room strobes. However, in very long corridors, Table 4-4.4.2.1 requires spacings of 100 ft (30.4 m) between appliances. This may preclude the need for synchronization.

4-4.4.3   Sleeping Areas.

4-4.4.3.1 Combination smoke detectors and visible notification appliances shall be installed in accordance with the applicable requirements of Chapter 2, Chapter 4, and Chapter 8.

This requirement reinforces the detector coverage requirements of Chapters 2 and 8, as well as those of Chapter 4.

4-4.4.3.2* Table 4-4.4.3.2 shall apply to sleeping areas that have no linear dimension greater than 16 ft (4.87 m). For larger rooms, the visible notification appliance shall be located within 16 ft (4.87 m) of the pillow.

Table 4-4.4.3.2 ***Effective Intensity Requirements for Sleeping Area Visible Notification Appliances***

Distance from Ceiling to Top of Lens		Intensity (cd)
in.	mm	
≥ 24	610	110
< 24	610	177

In order to offset the obscuration effect of smoke near the ceiling, it was determined by testing that a value of 177 cd eff. is required when the notification appliance is located within 24 in. (610 mm) of the ceiling, and a value of 110 cd eff. is sufficient when the distance is greater than 24 in. (610 mm) (UL 1971).

A-4-4.3.2 For sleeping areas, the use of lights with other intensities at distances different than within 16 ft (4.9 m) has not been researched and is not addressed in this code.

4-4.4.4* If visible notification appliances are required, a minimum of one appliance shall be installed in the concentrated viewing path.

A-4-4.4 Examples of rooms where there are concentrated viewing areas include classrooms and theater stages.

4-5 Visible Characteristics, Private Mode

Visible notification appliances used in the private mode shall be of a sufficient quantity and intensity, and located so as to meet the intent of the user and the authority having jurisdiction.

Visible notification appliances in the private mode are usually used in conjunction with an audible notification appliance to call the viewer's attention to the visible appliance. Many visible appliances in the private mode provide annunciated information that helps the viewer to locate the source of an alarm, supervisory, or trouble signal. A remote annunciator is an example of this usage.

4-6 Supplementary Visible Signaling Method

A supplementary visible notification appliance shall be intended to augment an audible or visible signal.

A supplementary visible notification appliance is not intended to serve as one of the required visible notification appliances. Examples include non-required remote annunciators, or non-required flashing lights located in the security or maintenance office. See the definition of Supplementary in Chapter 1.

4-6.1 A supplementary visible notification appliance shall comply with its marked rated performance.

Recognizing that this appliance is not satisfying a requirement, but is providing a supplemental function, subsection 4-6.1 makes it mandatory that the appliance function as marked and rated. This requirement discourages manufacturers from overrating the marking, which might not be detected because the appliances are supplementary, and

gives the authority having jurisdiction a basis for verifying the performance of such appliances.

4-6.2 Supplementary visible notification appliances shall be permitted to be located less than 80 in. (2.03 m) above the floor.

Because such an appliance is supplementary, it need not meet the mandatory height requirement for visible appliances.

4-7 Textual Audible Appliances

4-7.1 Fire Alarm Speaker Appliances.

4-7.1.1 Fire alarm speaker appliances shall comply with Section 4-3.

Speakers are audible appliances and must comply with the audibility and mounting requirements of Section 4-3. This includes the intelligibility requirements of 4-3.1.5.

4-7.1.2* The sound pressure level, in dBA, of the fire alarm speaker appliance evacuation tone signals of the particular mode installed (public or private) shall comply with all the requirements in 4-3.2 (public) or 4-3.3 (private).

In addition to conveying textual information, textual audible appliances are also used to produce tones used to warn occupants to evacuate the protected premises. This requirement ensures that the sound level requirements of Section 4-3 will be met by the textual audible appliances. See Exhibit 4.12 for an example of typical textual appliances (speakers).

Exhibit 4.12 *Typical loudspeaker used as textual notification appliances. (Source: Wheelock; photo courtesy of Mammoth Fire Alarms, Inc., Lowell, MA)*

A-4-7.1.2 The evacuation tone signal is used to evaluate the audibility produced by fire alarm speaker appliances because of the fluctuating sound pressure level of voice or recorded messages.

4-7.2 Telephone Appliances.

4-7.2.1 Telephone appliances shall be in accordance with EIA Tr 41.3, *Telephones*.

The Electronic Industries Alliance standard helps to ensure the quality and technical suitability of a telephone handset. EIA Tr 4.3, *Telephones*, is available from the Electronics Industries Alliance, 2500 Wilson Blvd., Arlington, VA 22201-3834; www.eia.org.

4-7.2.2 Wall-mounted telephone appliances or related jacks shall be not less than 36 in. (914 mm) and not more than 66 in. (1676 mm) above floor level with clear access to the appliance that is at least 30 in. (762 mm) wide.

Exception: If accessible to the general public, one telephone appliance per location shall be not more than 48 in. (1219 mm) above floor level.

The term *accessible* in this context means "available to and intended to be used by the general public." This includes use by floor or section fire wardens who might be required to communicate with the building emergency communication center by means of the fire alarm system telephones.

4-8* Textual Visible Appliances

Examples of textual visible appliances include annunciators, panel displays (LED and LCD), CRTs, screens, and signs. See Exhibit 4.13 for an annunciator representing a typical textual visible appliance.

Exhibit 4.13 *Typical textual visible appliance (annunciator).* *(Source: Radionics, Inc., Salinas, CA)*

A-4-8 Textual visible appliances are selected and installed to provide temporary text, permanent text, or symbols. Textual visible appliances are most commonly used in the private mode. The use of microprocessors with computer monitors and printers has resulted in the ability to provide detailed fire alarm system information in the form of text and graphics to persons charged with directing emergency response and evacuation. Textual visible appliances are also used in the public mode to communicate emergency response and evacuation information directly to the occupants or inhabitants of the area protected by the fire alarm system. Because textual visible appliances do not necessarily have the ability to alert, they should only be used to supplement audible or visible notification appliances.

Textual visible information should be of a size and visual quality that is easily read. Many factors influence the readability of textual visible appliances, including the following:

(1) Size and color of the text or graphic
(2) Distance from the point of observation
(3) Observation time
(4) Contrast
(5) Background luminance
(6) Lighting
(7) Stray lighting (glare)
(8) Shadows
(9) Physiological factors

While many of these factors can be influenced by the fire alarm equipment manufacturer and by the building designers, there is no readily available method to measure readability.

4-8.1 Application.

Textual visible appliances shall be permitted if used in addition to audible or visible, or both, notification appliances.

4-8.2 Performance.

The information produced by textual visible appliances shall be legible.

4-8.3 Location.

4-8.3.1 Private Mode. All textual visible notification appliances in the private mode shall be located in rooms that are accessible only to those persons directly concerned with the implementation and direction of emergency action initiation and procedure in the areas protected by the fire alarm system.

Exception: In locations where required by the authority having jurisdiction.

Paragraph 4-8.3.1 intends to limit access to private mode textual visible displays to only those persons authorized to obtain such information. The exception permits the authority having jurisdiction to specify a more public location for

the textual visible appliance, presumably for use by responding emergency personnel.

4-8.3.2 Public Mode. Textual visible notification appliances used in the public mode shall be located to ensure readability by the occupants or inhabitants of the protected area.

References Cited in Commentary

American National Standards Institute, Inc, 1998, ANSI 117.1, *Accessible and Usable Buildings and Facilities*, American National Standards Institute, Inc., Washington, DC.

American National Standards Institute, Inc. R-1996, ANSI S3.41-1990, *American National Standard Audible Emergency Evacuation Signal*, American National Standards Institute, Inc., Washington, DC.

Americans with Disabilities Act, *Accessibility Guidelines*, U.S. Government Printing Office, Washington, DC 20402.

Butler, Boyer, et al, 1981, *Locating Fire Alarm Sounders for Audibility*, Building Services Research and Information Association, UK.

Electronic Industries Alliance, EIA TR 4.3, *Telephones*, Electronic Industries Alliance, Arlington, VA.

Myles, M., 1979, *Analysis of Acoustic Signals Produced by Residential Fire Alarms*, National Fire Protection Association (May Annual Meeting), Quincy, MA.

National Fire Protection Association, 1995, *SFPE Handbook of Fire Protection Engineering*, National Fire Protection Association, Quincy, MA, pp. 4-20–4-26.

Schifiliti, R.P., 1988, "Designing Fire Alarm Audibility," *Fire Technology* (May).

Underwriters Laboratories Inc., 1991, UL 1971, *Report of Research on Emergency Signaling Devices for Use by the Hearing Impaired*, Underwriters Laboratories, Northbrook, IL.

Underwriters Laboratories Inc., 1996, *Standard for Safety Control Units for Fire Signalling Systems*.

CHAPTER 5

Supervising Station Fire Alarm Systems

Chapter 5 presents the requirements for three supervising station services: central station, proprietary, and remote station. It also presents the requirements for various transmission technologies. Public fire reporting systems and auxiliary systems are now found in Chapter 6.

5-1* Scope

Chapter 5 shall cover the requirements for the performance, installation, and operation of fire alarm systems at a continuously attended supervising station and between the protected premises and the continuously attended supervising station.

Chapter 5 of the code covers the requirements for a protected premises fire alarm system connected to, and monitored by, a continuously attended supervising station. This supervising station may be either a central station, proprietary supervising station, or a remote supervising station. See Exhibit 5.1 for an illustration of Chapter 5 organization.

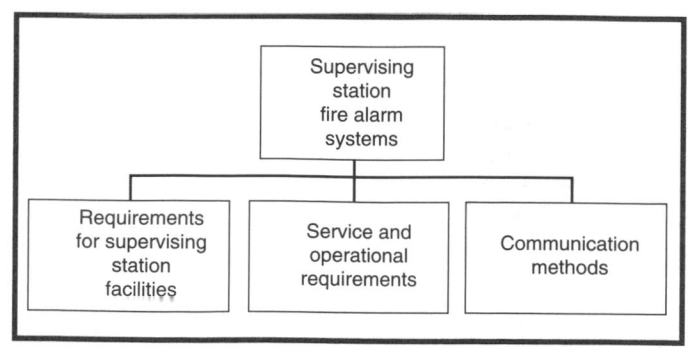

Exhibit 5.1 *Organization of Chapter 5, covering supervising fire alarm systems. (Source: R.P. Schifiliti Associates, Inc., Reading, MA)*

A-5-1 Table A-5-1 provides a tool for users of the code to easily and systematically look up requirements for protected premises, central station service, remote supervising station, and proprietary supervising station fire alarm systems.

5-2 Fire Alarm Systems for Central Station Service

The requirements of Chapters 1 and 7 and Section 5-5 shall apply to central station fire alarm systems, unless they conflict with the requirements of this section.

5-2.1 Scope.

Section 5-2 shall describe the general requirements and use of fire alarm systems to provide central station service as defined in Section 1-4.

5-2.2 General.

5-2.2.1 Fire alarm systems for central station service shall include the central station physical plant, exterior communications channels, subsidiary stations, and signaling equipment located at the protected premises.

See Exhibit 5.2 for an example of a central station and Exhibit 5.3 for the three methods of contracting central station service.

5-2.2.2* Section 5-2 shall apply to central station service, which consists of the following elements:

(1) Installation of fire alarm transmitters
(2) Alarm, guard, supervisory, and trouble signal monitoring
(3) Retransmission
(4) Associated record keeping and reporting
(5) Testing and maintenance
(6) Runner service

Table A-5-1 Supervising Station Performance Criteria

System	Protected Premises	Central Station Service	Remote Supervising Station	Proprietary Supervising Station
Qualify	All fire alarm systems	Supervising station service provided by a prime contractor. There is a subscriber (5-2.2.2 and 5-2.2.3).	Where central station service is neither required nor elected, properties under various ownership monitored by a remote supervising station (5-4.1)	Supervising station service monitoring contiguous or non-contiguous properties under one ownership and responsible to the owner of the protected property (5-3.2 et al)
Listed	Equipment listed for the use intended (1-5.1.2)	Service listed or placarded as well as local equipment for intended use (5-1.2.1 and 5-1.2.2)	Equipment listed for use intended (1-5.1.2)	Equipment listed for use intended (1-5.1.2)
Design	According to code by experienced persons (1-5.1.3)	According to code by experienced persons (1-5.1.3)	According to code by experienced persons (1-5.1.3)	According to code by experienced persons (1-5.1.3)
Compatibility	Detector devices pulling power from initiating or signaling circuits listed for control panel (1-5.3)	Detector devices pulling power from initiating or signaling circuits listed for control panel (1-5.3)	Detector devices pulling power from initiating or signaling circuits listed for control panel (1-5.3)	Detector devices pulling power from initiating or signaling circuits listed for control panel (1-5.3)
Performance and Limitations	85 percent and 110 percent of the nameplate rated input voltage, 32°F (0°C) and 120°F (49°C) ambient temperature, 85 percent relative humidity at 85°F (29.4°C)	85 percent and 110 percent of the nameplate rated input voltage, 32°F (0°C) and 120°F (49°C) ambient temperature, 85 percent relative humidity at 85°F (29.4°C)	85 percent and 110 percent of the nameplate rated input voltage, 32°F (0°C) and 120°F (49°C) ambient temperature, 85 percent relative humidity at 85°F (29.4°C)	85 percent and 110 percent of the nameplate rated input voltage, 32°F (0°C) and 120°F (49°C) ambient temperature, 85 percent relative humidity at 85°F (29.4°C)
Documentation	Authority having jurisdiction notified of new or changed specifications, wiring diagrams, battery calculations. Floor plans approval statement from contractor meets manufacturer's specifications and NFPA requirements. Record of completion (1-6.2.1 and 1-6.2.2). Results of evaluation required in 3-4.3.3.	Authority having jurisdiction notified of new or changed specifications, wiring diagrams, battery calculations. Floor plans approval statement from contractor meets manufacturer's specifications and NFPA requirements. Record of completion (1-6.2.1 and 1-6.2.2). Results of evaluation required in 3-4.3.3.	Authority having jurisdiction notified of new or changed specifications, wiring diagrams, battery calculations. Floor plans approval statement from contractor meets manufacturer's specifications and NFPA requirements. Record of completion (1-6.2.1 and 1-6.2.2). Results of evaluation required in 3-4.3.3.	Authority having jurisdiction notified of new or changed specifications, wiring diagrams, battery calculations. Floor plans approval statement from contractor meets manufacturer's specifications and NFPA requirements. Record of completion (1-6.2.1 and 1-6.2.2). Results of evaluation required in 3-4.3.3.
Supervising Station Facilities	None	UL 827 compliant for both the supervising station and subsidiary station (5-1.2.1 and 5-1.2.2)	Access restricted, retransmit to public fire or governmental agency. Trouble and supervisory can be received elsewhere at a continuously attended station (5-4.3 et al).	Fire resistive, detached building or cut-off room not near or exposed to hazards. Access restricted, NFPA 10, 26-hour emergency lighting (5-3.3 et al)
Testing and Maintenance	Chapter 7	Chapter 7	Chapter 7	Chapter 7

Table A-5-1 Continued

System	Protected Premises	Central Station Service	Remote Supervising Station	Proprietary Supervising Station
Runner Service	No	Yes—Alarm arrive at the protected premises within 1 hour where equipment needs to be reset, guard signal —30 minutes, supervisory—1 hour, trouble — 4 hour *(5-2.6.1 et al)*	No	Yes—alarm—1 hour, guard alarm—30 minutes, supervisory —1 hour, trouble — 1 hour *(5-3.6.6 et al)*
Operations and Management Requirements	None	Central station provides all but test, maintenance, installation and runner service provided by local alarm service. Local alarm service prime contractor provides above and central station provides remainder *(5-2.2.2)*.	None	See Qualify
Staff	None	Minimum of two persons on duty at supervising station. Operation and supervision primary task *(5-2.5)*.	Sufficient to receive alarms. Other duties permitted per the authority having jurisdiction.	Two operators of which one may be the runner, when runner is not in attendance at station contact not to exceed 15 minutes. Primary duties are monitoring alarms and operations of station *(5-3.5 et al)*.
Monitor Signals	Control unit, fire command center and supervising station if sent to supervising station *(1-5.4.3.2.2)*	Control unit, fire command center and supervising station if sent to supervising station *(1-5.4.3.2.2)*	Control unit, fire command center and supervising station if sent to supervising station *(1-5.4.3.2.2)*	Control unit, fire command center and supervising station if sent to supervising station *(1-5.4.3.2.2)*
Retransmission	None—Local	Alarm—public fire service communications center, subscriber, dispatch runner provide notice to subscriber and authority having jurisdiction if required. Supervisory, trouble, and guard service similar but not same *(5-2.6.1.2 –5-2.6.1.5)*.	(1) Alarm to fire department or governmental agency, if not available to another approved location. By dedicated phone circuit or one-way phone, private radio, or other methods acceptable. (2) Supervisory and trouble can go elsewhere *(5-4.4.4)*.	Alarm—public fire department or plant brigade or other parties as required. Guard-protected premises, supervisory-designated person to find problem, others as required. Trouble-designated person to find problem and others as required *(5-3.6.7.2 –5-3.6.7.4)*.
Records	Current year and 1 year after *(1-6.3)*	Complete records of all signals received must be retained for at least 1 year. Reports provided of signals received to authority having jurisdiction in a form it finds acceptable *(5-2.6.2)*.	At least 1 year *(5-4.6.3)*.	Complete records of all signals received shall be retained for at least 1 year. Reports provided of signals received to authority having jurisdiction in a form it finds acceptable *(5-3.6.7)*.
Alarm Type	Audible—fire zone. Audible and visual control unit and fire command center *(1-5.7.1)*	Audible—fire zone. Audible and visual control unit and fire command center *(1-5.7.1)*	Audible—fire zone. Audible and visual control unit and fire command center *(1-5.7.1)*	Audible—fire zone. Audible and visual control unit and fire command center *(1-5.7.1)*

(continues)

Table A-5-1 Continued

System	Protected Premises	Central Station Service	Remote Supervising Station	Proprietary Supervising Station
Time/Retransmit	20 seconds *(1-5.4.2.2/None)*	20 seconds *(1-5.4.2.2)* Immediate—public fire "maximum" 90 seconds *(5-2.6.1.1)*	Same/immediate—public fire or owner designate *(5-4.6.2)*	90 seconds *(5-3.4.7)* Immediate—public fire plant brigade or others *(5-3.6.6.1)*
Supervisory Type	Audible—covered zone. Audible and visual at the control panel and fire control center *(1-5.7.1)*	Audible—covered zone. Audible and visual at the control panel and fire control center *(1-5.7.1)*	Audible—covered zone. Audible and visual at the control panel and fire control center *(1-5.7.1)*	Audible—covered zone. Audible and visual at the control panel and fire control center *(1-5.7.1)*
Time/Retransmit	20 seconds *(1-5.4.2.2/None)*	20 seconds *(1-5.4.2.2)* Immediate person designated by subscriber maximum 4 minimum *(5-2.6.1.3)*	20 seconds *(1-5.4.2.2)* Immediate—public fire or owner designate *(5-4.6.2)*	90 seconds *(5-3.4.7)* Where required immediate person designated *(5-3.6.6.3)*
Trouble Type	Audible and visual at the control panel and fire control panel *(1-5.7.1)*	Audible and visual at the control panel and fire control panel *(1-5.7.1)*	Audible and visual at the control panel and fire control panel *(1-5.7.1)*	Audible and visual at the control panel and fire control panel *(1-5.7.1)*
Silence/Reset	Can be silenced, can be intermittent every 10 seconds for $^1/_2$ second, can have a switch secured. Re-sound every 24 hours if silenced, if restored and still silenced tone resounds *(1-5.4.8)*.	Can be silenced, can be intermittent every 10 seconds for $^1/_2$ second, can have a switch secured. Re-sound every 24 hours if silenced, if restored and still silenced tone resounds *(1-5.4.8)*.	Can be silenced, can be intermittent every 10 seconds for $^1/_2$ second, can have a switch secured. Re-sound every 24 hours if silenced, if restored and still silenced tone resounds *(1-5.4.8)*.	Can be silenced, can be intermittent every 10 seconds for $^1/_2$ second, can have a switch secured. Re-sound every 24 hours if silenced, if restored and still silenced tone resounds *(1-5.4.8)*.
Time/Retransmit	200 seconds *(1-5.4.6/None)*	200 seconds *(1-5.4.6)* Immediate—person designated by subscriber. Maximum 4 minutes *(5-2.6.1.4)*	200 seconds *(1-5.4.6)* Immediate—public fire or owner designate *(5-4.6.2)*	200 seconds *(5-3.4.7)* Where required immediate person designated *(5-3.6.6.4)*
Primary Source	Light and power service or engine-driven generator with trained operator *(1-5.2.4)*. Low-power radio and DACT increased.	Same low-power radio and DACT increased	Same low-power radio and DACT increased	Same low-power radio and DACT increased
Capacity	Direct-current voltages not to exceed 350 volts above earth ground *(1-5.2.3)*	Direct-current voltages not to exceed 350 volts above earth ground *(1-5.2.3)*	Direct-current voltages not to exceed 350 volts above earth ground *(1-5.2.3)*	Direct-current voltages not to exceed 350 volts above earth ground *(1-5.2.3)*
Duration	Constant	Constant	Constant	Constant
Type	2-wire ac, 3-wire ac or dc with continuous unfused neutral conductor or polyphase ac with unfused neutral, dedicated branch circuits marked and overcurrent protection *(1-5.2.4)*	2-wire ac, 3-wire ac or dc with continuous unfused neutral conductor or polyphase ac with unfused neutral, dedicated branch circuits marked and overcurrent protection *(1-5.2.4)*	2-wire ac, 3-wire ac or dc with continuous unfused neutral conductor or polyphase ac with unfused neutral, dedicated branch circuits marked and overcurrent protection *(1-5.2.4)*	2-wire ac, 3-wire ac or dc with continuous unfused neutral conductor or polyphase ac with unfused neutral, dedicated branch circuits marked and overcurrent protection *(1-5.2.4)*

Table A-5-1 Continued

System	Protected Premises	Central Station Service	Remote Supervising Station	Proprietary Supervising Station
Secondary	(1) Storage battery. (2) Auto-starting engine generator and storage batteries with 4-hour capacity. (3) Multiple engine-driven generators, one auto starting and capable of largest generator being out of service. Does not cause loss of signal *(1-5.2.5)*	(1) Storage battery. (2) Auto-starting engine generator and storage batteries with 4-hour capacity. (3) Multiple engine-driven generators, one auto starting and capable of largest generator being out of service. Does not cause loss of signal *(1-5.2.5)*	(1) Storage battery. (2) Auto-starting engine generator and storage batteries with 4-hour capacity. (3) Multiple engine-driven generators, one auto starting and capable of largest generator being out of service. Does not cause loss of signal *(1-5.2.5)*	(1) Storage battery. (2) Auto-starting engine generator and storage batteries with 4-hour capacity. (3) Multiple engine-driven generators, one auto starting and capable of largest generator being out of service. Does not cause loss of signal *(1-5.2.5)*
Capacity	Direct-current voltages not to exceed 360 volts above earth ground *(1-5.2.4)*	Direct-current voltages not to exceed 360 volts above earth ground *(1-5.2.4)*	Direct-current voltages not to exceed 360 volts above earth ground *(1-5.2.4)*	Direct-current voltages not to exceed 360 volts above earth ground *(1-5.2.4)*
Duration	Automatic without signal loss for 30 seconds and then for 24 hours on protected premises *(1-5.2.5)*	Automatic without signal loss for 30 seconds and then for 24 hours on protected premises and central station *(1-5.2.5)*	Automatic without signal loss for 30 seconds and then for 60 hours on remote station *(1-5.2.5)*	Automatic without signal loss for 30 seconds and then for 24 hours on protected premises and proprietary station *(1-5.2.5)*
Switch	Automatic but see the different configurations for exception *(1-5.2.6)*	Automatic but see the different configurations for exception *(1-5.2.6)*	Automatic but see the different configurations for exception *(1-5.2.6)*	Automatic but see the different configurations for exception *(1-5.2.6)*
Type	Automatic *(1-5.2.6)*	Automatic *(1-5.2.6)*	Automatic *(1-5.2.6)*	Automatic *(1-5.2.6)*

The central station service elements shall be provided under contract to a subscriber by one of the following:

(a) A listed central station that provides all of the elements of central station service with its own facilities and personnel.

(b) A listed central station that provides, as a minimum, the signal monitoring, retransmission, and associated record keeping and reporting with its own facilities and personnel and that shall be permitted to subcontract all or any part of the installation, testing, and maintenance and runner service.

(c) A listed fire alarm service–local company that provides the installation and testing and maintenance with its own facilities and personnel and that subcontracts the monitoring, retransmission, and associated record keeping and reporting to a listed central station. The required runner service shall be provided by the listed fire alarm service–local company with its own personnel or the listed central station with its own personnel.

A-5-2.2.2 There are related types of contract service that often are provided from, or controlled by, a central station but that are neither anticipated by, nor consistent with, the provisions of 5-2.2.2. Although 5-2.2.2 does not preclude such arrangements, a central station company is expected to recognize, provide for, and preserve the reliability, adequacy, and integrity of those supervisory and alarm services intended to be in accordance with the provisions of 5-2.2.2.

Central station service consists of eight distinct elements: installation, testing, maintenance, and runner service at the protected premises; management or operation of the system; monitoring of signals from the protected premises; retransmission of signals; and record keeping at the supervising station. Chapter 5 provides the guidelines for the following three ways of providing central station service:

(1) A central station may provide all eight elements.
(2) A central station may provide the four elements at the supervising station and subcontract one or more of the four elements at the protected premises.

Exhibit 5.2 Central station fire alarm system. (Source: Simplex Time Recorder Co., Gardner, MA)

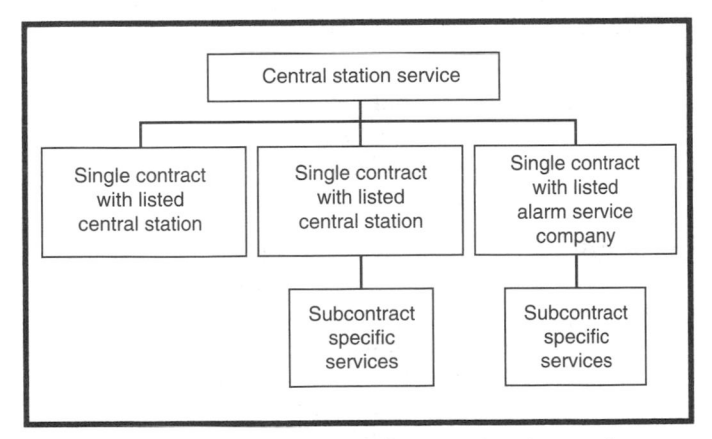

Exhibit 5.3 Subscriber contracts for central station services. (Source: R.P. Schifiliti Associates, Inc., Reading, MA)

(3) A listed fire alarm service–local company may provide the four elements at the protected premises and subcontract the supervising station duties to a central station.

In arrangement (3), either the listed fire alarm service–local company, or the listed central station must provide the runner service.

Typically, the listed central station provides the four elements at the supervising station and subcontracts one or more of the elements at the protected premises. Most commonly, the central station subcontracts part of the installation, notably the installation of waterflow alarm and sprinkler supervisory devices. Under this type of arrangement, the central station subcontracts these to a sprinkler system contractor and provides all the other elements at the protected premises.

Many fire alarm system installers connect protected premises fire alarm systems to a location remote from the protected premises that monitors signals. Relatively few such arrangements meet the requirements of Section 5-2 and should not be called central station service. Only service that incorporates all eight elements of central station service provided by listed alarm service providers who design, specify, install, test, maintain, and use the system in accordance with the requirements of Section 5-2 should be called central station service.

5-2.2.3 The prime contractor shall conspicuously indicate that the fire alarm system providing service at a protected premises complies with all the requirements of this code by providing a means of third party verification, as specified in 5-2.2.3.1 or 5-2.2.3.2.

To help ensure the inherent higher level of protection that a central station fire alarm system provides, the code requires the prime contractor to conspicuously indicate that the entire fire alarm system meets the requirements of the code, by providing a means of third party verification. This indication does not mean that the organization providing third party verification will actually inspect every supervising station fire alarm system. Nor does it mean that when the organization providing third party verification does inspect a supervising station fire alarm system that such an organization will inspect every aspect of that system. However, by providing a means of third party verification, a prime contractor makes provision for an additional level of oversight. The requirement in 5-2.2.3 tends to promote and encourage installation, testing, and maintenance procedures that will help ensure the overall quality of the supervising station fire alarm system.

The prime contractor may conspicuously post a certificate issued by the organization that has listed the central station. Or, the prime contractor may post a placard that indicates compliance. By intent, the code does not provide details of the process by which the listing organization issues the required certificate to the listed prime contractor. Rather, the code leaves these details up to the procedures and practices of the listing organization.

Unless an authority having jurisdiction specifies one of these two methods, the prime contractor—either the central station or the listed fire alarm service (local company)—may choose the method of verification.

Formal Interpretations 72-96-1 and 93-2 provide further clarification of 5-2.3.3.

Formal Interpretation 72-96-1

Reference: 5-2.2.3 through 5-2.2.5

Question No. 1: Do paragraphs 5-2.2.3 through 5-2.2.5 require that compliance of central station fire alarm systems with the requirements of NFPA 72 be verified by a third party inspection?

Answer: No

Question No. 2: Does the following sentence express the intent of the Technical Committee? "Although inspection may be a part of a listing organization's verification procedure, the Code requires only that such third party verification be accomplished by certification or by placarding."

Answer: Yes

Issue Edition: 1996

Reference: 4-3.2.3 through 4-3.2.5

Issue Date: December 20, 1996

Effective Date: January 9, 1997

Formal Interpretation 93-2

Reference: 5-2.2.3

Question No. 1: Is the verification of compliance to "all the requirements of this code" in 5-2.2.3 intended to be limited to the components of a central station fire alarm system as listed in 5-2.2.1?

Answer: No

Question No. 2: Is the verification of compliance to "all the requirements of this code" in 4-2.2.3 intended to include the entire protected premises fire alarm system, as well as the components comprising the central station fire alarm system as listed in 5-2.2.1?

Answer: Yes

Issue Edition: 1993

Reference: 4-3.2.3

Issue Date: August 29, 1996

Effective Date: September 18, 1996

5-2.2.3.1 The installation shall be certificated.

5-2.2.3.1.1 Fire alarm systems providing service that complies with all the requirements of this code shall be certified by the organization that has listed the central station, and a document attesting to this certification shall be located on or within 36 in. (1 m) of the fire alarm system control unit or, if no control unit exists, on or within 36 in. (1 m) of a fire alarm system component.

The organization that has listed the central station determines the detailed procedures that result in the issuing of a certificate. This organization produces a document, or certificate, for a specific protected premises. The prime contractor then conspicuously posts this certificate to indicate that the supervising station fire alarm system complies with the requirements of the code, and that the prime contractor has provided a means of third party verification.

As mentioned previously, the code does not provide details of the process by which the listing organization issues the required certificate to the listed prime contractor. Rather, the code leaves these details up to the procedures and practices of the listing organization. For example, Underwriters Laboratories Inc. (UL) has a fire alarm certificate program that meets the intent of 5-2.2.3.1.1 of the code.

In UL's program, a UL-listed prime contractor submits an application form to UL asking it to issue a certificate for the specific installation. UL reviews the details supplied on the application form, and if it judges that the installation described on the application form meets the requirements of the code, and UL's own requirements, it issues the certificate. To help maintain the integrity of the certification process, UL annually inspects a statistically significant sampling of certified installations for each listed prime contractor. To comply with the requirements of 5-2.2.3.1.2, UL maintains a computer database of certified installations at its headquarters in Northbrook, IL. Authorities having jurisdiction may access this database by means of its certificate verification service (ULCVS).

It should be noted that this process bears close resemblance to listing procedures for fire alarm equipment. The listing organization does not inspect every smoke detector for compliance, but rather conducts field inspections and random audits to verify compliance.

Any other organization that lists prime contractors (see the definition of *Listed* in 1-4) could also develop a supervising station fire alarm system certificate program.

5-2.2.3.1.2 A central repository of issued certification documents, accessible to the authority having jurisdiction, shall be maintained by the organization that has listed the central station.

5-2.2.3.2 The installation shall be placarded.

5-2.2.3.2.1 Fire alarm systems providing service that complies with all the requirements of this code shall be conspicuously marked by the central station to indicate compliance. The marking shall be by one or more placards that meet the requirements of the organization that has listed the central station and requires the placard.

5-2.2.3.2.2 The placard(s) shall be 20 in.^{2} (130 cm^2) or larger, shall be located on or within 36 in. (1 m) of the fire alarm system control unit or, if no control unit exists, on or within 36 in. (1 m) of a fire alarm system component, and shall identify the central station by name and telephone number.

5-2.2.4* Fire alarm system service that does not comply with all the requirements of Section 5-2 shall not be designated as central station service.

A-5-2.2.4 It is the prime contractor's responsibility to remove all compliance markings (certification markings or placards) when a service contract goes into effect that conflicts in any way with the requirements of 5-2.2.3.

If someone makes a change that invalidates the designation, central station service, the prime contractor must remove the certificate, or placard, as well as any other means that designates central station service. The authority having jurisdiction should enforce this requirement. This enforcement will ensure that only those systems meeting all eight elements of central station service will have this designation.

5-2.2.5* For the purpose of Section 5-2, the subscriber shall notify the prime contractor, in writing, of the identity of the authority(ies) having jurisdiction.

A-5-2.2.5 The prime contractor should be aware of statutes, public agency regulations, or certifications regarding fire alarm systems that might be binding on the subscriber. The prime contractor should identify for the subscriber which agencies could be an authority having jurisdiction and, if possible, advise the subscriber of any requirements or approvals being mandated by these agencies.

The subscriber has the responsibility for notifying the prime contractor of those private organizations that are being designated as an authority having jurisdiction. The subscriber also has the responsibility to notify the prime contractor of changes in the authority having jurisdiction, such as where there is a change in insurance companies. Although the responsibility is primarily the subscriber's, the prime contractor should also take responsibility for seeking out these private authorities having jurisdiction through the subscriber. The prime contractor is responsible for maintaining current records on the authority(ies) having jurisdiction for each protected premises.

The most prevalent public agency involved as an authority having jurisdiction with regard to fire alarm sys-

tems is the local fire department or fire prevention bureau. These are normally city or county agencies with statutory authority, and their approval of fire alarm system installations might be required. At the state level, the fire marshal's office is most likely to serve as the public regulatory agency.

The most prevalent private organizations involved as authorities having jurisdiction are insurance companies. Others include insurance rating bureaus, insurance brokers and agents, and private consultants. It is important to note that these organizations have no statutory authority and become authorities having jurisdiction only when designated by the subscriber.

With both public and private concerns to satisfy, it is not uncommon to find multiple authorities having jurisdiction involved with a particular protected premises. It is necessary to identify all authorities having jurisdiction in order to obtain all the necessary approvals for a central station fire alarm system installation.

The subscriber and the prime contractor must identify all of the authorities having jurisdiction involved at the protected premises. Although this responsibility rests primarily with the subscriber, the subscriber would normally only know the private authorities having jurisdiction. The prime contractor would most often know any additional public authorities having jurisdiction. Thus, a joint effort most effectively resolves this important requirement. Exhibit 5.4 illustrates a typical certificate and a typical placard issued by two organizations for central station service.

5-2.3 Facilities.

5-2.3.1 The central station building or that portion of a building occupied by a central station shall conform to the construction, fire protection, restricted access, emergency lighting, and power facilities requirements of the latest edition of ANSI/UL 827, *Standard for Safety Central-Station for Watchman, Fire-Alarm and Supervisory Services*.

The ANSI/UL standard referenced in 5-2.3.1 details protection features that help to maintain the integrity and continuity of the physical central station. In 5-2.2.2, the code requires that a qualified testing laboratory—acceptable to the authority having jurisdiction—must list the central station and must examine the protection features required by ANSI/UL 827, *Standard for Safety Central-Station for Watchman, Fire-Alarm and Supervisory Services* for compliance.

5-2.3.2 Subsidiary station buildings or those portions of buildings occupied by subsidiary stations shall conform to the construction, fire protection, restricted access, emergency lighting, and power facilities requirements of the latest edition of ANSI/UL 827, *Standard for Safety Central-Station for Watchman, Fire-Alarm and Supervisory Services*.

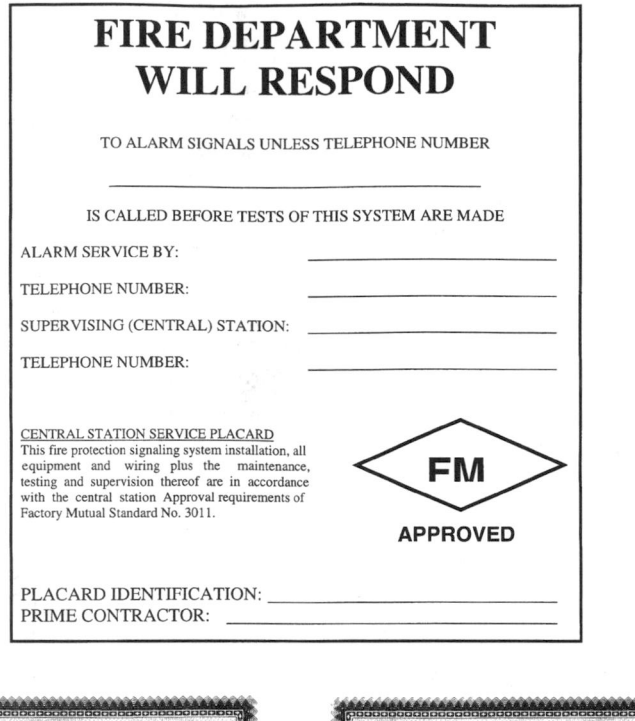

FIRE DEPARTMENT WILL RESPOND

TO ALARM SIGNALS UNLESS TELEPHONE NUMBER

_____

IS CALLED BEFORE TESTS OF THIS SYSTEM ARE MADE

ALARM SERVICE BY: _____

TELEPHONE NUMBER: _____

SUPERVISING (CENTRAL) STATION: _____

TELEPHONE NUMBER: _____

CENTRAL STATION SERVICE PLACARD
This fire protection signaling system, all equipment and wiring plus the maintenance, testing and supervision thereof are in accordance with the central station Approval requirements of Factory Mutual Standard No. 3011.

FM
APPROVED

PLACARD IDENTIFICATION: _____
PRIME CONTRACTOR: _____

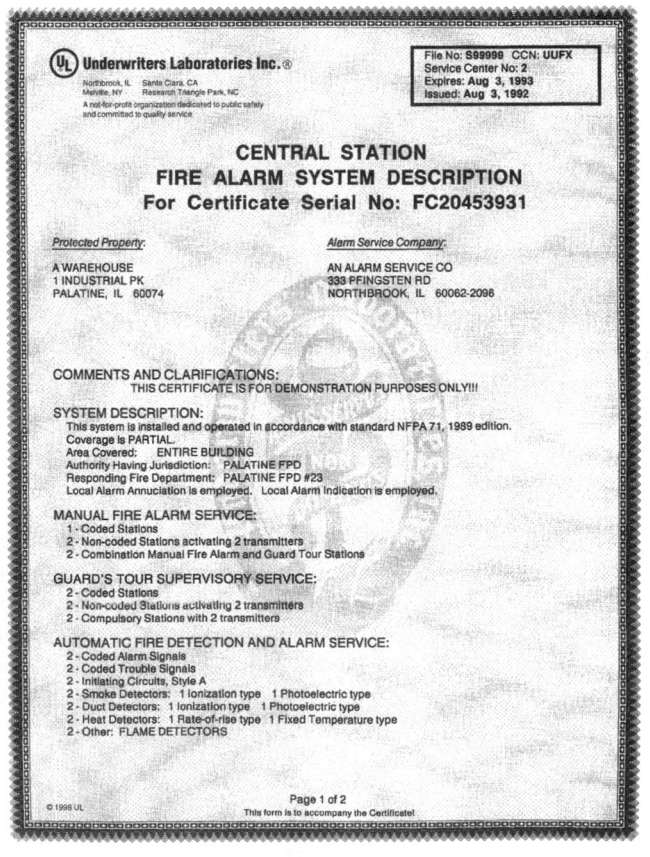

Underwriters Laboratories Inc.®
Northbrook, IL   Santa Clara, CA
Melville, NY   Research Triangle Park, NC
A not-for-profit organization dedicated to public safety and committed to quality service

File No: S99998 CCN: UUFX
Service Center No: 2
Expires: Aug 3, 1993
Issued: Aug 3, 1992

CENTRAL STATION
FIRE ALARM SYSTEM DESCRIPTION
For Certificate Serial No: FC20453931

Protected Property:
A WAREHOUSE
1 INDUSTRIAL PK
PALATINE, IL 60074

Alarm Service Company:
AN ALARM SERVICE CO
333 PFINGSTEN RD
NORTHBROOK, IL 60062-2096

COMMENTS AND CLARIFICATIONS:
        THIS CERTIFICATE IS FOR DEMONSTRATION PURPOSES ONLY!!!

SYSTEM DESCRIPTION:
    This system is installed and operated in accordance with standard NFPA 71, 1989 edition.
    Coverage is PARTIAL.
    Area Covered: ENTIRE BUILDING
    Authority Having Jurisdiction: PALATINE FPD
    Responding Fire Department: PALATINE FPD #23
    Local Alarm Annunciation is employed.   Local Alarm Indication is employed.

MANUAL FIRE ALARM SERVICE:
    1 - Coded Stations
    2 - Non-coded Stations activating 2 transmitters
    2 - Combination Manual Fire Alarm and Guard Tour Stations

GUARD'S TOUR SUPERVISORY SERVICE:
    2 - Coded Stations
    2 - Non-coded Stations activating 2 transmitters
    2 - Compulsory Stations with 2 transmitters

AUTOMATIC FIRE DETECTION AND ALARM SERVICE:
    2 - Coded Alarm Signals
    2 - Coded Trouble Signals
    2 - Initiating Circuits, Style A
    2 - Smoke Detectors: 1 Ionization type   1 Photoelectric type
    2 - Duct Detectors: 1 Ionization type   1 Photoelectric type
    2 - Heat Detectors: 1 Rate-of-rise type 1 Fixed Temperature type
    2 - Other: FLAME DETECTORS

Page 1 of 2
© 1998 UL   This form is to accompany the Certificate!

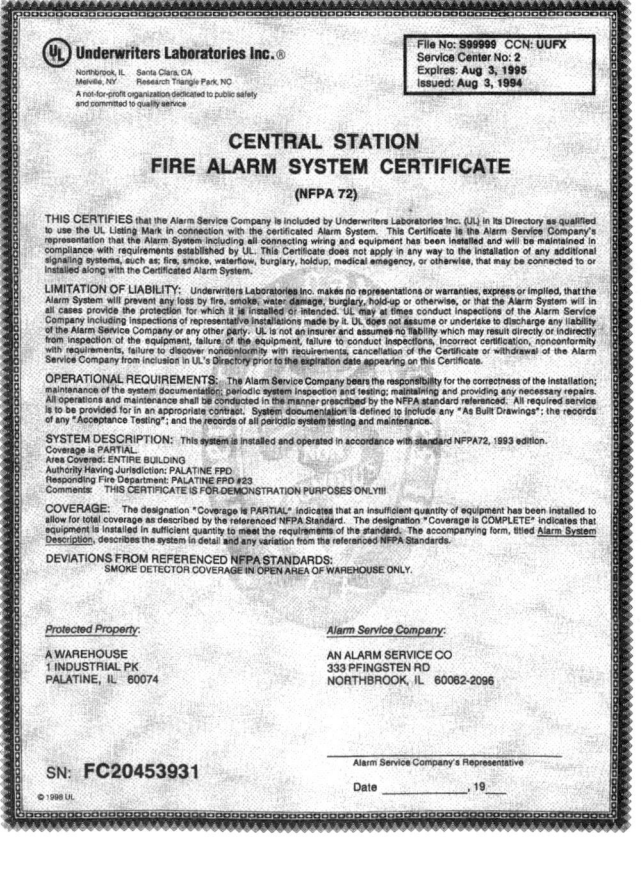

Underwriters Laboratories Inc.®
Northbrook, IL   Santa Clara, CA
Melville, NY   Research Triangle Park, NC
A not-for-profit organization dedicated to public safety and committed to quality service

File No: S99999 CCN: UUFX
Service Center No: 2
Expires: Aug 3, 1995
Issued: Aug 3, 1994

CENTRAL STATION
FIRE ALARM SYSTEM CERTIFICATE
(NFPA 72)

THIS CERTIFIES that the Alarm Service Company is included by Underwriters Laboratories Inc. (UL) in its Directory as qualified to use the UL Listing Mark in connection with the certificated Alarm System. This Certificate is the Alarm Service Company's representation that the Alarm System including all connecting wiring and equipment has been installed and will be maintained in compliance with requirements established by UL. This Certificate does not apply in any way to the installation of any additional signaling systems, such as; fire, smoke, waterflow, burglary, holdup, medical emergency, or otherwise, that may be connected to or installed along with the Certificated Alarm System.

LIMITATION OF LIABILITY: Underwriters Laboratories Inc. makes no representations or warranties, express or implied, that the Alarm System will prevent any loss by fire, smoke, water damage, burglary, hold-up or otherwise, or that the Alarm System will in all cases provide the protection for which it is intended. UL may at times conduct inspections of the Alarm Service Company including inspections of representative installations made by it. UL does not assume or undertake to discharge any liability of the Alarm Service Company or any other party. UL is not an insurer and assumes no liability which may result directly or indirectly from inspection of the equipment, failure of the equipment, failure to conduct inspections, incorrect certification, nonconformity with requirements, failure to discover nonconformity with requirements, cancellation of the Certificate or withdrawal of the Alarm Service Company from inclusion in UL's Directory prior to the expiration date appearing on this Certificate.

OPERATIONAL REQUIREMENTS: The Alarm Service Company bears the responsibility for the correctness of the installation; maintenance of the system documentation; periodic system inspection and testing; maintaining and providing any necessary repairs. All operations and maintenance shall be conducted in the manner prescribed by the NFPA standard referenced. All required service is to be provided for in an appropriate contract. System documentation is defined to include any "As Built Drawings"; the records of any "Acceptance Testing"; and the records of all periodic system testing and maintenance.

SYSTEM DESCRIPTION: This system is installed and operated in accordance with standard NFPA72, 1993 edition.
    Coverage is PARTIAL.
    Area Covered: ENTIRE BUILDING
    Authority Having Jurisdiction: PALATINE FPD
    Responding Fire Department: PALATINE FPD #23
    Comments: THIS CERTIFICATE IS FOR DEMONSTRATION PURPOSES ONLY!!!

COVERAGE: The designation "Coverage is PARTIAL" indicates that an insufficient quantity of equipment has been installed to allow for total coverage as defined by the referenced NFPA Standard. The designation "Coverage is COMPLETE" indicates that equipment is installed in sufficient quantity to meet the requirements of the standard. The accompanying form, titled Alarm System Description, describes the system in detail and any variation from the referenced NFPA Standards.

DEVIATIONS FROM REFERENCED NFPA STANDARDS:
    SMOKE DETECTOR COVERAGE IN OPEN AREA OF WAREHOUSE ONLY.

Protected Property:
A WAREHOUSE
1 INDUSTRIAL PK
PALATINE, IL 60074

Alarm Service Company:
AN ALARM SERVICE CO
333 PFINGSTEN RD
NORTHBROOK, IL 60062-2096

SN: **FC20453931**

_____
Alarm Service Company's Representative

Date _____, 19___

© 1998 UL

Exhibit 5.4 _A placard and a certificate for central station systems. (Sources: Underwriters Laboratories Inc., Northbrook, IL, and Factory Mutual Research Corporation, Norwood, MA)_

The requirement detailed in 5-2.3.2, and those that follow in 5-2.3.2.1 through 5-2.3.2.7, reflect the fact that under normal operating conditions, no one staffs a subsidiary station. Usually, a subsidiary station serves a particular geographic area. It concentrates signals from many protected premises and transmits those concentrated signals to a supervising station. A malfunction at a subsidiary station can substantially impair the successful transmission of signals from the properties it serves. Thus, these requirements help ensure the overall operational reliability of the subsidiary station. They also help ensure the integrity of the transmission path between the subsidiary station and the supervising station.

5-2.3.2.1 All intrusion, fire, power, and environmental control systems for subsidiary station buildings shall be monitored by the central station in accordance with 5-2.3.

The central station staff manages the subsidiary station by monitoring critical building systems, helping to ensure the operational integrity of the subsidiary station.

5-2.3.2.2 The subsidiary facility shall be inspected at least monthly by central station personnel for the purpose of verifying the operation of all supervised equipment, all telephones, all battery conditions, and all fluid levels of batteries and generators.

The central station staff also manages the integrity of the subsidiary station by inspecting it monthly. Not simply a stop-by visit, this inspection verifies the continuity of the systems installed at the subsidiary station.

5-2.3.2.3 In the event of the failure of equipment at the subsidiary station or the communications channel to the central station, a backup shall be operational within 90 seconds. Restoration of a failed unit shall be accomplished within 5 days.

The subsidiary station must have backup equipment and, in the event of a failure, the central station must be able to place the backup equipment into operation within 90 seconds. A technician must repair or replace any defective equipment within 5 days.

5-2.3.2.4 There shall be continuous supervision of each communications channel between the subsidiary station and the central station.

The equipment connected to each communications channel between the subsidiary station and the central station must continuously monitor for channel integrity. In most cases, the equipment uses some form of continuous multiplex transmission technology. Equipment using T1 network technology is quite commonly used for this purpose.

5-2.3.2.5 When the communications channel between the subsidiary station and the supervising station fails, the com-

munications shall be switched to an alternate path. Public switched telephone network facilities shall be used only as an alternate path.

In addition to a fully redundant communications channel, the central station may use "dial up, make good" service provided by the public telephone utility. If the hard-wired trunk between the central station and the subsidiary station fails, "dial up, make good" service allows the central station to access a substitute trunk line using the normal voice network. Use of this substitute trunk line provides an emergency path until the primary trunk lines can be restored. The central station initiates the "dial up, make good" service to re-establish the communications channel between the subsidiary station and the central station.

5-2.3.2.6 In the subsidiary station, there shall be a communications path, such as a cellular telephone, that is independent of the telephone cable between the subsidiary station and the serving wire center.

The requirement in 5-2.3.2.6 ensures that service personnel can establish communication with the central station upon arrival at a totally impaired subsidiary station.

5-2.3.2.7 A plan of action to provide for restoration of services specified by this code shall exist for each subsidiary station.

5-2.3.2.7.1 This plan shall provide for restoration of services within 4 hours of any impairment that causes loss of signals from the subsidiary station to the central station.

The central station must formulate a written plan for restoring service from a subsidiary station. This plan must encompass all services not already covered by 5-2.3.2.3. Such restoration must occur within 4 hours. Commonly, the organization listing a central station that uses one or more subsidiary stations will review such a plan.

5-2.3.2.7.2 There shall be an exercise to demonstrate the adequacy of the plan at least annually.

As with all emergency plans, the central station must test the plan's accuracy and validity. By performing this annual exercise, the implementing personnel have an opportunity to become thoroughly familiar with the procedure. Such an exercise also helps keep the plan up-to-date, and discloses changes at either the subsidiary station or the central station that may affect the integrity of the plan.

5-2.4 Equipment.

5-2.4.1 The central station and all subsidiary stations shall be equipped so as to receive and record all signals in accordance with 5-5.5. Circuit-adjusting means for emergency operation shall be permitted to be automatic or to be provided through manual operation upon receipt of a trouble signal. Computer-

aided alarm and supervisory signal processing hardware and software shall be listed for the specific application.

Paragraph 5-2.4.1 permits specially trained central station operators (see 5-2.5 for training requirements) to manually operate circuit-adjusting means. Equipment may also automatically operate circuit-adjusting means. The organization listing the central station must also specifically list any computer-aided alarm and supervisory signal processing hardware and software for central station service.

5-2.4.2 Power supplies shall comply with the requirements of Chapter 1.

5-2.4.3 Transmission means shall comply with the requirements of Section 5-5.

5-2.4.4* Two independent means shall be provided to retransmit a fire alarm signal to the designated public fire service communications center.

A-5-2.4.4 Two telephone lines (numbers) at the central station connected to the public switched telephone network, each having its own telephone instrument connected, and two telephone lines (numbers) available at the public fire service communications center to which a central station operator can retransmit an alarm meet the intent of this requirement.

5-2.4.4.1 The use of a universal emergency number, for example 911 public safety answering point, shall not meet the intent of this code for the principal means of retransmission.

In 5-2.2.2, the code declares that one of the eight elements of central station service includes retransmitting fire alarm signals to the public fire service communications center. The central station must have a reasonably secure means to retransmit.

In most cases the central station will use the public switched telephone network to dial the seven-digit fire reporting number (or ten digits with a dialing sequence that includes an area code). The code states that the central station may not use 911 as the primary means of retransmission because (1) the central station may not be located in the same community as the protected premises, (2) a public service answering point (PSAP) staffed with personnel who are not a part of the public fire department answer most 911 and enhanced 911 emergency telephone calls, and (3) the vast majority of 911 calls concern non-fire emergencies. Following this requirement helps to avoid the bottleneck that sometimes occurs at the PSAP.

5-2.4.4.2 If the principal means of retransmission is not equipped to allow the communications center to acknowledge receipt of each fire alarm report, both means shall be used to retransmit.

5-2.4.4.3* If required by the authority having jurisdiction, one of the means of retransmission shall be supervised so that

interruption of retransmission circuit (channel) communications integrity results in a trouble signal at the central station.

A-5-2.4.4.3 The following methods have been used successfully for supervising retransmission circuits (channels).

(a) An electrically supervised circuit (channel) provided with suitable code sending and automatic recording equipment.

(b) A supervised circuit (channel) providing suitable voice transmitting, receiving, and automatic recording equipment. The circuit can be permitted to be a telephone circuit with the following stipulations:

(1) It cannot be used for any other purpose.
(2) It is provided with a two-way ring-down feature for supervision between the fire department communications center and the central station.
(3) It is provided with terminal equipment located on the premises at each end.
(4) It is provided with 24-hour standby power.

> Note: Local on-premises circuits are not required to be supervised.

(c) Radio facilities using transmissions over a supervised channel with supervised transmitting and receiving equipment. Circuit continuity ensured by any means at intervals not exceeding 8 hours is satisfactory.

In certain cases where the authority having jurisdiction does not accept the reliability of the public switched telephone network, the authority having jurisdiction may require the central station to have a retransmission channel that is monitored for integrity for each public fire service communications center that serves the central station's customers.

5-2.4.4.4 The retransmission means shall be tested in accordance with Chapter 7.

5-2.4.4.5 The retransmission signal and the time and date of retransmission shall be recorded at the central station.

The requirement in 5-2.4.4.5 does not mandate the central station to record the time and date automatically. However, when computer-based automation systems manage the receipt and retransmission of signals, the central station can and should automatically record the time and date.

Further, based on a somewhat broad interpretation of 5-2.4.1, most central stations record the actual telephone call to the public fire service communications center. This tape recording, along with records of signals received at the central station, often helps investigators reconstruct the sequence of events that occurred during a major fire.

5-2.5 Personnel.

5-2.5.1 The central station shall have sufficient personnel, but not less than two persons, on duty at the central station

at all times to ensure disposition of signals in accordance with the requirements of 5-2.6.1.

The central station must have two operators on duty at all times. By mandating the presence of two operators, the code maximizes the likelihood that at least one of the operators will always remain fully alert to incoming signals.

5-2.5.2 Operation and supervision shall be the primary functions of the operators, and no other interest or activity shall take precedence over the protective service.

The code requires that the operators have no other duties that would distract them from the prompt, effective handling of signals.

5-2.6 Operations.

5-2.6.1 Disposition of Signals.

5-2.6.1.1 Alarm signals initiated by manual fire alarm boxes, automatic fire detectors, waterflow from the automatic sprinkler system, or actuation of other fire suppression system(s) or equipment shall be treated as fire alarms.
The central station shall perform the following actions:

It is required to perform the actions in the order in which they appear in the code.

(1)* Immediately retransmit the alarm to the public fire service communications center

A-5-2.6.1.1(1) The term *immediately* in this context is intended to mean "without unreasonable delay." Routine handling should take a maximum of 90 seconds from receipt of an alarm signal by the central station until the initiation of retransmission to the public fire service communications center.

A central station must give highest priority to the prompt handling and retransmission of fire alarm signals. Under the most adverse circumstances, some transmission technologies may have already taken up to 15 minutes to complete the transmission of a signal from the protected premises to the central station. (See 5-5.3.2.1.4 and 5-5.3.2.1.5.)

(2) Dispatch a runner or technician to the protected premises to arrive within 1 hour after receipt of a signal if equipment needs to be manually reset by the prime contractor

The runner or technician only needs to respond when the prime contractor must manually reset equipment at the protected premises. Where permitted by the authority having jurisdiction in accordance with 3-8.1.3, some supervising station fire alarm systems automatically reset when the specific initiating device that initiated the alarm signal no longer senses a fire. For example, when water stops flowing past a waterflow alarm-initiating device installed in the riser of a wet pipe sprinkler system, or when water pressure act-

ing on a waterflow alarm-initiating device connected to the alarm line of a sprinkler system alarm check valve drops to zero, the contacts within the device return to normal. In this case, a prime contractor would not need to reset the system. Thus, a runner does not need to respond.

For other central station fire alarm systems, the authority having jurisdiction may permit the subscriber or some other trained individual to reset the equipment.

(3) Immediately notify the subscriber

The central station will usually notify the subscriber by telephone, which is the quickest available method of notification.

(4) Provide notice to the subscriber or authority having jurisdiction, or both, if required

Written notice to the subscriber and the authority having jurisdiction should follow a format useful to each recipient. The subscriber can use such notice to document system operations. The authority having jurisdiction can use the notice to help document response to system operations at the location.

Exception: If the alarm signal results from a prearranged test, the actions specified by 5-2.6.1.1(1) and (3) shall not be required.

5-2.6.1.2 Guard's Tour Supervisory Signal.

5-2.6.1.2.1 Upon failure to receive a guard's tour supervisory signal within a 15-minute maximum grace period, the central station shall perform the following actions:

(1) Communicate without unreasonable delay with personnel at the protected premises
(2) Dispatch a runner to the protected premises to arrive within 30 minutes of the delinquency if communications cannot be established

If the central station cannot promptly contact personnel at the protected premises, then it should dispatch a runner to investigate why the guard missed a signal. Once dispatched, the runner must arrive at the protected premises within 30 minutes. This means the runner may actually arrive 45 minutes after the guard missed the signal. Even so, in actual cases, a responding runner has found the guard injured or ill and, by summoning medical assistance, has saved the guard's life.

(3) Report all delinquencies to the subscriber or authority having jurisdiction, or both, if required

5-2.6.1.2.2 Failure of the guard to follow a prescribed route in transmitting signals shall be handled as a delinquency.

Guard's tour supervision by a central station mandates a compulsory tour arrangement. The central station can monitor every reporting station along a route. Alternatively, the central station can monitor only a few of the stations along a route of stations. But in either case, the guard must follow a prescribed route, proceeding from station to station in a

fixed sequence. The guard incurs a delinquency if he or she fails to follow the prescribed route.

5-2.6.1.3* Upon receipt of a supervisory signal from a sprinkler system, other fire suppression system(s), or other equipment, the central station shall perform the following actions:

A-5-2.6.1.3 It is anticipated that the central station will first attempt to notify designated personnel at the protected premises. When such notification cannot be made, it might be appropriate to notify law enforcement or the fire department, or both. For example, if a valve supervisory signal is received where protected premises are not occupied, it is appropriate to notify the police.

(1)* Communicate immediately with the person(s) designated by the subscriber

The central station must handle supervisory signals promptly and accurately. These signals may indicate that something or someone has impaired a vital protection system.

A-5-2.6.1.3(1) The term *immediately* in this context is intended to mean "without unreasonable delay." Routine handling should take a maximum of 4 minutes from receipt of a supervisory signal by the central station until the initiation of communications with a person(s) designated by the subscriber.

(2) Dispatch a runner or maintenance person to arrive within 1 hour to investigate

Exception: Where the supervisory signal is cleared in accordance with a scheduled procedure determined by 5-2.6.1.3(1).

The runner or technician only needs to respond when the central station operator cannot resolve the restoration of the supervisory signal to normal by contacting designated personnel. Typically, upon receipt of a supervisory signal, a central station operator will telephone the premises. If he or she receives no answer, then the operator will telephone the individuals on a calling list provided by the subscriber. If the operator cannot reach someone on the calling list who will promptly respond to investigate the signal, then the operator must dispatch a runner. When dispatched, the runner must arrive at the protected premises within 1 hour.

(3) Notify the fire department or law enforcement agency, or both, if required
(4) Notify the authority having jurisdiction when sprinkler systems or other fire suppression systems or equipment has been wholly or partially out of service for 8 hours
(5) When service has been restored, provide notice, if required, to the subscriber or the authority having jurisdiction, or both, as to the nature of the signal, the time of occurrence, and the restoration of service when equipment has been out of service for 8 hours or more

Exception: If the supervisory signal results from a prearranged test, the actions specified by 5-2.6.1.3 (1), (3), and (5) shall not be required.

5-2.6.1.4 Upon receipt of trouble signals or other signals pertaining solely to matters of equipment maintenance of the fire alarm systems, the central station shall perform the following actions:

(1)* Communicate immediately with persons designated by the subscriber

The central station must handle trouble signals promptly and accurately. These signals indicate that the fire alarm system is wholly or partly out of service.

A-5-2.6.1.4(1) The term *immediately* in this context is intended to mean "without unreasonable delay." Routine handling should take a maximum of 4 minutes from receipt of a trouble signal by the central station until initiation of the investigation by telephone.

(2) Dispatch personnel to arrive within 4 hours to initiate maintenance, if necessary

The personnel, dispatched to arrive within 4 hours, must initiate repairs. This requirement generally means that a technician, rather than a runner, must respond.

(3) Provide notice, if required, to the subscriber or the authority having jurisdiction, or both, as to the nature of the interruption, the time of occurrence, and the restoration of service, when the interruption is more than 8 hours

5-2.6.1.5 All test signals received shall be recorded to indicate date, time, and type.

5-2.6.1.5.1 Test signals initiated by the subscriber, including those for the benefit of an authority having jurisdiction, shall be acknowledged by central station personnel whenever the subscriber or authority inquires.

5-2.6.1.5.2* Any test signal not received by the central station shall be investigated immediately and action shall be taken to reestablish system integrity.

A-5-2.6.1.5.2 The term *immediately* in this context is intended to mean "without unreasonable delay." Routine handling should take a maximum of 4 minutes from receipt of a trouble signal by the central station until initiation of the investigation by telephone.

The central station must handle test signals immediately. The code recommends accomplishing this within 4 minutes. These signals help to ensure that the fire alarm system continues to function properly. The central station must cooperate with any authority having jurisdiction that inquires regarding test signals. If a subscriber initiates a test signal, then calls the central station and determines that the central station did not receive the signal, the central station should

treat this occurrence as a trouble signal. The central station should dispatch a service technician to arrive within 4 hours to begin repairs.

5-2.6.1.5.3 The central station shall dispatch personnel to arrive within 1 hour if protected premises equipment needs to be manually reset after testing.

5-2.6.2 Record Keeping and Reporting.

5-2.6.2.1 Complete records of all signals received shall be retained for at least 1 year.

5-2.6.2.2 Testing and maintenance records shall be retained as required by 7-5.3.

5-2.6.2.3 The central station shall make arrangements to furnish reports of signals received to the authority having jurisdiction in a manner approved by the authority having jurisdiction.

When an authority having jurisdiction requests reports from a central station, the central station must provide the reports in a useful and usable form.

5-2.7 Testing and Maintenance.

5-2.7.1 Testing and maintenance for central station service shall be performed in accordance with Chapter 7.

5-2.7.2 The prime contractor shall provide each of its representatives and each alarm system user with a unique personal identification code.

5-2.7.3 In order to authorize the placing of an alarm system into test status, a representative of the prime contractor or an alarm system user shall first provide the central station with his or her personal identification code.

The prime contractor issues each of its representatives and each alarm system user a unique personal identification code (see 5-2.7.2) and requires its use (see 5-2.7.3) in order to carefully control those who may place the system into a test mode. This requirement helps to maintain the security and operational integrity of the system. Without this precaution, the central station has no way of verifying that the person placing the fire alarm system into test status has authorization to do so.

5-3 Proprietary Supervising Station Systems

The requirements of Chapters 1 and 7 and Section 5-5 shall apply to proprietary fire alarm systems, unless they conflict with the requirements of this section.

5-3.1 Scope.

Section 5-3 describes the operational procedures for the supervising facilities of proprietary fire alarm systems. It provides the minimum requirements for the facilities, equipment, personnel, operation, and testing and maintenance of the proprietary supervising station.

5-3.2 General.

5-3.2.1 Proprietary supervising stations shall be operated by trained, competent personnel in constant attendance who are responsible to the owner of the protected property. The requirements of 5-3.5.3 shall apply.

The management of a facility protected by a proprietary fire alarm system uses that system to oversee the built-in fire protection systems. Used as a management tool, the proprietary fire alarm system can help ensure that all other fire protection systems remain in service.

5-3.2.2 The protected property shall be either a contiguous property or noncontiguous properties under one ownership.

From a single proprietary supervising station, an owner may oversee the protection features at one or more properties. These properties may contiguously occupy a single piece of land or may occupy noncontiguous portions of land.

5-3.2.3 If a protected premises master control unit is integral to or co-located with the supervising station equipment, the requirements of Section 5-5 shall not apply.

Paragraph 5-3.2.3 recognizes that in some cases the proprietary fire alarm system may have a Master Control Unit, as defined in 1-4, co-located in the proprietary supervising station. (See Figure 5-5-1.) Where this occurs, the transmission technology requirements described in Section 5-5 would not apply. Rather, the system would use initiating device circuits and signaling line circuits, as described in Chapter 3, to transmit signals to the master fire alarm control unit co-located in the proprietary supervising station. Section 5-3 provides the requirements for all other aspects of such a proprietary fire alarm system.

5-3.2.4* The systems of Section 5-3 shall be permitted to be interconnected to other systems intended to make the premises safer in the event of fire or other emergencies indicative of hazards to life or property.

In addition to the information contained in Section 5-3, refer to 3-8.4.3 and Section 3-9.

A-5-3.2.4 The following functions are included in Appendix A to provide guidelines for utilizing building systems and equipment in addition to proprietary fire alarm equipment in order to provide life safety and property protection.

Building functions that should be initiated or controlled during a fire alarm condition include, but should not be limited to, the following:

(1) Elevator operation consistent with ANSI A17.1, *Safety Code for Elevators and Escalators*
(2) Unlocking of stairwell and exit doors *(refer to NFPA 80, Standard for Fire Doors and Fire Windows, and NFPA 101, Life Safety Code)*
(3) Release of fire and smoke dampers *(refer to NFPA 90A, Standard for the Installation of Air Conditioning and Ventilating Systems, and NFPA 90B, Standard for the Installation of Warm Air Heating and Air Conditioning Systems)*
(4) Monitoring and initiating of self-contained automatic fire extinguishing system(s) or suppression system(s) and equipment *(refer to NFPA 11, Standard for Low-Expansion Foam; NFPA 11A, Standard for Medium- and High-Expansion Foam Systems; NFPA 12, Standard on Carbon Dioxide Extinguishing Systems; NFPA 12A, Standard on Halon 1301 Fire Extinguishing Systems; NFPA 13, Standard for the Installation of Sprinkler Systems; NFPA 14, Standard for the Installation of Standpipe and Hose Systems; NFPA 15, Standard for Water Spray Fixed Systems for Fire Protection; and NFPA 17, Standard for Dry Chemical Extinguishing Systems)*
(5) Lighting control necessary to provide essential illumination during fire alarm conditions *(refer to NFPA 70, National Electrical Code, and NFPA 101, Life Safety Code)*
(6) Emergency shutoff of hazardous gas
(7) Control of building environmental heating, ventilating, and air-conditioning equipment to provide smoke control *(refer to NFPA 90A, Standard for the Installation of Air Conditioning and Ventilating Systems)*
(8) Control of process, data processing, and similar equipment as necessary during fire alarm conditions

5-3.3 Facilities.

5-3.3.1 The proprietary supervising station shall be located in a fire-resistive, detached building or in a cutoff room and shall not be exposed to the hazardous parts of the premises that are protected.

The requirements of 5-3.3.1 help to maintain a high degree of physical integrity for the proprietary supervising station. A "cutoff room" is intended to be a fire-resistive room separated from the normal operations of the facility. Although a specific fire resistance is not specified, the requirements of 3-8.4.1.3.3 should be applicable.

5-3.3.2 Access to the proprietary supervising station shall be restricted to those persons directly concerned with the

implementation and direction of emergency action and procedure.

The proprietary supervising station must not become a congregating place for guards, fire fighters, or other personnel. The presence of such persons may interfere with the operators and distract them from giving proper attention to signal traffic. If management locates the proprietary supervising station within a guard house where guards admit vehicles and personnel to the premises, management should provide some means of segregation to separate the operators of the proprietary supervising station from other incidental employees. This segregation will help to ensure that operators may effectively and efficiently handle the signal traffic without distraction.

5-3.3.3 The proprietary supervising station, as well as remotely located power rooms for batteries or engine-driven generators, shall be provided with portable fire extinguishers that comply with the requirements of NFPA 10, *Standard for Portable Fire Extinguishers*.

Personnel in a proprietary supervising station must have the means to handle a small fire in the supervising station or in the power rooms for batteries or engine-driven generators. Management should refer to the requirements of NFPA 600, *Standard on Industrial Fire Brigades*. These requirements ensure that management has properly organized and trained personnel to safely use the fire extinguishers provided.

5-3.3.4 Emergency Lighting System.

5-3.3.4.1 The proprietary supervising station shall be provided with an automatic emergency lighting system. The emergency source shall be independent of the primary lighting source.

5-3.3.4.2 In the event of a loss of the primary lighting for the supervising station, the emergency lighting system shall provide illumination for a period of not less than 26 hours to permit the operators to carry on operations, and shall be tested in accordance with the requirements of Chapter 7.

5-3.3.5 If 25 or more protected buildings or premises are connected to a subsidiary station, both of the following shall be provided at the subsidiary station:

(1) Automatic means for receiving and recording signals under emergency-staffing conditions
(2) A telephone

A proprietary supervising station may receive signals from many buildings on a very large premises or from several noncontiguous premises through one or more subsidiary stations. Where 25 or more protected buildings or premises

transmit through a subsidiary station, management must equip that subsidiary station so that it can be staffed by operators in an emergency. For example, if the signaling path between the subsidiary station and the proprietary supervising station fails, operators will travel to the subsidiary station, staff it, and operate it independently from the proprietary supervising station. See Exhibit 5.5 for an illustration of a proprietary supervisory station that uses subsidiary stations.

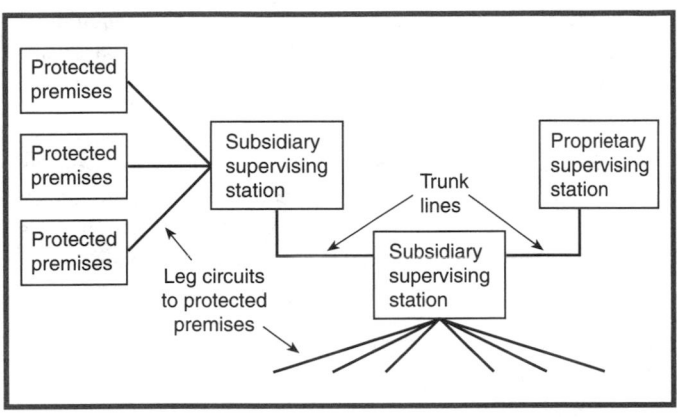

Exhibit 5.5 *Proprietary supervising station facilities—subsidiary stations. (Source: R.P. Schifiliti Associates, Inc., Reading, MA)*

5-3.4 Equipment.

5-3.4.1 Signal-receiving equipment in a proprietary supervising station shall comply with 5-3.4.

5-3.4.2 Provision shall be made to designate the building in which a signal originates. The floor, section, or other subdivision of the building shall be designated at the proprietary supervising station or at the building that is protected.

Exception: Where the area, height, or special conditions of occupancy make detailed designation unessential as approved by the authority having jurisdiction. This detailed designation shall use indicating appliances accepted by the authority having jurisdiction.

To effectively manage the built-in fire protection features of the protected premises, the code requires that the signals received by the proprietary fire alarm system contain sufficient detail to allow operators to quickly and accurately locate the source of the signals. If the nature of the protected property eliminates the need for such detail, the authority having jurisdiction may waive this requirement. Graphic annunciators, video displays, and addressable systems are commonly used to meet this requirement.

5-3.4.3 The proprietary supervising station shall have, in addition to a recording device, two different means for alert-

ing the operator when each signal is received that indicates a change of state of any connected initiating device circuit. One of these means shall be an audible signal, which shall persist until manually acknowledged. Means shall include the receipt of alarm signals, supervisory signals, and trouble signals, including signals indicating restoration.

In 5-3.4.3, the code requires two means of notifying the operators of the receipt of a signal. Paragraph 5-3.5.3 requires that the operators must have no other duties that would impair their ability to process signals from the proprietary supervising station fire alarm system. However, by implication the code accepts that the operators in the supervising station may attend to other duties, as long as those duties do not interfere with the operation of the protective service. The two means of notification make certain the operators attend to the incoming signals.

5-3.4.4 If means is provided in the proprietary supervising station to identify the type of signal received, a common audible indicating appliance shall be permitted to be used for alarm, supervisory, and trouble indication.

Paragraph 1-5.4.7 requires distinctive signals for alarm, supervisory, and trouble signals. Paragraph 5-3.4.4 modifies this requirement to permit a common audible notification appliance in the supervising station, as long as other means readily identify the type of signal.

5-3.4.5 At a proprietary supervising station, an audible trouble signal shall be permitted to be silenced, provided the act of silencing does not prevent the signal from operating immediately upon receipt of a subsequent trouble signal.

5-3.4.6 All signals required to be received by the proprietary supervising station that show a change in status shall be automatically and permanently recorded, including time and date of occurrence. This record shall be in a form that expedites operator interpretation in accordance with any one of the following.

(a) If a visual display is used that automatically provides change of status information for each required signal, including type and location of occurrence, any form of automatic permanent visual record shall be permitted. The recorded information shall include the content described above. The visual display shall show status information content at all times and shall be distinctly different after the operator has manually acknowledged each signal. Acknowledgment shall produce recorded information indicating the time and date of acknowledgment.

Paragraph 5-3.4.6(a) describes an annunciator that continuously shows the status of every point in the system that generates a signal. At a glance, the operator can see the status of every point. With this type of visual display, the proprietary

fire alarm system may use any type of permanent visual record. Such systems most often use a logging-type printer. Such a printer keeps a running list of signals received as a back-up to the visual display. The printed information will not necessarily have a format that would allow an operator to easily locate information. However, it does provide a running summary of signals as they occur with respect to date and time.

(b) If a visual display is not provided, required signal content information shall be automatically recorded on duplicate permanent visual recording instruments.

One recording instrument shall be used for recording all incoming signals, while the other shall be used for required fire, supervisory, and trouble signals only. Failure to acknowledge a signal shall not prevent subsequent signals from recording. Restoration of the signal to its prior condition shall be recorded.

When the proprietary fire alarm system does not provide a visual display, it must use two printers. Where a proprietary supervising station system receives signals from systems other than the fire alarm system, these printers will assist the operators in giving priority to signals from the fire alarm system. One printer will record all signals that the system receives. The other printer will record only fire alarm, supervisory, and trouble signals. Both printers must format the output to allow the operator to easily read, interpret, and act on the information provided.

(c) In the event that a system combines the use of a sequential visual display and recorded permanent visual presentation, the required signal content information shall be displayed and recorded. The visual information component shall be retained either on the display until manually acknowledged or repeated at intervals not greater than 5 seconds, for durations of 2 seconds each, until manually acknowledged. Each new displayed status change shall be accompanied by an audible indication that shall persist until manual acknowledgment of the signal is performed.

A means shall be provided for the operator to redisplay the status of required signal initiating inputs that have been acknowledged but not yet restored. If the system retains the signal on the visual display until manually acknowledged, subsequent recorded presentations shall not be inhibited upon failure to acknowledge. Fire alarm signals shall be segregated on a separate visual display in this configuration.

The visual display unit described in 5-3.4.6(c) presents one or more lines of information at a time, but does not simultaneously display the status of all points covered by the proprietary supervising station fire alarm system. The operator must scroll through the display once he or she has acknowledged each signal. To help the operator give proper precedence to fire alarm signals, the signals must either appear on a separate display, or the system must give them priority status on a common display. The system must still provide a permanent visual record, but the code does not specify the type of printer. Such a system most often uses a logging-type printer, as described in the commentary for 5-3.4.6(a).

Exception: Fire alarm signals shall not be required to be segregated on a separate display if given priority status on the common visual display.

5-3.4.7 The maximum elapsed time from sensing a fire alarm at an initiating device or initiating device circuit until it is recorded or displayed at the proprietary supervising station shall not exceed 90 seconds.

It is not the intent of 5-3.4.7 to severely limit the type of transmission technology used for a proprietary supervising station fire alarm system. Paragraph 5-3.4.7 requires a much higher level of performance than do the requirements of 5-5.3.2.1.5, 5-5.3.2.3.3, or 5-5.3.5.2(1), (2), and (3). Paragraph 5-3.4.7 further reinforces the requirements in 5-5.3.1.2.3(a) and 5-5.3.4.1(a). The requirements conclusively state that, whatever transmission technology the proprietary supervising station fire alarm system uses, that technology must guarantee delivery of a fire alarm signal within 90 seconds of actuation of the initiating device. However, the code has no provision that clearly requires a 90-second response in all cases. Rather, the code states such requirements assuming a normal transmission pathway. Such a requirement would not apply to an impaired transmission pathway. As long as the transmission pathway meets the requirements under normal circumstances, the pathway complies with the requirement of 5-3.4.7.

5-3.4.8 To facilitate the prompt receipt of fire alarm signals from systems handling other types of signals that are able to produce multiple simultaneous status changes, the requirements of either of the following shall be met:

(1) In addition to the maximum processing time for a single alarm, the system shall record simultaneous status changes at a rate not slower than either a quantity of 50, or 10 percent of the total number of initiating device circuits connected, within 90 seconds, whichever number is smaller, without loss of any signal.

(2) In addition to the maximum processing time, the system shall either display or record fire alarm signals at a rate not slower than one every 10 seconds, regardless of the rate or number of status changes occurring, without loss of any signals.

Exception: If fire alarm, waterflow alarm, and sprinkler supervisory signals and their associated trouble signals are the only signals processed by the system, the rate of recording shall not be slower than one signal every 30 seconds.

The requirements of 5-3.4.8 help to ensure the prompt receipt of fire alarm signals when a proprietary fire alarm system receives other types of signals from initiating devices that may produce many status changes at the same time. Paragraph 5-3.4.8 reinforces the requirement for active multiplex transmission systems in 5-5.3.1.2.3(c). Subparagraph 5-5.3.1.2.3 also applies that requirement to other transmission technologies.

However, when a proprietary fire alarm system handles only fire alarm signals, waterflow alarm signals, sprinkler supervisory signals, and their associated trouble signals, the exception relieves such a system from meeting the requirements of 5-3.4.8.

5-3.4.9 Trouble signals required by 1-5.8 and their restoration shall be automatically indicated and recorded at the proprietary supervising station within 200 seconds.

Paragraph 5-3.4.9 also limits the transmission technologies used for a proprietary fire alarm system. Paragraph 5-3.4.9 modifies, in part, the requirements in 5-5.3.1.2.3(b) for active multiplex transmission systems. Subparagraph 5-5.3.1.2.3(b) requires Type 1 active multiplex systems to transmit all signals, including trouble signals, within 90 seconds.

The requirements of 5-3.4.9 also impose the 200-second limit on the receipt of trouble signals on the other transmission technologies when the transmission pathway functions normally.

5-3.4.10 The recorded information for the occurrence of any trouble condition of signaling line circuit, leg facility, or trunk facility that prevents receipt of alarm signals at the proprietary supervising station shall be such that the operator is able to determine the presence of the trouble condition. Trouble conditions in a leg facility shall not affect or delay receipt of signals at the proprietary supervising station from other leg facilities on the same trunk facility.

The last sentence of 5-3.4.10 effectively mandates that the transmission technology preserve the signals from other leg facilities when one leg facility experiences trouble. Management of a facility equipped with a proprietary supervising station fire alarm system would have to analyze each transmission technology chosen to determine if it could meet the requirements of 5-3.4.10. For example, if management chooses to use an active multiplex transmission technology, 5-3.4.10 dictates that the system meet the requirements of a Type 1 or Type 2 active multiplex system.

5-3.5 Personnel.

5-3.5.1 At least two operators shall be on duty at all times. One of the two operators shall be permitted to be a runner.

Exception: If the means for transmitting alarms to the fire department is automatic, at least one operator shall be on duty at all times.

The exception implies that where management provides automatic retransmission of fire alarm signals, the sole operator could respond as a runner.

5-3.5.2 When the runner is not in attendance at the proprietary supervising station, the runner shall establish two-way communications with the station at intervals not exceeding 15 minutes.

5-3.5.3 The primary duties of the operator(s) shall be to monitor signals, operate the system, and take such action as shall be required by the authority having jurisdiction. The operator(s) shall not be assigned any additional duties that would take precedence over the primary duties.

Although it does not prohibit the operators from performing other duties, the code expects the operators to have no duties that would distract them from the prompt, effective handling of signals.

5-3.6 Operations.

5-3.6.1 Communications and Transmission Channels.

5-3.6.1.1 All communications and transmission channels between the proprietary supervising station and the protected premises master control unit (panel) shall be operated manually or automatically once every 24 hours to verify operation.

5-3.6.1.2 If a communications or transmission channel fails to operate, the operator shall immediately notify the person(s) identified by the owner or authority having jurisdiction.

5-3.6.2 Operator Controls.

5-3.6.2.1 All operator controls at the proprietary supervising station(s) designated by the authority having jurisdiction shall be operated at each change of shift.

5-3.6.2.2 If operator controls fail, the operator shall immediately notify the person(s) identified by the owner or authority having jurisdiction.

5-3.6.3 Indication of a fire shall be promptly retransmitted to the public fire service communications center or other locations accepted by the authority having jurisdiction, indicating the building or group of buildings from which the alarm has been received.

The requirements of 5-3.6.1, 5-3.6.1.1, 5-3.6.1.2, 5-3.6.2, and 5-3.6.3 help the proprietary supervising station to continue operating. Exercising the communications channels and operating controls allows the operator to more quickly identify potential failures. Also, by promptly contacting designated persons when a failure occurs, the operators help ensure that repairs begin as soon as possible.

5-3.6.4* The means of retransmission shall be accepted by the authority having jurisdiction and shall be in accordance with Sections 5-2 or 5-4, or Chapter 6.

Exception: Secondary power supply capacity shall be as required in Chapter 1.

A-5-3.6.4 It is the intent of this code that the operator within the proprietary supervising station should have a secure means of immediately retransmitting any signal indicative of a fire to the public fire department communications center. Automatic retransmission using an approved method installed in accordance with Sections 5-2, 5-3, 5-4, and Chapter 6 is the best method for proper retransmission. However, a manual means can be permitted to be used, consisting of either a manual connection following the requirements of Sections 5-2, 5-4, and Chapter 6, or, for proprietary supervising stations serving only contiguous properties, a means in the form of a municipal fire alarm box installed within 50 ft (15 m) of the proprietary supervising station in accordance with Chapter 6 can be permitted.

Paragraph 5-3.6.4 requires that the proprietary supervising station retransmit signals to the public fire service communications center by means of a central station fire alarm system, a remote supervising station fire alarm system, or an auxiliary fire alarm system. It also permits a proprietary supervising station covering a contiguous property to retransmit signals to the public fire service communications center by means of manually actuating a municipal fire alarm box.

The authority having jurisdiction would have to make an exception to the requirements of 5-3.6.4 if management desired to retransmit alarms solely by using an ordinary telephone. (See commentary following A-5-3.6.5.)

5-3.6.5* Retransmission by coded signals shall be confirmed by two-way voice communications indicating the nature of the alarm.

A-5-3.6.5 Regardless of the type of retransmission facility used, telephone communications between the proprietary supervising station and the fire department should be available at all times and should not depend on a switchboard operator.

Management should provide the proprietary supervising station with a connection to the public switched telephone network that does not require the operator of a private branch exchange (PBX) switchboard to manually intervene to obtain access to the network. In fact, management should provide a direct connection to the network that completely bypasses the PBX. This set-up will allow operators in the proprietary supervising station to make a telephone call, even when the PBX fails.

5-3.6.6 Dispositions of Signals.

5-3.6.6.1 Alarms. Upon receipt of a fire alarm signal, the proprietary supervising station operator shall initiate action to perform the following:

(1) Immediately notify the fire department, the plant fire brigade, and such other parties as the authority having jurisdiction requires.
(2) Promptly dispatch a runner to the alarm location (travel time shall not exceed 1 hour).
(3) Restore the system as soon as possible after disposition of the cause of the alarm signal.

5-3.6.6.2 Guard's Tour Supervisory Signal. If a guard's tour supervisory signal is not received from a guard within a 15-minute maximum grace period, or if a guard fails to follow a prescribed route in transmitting the signals (where a prescribed route has been established), the supervisory signal shall be treated as a delinquency signal. If a guard's tour supervisory signal is delinquent, the proprietary supervising station operator shall initiate action to perform the following:

(1) Communicate at once with the protected areas or premises by telephone, radio, calling back over the system circuit, or other means accepted by the authority having jurisdiction.
(2) Dispatch a runner to investigate the delinquency, if communications with the guard cannot be promptly established. (Travel time shall not exceed $^1/_2$ hour.)

5-3.6.6.3 Supervisory Signals. Upon receipt of sprinkler system and other supervisory signals, the proprietary supervising station operator shall initiate action to perform the following, if required:

(1) Communicate immediately with the designated person(s) to ascertain the reason for the signal
(2) Dispatch a runner or maintenance person (travel time not to exceed 1 hour) to investigate, unless supervisory conditions are promptly restored
(3) Notify the fire department
(4) Notify the authority having jurisdiction when sprinkler systems are wholly or partially out of service for 8 hours or more
(5) Provide written notice to the authority having jurisdiction as to the nature of the signal, time of occurrence, and restoration of service, when equipment has been out of service for 8 hours or more

5-3.6.6.4 Trouble Signals. Upon receipt of trouble signals or other signals pertaining solely to matters of equipment maintenance of the fire alarm system, the proprietary supervising station operator shall initiate action to perform the following, if required:

(1) Communicate immediately with the designated person(s) to ascertain reason for the signal

(2) Dispatch a runner or maintenance person (travel time not to exceed 1 hour) to investigate

(3) Notify the fire department

(4) Notify the authority having jurisdiction when interruption of service exists for 4 hours or more

(5) Provide written notice to the authority having jurisdiction as to the nature of the signal, time of occurrence, and restoration of service, when equipment has been out of service for 8 hours or more

The requirements in 5-3.6.6 almost match those in 5-2.6.1 for central station fire alarm systems with the following notable differences:

- The term *immediately* has no specific definition.
- Upon receipt of an alarm, the proprietary supervising station must always dispatch a runner.
- The authority having jurisdiction may require the proprietary supervising station to notify the fire department upon receipt of a supervisory signal or a trouble signal. This notification will alert the fire department to impaired protection.
- The runner must respond to a trouble signal within 1 hour.
- The proprietary supervising station must notify the authority having jurisdiction if an interruption to service producing a trouble signal persists for 4 hours.

5-3.6.7 Record Keeping and Reporting.

5-3.6.7.1 Complete records of all signals received shall be retained for at least 1 year.

5-3.6.7.2 Testing and maintenance records shall be retained as required by 7-5.3.

5-3.6.7.3 The proprietary supervising station shall make arrangements to furnish reports of signals received to the authority having jurisdiction in a form the authority will accept.

When an authority having jurisdiction requests reports from a proprietary supervising station, the supervising station must provide the reports in a useful and usable form.

5-3.7 Testing and Maintenance.

Testing and maintenance of proprietary fire alarm systems shall be performed in accordance with Chapter 7.

5-4 Remote Supervising Station Fire Alarm Systems

The requirements of Chapters 1 and 7 and Section 5-5 shall apply to remote supervising station fire alarm systems, unless they conflict with the requirements of this section.

5-4.1 Scope.

Section 5-4 shall apply where central station service is neither required nor elected. Section 5-4 shall describe the installation, maintenance, testing, and use of a remote supervising station fire alarm system that serves properties under various ownership from a remote supervising station where trained, competent personnel are in constant attendance. Section 5-4 shall cover the minimum requirements for the remote supervising station physical facilities, equipment, operating personnel, response, retransmission, signals, reports, and testing.

An authority having jurisdiction may require a remote supervising station fire alarm system when that authority does not need the level of protection offered by a central station fire alarm system. Alternatively, the management of a facility may choose to provide a remote station fire alarm system when management does not believe it needs the level of protection offered by a central station fire alarm system or a proprietary supervising station fire alarm system.

When first appearing in the code in 1961, the requirements for a remote supervising station fire alarm system provided a means of transmitting fire alarm, supervisory, and trouble signals from a protected premises to the public fire service communications center. In most of these cases, the municipality did not have a public fire alarm reporting system, and no one could provide central station service to that locale.

Sometimes the municipality did not have a constantly attended public fire service communications center. Rather, officials relied on a multiple-location fire telephone system. When an individual placed a telephone call to the fire reporting number, telephones in several locations throughout the community rang. These locations included local businesses, as well as the homes of the fire chief and other fire officers. A switch at each telephone could actuate sirens throughout the community, or transmit signals to radio receivers or pagers that would summon the volunteer fire fighters.

In these communities, officials had to find an alternate location to receive signals from remote station fire alarm systems. Often, officials chose a local 24-hour telephone answering service used by doctors, dentists, or tradespeople to receive the remote supervising station fire alarm signals. In some cases, the officials chose a gasoline service station or local restaurant that remained open around the clock to receive the signals.

In later years, some alarm system installers who chose not to provide listed central station service set up monitoring centers to receive remote station fire alarm signals. In turn, some listed central station operating companies began to provide equipment that would meet the requirements for remote supervising station fire alarm systems in order to legitimately receive such signals.

5-4.2 General.

5-4.2.1 Remote supervising station fire alarm systems shall provide an automatic audible and visible indication of alarm and, if required, of supervisory and trouble conditions at a location remote from the protected premises. A manual or automatic permanent record of these conditions shall be provided.

Unlike central station and proprietary fire alarm systems that keep records automatically, remote station fire alarm systems may record signals in a manually written or type-written log.

5-4.2.2 Section 5-4 shall not require the use of audible or visible notification appliances other than those required at the remote supervising station. If it is desired to provide fire alarm evacuation signals in the protected premises, the alarm signals, circuits, and controls shall comply with the provisions of Chapter 3 and Chapter 4 in addition to the provisions of Section 5-4.

A remote station fire alarm system only needs to provide audible and visible notification at the remote supervising station. If either an authority having jurisdiction or a member of management desires to provide audible and visible notification appliances throughout a protected premises, he or she should refer to the requirements in Chapter 3 and Chapter 4 of the code.

5-4.2.3 The loading capacities of the remote supervising station equipment for any approved method of transmission shall be as designated in Section 5-5.

Remote station fire alarm systems have the full range of transmission technologies available, if those technologies meet the requirements of Section 5-5.

5-4.3* Facilities.

A-5-4.3 As a minimum, the room or rooms containing the remote supervising station equipment should have a 1-hour fire rating, and the entire structure should be protected by an alarm system complying with Chapter 3.

5-4.3.1 If a remote supervising station connection is used to transmit an alarm signal, the signal shall be received at the public fire service communications center, at a fire station, or at the governmental agency that has a public responsibility for taking prescribed action to ensure response upon receipt of a fire alarm signal.

Exception: If such an agency is unwilling to receive alarm signals or permits the acceptance of another location by the authority having jurisdiction, such alternate location shall have personnel on duty at all times who are trained to receive the alarm signal and immediately retransmit it to the fire department.

When conditions meet the requirements of the exception, the authority having jurisdiction may accept any suitable location as the remote supervising station. For example, an authority having jurisdiction could permit a listed central station to receive these signals. Accepting the use of a listed central station would constitute remote supervising station service, but not central station service. The equipment at the protected premises and at the central station must meet the requirements for remote supervising station fire alarm systems. According to the requirements, the equipment must be specifically listed for use as part of a remote supervising station fire alarm system. Also, the equipment must provide 60 hours standby power followed by power to operate for 5 minutes in alarm.

5-4.3.2 Supervisory and trouble signals shall be handled at a constantly attended location that has personnel on duty who are trained to recognize the type of signal received and to take prescribed action. The location shall be permitted to be other than that at which alarm signals are received.

In some installations, a remote station fire alarm system transmits fire alarm signals to the public fire service communications center, and transmits supervisory signals and trouble signals to another location acceptable to the authority having jurisdiction.

5-4.3.3 If locations other than the public fire service communications center are used for the receipt of signals, access to receiving equipment shall be restricted in accordance with the requirements of the authority having jurisdiction.

The requirement of 5-4.3.3 helps to ensure the security and operational integrity of the remote station receiving equipment.

5-4.4 Equipment.

5-4.4.1 Signal-receiving equipment shall indicate receipt of each signal both audibly and visibly.

5-4.4.1.1 Audible signals shall meet the requirements of Chapter 4 for the private operating mode.

See 4-3.3 for requirements for private mode audible signaling.

5-4.4.1.2 Means for silencing alarm, supervisory, and trouble signals shall be provided and shall be arranged so that subsequent signals shall re-sound.

Silencing one signal must not prevent a subsequent signal from causing the audible notification appliance to sound.

5-4.4.1.3 A trouble signal shall be received when the system or any portion of the system at the protected premises is placed in a bypass or test mode.

The requirements in 5-4.4.1.3 prevent any type of so-called silent disconnect switch at the protected premises. Opera-

tion of any disconnect switch must produce a trouble signal at the remote supervising station.

5-4.4.1.4 An audible and visible indication shall be provided upon restoration of the system after receipt of any signal.

In contrast with the requirements of 3-8.3.3.1.3, in 5-4.4.1.4 the code does not permit the indication of restoration to normal by merely extinguishing a lamp. The audible notification appliance at the remote supervising station must also sound.

5-4.4.1.5 If visible means are provided in the remote supervising station to identify the type of signal received, a common audible notification appliance shall be permitted to be used.

5-4.4.2 Power supplies shall comply with the requirements of Chapter 1.

Exception: In a remote supervising station fire alarm system where the alarm and supervisory signals are transmitted over a listed supervised one-way radio system, 24 hours of secondary (standby) power shall be permitted in lieu of 60 hours, as required in 1-5.2.5.3, at the radio alarm repeater station receivers (RARSR), provided that personnel are dispatched to arrive within 4 hours after detection of failure to initiate maintenance.

Many of the one-way radio alarm system sites that house a radio alarm repeater station receiver (RARSR) have a minimal footprint in equipment rooms. For example, at the top of a high-rise building, alarm service providers lease such space at a very high cost. This small footprint does not give adequate room for 60 hours of standby batteries. The exception permits the use of 24 hours of standby power as long as a technician will arrive within 4 hours of the receipt of a trouble signal from the RARSR.

5-4.4.3 Transmission means shall comply with the requirements of Section 5-5.

Remote station fire alarm systems have the full range of transmission technologies available from Section 5-5, as long as they meet any special requirements of Section 5-4.

5-4.4.4 Retransmission of an alarm signal, if required, shall be by one of the following methods, which appear in descending order of preference as follows.

(a) A dedicated circuit that is independent of any switched telephone network. This circuit shall be permitted to be used for voice or data communications.

(b) A one-way (outgoing only) telephone at the remote supervising station that utilizes the public switched telephone network. This telephone shall be used primarily for voice transmission of alarms to a telephone at the public fire service communications center that cannot be used for outgoing calls.

(c) A private radio system using the fire department frequency, where permitted by the fire department.

(d) Other methods accepted by the authority having jurisdiction.

The vast majority of remote supervising stations will use a retransmission method that complies with the requirements of 5-4.4.4(b). The telephone utility switch blocks incoming telephone calls to the number, allowing operators at the remote supervising station to make outgoing calls on this line, but not to receive incoming calls on the line.

5-4.5 Personnel.

The authority having jurisdiction may permit operators at the remote supervising station to perform other duties not related to the operation of the system. However, the code limits the extent to which these other duties may interfere with the proper handling of signals.

5-4.5.1 The remote supervising station shall have sufficient personnel, but not less than two persons, on duty at the remote supervising station at all times to ensure disposition of signals in accordance with the requirements of 5-4.6.

5-4.5.2 Duties pertaining to other than operation of the remote supervising station receiving and transmitting equipment shall be permitted subject to the approval of the authority having jurisdiction.

5-4.6 Operations.

5-4.6.1 If the remote supervising station is at a location other than the public fire service communications center, alarm signals shall be immediately retransmitted to the public fire service communications center.

5-4.6.2 Upon receipt of an alarm, supervisory, or trouble signal by the remote supervising station other than the public fire service communications center, the operator on duty shall be responsible for notifying the owner or the owner's designated representative immediately.

Promptly contacting designated persons when a failure occurs helps to ensure that repairs will begin as soon as possible.

5-4.6.3 All operator controls at the remote supervising station shall be operated at the beginning of each shift or change in personnel, and the status of all alarm, supervisory, and trouble signals shall be noted and recorded.

The requirements of subsection 5-4.6.3 provide for the operational continuity of the supervising station. By exercising operating controls, operators can more quickly identify potential failures.

5-4.7 Record Keeping and Reporting.

5-4.7.1 A permanent record of the time, date, and location of all signals and restorations received and the action taken shall be maintained for at least 1 year and shall be made available to the authority having jurisdiction. These records shall be permitted to be created by manual means.

Most often, a manually maintained, either handwritten or typewritten log book will contain the required records.

5-4.7.2 Testing and maintenance records shall be retained as required in 7-5.3. These records shall be permitted to be created by manual means.

5-4.8 Testing and Maintenance.

Testing and maintenance for remote supervising stations shall be performed in accordance with Chapter 7.

5-5 Communications Methods for Supervising Station Fire Alarm Systems

The requirements of Chapters 1 and 7 shall apply to continuously attended supervising station fire alarm systems, unless they conflict with the requirements of this section.

5-5.1* Scope.

Section 5-5 and Figure 5-5.1 shall describe the requirements for the methods of communications between the protected premises and the supervising station. These requirements shall include the following:

(1) The transmitter located at the protected premises

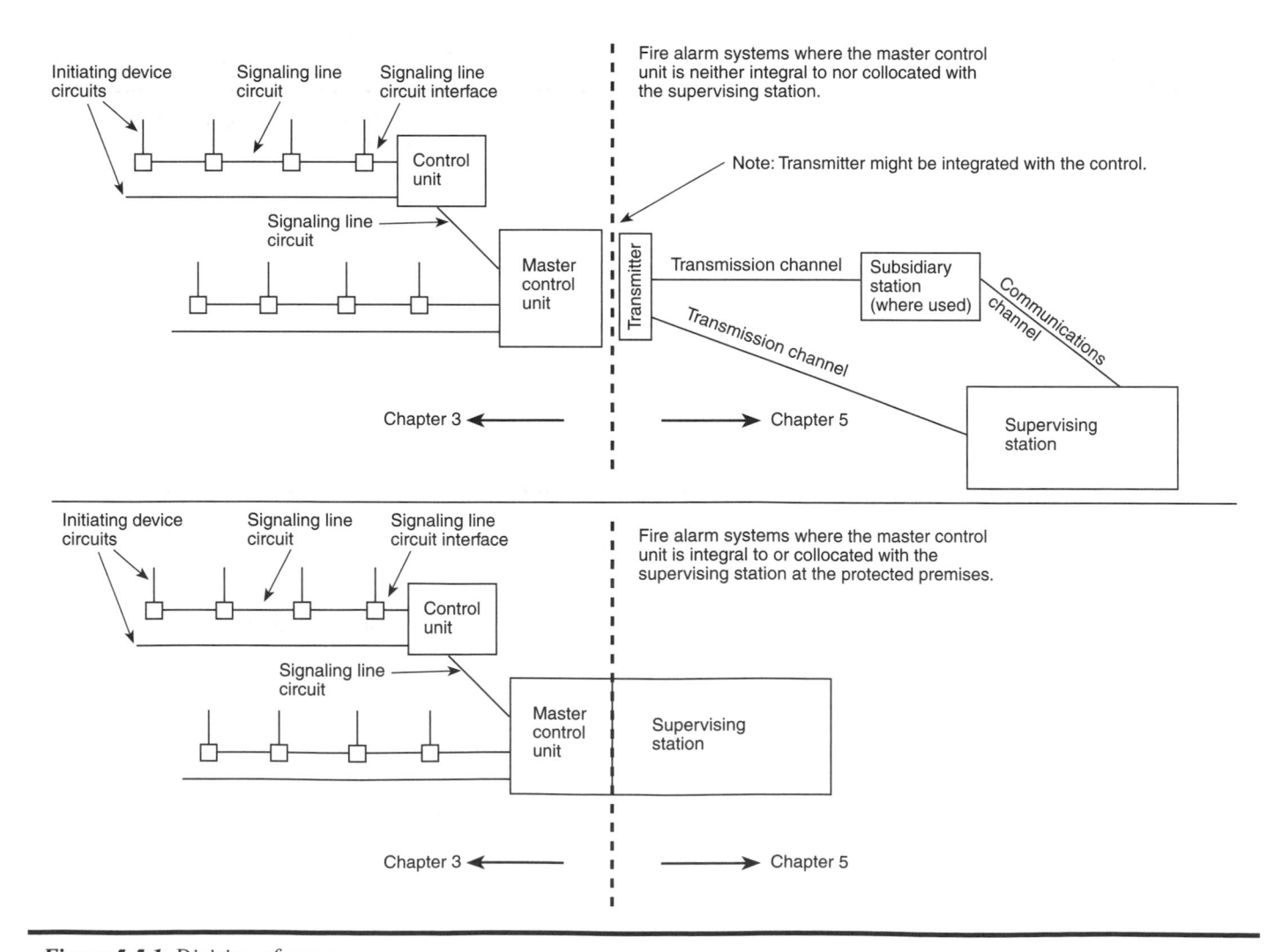

Figure 5-5.1 *Division of scope.*

(2) The transmission channel between the protected premises and the supervising station or subsidiary station

(3) If used, any subsidiary station and its communications channel

(4) The signal receiving, processing, display, and recording equipment at the supervising station

Exception: Transmission channels owned by and under the control of the protected premises owner, that are not facilities leased from a supplier of communications service capabilities, such as video cable, telephone, or other communications services that are also offered to other customers.

The code makes a full range of transmission technologies available to all of the supervising station services. This range of transmission technologies gives designers maximum flexibility in choosing the transmission technology most appropriate for the particular application. Available technologies include the following:

- Active multiplex, including derived local channel
- Digital alarm communicator systems; digital alarm radio systems
- McCulloh systems
- Two-way radio frequency multiplex systems
- One-way radio alarm systems
- Directly connected, noncoded systems
- Private microwave radio systems

However, specific requirements of each section of Chapter 5 may limit the use of a particular transmission technology. The code has no jurisdiction over public utilities such as telephone service. See Exhibit 5.6 for a graphic representation of communications methods used for supervising stations.

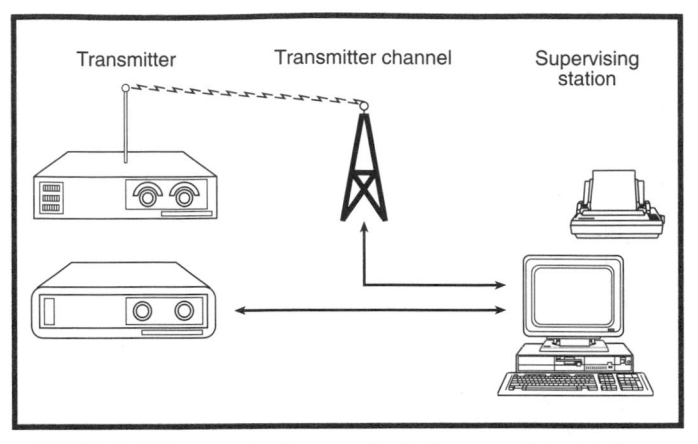

Exhibit 5.6 *Communications methods for supervising station fire alarm systems. (Source: R.P. Schifiliti Associates, Inc., Reading, MA)*

A-5-5.1 Refer to Table A-5-5.1 for communications methods.

Table A-5-5.1 Communications Methods for Supervising Stations

Criteria	5-5.3.1 Active Multiplex	5-5.3.2 Digital Alarm Communicator Systems	5-5.3.3 McCulloh Systems	5-5.3.4 Two-Way Radio Frequency (RF) Multiplex Systems	5-5.3.5 One-Way Private Radio Alarm Systems	5-5.3.6 Directly Connected Noncoded Systems	5-5.3.7 Private Microwave Radio Systems	5-5.4 Other Transmission Technologies
FCC approval when applicable	Yes	Yes	Yes	Yes	Yes	Yes	Yes	Yes
Conform to NFPA 70, *National Electrical Code*	Yes	Yes	Yes	Yes	Yes	Yes	Yes	Yes

Table A-5-5-1 Continued

Criteria	5-5.3.1 Active Multiplex	5-5.3.2 Digital Alarm Communicator Systems	5-5.3.3 McCulloh Systems	5-5.3.4 Two-Way Radio Frequency (RF) Multiplex Systems	5-5.3.5 One-Way Private Radio Alarm Systems	5-5.3.6 Directly Connected Noncoded Systems	5-5.3.7 Private Microwave Radio Systems	5-5.4 Other Transmission Technologies
Monitoring for integrity of the transmission and communications channel	Systems are periodically polled for end-to-end communications integrity	Both the premises unit and the system unit monitor for integrity in a manner approved for the means of transmission employed. A single signal received on each incoming DACR line once every 24 hours	Continuous dc supervision	Systems are periodically polled for end-to-end communications integrity	Test signal from every transmitter once every 24 hours	Continuous dc supervision	Used as a portion of another type of transmission technology. End-to-end integrity monitored by the main transmission technology. Microwave portion is continuously monitored	Monitor for integrity or provide back-up channel tested as below
Annunciate, at the supervising station, the degradation and restoration of the transmission or communications channel	Within 200 seconds for Type III multiplex. Within 90 seconds for Type I and II multiplex.	Within 4 minutes using alternate phone line to report the trouble	Indicate automatically and operation under fault condition achieved either manually or automatically	Not exceed 90 seconds from the time of the actual failure	Only monitor the quality of signal received and indicate if the signal falls below minimum signal quality specified in code	Presented in a form to expedite prompt operator interpretation	Presented in a form to expedite prompt operator interpretation	Within 5 minutes (may use a second separate path to report failure)
Redundant communication path where a portion of the transmission or communications channel cannot be monitored for integrity	Redundant path not required—supervising station always indicates a communications failure	Employ a combination of two separate transmission channels alternately tested at intervals not exceeding 24 hours	Redundant path not required—supervising station always indicates a communications failure. Exception is the use of nonmetallic channels that require two channels or an immediate transfer to a standby channel	Redundant path not required—supervising station always indicates a communications failure	Minimum of two independent RF paths must be simultaneously employed	None required	Dual transmitters required if more than 5 buildings or premises or 50 initiating devices circuits	Provide a redundant path if communication failure not annunciated at supervising station

(continued)

Table A-5-5-1 Continued

Criteria	5-5.3.1 Active Multiplex	5-5.3.2 Digital Alarm Communicator Systems	5-5.3.3 McCulloh Systems	5-5.3.4 Two-Way Radio Frequency (RF) Multiplex Systems	5-5.3.5 One-Way Private Radio Alarm Systems	5-5.3.6 Directly Connected Noncoded Systems	5-5.3.7 Private Microwave Radio Systems	5-5.4 Other Transmission Technologies
Interval testing of the back-up path(s)	For Type I, 1-hour testing for dedicated lines and 24-hour testing for dial-up. No requirement for Type II and III.	When two phone lines are used, test alternately every 24 hours. Testing for other back-up technologies, see 5-5.3.2.1.6.2	No testing requirement	Back-up path not required	No requirement because the quality of the signal is continuously monitored	Back-up path not required	Dual transmitters shall be operated on time ratio of 2:1 within each 24-hours	If back-up path required, test path once every 24-hours on alternating channels testing each channel every 48-hours
Annunciation of communication failure or ability to communicate at the protected premises	Not required—always annunciated at the supervising station that initiates corrective action	Indication of failure at premises due to line failure or failure to communicate after from 5 to 10 dialing attempts	Not required—always annunciated at the supervising station that initiates corrective action	Not required—always annunciated at the supervising station that initiates corrective action	Monitor the interconnection of the premises unit elements of transmitting equipment and indicate a failure at the premises or transmit a trouble signal to the supervising station	None required	None required	Systems where the transmitter at the local premises unit detects a communication failure before the supervising station, the premises unit will annunciate the failure within 5 minutes of detecting the failure
Time to restore signal receiving, processing, display, and recording equipment	Where duplicate equipment not provided, spare hardware required so a repair can be effected within 30 minutes	Spare digital alarm communicator receivers required for switchover to back-up receiver in 30 seconds. One back-up system unit for every 5 system units.	Where duplicate equipment not provided, spare hardware required so a repair can be effected within 30 minutes	Where duplicate equipment not provided, spare hardware required so a repair can be effected within 30 minutes	Where duplicate equipment not provided, spare hardware required so a repair can be effected within 30 minutes	Where duplicate equipment not provided, spare hardware required so a repair can be effected within 30 minutes	Where duplicate equipment not provided, spare hardware required so a repair can be effected within 30 minutes	Where duplicate equipment not provided, spare hardware required so a repair can be effected within 30 minutes. Complete set of critical spare parts on a 1 to 5 ratio of parts to system units or a duplicate functionally equivalent system unit for every 5 system units

Table A-5-5-1 Continued

Criteria	5-5.3.1 Active Multiplex	5-5.3.2 Digital Alarm Communicator Systems	5-5.3.3 McCulloh Systems	5-5.3.4 Two-Way Radio Frequency (RF) Multiplex Systems	5-5.3.5 One-Way Private Radio Alarm Systems	5-5.3.6 Directly Connected Noncoded Systems	5-5.3.7 Private Microwave Radio Systems	5-5.4 Other Transmission Technologies
Loading capacities for system units and transmission and communications channels	512 buildings and premises on one system unit. Unlimited if you can switch to duplicate system unit within 30 seconds. Loading capacity for transmission and communications channels (trunks) is listed in Table 5-5.3.1.4.	See Table 5-5.3.2.2.2.3 for the maximum number of transmitters on a hunt group in a system unit	Alarm and sprinkler supervisory limited to 25 plants and 250 code wheels on one circuit. 60 scheduled guard reports per hour	512 buildings and premises on a system unit with no back-up. Unlimited if you can switch to a back-up in 30 seconds.	512 buildings and premises on a system unit with no back-up. Unlimited if you can switch to a back-up in 30 seconds.	A single circuit must not serve more than one plant.	Up to 5 buildings or premises or 50 initiating device circuits on one transmitter. Unlimited if dual transmitters are used with automatic switchover or manual switchover in 30 seconds	512 independent fire alarm systems on a system unit with no back-up. Unlimited if you can switch to a back-up in 30 seconds. The system shall be designed such that a failure of a transmission channel serving a system unit shall not result in the loss in the ability to monitor more than 3000 transmitters
End-to-end communication time for an alarm	90 seconds from initiation until it is recorded	Off-hook to on-hook not to exceed 90 seconds per attempt. 10 attempts maximum. 900 seconds maximum for all attempts.	Not addressed	90 seconds from initiation until it is recorded	90% probability to receive an alarm in 90 seconds, 99% probability in 180 seconds, 99.999% probability in 450 seconds	Not addressed	Not addressed	90 seconds from initiation of alarm until displayed to the operator and recorded on a media from which the information can be retrieved
Record and display rate of subsequent alarms at supervising station	Not slower than one every 10 additional seconds	Not addressed	Not addressed	When any number of subsequent alarms come in, record at a rate not slower than one every additional 10 seconds	When any number of subsequent alarms come in, record at a rate not slower than one every additional 10 seconds	Not addressed	Not addressed	No slower than one every 10 additional seconds

(continues)

Table A-5-5-1 Continued

Criteria	5-5.3.1 Active Multiplex	5-5.3.2 Digital Alarm Communicator Systems	5-5.3.3 McCulloh Systems	5-5.3.4 Two-Way Radio Frequency (RF) Multiplex Systems	5-5.3.5 One-Way Private Radio Alarm Systems	5-5.3.6 Directly Connected Noncoded Systems	5-5.3.7 Private Microwave Radio Systems	5-5.4 Other Transmission Technologies
Signal error detection and correction	Not addressed	Signal repetition, digital parity check, or some equivalent means of signal verification must be used	Not addressed	Not addressed	Not addressed	Not applicable	Not addressed	Signal repetition, parity check, or some equivalent means of error detection and correction shall be used
Path sequence priority	Not addressed	The first transmission attempt uses the primary channel	Not addressed	Not addressed	Not addressed	Not addressed	Not addressed	No need for prioritization of paths. The requirement is that both paths are equivalent
Carrier diversity	None required	Where long distance service (including WATS) is used, the second telephone number shall be provided by a different long distance service provider where there are multiple providers	Not addressed	Not addressed	Not addressed	Not addressed	Not addressed	When a redundant path is required, the alternate path shall be provided by a public communication service provider different from the primary path where available

Table A-5-5-1 Continued

Criteria	5-5.3.1 Active Multiplex	5-5.3.2 Digital Alarm Communicator Systems	5-5.3.3 McCulloh Systems	5-5.3.4 Two-Way Radio Frequency (RF) Multiplex Systems	5-5.3.5 One-Way Private Radio Alarm Systems	5-5.3.6 Directly Connected Noncoded Systems	5-5.3.7 Private Microwave Radio Systems	5-5.4 Other Transmission Technologies
Throughput probability	Not addressed	Demonstrate 90% probability of a system unit immediately answering a call or follow the loading Table 5-5.3.2.2.2.3. One-way radio back-up demonstrates 90% probability of transmission.	Not addressed	Not addressed	90% probability to receive an alarm in 90 seconds. 99% probability in 180 seconds. 99.999% in probability 450 seconds	Not addressed	Not addressed	When the supervising station does not regularly communicate with the transmitter at least once every 200 seconds, then the throughput probability of the alarm transmission must be at least 90% in 90 seconds, 99% in 180 seconds, 99.999% in 450 seconds
Unique premises identifier	Yes	Yes	Yes	Yes	Yes	Yes	Yes	If a transmitter shares a transmission or communication channel with other transmitters, it shall have a unique transmitter identifier
Unique flaws	None addressed	If call forwarding is used to communicate to the supervising station, verify the integrity of this feature every 4 hours	None addressed	None addressed	None addressed	None addressed	None addressed	From time to time, there may be unique flaws in a communication system. Unique requirements shall be written for these unique flaws

(continues)

Table A-5-5-1 Continued

Criteria	5-5.3.1 Active Multiplex	5-5.3.2 Digital Alarm Communicator Systems	5-5.3.3 McCulloh Systems	5-5.3.4 Two-Way Radio Frequency (RF) Multiplex Systems	5-5.3.5 One-Way Private Radio Alarm Systems	5-5.3.6 Directly Connected Noncoded Systems	5-5.3.7 Private Microwave Radio Systems	5-5.4 Other Transmission Technologies
Signal priority	Fire alarm, supervisory, and trouble signals shall take precedence, in that respective order of priority, over all other signals (except life threatening signals over supervisory and trouble)	Chapter 1 on fundamentals requires that alarm signals take priority over supervisory signals unless there is sufficient repetition of the alarm signal to prevent the loss of an alarm signal	Chapter 1 on fundamentals requires that alarm signals take priority over supervisory signals unless there is sufficient repetition of the alarm signal to prevent the loss of an alarm signal	Chapter 1 on fundamentals requires that alarm signals take priority over supervisory signals unless there is sufficient repetition of the alarm signal to prevent the loss of an alarm signal	Chapter 1 on fundamentals requires that alarm signals take priority over supervisory signals unless there is sufficient repetition of the alarm signal to prevent the loss of an alarm signal	Chapter 1 on fundamentals requires that alarm signals take priority over supervisory signals unless there is sufficient repetition of the alarm signal to prevent the loss of an alarm signal	Chapter 1 on fundamentals requires that alarm signals take priority over supervisory signals unless there is sufficient repetition of the alarm signal to prevent the loss of an alarm signal	If the communication methodology is shared with any other usage, all fire alarm transmissions shall preempt and take precedence over any other usage. Fire alarm signals take precedent over supervisory signals
Sharing communications equipment on premises	Not addressed	Disconnect outgoing or incoming telephone call and prevent its use for outgoing telephone calls until signal transmission has been completed	Not addressed	Not addressed	Not addressed	Not addressed	Not addressed	If the transmitter is sharing on-premises communications equipment, the shared equipment shall be listed for the purpose (otherwise the transmitter must be installed ahead of the unlisted equipment)

5-5.2 General.

5-5.2.1 Applicable Requirements.

Over the years, as the code encompassed new transmission technologies, it most often compared them to the performance capabilities of the McCulloh system, which was the first transmission technology used by supervising station systems. This comparison allowed an intermixing of certain operational requirements among the various technologies. Therefore, 5-5.2.1 becomes a caveat to make certain that when applying a particular transmission technology, authorities having jurisdiction, as well as system designers, installers, and users, do not ignore critical operational requirements that appear in only one section of the code.

5-5.2.1.1 If the protected premises master control unit is neither integral to nor co-located with the supervising station, the communications methods of Section 5-5 shall be

used to connect the protected premises to either a subsidiary station, if used, or a supervising station providing central station service in accordance with Section 5-2, proprietary service in accordance with Section 5-3, or remote station service in accordance with Section 5-4. These communications methods shall be permitted to include the following:

(1) Active multiplex circuits that are part of a supervising station, including systems derived channels
(2) Digital alarm communicator systems, including digital alarm radio systems
(3) McCulloh systems
(4) Two-way radio frequency (RF) multiplex systems
(5) One-way radio alarm systems
(6) Directly-connected noncoded systems

5-5.2.1.2* Nothing in Chapter 5 shall be interpreted as prohibiting the use of listed equipment using alternate communications methods that provide a level of reliability and supervision consistent with the requirements of Chapter 1 and the intended level of protection.

A-5-5.2.1.2 It is not the intent of Section 5-5 to limit the use of listed equipment using alternate communications methods, provided these methods demonstrate performance characteristics that are equal to or superior to those technologies described in Section 5-5. Such demonstration of equivalency is to be evidenced by the equipment using the alternate communications methods meeting all the requirements of Chapter 1, including those that deal with such factors as reliability, monitoring for integrity, and listing. It is further expected that suitable proposals stating the requirements for such technology will be submitted for inclusion in subsequent editions of this code.

5-5.2.1.3 For multiple building premises, the requirements of 1-5.7.4 shall apply to the alarm, supervisory, and trouble signals transmitted to the supervising station.

5-5.2.2 Equipment.

5-5.2.2.1 Fire alarm system equipment and installations shall comply with Federal Communication Commission (FCC) rules and regulations, as applicable, concerning the following:

(1) Electromagnetic radiation
(2) Use of radio frequencies
(3) Connection to the public switched telephone network of telephone equipment, systems, and protection apparatus

Subparagraph 5-5.2.2.1 recognizes that the Federal Communications Commission (FCC) has jurisdiction over the installation requirements for certain communications equipment used to transmit signals from a protected premises to a supervising station.

5-5.2.2.2 Radio receiving equipment shall be installed in compliance with NFPA 70, *National Electrical Code*®, Article 810.

When a particular transmission technology uses television or radio equipment, that equipment, and in particular its antenna, must be installed in compliance with the appropriate articles of NFPA 70, *National Electrical Code*®.

5-5.2.2.3 The external antennas of all radio transmitting and receiving equipment shall be protected in order to minimize the possibility of damage by static discharge or lightning.

5-5.2.3 Adverse Conditions.

5-5.2.3.1 For active and two-way RF multiplex systems, the occurrence of an adverse condition on the transmission channel between a protected premises and the supervising station that prevents the transmission of any status change signal shall be automatically indicated and recorded at the supervising station. This indication and record shall identify the affected portions of the system so that the supervising station operator will be able to determine the location of the adverse condition by trunk or leg facility, or both.

Interrogation and response transmission back and forth along the communication path monitors the integrity of active multiplex transmission technology. The satisfactory exchange of data ensures that all trunks and legs remain operational. If the system does not successfully complete an interrogation and response sequence, the possible failure of a trunk or a leg is indicated. In such a case, 5-5.2.3.1 requires the system to notify the supervising station and provide sufficient detail to allow prompt troubleshooting and repair of the trunk or leg. This adverse condition must be indicated as a trouble signal as per the requirements of 1-5.8.1.

5-5.2.3.2 For a one-way radio alarm system, the system shall be supervised to ensure that at least two independent radio alarm repeater station receivers (RARSR) are receiving signals for each radio alarm transmitter (RAT) during each 24-hour period. The occurrence of a failure to receive a signal by either RARSR shall be automatically indicated and recorded at the supervising station. The indication shall identify which RARSR failed to receive such supervisory signals. Received test signals shall not be required to be indicated at the supervising station.

The satisfactory receipt of at least one transmission every 24 hours by at least two independent RARSRs monitors the integrity of one-way radio transmission technology. If receivers do not receive such a signal, then 5-5.2.3.2 requires the system to notify the supervising station and provide sufficient detail to allow prompt troubleshooting and repair of the trunk or leg. This adverse condition must be indicated as a trouble signal as per the requirements of 1-5.8.1.

5-5.2.3.3 For active and two-way RF multiplex systems that are part of a central station fire alarm system, restoration of service to the affected portions of the system shall be automatically recorded. When service is restored, the first status change of any initiating device circuit, or any initiating device directly connected to a signaling line circuit, or any combination thereof that occurred at any of the affected premises during the service interruption, also shall be recorded.

Central station equipment must automatically record restoration of interrupted service and report the first status change on any initiating device circuit. This reporting procedure means that for each initiating device circuit, the equipment at the protected premises must retain and later report the first status change that occurs during the transmission interruption.

5-5.2.4 Dual Control.

5-5.2.4.1 Dual control, if required, shall provide for redundancy in the form of a standby circuit or other alternate means of transmitting signals over the primary trunk portion of a transmission channel. The same method of signal transmission shall be permitted to be used over separate routes, or alternate methods of signal transmission shall be permitted to be used. Public switched telephone network facilities shall be used only as an alternate method of transmitting signals.

Although dual control does not provide full redundancy for every trunk and leg, it does offer an option that an authority having jurisdiction or system designer can choose to help ensure the receipt of signals during interruptions to the primary trunk. Most often, technology called DataPhone Select-A-Station (DSAS) that is offered by the public telephone utility provides this redundancy. When the primary trunk fails, this technology allows the supervising station to either automatically or manually dial into the public switched telephone network and establish an alternate path for the signals that would normally transmit over the primary trunk. Telephone technicians sometimes refer to this arrangement as "dial up, make good."

5-5.2.4.2 If using facilities leased from a telephone company, that portion of the primary trunk facility between the supervising station and its serving wire center shall not be required to comply with the separate routing requirement of the primary trunk facility. Dual control, if used, shall require supervision as follows.

(1) Dedicated facilities that are able to be used on a full-time basis, and whose use is limited to signaling purposes as defined in this code, shall be exercised at least once every hour.

(2) Public switched telephone network facilities shall be exercised at least once every 24 hours.

To ensure that the dual control system can use the alternate path, operators must exercise the path. Operators must exercise a dedicated path once each hour and they must exercise a path provided as part of the public switched telephone network at least once each day.

5-5.3 Communications Methods.

5-5.3.1 Active Multiplex Transmission Systems.

5-5.3.1.1 The multiplex transmission channel shall terminate in a transmitter at the protected premises and in a system unit at the supervising station. The derived channel shall terminate in a transmitter at the protected premises and in derived channel equipment at a subsidiary station location or a telephone company wire center. The derived channel equipment at the subsidiary station location or a telephone company wire center shall select or establish the communications with the supervising station.

The protected facility may own its own equipment or it may lease some or all of the equipment from an alarm service provider. In the case of a derived local channel the protected facility or the alarm service provider may lease some of the equipment from the public telephone utility.

See Exhibit 5.7 for an illustration of active multiplex transmission methods.

Exhibit 5.7 *Communications methods—active multiplex transmission systems. (Source: R.P. Schifiliti Associates, Inc., Reading, MA)*

5-5.3.1.2* Operation of the transmission channel shall conform to the requirements of this code whether channels are private facilities, such as microwave, or leased facilities furnished by a communications utility company. If private signal transmission facilities are used, the equipment necessary to transmit signals shall also comply with the requirements for duplicate equipment or replacement of critical compo-

nents, as described in 5-5.5.2. The trunk transmission channels shall be dedicated facilities for the main channel. For Type 1 multiplex systems, the public switched telephone network facilities shall be permitted to be used for the alternate channel.

Exception: Derived channel scanners with no more than 32 legs shall be permitted to use the public switched telephone network for the main channel.

One manufacturer of derived local channel equipment offers an option of connecting the scanner in the telephone utility wire center to the supervising station by means of a dial-up modem that essentially performs as if it were a digital alarm communicator transmitter (DACT). The exception to 5-5.3.1.2 limits the loading of such a scanner to no more than 32 legs (protected premises).

A-5-5.3.1.2 Where derived channels are used, normal operating conditions of the telephone equipment are not to inhibit or impair the successful transmission of signals. These normal conditions include, but are not limited to, the following:

(1) Intraoffice calls with a transponder on the originating end
(2) Intraoffice calls with a transponder on the terminating end
(3) Intraoffice calls with transponders on both ends
(4) Receipt and origination of long distance calls
(5) Calls to announcement circuits
(6) Permanent signal receiver off-hook tone
(7) Ringing with no answer, with transponder on either the originating or the receiving end
(8) Calls to tone circuits (that is, service tone, test tone, busy, or reorder)
(9) Simultaneous signal with voice source
(10) Simultaneous signal with data source
(11) Tip and ring reversal
(12) Cable identification equipment

5-5.3.1.2.1 Derived channel signals shall be permitted to be transmitted over the leg facility, which shall be permitted to be shared by the telephone equipment under all on-hook and off-hook operating conditions.

Subparagraph 5-5.3.1.2.1 embodies the whole concept of derived local channel: it shares the leg facility used by the plain old telephone service (POTS) for a particular protected premises. The derived local channel system permits the transmission of alarm, supervisory, and trouble signals, even when normal telephone communications use the leg. An interrogation and response sequence, similar to that used by other active multiplex systems, monitors the integrity of the leg.

5-5.3.1.2.2 If derived channel equipment uses the public switched telephone network to communicate with a super-

vising station, such equipment shall meet the requirements of 5-5.3.2.

5-5.3.1.2.3 The maximum end-to-end operating time parameters allowed for an active multiplex system are as follows.

(a) The maximum allowable time lapse from the initiation of a single fire alarm signal until it is recorded at the supervising station shall not exceed 90 seconds. When any number of subsequent fire alarm signals occur at any rate, they shall be recorded at a rate no slower than one every 10 additional seconds.

Subparagraph 5-5.3.1.2.3(a) was designed to ensure that the system will complete an interrogation and response sequence at least every 90 seconds. As an alternative, the system may provide some other means to ensure alarm receipt in that time frame. For example, a designer could devise equipment that immediately transmits alarm signals from any multiplex interface at the protected premises, regardless of what point the system has reached in its normal 90-second interrogation and response sequence. However, most systems complete the interrogation and response sequence within 90 seconds.

(b)* The maximum allowable time lapse from the occurrence of an adverse condition in any transmission channel until recording of the adverse condition is started shall not exceed 90 seconds for Type 1 and Type 2 systems and 200 seconds for Type 3 systems. The requirements of 5-5.3.1.3 shall apply.

The reporting of an adverse condition on a Type 3 system within 200 seconds allows for a system, as described in 5-5.3.1.2.3, in which the equipment at the protected premises transmits alarm signals from any multiplex interface at the protected premises, regardless of what point the system has reached in its normal 200-second interrogation and response sequence. This requirement ensures that the interrogation and response sequence will occur at least every 200 seconds.

A-5-5.3.1.2.3(b) Derived channel systems comprise Type 1 and Type 2 systems only.

(c) In addition to the maximum operating time allowed for fire alarm signals, the requirements of one of the following shall be met:

(1) A system unit that has more than 500 initiating device circuits shall be able to record not less than 50 simultaneous status changes in 90 seconds.
(2) A system unit having fewer than 500 initiating device circuits shall be able to record not less than 10 percent of the total number of simultaneous status changes within 90 seconds.

Exception: Proprietary supervising station systems that have operating time requirements specified in 5-3.4.7 through 5-3.4.9.

These requirements ensure that the portion of the multiplex system that processes and records status changes can do so with sufficient speed to handle a reasonable volume of signal traffic, based on the system's signal capacity.

5-5.3.1.3 System Classification. Active multiplex systems shall be divided into three categories based on their ability to perform under adverse conditions of their transmission channels. The system classifications shall be as follows.

(a) A Type 1 system shall have dual control as described in 5-5.2.4. An adverse condition on a trunk or leg facility shall not prevent the transmission of signals from any other trunk or leg facility, except those signals dependent on the portion of the transmission channel in which the adverse condition has occurred. An adverse condition limited to a leg facility shall not interrupt service on any trunk or other leg facility. The requirements of 5-5.2.1, 5-5.2.2, and 5-5.2.3 shall be met by Type 1 systems.

To meet the requirements for a Type 1 system, an active multiplex system must isolate each system leg and trunk from other legs and trunks on the system. To do this wherever two or more trunks or two or more legs converge, the equipment uses a device called a "closed window bridge." Coupling circuitry within the closed window bridge allows signals to pass, but keeps a fault on one leg or trunk from interfering with signals from another leg or trunk. The public telephone utility may supply the bridge and locate it in a public telephone wire center or at a protected premises. The protected property or the alarm service provider may own or lease the bridge and determine its location. For example, equipment could multiplex the fire alarm systems for the individual stores in a shopping mall and for the mall common areas through a bridge located at an equipment room in the mall.

Type 1 systems also employ dual control, as described in 5-5.2.4. This gives the system an alternate transmission path should the primary trunk fail.

(b) A Type 2 system shall have the same requirements as a Type 1 system.

Exception: Dual control of the primary trunk facility shall not be required.

A Type 2 system also uses a closed window bridge to provide isolation between trunks and legs. Type 2 systems do not need to provide dual control. Thus, a Type 2 system has no alternate transmission path if the primary trunk fails.

(c) A Type 3 system shall automatically indicate and record at the supervising station the occurrence of an adverse condition on the transmission channel between a protected premises and the supervising station. The requirements of 5-5.2 shall be met.

Exception: The requirements of 5-5.2.4 shall not apply.

Type 3 systems have no requirement for isolation between legs and trunks. They commonly employ an "open window bridge." Although this device allows the coupling of signals wherever two or more legs or two or more trunks converge, an adverse condition on one leg or trunk may affect the operation of other legs or trunks. As the exception points out, Type 3 systems do not need to provide dual control. Thus, a Type 3 system has no alternate transmission path if the primary trunk fails.

5-5.3.1.4 System Loading Capacities. The capacities of active multiplex systems are based on the overall reliability of the signal receiving, processing, display, and recording equipment at the supervising and subsidiary stations, and the capability to transmit signals during adverse conditions of the signal transmission facilities. Allowable capacities of active multiplex systems shall be in accordance with Table 5-5.3.1.4.

The loading of trunks depends on the capability of the type of system. Since a Type 1 system has a redundant primary trunk (dual control) and isolation between legs and trunks, it has the greatest permitted trunk capacity. A Type 2 system does not have dual control, but does have isolation between legs and trunks. It has less trunk capacity than a Type 1 system, but more than a Type 3 system. A Type 3 system has neither dual control nor isolation between legs and trunks. It has the least trunk capacity of the three types.

5-5.3.1.5 Exceptions to Loading Capacities Listed in Table 5-5.3.1.4. If the signal receiving, processing, display, and recording equipment are duplicated at the supervising station and a switchover is able to be accomplished in not more than 30 seconds with no loss of signals during this period, the capacity of a system unit shall be unlimited.

Subparagraph 5-5.3.1.5 modifies requirements for system units at the supervising station in Table 5-5.3.1.4. However, to meet this requirement an active multiplex system has to employ complete redundancy of all critical components, and, when a failure occurs, complete a switchover in 30 seconds with no loss of signals. Those systems that meet this requirement generally process all incoming signals in tandem. That is, both the main unit and the standby unit process incoming signals at all times. When the main unit fails, the standby unit continues to function normally. Operators would actually change-over only those incidental peripheral devices that have no required redundancy.

Table 5-5.3.1.4 Loading Capacities for Active Multiplex Systems

Trunks	Type 1	Type 2	Type 3
	System Type		
	Type 1	Type 2	Type 3
Maximum number of fire alarm service initiating device circuits per primary trunk facility	5,120	1,280	256
Maximum number of leg facilities for fire alarm service per primary trunk facility	512	128	64
Maximum number of leg facilities for all types of fire alarm service per secondary trunk facility[a]	128	128	128
Maximum number of all types of initiating device circuits per primary trunk facility in any combination[a]	10,240	2,560	512
Maximum number of leg facilities for all types of fire alarm service per primary trunk facility in any combination[a]	1,024	256	128
System Units at the Supervising Station			
Maximum number of all types of initiating device circuits per system unit[a]	10,240[b]	10,240[b]	10,240[b]
Maximum number of fire protecting buildings and premises per system unit	512[b]	512[b]	512[b]
Maximum number of fire alarm service initiating device circuits per system unit	5,120[b]	5,120[b]	5,120[b]
Systems Emitting from Subsidiary Station[c]	—	—	—

[a]Includes every initiating device circuit (for example, waterflow, fire alarm, supervisory, guard, burglary, hold-up).
[b]Paragraph 5-5.3.1.5 shall apply.
[c]Same as system units at the supervising station.

5-5.3.2 Digital Alarm Communicator Systems.

5-5.3.2.1 Digital Alarm Communicator Transmitter (DACT).

5-5.3.2.1.1 A DACT shall be connected to the public switched telephone network upstream of any private telephone system at the protected premises. In addition, the connections to the public switched telephone network shall be under the control of the subscriber for whom service is being provided by the supervising station fire alarm system. Special attention shall be required to ensure that this connection is made only to a loop start telephone circuit and not to a ground start telephone circuit.

Exception: If public cellular telephone service is used as a secondary means of transmission, the requirements of 5-5.3.2.1.1 shall not apply to the cellular telephone service.

The DACT connects to the public switched telephone network so that it may seize the line to which it is connected. This seizure disconnects any private telephone equipment beyond the DACT's point of connection. This arrangement gives the DACT control over the line at all times.

On a loop start telephone line, the public telephone utility continuously supplies voltage from the first telephone utility wire center. The vast majority of residential telephone connections use loop start lines. In contrast, almost all busi-

ness telephone connections, particularly those employing private branch exchange (PBX) connections, use ground start lines. In order to obtain dial tone and operating power on a ground start line, the user equipment momentarily connects one side of the line to earth ground. Since the public telephone utility does not supply voltage to an idle ground start line, the DACT cannot use the presence of voltage to monitor the integrity of the ground start line as it can with a loop start line.

Functionally, a DACT can signal over a ground start line and frequently does so when used as part of a burglar alarm system. However, the DACT can only monitor a loop start line for integrity.

The exception is necessary since public cellular telephone systems do not use telephone lines. Thus, when the public cellular telephone system is used as a secondary means of signal transmission, the requirements of 5-5.3.2.1.1 do not apply to the cellular portion of the system.

Each DACT must be connected to two loop start telephone lines. Use of two telephone lines for several DACTs in a campus-style arrangement is not acceptable.

5-5.3.2.1.2 All information exchanged between the DACT at the protected premises and the digital alarm communicator receiver (DACR) at the supervising or subsidiary station shall be by digital code or some other approved means. Signal repetition, digital parity check, or some other approved means of signal verification shall be used.

The functional requirements of 5-5.3.2.1.2 rule out the use of an analog or digital voice tape dialer to transmit fire alarm signals. Such a device dials a predetermined telephone number and then plays a voice message such as, "There is a fire at 402 Spruce Street." Over the years, officials have reported many cases where a voice tape dialer malfunctions and endlessly repeats its message, tying up a vital emergency telephone line in a public fire service communications center. The code strictly forbids the use of analog or digital voice tape dialers, which do not verify the transmission of data.

5-5.3.2.1.3* A DACT shall be configured so that when it is required to transmit a signal to the supervising station, it shall seize the telephone line (going off-hook) at the protected premises and disconnect an outgoing or incoming telephone call and prevent use of the telephone line for outgoing telephone calls until signal transmission has been completed. A DACT shall not be connected to a party line telephone facility.

A-5-5.3.2.1.3 In order to give the DACT the ability to disconnect an incoming call to the protected premises, telephone service should be of the type that provides for timed-release disconnect. In some telephone systems (step-by-step offices), timed-release disconnect is not provided.

To ensure reliability for transmission of fire alarm, supervisory, and trouble signals, 5-5.3.2.1.3 and its related appendix recommendation give the DACT exclusive control over the telephone service to which it is connected.

5-5.3.2.1.4 A DACT shall have the means to satisfactorily obtain a dial tone, dial the number(s) of the DACR, obtain verification that the DACR is able to receive signals, transmit the signal, and receive acknowledgment that the DACR has accepted that signal. In no event shall the time from going off-hook to on-hook exceed 90 seconds per attempt.

Subparagraph 5-5.3.2.1.4 describes the normal sequence of operation for a DACT. Upon initiation of an alarm, a supervisory, or a trouble signal, the DACT seizes the line, obtains a dial tone, dials the number of the DACR, receives a "handshake" signal from the DACR, transmits its data, receives an acknowledgment signal—sometimes called the "kiss-off signal"—from the DACR, and hangs up. Each attempt of this calling and verification sequence must take no longer than 90 seconds to complete. It should be noted that 5-5-3.2.1.4 and 5-5.3.2.1.5 do not necessarily mean that the signal must be transmitted within 90 seconds. Each attempt is permitted to take a maximum of 90 seconds, and in some cases, the alarm signal could be received 900 seconds after the signal is generated at the protected premises.

5-5.3.2.1.5* A DACT shall have means to reset and retry if the first attempt to complete a signal transmission sequence is unsuccessful. A failure to complete connection shall not prevent subsequent attempts to transmit an alarm where such alarm is generated from any other initiating device circuit or signaling line circuit, or both. Additional attempts shall be made until the signal transmission sequence has been completed, up to a minimum of 5 and a maximum of 10 attempts.

If the maximum number of attempts to complete the sequence is reached, an indication of the failure shall be made at the premises.

A-5-5.3.2.1.5 A DACT can be programmed to originate calls to the DACR telephone lines (numbers) in any alternating sequence. The sequence can consist of single or multiple calls to one DACR telephone line (number), followed by single or multiple calls to a second DACR telephone line (number), or any combination thereof that is consistent with the minimum/maximum attempt requirements in 5-5.3.2.1.5.

The DACT, as described in 5-5.3.2.1.5, must make at least five attempts to complete the sequence. However, it must not make more than ten attempts in order to prevent a malfunctioning DACT from tying up the DACR.

Under the most adverse circumstances, where the DACT finally completes a transmission on the last, or tenth, attempt, at a maximum of 90 seconds per attempt (see subsection 5-5.3.2.1.4) nearly 900 seconds or fifteen minutes could have elapsed. Based on this potential delay, some authorities having jurisdiction may not accept digital alarm communicator systems (DACS) for proprietary fire alarm systems (see 5-3.4.7 and 5-3.4.9).

5-5.3.2.1.6 DACT Transmission Channels.

5-5.3.2.1.6.1 A DACT shall employ one of the following combinations of transmission channels:

(1) Two telephone lines (numbers)

It should be noted that the code has no jurisdiction over utility provided services such as telephone services.

(2) One telephone line (number) and one cellular telephone connection
(3) One telephone line (number) and a one-way radio system
(4) One telephone line (number) equipped with a derived local channel

Some public telephone companies offer so-called "cut line" detection to supervising station alarm system providers. This service uses derived local channel equipment to detect adverse conditions on a telephone line. When a supervising

station uses cut line supervision, it may operate its DACS with a single telephone line connected to each DACT.

(5) One telephone line (number) and a one-way private radio alarm system

(6) One telephone line (number) and a private microwave radio system

(7) One telephone line (number) and a two-way RF multiplex system

(8)*A single integrated services digital network (ISDN) telephone line using a terminal adapter specifically listed for supervising station fire alarm service, where the path between the transmitter and the switched telephone network serving central office is monitored for integrity so that the occurrence of an adverse condition in the path shall be annunciated at the supervising station within 200 seconds

A-5-5.3.2.1.6.1(8) A two-number ISDN line is not a substitute for the requirement to monitor the integrity of the path.

5-5.3.2.1.6.2 The following requirements shall apply to all combinations in 5-5.3.2.1.6.1:

(1) Both channels shall be supervised in a manner approved for the means of transmission employed.

(2) Both channels shall be tested at intervals not exceeding 24 hours.

The additional testing and reporting requirements in 5-5.3.2.1.10 should be noted.

Exception 1: For public cellular telephone service, a verification (test) signal shall be transmitted at least monthly.

Exception 2: Where two telephone lines (numbers) are used, it shall be permitted to test each telephone line (number) at alternating 24-hour intervals.

(3) The failure of either channel shall send a trouble signal on the other channel within 4 minutes.

As important as it is to monitor the integrity of the transmission means, it is equally important to avoid nuisance trouble signals. The permissible 4-minute delay in transmitting a trouble signal allows for momentary, or even somewhat longer, interruptions in the transmission path, such as might occur during a storm.

(4) When one transmission channel has failed, all status change signals shall be sent over the other channel.

Exception: Where used in combination with a DACT, a derived local channel shall not be required to send status change signals other than those indicating that adverse conditions exist on the telephone line (number).

(5) The primary channel shall be capable of delivering an indication to the DACT that the message has been received by the supervising station.

A one-way radio alarm system could not meet this requirement. Thus, it could not serve as the primary transmission means.

(6) The first attempt to send a status change signal shall use the primary channel.

Exception: Where the primary channel is known to have failed.

(7) Simultaneous transmission over both channels shall be permitted.

(8) Failure of telephone lines (numbers) or cellular service shall be annunciated locally.

5-5.3.2.1.7 DACT Transmission Means.

5-5.3.2.1.7.1 A DACT shall be connected to two separate means of transmission at the protected premises. The DACT shall be capable of selecting the operable means of transmission in the event of failure of the other means. The primary means of transmission shall be a telephone line (number) connected to the public switched network.

If the DACT detects that one of the two transmission means has failed (loss of voltage on a wire line, loss of one-way radio alarm service, or loss of cellular telephone service), it must switch to the other operable means. A trouble signal must also be transmitted over the other communications means. See 5-5.3.2.1.6.2 and Exhibits 5.8 and 5.9 for typical connections methods to a DACT.

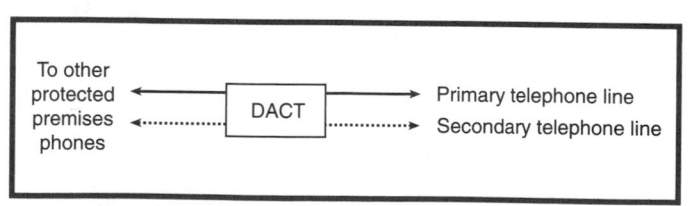

Exhibit 5.8 *Connections to a DACT. (Source: R.P. Schifiliti Associates, Inc., Reading, MA)*

5-5.3.2.1.7.2 The first transmission attempt shall utilize the primary means of transmission.

5-5.3.2.1.8 Each DACT shall be programmed to call a second DACR line (number) when the signal transmission sequence to the first called line (number) is unsuccessful.

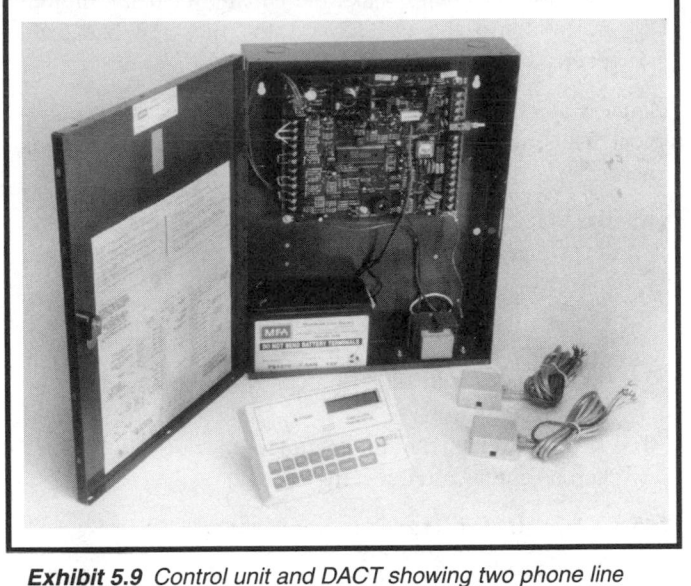

Exhibit 5.9 *Control unit and DACT showing two phone line connections. [Source: Mammoth Fire Alarms (Silent Knight), Lowell, MA]*

To help avoid a possible disarrangement of the transmission path on the receiving end of the digital alarm communicator system, the requirement in 5-5.3.2.1.8 specifies that the DACT must call a second number if calling the first number does not result in completion of the transmission.

5-5.3.2.1.9* If long distance telephone service, including WATS, is used, the second telephone number shall be provided by a different long distance service provider if there are multiple providers.

Formal Interpretation 72-96-2 provides further clarification of the requirements of 5-5.3.2.1.9.

Formal Interpretation 72-96-2

Reference: 5-5.3.2.1.9

Question No. 1: Is a "local call" one which is placed from and terminating within the same "LATA" or an "intra-LATA" call?

Answer: Yes.

Question No. 2: Is a long distance call one which is placed from one "LATA" to a different "LATA" or an "inter-LATA"?

Answer: Yes

Issue Edition: 1996

Reference: 4-5.3.2.1.9

Issue Date: November 2, 1998

Effective Date: November 22, 1998

A-5-5.3.2.1.9 The requirement for use of two different long distance providers is to prevent a lost signal due to a fault in one long distance provider's network. This requirement is not meant to apply in local situations where signal traffic is strictly within the area covered by one local telephone company.

Because it is never certain whether a subscriber has changed long distance providers, it is recommended that, if direct dialer service is used, a telephone call should be forced onto a specific long distance provider's network by using the dialing prefix carrier identification code (CIC) specific to each long distance provider.

5-5.3.2.1.10 Each DACT shall automatically initiate and complete a test signal transmission sequence to its associated DACR at least once every 24 hours. A successful signal transmission sequence of any other type within the same 24-hour period shall fulfill the requirement to verify the integrity of the reporting system, provided signal processing is automated so that 24-hour delinquencies are individually acknowledged by supervising station personnel.

At least once every 24 hours each DACT must initiate a signal to verify the end-to-end integrity of the digital alarm communicator system. If the receiving or processing equipment at the supervising station has sufficient intelligence to automatically keep track of signal traffic, any incoming signal from a particular DACT may serve to satisfy this requirement, as long as the receiver receives one signal during every 24-hour period.

The additional requirements in 5-5.3.2.1.6.2(b) including Exception Nos. 1 and 2 should be noted. The requirement that the daily test must alternate between telephone lines should be particularly noted.

5-5.3.2.1.11* If a DACT is programmed to call a telephone line (number) that is call forwarded to the line (number) of the DACR, a means shall be implemented to verify the integrity of the call forwarding feature every 4 hours.

A-5-5.3.2.1.11 Because call forwarding requires equipment at a telephone company central office that might occasionally interrupt the call forwarding feature, a signal should be initiated whereby the integrity of the forwarded telephone line (number) that is being called by DACTs is

verified every 4 hours. This can be accomplished by a single DACT, either in service or used solely for verification, that automatically initiates and completes a transmission sequence to its associated DACR every 4 hours. A successful signal transmission sequence of any other type within the same 4-hour period should be considered sufficient to fulfill this requirement.

Call forwarding should not be confused with WATS or 800 service. The latter, differentiated from the former by dialing the 800 prefix, is a dedicated service used mainly for its toll-free feature; all calls are preprogrammed to terminate at a fixed telephone line (number) or to a dedicated line.

Occasionally, a supervising station will maintain one or more telephone numbers in a local calling area that the telephone equipment will call forward to another number connected to the DACR. When the supervising station employs this practice, it must verify the integrity of the call-forward instruction every 4 hours. Most often the supervising station equipment does this by having the technicians program the automatic test signal of six of the DACTs in the service area that use the call forwarded number so they each initiate their test signals at 4 hour intervals during a 24-hour period.

5-5.3.2.2 Digital Alarm Communicator Receiver (DACR).

5-5.3.2.2.1 Equipment.

5-5.3.2.2.1.1 Spare DACRs shall be provided in the supervising or subsidiary station. The spare DACRs shall be able to be switched into the place of a failed unit within 30 seconds after detection of failure.

One spare DACR shall be permitted to serve as a backup for up to five DACRs in use.

The presence of a spare DACR does not by itself satisfy the requirements of 5-5.3.2.2.1.1. To meet the switching requirement in 5-5.3.2.2.1.1, someone must provide adequate written instructions and train the personnel on duty in the supervising station to accomplish the switch-over. Preferably, the connections to the unit should end in a manner that permits rapid, error-free reconnection to the spare unit. For example, multiple telephone line connections could terminate in a plug and jack assembly that would permit rapid disconnection and rapid reconnection to the second unit.

5-5.3.2.2.1.2 The number of incoming telephone lines to a DACR shall be limited to eight lines.

Exception: If the signal receiving, processing, display, and recording equipment at the supervising or subsidiary station is duplicated and a switchover is able to be accomplished in less than 30 seconds with no loss of signal during this period, the number of incoming lines to the unit shall be permitted to be unlimited.

The code allows a maximum of eight incoming lines to be connected to a single DACR. This helps to prevent overloading a DACR's ability to receive and process signals promptly. However, some fully automated supervising station facilities may exist to take advantage of the exception.

5-5.3.2.2.2 Transmission Channel.

5-5.3.2.2.2.1* The DACR equipment at the supervising or subsidiary station shall be connected to a minimum of two separate incoming telephone lines (numbers). If the lines (numbers) are in a single hunt group, they shall be individually accessible; otherwise, separate hunt groups shall be required. These lines (numbers) shall be used for no other purpose than receiving signals from a DACT. These lines (numbers) shall be unlisted.

A-5.3.2.2.2.1 The timed-release disconnect considerations as outlined in A-5-5.3.2.1.3 apply to the telephone lines (numbers) connected to a DACR at the supervising station.

It might be necessary to consult with appropriate telephone service personnel to ensure that numbers assigned to the DACR can be individually accessed even where they are connected in rotary (a hunt group).

Hunt groups provided by some older public telephone utility central office equipment may have the potential for locking onto a defective line. This disables all lines in the hunt group. The requirements in 5-5.3.2.2.2.1 help to ensure that the design of the digital alarm communicator system receiving network has a high degree of reliability.

5-5.3.2.2.2.2 The failure of any telephone line (number) connected to a DACR due to loss of line voltage shall be annunciated visually and audibly in the supervising station.

The DACR must connect to loop start telephone lines with voltage normally present. The DACR will monitor this voltage to ensure an operable line up to the first public telephone utility wire center.

5-5.3.2.2.2.3* The loading capacity for a hunt group shall be in accordance with Table 5-5.3.2.2.2.3 or be capable of demonstrating a 90 percent probability of immediately answering an incoming call.

Each supervised burglar alarm (open/close) or each suppressed guard tour transmitter shall reduce the allowable DACTs as follows:

(1) Up to a four-line hunt group, by 10
(2) Up to a five-line hunt group, by 7
(3) Up to a six-line hunt group, by 6
(4) Up to a seven-line hunt group, by 5
(5) Up to an eight-line hunt group, by 4

Each guard tour transmitter shall reduce the allowable DACTs as follows:

(1) Up to a four-line hunt group, by 30
(2) Up to a five-line hunt group, by 21
(3) Up to a six-line hunt group, by 18
(4) Up to a seven-line hunt group, by 15
(5) Up to an eight-line hunt group, by 12

Table 5-5.3.2.2.2.3 Loading Capacities for Hunt Groups

System Loading at the Supervising Station	Number of Lines in Hunt Group				
	1	2	3	4	5–8
With DACR lines processed in parallel					
Number of initiating circuits	NA	5,000	10,000	20,000	20,000
Number of DACTs*	NA	500	1,500	3,000	3,000
With DACR lines processed serially (put on hold, then answered one at a time)					
Number of initiating circuits	NA	3,000	5,000	6,000	6,000
Number of DACTs*	NA	300	800	1,000	1,000

NA: Not allowed.
*Table 5-5.3.2.2.2.3 shall be based on an average distribution of calls and an average connected time of 30 seconds for a message. The loading figures in the table shall presume that the lines are in a hunt group, that is, DACT is able to access any line not in use. A single-line DACR shall not be allowed (NA) for any of the configurations shown.

A-5-5.3.2.2.2.3 In determining system loading, Table 5-5.3.2.2.2.3 can be used, or it should be demonstrated that there is a 90 percent probability of incoming line availability. Table 5-5.3.2.2.2.3 is based on an average distribution of calls and an average connected time of 30 seconds per message. Therefore, where it is proposed to use Table 5-5.3.2.2.2.3 to determine system loading, if any factors are disclosed that could extend DACR connect time so as to increase the average connect time, the alternate method of determining system loading should be used. Higher (or possibly lower) loadings might be appropriate in some applications.

(a) Some factors that could increase (or decrease) the capacity of a hunt group follow.

(1) Shorter (or longer) average message transmission time can influence hunt group capacity.
(2) The use of audio monitoring (listen-in) slow-scan video or other similar equipment can significantly increase the connected time for a signal and reduce effective hunt group capacity.
(3) The clustering of active burglar alarm signals can generate high peak loads at certain hours.
(4) Inappropriate scheduling of 24-hour test signals can generate excessive peak loads.

(b) Demonstration of a 90 percent probability of incoming line availability can be accomplished by the following in-service monitoring of line activity.

(1) Incoming lines are assigned to telephone hunt groups. When a DACT calls the main number of a hunt group, it can connect to any currently available line in that hunt group.
(2) The receiver continuously monitors the "available" status of each line. A line is available when it is waiting for an incoming call. A line is unavailable for any of the following reasons:

   a. Currently processing a call
   b. Line in trouble
   c. Audio monitoring (listen-in) in progress
   d. Any other condition that makes the line input unable to accept calls

(3) The receiver monitors the "available" status of the hunt group. A hunt group is available when any line in it is available.
(4) A message is emitted by the receiver when a hunt group is unavailable for more than 1 minute out of 10 minutes. This message references the hunt group and the degree of overload.

The loading of a DACR helps to determine the overall reliability of a digital alarm communicator system. System designers may use one of two options to determine loading capacity. They may use Table 5-5.3.2.2.2.3 or they may ensure 90 percent availability. Larger capacity supervising stations that employ a computer-based automation system to oversee the handling of signals normally use the second option. Such a system can monitor traffic and report to management so that management may add equipment or take other action to maintain the necessary reliability as loading increases due to the addition of new customers.

5-5.3.2.2.2.4* A signal shall be received on each individual incoming DACR line at least once every 24 hours.

A-5-5.3.2.2.2.4 The verification of the 24-hour DACR line test should be done early enough in the day to allow repairs to be made by the telephone company.

Depending on the number of lines involved and the design and complexity of the particular hunt group arrangements, the supervising station automation system may perform these tests automatically. Alternatively, the supervising station operators may initiate the test signals manually while sequentially creating a busy signal on each line in a hunt group.

5-5.3.2.2.2.5 The failure to receive a test signal from the protected premises shall be treated as a trouble signal. The requirements of 5-2.6.1.4 shall apply.

The daily test signal serves to verify the end-to-end functioning of the system. It monitors the integrity of the system and guards against the loss of both telephone lines connected to the DACT. It may also detect the malfunctioning of an entire hunt group at the DACR. In large supervising stations, the computer-based automation system often oversees the test signals. Small supervising stations may use a manual logging system to keep track of the test signals.

5-5.3.2.3 Digital Alarm Radio System (DARS).

See 5-5.3.2.1.6.1(3) and (5) and 5-5.3.2.1.6.2 for additional requirements when DACTS are used with DARS.

5-5.3.2.3.1 If any DACT signal transmission is unsuccessful, the information shall be transmitted by means of the digital alarm radio transmitter (DART). The DACT shall continue its transmission sequence as required by 5-5.3.2.1.5.

When a digital alarm radio system (DARS) provides the secondary transmission path for a DACT, the DACT must continue to attempt to complete the call to the DACR. Subparagraph 5-5.3.2.1.6.2(7) permits simultaneous transmission by both the DACT and DART.

5-5.3.2.3.2 The DARS shall be capable of demonstrating a minimum of 90 percent probability of successfully completing each transmission sequence.

To fulfill the requirement in 5-5.3.2.3.2, engineers must complete radio propagation studies that satisfy the specified 90 percent reliability factor. The code requires similar studies for a one-way radio alarm system. See 5-5.3.5.2 and A-5-5.3.5.2.

5-5.3.2.3.3 Transmission sequences shall be repeated a minimum of five times. The digital alarm radio transmitter (DART) transmission shall be permitted to be terminated in less than five sequences if the DACT successfully communicates to the DACR.

To provide overall system reliability, the system must make a sufficient number of attempts to complete the signal transmission.

5-5.3.2.3.4 Each DART shall automatically initiate and complete a test signal transmission sequence to its associ-

ated digital alarm radio receiver (DARR) at least once every 24 hours. A successful DART signal transmission sequence of any other type within the same 24-hour period shall fulfill the requirement to test the integrity of the reporting system, provided signal processing is automated so that 24-hour delinquencies are individually acknowledged by supervising station personnel.

The requirement of 5-5.3.2.3.4 dovetails with 5-5.3.2.1.6.2(2). When a DACT connects to a single telephone line as the primary transmission path, and through a digital alarm radio system (DARS) as the secondary transmission path, the system must conduct a test at least once every 24 hours for each transmission path.

5-5.3.2.4 Digital Alarm Radio Transmitter (DART). A DART shall transmit a digital code or another approved signal by use of radio transmission to its associated digital alarm radio receiver (DARR). Signal repetition, digital parity check, or another approved means of signal verification shall be used. The DART shall comply with applicable FCC rules consistent with its operating frequency.

The requirement of 5-5.3.2.4 ensures that a DART uses digital information or other coded signal radio transmission to communicate the status of the fire alarm system at the protected premises. This requirement precludes the use of voice information transmission.

5-5.3.2.5 Digital Alarm Radio Receiver (DARR) Equipment.

5-5.3.2.5.1 A spare DARR shall be provided in the supervising station and shall be able to be switched into the place of a failed unit within 30 seconds after detection of failure.

5-5.3.2.5.2 Facilities shall be provided at the supervising station for supervisory and control functions of subsidiary and repeater station radio receiving equipment. This shall be accomplished via a supervised circuit where the radio equipment is remotely located from the supervising or subsidiary station. The following conditions shall be supervised at the supervising station:

(1) Failure of ac power supplying the radio equipment
(2) Malfunction of receiver
(3) Malfunction of antenna and interconnecting cable
(4) Indication of automatic switchover of the DARR
(5) Malfunction of data transmission line between the DARR and the supervising or subsidiary station

Monitoring the integrity of these functions helps to ensure overall system reliability. A large supervising station equipped with a computer-based automation system uses that system to perform most or all of these functions.

5-5.3.3 McCulloh Systems.

McCulloh systems provide the oldest form of transmission between a protected premises and a supervising station. Coded transmitters at a protected premises connect in series with transmitters at other protected premises and with receiving equipment at the supervising station. The interconnected circuits must maintain continuous metallic circuit continuity. This circuit continuity allows the dc current to flow from the power supply at the supervising station, out over the series circuit, and through the coded contacts of the transmitters at the protected premises. Where the public telephone utility does not wish to maintain circuits that will offer continuous metallic circuit continuity, an alternative exists (see 5-5.3.3.2.6). Between public telephone utility company wire centers, this alternative system converts the McCulloh-type system into a multiplex system. Then, the alternative system reconverts the multiplex system to a McCulloh system in order to deliver the signals to the supervising station.

Initiation of an alarm, supervisory, or trouble signal at the protected premises actuates the associated transmitter. As the code wheel of the actuated transmitter turns, it alternately breaks the circuit and connects the circuit to earth ground.

Under normal circumstances, the breaking of the circuit operates receiving equipment that records the coded pulses at the supervising station. Operators, either manually or by using a computer-based automation system at the supervising station, convert these pulses to information that gives the location of the protected premises.

If a single open fault or single ground fault impairs the circuit between the protected premises and the supervising station, then the signal produced by the turning of the code wheel transmits through earth ground. Operators must respond to a trouble signal generated by the fault. They then manually, or by using a computer-based automation system, recondition the circuit to receive the signals through earth ground.

5-5.3.3.1 Transmitters.

5-5.3.3.1.1 A coded alarm signal from a transmitter shall consist of not less than three complete rounds of the number or code transmitted.

5-5.3.3.1.2* A coded fire alarm box shall produce not less than three signal impulses for each revolution of the coded signal wheel or another approved device.

A-5-5.3.3.1.2 The recommended coded signal designations for a building having four floors and basements are provided in Table A-5-5.3.3.1.2.

5-5.3.3.1.3 Circuit-adjusting means for emergency operating shall be permitted to be either automatic or be provided through manual operation upon receipt of a trouble signal.

Table A-5-5.3.3.1.2 Recommended Coded Signal Designations

Location	Coded Signal
Fourth floor	2–4
Third floor	2–3
Second floor	2–2
First floor	2–1
Basement	3–1
Sub-basement	3–2

When a circuit became impaired by either an open fault or a ground fault, original McCulloh supervising stations have required operator action to condition a circuit to receive subsequent signals. As these stations have become equipped with computer-based automation systems, the interface equipment now often conditions the circuits automatically.

5-5.3.3.1.4* Equipment shall be provided at the supervising or subsidiary station on all circuits extending from the supervising or subsidiary station that is utilized for McCulloh systems for performing the following:

(1) Tests on current on each circuit under nontransmitting conditions
(2) Tests on current on each side of the circuit with the receiving equipment conditioned for an open circuit

A-5-5.3.3.1.4 The current readings, in accordance with 5-5.3.3.1.4(a), should be compared with the normal readings to determine if a change in the circuit condition has occurred. A zero current reading in accordance with 5-5.3.3.1.4(b) indicates that the circuit is clear of a foreign ground.

By taking a current reading on each McCulloh circuit, operators can sometimes detect a "strap" across the circuit (short circuit). Operators have less success at detecting short circuits when the strap exists close to or at the protected premises. If sufficient resistance exists between the location of the strap and the protected premises, the loss of that resistance when someone applies the strap will result in an increase in current. The operator taking the reading should detect the increase in current.

In most cases a strap alone does not disable the transmission of signals. The McCulloh transmitter also connects the circuit to ground with each pulse of the code wheel. However, the person placing the strap across the circuit may also disconnect both sides of the circuit beyond the strap. This action prevents any McCulloh transmitters, located beyond the strap and the open circuit fault, from transmitting signals to the supervising station.

A foreign ground—an unintentional connection of the current to earth ground—can adversely affect the circuit's ability to transmit a signal. These current readings also help

to detect the presence of foreign grounds. Once detected, foreign grounds can be located and cleared by a technician.

Typical foreign grounds include a tree branch rubbing against an aerial portion of the circuit, or water filling a conduit containing a portion of the circuit where the insulation has degraded.

5-5.3.3.2 Transmission Channels.

5-5.3.3.2.1 Circuits between the protected premises and the supervising or subsidiary station that are essential to the actuation or operation of devices that initiate a signal indicative of fire shall be arranged so that the occurrence of a single break or single ground fault does not prevent transmission of an alarm.

Exception 1: Circuits wholly within the supervising or subsidiary station.

Exception 2: The carrier system portion of circuits.

A McCulloh system can continue to function even with a single open fault or a single ground fault on the circuit. Once the supervising station receives a trouble signal that indicates a single open fault or single ground fault, operators must respond. They must manually, or by using a computer-based automation system, recondition the circuit to receive the signals through earth ground.

Exceptions No. 1 and No. 2 exclude this requirement from circuits completely within the supervising station and from the portion of the circuit, described in 5-5.3.3.2.6, that does not have continuous metallic continuity.

5-5.3.3.2.2 The occurrence of a single break or a single ground fault on any circuit shall not of itself cause a false signal that is able to be interpreted as an alarm of fire. If such a single fault prevents the functioning of any circuit, its occurrence shall be indicated automatically at the supervising station by a trouble signal that compels attention and that is distinguishable from signals other than those indicative of an abnormal condition of supervised parts of a fire suppression system(s).

The classic requirement of 5-5.3.3.2.2 dictates that the signals produced by a single open fault or a single ground fault on any circuit associated with the McCulloh system must not produce a false fire alarm signal. It also requires that such faults produce a trouble signal. Whereas 1-5.4.7(b) requires a distinct supervisory off-normal signal, 5-5.3.3.2.2 permits a trouble signal to indicate both a fault and a fire extinguishing system supervisory off-normal condition. The code limits this combining of trouble and supervisory off-normal signals to McCulloh systems.

5-5.3.3.2.3 The circuits and devices shall be arranged to receive and record a signal identifiable as to location of origin, and provisions shall be made for identifying transmission to the public fire service communications center.

5-5.3.3.2.4 Multipoint transmission channels between the protected premises and the supervising or subsidiary station and within the protected premises, consisting of one or more coded transmitters and an associated system unit(s), shall meet the requirements of either 5-5.3.3.2.5 or 5-5.3.3.2.6.

5-5.3.3.2.5 If end-to-end metallic continuity is present, signals shall be received from other points under any one of the following transmission channel fault conditions at one point on the line:

(1) Open
(2) Ground
(3)*Wire-to-wire short

A-5-5.3.3.2.5(3) Though rare, it is understood that the occurrence of a wire-to-wire short on the primary trunk facility near the supervising station could disable the transmission system without immediate detection.

(4) Open and ground

The traditional McCulloh system has end-to-end metallic continuity. Most often the subscriber leases the circuit between the protected premises and the supervising station from the public telephone utility. The telephone utility does not supply any power for the circuit. Rather, it simply provides a pair of wires. The telephone utility usually refers to such a circuit as a "PL circuit" (private line circuit).

The fact that the circuit must maintain end-to-end metallic continuity somewhat limits the electrical distance (resistance) between the protected premises and the supervising station. The resistance of the circuit at a specified current flow must not require more operating power than the supervising station can supply.

A typical central station McCulloh power supply might provide 115 vdc with a current flow of 15 mA, permitting a total McCulloh circuit resistance of 7,666 ohms. For No. 24 AWG, solid copper conductors [0.02567 ohms/ft (0.0806 ohms/m], permitting a total circuit length of 298,662.5 ft (95,115.5 m) translates to a two-way circuit length of 149,331.25 ft (47,557.7 m) or a little more than 28 1/4 miles (47.56 km). This circuit resistance would serve to limit the distance between the supervising station and the farthest protected premises.

5-5.3.3.2.6 If end-to-end metallic continuity is not present, the nonmetallic portion of transmission channels shall meet all of the following requirements.

(a) Two nonmetallic channels or one channel plus a means for immediate transfer to a standby channel shall be provided for each transmission channel, with a maximum of eight transmission channels being associated with each standby channel, or shall be furnished over one channel, provided that service is limited to one plant.

(b) The two nonmetallic channels (or one channel with standby arrangement) for each transmission channel shall be provided by one of the following means, shown in descending order of preference:

(1) Over separate facilities and separate routes
(2) Over separate facilities in the same route
(3) Over the same facilities in the same route

(c) Failure of a nonmetallic channel or any portion thereof shall be indicated immediately and automatically in the supervising station.

(d) Signals shall be received from other points under any one of the following fault conditions at one point on the metallic portion of the transmission channel:

(1) Open
(2) Ground
(3)*Wire-to-wire short

A-5-5.3.3.2.6(d)(3) Though rare, it is understood that the occurrence of a wire-to-wire short on the primary trunk facility near the supervising station could disable the transmission system without immediate detection.

As public telephone utilities have moved away from communications technology that uses end-to-end metallic continuity, the availability of PL circuits has significantly diminished. When the utility schedules the elimination of such circuits between certain telephone utility wire centers, the utility sometimes provides an alternative. Such an alternate circuit must have the features described in 5-5.3.3.2.6. Alarm service providers sometimes also use this method to transport signals from remote areas to the supervising station where the telephone utility does not have PL circuits generally available.

5-5.3.3.3 Loading Capacity of McCulloh Circuits.

The loading capacities discussed in 5-5.3.3.3 have been part of the NFPA signaling standards for almost 70 years. The capacities limit the number of signals lost under various adverse conditions. Such adverse conditions include open and ground faults and a clash of simultaneous signals coming from two or more protected premises on the same McCulloh circuit. Virtually every loading requirement contained in the code for other transmission technologies has its root in these numbers for McCulloh systems.

5-5.3.3.3.1 The number of transmitters connected to any transmission channel shall be limited to eliminate interference. The total number of code wheels or other approved devices connected to a single transmission channel shall not exceed 250. Alarm signal transmission channels shall be reserved exclusively for fire alarm signal transmitting service.

Exception: As provided in 5-5.3.3.3.4.

5-5.3.3.3.2 The number of waterflow switches permitted to be connected to actuate a single transmitter shall not exceed five switches.

5-5.3.3.3.3 The number of supervisory switches permitted to be connected to actuate a single transmitter shall not exceed 20 switches.

5-5.3.3.3.4 Combined alarm and supervisory transmission channels shall comply with the following:

(1) If both sprinkler supervisory signals and fire or waterflow alarm signals are transmitted over the same transmission channel, provision shall be made to obtain either alarm signal precedence or continuous repetition of the alarm signal to prevent the loss of any alarm signal.
(2) Other signal transmitters (for example, burglar, industrial processes) on an alarm transmission channel shall not exceed five.

5-5.3.3.3.5* If signals from manual fire alarm boxes and waterflow alarm transmitters within a building are transmitted over the same transmission channel and are operating at the same time, there shall be no interference with the fire box signals. Provision of the shunt noninterfering method of operation shall be permitted for this performance.

With a McCulloh coded-type manual fire alarm box, connected electrically first on the McCulloh circuit, operation of the box places a short circuit or shunt across the McCulloh circuit. It also disconnects the McCulloh transmitters connected electrically downstream from the manual fire alarm box. This arrangement effectively, if somewhat crudely, prevents another transmitter from interfering with the signal produced by the manual box.

A-5-5.3.3.3.5 At the time of system acceptance, verification should be made that manual fire alarm box signals are free of transmission channel interference.

5-5.3.3.3.6 One alarm transmission channel shall serve not more than 25 plants. A plant shall be permitted to consist of one or more buildings under the same ownership, and the circuit arrangement shall be such that an alarm signal cannot be received from more than one transmitter at a time within a plant. If such noninterference is not provided, each building shall be a plant.

The routing of the McCulloh circuit throughout a large, multiple-building facility could have a significant effect on the loading of the circuit. A designer can extend the number of buildings served by a single McCulloh circuit by using a noninterfering shunt arrangement and very carefully routing the circuit throughout the facility. The circuit should begin at the most important building and extend to the least important building.

5-5.3.3.3.7 One sprinkler supervisory transmission channel circuit shall serve not more than 25 plants. A plant shall be permitted to consist of one or more buildings under the same ownership.

5-5.3.3.3.8 Connections to a guard supervisory transmission channel or to a combination manual fire alarm and guard transmission channel shall be limited so that not more than 60 scheduled guard report signals are transmitted in any 1-hour period. Patrol scheduling shall be such as to eliminate interference between guard report signals.

The requirement in 5-5.3.3.3.8 presumes that an operator must manually record the guard supervisory signals.

5-5.3.4 Two-Way Radio Frequency (RF) Multiplex Systems.

A two-way radio frequency (RF) multiplex system consists of a traditional multiplex fire alarm system that uses a licensed two-way radio system to transmit signals from the protected premises to the supervising station. Essentially, the fire alarm system operates transparently over the radio portion of the system. The code states requirements for two-way RF multiplex systems identical to those for active multiplex systems.

5-5.3.4.1 The maximum end-to-end operating time parameters allowed for a two-way RF multiplex system are as follows.

(a) The maximum allowable time lapse from the initiation of a single fire alarm signal until it is recorded at the supervising station shall not exceed 90 seconds. When any number of subsequent fire alarm signals occur at any rate, they shall be recorded at a rate no slower than one every additional 10 seconds.

Subparagraph 5-5.3.4.1(a) was developed to ensure that the system will complete an interrogation and response sequence at least every 90 seconds. As an alternative, the system may provide some other means to ensure alarm receipt in that time frame. For example, a designer could devise equipment that immediately transmits alarm signals from any multiplex interface at the protected premises, regardless of what point the system has reached in its normal 90-second interrogation and response sequence. However, most systems complete the interrogation and response sequence within 90 seconds.

(b) The maximum allowable time lapse from the occurrence of an adverse condition in any transmission channel until recording of the adverse condition is started shall not exceed 90 seconds for Type 4 and Type 5 systems. The requirements of 5-5.3.4.4 shall apply.

As in 5-5.3.4.1(a), 5-5.3.4.1(b) also ensures that an interrogation and response sequence will be completed at least every 90 seconds for Type 4 and Type 5 systems.

(c) In addition to the maximum operating time allowed for fire alarm signals, the requirements of one of the following paragraphs shall be met:

(1) A system unit that has more than 500 initiating device circuits shall be able to record not less than 50 simultaneous status changes in 90 seconds.

(2) A system unit that has fewer than 500 initiating device circuits shall be able to record not less than 10 percent of the total number of simultaneous status changes within 90 seconds.

The requirements in 5-5.3.4.1 ensure that the portion of the multiplex system that processes and records status changes can do so with sufficient speed to handle a reasonable volume of signal traffic.

5-5.3.4.2 Facilities shall be provided at the supervising station for the following supervisory and control functions of the supervising or subsidiary station and the repeater station radio transmitting and receiving equipment. This shall be accomplished via a supervised circuit where the radio equipment is remotely located from the system unit.

The following conditions shall be supervised at the supervising station:

(1) RF transmitter in use (radiating)
(2) Failure of ac power supplying the radio equipment
(3) RF receiver malfunction
(4) Indication of automatic switchover

Independent deactivation of either RF transmitter shall be controlled from the supervising station.

These supervisory functions help to ensure the continuity of signal transmission between the protected premises and the supervising station. In addition, the system also transmits an interrogation and response sequence between the protected premises and the supervising station every 90 seconds.

5-5.3.4.3 Transmission Channel.

5-5.3.4.3.1 The RF multiplex transmission channel shall terminate in a RF transmitter/receiver at the protected premises and in a system unit at the supervising or subsidiary station.

With this system, each protected premises has its own radio frequency (RF) transmitter/receiver unit. The supervising station also has a RF transmitter/receiver unit. The interrogation and response sequence takes place between these units. This system is similar to an active multiplex system where each protected premises has a multiplex interface (transponder), and the supervising station has an active multiplex system control unit.

5-5.3.4.3.2 Operation of the transmission channel shall conform to the requirements of this code whether channels

are private facilities, such as microwave, or leased facilities furnished by a communications utility company. If private signal transmission facilities are used, the equipment necessary to transmit signals shall also comply with requirements for duplicate equipment or replacement of critical components, as described in 5-5.5.2.

The requirement in 5-5.3.4.3.2 ensures that the system complies with the requirements of the code, even if the facilities are leased from a communications utility company. It further ensures continuity of operations by requiring either redundant critical assemblies or replacement with on-premises spares. Either action must restore service within 30 minutes. (See 5-5.5.2.)

5-5.3.4.4* Two-way RF multiplex systems shall be divided into the following two categories based on their ability to perform under adverse conditions. System classifications shall be Type 4 or Type 5.

A Type 4 system shall have two or more control sites configured as follows:

(1) Each site shall have an RF receiver interconnected to the supervising or subsidiary station by a separate channel.
(2) The RF transmitter/receiver located at the protected premises shall be within transmission range of at least two RF receiving sites.
(3) The system shall contain two RF transmitters that are one of the following:
   a. Located at one site with the capability of interrogating all of the RF transmitters/receivers on the premises
   b. Dispersed with all of the RF transmitters/receivers on the premises having the capability to be interrogated by two different RF transmitters
(4) Each RF transmitter shall maintain a status that allows immediate use at all times. Facilities shall be provided in the supervising or subsidiary station to operate any off-line RF transmitter at least once every 8 hours.
(5) Any failure of one of the RF receivers shall in no way interfere with the operation of the system from the other RF receiver. Failure of any receiver shall be annunciated at the supervising station.
(6) A physically separate channel shall be required between each RF transmitter or RF receiver site, or both, and the system unit.

A Type 5 system shall have a single control site configured as follows:

(1) A minimum of one RF receiving site
(2) A minimum of one RF transmitting site

A-5-5.3.4.4 The intent of the plurality of control sites is to safeguard against damage caused by lightning and to mini-

mize the effect of interference on the receipt of signals. The control sites can be co-located.

A Type 4 two-way RF multiplex system must have a plurality of control sites. Each site contains an RF transmitter/receiver unit. The requirements for a Type 4 system essentially create a two-way RF multiplex system that has redundancy of critical components. An authority having jurisdiction, or system designer, expecting a high volume of traffic or unusual transient radio frequency propagation problems, would use such a system. The requirements for a Type 5 system provide for a minimum level of system integrity that would offer adequate service for most applications.

5-5.3.4.5 Loading Capacities.

5-5.3.4.5.1 The loading capacities of two-way RF multiplex systems are based on the overall reliability of the signal receiving, processing, display, and recording equipment at the supervising or subsidiary station and the capability to transmit signals during adverse conditions of the transmission channels. Allowable loading capacities shall comply with Table 5-5.3.4.5.1.

The loading of a two-way RF multiplex system depends on the capability of the type of system. Because a Type 4 system has a redundant transmitter/receiver exerting control over the interrogation and response sequence between the protected premises and the supervising station, it has the greatest permitted system loading. A Type 5 system does not have dual transmitters/receivers in control of the system, so it has a more limited trunk capacity.

5-5.3.4.5.2 Exceptions to Loading Capacities Listed in Table 5-5.3.4.5.1. The capacity of a system unit shall be unlimited if the signal receiving, processing, display, and recording equipment are duplicated at the supervising station and a switchover is able to be accomplished in not more than 30 seconds with no loss of signals during this period.

Subparagraph 5-5.3.4.5.2 modifies the lower half of Table 5-5.3.4.5.1. However, to meet this requirement a two-way RF multiplex system would have to employ complete redundancy of all critical components, and complete a switchover in 30 seconds with no loss of signals. Those systems that meet this requirement generally process all incoming signals in tandem. That is, both the main unit and the standby unit process incoming signals at all times. When the main unit fails, the standby unit continues to function normally. Operators would actually change over only those incidental peripheral devices that have no required redundancy.

Table 5-5.3.4.5.1 Loading Capacities for Two-Way RF Multiplex Systems

	System Type	
Trunks	**Type 4**	**Type 5**
Maximum number of fire alarm service initiating device circuits per primary trunk facility	5,120	1,280
Maximum number of leg facilities for fire alarm service per primary trunk facility	512	128
Maximum number of leg facilities for all types of fire alarm service per secondary trunk facility[a]	128	128
Maximum number of all types of initiating device circuits per primary trunk facility in any combination	10,240	2,560
Maximum number of leg facilities for types of fire alarm service per primary trunk facility in any combination[a]	1,024	256
System Units at the Supervising Station		
Maximum number of all types of initiating device circuits per system unit[a]	10,240[b]	10,240[b]
Maximum number of fire protected buildings and premises per system unit	512[b]	512[b]
Maximum number of fire alarm service initiating device circuits per system	5,120[b]	5,120[b]
Systems Emitting from Subsidiary Station[c]	—	—

[a]Includes every initiating device circuit (for example, waterflow, fire alarm supervisory, guard, burglary, hold-up).
[b]Paragraph 5-5.3.4.5.2 shall apply.
[c]Same as system units at the supervising station.

5-5.3.5 One-Way Private Radio Alarm Systems.

To create the requirements for a radio frequency transmission system that does not have an interrogation and response sequence to monitor the integrity of the transmission of signals between the protected premises and the supervising station, the Technical Committee on Supervising Station Fire Alarm Systems borrowed heavily from the requirements for digital alarm communicator systems.

5-5.3.5.1 The requirements of 5-5.3.5 for a radio alarm repeater station receiver (RARSR) shall be satisfied if signals from each radio alarm transmitter (RAT) are received and supervised, in accordance with Chapter 5, by at least two independently powered, independently operating, and separately located RARSRs.

The one-way radio alarm system consists of a radio frequency transmitter at the protected premises that connects to the protected premises control unit. This unit is capable of transmitting alarm, supervisory, and trouble signals to at least two receivers. The receivers relay the received signal to the supervising station by radio frequency or wired transmission means. This system allows the use of either a private system operated by a single alarm service provider, or a multi-user system operated by a one-way radio network provider. Most systems communicate through a multi-user network.

5-5.3.5.2* The end-to-end operating time parameters allowed for a one-way radio alarm system shall be as follows:

(1) There shall be a 90 percent probability that the time between the initiation of a single fire alarm signal until it is recorded at the supervising station will not exceed 90 seconds.

(2) There shall be a 99 percent probability that the time between the initiation of a single fire alarm signal until it is recorded at the supervising station will not exceed 180 seconds.

(3) There shall be a 99.999 percent probability that the time between the initiation of a single fire alarm signal until it is recorded at the supervising station will not exceed 7.5 minutes (450 seconds), at which time the RAT shall cease transmitting. When any number of subsequent fire alarm signals occur at any rate, they shall be recorded at an average rate no slower than one every additional 10 seconds.

(4) In addition to the maximum operating time allowed for fire signals, the system shall be able to record not less than 12 simultaneous status changes within 90 seconds at the supervising station.

A-5-5.3.5.2 It is intended that each RAT communicate with two or more independently located RARSRs. The location of such RARSRs should be such that they do not share common facilities.

Note: All probability calculations required for the purposes of Chapter 5 should be made in accordance with established communications procedures, should assume the maximum channel loading parameters

specified, and should further assume that 25 RATs are actively in alarm and are being received by each RARSR.

Because one-way private radio systems do not have an interrogation and response sequence to verify the operating capability of the communications channel and all equipment associated with it, the system must rely on other means to achieve an acceptable level of operational integrity. The probabilities specified in 5-5.3.5.2 help to ensure that level of integrity.

To achieve the probabilities, the system functions similarly to a digital alarm communicator transmitter. The system makes a given number of attempts to reach one or both of the two receivers. If it does not succeed, it stops transmitting, so it will not tie up the receiver.

5-5.3.5.3 Supervision.

5-5.3.5.3.1 Equipment shall be provided at the supervising station for the supervisory and control functions of the supervising or subsidiary station and for the repeater station radio transmitting and receiving equipment. This shall be accomplished via a supervised circuit where the radio equipment is remotely located from the system unit. The following conditions shall be supervised at the supervising station:

(1) Failure of ac power supplying the radio equipment
(2) Malfunction of RF receiver
(3) Indication of automatic switchover, if applicable

The specified supervisory functions help to ensure the continuity of signal transmission between the protected premises and the supervising station.

5-5.3.5.3.2 Protected Premises.

5-5.3.5.3.2.1 Interconnections between elements of transmitting equipment, including any antennas, shall be supervised either to cause an indication of failure at the protected premises or to transmit a trouble signal to the supervising station.

5-5.3.5.3.2.2 If elements of transmitting equipment are physically separated, the wiring or cabling between them shall be protected by conduit.

Subparagraphs 5-5.3.5.3.2.1 and 5-5.3.5.3.2.2 address two serious points of potential failure. Either the loss of the antenna, or the loss of connection between the transmitter and the antenna, would impair transmission.

In some systems, the transmitter connects directly to the antenna. In others, the installer locates the antenna at a point in the building more advantageous for successful transmission of a signal. The requirements in 5-5.3.5.3.2.1 and 5-5.3.5.3.2.2 ensure that a trouble signal resulting from the loss of the antenna or its connection will at least annun-

ciate locally. They further require mechanical protection by installing the conductors between the transmitter and remote antenna in conduit.

5-5.3.5.4 Transmission Channels.

5-5.3.5.4.1 The one-way RF transmission channel shall originate with a one-way RF transmitting device at the protected premises and shall terminate at the RF receiving system of an RARSR capable of receiving transmissions from such transmitting devices.

5-5.3.5.4.2 A receiving network transmission channel shall terminate at an RARSR at one end and with either another RARSR or a radio alarm supervising station receiver (RASSR) at the other end.

Subparagraph 5-5.3.5.4.2 permits the architecture necessary to develop a network to handle a large number of radio alarm transmitters. The network interconnections can use multiple radio alarm repeater station receivers (RARSRs) that, in turn, repeat the received signals to other RARSRs until the signals ultimately reach a radio alarm supervising station receiver (RASSR). Along each segment of the transmission path, at least two RARSRs must always receive the signal.

5-5.3.5.4.3 Operation of receiving network transmission channels shall conform to the requirements of this code whether channels are private facilities, such as microwave, or leased facilities furnished by a communications utility company. If private signal transmission facilities are used, the equipment necessary to transmit signals shall also comply with requirements for duplicate equipment or replacement of critical components as described in 5-5.5.2.

The requirement in 5-5.3.5.4.3 ensures that the system will comply with the code, even if the installer leases facilities from a communications utility company or other one-way radio network service provider. It further ensures continuity of operations by requiring either redundant critical assemblies, or that technicians can replace critical assemblies with on-premises spares and restore service within 30 minutes.

The system must also monitor the quality of the transmitted signal, including the various operating time parameters specified in 5-5.3.5.2. One design provides each Radio Alarm Transmitter (see 5-5.3.5.1) with a clock. Each transmitted signal includes the time of first transmission and the current time, along with the alarm, supervisory, or trouble data.

5-5.3.5.4.4 The system shall provide information that indicates the quality of the received signal for each RARSR supervising each RAT in accordance with 5-5.3.5 and shall provide information at the supervising station when such signal quality falls below the minimum signal quality levels set forth in 5-5.3.5.

5-5.3.5.4.5 Each RAT shall be installed in such a manner so as to provide a signal quality over at least two independent one-way RF transmission channels, of the minimum quality level specified, that satisfies the performance requirements in 5-5.2.2 and 5-5.5.

5-5.3.5.5 Nonpublic one-way radio alarm systems shall be divided into two categories based on the following number of RASSRs present in the system.

(1) A Type 6 system shall have one RASSR and at least two RARSRs.
(2) A Type 7 system shall have more than one RASSR and at least two RARSRs.

The Type 6 system serves a single supervising station. The Type 7 system serves more than one supervising station. A multi-user network most closely fits the Type 7 system description.

In a Type 7 system, when more than one RARSR is out of service and, as a result, any RATs are no longer being supervised, the affected supervising station shall be notified.

In a Type 6 system, when any RARSR is out of service, a trouble signal shall be annunciated at the supervising station.

5-5.3.5.6 The loading capacities of one-way radio alarm systems are based on the overall reliability of the signal receiving, processing, display, and recording equipment at the supervising or subsidiary station and the capability to transmit signals during adverse conditions of the transmission channels. Allowable loading capacities shall be in accordance with Table 5-5.3.5.6.

5-5.3.5.7 Exceptions to Loading Capacities Listed in Table 5-5.3.5.6. If the signal receiving, processing, display, and recording equipment is duplicated at the supervising station and a switchover is able to be accomplished in not more than 30 seconds with no loss of signals during this period, the capacity of a system unit shall be unlimited.

Subparagraph 5-5.3.5.7 modifies requirements for system units at the supervising station in Table 5-5.3.5.6. However, to meet this requirement a one-way radio alarm system would have to employ complete redundancy of all critical components, and complete a switchover in 30 seconds with no loss of signals. Those systems that meet this requirement generally process all incoming signals in tandem. That is, both the main unit and the standby unit process incoming signals at all times. When the main unit fails, the standby unit continues to function normally. Operators would actually change over only those incidental peripheral devices that have no required redundancy.

Table 5-5.3.5.6 Loading Capacities of One-Way Radio Alarm Systems

Radio Alarm Repeater Station Receiver (RARSR)	System Type	
	Type 6	Type 7
Maximum number of fire alarm service initiating device circuits per RARSR	5,120	5,120
Maximum number of RATs for fire	512	512
Maximum number of all types of initiating device circuits per RARSR in any combination[a]	10,240	10,240
Maximum number of RATs for all types of fire alarm service per RARSR in any combination[a,c]	1,024	1,024
System Units at the Supervising Station		
Maximum number of all types of initiating device circuits per system unit[a]	10,240[b]	10,240[b]
Maximum number of fire-protected buildings and premises per system unit	512[b]	512[b]
Maximum number of fire alarm service initiating device circuits per system unit	5,120[b]	5,120[b]

Notes:
1. Each guard tour transmitter shall reduce the allowable RATs by 15.
2. Each two-way protected premises radio transmitter shall reduce the allowable RATs by two.
[a]Includes every initiating device circuit (for example, waterflow, fire alarm, supervisory, guard, burglary, hold-up).
[b]Paragraph 5-5.3.5.7 shall apply.
[c]Each supervised BA (open/close) or each suppressed guard tour transmitter shall reduce the allowable RATs by five.

5-5.3.6 Directly Connected Noncoded Systems.

5-5.3.6.1 Circuits for transmission of alarm signals between the fire alarm control unit or the transmitter in the protected premises and the supervising station shall be arranged so as to comply with either of the following provisions.

(a) These circuits shall be arranged so that the occurrence of a single break or single ground fault does not pre-

vent the transmission of an alarm signal. Circuits complying with this paragraph shall be automatically self-adjusting in the event of either a single break or a single ground fault and shall be automatically self-restoring in the event that the break or fault is corrected.

Only one manufacturer offers a system that meets the requirements of 5-5.3.6.1(a). This system powers the circuit by means of a float-charged set of batteries with a center tap connection to earth ground. Two sets of alarm relays are connected across the circuit at the supervising station and reference to earth ground. Loss of one side of the circuit would not prevent the transmission of the signal over the remaining side of the circuit and earth ground.

(b) These circuits shall be arranged so that they are isolated from ground (except for reference ground detection) and so that a single ground fault does not prevent the transmission of an alarm signal. Circuits complying with this paragraph shall be provided with a ground reference circuit so as to detect and indicate automatically the existence of a single ground fault.

The vast majority of the circuits employed for transmitting signals from a protected premises to a supervising station over directly connected noncoded systems meet the requirements of 5-5.3.6.1(b). In most cases, the protected premises or alarm service provider leases the circuits for directly connected noncoded systems from the public telephone utility. The telephone utility does not supply any power for the circuit. Rather, it simply provides a pair of wires. The telephone utility usually refers to such a circuit as a "PL circuit" (private line circuit).

The fact that the circuit must maintain end-to-end metallic continuity somewhat limits the electrical distance (resistance) between the protected premises and the supervising station. The resistance of the circuit at a specified current flow must not require more operating power than either the protected premises or the supervising station can supply.

A typical directly connected noncoded system power supply might provide 24 vdc with a current flow of 15 mA, permitting a total McCulloh circuit resistance of 1,600 ohms. For No. 24 AWG solid copper conductors [0.02567 ohms/ft (0.0806 ohms/m)] circuit resistance would permit a total circuit length of 62,329.6 ft (19,850.2 m) total. This then translates to a two-way circuit length of 31,164.8 ft (9,925.1 m) or a little less than 6 miles (10 km). Circuit resistance would serve to limit the distance between the supervising station and the protected premises.

Exception: A multiple ground-fault condition that would prevent alarm operation shall be indicated by an alarm or by a trouble signal.

5-5.3.6.2 Circuits for transmission of supervisory signals shall be separate from alarm circuits. These circuits within

the protected premises and between the protected premises and the supervising station shall be arranged as described in 5-5.3.6.1(a) or (b).

Exception: If the reception of alarm signals and supervisory signals at the same supervising station is permitted by the authority having jurisdiction, the supervisory signals do not interfere with the alarm signals, and alarm signals have priority, the same circuit between the protected premises and the supervising station shall be permitted to be used for alarm and supervisory signals.

5-5.3.6.3 The occurrence of a single break or a single ground fault on any circuit shall not of itself cause a false signal that is able to be interpreted as an alarm of fire.

5-5.3.6.4 The requirements of 5-5.3.6.1 and 5-5.3.6.2 shall not apply to the following circuits:

(1) Circuits wholly within the supervising station
(2) Circuits wholly within the protected premises extending from one or more automatic fire detectors or other non-coded initiating devices other than waterflow devices to a transmitter or control unit
(3) Power supply leads wholly within the building or buildings protected

These requirements clarify that the named circuits do not need to have the operational capability of the circuits extending between the protected premises and the supervising station.

5-5.3.6.5 Loading Capacity of Circuits.

5-5.3.6.5.1 The number of initiating devices connected to any signaling circuit and the number of plants that shall be permitted to be served by a signal circuit shall be determined by the authority having jurisdiction and shall not exceed the limitations specified in 5-5.3.6.5.

A plant shall be permitted to consist of one or more buildings under the same ownership.

5-5.3.6.5.2* A single circuit shall not serve more than one plant.

A-5-5.3.6.5.2 If a single plant involves more than one gate entrance or involves a number of buildings, separate circuits might be required so that the alarm to the supervising station indicates the area to which the fire department is to be dispatched.

Unique among transmission technologies, a directly connected noncoded system may serve only one plant.

5-5.3.7 Private Microwave Radio Systems.

In developing the requirements for 5-5.3.7, the Technical Committee on Supervising Station Fire Alarm Systems originally consulted with AT&T. Based on the requirements for

a standard microwave relay link, these requirements help to ensure normal network reliability.

In most cases, a private microwave radio system would transport a rather high volume of signal traffic from a subsidiary station to a supervising station.

5-5.3.7.1* If a private microwave radio is used as the transmission channel and communications channel, supervised transmitting and receiving equipment shall be provided at supervising, subsidiary, and repeater stations.

A-5-5.3.7.1 A private microwave radio can be used either as a transmission channel, to connect a transmitter to a supervising station or subsidiary station, or as a communications channel to connect a subsidiary station(s) to a supervising station(s). This can be done independently or in conjunction with wireline facilities.

5-5.3.7.2 If more than five protected buildings or premises or 50 initiating devices or initiating device circuits are being serviced by a private radio carrier, the supervising, subsidiary, and repeater station radio facilities shall meet all of the following criteria.

(a) Dual supervised transmitters, arranged for automatic switching from one to the other in case of trouble, shall be installed. If the transmitters are located where someone is always on duty, switchboard facilities shall be permitted to be manually operated, provided the switching is able to be carried out within 30 seconds. If the transmitters are located where no one is continuously on duty, the circuit extending between the supervising station and the transmitters shall be a supervised circuit.

(b)* Transmitters shall be operated on a time ratio of 2:1 within each 24 hours.

A-5-5.3.7.2(b) Transmitters should be operated alternately, 16 hours on and 16 hours off.

(c) Dual receivers shall be installed with a means for selecting a usable output from one of the two receivers. The failure of one shall in no way interfere with the operation of the other. Failure of either receiver shall be annunciated.

5-5.3.7.3 Means shall be provided at the supervising station for the supervision and control of supervising, subsidiary, and repeater station radio transmitting and receiving equipment. If the radio equipment is remote from the supervising station, this shall be accomplished via a supervised circuit.

The following conditions shall be supervised at the supervising station:

(1) Transmitter in use (radiating)
(2) Failure of ac power supplying the radio equipment
(3) Receiver malfunction
(4) Indication of automatic switchover

It shall be possible to independently deactivate either transmitter from the supervising station.

5-5.4 Other Transmission Technologies.

Subsection 5-5.4 provides guidance for those who might develop some transmission technology not currently covered by the requirements of the code. It parallels the requirements of the other existing transmission technologies.

5-5.4.1 Other transmission technologies shall include those transmission technologies that operate on principles different from transmission technologies covered by 5-5.3.1 through 5-5.3.7. Such transmission technologies shall be permitted to be installed if they conform to the requirements of 5-5.4 and to all other applicable requirements of this code.

5-5.4.2 Fire alarm system equipment and installations shall comply with the Federal Communications Commission (FCC) rules and regulations, as applicable, concerning electromagnetic radiation, use of radio frequencies, and connections to the public switched telephone network of telephone equipment, systems, and protection apparatus.

5-5.4.3 Equipment shall be installed in compliance with NFPA 70, *National Electrical Code.*

5-5.4.4 Provision shall be made to monitor the integrity of the transmission technology and its communications path.

5-5.4.4.1 Any failure shall be annunciated at the supervising station within 5 minutes of the failure.

5-5.4.4.2 If communications cannot be established with the supervising station, an indication of this failure to communicate shall be annunciated at the protected premises.

5-5.4.4.3 If a portion of the communications path cannot be monitored for integrity, a redundant communications path shall be provided.

5-5.4.4.3.1 Provision shall be made to monitor the integrity of the redundant communications path.

5-5.4.4.3.2 Failure of both the primary and redundant communications paths shall be annunciated at the supervising station within not more than 24 hours of the failure.

5-5.4.4.4 System units at the supervising station shall be restored to service within 30 minutes of a failure.

5-5.4.4.5 The transmission technology shall be designed so that upon failure of a transmission channel serving a system unit at the supervising station, the loss of the ability to monitor shall not affect more than 3000 transmitters.

5-5.4.5 Spare System Unit Equipment.

5-5.4.5.1 Sufficient spare equipment shall be maintained at the supervising station such that any failed piece of equip-

ment can be replaced and the systems unit restored to full operation.

5-5.4.5.2 The complement of spare parts shall be sufficient to replace any failed critical component.

Exception: If multiples of the same components are used at the supervising station, the ratio of available spare parts to units in service shall be 1 to 5 minimum.

5-5.4.6 Loading Capacity of a System Unit.

5-5.4.6.1 The maximum number of independent fire alarm systems connected to a single system unit shall be limited to 512.

5-5.4.6.2 If duplicate spare system units are maintained at the supervising station and switchover can be achieved in 30 seconds, then the system capacity is unlimited.

5-5.4.7 End-to-End Communication Time for an Alarm.
The maximum duration between the initiation of an alarm signal at the protected premises, transmission of the signal, and subsequent display and recording of the alarm signal at the supervising station shall not exceed 90 seconds.

5-5.4.8 Unique Premises Identifier.
If a transmitter shares a transmission or communications channel with other transmitters, it shall have a unique transmitter identifier.

5-5.4.9 Recording and Display Rate of Subsequent Alarms.
Recording and display of alarms at the supervising station shall be at a rate no slower than one complete signal every 10 seconds.

5-5.4.10 Signal Error Detection and Correction.

5-5.4.10.1 Communication of alarm, supervisory, and trouble signals shall be in a highly reliable manner to prevent degradation of the signal in transit, which in turn would result in either of the following:

(1) Failure of the signal to be displayed and recorded at the supervising station
(2) An incorrect corrupted signal displayed and recorded at the supervising station

5-5.4.10.2 Enhanced reliability of the signal shall be achieved by any of the following:

(1) Signal repetition—multiple transmissions repeating the same signal
(2) Parity check—a mathematically check sum algorithm of a digital message that verifies correlation between transmitted and received message
(3) An equivalent means to (1) or (2) that provides a certainty of 99.99 percent that the received message is identical to the transmitted message.

5-5.4.11 Signal Priority.
If the communications methodology is shared with any other usage, all fire alarm, supervisory, and trouble signals shall take precedence, in that respective order of priority, over all other signals.

Exception: Signals from hold-up alarms or other signals indicating life-threatening situations shall be permitted to take precedence over supervisory and trouble signals if acceptable to the authority having jurisdiction.

5-5.4.12 Sharing Communications Equipment On-Premises.
If the fire alarm transmitter is sharing on-premises communications equipment, the shared equipment shall be listed for the purpose. If on-premises communications equipment is not listed for the purpose, the fire alarm transmitter shall be installed ahead of the unlisted communications equipment.

5-5.4.13 Service Provider Diversity.
When a redundant path is required, the alternate path shall be provided by a public communications service provider different from the primary path, if available.

5-5.4.14 Throughput Probability.
When the supervising station does not regularly communicate with the transmitter at least once every 200 seconds, then the throughput probability of the alarm transmission must be at least 90 percent in 90 seconds, 99 percent in 180 seconds, or 99.999 percent in 450 seconds.

5-5.4.15 Unique Flaws Not Covered by This Code.
If a communications technology has a unique flaw that could result in the failure to communicate a signal, the implementation of that technology for fire alarm signaling shall compensate for that flaw so as to eliminate the risk of missing a fire alarm signal.

5-5.5 Display and Recording Requirements for All Transmission Technologies.

Subsection 5-5.5 specifies the content and nature of the display and recording of signals received at a supervising station. The requirements take into account a reasonable quantity of signal traffic. They also consider certain ergonomic necessities for interfacing electronically reproduced signals with one or more human operators.

5-5.5.1* Any status changes, including the initiation or restoration to normal of a trouble condition, that occur in an initiating device or in any interconnecting circuits or equipment, including the local protected premises controls from the location of the initiating device(s) to the supervising station, shall be presented in a form to expedite prompt operator interpretation. Status change signals shall provide the following information.

(a) *Type of Signal.* Identification of the type of signal to show whether it is an alarm, supervisory, delinquency, or trouble signal

(b) *Condition.* Identification of the signal to differentiate between an initiation of an alarm, supervisory, delinquency, or trouble signal and a clearing from one or more of these conditions

(c) *Location.* Identification of the point of origin of each status change signal

A-5-5.5.1 The signal information can be permitted to be provided in coded form. Records can be permitted to be used to interpret these codes.

5-5.5.2* If duplicate equipment for signal receiving, processing, display, and recording is not provided, the installed equipment shall be designed so that any critical assembly is able to be replaced from on-premises spares and the system is able to be restored to service within 30 minutes. A critical assembly shall be an assembly in which a malfunction prevents the receipt and interpretation of signals by the supervising station operator.

Exception: Proprietary and remote station systems.

The requirement in 5-5.5.2 ensures that a technician will properly repair any malfunction in a critical assembly. For a description of a critical assembly, see 5-5.5.2. The technician may repair the defective assembly but, most often, the technician will replace the defective assembly with an on-premises spare. Any assembly too complex for a technician to readily repair requires a duplicate. It should be noted that the exception limits the application of 5-5.5.2 to central station supervising station fire alarm systems.

A-5-5.5.2 In order to expedite repairs, it is recommended that spare modules, such as printed circuit boards, CRT displays, or printers, be stocked at the supervising station.

5-5.5.3* Any method of recording and display or indication of change of status signals shall be permitted, provided all of the following conditions are met.

(a) Each change of status signal requiring action to be taken by the operator shall result in an audible signal and not less than two independent methods of identifying the type, condition, and location of the status change.

(b) Each change of status signal shall be automatically recorded. The record shall provide the type of signal, condition, and location as required by 5-5.5.1 in addition to the time and date the signal was received.

(c) Failure of an operator to acknowledge or act upon a change of status signal shall not prevent subsequent alarm signals from being received, indicated or displayed, and recorded.

(d) Change of status signals requiring action to be taken by the operator shall be displayed or indicated in a manner that clearly differentiates them from those that have been acted upon and acknowledged.

(e) Each incoming signal to a DACR or DARR shall cause an audible signal that persists until manually acknowledged.

Exception: Test signals required by 5-5.3.2.1.10 received at a DACR or a DARR.

A-5-5.5.3 For all forms of transmission, the maximum time to process an alarm signal should be 90 seconds. The maximum time to process a supervisory signal should be 4 minutes. The time to process an alarm or supervisory signal is defined as that time measured from receipt of a signal until retransmission or subscriber contact is initiated.

When the level of traffic in a supervising station system reaches a magnitude such that delayed response is possible, even if the loading tables or loading formulas of this code are not exceeded, it is envisioned that it will be necessary to employ an enhanced method of processing.

For example, in a system where a single DACR instrument provided with fire and burglar alarm service is connected to multiple telephone lines, it is conceivable that, during certain periods of the day, fire alarm signals could be delayed by the security signaling traffic, such as opening and closing signals. Such an enhanced system would perform as follows, upon receipt of a signal:

(1) Automatically process the signals, differentiating between those that require immediate response by supervising station personnel and those that need only be logged

(2) Automatically provide relevant subscriber information to assist supervising station personnel in their response

(3) Maintain a timed, unalterable log of the signals received and the response of supervising station personnel to such signals

5-5.6 Testing and Maintenance Requirements for All Transmission Technologies.

Testing and maintenance of communications methods shall be in accordance with the requirements of Chapter 7.

References Cited in Commentary

ANSI/UL 827, *Standard for Safety Central-Station for Watchman, Fire-Alarm and Supervisory Services*, 1996.

NFPA 70, *National Electrical Code*®, National Fire Protection Association, Quincy, MA, 1999.

NFPA 600, *Standard on Industrial Fire Brigades*, National Fire Protection Association, Quincy, MA, 2000.

Factory Mutual Standard No. 3011, *Central Station Service for Fire Alarms and Protective Equipment Supervision*, FMRC 3011, 1999.

CHAPTER 6

Public Fire Alarm Reporting Systems

Chapter 6 contains requirements for public fire alarm reporting systems. These systems allow the public to initiate a fire alarm signal that is transmitted to a public fire service communications center.

Chapter 6 covers only the transmission of the signals to the public fire service communications center, whereas NFPA 1221, *Standard for the Installation, Maintenance, and Use of Emergency Services Communications,* contains the requirements for the equipment, operations, and maintenance of a public fire service communications center.

The requirements of Chapter 6 only permit transmission of fire alarm signals or trouble signals relating to the reporting system itself. These requirements do not consider the transmission of supervisory signals or trouble signals from the protected premises.

6-1 Public Fire Alarm Reporting Systems

If permitted by the authority having jurisdiction, use of systems described in Chapter 6 shall be permitted to provide defined reporting functions from or within private premises.

In rare cases, a municipality or other governmental entity may exercise its right as an authority having jurisdiction and permit the transmission of supervisory or trouble signals from buildings the entity owns. However, Section 6-1 of the code does not contain requirements for transmission of supervisory or trouble signals from a protected premises.

A public fire alarm reporting system allows the public to initiate a fire alarm signal from one or more publicly accessible manual fire alarm boxes. Protected premises located throughout the area served by the system may also connect their protected premises fire alarm systems to the public fire alarm reporting system. Section 6-16 terms this connection an *auxiliary fire alarm system.* A publicly accessible manual fire alarm box that contains interface equipment to allow connection of a protected premises fire alarm system is called a *master box.*

Of course, the protected premises fire alarm systems must meet the requirements of Chapters 1, 2, 3, 4, 5, and 7.

6-1.1 The requirements of Chapters 1 and 7 shall apply to public and auxiliary fire alarm reporting systems, unless they conflict with the requirements of this chapter.

6-1.2 Scope.

This chapter covers the general requirements and use of public and auxiliary fire alarm reporting systems. These systems include the equipment necessary to effect the transmission and reception of fire alarms or other emergency calls connected to the public fire alarm reporting system.

6-2 General Fundamentals

6-2.1* If implemented at the option of the authority having jurisdiction, a public fire alarm reporting system shall be designed, installed, operated, and maintained to provide the maximum practicable reliability for transmission and receipt of fire alarms.

A-6-2.1 When choosing from available options to implement a public fire alarm reporting system, the operating agency should consider which of the choices would facilitate the maximum reliability of the system, where such a choice is not cost prohibitive.

6-2.2 A public fire alarm reporting system, as described herein, shall be permitted to be used for the transmission of other signals or calls of a public emergency nature, provided such transmission does not interfere with the transmission and receipt of fire alarms.

Subsection 6-2.2 permits a system to transmit other signals of a public emergency nature, such as a request for emergency medical response or police response. The code permits these non-fire transmissions as long as they do not interfere with the transmission of fire alarm signals.

6-2.3 Alarm systems shall be Type A or Type B. A Type A system shall be provided where the number of all fire alarms required to be transmitted over the dispatch circuits exceeds 2500 per year.

6-2.3.1 If a Type A system is required, automatic retransmission of alarms from boxes by use of electronic equipment shall be permitted, provided the following conditions are met:

(1) Approved facilities are provided for the automatic receipt, storage, retrieval, and retransmission of alarms in the order received.
(2) Override capability is provided to the operator(s) so that manual transmission at the dispatch facilities is immediately able to be used.

In a Type A system, operators at the public fire service communications center receive signals from the public fire reporting system. They then manually retransmit these signals to the fire stations designated to respond.

In a Type B system, equipment at the public fire service communications center automatically retransmits the signals to all fire stations and other locations connected to the system.

6-2.4 If applicable, electronic computer/data processing equipment shall be protected in accordance with NFPA 75, *Standard for the Protection of Electronic Computer/Data Processing Equipment.*

6-3 Management and Maintenance

The requirements of Chapter 7 shall apply.

6-3.1 All systems shall be under the control of a responsible jurisdictional employee. If maintenance is provided by an organization or person other than the municipality or its employees, complete written records of the installation, maintenance, test, and extension of the system shall be forwarded to the responsible employee as soon as possible.

Subsection 6-3.1 is new, and it requires those operating public fire alarm reporting systems to test and maintain the systems in accordance with the requirements of Chapter 7. A single employee should have responsibility for controlling the system. In many communities this individual has the title Fire Alarm Superintendent, Superintendent of Fire Alarms, or Deputy Chief of Communications, Director of Signals, and so forth. Sometimes, this individual has

responsibility for both the public fire alarm reporting system and the traffic signals. The International Municipal Signal Association (IMSA) serves as the professional membership association of such individuals. IMSA provides certification programs, technical literature, and other professional services to assist in the continuing education and professional development of those individuals responsible for overseeing and operating pubic fire alarm reporting systems.

Where the jurisdiction does not have adequate staff or knowledge of the system to perform testing and maintenance, 6-3.1 and 6-3.2 permit a contract with a maintenance organization.

6-3.2 Maintenance by an organization or person other than from the jurisdiction or an employee of the jurisdiction shall be by written contract, guaranteeing performance acceptable to the authority having jurisdiction.

6-3.3 All equipment shall be accessible to the authority having jurisdiction for the purpose of maintenance.

6-4 Equipment and Installation

6-4.1 Means for actuation of alarms by the public shall be located where they are conspicuous and accessible for operation.

6-4.2 Records of wired public fire alarm reporting system circuits shall include the following:

(1) Outline plans showing terminals and box sequence
(2) Diagrams of applicable office wiring
(3) List of materials used, including trade name, manufacturer, and year of purchase or installation

Proper plans, material specification sheets, and diagrams allow for ease of repair, maintenance, and testing of the system.

6-4.2.1 Public fire alarm reporting systems as defined in this chapter, shall, in their entirety, be subject to a complete operational acceptance test upon completion of system installation. This test(s) shall be made in accordance with the requirements of the authority having jurisdiction. However, in no case shall the operational functions tested be less than those stipulated in Chapter 7. Acceptance tests shall also be performed on any alarm reporting devices as identified in this chapter that are added subsequent to the installation of the initial system.

The requirements for testing and maintaining public fire reporting systems are found in Chapter 7. Chapter 6 requires a complete acceptance test for all public fire alarm reporting systems, just as for any other fire alarm system.

6-4.3 Publicly accessible boxes shall be recognizable as such. Boxes shall have operating instructions plainly marked on the exterior surface.

6-4.4 The actuating device shall be of such design and so located as to make the method of its use apparent.

6-4.5 Publicly accessible boxes shall be conspicuous. Their color shall be distinctive.

6-4.6 All publicly accessible boxes mounted on support poles shall be identified by a wide band of distinctive colors or signs placed 8 ft (2.44 m) above the ground and visible from all directions wherever possible.

6-4.7* Indicating lights of a distinctive color, visible for at least 1500 ft (460 m), shall be installed over publicly accessible boxes in mercantile and manufacturing areas. Equipping the street light nearest the box with a distinctively colored light shall meet this requirement.

A-6-4.7 The current supply for designating lamps at street boxes should be secured at lamp locations from the local electric utility company.

Alternating-current power can be permitted to be superimposed on metallic fire alarm circuits for supplying designating lamps or for control or actuation of equipment devices for fire alarm or other emergency signals, provided the following conditions exist.

(a) Voltages between any wire and ground or between one wire and any other wire of the system should not exceed 150 volts, and the total resultant current in any line circuit should not exceed $^1/_4$ ampere.

(b) Components such as coupling capacitors, transformers, chokes, or coils are rated for 600-volt working voltage and have a breakdown voltage of at least twice the working voltage plus 1000 volts.

(c) There is no interference with fire alarm service under any conditions.

Superimposing box light power on the fire alarm circuit was popular in the 1930s and 1940s, but system operators seldom use this method today.

6-4.8 Boxes shall be mounted in an approved manner on poles, pedestals, or structural surfaces as directed by the authority having jurisdiction.

Some jurisdictions require a pedestal-mounted box, whereas others permit the box to be mounted directly on a utility pole or on the exterior of the protected premises.

6-4.9 Concurrent operation of at least four boxes shall not result in the loss of an alarm.

To meet the requirement of 6-4.9, each box installed on a circuit must sense that another box has begun to transmit a signal over the common box circuit. The first box withholds

transmitting its signal until it senses a clear circuit, and then it transmits the signal. Manufacturers describe this box design as non-interfering and successive.

6-5 Publicly Accessible Fire Service Boxes (Street Boxes)

6-5.1 Street boxes, when in an abnormal condition, shall leave the circuit usable.

Locating publicly accessible manual fire alarm boxes along public streets and thoroughfares subjects them to possible damage from a variety of sources. Vandals, vehicular accidents, and street repair and maintenance operations could all damage a box. When something damages a box, however, the remainder of the system must continue to operate normally.

6-5.2 Street boxes shall be designed so that recycling does not occur if a box actuating device is held in the actuating position. Street boxes shall be ready to accept a new signal as soon as the actuating device is released.

The requirement in 6-5.2 ensures that if a person in the panic of an emergency continues to hold the box actuating lever in the actuated position, the box will not recycle. This feature prevents a box from tying up the circuit. Exhibit 6.1 shows a typical street box.

Exhibit 6.1 *Typical street box. [Source: Mammoth Fire Alarms, Inc. (Gamewell) Lowell, MA]*

6-5.3* Street boxes, when actuated, shall give a visible or audible indication to the user that the box is operating or that the signal has been transmitted to the communications center.

Most coded wired boxes provide an audible indication of actuation because of the noise created by the mechanism that drives the mechanical code wheel. Series telephone boxes indicate actuation by means of a sound generated in the handset or speaker. Coded radio street boxes usually provide a visible means to indicate actuation.

A-6-5.3 If the operating mechanism of a box creates sufficient sound to be heard by the user, the requirements are satisfied.

6-5.4 The street box housing shall protect the internal components from the weather.

6-5.5 Doors on street boxes shall remain operable under adverse climatic conditions, including icing and salt spray.

6-5.6 Street boxes shall be recognizable as such. Street boxes shall have instructions for use plainly marked on their exterior surfaces.

6-5.7 Street boxes shall be securely mounted on poles, pedestals, or structural surfaces as directed by the authority having jurisdiction.

6-5.8 Street boxes shall be as conspicuous as possible. Their color shall be distinctive, and they shall be visible from as many directions as possible. A wide band of distinctive colors visible over the tops of parked cars or adequate signs completely visible from all directions shall be applied to supporting poles.

6-5.9* Location-designating lights of distinctive color, visible for at least 1500 ft (460 m) in all directions, shall be installed over street boxes. The street light nearest the street box, if equipped with a distinctively colored light, shall be considered as meeting this requirement.

A-6-5.9 The current supply for designating lights at street boxes should be secured at lamp locations from the local electric utility company.

Alternating-current power can be permitted to be superimposed on metallic fire alarm circuits for supplying designating lamps or for control or actuation of equipment devices for fire alarm or other emergency signals, provided the following conditions exist.

(a) Voltage between any wire and ground or between one wire and any other wire of the system does not exceed 150 volts, and the total resultant current in any line circuit does not exceed $^1/_4$ ampere.

(b) Components such as coupling capacitors, transformers, chokes, or coils are rated for 600-volt working voltage and have a breakdown voltage of at least twice the working voltage plus 1000 volts.

(c) There is no interference with fire alarm service under any conditions.

6-5.10 Street box cases and parts that are, at any time, accessible to users shall be of insulating materials or permanently and effectively grounded. All ground connections to street boxes shall comply with the requirements of NFPA 70, *National Electrical Code*®, Article 250.

6-5.11 If a street box is installed inside a structure, it shall be placed as close as is practicable to the point of entrance of the circuit, and the exterior wire shall be installed in conduit or electrical metallic tubing in accordance with Chapter 3 of NFPA 70, *National Electrical Code*.

6-5.12 Coded Radio Street Boxes.

6-5.12.1 Coded radio street boxes shall be designed and operated in compliance with all applicable rules and regulations of the FCC, as well as with the requirements established herein.

6-5.12.2 Coded radio street boxes shall provide no less than three specific and individually identifiable functions to the communications center, in addition to the street box number, as follows:

(1) Test
(2) Tamper
(3) Fire

6-5.12.3* Coded radio street boxes shall transmit to the communications center no less than one repetition for "test," no less than one repetition for "tamper," and no less than three repetitions for "fire."

A-6-5.12.3 The following is an excerpt from the FCC *Rules and Regulations*, Vol. V, Part 90, March 1979:

> Except for test purposes, each transmission must be limited to a maximum of 2 seconds and may be automatically repeated not more than two times at spaced intervals within the following 30 seconds; thereafter, the authorized cycle may not be reactivated for 1 minute.

6-5.12.4 If multifunction coded radio street boxes are used to transmit to the communications center a request(s) for emergency service or assistance in addition to those stipulated in 6-5.12.2, each such additional message function shall be individually identifiable.

6-5.12.5 Multifunction coded radio street boxes shall be designed so as to prevent the loss of supplemental or concurrently actuated messages.

6-5.12.6 An actuating device held or locked in the activating position shall not prevent the activation and transmission of other messages.

6-5.12.7 Power Source.

6-5.12.7.1 Box primary power shall be permitted to be from a utility distribution system, a photovoltaic power system, or user power, or shall be self-powered using either an integral battery or other stored energy source, as approved by the authority having jurisdiction.

6-5.12.7.2 Self-powered boxes shall have sufficient power for uninterrupted operation for a period of not less than 6 months. Self-powered boxes shall transmit a low power warning message to the communications center for at least 15 days prior to the time the power source will fail to operate the box. This message shall be part of all subsequent transmissions.

Use of a charger to extend the life of a self-powered box shall be permitted if the charger does not interfere with box operation. The box shall be capable of operation for not less than 6 months with the charger disconnected.

6-5.12.7.3 Boxes powered by a utility distribution system shall have an integral standby, sealed, rechargeable battery that is capable of powering box functions for at least 60 hours in the event of primary power failure. Transfer to standby battery power shall be automatic and without interruption to box operation. If operating from primary power, the box shall be capable of operation with a dead or disconnected battery. A local trouble indication shall activate upon primary power failure. A battery charger shall be provided in compliance with 1-5.2.8.2, except as modified herein.

If the primary power has failed, boxes shall transmit a power failure message to the communications center as part of subsequent test messages until primary power is restored. A low-battery message shall be transmitted to the communications center if the remaining battery standby time is less than 54 hours.

6-5.12.7.4 Photovoltaic power systems shall provide box operation for not less than 6 months.

Photovoltaic power systems shall be supervised. The battery shall have power to sustain operation for a minimum period of 15 days without recharging. The box shall transmit a trouble message to the communications center when the charger has failed for more than 24 hours. This message shall be part of all subsequent transmissions. If the remaining battery standby duration is less than 10 days, a low-battery message shall be transmitted to the communications center.

6-5.12.7.5 User-powered boxes shall have an automatic self-test feature.

6-5.13 Telephone Street Boxes.

6-5.13.1 If a handset is used, the caps on the transmitter and receiver shall be secured to reduce the probability of the telephone street box being disabled due to vandalism.

6-5.13.2 Telephone street boxes shall be designed to allow the communications center operator to determine whether or not the telephone street box has been restored to normal condition after use.

6-6* Location

A-6-6 Where the intent is for complete coverage, it should not be necessary to travel in excess of one block or 500 ft (150 m) to reach a box. In residential areas, it should not be necessary to travel in excess of two blocks or 800 ft (240 m) to reach a box.

6-6.1 The location of publicly accessible boxes shall be designated by the authority having jurisdiction.

In most cases, the municipal fire officials serve as the authority having jurisdiction. The municipal grading schedule of the Insurance Services Office may also influence their decisions regarding box location.

6-6.2 Schools, hospitals, nursing homes, and places of public assembly shall have a box located at the main entrance, as directed by the authority having jurisdiction.

6-7 Power Supply

6-7.1 General.

6-7.1.1 Batteries, motor-generators, or rectifiers shall be able to supply all connected circuits without exceeding the capacity of any battery or overloading any generator or rectifier, so that circuits developing grounds or crosses with other circuits each shall be able to be supplied by an independent source to the extent required by 6-7.1.8.

6-7.1.2 Provision shall be made in the operating room for supplying any circuit from any battery, generator, or rectifier. Enclosed fuses shall be provided at points where supplies for individual circuits are taken from common leads. Necessary switches, testing, and signal transmitting and receiving devices shall be provided to allow the isolation, control, and test of each circuit up to at least 10 percent of

the total number of box and dispatch circuits, but never less than two circuits.

The requirements in 6-7.1.2 ensure maximum reliability for the public fire alarm reporting system. The last phrase intends that the system must have enough equipment, such as transmitting and receiving equipment, so that no one device or appliance serves more than 10 percent of the circuits. The stock of spare parts must always have at least two of every device or appliance on hand.

6-7.1.3 If common-current source systems are grounded, the ground shall not exceed 10 percent of resistance of any connected circuit and shall be located at one side of the battery. Visual and audible indicating devices shall be provided for each box and dispatch circuit to give immediate warning of ground leakage endangering operability.

Designers and installers use the term *cable plant* to refer to the wires and cables that interconnect the components of a public fire alarm reporting system. Most system cable plants include a combination of aerial and underground cables. Over time, a circuit may become accidentally connected to earth ground. Usually this connection occurs when some foreign object rubs through the insulation on the conductors. Such foreign grounds on a circuit will render a portion of the circuit inoperable. For this reason, operators must give attention to the prompt discovery of excess current leakage to ground.

6-7.1.4 Local circuits at communications centers shall be supplied either in common with box circuits or coded radio-receiving system circuits or by a separate power source. The source of power for local circuits required to operate the essential features of the system shall be supervised.

The system must monitor the integrity of the power for circuits and equipment within the public fire service communications center. The loss of this power must cause a trouble signal.

6-7.1.5 Visual and audible means to indicate a 15 percent or greater reduction of normal power supply (rated voltage) shall be provided.

When power for the public fire alarm reporting system or for local circuits drops 15 percent or more below the normal rated voltage, such reduction must initiate a trouble signal.

6-7.1.6 If the electrical service/capacity of the equipment required under 2-1.6 of NFPA 1221, *Standard for the Installation, Maintenance, and Use of Emergency Services Communications Systems*, is adequate to satisfy the needs of equipment in this chapter, such equipment shall not be required to be duplicated.

6-7.1.7 The forms and arrangements of power supply shall be as classified in 6-7.1.7.

6-7.1.8 Form 4.

6-7.1.8.1 Each box circuit or coded radio receiving system shall be served by the following:

(a)* *Form 4A.* An inverter, powered from a common rectifier, receiving power by a single source of alternating current with a floating storage battery having a 24-hour standby capacity.

A-6-7.1.8.1(a) Figure A-6-7.1.8.1(a) illustrates a Form 4A arrangement.

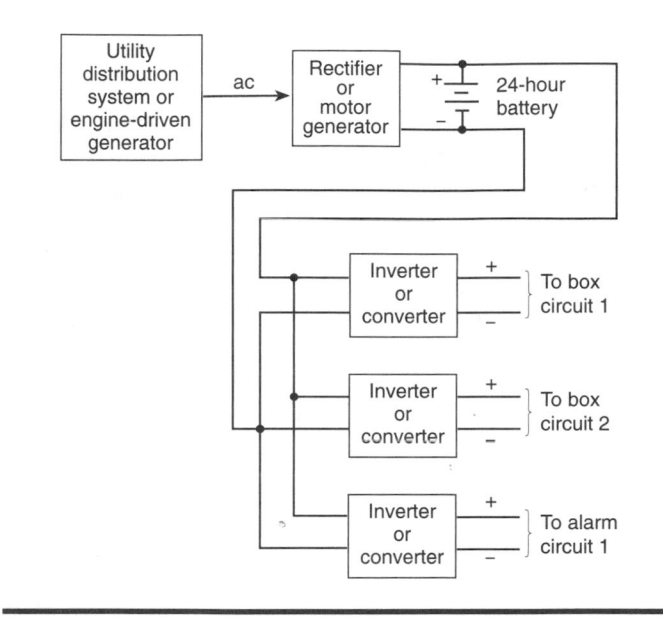

Figure A-6-7.1.8.1(a) Form 4A.

(b)* *Form 4B.* An inverter, powered from a common rectifier, receiving power from two sources of alternating current with a floating storage battery having a 4-hour standby capacity.

A-6-7.1.8.1(b) Figure A-6-7.1.8.1(b) illustrates a Form 4B arrangement.

(c) *Form 4A and Form 4B.* It shall be permitted to distribute the system load between two or more common rectifiers and batteries.

(d)* *Form 4C.* A rectifier, converter, or motor-generator receiving power from two sources of alternating current with transfer facilities to apply power from the secondary source to the system within 30 seconds.

A-6-7.1.8.1(d) Figure A-6-7.1.8.1(d) illustrates a Form 4C arrangement. Refer to NFPA 1221, *Standard for the Installation, Maintenance, and Use of Emergency Services Communications Systems*.

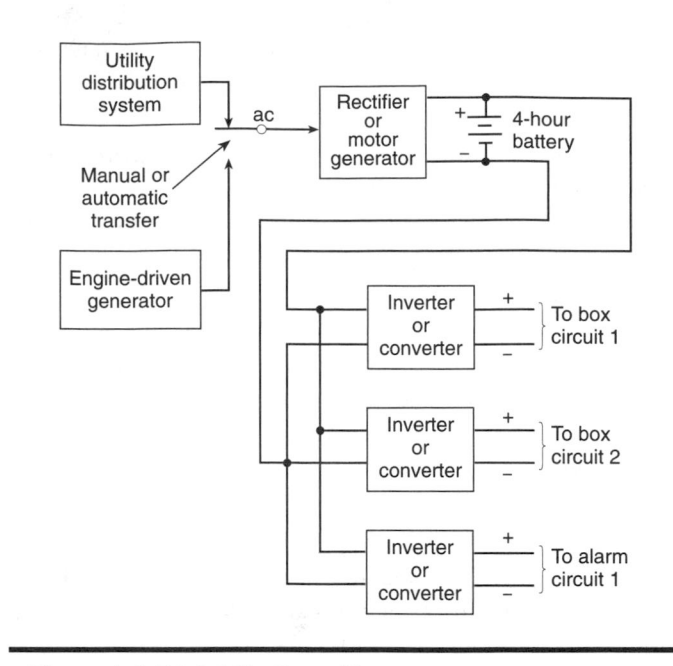

Figure A-6-7.1.8.1(b) Form 4B.

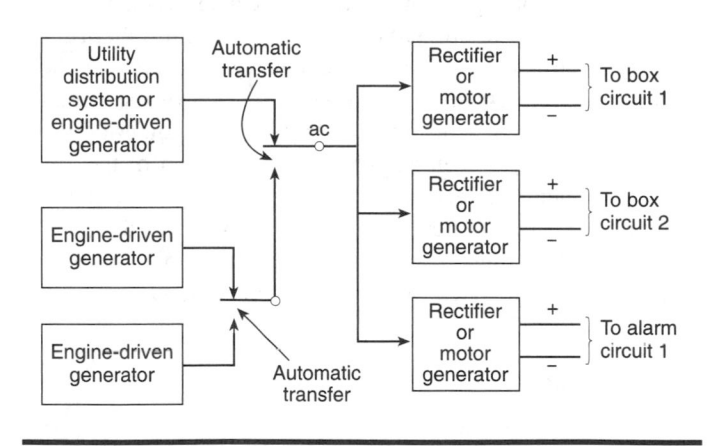

Figure A-6-7-1.8.1(d) Form 4C.

6-7.1.8.2 Form 4A and Form 4B shall be permitted to distribute the system load between two or more common rectifiers and batteries.

6-7.2 Rectifiers, Converters, Inverters, and Motor-Generators.

6-7.2.1 Rectifiers shall be supplied through an isolating transformer that takes energy from a circuit not to exceed 250 volts.

6-7.2.2 Complete spare units or spare parts shall be in reserve.

6-7.2.3 One spare rectifier shall be provided for each ten required for operation, but in no case shall less than one be provided.

6-7.2.4 Leads from rectifiers or motor-generators, with storage battery floating, shall have fuses rated at not less than 1 ampere and not more than 200 percent of maximum connected load. If not provided with battery floating, the fuse shall be not less than 3 amperes.

The requirements in 6-7.2.4 for sizing overcurrent protection provide circuits with sufficient protection without causing the protective fuses to be overly sensitive. Too frequent operation of the fuses reduces the overall reliability of the public fire alarm reporting system.

6-7.3 Engine-Driven Generator Sets.

To maintain reliability, the system must ensure the continuity of supplied power. The requirements in 6-7.3 ensure this continuity.

6-7.3.1 The provisions of 6-7.3 shall apply to generators driven by internal combustion engines.

6-7.3.2 The installation of engine-driven generator sets shall conform to the provisions of NFPA 37, *Standard for the Installation and Use of Stationary Combustion Engines and Gas Turbines*, and NFPA 110, *Standard for Emergency and Standby Power Systems*.

Exception: Where restricted by the provisions of 6-7.3.

6-7.3.3 The engine-driven generator shall be located in a ventilated cutoff area of the building that houses the alarm-receiving equipment.

6-7.3.3.1 The area that houses the unit shall be used for no other purpose than for storage of spare parts or equipment.

6-7.3.3.2 Exhaust fumes shall be discharged directly outside the building.

6-7.3.4 Liquid fuel shall be stored in outside underground tanks and gravity feed shall not be used. Fuel shall be provided for 24 hours of operation at full load where a reliable source of fuel supply is able to be provided, at any time, on a 2-hour notice. If a source of supply is not reliable or able to be provided on a 2-hour notice, special arrangements shall to be made for refueling as necessary. A supply for 48 hours of operation at full load shall be maintained.

Underground tanks reduce fire hazards associated with fuel leaks. To minimize the potential for leaks, the code does not permit gravity feed.

6-7.3.5 Liquefied petroleum gas and natural gas installations shall meet the requirements of NFPA 54, *National Fuel Gas Code*, and NFPA 58, *Liquefied Petroleum Gas Code*.

6-7.3.6 The unit, as a minimum, shall have the capacity to supply power to operate all fire alarm facilities and emergency lighting of the operating rooms or communications building.

6-7.3.7 A separate storage battery on automatic float charger shall be provided for starting the engine-driven generator.

6-7.3.8 If more than one engine-driven generator is provided, each shall be provided with a separate fuel line and transfer pump.

6-7.4 Float-Charged Batteries.

To maintain reliability, the system must ensure the continuity of supplied power. The requirements in 6-7.4 ensure this continuity.

6-7.4.1 Float-charged batteries shall be of the storage type. Primary batteries (dry cells) shall not be used. Lead-acid batteries shall be in jars of glass or other approved transparent materials; other types of batteries shall be in containers approved for the purpose.

6-7.4.2 Float-charged batteries shall be located on the same floor of the building as the operating equipment and shall be available for maintenance and inspection. The battery room shall be above ground and shall be ventilated to prevent accumulation of explosive gas mixtures; special ventilation shall be required only for unsealed cells.

Paragraph 6-7.4.2 was revised in 1999 to require the storage batteries to be on the same floor or level as the operating equipment. The installer must select an above-grade location, unless he or she constructs a location specifically for use below grade. System designers generally avoid below-grade locations to minimize the potential for flooding.

6-7.4.3 Batteries shall be mounted to provide effective insulation from the ground and from other batteries. The mounting shall be suitably protected against deterioration, and shall provide stability, especially in geographic areas subject to seismic disturbance.

6-8 Requirements for Metallic Systems and Metallic Interconnections

6-8.1 Circuit Conductors.

6-8.1.1 Wires shall be terminated so as to prevent breaking from vibration or stress.

6-8.1.2 Circuit conductors on terminal racks shall be identified and isolated from conductors of other systems wherever possible and shall be suitably protected from mechanical injury.

6-8.1.3 Exterior cable and wire shall conform to International Municipal Signal Association (IMSA) specifications or an approved equivalent.

The International Municipal Signal Association (IMSA) publishes wire and cable specifications for use in the installation of public fire alarm reporting systems. IMSA can be reached at P.O. Box 539, Newark, NY 14513, or www.imsasafety.org.

Exception: If circuit conductors are provided by a public utility on a lease basis, IMSA specifications shall not apply.

Some jurisdictions lease conductors from utilities, such as a local telephone company, rather than install their own conductors. The code has no authority over utilities; therefore, conductors owned by a utility are beyond the scope of code requirements.

6-8.1.4 If a public box is installed inside a building, the circuit from the point of entrance to the public box shall be installed in rigid metal, intermediate metal conduit, or electrical metallic tubing in accordance with NFPA 70, *National Electrical Code.*

Exception: This requirement shall not apply to coded radio box systems.

Paragraph 6-8.1.4 intends to limit the exposure of the public fire alarm reporting system circuit to mechanical damage or heat from a hostile fire within a building. If an installer runs the fire alarm reporting circuit extensively throughout the building, a fire could burn through a portion of the circuit before the system transmits an alarm signal, rendering the protection system useless. However, installation of the circuit in conduit or raceway alone would not prevent a fire from damaging the circuit. Though not required by the code, where a circuit must be protected against possible fire damage, a cable type listed for fire survivability should be used. Also see 6-9.1.1.2.

6-8.2 Cables.

6-8.2.1 General.

6-8.2.1.1 Exterior cable and wire shall conform to IMSA specifications or an approved equivalent.

6-8.2.1.1.1 Overhead, underground, or direct burial cables shall be specifically approved for the purpose.

For example, underground conductors must be listed for wet locations even if installed in raceway. Cables must be listed for direct burial if run underground without a raceway.

6-8.2.1.1.2 Cables used in interior installations shall comply with NFPA 70, *National Electrical Code.*

6-8.2.1.2 Paper or pressed pulp insulation shall not be permitted.

Exception: Cables containing conductors with such insulation shall be permitted if pressurized with dry air or nitrogen.

Certain cable constructions installed in the 1940s, particularly for telephone and other communications purposes, used copper wire insulated with paper or pressed pulp materials. These cables must remain dry or the insulation materials will degrade. To ensure dryness, installers commonly pressurize the cable with dry nitrogen and monitor the cable pressure for leakage.

6-8.2.1.2.1 Loss of pressure in cables shall be indicated by a visual or audible warning system, located where an individual is in constant attendance who is able to interpret the pressure readings and who has authority to have the indicated abnormal condition corrected.

6-8.2.1.3 Natural rubber-sheathed cable shall not be used if it is able to be exposed to oil, grease, or other substances or conditions that tend to deteriorate the cable sheath. Braided-sheathed cable shall be used only inside of buildings if it is run in conduit or metal raceways.

6-8.2.1.4 Other municipally controlled signal wires shall be permitted to be installed in the same cable with fire alarm wires. Cables controlled by or containing wires of private signaling organizations shall be permitted to be used for fire alarm purposes only by permission of the authority having jurisdiction.

Occasionally, municipalities that maintain their own governmental service telephone system run the wiring for that service in the same cable as the public fire alarm reporting system. Conversely, on rare occasions public fire reporting systems may consist of interconnecting wiring leased from the public telephone utility or even from a private organization such as Western Union. The requirements of 6-8.2.1.4 apply to such cases.

6-8.2.1.5 Signaling wires that are able to introduce a hazard, because of the source of current supply, shall be protected in accordance with NFPA 70, *National Electrical Code.*

See NFPA 70, *National Electrical Code®*, Articles 760 and 800 for protection requirements.

6-8.2.1.6 All cables with all taps and splices made shall be tested for insulation resistance when installed, but before connection to terminals. Such tests shall indicate an insulation resistance of at least 200 megohms per mile between any one conductor and all other conductors, the sheath, and the ground.

Subparagraph 6-8.2.1.6 requires installers to test cables and splices with a megohm meter (megger) to ensure the dielectric strength of the insulation. Installers must conduct this test before they connect any devices or appliances to the cable plant.

6-8.2.2 Underground Cables.

To maintain the overall operational integrity of a public fire alarm reporting system, the requirements of 6-8.2.2 protect underground cables from exposure to potential mechanical injury.

6-8.2.2.1 Underground cables in duct or direct burial shall be brought aboveground only at points where liability of mechanical injury or of disablement from heat incidental to fires in adjacent buildings is minimized.

6-8.2.2.2 Cables shall be permitted in duct systems and manholes containing low-tension fire alarm system conductors only, except low-tension secondary power cables shall be permitted. If in duct systems or manholes that contain power circuit conductors in excess of 250 volts to ground, fire alarm cables shall be located as far as possible from such power cables and shall be separated from them by a noncombustible barrier or by such other means as is practicable to protect the fire alarm cables from injury.

6-8.2.2.3 All cables installed in manholes shall be racked and marked for identification.

6-8.2.2.4 All conduits or ducts entering buildings from underground duct systems shall be effectively sealed against moisture or gases entering the building.

6-8.2.2.5 Cable joints shall be located only in manholes, fire stations, and other locations where accessibility is provided and where there is little liability of injury to the cable due to either falling walls or operations in the buildings. Cable joints shall be made to provide and maintain conductivity, insulation, and protection at least equal to that afforded by the cables that are joined. Open cable ends shall be sealed against moisture.

6-8.2.2.6 Direct-burial cable, without enclosure in ducts, shall be laid in grass plots, under sidewalks, or in other places where the ground is not likely to be opened for other underground construction. If splices are made, such splices shall, if practicable, be accessible for inspection and tests. Such cables shall be buried at least 18 in. (0.5 m) deep and, where crossing streets or other areas likely to be opened for other underground construction, shall be in duct or conduit or be covered by creosoted planking of at least 2 in. × 4 in. (50 mm × 100 mm) with half-round grooves, spiked or banded together after the cable is installed.

6-8.2.3 Aerial Construction.

To maintain the overall operational integrity of a public fire alarm reporting system, the requirements of 6-8.2.3 and 6-8.2.4 protect aerial cables and leads down poles to reduce the risk of both mechanical injury and electrical failure.

6-8.2.3.1 Fire alarm wires shall be run under all other wires except communications wires. Suitable precautions shall be provided if passing through trees, under bridges, over railroads, and at other places where injury or deterioration is possible. Wires and cables shall not be attached to a

crossarm that carries electric light and power wires, except circuits carrying up to 220 volts for public communications use, and then only if the 220-volt circuits are tagged or otherwise identified.

6-8.2.3.2 Aerial cable shall be supported by messenger wire of approved tensile strength.

Exception: Two-conductor cable that has conductors of No. 20 AWG or larger size and has mechanical strength equal to No. 10 AWG hard-drawn copper.

6-8.2.3.3 Single wire shall meet IMSA specifications and shall not be smaller than No. 10 Roebling gauge if of galvanized iron or steel, No. 10 AWG if of hard-drawn copper, No. 12 AWG if of approved copper-covered steel, or No. 6 AWG if of aluminum. Span lengths shall not exceed the manufacturer's recommendations.

6-8.2.3.4 Wires to buildings shall contact only intended supports and shall enter through an approved weatherhead or sleeves slanting upward and inward. Drip loops shall be formed on wires outside of buildings.

6-8.2.4 Leads Down Poles.

6-8.2.4.1 Leads down poles shall be protected against mechanical injury. Any metallic covering shall form a continuous conducting path to ground. Installation, in all cases, shall prevent water from entering the conduit or box.

6-8.2.4.2 Leads to boxes shall have 600-volt insulation approved for wet locations, as defined in NFPA 70, *National Electrical Code.*

6-8.2.5 Wiring Inside Buildings.

To maintain the overall operational integrity of a public fire alarm reporting system, the requirements of 6-8.2.5 protect exposed wiring inside a building. This protection limits the risk of mechanical injury or electrical failure. The requirements also make certain that the wiring does not contribute to the spread of a fire in a building.

6-8.2.5.1 At the communications center, conductors shall extend as directly as possible to the operating room in conduits, ducts, shafts, raceways, or overhead racks and troughs of a type of construction affording protection against fire and mechanical injury.

6-8.2.5.2 All conductors inside buildings shall be in conduit, electrical tubing, metal molding, or raceways. Installation shall be in accordance with NFPA 70, *National Electrical Code.*

6-8.2.5.3 Conductors shall have an approved insulation. The insulation or other outer covering shall be flame retardant and moisture resistant.

6-8.2.5.4 Conductors shall be installed as far as possible without joints. Splices shall be permitted only in listed junction or terminal boxes. Fire alarm circuits shall be identified by the use of red covers or doors and the words "municipal fire alarm circuit" shall be clearly marked on all terminal and junction locations to prevent unintentional interference. Wire terminals, terminal boxes, splices, and joints shall conform to NFPA 70, *National Electrical Code.*

6-8.2.5.5 Conductors bunched together in a vertical run that connects two or more floors shall have a flame-retardant covering that is able to prevent the carrying of fire from floor to floor.

Exception: This requirement shall not apply where the conductors are encased in a metallic conduit or located in a fire-resistive shaft having fire stops at each floor.

6-8.2.5.6 Cable and wiring exposed to a fire hazard shall be protected in an approved manner.

6-8.2.5.7 Cable terminals and cross-connecting facilities shall be located in or adjacent to the operations room.

6-8.2.5.8 If signal conductors and electric light and power wires are run in the same shaft, they shall be separated by at least 2 in. (51 mm), or either system shall be encased in a noncombustible enclosure.

6-9 Facilities for Signal Transmission

6-9.1 Circuits.

6-9.1.1 General.

6-9.1.1.1 ANSI/IEEE C2, *National Electrical Safety Code,* shall be used as a guide for the installation of outdoor circuitry.

Public and private electric company utilities, public and private telephone utilities, and public and private community antenna television company utilities use ANSI/IEEE C2, *National Electrical Safety Code* (NESC). A committee from the Institute of Electrical and Electronic Engineers developed the NESC to describe the placement and spacing of outdoor aerial cable installations. The NESC ensures the safe operation of the associated systems.

6-9.1.1.2 Installation shall provide for the following:

(1) Continuity of service
(2) Protection from mechanical damage
(3) Disablement from heat that is incidental to fire
(4) Protection from falling walls
(5) Damage by floods, corrosive vapors, or other causes

6-9.1.1.3 Open local circuits within single buildings shall be permitted in accordance with Chapter 3.

Protected premises fire alarm system circuits that are not a part of the public fire alarm reporting system must be installed in accordance with Chapter 3 of the code.

6-9.1.1.4 All circuits shall be routed so as to allow tracing of circuits for trouble.

6-9.1.1.5 Circuits shall not pass over, under, through, or be attached to buildings or property not owned by or under the control of the authority having jurisdiction or the agency responsible for maintaining the system.

Exception: Where the circuit is terminated at a public fire alarm reporting system initiating device on the premises and where a means, approved by the authority having jurisdiction, is provided to disconnect the circuit from the building or property.

Installers strung the circuits for many of the original public fire alarm reporting systems in various east coast municipalities throughout a city from building to building. Engineers soon learned that fires in those buildings would damage the circuits, thereby placing the operational integrity of the public fire alarm reporting system in jeopardy.

The exception was revised for the 1999 code to permit the circuit to be terminated at a public fire alarm reporting system initiating device on the protected premises. However, a means must exist to disconnect the circuit from the building. This disconnecting means allows isolation of the device inside the protected premises in the event of a fault on the conductors that run through the protected premises.

6-9.1.2 Box Circuits. A means provided only to the authority having jurisdiction or the agency responsible for maintaining the public fire alarm reporting system shall be provided for disconnecting the circuit inside the building. Definite notification shall be given to the designated building representative when the interior box(es) is out of service.

If an installer makes a connection to the public fire alarm reporting system in accordance with the requirements of Section 6-16, he or she must provide a means to disconnect the protected premises connection. Only the authority having jurisdiction over the public fire alarm reporting system may have access to this disconnecting means.

6-9.1.3 Tie Circuits.

Tie circuits connect the public fire service communications center with a subsidiary communications center. For example, in a large municipality, a subsidiary communications center concentrates signals from a particular neighborhood before transmitting them to the public fire service communi-

cations center. Also, in some cities where several boroughs have their own public fire service communications center (New York City, for example), the system may use tie circuits to interconnect the centers. This interconnection allows the centers to handle signals from all of the boroughs even if one of the centers is impaired.

6-9.1.3.1 A separate tie circuit shall be provided from the communications center to each subsidiary communications center.

6-9.1.3.2 The tie circuit between the communications center and the subsidiary communications center shall not be used for any other purpose.

6-9.1.3.3 In a Type B wire system, if all boxes in the system are of the succession type, it shall be permitted to use the tie circuit as a dispatch circuit to the extent permitted by NFPA 1221, *Standard for the Installation, Maintenance, and Use of Emergency Services Communications Systems.*

6-9.1.4* Circuit Protection.

A-6-9.1.4 All requirements for circuit protection do not apply to coded radio reporting systems. These systems do not use metallic circuits.

Circuit protection limits equipment damage caused when an incident applies transient currents to the circuits of the public fire alarm reporting system. Lightning is one source of such transients.

Article 800 of the *National Electrical Code* covers protection of communications circuits, including surge suppressors and lightning arresters.

6-9.1.4.1 General.

6-9.1.4.1.1 The protective devices shall be located close to or be combined with the cable terminals.

6-9.1.4.1.2 Surge arresters approved for the purpose shall be provided. Surge arresters shall be marked with the name of the manufacturer and model designation.

6-9.1.4.1.3 All surge arresters shall be connected to a ground in accordance with NFPA 70, *National Electrical Code.*

6-9.1.4.1.4 All fuses shall be plainly marked with their rated ampere capacity. All fuses rated over 2 amperes shall be of the enclosed type.

6-9.1.4.1.5 Circuit protection required at the communications center shall be provided in every building that houses communications center equipment.

6-9.1.4.1.6 Each conductor entering a fire station from partially or entirely aerial lines shall be protected by a lightning arrester.

6-9.1.4.2 Communications Center.

6-9.1.4.2.1 All conductors entering the communications center shall be protected by the following devices, in the order named, starting from the exterior circuit:

(1) A fuse rated at 3 amperes minimum to 7 amperes maximum and not less than 2000 volts
(2) A surge arrester(s)
(3) A fuse or circuit breaker rated at $^1/_2$ ampere

6-9.1.4.2.2 The $^1/_2$-ampere protection on the tie circuits shall be omitted at subsidiary communications centers.

6-9.1.4.3 Protection on Aerial Construction.

6-9.1.4.3.1 At junction points of open aerial conductors and cable, each conductor shall be protected by a surge arrester(s) of the weatherproof type. There also shall be a connection between the surge arrester ground, any metallic sheath, and messenger wire.

6-9.1.4.3.2 Aerial open-wire and non–messenger-supported, two-conductor cable circuits shall be protected by a surge arrester(s) at intervals not to exceed 2000 ft (610 m).

6-9.1.4.3.3 Surge arresters, other than of the air-gap or self-restoring type, shall not be installed in fire alarm circuits.

6-9.1.4.3.4 All protective devices shall be accessible for maintenance and inspection.

6-10 Power

6-10.1 Requirements for Constant-Current Systems.

The coded wired public fire alarm reporting system operates at a constant current of nominally 100 mA. The requirements of 6-10.1 regulate and maintain the current; limit the voltage; provide a visual indication of current reduction; and provide meters to allow operators to measure current. These features all ensure that such a system maintains a high level of operational integrity.

6-10.1.1 Means shall be provided for manually regulating the current in box circuits so that the operating current is maintained within 10 percent of normal throughout changes in external circuit resistance from 20 percent above to 50 percent below normal.

6-10.1.2 The voltage supplied to maintain normal line current on box circuits shall not exceed 150 volts, measured under no-load conditions, and shall be such that the line current cannot be reduced below the approved operating value by the simultaneous operation of four boxes.

6-10.1.3 Visual and audible means to indicate a 20 percent or greater reduction in the normal current in any alarm circuit shall be provided. All devices connected in series with any alarm circuit shall function when the alarm circuit current is reduced to 70 percent of normal.

6-10.1.4 Meters shall be provided to indicate the current in any box circuit and the voltage of any power source. Meters used in common for two or more circuits shall be provided with cut-in devices designed to reduce the probability of cross-connecting circuits.

6-11 Receiving Equipment—Facilities for Receipt of Box Alarms

Meeting the requirements of 6-11 ensures that the public fire service communications center will receive the signals transmitted over the public fire alarm reporting system and will automatically record them in a manner that provides a permanent visual record of the signals. At the same time, an audible notification appliance will alert the operators to incoming signals.

The signal indicates the exact location of its origin. This indication comes from a unique number assigned to each public fire alarm reporting box. An operator reading a manual chart, or an interface to a computer-aided dispatching system then translates the box number to an exact location. It is important to note that NFPA 1221, *Standard for Public Fire Service Communications,* covers the requirements for computer-aided dispatching systems.

6-11.1 General.

6-11.1.1 Alarms from boxes shall be automatically received and recorded at the communications center.

6-11.1.2 A permanent visual record and an audible signal shall be required to indicate the receipt of an alarm. The permanent record shall indicate the exact location from which the alarm is being transmitted.

The audible signal device shall be permitted to be common to two or more box circuits and arranged so that the fire alarm operator is able to manually silence the signal temporarily by a self-restoring switch.

6-11.1.3 Facilities shall be provided that automatically record the date and time of receipt of each alarm.

Exception: Only the time shall be required to be automatically recorded for voice recordings.

6-11.2 Visual Recording Devices.

6-11.2.1 A device for producing a permanent graphic recording of all alarm, supervisory, trouble, and test signals

received or retransmitted, or both, shall be provided at each communications center for each alarm circuit and tie circuit. If each circuit is served by a dedicated recording device, the number of reserve recording devices required on site shall be equal to at least 5 percent of the circuits in service and in no case less than 1 percent. If two or more circuits are served by a common recording device, a reserve recording device shall be provided on site for each circuit connected to a common recorder.

6-11.2.2 In a Type B wire system, one such recording device shall be installed in each fire station and at least one shall be installed in the communications center.

6-11.3 System Integrity.

6-11.3.1 Wired circuits upon which transmission and receipt of alarms depend, shall be constantly monitored for integrity to provide prompt warning of conditions adversely affecting reliability.

6-11.3.2 The power supplied to all required circuits and devices of the system shall be constantly monitored for integrity.

6-11.4 Trouble Signals.

6-11.4.1 Trouble signals shall actuate a sounding device located where there is a responsible person on duty at all times.

6-11.4.2 Trouble signals shall be distinct from alarm signals and shall be indicated by both a visual light and an audible signal.

6-11.4.3 The audible signal shall be permitted to be common to more than one circuit that is monitored for integrity.

6-11.4.4 A switch for silencing the audible trouble signal shall be permitted, provided the visual signal remains operating until the silencing switch is restored to its normal position.

6-11.4.5 The audible signal shall be responsive to faults on any other circuits that occur prior to restoration of the silencing switch to its normal position.

6-12 Remote Receiving Equipment— Facilities for Receipt of Box Alarms at a Remote Communications Center

When the alarm receiving equipment is located at a location other than where the box circuit protection, controls, and power supplies are located, the following requirements, in addition to all of the requirements of Section 6-11, shall apply. All equipment used to provide the primary and remote receiving facilities shall be listed for its intended use and shall be installed in accordance with NFPA 70, *National Electrical Code.*

6-12.1 The monitoring for integrity of all box circuits shall be provided with a visual and audible means to indicate a 20 percent or greater reduction or increase in the normal current in any box alarm circuit. The visual means shall identify the exact circuit affected.

6-12.2 Monitoring for integrity of all power supplies shall be provided with visual and audible means to indicate a loss of primary or standby power supplies at both the primary and remote communications centers.

6-12.3 A minimum of two separate means of interconnection shall be provided between the primary and remote communications center receiving equipment. This interconnection shall be dedicated and shall not be used for any other purpose.

6-12.4 When data transmission or multiplexing equipment is used that is not an integral part of the alarm-receiving equipment, a visual and audible means shall be provided to monitor the integrity of the external equipment. This shall include monitoring all primary and stand-by power supplies as well as the transmission of data.

6-12.5 Power shall be provided in accordance with Section 6-7. The use of an uninterruptible power supply (UPS) to comply with standby power requirements shall not be permitted.

6-13 Coded Wired Reporting Systems

Sections 6-13, 6-14, and 6-15 establish requirements for each of the three types of public fire alarm reporting systems: coded wired, coded radio, and telephone (series).

6-13.1 For a Type B system, the effectiveness of noninterference and succession functions between box circuits shall be no less than between boxes in any one circuit. The disablement of any metallic box circuit shall cause a warning signal in all other circuits, and, thereafter, the circuit or circuits not otherwise broken shall be automatically restored to operative condition.

In a Type B coded wired system, the system repeats signals from one box circuit or alarm circuit onto the other box and alarm circuits. The repetition of these signals causes other boxes to sense a busy circuit and wait for a clear circuit before transmitting. This approach provides for the proper functioning of the non-interfering and successive features.

6-13.2 Box circuits shall be sufficient in number and laid out so that the areas that would be left without box protec-

tion in case of disruption of a circuit do not exceed those covered by 20 properly spaced boxes if all or any part of the circuit is of aerial open-wire, or 30 properly spaced boxes if the circuit is entirely in underground or messenger-supported cable.

6-13.3 If all boxes on any individual circuit and associated equipment are designed and installed to provide for receipt of alarms through the ground in the event of a break in the circuit, the circuit shall be permitted to serve twice the number of aerial open-wire and cable circuits, respectively, as are specified in 6-13.2.

In most coded wired systems, when an actuated fire alarm box senses that the circuit has an open fault, it idles for one round and then connects the box to earth ground. Sensing an open circuit, the receiving equipment at the public fire service communications center also connects itself to earth ground. This conditioning of the circuit allows the box to transmit its signal through earth ground. If two open faults occur on the circuit, the boxes isolated between the faults cannot transmit a signal.

6-13.4 The installation of additional boxes in an area served by the number of properly spaced boxes indicated in 6-13.1 through 6-13.3 shall not constitute geographical overloading of a circuit.

Once an installer has properly spaced boxes throughout an area, additional boxes must not be added to overload the circuit. A circuit overload might occur when an installer adds master fire alarm boxes to connect protected premises fire alarm systems to the public fire alarm reporting system.

6-13.5 Sounding devices for signals shall be provided for box circuits.

6-13.5.1 A common sounding device for more than one circuit shall be permitted to be used in a Type A system and shall be installed at the communications center.

6-13.5.2 In a Type B system, a sounding device shall be installed in each fire station at the same location as the recording device for that circuit, unless installed at the communications center, where a common sounding device shall be permitted.

6-14 Coded Radio Reporting Systems

6-14.1 Radio Box Channel (Frequency).

6-14.1.1 The number of boxes permitted on a single frequency shall be governed by the following.

(a) For systems that use one-way transmission in which the individual box automatically initiates the required message *(refer to* 6-14.6.3*)* using circuitry integral to the boxes, not more than 500 boxes shall be permitted on a single frequency.

(b) For systems that use a two-way concept in which interrogation signals *(refer to* 6-14.6.3*)* are transmitted to the individual boxes from the communications center on the same frequency used for receipt of alarms, not more than 250 boxes shall be permitted on a single frequency. If interrogation signals are transmitted on a frequency that differs from that used for receipt of alarms, not more than 500 boxes shall be permitted on a single frequency.

(c) A specific frequency shall be designated for both fire and other fire-related or public safety alarm signals and for monitoring for integrity signals.

Paragraph 6-14.1.1(c) prevents the public fire service communications center from using the radio frequency assigned to the boxes for normal two-way or one-way radio communications. Such use might inadvertently interfere with receipt of signals from the boxes.

6-14.1.2 If box message signals to the communications center or acknowledgment of message receipt signals from the communications center to the box are repeated, associated repeating facilities shall conform to the requirements indicated in 7-1.1.4(d) of NFPA 1221, *Standard for the Installation, Maintenance, and Use of Emergency Services Communications Systems.*

6-14.2 Metallic Interconnections.

A means that is available only to the agency responsible for maintaining the public fire alarm reporting system shall be provided for disconnecting the auxiliary loop to the connected property. Notification shall be given to the designated representative of the property when the auxiliary box is not in service.

If an installer makes a connection to the public fire alarm reporting system in accordance with the requirements of Section 6-16, he or she must provide a means to disconnect the protected premises connection. Only the authority having jurisdiction over the public fire alarm reporting system may have access to this disconnecting means.

6-14.3 The antenna transmission line between the transmitter and the antenna shall be installed in rigid metal, intermediate metal conduit, or electrical metallic tubing in accordance with NFPA 70, *National Electrical Code.*

6-14.3.1 Type A System.

6-14.3.1.1* For each frequency used, two separate receiving networks, each including an antenna, an audible alerting

device, a receiver, a power supply, signal processing equipment, a means of providing a permanent graphic recording of the incoming message that is both timed and dated, and other associated equipment shall be provided and shall be installed at the communications center. Facilities shall be arranged so that a failure of either receiving network cannot affect the receipt of messages from boxes.

Redundant equipment increases the overall reliability of the coded radio public fire alarm reporting system.

A-6-14.3.1.1 Figure A-6-14.3.1.1 illustrates a Type A receiving network.

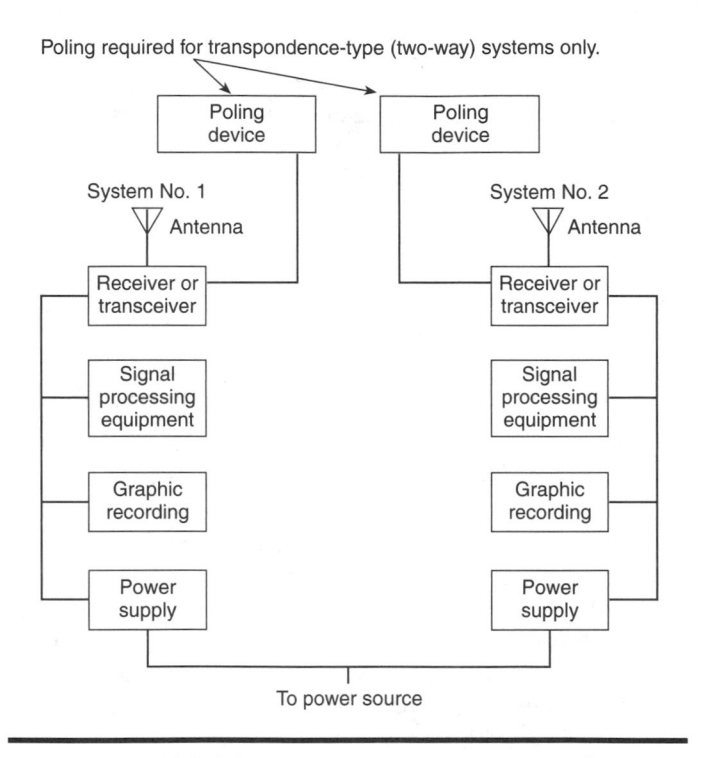

Figure A-6-14.3.1.1 Type A system receiving networks.

6-14.3.1.2 If the system configuration is such that a polling device is incorporated into the receiving network to allow remote or selective initiation of box tests, a separate such device shall be included in each of the two required receiving networks. The requirements of Chapter 7 shall apply. Furthermore, the polling devices shall be configured for automatic cycle initiation in their primary operating mode, shall be capable of continuous self-monitoring, and shall be integrated into the network(s) to provide automatic switchover and operational continuity in the event of failure of either device.

Some coded radio systems provide for an interrogation and response sequence initiated from the public fire service

communications center. This interrogation and response sequence monitors the integrity of the radio channel signaling pathway. Redundant equipment increases the overall reliability of the system.

6-14.3.1.3 Test signals from boxes shall not be required to include the date as part of their permanent recording, provided that the date is automatically printed on the recording tape at the beginning of each calendar day.

6-14.3.2 Type B System.

6-14.3.2.1 For each frequency used, a single, complete receiving network shall be permitted in each fire station, provided the communications center conforms to 6-14.3.1.1. If the jurisdiction maintains two or more alarm reception points in operation, one receiving network shall be permitted to be at each alarm reception point.

6-14.3.2.2 If alarm signals are transmitted to a fire station from the communications center using the coded radio-type receiving equipment in the fire station to receive and record the alarm message, a second receiving network conforming to 6-14.3.2.1 shall be provided at each fire station, and that receiving network shall employ a frequency other than that used for the receipt of box messages.

6-14.4 Power.

Power shall be provided in accordance with Section 6-7.

6-14.5 Testing.

The requirements of Chapter 7 shall apply.

6-14.6 Supervision.

6-14.6.1 All coded radio box systems shall provide constant monitoring of the frequency in use. Both an audible and a visual indication of any sustained carrier signal, if in excess of 15-second duration, shall be provided for each receiving system at the communications center.

An open fault or ground fault on a coded wired public fire alarm reporting system interferes with the transmission of signals. Similarly, the sustained transmission of a radio carrier signal interferes with the transmission of signals from the boxes on a coded radio public fire alarm reporting system.

6-14.6.2 The power supplied to all required circuits and devices of the system shall be supervised.

6-14.6.3* Each coded radio box shall automatically transmit a test message at least once in each 24-hour period.

The 24-hour test signal safeguards against the catastrophic failure of a single fire alarm box.

A-6-14.6.3 The transmission of an actual emergency-related message, initiated at the same time it is preselected for a test message, and, in turn, preempts said test message, must satisfy the intent of this requirement.

6-14.6.4 Receiving equipment associated with coded radio-type systems, including any related repeater(s), shall be tested at least hourly. The receipt of test messages that do not exceed 60-minute intervals shall meet this requirement.

6-14.6.5 Radio repeaters upon which receipt of alarms depend shall be provided with dual receivers, transmitters, and power supplies. Failure of the primary receiver, transmitter, or power supply shall cause an automatic switchover to the secondary receiver, transmitter, or power supply.

Exception: Manual switchover shall be permitted provided it is completed within 30 seconds.

6-14.6.6 Trouble signals shall actuate a sounding device located where there is always a responsible person on duty.

6-14.6.7 rouble signals shall be distinct from alarm signals and shall be indicated by both a visual light and an audible signal.

The audible signal shall be permitted to be common to two or more supervised circuits.

A switch for silencing the audible trouble signal shall be permitted if the visual signal remains operating until the silencing switch is restored to its normal position.

6-14.6.8 The audible signal shall be responsive to subsequent faults in other monitored functions prior to restoration of the silencing switch.

6-15 Telephone (Series) Reporting Systems

Sometimes installers add components to all or a portion of an existing coded wired reporting system to give that system the capability of transmitting and receiving voice alarms. In these cases, the telephone (series) system uses the same cable plant as the coded wired system.

6-15.1 A permanent visual recording device installed in the communications center shall be provided to record all incoming box signals. A spare recording device shall be provided for five or more box circuits.

This permanent visual recording device records the date, time, and box number, but not the content of the voice message. See 6-15.4.

6-15.2 A second visual means of identifying the calling box shall be provided.

6-15.3 Audible signals shall indicate all incoming calls from box circuits.

6-15.4 All voice transmissions from boxes for emergencies shall be recorded with the capability of instant playback.

Specially designed audio recording equipment not only provides an audio log of signal content from the boxes, but also allows an operator to instantly recycle to the beginning of each message. In this way, operators at the public fire service communications center can review messages whose content is unclear.

6-15.5 A voice recording facility shall be provided for each operator handling incoming alarms to eliminate the possibility of interference.

6-15.6 Box circuits shall be sufficient in number and laid out so that the areas that would be left without box protection in case of disruption of a circuit do not exceed those covered by 20 properly spaced boxes if all or any part of the circuit is of aerial open-wire, or 30 properly spaced boxes if the circuit is entirely in underground or messenger-supported cable.

6-15.7 If all boxes on any individual circuit and associated equipment are designed and installed to provide for receipt of alarms through the ground in the event of a break in the circuit, the circuit shall be permitted to serve twice the number of aerial open-wire and cable circuits, respectively, as is specified in 6-15.6.

In the coded wired portion of some telephone (series) systems, when an actuated fire alarm box senses that the circuit has an open fault, it idles for one round and then connects the box to earth ground. Sensing an open circuit, the receiving equipment at the public fire service communications center also connects itself to earth ground. This conditioning of the circuit allows the box to transmit its signal through earth ground. If two open faults occur on the circuit, the boxes isolated between the faults cannot transmit a signal.

6-15.8 The installation of additional boxes in an area served by the number of properly spaced boxes indicated in 6-15.6 shall not constitute geographical overloading of a circuit.

Once an installer has properly spaced boxes throughout an area, additional boxes must not be added to overload the circuit. A circuit overload might occur when an installer adds master fire alarm boxes to connect protected premises fire alarm systems to the public fire alarm reporting system.

6-16 Auxiliary Fire Alarm Systems

The requirements of Chapter 1, 3, and 7 shall apply to auxiliary fire alarm systems, unless they conflict with the requirements of Section 6-16. If permitted by the authority having jurisdiction, the use of systems described in Chapter 6 shall

be permitted to provide defined reporting functions from or within private premises.

6-16.1 Scope.

Section 6-16 describes the equipment and circuits necessary to connect a protected premises.

6-16.2 General.

6-16.2.1 An auxiliary fire alarm system shall be used only in connection with a public fire alarm reporting system that is approved for the service. A system approved by the authority having jurisdiction shall meet this requirement.

If a community has not provided a public fire alarm reporting system, no auxiliary fire alarm system can exist. An auxiliary system depends on the public fire alarm reporting system to transmit signals from the protected premises to the public fire service communications center.

6-16.2.2 Permission for the connection of an auxiliary fire alarm system to a public fire alarm reporting system, and acceptance of the type of auxiliary transmitter and its actuating mechanism, circuits, and components connected thereto, shall be obtained from the authority having jurisdiction.

6-16.2.3 An auxiliary fire alarm system shall be maintained and supervised by a responsible person or corporation.

Proper maintenance of an auxiliary system requires careful coordination with those who operate and maintain the public fire alarm reporting system.

6-16.2.4 Section 6-16 shall not require the use of audible alarm signals other than those necessary to operate the auxiliary fire alarm system. If it is desired to provide fire alarm evacuation signals in the protected property, the alarms, circuits, and controls shall comply with the provisions of Chapter 3 in addition to the provisions of Section 6-16.

An auxiliary system does not, itself, notify occupants of a fire alarm. If the authority having jurisdiction requires such notification, then a protected premises fire alarm system with notification appliances must be installed in accordance with Chapters 1, 3, 5, 6, and 7. A building fire alarm scheme fairly popular in the past imposed alternating current for audible fire alarm notification appliances on the direct current manual fire alarm-initiating device circuit of a shunt-type master fire alarm box. The Gamewell Company marketed this system under the trade name of "Dual Alarm." Such a system would no longer meet the requirements of Chapters 1, 2, 3, 4, 5, and 7.

6-16.3 Communications Center Facilities.

The communications center facilities shall be in accordance with the requirements of Sections 6-1 through 6-15.

6-16.4 Equipment.

6-16.4.1 Types of Systems. Auxiliary fire alarm systems shall be of the following two types.

(a)* *Local Energy Type.*

A-6-16.4.1(a) The local energy-type system *[refer to Figures A-6-16.4.1(a)(1) and A-6-16.4.1(a)(2)]* is electrically isolated from the public fire alarm reporting system and has its own power supply. The tripping of the transmitting device does not depend on the current in the system. In a wired circuit, receipt of the alarm by the communications center when the circuit is accidentally opened depends on the design of the transmitting device and the associated communications center equipment (in other words, whether or not the system is designed to receive alarms through manual or automatic ground operational facilities). In a radio box-type system, receipt of the alarm by the communications center depends on the proper operation of the radio transmitting and receiving equipment.

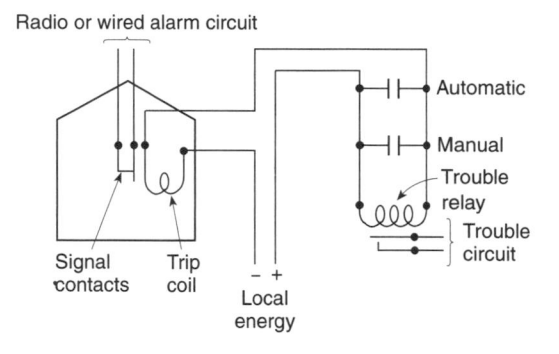

Figure A-6-16.4.1(a)(1) *Local energy-type auxiliary fire alarm system.*

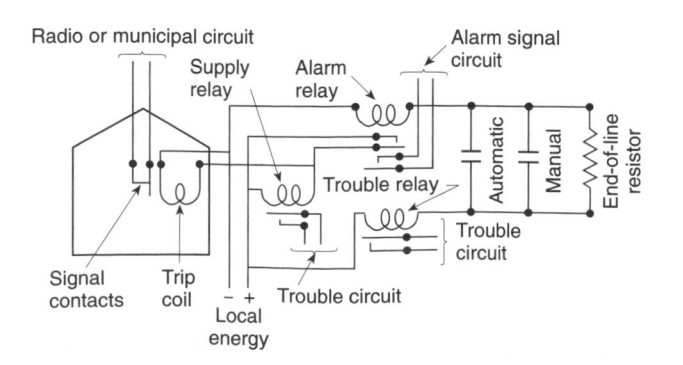

Figure A-6-16.4.1(a)(2) *Local energy-type auxiliary fire alarm system.*

(1) Local energy systems shall be permitted to be of the coded or noncoded type.

(2) Power supply sources for local energy systems shall conform to Chapter 1.

(b)* *Shunt Type.*

A-6-16.4.1(b) The shunt-type system *[refer to Figures A-6-16.4.1(b)(1) and A-6-16.4.1(b)(2)]* is electrically connected to, and is an integral part of, the public fire alarm reporting system. A ground fault on the auxiliary circuit is a fault on the public fire alarm reporting system circuit, and an accidental opening of the auxiliary circuit sends a needless (or false) alarm to the communications center. An open circuit in the transmitting device trip coil is not indicated either at the protected property or at the communications center. Also, if an initiating device is operated, an alarm is not transmitted, but an open circuit indication is given at the communications center. If a public fire alarm reporting system circuit is open when a connected shunt-type system is operated, the transmitting device does not trip until the public fire alarm reporting system circuit returns to normal, at which time the alarm is transmitted, unless the auxiliary circuit is first returned to a normal condition.

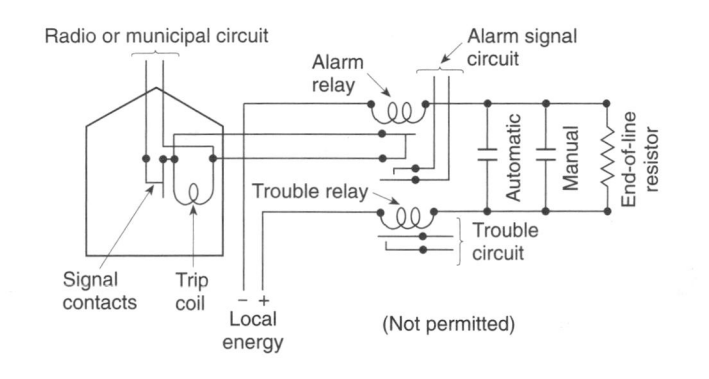

Figure A-6-16.4.1(b)(1) Shunt-type auxiliary fire alarm system.

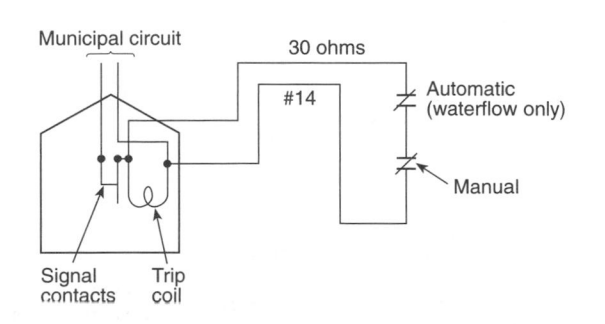

Figure A-6-16.4.1(b)(2) Shunt-type auxiliary fire alarm system.

Additional design restrictions for shunt-type systems are found in laws or ordinances.

The detailed requirements of 6-16.4.1(a) and (b) describe the two types of auxiliary systems: local energy and shunt type. Most authorities having jurisdiction consider the shunt system the least desirable. Also, see definitions and commentary for each type of system in Section 1-4.

Public telephone company utilities no longer install parallel telephone systems. Even though some communities still use these systems, they will only be able to do so as long as the telephone utility continues to maintain them. For these reasons, references to parallel telephone systems were removed from the 1999 code.

(1) Shunt systems shall be noncoded with respect to any remote electrical tripping or actuating devices.

(2) All conductors of the shunt circuit shall be installed in accordance with NFPA 70, *National Electrical Code*, Article 346, for rigid conduit, or Article 348, for electrical metallic tubing.

(3) Both sides of the shunt circuit shall be in the same conduit.

(4) If an auxiliary transmitter is located within a private premises, it shall be installed in accordance with 6-9.1.

(5) If a shunt loop is used, it shall not exceed a length of 750 ft (230 m) and shall be in conduit.

(6) Conductors of the shunt circuits shall not be smaller than No. 14 AWG and shall be insulated as prescribed in NFPA 70, *National Electrical Code*, Article 310.

(7) The power for shunt-type systems shall be provided by the public fire alarm reporting system.

(8) A local system made to an auxiliary system by the addition of a relay whose coil is energized by a local power supply and whose normally closed contacts trip a shunt-type master box shall not be permitted.

6-16.4.2 The interface of the two types of auxiliary fire alarm systems with the three types of public fire alarm reporting systems shall be in accordance with Table 6-16.4.2.

Table 6-16.4.2 Application of Public Fire Alarm Reporting Systems with Auxiliary Fire Alarm Systems

Reporting Systems	Local Energy Type	Shunt Type	Parallel Type
Coded wired	Yes	Yes	No
Coded radio	Yes	No	No
Telephone series	Yes	No	No

6-16.4.3 The application of the two types of auxiliary fire alarm systems shall be limited to the initiating devices specified in Table 6-16.4.3.

6-16.4.4 Location of Transmitting Devices.

6-16.4.4.1 Shunt-type auxiliary systems shall be arranged so that one auxiliary transmitter does not serve more than 100,000 ft² (9290 m²) total area.

Exception: Where otherwise permitted by the authority having jurisdiction.

6-16.4.4.2 A separate auxiliary transmitter shall be provided for each building, or where permitted by the authority having jurisdiction, for each group of buildings of single ownership or occupancy.

6-16.4.4.3 The same box shall be permitted to be used as a public fire alarm reporting system box and as a transmitting device for an auxiliary system if permitted by the authority having jurisdiction, provided that the box is located at the outside of the entrance to the protected property.

The fire department shall be permitted to require the box to be equipped with a signal light to differentiate between automatic and manual operation, unless local outside alarms at the protected property serve the same purpose.

Table 6-16.4.3 Application of Initiating Device with Auxiliary Fire Alarm Systems

Initiating Devices	Local Energy Type	Shunt Type	Parallel Type
Manual fire alarm	Yes	Yes	Yes
Waterflow or actuation of the fire extinguishing system(s) or suppression system(s)	Yes	Yes	Yes
Automatic detection devices	Yes	No	Yes

6-16.4.4.4 The transmitting device shall be located as required by the authority having jurisdiction.

6-16.4.4.5 The system shall be designed and arranged so that a single fault on the auxiliary system shall not jeopardize operation of the public fire alarm reporting system and shall not, in case of a single fault on either the auxiliary or public fire alarm reporting system, transmit a false alarm on either system.

Exception: Shunt systems complying with 6-16.4.1(b).

6-16.5 Personnel.

Personnel necessary to receive and act on signals from auxiliary fire alarm systems shall be in accordance with the requirements of Sections 6-1 through 6-16 and NFPA 1221, *Standard for the Installation, Maintenance, and Use of Emergency Services Communications Systems.*

6-16.6 Operations.

Operations for auxiliary fire alarm systems shall be in accordance with the requirements of Sections 6-1 through 6-16 and NFPA 1221, *Standard for the Installation, Maintenance, and Use of Emergency Services Communications Systems.*

6-16.7 Testing and Maintenance.

Testing and maintenance of auxiliary fire alarm systems shall be in accordance with the requirements of Chapter 7.

References Cited in Commentary

ANSI/IEEE C2, *National Electrical Safety Code,* 1997.
NFPA 70, *National Electrical Code®*, National Fire Protection Association, Quincy, MA, 1999.
NFPA 1221, *Standard for the Installation, Maintenance, and Use of Emergency Services Communications*, National Fire Protection Association, Quincy, MA, 1999.

CHAPTER 7

Inspection, Testing, and Maintenance

Chapter 7 covers minimum requirements for inspection, testing, and maintenance of fire alarm systems. More specifically, Chapter 7 includes requirements for test methods, the frequency of inspection and testing, maintenance requirements, and record keeping.

7-1 General

7-1.1 Scope.

Chapter 7 shall cover the minimum requirements for the inspection, testing, and maintenance of the fire alarm systems described in Chapter 1, 3, and 5 and for their initiation and notification components described in Chapter 2 and 4. The testing and maintenance requirements for one- and two-family dwelling units shall be located in Chapter 8. Single station detectors used for other than one- and two-family dwelling units shall be tested and maintained in accordance with Chapter 7. More stringent inspection, testing, or maintenance procedures that are required by other parties shall be permitted.

Chapter 7 addresses testing and maintenance requirements for protected premises fire alarm systems and for supervising station fire alarm systems. It does not include the testing and maintenance requirements for fire warning equipment for dwelling units, which are contained in Chapter 8.

Single station smoke detectors used for other than one- and two-family dwelling units are often found in a variety of residential occupancies such as apartments, hotel and motel rooms, and dormitory living units. Non-system connected detection devices (sometimes called stand alone detectors) are sometimes found in HVAC systems, door releasing applications, and special hazard releasing devices. The requirements in Chapter 7, including sensitivity testing, apply to these types of detectors.

7-1.1.1 Inspection, testing, and maintenance programs shall satisfy the requirements of this code, shall conform to the equipment manufacturer's recommendations, and shall verify correct operation of the fire alarm system.

Paragraph 7-1.1.1 incorporates any manufacturer's instructions into the requirements of the code. Therefore, these instructions should be enforced as code. Verifying the correct operation of the fire alarm system includes conformance with the code and also with the owner's fire protection goals and the designer's specifications. Owners should not expect the authority having jurisdiction to do any more than enforce compliance with the *National Fire Alarm Code*. The fire alarm system designer should be retained to ensure that the system goals and owner's fire protection goals are met.

7-1.1.2 System defects and malfunctions shall be corrected. If a defect or malfunction is not corrected at the conclusion of system inspection, testing, or maintenance, the system owner or the owner's designated representative shall be informed of the impairment in writing within 24 hours.

The requirement to notify the owner or the owner's designated representative (usually the fire alarm system designer) in writing is to ensure that the defects and malfunctions will be corrected.

7-1.1.3 Nothing in Chapter 7 shall be intended to prevent the use of alternate test methods or testing devices. Such methods or devices shall provide the same level of effectiveness, and safety, and shall meet the intent of the requirements of Chapter 7.

The authority having jurisdiction has the responsibility of ensuring that the alternative methods are equivalent and will ultimately determine the acceptability of the proposed method.

7-1.1.4 The requirements of Chapter 7 shall apply to both new and existing systems.

Because the requirements of Chapter 7 apply to both new and existing systems, they are retroactive. The committee intends that the most current edition of the code be used for testing, inspection, and maintenance of both new and existing fire alarm systems. The requirements of the other chapters are not retroactive and only apply to new installations. See 1-2.3 for further details on these requirements.

7-1.2 The owner or the owner's designated representative shall be responsible for inspection, testing, and maintenance of the system and alterations or additions to this system. The delegation of responsibility shall be in writing, with a copy of such delegation provided to the authority having jurisdiction upon request.

The owner of the system is responsible for testing and maintaining the system. If the owner chooses to appoint a representative to ensure this responsibility, the delegation must be confirmed in writing. This written delegation may take the form of a testing and maintenance contract with a qualified contractor or delegation to a qualified staff specialist. The technicians delegated must be qualified. In most cases the authority having jurisdiction determines whether or not the technician is qualified. An obvious question to be asked is whether or not the technician has received training from the equipment manufacturer for the equipment he or she is supposed to maintain. See 7-1.2.2 for qualifications of testing personnel and Exhibit 7.1 for an illustration of a technician performing a test.

Exhibit 7.1 *Technician testing fire alarm system. (Source: Simplex Time Recorder Co., Gardner, MA)*

7-1.2.1 Inspection, testing, or maintenance shall be permitted to be done by a person or organization other than the owner if conducted under a written contract. Testing and maintenance of central station service systems shall be performed under the contractual arrangements specified in 5-2.2.2.

Paragraph 7-1.2.1 clarifies the requirement that a contractual agreement is required with the central station providing the service for testing of central station service systems. However, a central station system contract may still allow the owner or the owner's designated representative to perform these duties if qualified and if the contract contains the delegation of duties.

7-1.2.2 Service personnel shall be qualified and experienced in the inspection, testing, and maintenance of fire alarm systems. Examples of qualified personnel shall be permitted to include, but shall not be limited to, individuals with the following qualifications:

(1) Factory trained and certified
(2) National Institute for Certification in Engineering Technologies fire alarm certified
(3) International Municipal Signal Association fire alarm certified
(4) Certified by a state or local authority
(5) Trained and qualified personnel employed by an organization listed by a national testing laboratory for the servicing of fire alarm systems

The qualification requirement stated in 7-1.2.2 is to ensure that the persons testing and maintaining fire alarm systems have an appropriate level of knowledge, skill, and understanding of the systems and equipment.

Because each manufacturer's equipment is different, personnel who are not factory trained and certified should not be allowed to maintain equipment they are not qualified to work on. Technicians must receive training from all manufacturers of equipment encountered on any project that they are required to maintain. Fire alarm installation and testing personnel should also be trained to know and understand the concepts of the code.

The state or local authority having jurisdiction may have specific certification or licensing tests or other requirements that must be met. The National Institute for Certification in Engineering Technologies (NICET) offers a program of certification for fire alarm technicians. NICET can be reached at the following address: 1420 King Street, Alexandria, VA 22314-2715.

The International Municipal Signal Association (IMSA) is the professional association of those individuals who oversee public fire communications systems and traffic signaling systems. This organization offers educational programs for technicians and authorities having jurisdiction.

IMSA also offers interior fire alarm certification programs and publishes cable requirements for public fire reporting systems. IMSA can be reached at P.O. Box 539, Newark, NY 14513. See 6-8.1.3 and 6-8.2.1.1.

7-1.3* Notification.

A-7-1.3 Prior to any scheduled inspection or testing, the service company should consult with the building owner or the owner's designated representative. Issues of advance notification in certain occupancies, including advance notification time, building posting, systems interruption and restoration, evacuation procedures, accommodation for evacuees, and other related issues, should be agreed upon by all parties prior to any inspection or testing.

7-1.3.1 Before proceeding with any testing, all persons and facilities receiving alarm, supervisory, or trouble signals, and all building occupants, shall be notified of the testing to prevent unnecessary response. At the conclusion of testing, those previously notified (and others, as necessary) shall be notified that testing has been concluded.

Everyone who may be affected by testing of a fire alarm system at a protected premises must be notified that the testing will take place. Those notified should include, but not be limited to, the building owner, building manager, and switchboard operator, building engineer, building or floor fire wardens, and building maintenance personnel. In addition, the fire service communications center, the supervising station, and building occupants should be notified prior to testing. Methods of notification include bulletin board postings, electronic mail, public address announcements, and lobby signs.

A fire emergency plan should be established for each protected premises that provides for notifying occupants, the fire service communications center, and the supervising station in case a fire occurs during testing.

Paragraph 5-2.7.2 of the code requires the prime contractor (or designated representative) to provide a unique identification code to the central station before placing the central station fire alarm system into test status. The requirement of 7-1.3.1 is intended to prevent unauthorized tampering with the fire alarm system.

7-1.3.2 The owner or the owner's designated representative and service personnel shall coordinate system testing to prevent interruption of critical building systems or equipment.

If the fire alarm system has interface connections to other building systems, such as elevators and HVAC systems, the interfaces must be managed so that testing does not disrupt building systems or equipment that may be critical to the continuity of building operations. Interfaced equipment that is part of the overall fire protection system, such as smoke control or HVAC system shutdown, must be tested. The testing must be coordinated with the specialists maintaining these interfaced systems.

7-1.4 Prior to system maintenance or testing, the system certificate and the information regarding the system and system alterations, including specifications, wiring diagrams, and floor plans, shall be provided by the owner or a designated representative to the service personnel upon request.

At the time of an acceptance test, the authority having jurisdiction and the fire alarm system designer must ensure that all documentation for the fire alarm system installation has been completed and is presented to the owner or the owner's designated representative in a usable format. The required documentation includes all documents outlined in both Chapter 1 (Section 1-6) and Chapter 7. Maintenance of a fire alarm system directly impacts the mission effectiveness of the system. Service personnel cannot effectively maintain or test a system without full access to all of the required fire alarm system documentation.

Exhibit 7.2 is a suggested checklist of documents needed to test and maintain a system. All items on this checklist should be provided prior to testing or maintaining the system.

Document Checklist:

- ❏ Fire Alarm System Record of Completion
- ❏ Point-to-Point Wiring Diagrams
- ❏ Individual Device Interconnection Drawings
- ❏ As-Built (Record) Drawings
- ❏ Copy of Original Equipment Submittals
- ❏ Operational Manuals
- ❏ Manufacturer's Proper Testing and Maintenance Requirements

---

Exhibit 7.2 *Checklist for required system testing documentation.*

7-1.5 Releasing Systems.

Requirements pertinent to testing the fire alarm systems initiating fire suppression system releasing functions shall be covered by 7-1.5.

Testing of special hazard fire protection systems that are equipped with their own fire alarm control unit should be conducted as a separate series of tests. These systems are typically covered by standards that provide testing requirements for the special hazard system. The code covers the fire

alarm equipment up to the point of connection to the special hazard system. Only the interface functions between the separate control unit and the building fire alarm system should be tested as part of the building fire alarm system testing.

7-1.5.1 Testing personnel shall be qualified and experienced in the specific arrangement and operation of a suppression system(s) and a releasing function(s) and cognizant of the hazards associated with inadvertent system discharge.

Again, if the technician testing the fire alarm system is not qualified to test the special hazard system, then additional technicians trained and qualified to work on the special hazard system should be present to assist in conducting the tests. (See 7-1.2.2.) Having trained and qualified personnel present during testing of special hazard systems is necessary to prevent unwanted discharge of the fire suppression agent that can cause significant property damage or cause accidental injury or death to the building occupants.

7-1.5.2 Occupant notification shall be required whenever a fire alarm system configured for releasing service is being serviced or tested.

Notification allows the occupants to either evacuate the area being tested or prepare for the possible interruption of their work caused by the test.

7-1.5.3 Discharge testing of suppression systems shall not be required by this code. Suppression systems shall be secured from inadvertent actuation, including disconnection of releasing solenoids or electric actuators, closing of valves, other actions, or combinations thereof, for the specific system, for the duration of the fire alarm system testing.

7-1.5.4 Testing shall include verification that the releasing circuits and components energized or actuated by the fire alarm system are electrically supervised and operate as intended on alarm.

If the building fire alarm system also controls the special hazard fire protection system, the special hazard system operation (without the discharge of suppression agent) must be tested as part of the building fire alarm system testing procedures. Care must be taken to ensure that the special hazard system is not inadvertently actuated.

7-1.5.5 Suppression systems and releasing components shall be returned to their functional operating condition upon completion of system testing.

7-1.6 System Testing.

7-1.6.1 Initial Acceptance Testing. All new systems shall be inspected and tested in accordance with the requirements of Chapter 7.

7-1.6.2* Reacceptance Testing.

A-7-1.6.2 The additional devices to be tested should be a sample representation of the types of devices and locations on the system.

7-1.6.2.1 Reacceptance testing shall be performed after any of the following:

(1) Added or deleted system components
(2) Any modification, repair, or adjustment to system hardware or wiring
(3) Any change to site-specific software.

All components, circuits, systems operations, or site-specific software functions known to be affected by the change or identified by a means that indicates the system operational changes shall be 100 percent tested. In addition, 10 percent of initiating devices that are not directly affected by the change, up to a maximum of 50 devices, also shall be tested, and correct system operation shall be verified. A revised record of completion in accordance with 1-6.2.1 shall be prepared to reflect any changes.

Any modification to the fire alarm system, no matter how small it may seem, may affect the overall operation of the modified portion of the system. Seemingly harmless changes in software have caused tremendous changes in operation and have sometimes resulted in disastrous events. Systems have been found to have large portions inoperable due to faulty reprogramming during a repair or test. The requirement of 7-1.6.2.1 ensures that the affected portion of the system will be completely reacceptance tested. It further requires that a random sampling of other portions of the fire alarm system be tested to determine that other, seemingly unrelated, portion(s) of the system have not been adversely affected by the modification. See Section 1-4 of the code for definitions of software types.

The 10 percent sample should be randomly selected, and it must include at least one device per initiating device circuit or signaling line circuit to ensure correct operation. If all of the devices are installed on one signaling line circuit, multiple devices (10 percent sample) should be tested at different sections of the single circuit. Use of software comparison algorithms can also help determine where changes may have occurred.

Formal Interpretation 93-1 further clarifies the requirements of this section.

Formal Interpretation 93-1

Reference: 7-1.6.2

Question: Is it the intent of the committee to mandate a complete retesting of an entire system including all devices, circuits, and connections when only a single device or circuit has been modified?

Answer: No.

Issue Edition: 1993

Reference: 7-1.6

Issue Date: September 9, 1993

Effective Date: September 29, 1993

7-1.6.2.2 Changes to all control units connected or controlled by the system executive software shall require a 10 percent functional test of the system, including a test of at least one device on each input and output circuit to verify critical system functions such as notification appliances, control functions, and off-premises reporting.

Changes to software are frequently made in the field using laptop computers and manufacturer's software. See Exhibit 7.3 for an illustration of a technician programming a fire alarm control unit.

7-2 Test Methods

7-2.1* Central Stations.

At the request of the authority having jurisdiction, the installation shall be inspected for complete information regarding the system, including specifications, wiring diagrams, and floor plans that have been submitted for approval prior to installation of equipment and wiring.

A-7-2.1 If the authority having jurisdiction strongly suspects significant deterioration or otherwise improper operation by a central station, a surprise inspection to test the operation of the central station should be made, but extreme caution should be exercised. This test is to be conducted without advising the central station. However, the public fire service communications center must be contacted when manual alarms, waterflow alarms, or automatic fire detection systems are tested so that the fire department will not respond. In addition, persons normally receiving calls for supervisory alarms should be notified when items such as gate valves and functions such as pump power are tested. Confirmation of the authenticity of the test procedure should be obtained and should be a matter for resolution between plant management and the central station.

7-2.2* Fire alarm systems and other systems and equipment that are associated with fire alarm systems and accessory equipment shall be tested according to Table 7-2.2.

A-7-2.2 The wiring diagrams depicted in Figures A-7-2.2(a)(1) through A-7-2.2(a)(17) are representative of typi-

Exhibit 7.3 *Technician using a laptop computer to modify software. (Source: Mammoth Fire Alarms, Inc., Lowell, MA)*

cal circuits encountered in the field and are not intended to be all-inclusive.

The noted styles are as indicated in Table 3-5, Table 3-6, Table 3-7, and Table 5-5.3.2.2.2.3.

The noted systems are as indicated in NFPA 170, *Standard for Fire Safety Symbols*.

Because ground-fault detection is not required for all circuits, tests for ground-fault detection should be limited to those circuits equipped with ground-fault detection.

An individual point-identifying (addressable) initiating device operates on a signaling line circuit and not on a Style A, B, C, D, or E (Class B and Class A) initiating device circuit.

All of the following initiating device circuits are illustrative of either alarm or supervisory signaling. Alarm-initiating devices and supervisory initiating devices are not permitted on the same initiating device circuit.

Table 7-2.2 Test Methods

Device	Method
1. Control Equipment	
a. Functions	At a minimum, control equipment shall be tested to verify correct receipt of alarm, supervisory, and trouble signals (inputs), operation of evacuation signals and auxiliary functions (outputs), circuit supervision including detection of open circuits and ground faults, and power supply supervision for detection of loss of ac power and disconnection of secondary batteries.
	The input/output control equipment functions must be tested to ensure proper operation. The checking of functions internal to the equipment, such as software algorithms and communications protocols (sometimes called firmware), is not intended. Verifying that Class A circuits transmit an alarm in either direction under a single fault condition is an example of a functional test.
b. Fuses	The rating and supervision shall be verified.
	Verifying the rating and supervision of fuses is important because an incorrect fuse rating can lead to equipment damage or unnecessary loss of power.
c. Interfaced Equipment	Integrity of single or multiple circuits providing interface between two or more control panels shall be verified. Interfaced equipment connections shall be tested by operating or simulating operation of the equipment being supervised. Signals required to be transmitted shall be verified at the control panel.
	The wiring connections must be tested by simulating a single open and a single ground to verify proper indications for the monitoring of the interfaced equipment wiring integrity. In addition, the interfaced equipment must be placed in a simulated trouble condition to test for proper supervisory signal receipt and reaction at the main control unit. See Exhibit 7.4, page 278, for an illustration of one method of checking electrical supervision.
d. Lamps and LEDs	Lamps and LEDs shall be illuminated.
e. Primary (Main) Power Supply	All secondary (standby) power shall be disconnected and tested under maximum load, including all alarm appliances requiring simultaneous operation. All secondary (standby) power shall be reconnected at end of test. For redundant power supplies, each shall be tested separately.
2. Engine-Driven Generator	If an engine-driven generator dedicated to the fire alarm system is used as a required power source, operation of the generator shall be verified in accordance with NFPA 110, *Standard for Emergency and Standby Power Systems*, by the building owner.
3. Secondary (Standby) Power Supply	All primary (main) power supplies shall be disconnected and the occurrence of required trouble indication for loss of primary power shall be verified. The system's standby and alarm current demand shall be measured or verified and, using manufacturer's data, the ability of batteries to meet standby and alarm requirements shall be verified. General alarm systems shall be operated for a minimum of 5 minutes and emergency voice communications systems for a minimum of 15 minutes. Primary (main) power supply shall be reconnected at end of test.
	Determining the correct amount of battery standby needed is usually accomplished by calculating the supervisory and alarm loads on the system. Although not required by the code, it is considered good engineering practice to include a safety factor in the battery standby calculations to ensure correct operation as the batteries age or are subjected to extreme ambient conditions.

Table 7-2.2 Continued

Device	Method
4. Uninterrupted Power Supply (UPS)	If a UPS system dedicated to the fire alarm system is used as a required power source, operation of the UPS system shall be verified by the building owner in accordance with NFPA 111, *Standard on Stored Electrical Energy Emergency and Standby Power Systems.*
5. Batteries—General Tests	
a. Visual Inspection	Prior to conducting any battery testing, the person conducting the test shall ensure that all system software stored in volatile memory is protected from loss. Batteries shall be inspected for corrosion or leakage. Tightness of connections shall be checked and ensured. If necessary, battery terminals or connections shall be cleaned and coated. Electrolyte level in lead-acid batteries shall be visually inspected.
b. Battery Replacement	Batteries shall be replaced in accordance with the recommendations of the alarm equipment manufacturer or when the recharged battery voltage or current falls below the manufacturer's recommendations.
c. Charger Test	Operation of battery charger shall be checked in accordance with charger test for the specific type of battery.
d. Discharge Test	With the battery charger disconnected, the batteries shall be load tested following the manufacturer's recommendations. The voltage level shall not fall below the levels specified. *Exception: An artificial load equal to the full fire alarm load connected to the battery shall be permitted to be used in conducting this test.*
e. Load Voltage Test	With the battery charger disconnected, the terminal voltage shall be measured while supplying the maximum load required by its application. The voltage level shall not fall below the levels specified for the specific type of battery. If the voltage falls below the level specified, corrective action shall be taken and the batteries shall be retested. *Exception: An artificial load equal to the full fire alarm load connected to the battery shall be permitted to be used in conducting this test.*
6. Battery Tests (Specific Types)	
a. Primary Battery Load Voltage Test	The maximum load for a No. 6 primary battery shall not be more than 2 amperes per cell. An individual (1.5-volt) cell shall be replaced when a load of 1 ohm reduces the voltage below 1 volt. A 6-volt assembly shall be replaced when a test load of 4 ohms reduces the voltage below 4 volts. It is not desirable to completely drain a battery during the discharge test because, in the event of a power failure shortly after the test, the system could be left without a power supply. A typical test places the battery under load for a shorter period (1 hour to 2 hours). However, the battery should be tested to ensure that it can deliver the required current at rated voltage under maximum expected load. Battery calculations must be relied on to ensure capacity. See Table 7-2.2, Item 3, and 1-5.2.6.
b. Lead-Acid Type	
1. Charger Test	With the batteries fully charged and connected to the charger, the voltage across the batteries shall be measured with a voltmeter. The voltage shall be 2.30 volts per cell ±0.02 volts at 25°C (77°F) or as specified by the equipment manufacturer.
2. Load Voltage Test	Under load, the battery shall not fall below 2.05 volts per cell.

(continues)

Table 7-2.2 Continued

Device	Method
3. Specific Gravity	The specific gravity of the liquid in the pilot cell or all of the cells shall be measured as required. The specific gravity shall be within the range specified by the manufacturer. Although the specified specific gravity varies from manufacturer to manufacturer, a range of 1.205–1.220 is typical for regular lead-acid batteries, while 1.240–1.260 is typical for high-performance batteries. A hydrometer that shows only a pass or fail condition of the battery and does not indicate the specific gravity shall not be used, because such a reading does not give a true indication of the battery condition. Caution: Adding acid to a lead-acid type battery cell may destroy the plates, affecting battery capacity.
c. Nickel-Cadmium Type 1. Charger Test[*]	With the batteries fully charged and connected to the charger, an ampere meter shall be placed in series with the battery under charge. The charging current shall be in accordance with the manufacturer's recommendations for the type of battery used. In the absence of specific information, $^1/_{30}$ to $^1/_{25}$ of the battery rating shall be used.
2. Load Voltage Test	Under load, the float voltage for the entire battery shall be 1.42 volts per cell, nominal. If possible, cells shall be measured individually.
d. Sealed Lead-Acid Type 1. Charger Test	With the batteries fully charged and connected to the charger, the voltage across the batteries shall be measured with a voltmeter. The voltage shall be 2.30 volts per cell ±0.02 volts at 25°C (77°F) or as specified by the equipment manufacturer.
2. Load Voltage Test	Under load, the battery shall perform in accordance with the battery manufacturer's specifications.
7. Public Reporting System Tests	In addition to the tests and inspection required above, the following requirements shall apply. Manual tests of the power supply for public reporting circuits shall be made and recorded at least once during each 24-hour period. Such tests shall include the following: (a) Current strength of each circuit. Changes in current of any circuit exceeding 10 percent shall be investigated immediately. (b) Voltage across terminals of each circuit inside of terminals of protective devices. Changes in voltage of any circuit exceeding 10 percent shall be investigated immediately. (c)* Voltage between ground and circuits. If this test shows a reading in excess of 50 percent of that shown in the test specified in (b), the trouble shall be immediately located and cleared. Readings in excess of 25 percent shall be given early attention. These readings shall be taken with a calibrated voltmeter of not more than 100-ohms resistance per volt. Systems in which each circuit is supplied by an independent current source (Forms 3 and 4) require tests between ground and each side of each circuit. Common current source systems (Form 2) require voltage tests between ground and each terminal of each battery and other current source. (d) Ground current reading shall be permitted in lieu of (c). If this method of testing is used, all grounds showing a current reading in excess of 5 percent of the supplied line current shall be given immediate attention. (e) Voltage across terminals of common battery, on switchboard side of fuses. (f) Voltage between common battery terminals and ground. Abnormal ground readings shall be investigated immediately. Tests specified in (e) and (f) shall apply only to those systems using a common battery. If more than one common battery is used, each common battery shall be tested.
8. Transient Suppressors	Lightning protection equipment shall be inspected and maintained per the manufacturer's specifications.

Table 7-2.2 Continued

Device	Method
	Equipment located in moderate to severe areas outlined in NFPA 780, *Standard for the Installation of Lightning Protection Systems*, Appendix H, shall be inspected semiannually and after any lightning strikes. Additional inspections shall be required after any lightning strikes.
	In areas prone to lightning storms, the owner should be advised to notify the fire alarm service company when a storm has occurred so that all original lightning protection can be checked.

9. Control Unit Trouble Signals

Device	Method
a. Audible and Visual	Operation of panel trouble signals shall be verified as well as ring-back feature for systems using a trouble-silencing switch that requires resetting.
b. Disconnect Switches	If control unit has disconnect or isolating switches, performance of intended function of each switch shall be verified and receipt of trouble signal when a supervised function is disconnected shall also be verified.
c. Ground-Fault Monitoring Circuit	If the system has a ground detection feature, the occurrence of ground-fault indication shall be verified whenever any installation conductor is grounded.
	Each conductor should be grounded temporarily to ensure proper ground detection circuit operation. The results of these tests should be recorded on the acceptance test report for future troubleshooting information. Note that the control equipment may take up to 200 seconds to indicate a trouble/ground condition. See Section 1-5.8.1.
d. Transmission of Signals to Off-Premises Location	Initiating device shall be actuated and receipt of alarm signal at the off-premises location shall be verified. A trouble condition shall be created and receipt of a trouble signal at the off-premises location shall be verified. A supervisory device shall be actuated and receipt of a supervisory signal at the off-premises location shall be verified. If a transmission carrier is capable of operation under a single or multiple fault condition, an initiating device shall be activated during such fault condition and receipt of a trouble signal at the off-premises location shall be verified, in addition to the alarm signal.

10. Remote Annunciators

Device	Method
	The correct operation and identification of annunciators shall be verified. If provided, the correct operation of annunciator under a fault condition shall be verified.
	Remote annunciation is very important to the fire department personnel responding to the alarm. The intent of remote annunciation is to reduce the time spent finding the source of the alarm by providing clear and accurate information to the responding fire service. For that reason, remote annunciation information should be given to the fire department that is assigned to respond to the protected premises for review and input. Too much detail is as bad as too little information, so a balance must be found to help locate the fire as quickly as possible.

11. Conductors/Metallic

Device	Method
a. Stray Voltage	All installation conductors shall be tested with a volt/ohmmeter to verify that there are no stray (unwanted) voltages between installation conductors or between installation conductors and ground. Unless a different threshold is specified in the system installed equipment manufacturer's specifications, the maximum allowable stray voltage shall not exceed 1 volt ac/dc.
b. Ground Faults	All installation conductors other than those intentionally and permanently grounded shall be tested for isolation from ground per the installed equipment manufacturer's specifications.
c. Short-Circuit Faults	All installation conductors other than those intentionally connected together shall be tested for conductor-to-conductor isolation per the installed equipment manufacturer's specifications. These same circuits also shall be tested conductor-to-ground.

(continues)

Table 7-2.2 Continued

Device	Method
d. Loop Resistance	With each initiating and indicating circuit installation conductor pair short-circuited at the far end, the resistance of each circuit shall be measured and recorded. It shall be verified that the loop resistance does not exceed the installed equipment manufacturer's specified limits.
	If the loop resistance exceeds the installed equipment manufacturer's specified limits, the wiring must be changed. In this instance, the fire alarm control equipment manufacturer, rather than the cable manufacturer, would be consulted because the equipment manufacturer generally has more stringent operational requirements.
12. Conductors/Nonmetallic	
a. Circuit Integrity	Each initiating device, notification appliance, and signaling line circuit shall be tested to confirm that the installation conductors are monitored for integrity in accordance with the requirements of Chapters 1 and 3.
b. Fiber Optics	The fiber-optic transmission line shall be tested in accordance with the manufacturer's instructions by the use of an optical power meter or by an optical time domain reflectometer used to measure the relative power loss of the line. This relative figure for each fiber-optic line shall be recorded in the fire alarm control panel. If the power level drops 2 percent or more from the value recorded during the initial acceptance test, the transmission line, section thereof, or connectors shall be repaired or replaced by a qualified technician to bring the line back into compliance with the accepted transmission level per the manufacturer's recommendations.
c. Supervision	Introduction of a fault in any supervised circuit shall result in a trouble indication at the control unit. One connection shall be opened at not less than 10 percent of the initiating device, notification appliance, and signaling line circuit.
	The term *supervision* as used in Table 7-2.2, part 12, item c, means the monitoring of the circuit conductor integrity.
	Each initiating device, notification appliance, and signaling line circuit shall be tested for correct indication at the control unit. All circuits shall perform as indicated in Table 3-5, Table 3-6, or Table 3-7.
13. Initiating Devices	
a. Electromechanical Releasing Device	
1. Nonrestorable-Type Link	Correct operation shall be verified by removal of the fusible link and operation of the associated device. Any moving parts shall be lubricated as necessary.
2. Restorable-Type Link*	Correct operation shall be verified by removal of the fusible link and operation of the associated device. Any moving parts shall be lubricated as necessary.
b. Fire Extinguishing System(s) or Suppression System(s) Alarm Switch	The switch shall be mechanically or electrically operated and receipt of signal by the control panel shall be verified.
c. Fire–Gas and Other Detectors	Fire–gas detectors and other fire detectors shall be tested as prescribed by the manufacturer and as necessary for the application.
d. Heat Detectors	
1. Fixed-Temperature, Rate-of-Rise, Rate-of Compensation, Restorable Line, Spot Type (excluding Pneumatic Tube Type)	Heat test shall be performed with a heat source per the manufacturer's recommendations for response within 1 minute. A test method shall be used that is recommended by the manufacturer or other method shall be used that will not damage the nonrestorable fixed-temperature element of a combination rate-of-rise/fixed-temperature element.
	Extreme caution must be used in hazardous locations (those containing explosive vapors or dusts) when heat testing these types of detectors. The use of a bucket of hot water or hot towels is recommended. Heat detectors must not be tested with live flame.

Table 7-2.2 Continued

Device	Method
2. Fixed-Temperature, Nonrestorable Line Type	Heat test shall not be performed. Functionality shall be tested mechanically and electrically. Loop resistance shall be measured and recorded. Changes from acceptance test shall be investigated.
3. Fixed-Temperature, Nonrestorable Spot Type	After 15 years from initial installation, all devices shall be replaced or two detectors per 100 shall be laboratory tested. The two detectors shall be replaced with new devices. If a failure occurs on any of the detectors removed, additional detectors shall be removed and tested to determine either a general problem involving faulty detectors or a localized problem involving one or two defective detectors.
	If detectors are tested instead of replaced, tests shall be repeated at intervals of 5 years.
	The laboratory test referenced is conducted by an independent testing laboratory engaged in the listing or approval of heat detectors.
4. Nonrestorable (General)	Heat tests shall not be performed. Functionality shall be tested mechanically and electrically.
	Contacts may be operated by hand or electrically shorted to ensure alarm response.
5. Restorable Line Type, Pneumatic Tube Only	Heat tests shall be performed (where test chambers are in circuit) or a test with pressure pump shall be conducted.
e. Fire Alarm Boxes	Manual fire alarm boxes shall be operated per the manufacturer's instructions. Key-operated presignal and general alarm manual fire alarm boxes shall both be tested.
f. Radiant Energy Fire Detectors	Flame detectors and spark/ember detectors shall be tested in accordance with the manufacturer's instructions to determine that each detector is operative.
	Flame detector and spark/ember detector sensitivity shall be determined using any of the following:
	(a) Calibrated test method
	(b) Manufacturer's calibrated sensitivity test instrument
	(c) Listed control unit arranged for the purpose
	(d) Other approved calibrated sensitivity test method that is directly proportional to the input signal from a fire, consistent with the detector listing or approval
	If designed to be field adjustable, detectors found to be outside of the approved range of sensitivity shall be replaced or adjusted to bring them into the approved range.
	Flame detector and spark/ember detector sensitivity shall not be determined using a light source that administers an unmeasured quantity of radiation at an undefined distance from the detector.
g. Smoke Detectors 1. Systems Detectors	The detectors shall be tested in place to ensure smoke entry into the sensing chamber and an alarm response. Testing with smoke or listed aerosol approved by the manufacturer shall be permitted as acceptable test methods. Other methods approved by the manufacturer that ensure smoke entry into the sensing chamber shall be permitted.
	The test described in the first paragraph of Table 7-2.2, part 13, item g(1), is a "go, no-go" type of test to ensure smoke entry into the chamber and alarm response. It does not test the detector's sensitivity. Table 7-2.2, part 13, item g(1), requires more than a visual test. See Exhibits 7.5 and 7.6, pages 278 and 279, for examples of equipment used in a functional test of a fire alarm system. Because smoke entry must be part of the test, using a test button or a test magnet does not meet the requirements of Table 7-2.2, part 13, item g(1).
	Any of the following tests shall be performed to ensure that each smoke detector is within its listed and marked sensitivity range:
	(a) Calibrated test method
	(b) Manufacturer's calibrated sensitivity test instrument
	(c) Listed control equipment arranged for the purpose

(continues)

Table 7-2.2 Continued

Device	Method
	(d) Smoke detector/control unit arrangement whereby the detector causes a signal at the control unit when its sensitivity is outside its listed sensitivity range
	(e) Other calibrated sensitivity test method approved by the authority having jurisdiction
2. Single Station Detectors	The detectors shall be tested in place to ensure smoke entry into the sensing chamber and an alarm response. Testing with smoke or listed aerosol approved by the manufacturer shall be permitted as acceptable test methods. Other methods approved by the manufacturer that ensure smoke entry into the sensing chamber shall be permitted.
	Single station smoke detectors (smoke alarms) other than those used in one- and two-family dwellings are often found in other residential occupancies, such as apartments, hotel and motel rooms, and dormitory living units. Other non-system connected detection devices (sometimes called stand-alone detectors) are sometimes found in HVAC systems, door releasing applications, and special hazards-releasing devices. The requirements in Chapter 7 of the code, including sensitivity testing, apply to these types of detectors. Single station smoke detectors used in one- and two-family dwellings, are required to be functionally tested, but are not required to be sensitivity tested. See 7-3.2.1 for additional information on this requirement.
3. Air Sampling	Per manufacturer's recommended test methods, detector alarm response shall be verified through the end sampling port on each pipe run; airflow through all other ports shall be verified as well.
4. Duct Type	Air duct detectors shall be tested or inspected to ensure that the device will sample the airstream. The test shall be made in accordance with the manufacturer's instructions.
	Often duct-type smoke detectors are installed by a mechanical contractor and connected to the fire alarm system by the alarm system contractor. The final responsibility for ensuring the sampling tubes and the smoke detector have been installed correctly rests with the alarm system contractor. One method of determining that air is being sampled is to use a manometer, as shown in Exhibit 7.7, page 279.
5. Projected Beam Type	The detector shall be tested by introducing smoke, other aerosol, or an optical filter into the beam path.
6. Smoke Detector with Built-in Thermal Element	Both portions of the detector shall be operated independently as described for the respective devices.
	Table 7-2.2, part 13, item g(6) requires a test of both portions of a combination unit if possible. The code does not explicitly address the issue of the failure of one feature. However, it is assumed that if one feature of a combination smoke/heat detector fails a test, the entire unit should be removed and replaced.
7. Smoke Detectors with Control Output Functions	It shall be verified that the control capability shall remain operable even if all of the initiating devices connected to the same initiating device circuit or signaling line circuit are in an alarm state.
	The requirement of Table 7-2.2, part 13, item g(7) forbids the use of two-wire smoke detectors for controlling operations on an initiating device circuit (e.g., fan shutdown) where other devices are installed on the same circuit. If, for instance, the smoke detector tries to actuate after a manual fire alarm box has been actuated, the smoke detector may not actuate. See 3-8.2.4 of the code and related commentary for more detailed information.
h. Initiating Devices, Supervisory	
1. Control Valve Switch	Valve shall be operated and signal receipt shall be verified to be within the first two revolutions of the hand wheel or within one-fifth of the travel distance, or per the manufacturer's specifications.

Table 7-2.2 Continued

Device	Method
2. High- or Low-Air Pressure Switch	Switch shall be operated. Receipt of signal obtained where the required pressure is increased or decreased a maximum 10 psi (70 kPa) from the required pressure level shall be verified.
3. Room Temperature Switch	Switch shall be operated. Receipt of signal to indicate the decrease in room temperature to 40°F (4.4°C) and its restoration to above 40°F (4.4°C) shall be verified.
4. Water Level Switch	Switch shall be operated. Receipt of signal indicating the water level raised or lowered 3 in. (76.2 mm) from the required level within a pressure tank, or 12 in. (305 mm) from the required level of a nonpressure tank, shall be verified, as shall its restoral to required level.
5. Water Temperature Switch	Switch shall be operated. Receipt of signal to indicate the decrease in water temperature to 40°F (4.4°C) and its restoration to above 40°F (4.4°C) shall be verified.
i. Mechanical, Electrosonic, or Pressure-Type Waterflow Device	Water shall be flowed through an inspector's test connection indicating the flow of water equal to that from a single sprinkler of the smallest orifice size installed in the system for wet-pipe systems, or an alarm test bypass connection for dry-pipe, pre-action, or deluge systems in accordance with NFPA 25, *Standard for the Inspection, Testing, and Maintenance of Water-Based Fire Protection Systems*. It is unacceptable to electrically or mechanically (without waterflow) operate the waterflow switch. The flow test ensures that when the automatic sprinkler system is operated, an alarm signal is generated on the fire alarm system.
14. Alarm Notification Appliances a. Audible	Sound pressure level shall be measured with sound level meter meeting ANSI S1.4a, *Specifications for Sound Level Meters*, Type 2 requirements. Levels throughout protected area shall be measured and recorded. During the acceptance test, and with the authority having jurisdiction's approval, areas that are physically remote from the audible notification appliances should be chosen to measure and record sound pressure levels. If these areas comply with code requirements, then the authority having jurisdiction may deem further measurements unnecessary. However, areas that fail to meet the requirements of Section 4-3 of the code will require additional appliances.
b. Audible Textural Notification Appliances (Speakers and other Appliances to Convey Voice Messages)	Sound pressure level shall be measured with sound level meter meeting ANSI S1.4a, *Specifications for Sound Level Meters*, Type 2 requirements. Levels throughout protected area shall be measured and recorded. Audible information shall be verified to be distinguishable and understandable. To comply with the sound level requirements of the code, many installers attempt to tap speakers at a higher wattage, rather than increase the number of speakers in an area. This incorrect approach to sound level compliance leads to distortion of voice messages through the speakers. Therefore, in addition to meeting the sound level requirements of the code, the clarity or intelligibility of the voice message must be verified where speakers are used for voice communication. Sound-reflective ambient environments (such as walls, floors, and concrete and steel stair towers) have a major effect on the clarity of voice messages. See 4-3.1.5 and the related appendix material.
c. Visible	Test shall be performed in accordance with the manufacturer's instructions. Device locations shall be verified to be per approved layout and it shall be confirmed that no floor plan changes affect the approved layout. The tests must ensure that visible notification appliances operate and are not blocked by shelving, furniture, ceiling-mounted light fixtures, or movable partitions. The owner also should be advised to keep all viewing paths to visible notification appliances clear.

(continues)

Table 7-2.2 Continued

Device	Method
15. Special Hazard Equipment	
a. Abort Switch (IRI Type)	Abort switch shall be operated. Correct sequence and operation shall be verified.
b. Abort Switch (Recycle Type)	Abort switch shall be operated. Development of correct matrix with each sensor operated shall be verified.
c. Abort Switch (Special Type)	Abort switch shall be operated. Correct sequence and operation in accordance with authority having jurisdiction shall be verified. Sequencing on as-built drawings or in owner's manual shall be observed.
d. Cross Zone Detection Circuit	One sensor or detector on each zone shall be operated. Occurrence of correct sequence with operation of first zone and then with operation of second zone shall be verified.
e. Matrix-Type Circuit	All sensors in system shall be operated. Development of correct matrix with each sensor operated shall be verified.
f. Release Solenoid Circuit	Solenoid shall be used with equal current requirements. Operation of solenoid shall be verified.
g. Squibb Release Circuit	AGI flashbulb or other test light approved by the manufacturer shall be used. Operation of flashbulb or light shall be verified.
h. Verified, Sequential, or Counting Zone Circuit	Required sensors at a minimum of four locations in circuit shall be operated. Correct sequence with both the first and second detector in alarm shall be verified.
i. All Above Devices or Circuits or Combinations Thereof	Supervision of circuits by creating an open circuit shall be verified.
	Caution should be used when testing the interfaced special hazard equipment to avoid unnecessary actuation. When testing these interconnections, it should never be assumed that the previous tests were conducted properly. The manufacturer's test procedures should always be reviewed prior to conducting these tests. After all equipment has been tested independently, it should be ensured that all connections or test switches are returned to their normal positions.
16. Supervising Station Fire Alarm Systems—Transmission Equipment	
a. All Equipment	Test shall be performed on all system functions and features in accordance with the equipment manufacturer's instructions for correct operation in conformance with the applicable sections of Chapter 5.
	Initiating device shall be actuated. Receipt of the correct initiating device signal at the supervising station within 90 seconds shall be verified. Upon completion of the test, the system shall be restored to its functional operating condition.
	To ensure the proper compliance to the time limits for reporting, the synchronization of all timing devices should be checked.
	If test jacks are used, the first and last tests shall be made without the use of the test jack.
b. Digital Alarm Communicator Transmitter (DACT)	Connection of the DACT to two separate means of transmission shall be ensured.
	The primary line from the DACT must be a loop start telephone line (number). Where two telephone lines are used, both lines must be loop start lines. See 5-5.3.2.1.1 and 5-5.3.2.1.6 for more information on this requirement.
	Exception: DACTs that are connected to a telephone line (number) that is also supervised for adverse conditions by a derived local channel.
	DACT shall be tested for line seizure capability by initiating a signal while using the primary line for a telephone call. Receipt of the correct signal at the supervising station shall be verified. Completion of the transmission attempt within 90 seconds from going off-hook to on-hook shall be verified.

Table 7-2.2 Continued

Device	Method
	The primary line from the DACT shall be disconnected. Indication of the DACT trouble signal at the premises shall be verified as well as transmission to the supervising station within 4 minutes of detection of the fault.
	The secondary means of transmission from the DACT shall be disconnected. Indication of the DACT trouble signal at the premises shall be verified as well as transmission to the supervising station within 4 minutes of detection of the fault.
	To ensure the proper compliance to the time limits for reporting, the synchronization of all timing devices should be checked.
c. Digital Alarm Radio Transmitter (DART)	The DACT shall be caused to transmit a signal to the DACR while a fault in the primary telephone number is simulated. Utilization of the secondary telephone number by the DACT to complete the transmission to the DACR shall be verified.
	The primary telephone line shall be disconnected. Transmission of a trouble signal to the supervising station by the DART within 4 minutes shall be verified.
d. McCulloh Transmitter	Initiating device shall be actuated. Production of not less than three complete rounds of not less than three signal impulses each by the McCulloh transmitter shall be verified.
	If end-to-end metallic continuity is present and with a balanced circuit, each of the following four transmission channel fault conditions shall be caused in turn, and receipt of correct signals at the supervising station shall be verified: (a) Open (b) Ground (c) Wire-to-wire short (d) Open and ground If end-to-end metallic continuity is not present and with a properly balanced circuit, each of the following three transmission channel fault conditions shall be caused in turn, and receipt of correct signals at the supervising station shall be verified (a) Open (b) Ground (c) Wire-to-wire short
e. Radio Alarm Transmitter (RAT)	A fault between elements of the transmitting equipment shall be caused. Indication of the fault at the protected premises shall be verified or it shall be verified that a trouble signal is transmitted to the supervising station.
	The tests required in items 16 and 17 provide more comprehensive methods for testing supervising station transmission and receiving equipment. The transmission and receipt of fire alarm signals are no less important than the detection of the fire. The goal of fire alarm systems with off-premises transmission is to ensure that the fire department response is not delayed in any way.
17. Supervising Station Fire Alarm Systems—Receiving Equipment	
a. All Equipment	Tests shall be performed on all system functions and features in accordance with the equipment manufacturer's instructions for correct operation in conformance with the applicable sections of Chapter 5.
	Initiating device shall be actuated. Receipt of the correct initiating device signal at the supervising station within 90 seconds shall be verified. Upon completion of the test, the system shall be restored to its functional operating condition.
	If test jacks are used, the first and last tests shall be made without the use of the test jack.
b. Digital Alarm Communicator Receiver (DACR)	Each telephone line (number) shall be disconnected in turn from the DACR and audible and visual annunciation of a trouble signal in the supervising station shall be verified.

(continues)

Table 7-2.2 Continued

Device	Method
c. Digital Alarm Radio Receiver (DARR)	A signal shall be caused to be transmitted on each individual incoming DACR line at least once every 24 hours. Receipt of these signals shall be verified. The following conditions of all DARRs on all subsidiary and repeater station receiving equipment shall be caused. Receipt at the supervising station of correct signals for each of the following conditions shall be verified: (a) AC power failure of the radio equipment (b) Receiver malfunction (c) Antenna and interconnecting cable failure (d) Indication of automatic switchover of the DARR (e) Data transmission line failure between the DARR and the supervising or subsidiary station
d. McCulloh Systems	The current on each circuit at each supervising and subsidiary station under the following conditions shall be tested and recorded: (a) During functional operation (b) On each side of the circuit with the receiving equipment conditioned for an open circuit A single break or ground condition shall be caused on each transmission channel. If such a fault prevents the functioning of the circuit, receipt of a trouble signal shall be verified. Each of the following conditions at each of the supervising or subsidiary stations and all repeater station radio transmitting and receiving equipment shall be caused; receipt of correct signals at the supervising station shall be verified: (a) RF transmitter in use (radiating) (b) AC power failure supplying the radio equipment (c) RF receiver malfunction (d) Indication of automatic switchover
e. Radio Alarm Supervising Station Receiver (RASSR) and Radio Alarm Repeater Station Receiver (RARSR)	Each of the following conditions at each of the supervising or subsidiary stations and all repeater station radio transmitting and receiving equipment shall be caused; receipt of correct signals at the supervising station shall be verified: (a) AC power failure supplying the radio equipment (b) RF receiver malfunction (c) Indication of automatic switchover, if applicable
f. Private Microwave Radio Systems	Each of the following conditions at each of the supervising or subsidiary stations and all repeater station radio transmitting and receiving equipment shall be caused; receipt of correct signals at the supervising station shall be verified: (a) RF transmitter in use (radiating) (b) AC power failure supplying the radio equipment (c) RF receiver malfunction (d) Indication of automatic switchover
18. Emergency Communications Equipment	
a. Amplifier/Tone Generators	Correct switching and operation of backup equipment shall be verified.
b. Call-in Signal Silence	Function shall be operated and receipt of correct visual and audible signals at control panel shall be verified.
c. Off-hook Indicator (Ring Down)	Phone set shall be installed or phone shall be removed from hook and receipt of signal at control panel shall be verified.
d. Phone Jacks	Phone jack shall be visually inspected and communications path through jack shall be initiated. During an acceptance test, all phone jacks on each floor or zone must be checked for proper operation.
e. Phone Set	Each phone set shall be activated and correct operation shall be verified.

Table 7-2.2 Continued

Device	Method
f. System Performance	System shall be operated with a minimum of any five handsets simultaneously. Voice quality and clarity shall be verified.
19. Interface Equipment	Interface equipment connections shall be tested by operating or simulating the equipment being supervised. Signals required to be transmitted shall be verified at the control panel. Test frequency for interface equipment shall be the same as the frequency required by the applicable NFPA standard(s) for the equipment being supervised.
	The signals being verified include the status (i.e., alarm, trouble, supervisory conditions) of the interfaced equipment. The main fire alarm control unit indicates a supervisory signal for any trouble or supervisory conditions at the interfaced equipment.
20. Guard's Tour Equipment	The device shall be tested in accordance with the manufacturer's specifications.
21. Special Procedures a. Alarm Verification	Time delay and alarm response for smoke detector circuits identified as having alarm verification shall be verified.
	Alarm verification may be disconnected during a system test and then reconnected after the initial test. If this happens, all circuits with devices to be verified must be tested again to ensure that the alarm verification feature is operable.
b. Multiplex Systems	Communications between sending and receiving units under both primary and secondary power shall be verified. Communications between sending and receiving units under open circuit and short circuit trouble conditions shall be verified. Communications between sending and receiving units in all directions where multiple communications pathways are provided shall be verified. If redundant central control equipment is provided, switchover and all required functions and operations of secondary control equipment shall be verified. All system functions and features shall be verified in accordance with manufacturer's instructions.
	System functions and features should be verified in accordance with the circuit styles as designed, as well as the manufacturer's specifications. The authority having jurisdiction is not responsible for ensuring that the fire alarm system is tested in accordance with the fire alarm system designer's specifications. The authority having jurisdiction requires tests to the minimum requirements of the code. It is the contractor's responsibility to test the entire system for its operation as described in the design documents.
22. Low-Power Radio (Wireless Systems)	The following procedures describe additional acceptance and reacceptance test methods to verify wireless protection system operation: (a) The manufacturer's manual and the as-built drawings provided by the system supplier shall be used to verify correct operation after the initial testing phase has been performed by the supplier or by the supplier's designated representative. (b) Starting from the functional operating condition, the system shall be initialized in accordance with the manufacturer's manual. A test shall be conducted to verify the alternative path, or paths, by turning off or disconnecting the primary wireless repeater. The alternative communications path shall exist between the wireless control panel and peripheral devices used to establish initiation, indicating, control, and annunciation. The system shall be tested for both alarm and trouble conditions. (c) Batteries for all components in the system shall be checked monthly. If the control panel checks all batteries and all components daily, the system shall not require monthly testing of the batteries.
	The requirement of Table 7-2.2, part 22(c) applies to low-power wireless systems covered by Section 3-11, not radio-type public fire reporting systems.

Exhibit 7.4 *Technician removing wire from device to check electrical supervision. (Source: Simplex Time Recorder Co., Gardner, MA)*

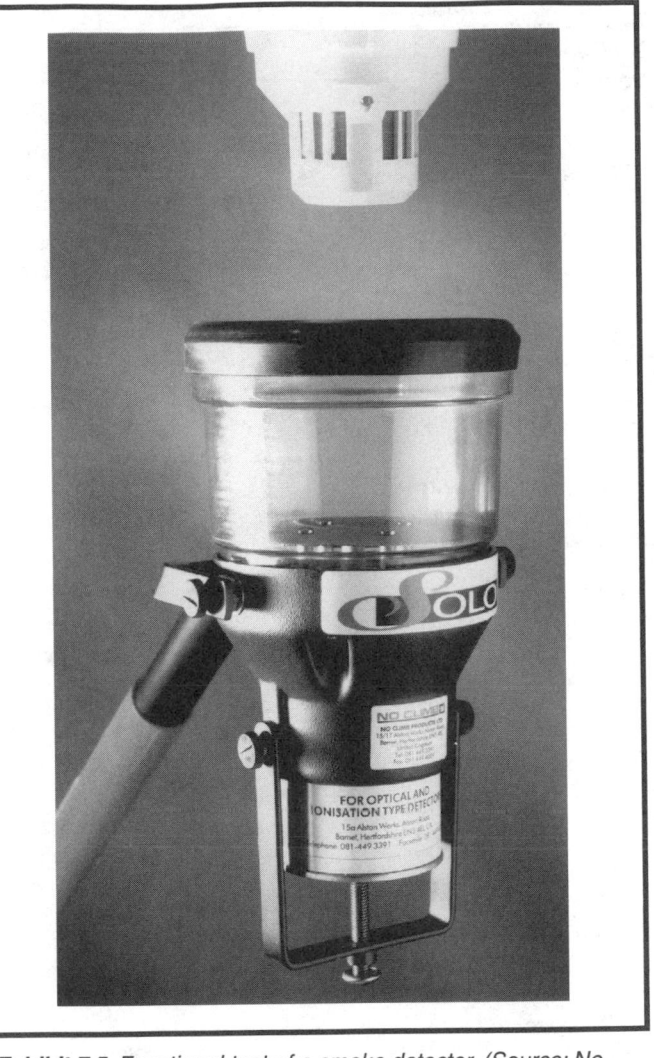

Exhibit 7.5 *Functional test of a smoke detector. (Source: No Climb Products Ltd., Hertfordshire, UK)*

In addition to losing its ability to receive an alarm from an initiating device located beyond an open fault, a Style A (Class B) initiating device circuit also loses its ability to receive an alarm when a single ground fault is present.

Style C and Style E (Class B and Class A) initiating device circuits can discriminate between an alarm condition and a wire-to-wire short. In these circuits, a wire-to-wire short provides a trouble indication. However, a wire-to-wire short prevents alarm operation. Shorting-type initiating devices cannot be used without an additional current or voltage limiting element.

Directly-connected system smoke detectors, commonly referred to as two-wire detectors, should be listed as being electrically and functionally compatible with the control unit and the specific subunit or module to which they are connected. If the detectors and the units or modules are not compatible, it is possible that, during an alarm condition, the detector's visible indicator will illuminate, but no change of state to the alarm condition will occur at the control unit. Incompatibility can also prevent proper system operation at extremes of operating voltage, temperature, and other environmental conditions.

Where two or more two-wire detectors with integral relays are connected to a single initiating device circuit and their relay contacts are used to control essential building functions (for example, fan shutdown, elevator recall), it should be clearly noted that the circuit might be capable of supplying only enough energy to support one detector/relay combination in an alarm mode. If control of more than one building function is required, each detector/relay combination used to control separate functions should be connected to separate initiating device circuits, or they should be connected to an initiating device circuit that provides adequate power to allow all the detectors connected to the circuit to be in the alarm mode simultaneously. During acceptance and reacceptance testing, this feature should always be tested and verified.

A speaker is an alarm-indicating appliance, and, if used as shown in the following diagrams, the principle of operation and supervision is the same as for other audible alarm indicating appliances (for example, bells and horns).

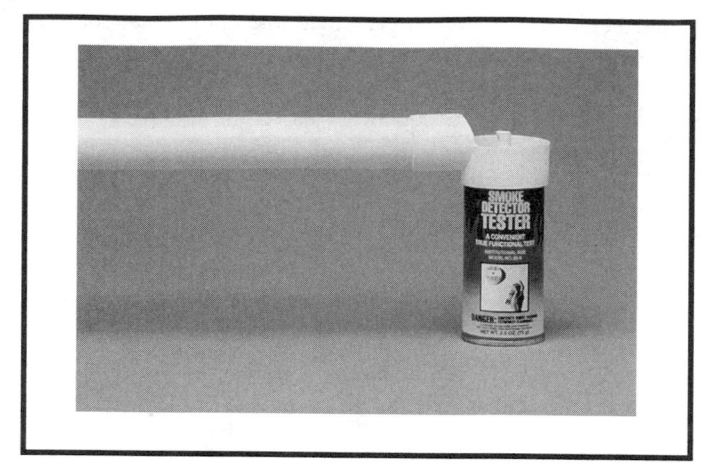

Exhibit 7.6 *Aerosol smoke product for functional test of a smoke detector. (Source: R.P. Schifiliti & Associates and Home Safeguard Industries, Inc.)*

The testing of supervised remote relays is to be conducted in the same manner as for indicating appliances.

(a) Wiring Diagrams.

NOTE: When testing circuits, the correct wiring size, insulation type, and conductor fill should be verified in accordance with the requirements of NFPA 70, *National Electrical Code.*

(1) *Nonpowered Alarm-Initiating or Supervisory-Initiating Devices (e.g., Manual Station or Valve Supervisory Switch) Connected to Style A, B, or C Initiating Device Circuits.* Disconnect conductor at device or control unit, then reconnect. Temporarily connect a ground to either leg of conductors, then remove ground. Both operations should indicate audible and visual trouble with subsequent restoration at control unit. Conductor-to-conductor short should initiate alarm. Style A and Style B (Class B) indicate trouble Style C (Class B). Style A (Class B) does not initiate alarm while in trouble condition.

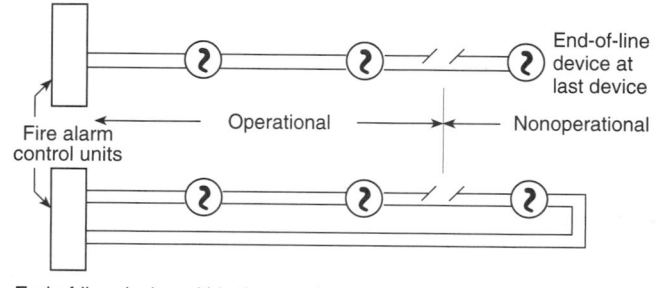

Figure A-7-2.2(a)(1) *Nonpowered alarm-initiating or supervisory-initiating devices connected to Style A, B, or C initiating device circuits.*

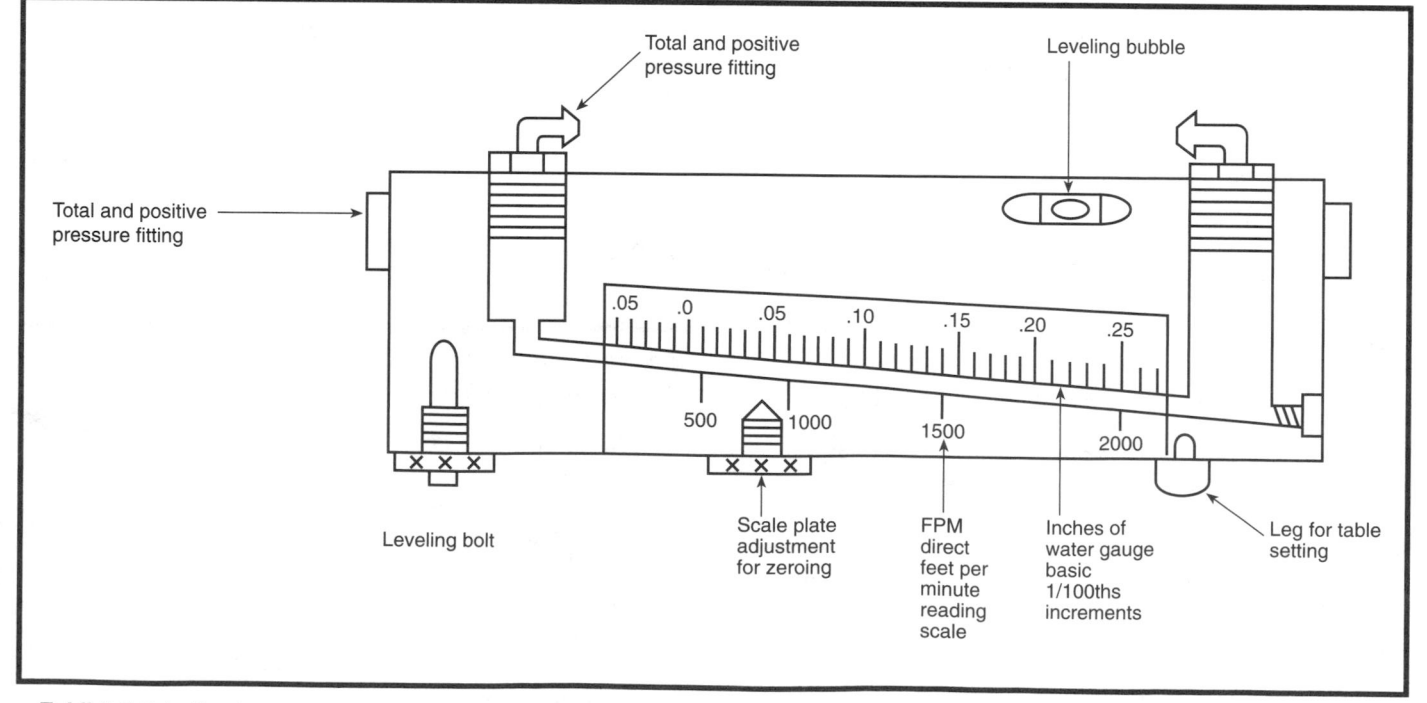

Exhibit 7.7 *Inclined manometer for velocity pressure readings in low velocity ducts, 400 fpm to 2000 fpm (2.0 m/sec to 10.2 m/sec). (Source: System Sensor, St. Charles, IL)*

(2) *Nonpowered Alarm-Initiating or Supervisory-Initiating Devices Connected to Style D or E Initiating Device Circuits.* Disconnect a conductor at a device at midpoint in the circuit. Operate a device on either side of the device with the disconnected conductor. Reset control unit and reconnect conductor. Repeat test with a ground applied to either conductor in place of the disconnected conductor. Both operations should indicate audible and visual trouble, then alarm or supervisory indication with subsequent restoration.

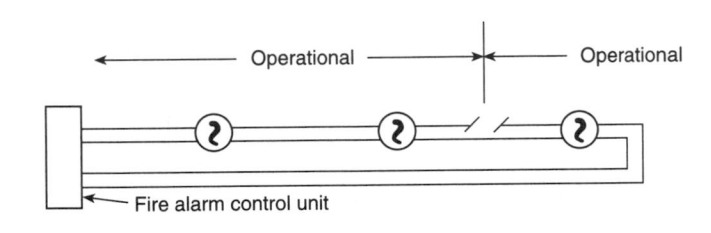

Figure A-7-2.2(a)(2) *Nonpowered alarm-initiating or supervisory-initiating devices connected to Style D or E initiating device circuits.*

(3) *Circuit-Powered (Two-Wire) Smoke Detectors for Style A, B, or C Initiating Device Circuits.* Remove smoke detector where installed with plug-in base or disconnect conductor from control unit beyond first device. Activate smoke detector per manufacturer's recommendations between control unit and circuit break. Restore detector or circuit, or both. Control unit should indicate trouble where fault occurs and alarm where detectors are activated between the break and the control unit.

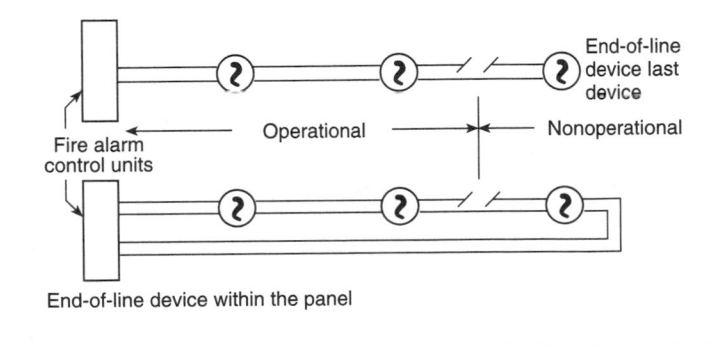

Figure A-7-2.2(a)(3) *Circuit-powered (two-wire) smoke detectors for Style A, B, or C initiating device circuits.*

(4) *Circuit-Powered (Two-Wire) Smoke Detectors for Style D or E Initiating Device Circuits.* Disconnect conductor at a smoke detector or remove where installed with a plug-in base at midpoint in the circuit. Operate a device

on either side of the device with the fault. Reset control unit and reconnect conductor or detector. Repeat test with a ground applied to either conductor in place of the disconnected conductor or removed device. Both operations should indicate audible and visual trouble, then alarm indication with subsequent restoration.

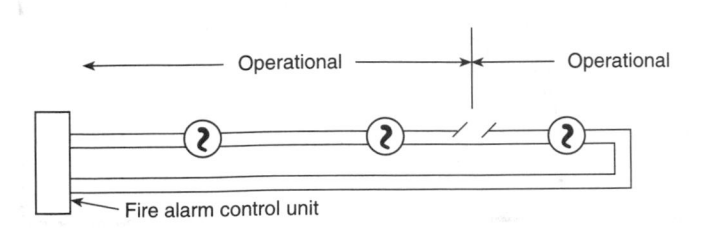

Figure A-7-2.2(a)(4) *Circuit-powered (two-wire) smoke detectors for Styles D or E initiating device circuits.*

(5) *Combination Alarm-Initiating Device and Indicating Appliance Circuits.* Disconnect a conductor either at indicating or initiating device. Activate initiating device between the fault and the control unit. Activate additional smoke detectors between the device first activated and the control unit. Restore circuit, initiating devices, and control unit. Confirm that all indicating appliances on the circuit operate from the control unit up to the fault and that all smoke detectors tested and their associated ancillary functions, if any, operate.

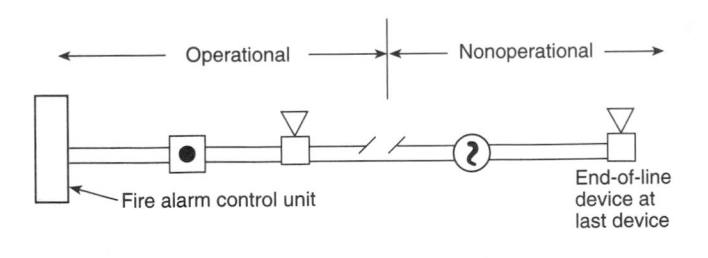

Figure A-7-2.2(a)(5) *Combination alarm-initiating device and indicating appliance circuits.*

(6) *Combination Alarm-Initiating Device and Indicating Appliance Circuits Arranged for Operation with a Single Open or Ground Fault.* Testing of the circuit is similar to that described in (e). Confirm that all indicating appliances operate on either side of fault.

(7) *Style A, B, or C Circuits with Four-Wire Smoke Detectors and an End-of-Line Power Supervision Relay.* Testing of the circuit is similar to that described in (c) and (d). Disconnect a leg of the power supply circuit beyond the first device on the circuit. Activate initiating device between the fault and the control unit. Restore circuits, initiating devices, and control unit. Audible and visual

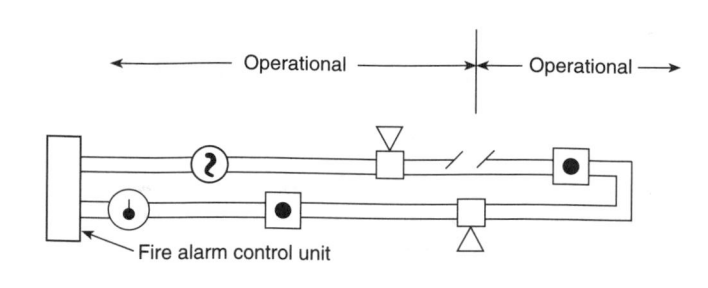

Figure A-7-2.2(a)(6) *Combination alarm-initiating device and indicating appliance circuits arranged for operation with a single open or ground fault.*

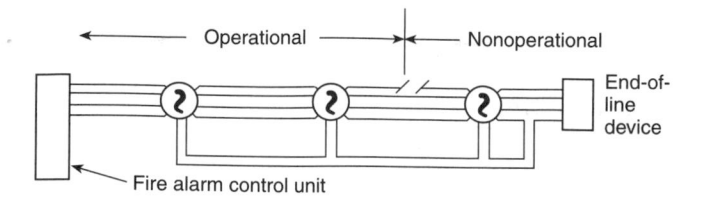

Figure A-7-2.2(a)(8) *Style A, B, or C initiating device circuits with four-wire smoke detectors that include integral individual supervision relays.*

trouble should indicate at the control unit where either the initiating or power circuit is faulted. All initiating devices between the circuit fault and the control unit should activate. In addition, removal of a smoke detector from a plug-in-type base can also break the power supply circuit. Where circuits contain various powered and nonpowered devices on the same initiating circuit, verify that the nonpowered devices beyond the power circuit fault can still initiate an alarm. A return loop should be brought back to the last powered device and the power supervisory relay to incorporate into the end-of-line device.

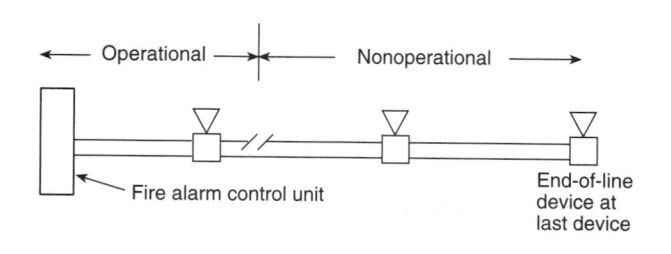

Figure A-7-2.2(a)(9) *Alarm indicating appliances connected to Styles W and Y (two-wire) circuits.*

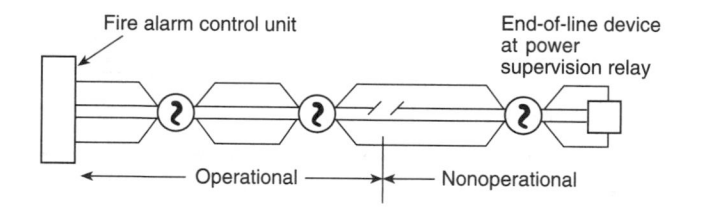

Figure A-7-2.2(a)(7) *Style A, B, or C circuits with four-wire smoke detectors and an end-of-line power supervision relay.*

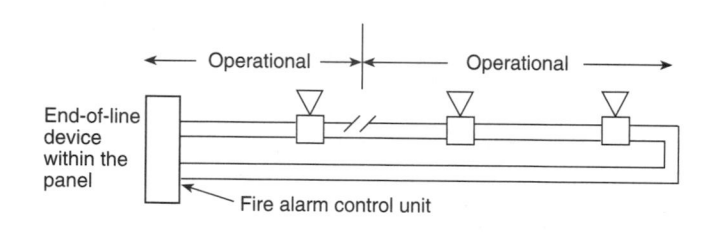

Figure A-7-2.2(a)(10) *Alarm indicating appliances connected to Styles X and Z (four-wire) circuits.*

(8) *Style A, B, or C Initiating Device Circuits with Four-Wire Smoke Detectors that Include Integral Individual Supervision Relays.* Testing of the circuit is similar to that described in (c) with the addition of a power circuit.

(9) *Alarm Indicating Appliances Connected to Styles W and Y (Two-Wire) Circuits.* Testing of the indicating appliances connected to Style W and Style Y (Class B) is similar to that described in (c).

(10) *Alarm Indicating Appliances Connected to Styles X and Z (Four-Wire) Circuits.* Testing of the indicating appliances connected to Style X and Style Z (Class B and Class A) is similar to that described in (d).

(11) *System with a Supervised Audible Indicating Appliance Circuit and an Unsupervised Visible Indicating Appliance Circuit.* Testing of the indicating appliances connected to Style X and Style Z (Class B and Class A) is similar to that described in (d).

(12) *System with Supervised Audible and Visible Indicating Appliance Circuits.* Testing of the indicating appliances connected to Style X and Style Z (Class B and Class A) is similar to that described in (d).

(13) *Series Indicating Appliance Circuit, which No Longer Meets the Requirements of NFPA 72.* An open fault in the circuit wiring should cause a trouble condition.

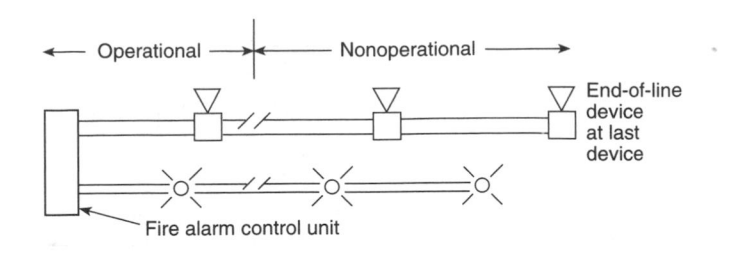

Figure A-7-2.2(a)(11) *Supervised audible indicating appliance circuit and an unsupervised visible indicating appliance circuit.*

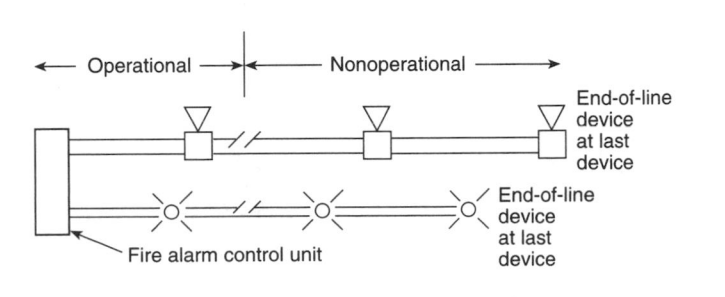

Figure A-7-2.2(a)(12) *Supervised audible and visible indicating appliance circuits.*

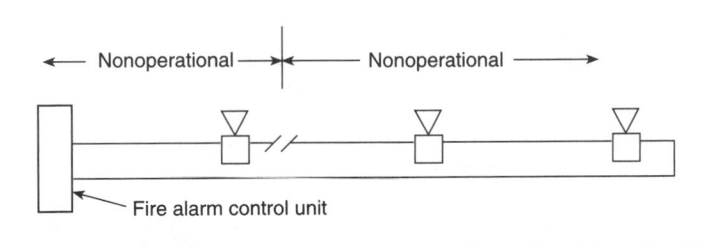

Figure A-7-2.2(a)(13) *Series indicating appliance circuit.*

(14) *Supervised Series Supervisory-Initiating Circuit with Sprinkler Supervisory Valve Switches Connected, which No Longer Meets the Requirements of NFPA 72.* An open fault in the circuit wiring of operation of the valve switch (or any supervisory signal device) should cause a trouble condition.

(15) *Initiating Device Circuit with Parallel Waterflow Alarm Switches and a Series Supervisory Valve Switch, which No Longer Meets the Requirements of NFPA 72.* An open fault in the circuit wiring or operation of the valve switch should cause a trouble signal.

(16) *System Connected to a Municipal Fire Alarm Box Circuit.* Disconnect a leg of municipal circuit at master box. Verify alarm sent to public communications

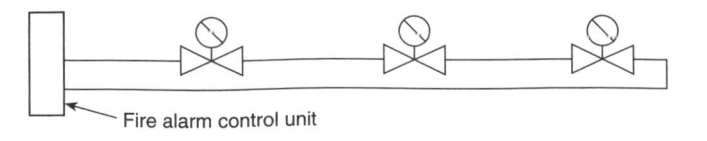

Figure A-7-2.2(a)(14) *Supervised series supervisory-initiating circuit with sprinkler supervisory valve switches connected.*

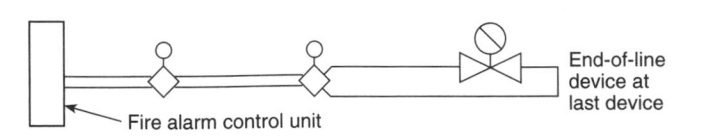

Figure A-7-2.2(a)(15) *Initiating device circuit with parallel waterflow alarm switches and a series supervisory-valve switch.*

center. Disconnect leg of auxiliary circuit. Verify trouble condition on control unit. Restore circuits. Activate control unit and send alarm signal to communications center. Verify control unit in trouble condition until master box reset.

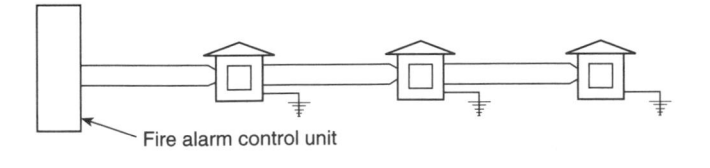

Figure A-7-2.2(a)(16) *System connected to a municipal fire alarm master box circuit.*

(17) *Auxiliary Circuit Connected to a Municipal Fire Alarm Master Box.* For operation with a master box, an open or ground fault (where ground detection is provided) on the circuit should result in a trouble condition at the control unit. A trouble signal at the control unit should persist until the master box is reset. For operation with a shunt trip master box, an open fault in the auxiliary circuit should cause an alarm on the municipal system.

(b) *Circuit Styles.*

NOTE: Some testing laboratories and authorities having jurisdiction permit systems to be classified as a Style 7 (Class A) by the application of two circuits of the same style operating in parallel. An example of this is to take two series circuits, either Style 0.5 or Style 1.0 (Class B), and operate them in parallel. The

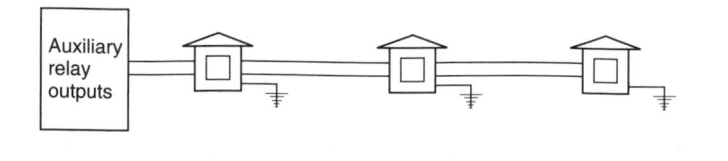

Figure A-7-2.2(a)(17) Auxiliary circuit connected to a municipal fire alarm master box.

logic is that if a condition occurs on one of the circuits, the other parallel circuit remains operative.

In order to understand the principles of the circuit, alarm receipt capability should be performed on a single circuit, and the style type, based on the performance, should be indicated on the record of completion.

(1) *Style 0.5*. This signaling circuit operates as a series circuit in performance. This is identical to the historical series audible signaling circuits. Any type of break or ground in one of the conductors, or the internal of the multiple interface device, and the total circuit is rendered inoperative.

To test and verify this type of circuit, either a conductor should be lifted or an earth ground should be placed on a conductor or a terminal point where the signaling circuit attaches to the multiplex interface device.

(2) *Style 0.5(a) (Class B) Series*. Style 0.5(a) functions so that, when a box is operated, the supervisory contacts open, making the succeeding devices nonoperative while the operating box sends a coded signal. Any alarms occurring in any successive devices will not be received at the receiving station during this period.

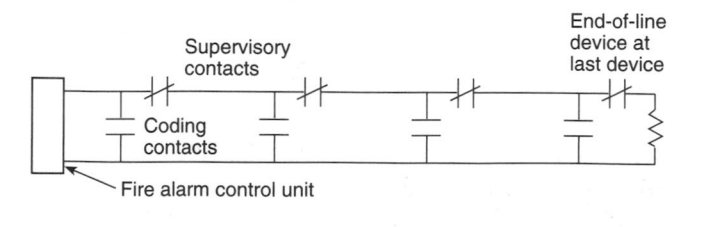

Figure A-7-2.2(b)(2) Style 0.5(a) (Class B) series.

(3) *Style 0.5(b) (Class B) Shunt*. The contact closes when the device is operated and remains closed to shunt out the remainder of the system until the code is complete.

(4) *Style 0.5(c) (Class B) Positive Supervised Successive*. An open or ground fault on the circuit should cause a trouble condition at the control unit.

(5) *Style 1.0 (Class B)*. This is a series circuit identical to the diagram for Style 0.5, except that the fire alarm sys-

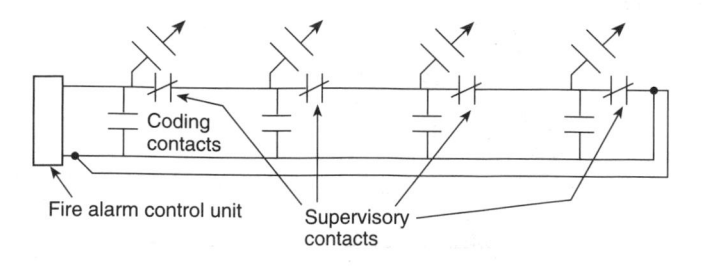

Figure A-7-2.2(b)(3) Style 0.5(b) (Class B) shunt.

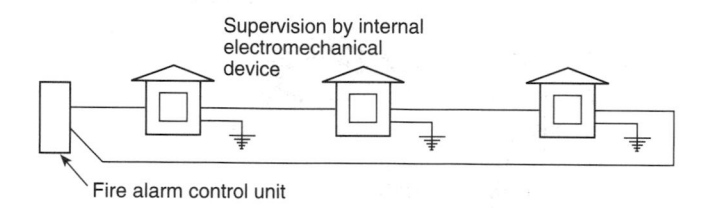

Figure A-7-2.2(b)(4) Style 0.5(c) (Class B) positive supervised successive.

tem hardware has enhanced performance. A single earth ground can be placed on a conductor or multiplex interface device, and the circuit and hardware still have alarm operability.

If a conductor break or an internal fault occurs in the pathway of the circuit conductors, the entire circuit becomes inoperative.

To verify alarm receipt capability and the resulting trouble signal, place an earth ground on one of the conductors or at the point where the signaling circuit attaches to the multiplex interface device. One of the transmitters or an initiating device should then be placed into alarm.

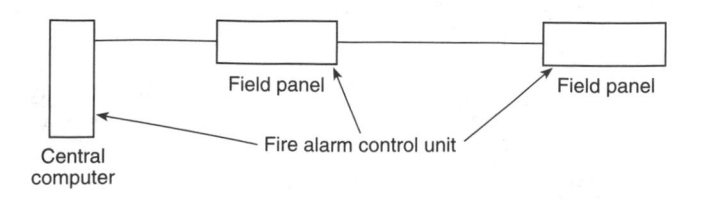

Figure A-7-2.2(b)(5)(a) Style 1.0 (Class B).

(6) *Typical McCulloh Loop*. This is the central station McCulloh redundant-type circuit and has alarm receipt capability on either side of a single break.

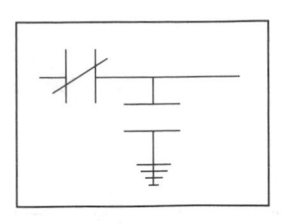

Figure A-7-2.2(b)(5)(b) *Typical transmitter layout.*

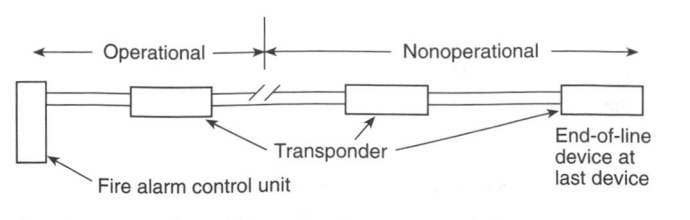

Figure A-7-2.2(b)(7) *Style 3.0 (Class B).*

a. To test, lift one of the conductors and operate a transmitter or initiating device on each side of the break. This activity should be repeated for each conductor.

b. Place an earth ground on a conductor and operate a single transmitter or initiating device to verify alarm receipt capability and trouble condition for each conductor.

c. Repeat the instructions of (a) and (b) at the same time and verify alarm receipt capability, and verify that a trouble condition results.

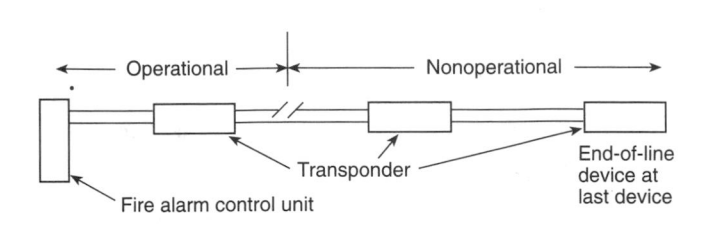

Figure A-7-2.2(b)(8) *Style 3.5 (Class B).*

(9) *Style 4.0 (Class B).* Repeat the instructions for Style 3.0 (Class B) and include a loss of carrier where the signal is being used.

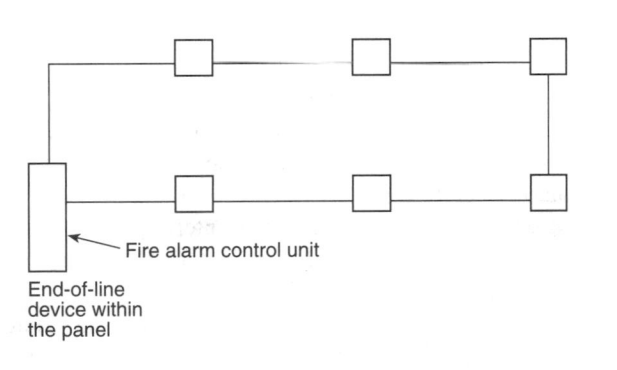

Figure A-7-2.2(b)(6) *Typical McCulloh loop.*

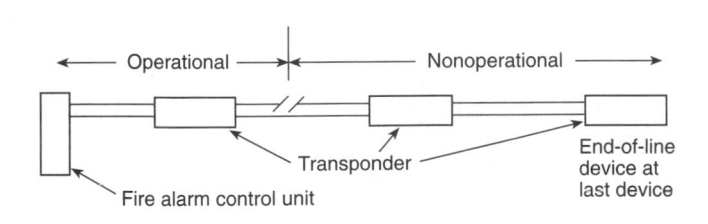

Figure A-7-2.2(b)(9) *Style 4.0 (Class B).*

(7) *Style 3.0 (Class B).* This is a parallel circuit in which multiplex interface devices transmit signal and operating power over the same conductors. The multiplex interface devices might be operable up to the point of a single break. Verify by lifting a conductor and causing an alarm condition on one of the units between the central alarm unit and the break. Either lift a conductor to verify the trouble condition or place an earth ground on the conductors. Test for all the valuations shown on the signaling table.

On ground-fault testing, verify alarm receipt capability by actuating a multiplex interface initiating device or a transmitter.

(8) *Style 3.5 (Class B).* Repeat the instructions for Style 3.0 (Class B) and verify the trouble conditions by either lifting a conductor or placing a ground on the conductor.

(10) *Style 4.5 (Class B).* Repeat the instructions for Style 3.5 (Class B). Verify alarm receipt capability while lifting a conductor by actuating a multiple interface device or transmitter on each side of the break.

(11) *Style 5.0 (Class A).* Verify the alarm receipt capability and trouble annunciation by lifting a conductor and actuating a multiplex interfacing device or a transmitter on each side of the break. For the earth ground verification, place an earth ground and certify alarm receipt capability and trouble annunciation by actuating a single multiplex interfacing device or a transmitter.

(12) *Style 6.0 (Class A).* Repeat the instructions for Style 2.0 [Class A (a) through (c)]. Verify the remaining steps for trouble annunciation for the various combinations.

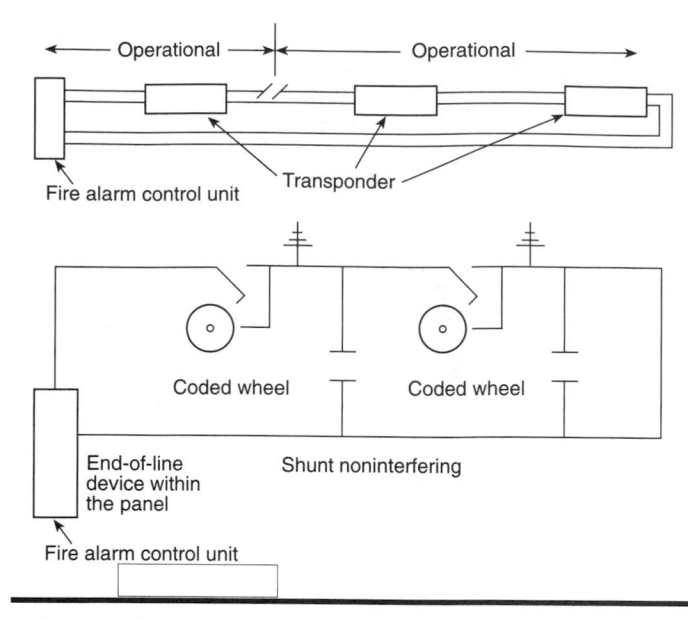

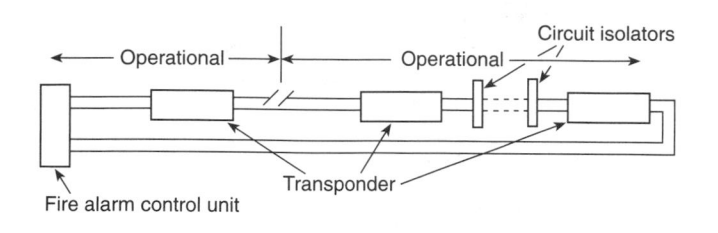

Figure A-7-2.2(b)(13) Style 6.0 (with circuit isolators) (Class A).

be clearly noted that the alarm receipt capability for remaining portions of the circuit protection isolators is not the capability of the circuit, but is permitted with enhanced system capabilities.

Figure A-7-2.2(b)(10) Style 4.5 (Class B).

(14) *Style 7.0 (Class A).* Repeat the instructions for testing of Style 6.0 (Class A) for alarm receipt capability and trouble annunciation.

NOTE 1: A portion of the circuit between the alarm processor or central supervising station and the first circuit isolator does not have alarm receipt capability in the presence of a wire-to-wire short. The same is true for the portion of the circuit from the last isolator to the alarm processor or the central supervising station.

NOTE 2: Some manufacturers of this type of equipment have isolators as part of the base assembly. Therefore, in the field, this component might not be readily observable without the assistance of the manufacturer's representative.

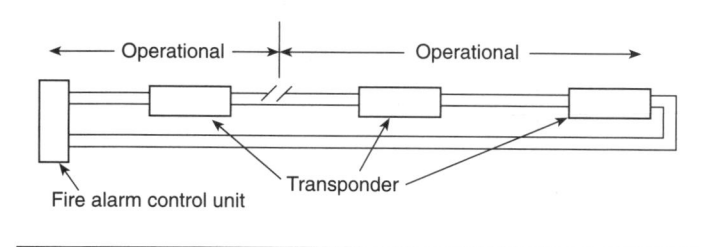

Figure A-7-2.2(b)(11) Style 5.0 (Class A).

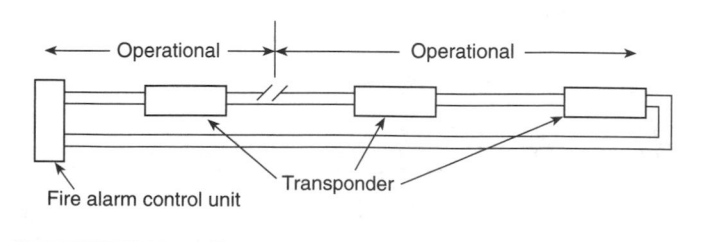

Figure A-7-2.2(b)(12) Style 6.0 (Class A).

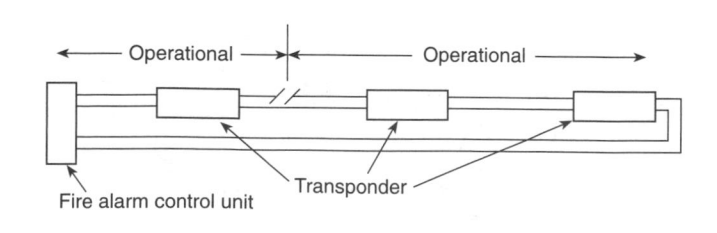

Figure A-7-2.2(b)(14)(a) Style 7.0 (Class A).

(13) *Style 6.0 (with Circuit Isolators) (Class A).* For the portions of the circuits electrically located between the monitoring points of circuit isolators, follow the instructions for a Style 7.0 (Class A) circuit. It should

Figures A-7-2.2(b)(14)(b) through A-7-2.2(b)(14)(e) depict block diagrams of radio fire alarm systems.

CP = Wireless control panel
(with power supply and standby power)

R = Wireless repeater
(with power supply and standby power)

D = Wireless initiating, indicating, and control device
(either primary battery or primary standby battery)

Figure A-7-2.2(b)(14)(b) *Low-power radio (wireless) fire alarm system.*

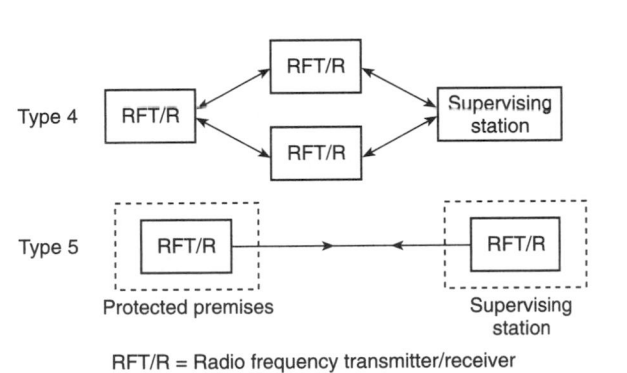

Type 4

Type 5

Protected premises

Supervising station

RFT/R = Radio frequency transmitter/receiver

Figure A-7-2.2(b)(14)(c) *Two-way RF multiplex systems.*

Optional unlimited

RAT

Protected premises

Physically separated

RAT = Radio alarm transmitter
RARSR = Radio alarm repeater station receiver
RASSR = Radio alarm supervising station receiver

Figure A-7-2.2(b)(14)(d) *One-way radio alarm system.*

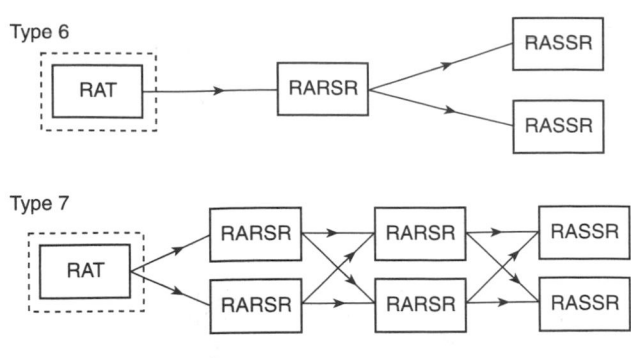

Type 6

Type 7

RAT = Radio alarm transmitter
RARSR = Radio alarm repeater station receiver
RASSR = Radio alarm supervising station receiver

Figure A-7-2.2(b)(14)(e) *One-way radio alarm system.*

Figures A-7-2.2(b)(14)(f) through A-7-2.2(b)(14)(k) show fiber optic network operation under various fault conditions.

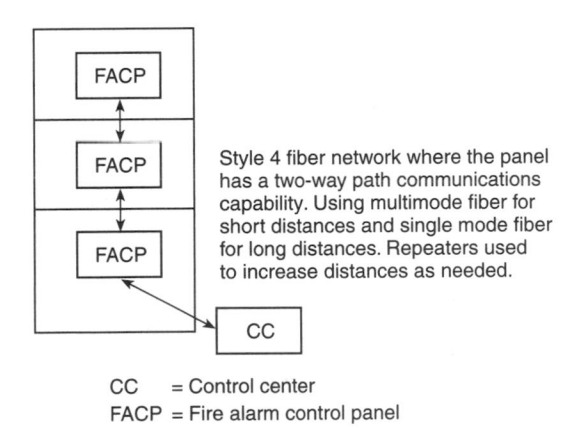

Style 4 fiber network where the panel has a two-way path communications capability. Using multimode fiber for short distances and single mode fiber for long distances. Repeaters used to increase distances as needed.

CC = Control center
FACP = Fire alarm control panel

Figure A-7-2.2(b)(14)(f) *Style 4 fiber network.*

(c) *Batteries.* To maximize battery life, nickel-cadmium batteries should be charged as in Table A-7-2.2(c)(1).

Table A-7-2.2(c)(1) *Voltage for Nickel-Cadmium Batteries*

Float voltage	1.42 volts/cell + 0.01 volt
High rate voltage	1.58 volts/cell + 0.07 volt − 0.00 volt

Note: High and low gravity voltages are (+) 0.07 volt and (−) 0.03 volt, respectively.

To maximize battery life, the battery voltage for lead-acid cells should be maintained within the limits shown in Table A-7-2.2(c)(2).

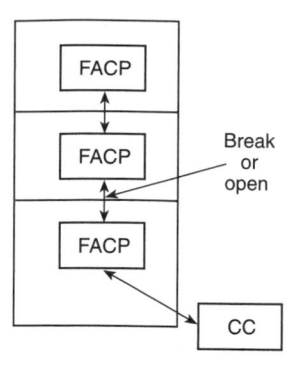

Style 4 fiber network where the panel has a two-way path communications capability. A single break separates the system into two LANs both with Style 4 capabilities.

CC   = Control center
FACP = Fire alarm control panel

Figure A-7-2.2(b)(14)(g) Style 4 fiber network.

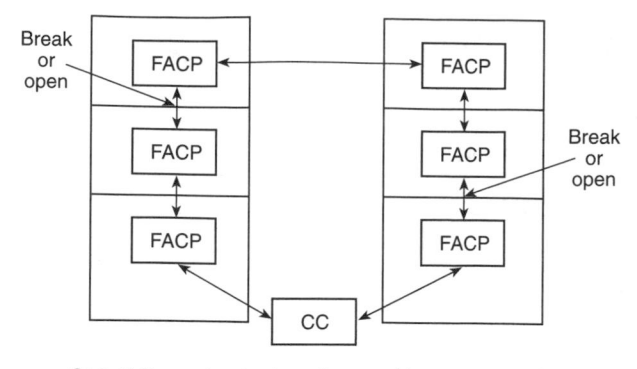

Style 7 fiber network where the panel has a two-way path communications capability with the two breaks now breaking into two LANs, both functioning as independent networks with the same Style 7 capabilities.

CC   = Control center
FACP = Fire alarm control panel

Figure A-7-2.2(b)(14)(i) Style 7 fiber network.

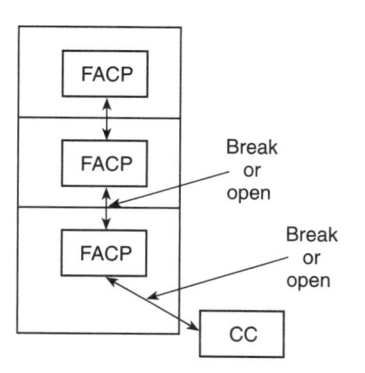

Style 4 fiber network where the panel has a two-way path communications capability. A double break isolates the panels and the control center in this case. In this case, there is one LAN and one isolated panel operating on its own. Control center is isolated completely with no communications with the network.

CC   = Control center
FACP = Fire alarm control panel

Figure A-7-2.2(b)(14)(h) Style 4 fiber network.

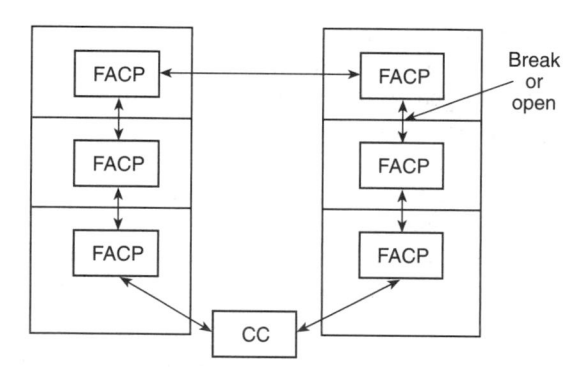

Style 7 fiber network where the panel has a two-way path communications capability, with one break. System remains as one LAN and meets Style 7.

CC   = Control center
FACP = Fire alarm control panel

Figure A-7-2.2(b)(14)(j) Style 7 fiber network.

Table A-7-2.2(c)(2) Voltage for Lead-Acid Batteries

Float Voltage	High-Gravity Battery (Lead Calcium)	Low-Gravity Battery (Lead Antimony)
Maximum	2.25 volts/cell	2.17 volts/cell
Minimum	2.20 volts/cell	2.13 volts/cell
High rate voltage	—	2.33 volts/cell

The following procedure is recommended for checking the state of charge for nickel-cadmium batteries:

(1) The battery charger should be switched from float to high-rate mode.

(2) The current, as indicated on the charger ammeter, will immediately rise to the maximum output of the charger, and the battery voltage, as shown on the charger voltmeter, will start to rise at the same time.

(3) The actual value of the voltage rise is unimportant, because it depends on many variables. The length of time it takes for the voltage to rise is the important factor.

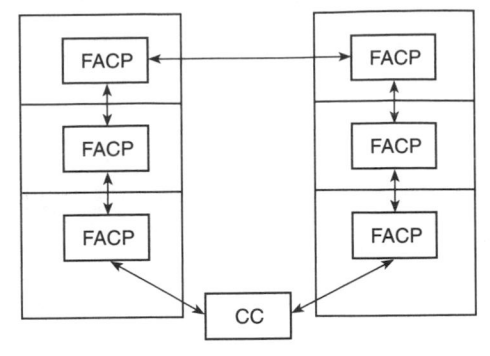

Style 7 fiber network where the panel has a two-way path communications capability.

CC = Control center
FACP = Fire alarm control panel

Figure A-7-2.2(b)(14)(k) Style 7 fiber network.

(4) If, for example, the voltage rises rapidly in a few minutes, then holds steady at the new value, the battery is fully charged. At the same time, the current will drop to slightly above its original value.

(5) In contrast, if the voltage rises slowly and the output current remains high, the high-rate charge should be continued until the voltage remains constant. Such a condition is an indication that the battery is not fully charged, and the float voltage should be increased slightly.

A-7-2.2 Table Line/Item 6.c.1. Example: 4000 mAh × $^1/_{25}$ = 160 mA charging current at 25°C (77°F).

A-7-2.2 Table Line/Item 7(c). The voltmeter sensitivity has been changed from 1000 ohms per volt to 100 ohms per volt so that false ground readings (caused by induced voltages) are minimized.

A-7-2.2 Table Line/Item 13.a.2. Fusible thermal link detectors are commonly used to close fire doors and fire dampers. They are actuated by the presence of external heat, which causes a solder element in the link to fuse, or by an electric thermal device, which, when energized, generates heat within the body of the link, causing the link to fuse and separate.

7-3 Inspection and Testing Frequency

7-3.1* Visual Inspection.

Visual inspection shall be performed in accordance with the schedules in Section 7-3 or more often if required by the

authority having jurisdiction. The visual inspection shall be made to ensure that there are no changes that affect equipment performance.

Where the authority having jurisdiction suspects building conditions are changing more rapidly than normal and that these changes are likely to affect the performance of the fire alarm system, the authority having jurisdiction may require more frequent visual inspections.

Table 7-3.1 displays how often visual inspections are to be performed on various components and subsystems of the fire alarm system.

A visual inspection should always be conducted prior to any testing. A copy of the as-built drawings and system documentation provides quantities and locations of devices. Improperly located, damaged, or non-functional equipment should be identified and corrected before tests begin.

Exception 1: Devices or equipment that is inaccessible for safety considerations (for example, continuous process operations, energized electrical equipment, radiation, and excessive height) shall be inspected during scheduled shutdowns if approved by the authority having jurisdiction. Extended intervals shall not exceed 18 months.

Exception No.1 clearly defines the intended safety considerations where special conditions are encountered.

Exception 2: If automatic inspection is performed at a frequency of not less than weekly by a remotely monitored fire alarm control unit specifically listed for such application, the visual inspection frequency shall be permitted to be annual. Table 7-3.1 shall apply.

Exception No. 2 was revised to allow the development of new technology that will be able to remotely inspect equipment. This technology is not yet available for use in the field, but may become available during the course of this code cycle

A-7-3.1 Equipment performance can be affected by building modifications, occupancy changes, changes in environmental conditions, device location, physical obstructions, device orientation, physical damage, improper installation, degree of cleanliness, or other obvious problems that might not be indicated through electrical supervision.

7-3.2* Testing.

Testing shall be performed in accordance with the schedules in Chapter 7 or more often if required by the authority having jurisdiction. If automatic testing is performed at least weekly by a remotely monitored fire alarm control unit specifically listed for the application, the manual testing frequency shall be permitted to be extended to annual. Table 7-3.2 shall apply.

Table 7-3.1 Visual Inspection Frequencies

Component	Initial/ Reacceptance	Monthly	Quarterly	Semiannually	Annually
1. Control Equipment: Fire Alarm Systems Monitored for Alarm, Supervisory, Trouble Signals					
a. Fuses	X	—	—	—	X
b. Interfaced Equipment	X	—	—	—	X
c. Lamps and LEDs	X	—	—	—	X
d. Primary (Main) Power Supply	X	—	—	—	X

The term *monitored* refers to systems connected to a supervising station that receives all three signals, that is, alarm, trouble, and supervisory. In unmonitored systems, signals are not transmitted to a supervising station for appropriate action to be taken. Therefore, weekly inspection of fuses, lamps, LEDs, interfaced equipment, and power supplies to ensure reliability is necessary. See item 2 below.

Component	Initial/ Reacceptance	Monthly	Quarterly	Semiannually	Annually
2. Control Equipment: Fire Alarm Systems Unmonitored for Alarm, Supervisory, Trouble Signals					
a. Fuses	X (weekly)	—	—	—	—
b. Interfaced Equipment	X (weekly)	—	—	—	—
c. Lamps and LEDs	X (weekly)	—	—	—	—
d. Primary (Main) Power Supply	X (weekly)	—	—	—	—
3. Batteries					
a. Lead-Acid	X	X	—	—	—
b. Nickel-Cadmium	X	—	—	X	—
c. Primary (Dry Cell)	X	X	—	—	—
d. Sealed Lead-Acid	X	—	—	X	—
4. Transient Suppressors	X	—	—	X	—
5. Control Panel Trouble Signals	X	—	—	X	—
6. Fiber-Optic Cable Connections	X	—	—	—	X
7. Emergency Voice/Alarm Communications Equipment	X	—	—	X	—
8. Remote Annunciators	X	—	—	X	—
9. Initiating Devices					
a. Air Sampling	X	—	—	X	—
b. Duct Detectors	X	—	—	X	—
c. Electromechanical Releasing Devices	X	—	—	X	—
d. Fire-Extinguishing System(s) or Suppression System(s) Switches	X	—	—	X	—
e. Fire Alarm Boxes	X	—	—	X	—

Often, after a system has been installed, an owner or occupant will place large plants, file cabinets, or other obstructions in front of manual fire alarm boxes. A visual inspection ensures that these conditions will be found and corrective measures can be taken.

Component	Initial/ Reacceptance	Monthly	Quarterly	Semiannually	Annually
f. Heat Detectors	X	—	—	X	—

Heat detectors should be inspected to ensure that there is no mechanical damage, that they are properly located, and that building conditions (such as the installation of a new wall) have not changed, possibly reducing the effectiveness of the devices.

Component	Initial/ Reacceptance	Monthly	Quarterly	Semiannually	Annually
g. Radiant Energy Fire Detectors	X	—	X	—	—

(continues)

Table 7-3.1 Continued

Component	Initial/ Reacceptance	Monthly	Quarterly	Semiannually	Annually
Radiant energy fire detectors should be inspected to ensure that there are no obstructions between the detector and the protected area, that the lenses are clear and free of contaminates, that there is no mechanical damage, and that the unit is directed toward the intended hazard.					
h. Smoke Detectors	X	—	—	X	—
Some smoke detectors may be monitored for contamination by the fire alarm control unit. However, without a visual inspection, covered detectors might not be found. Often during special cleaning or renovation projects, someone covers the smoke detector to avoid contamination and possible nuisance alarms. After the project is complete, the covers are inadvertently left in place. See Exhibit 7.8 for an example of dust covers for smoke detectors.					
i. Supervisory Signal Devices	X	—	X	—	—
j. Waterflow Devices	X	—	X	—	—
10. Guard's Tour Equipment	X	—	—	X	—
11. Interface Equipment	X	—	—	X	—
12. Alarm Notification Appliances—Supervised	X	—	—	X	—
Notification appliances should be inspected to ensure that there are no obstructions that impair effectiveness, that there is no mechanical damage, and that changing building conditions have not rendered the device ineffective. Where walls have been added to a space or floor or wall coverings have changed, additional sound pressure level measurements need to be performed. The term *supervised* refers to monitored for integrity as covered by 1-5.8.1.					
13. Supervising Station Fire Alarm Systems—Transmitters					
a. DACT	X	—	—	X	—
b. DART	X	—	—	X	—
c. McCulloh	X	—	—	X	—
d. RAT	X	—	—	X	—
14. Special Procedures	X	—	—	X	—
15. Supervising Station Fire Alarm Systems—Receivers					
a. DACR*	X	X	—	—	—
b. DARR*	X	—	—	X	—
c. McCulloh Systems*	X	—	—	X	—
d. Two-Way RF Multiplex*	X	—	—	X	—
e. RASSR*	X	—	—	X	—
f. RARS*	X	—	—	X	—
g. Private Microwave*	X	—	—	X	—

*Reports of automatic signal receipt shall be verified daily.

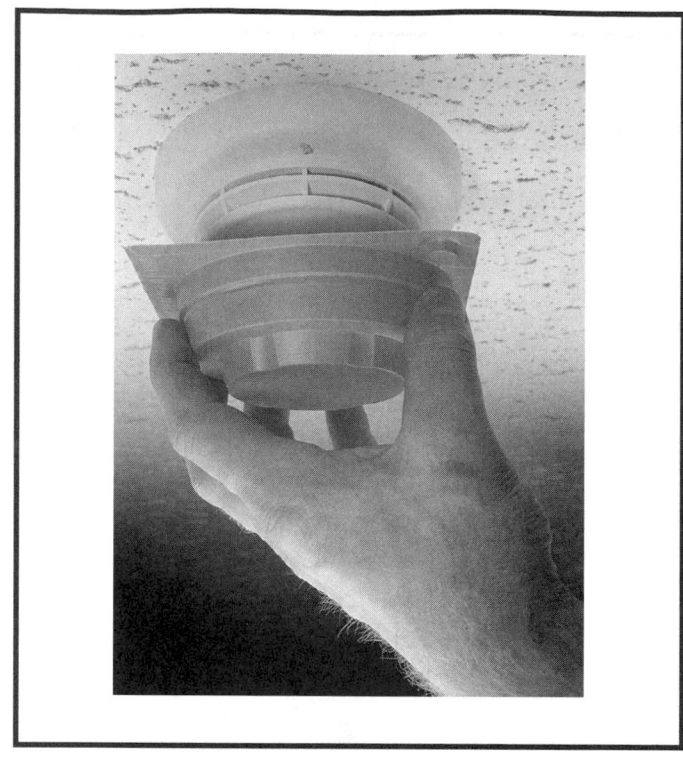

Exhibit 7.8 *Example of a typical smoke detector cover. (Source: Mammoth Fire Alarms, Inc., Lowell, MA)*

Exception: Devices or equipment that are inaccessible for safety considerations (for example, continuous process operations, energized electrical equipment, radiation, and excessive height) shall be tested during scheduled shutdowns if approved by the authority having jurisdiction but shall not be tested more than every 18 months.

The exception to 7-3.2 clearly defines the intended safety considerations. Table 7-3.2 indicates the frequencies for tests on the various components and subsystems of the fire alarm system.

A-7-3.2 It is suggested that the annual test be conducted in segments so that all devices are tested annually.

7-3.2.1* Detector sensitivity shall be checked within 1 year after installation and every alternate year thereafter. After the second required calibration test, if sensitivity tests indicate that the detector has remained within its listed and marked sensitivity range (or 4 percent obscuration light gray smoke, if not marked), the length of time between calibration tests shall be permitted to be extended to a maximum of 5 years. If the frequency is extended, records of detector-caused nuisance alarms and subsequent trends of these alarms shall be maintained. In zones or in areas where nuisance alarms show any increase over the previous year, calibration tests shall be performed.

Table 7-3.2 Testing Frequencies

Component	Initial/ Reacceptance	Monthly	Quarterly	Semiannually	Annually	Table 7-2.2 Reference
1. Control Equipment—Building Systems Connected to Supervising Station						1, 7, 16, 17
a. Functions	X	—	—	—	X	
b. Fuses	X	—	—	—	X	—
c. Interfaced Equipment	X	—	—	—	X	—
d. Lamps and LEDs	X	—	—	—	X	—
e. Primary (Main) Power Supply	X	—	—	—	X	—
f. Transponders	X	—	—	—	X	—

This section refers to systems connected to a supervising station that receives all three signals: alarm, trouble and supervisory. In unmonitored systems, signals are not transmitted to a supervising station for appropriate action to be taken. Therefore, quarterly inspection of these items to ensure reliability is necessary. See item 2 below.

Component	Initial/ Reacceptance	Monthly	Quarterly	Semiannually	Annually	Table 7-2.2 Reference
2. Control Equipment—Building Systems—Not Connected to a Supervising Station	—	—	—	—	—	1
a. Functions	X	—	X			—
b. Fuses	X	—	X	—	—	—
c. Interfaced Equipment	X	—	X	—	—	—
d. Lamps and LEDs	X	—	X	—	—	—

(continues)

Table 7-3.2 Continued

Component	Initial/ Reacceptance	Monthly	Quarterly	Semiannually	Annually	Table 7-2.2 Reference
e. Primary (Main) Power Supply	X	—	X	—	—	—
f. Transponders	X	—	X	—	—	—
3. Engine-Driven Generator—Central Station Facilities and Fire Alarm Systems	X	X	—	—	—	—
4. Engine-Driven Generator—Public Fire Alarm Reporting Systems	X (weekly)	—	—	—	—	—
5. Batteries—Central Station Facilities						
a. Lead-Acid Type	—	—	—	—	—	6b
1. Charger Test (Replace battery as needed.)	X	—	—	—	X	—
2. Discharge Test (30 minutes)	X	X	—	—	—	—
3. Load Voltage Test	X	X	—	—	—	—
4. Specific Gravity	X	—	—	X	—	—
b. Nickel-Cadmium Type	—	—	—	—	—	6c
1. Charger Test (Replace battery as needed.)	X	—	X	—	—	—
2. Discharge Test (30 minutes)	X	—	—	—	X	—
3. Load Voltage Test	X	—	—	—	X	—
c. Sealed Lead-Acid Type	X	X	—	—	—	6d
1. Charger Test (Replace battery as needed.)	—	X	X	—	—	—
2. Discharge Test (30 minutes)	X	X	—	—	—	—
3. Load Voltage Test	X	X	—	—	—	—
6. Batteries—Fire Alarm Systems						
a. Lead-Acid Type	—	—	—	—	—	6b
1. Charger Test (Replace battery as needed.)	X	—	—	—	X	—
2. Discharge Test (30 minutes)	X	—	—	X	—	—
3. Load Voltage Test	X	—	—	X	—	—
4. Specific Gravity	X	—	—	X	—	—
b. Nickel-Cadmium Type	—	—	—	—	—	6c
1. Charger Test (Replace battery as needed.)	X	—	—	—	X	—
2. Discharge Test (30 minutes)	X	—	—	—	X	—
3. Load Voltage Test	X	—	—	X	—	—
c. Primary Type (Dry Cell)	—	—	—	—	—	6a
1. Load Voltage Test	X	X	—	—	—	—
d. Sealed Lead-Acid Type	—	—	—	—	—	6d
1. Charger Test (Replace battery every 4 years.)	X	—	—	—	X	—
2. Discharge Test (30 minutes)	X	—	—	—	X	—
3. Load Voltage Test	X	—	—	X	—	—
7. Batteries—Public Fire Alarm Reporting Systems Voltage tests in accordance with Table 7-2.2, item 7(a)–(f)	X (daily)	—	—	—	—	—
a. Lead-Acid Type	—	—	—	—	—	6b
1. Charger Test (Replace battery as needed.)	X	—	—	—	X	—
2. Discharge Test (2 hours)	X	—	X	—	—	—
3. Load Voltage Test	X	—	X	—	—	—
4. Specific Gravity	X	—	—	X	—	—
b. Nickel-Cadmium Type	—	—	—	—	—	6c
1. Charger Test (Replace battery as needed.)	X	—	—	—	X	—
2. Discharge Test (2 hours)	X	—	—	—	X	—
3. Load Voltage Test	X	—	X	—	—	—
c. Sealed Lead-Acid Type	—	—	—	—	—	6d
1. Charger Test (Replace battery as needed.)	X	—	—	—	X	—

Table 7-3.2 Continued

Component	Initial/ Reacceptance	Monthly	Quarterly	Semiannually	Annually	Table 7-2.2 Reference
2. Discharge Test (2 hours)	X	—	—	—	X	—
3. Load Voltage Test	X	—	X	—	—	—
8. Fiber-Optic Cable Power	X	—	—	—	X	12b
9. Control Unit Trouble Signals	X	—	—	—	X	9
10. Conductors—Metallic	X	—	—	—	—	11
11. Conductors—Nonmetallic	X	—	—	—	—	12
12. Emergency Voice/Alarm Communications Equipment	X	—	—	—	X	18
13. Retransmission Equipment (The requirements of 7-3.4 shall apply.)	X	—	—	—	—	—
14. Remote Annunciators	X	—	—	—	X	10
15. Initiating Devices	—	—	—	—	—	13
a. Duct Detectors	X	—	—	—	X	—
b. Electromechanical Releasing Device	X	—	—	—	X	—
c. Fire-Extinguishing System(s) or Suppression System(s) Switches	X	—	—	—	X	—
d. Fire–Gas and Other Detectors	X	—	—	—	X	—
e. Heat Detectors (The requirements of 7-3.2.3 shall apply.)	X	—	—	—	X	—
f. Fire Alarm Boxes	X	—	—	—	X	—
g. Radiant Energy Fire Detectors	X	—	—	X	—	—
h. All Smoke Detectors—Functional	X	—	—	—	X	—
i. Smoke Detectors—Sensitivity (The requirements of 7-3.2.1 shall apply.)	—	—	—	—	—	—
j. Supervisory Signal Devices	X	—	X	—	—	—
k. Waterflow Devices (except valve tamper switches)	X	—	—	X	—	—
1. Valve Tamper Switches	X	—	—	X	—	—
16. Guard's Tour Equipment	X	—	—	—	X	—
17. Interface Equipment	X	—	—	—	X	19
18. Special Hazard Equipment	X	—	—	—	X	15
19. Alarm Notification Appliances	—	—	—	—	—	14
a. Audible Devices	X	—	—	—	X	—
b. Audible Textual Notification Appliances	X	—	—	—	X	—
c. Visible Devices	X	—	—	—	X	—
20. Off-Premises Transmission Equipment	X	—	X	—	—	—
21. Supervising Station Fire Alarm Systems— Transmitters	—	—	—	—	—	16
a. DACT	X	—	—	—	X	—
b. DART	X	—	—	—	X	—
c. McCulloh	X	—	—	—	X	—
d. RAT	X	—	—	—	X	—
22. Special Procedures	X	—	—	—	X	21
23. Supervising Station Fire Alarm Systems— Receivers	—	—	—	—	—	17
a. DACR	X	X	—	—	—	—
b. DARR	X	X	—	—	—	—
c. McCulloh Systems	X	X	—	—	—	—
d. Two-Way RF Multiplex	X	X	—	—	—	—
e. RASSR	X	X	—	—	—	—
f. RARSR	X	X	—	—	—	—
g. Private Microwave	X	X	—	—	—	—

To ensure that each smoke detector is within its listed and marked sensitivity range, it shall be tested using any of the following methods:

(1) Calibrated test method
(2) Manufacturer's calibrated sensitivity test instrument
(3) Listed control equipment arranged for the purpose
(4) Smoke detector/control unit arrangement whereby the detector causes a signal at the control unit where its sensitivity is outside its listed sensitivity range
(5) Other calibrated sensitivity test methods approved by the authority having jurisdiction

See Exhibits 7.9 and 7.10 for illustrations of calibrated test instruments for testing sensitivity of smoke detectors.

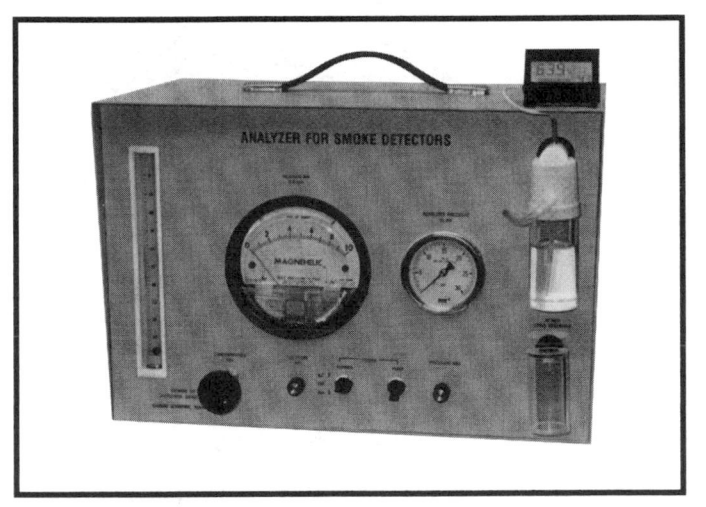

Exhibit 7.9 *Calibrated test instrument. (Source: Gemini Scientific, Sunnyvale, CA)*

Detectors found to have a sensitivity outside the listed and marked sensitivity range shall be cleaned and recalibrated or be replaced.

Exception 1: Detectors listed as field adjustable shall be permitted to be either adjusted within the listed and marked sensitivity range and cleaned and recalibrated, or they shall be replaced.

Exception 2: This requirement shall not apply to single station detectors referenced in 7-3.3 and Table 7-2.2.

The detector sensitivity shall not be tested or measured using any device that administers an unmeasured concentration of smoke or other aerosol into the detector.

Formal Interpretation 87-1 further clarifies the requirements of this section.

Formal Interpretation 87-1

Reference: 7-2.2

Background: There is a large population of installed smoke detectors which were manufactured in compliance with testing laboratory standards which are no longer used. It is not possible to test the listed sensitivity range of some of these detectors.

Question 1: Is it the Code's intent that these detectors need not be tested for sensitivity?

Answer: No.

Question 2: It is the Code's intent that these detectors be replaced with new detectors whose sensitivity can be tested?

Answer: No.

Issue Edition: 1987 of NFPA 72E

Reference: 8-2.4.2

Issue Date: July 11, 1989

Effective Date: July 31, 1989

Older detectors that were manufactured prior to current standards did not require a sensitivity range to be marked on the product. These detectors should have a sensitivity of between 0.5 and 4 percent per foot obscuration (light gray smoke). Sensitivities less than 0.5 percent obscuration per foot may lead to unwanted alarms, and sensitivities over 4 percent per foot may cause delays in, or failure of, detection.

The requirements of 7-3.2.1 give the possible options for testing the sensitivity of smoke detectors. It is important to note that each of the five options provides a measured means of ensuring sensitivity. Furthermore, after two successful tests in which sensitivity has remained stable, sensitivity testing extends to 5-year intervals, recognizing the apparent stability of the environment in which the detector is installed, as well as the apparent stability of the detector itself. When the frequency of sensitivity testing is extended, the required records of detector operation help to warn of possible changes in the environment or changes in the stability of the detector. Any such changes may warrant more frequent testing.

Detectors that are outside their marked sensitivity range must be recalibrated and retested or replaced. Removal tools can assist maintenance personnel in removal of smoke detectors. See Exhibit 7.11 for an example of a removal tool.

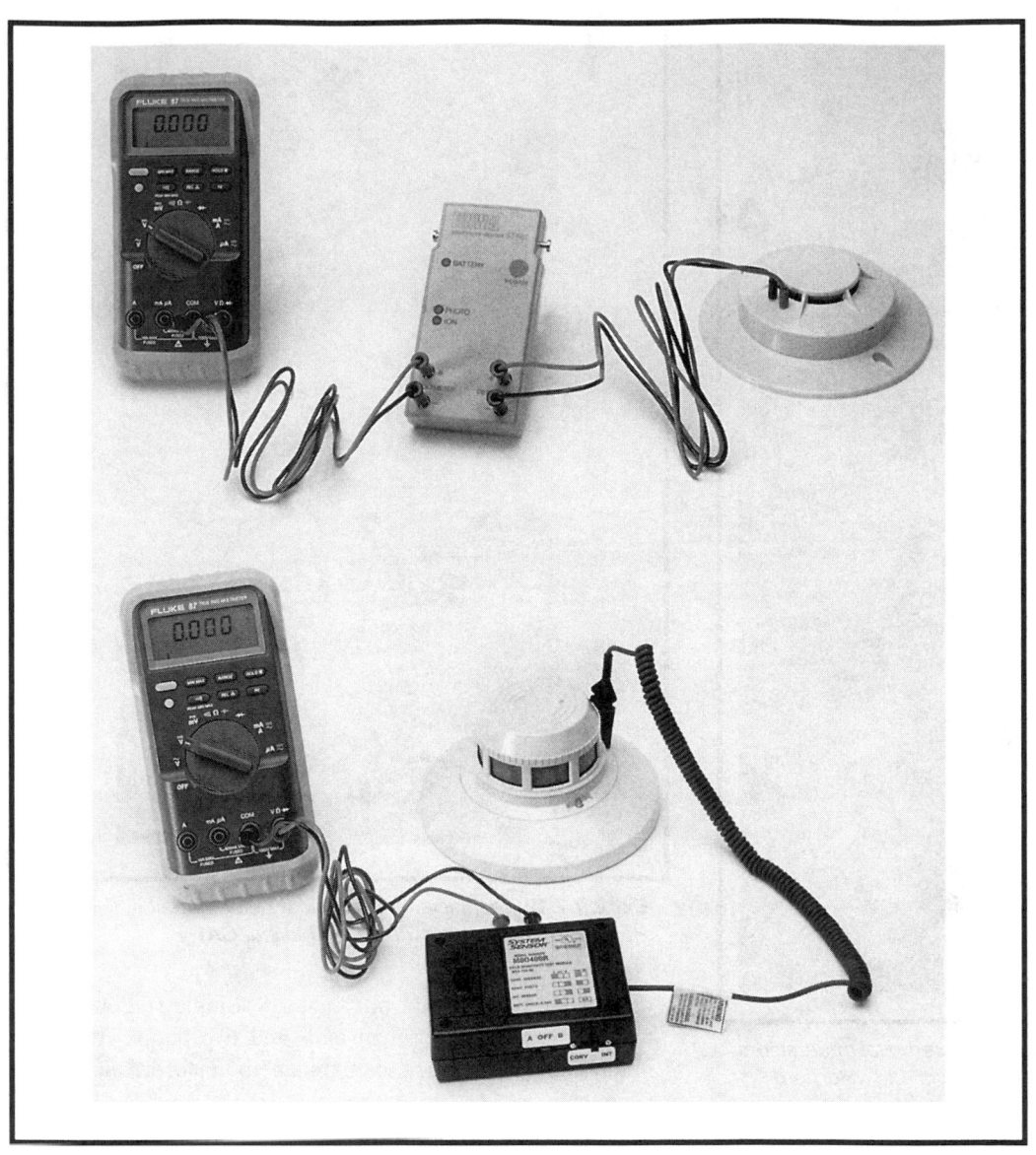

Exhibit 7.10 *Manufacturer's calibrated test instruments. (Source: Mammoth Fire Alarms, Inc., Lowell, MA)*

A-7-3.2.1 Detectors that cause unwanted alarms should be tested at their lower listed range (or at 0.5 percent obscuration if unmarked or unknown). Detectors that activate at less than this level should be replaced.

7-3.2.2 Test frequency of interfaced equipment shall be the same as specified by the applicable NFPA standards for the equipment being supervised.

For example, the test frequency for a carbon dioxide special hazard fire extinguishing system is specified in NFPA 12, *Standard on Carbon Dioxide Extinguishing Systems.*

7-3.2.3 For restorable fixed-temperature, spot-type heat detectors, two or more detectors shall be tested on each initiating circuit annually. Different detectors shall be tested each year, with records kept by the building owner specifying which detectors have been tested. Within 5 years, each detector shall have been tested.

The requirements of paragraph 7-3.2.3 represent a reduction in the testing frequency requirements from previous editions of the code. It is imperative that accurate records are kept so that the same detectors are not tested each year. Paragraph

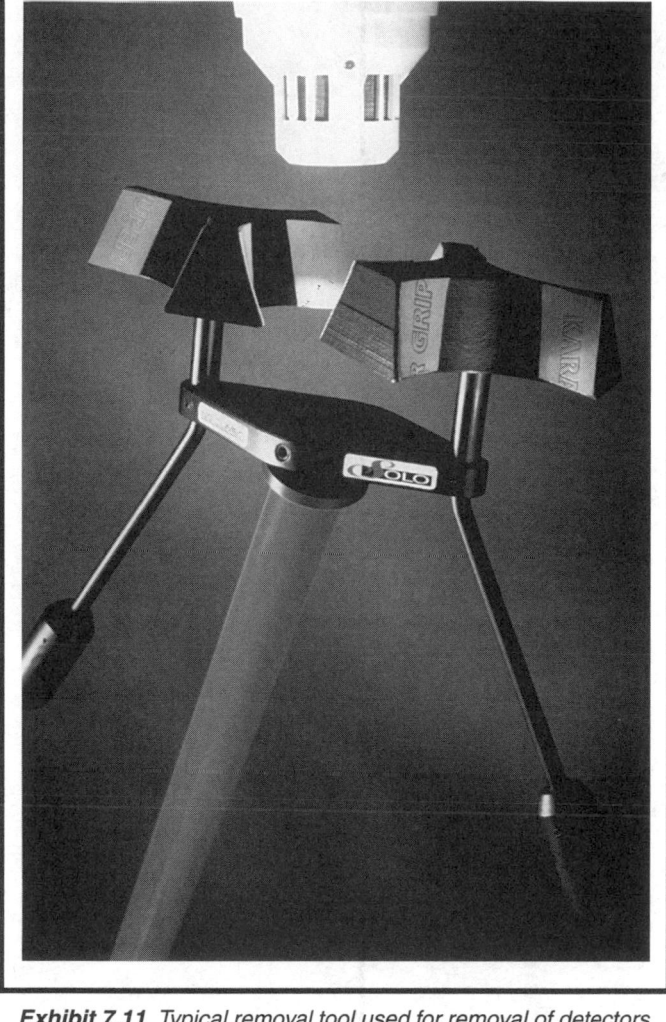

Exhibit 7.11 *Typical removal tool used for removal of detectors on high ceilings. (Source: No-Climb Products, Ltd., Hertfordshire, UK)*

Exhibit 7.12 *Technician using a heat detector tester. (Source: Home Safeguard Industries, Inc., Malibu, CA)*

7-3.2.3 only applies to restorable fixed-temperature type detectors. All other heat detectors still require annual tests. See Exhibit 7.12.

7-3.2.4* For testing addressable and analog-described devices, which are affixed to either a single, molded assembly or a twist-lock type affixed to a base, testing shall be conducted using the signaling style circuits (Styles 0.5 through 7).

Analog-type detectors shall be tested with the same criteria.

A-7-3.2.4 The addressable term was determined by the Technical Committee in Formal Interpretation 79-8 on NFPA 72D and Formal Interpretation 87-1 on NFPA 72A.

7-3.3 Single station smoke detectors installed in one- and two-family dwelling units shall be inspected, tested, and maintained as specified in Chapter 8. Single station detectors installed in other than one- and two-family dwelling units shall be tested and maintained in accordance with Chapter 7.

7-3.4 Test of all circuits extending from the central station shall be made at intervals of not more than 24 hours.

Operators at the central station initiate these tests to verify that all circuits are operational.

7-3.5 Public Fire Reporting Systems.

7-3.5.1 Emergency power sources other than batteries shall be operated to supply the system for a continuous period of 1 hour at least weekly. This test shall require simulated failure of the normal power source.

7-3.5.2 Testing facilities shall be installed at the communications center and each subsidiary communications center, if used.

Exception: If satisfactory to the authority having jurisdiction, those facilities for systems leased from a nonmunicipal organization that might be located elsewhere.

7-4 Maintenance

See Section 1-4 for definition of Maintenance.

7-4.1 Fire alarm system equipment shall be maintained in accordance with the manufacturer's instructions. The frequency of maintenance shall depend on the type of equipment and the local ambient conditions.

7-4.2 The frequency of cleaning shall depend on the type of equipment and the local ambient conditions.

Exhibit 7.13 shows an example of a product used to clean smoke detectors.

Exhibit 7.13 *A compressed air smoke detector cleaner. (Source: Home Safeguard Industries, Inc., Malibu, CA)*

Examples of areas subject to accumulations of dust and dirt are elevator hoistways and machine rooms, HVAC ducts, and boiler rooms.

7-4.3 All apparatus requiring rewinding or resetting to maintain normal operation shall be rewound or reset as promptly as possible after each test and alarm. All test signals received shall be recorded to indicate date, time, and type.

7-4.4 The retransmission means as defined in Section 5-2 shall be tested at intervals of not more than 12 hours. The retransmission signal and the time and date of the retransmission shall be recorded in the central station.

Exception: If the retransmission means is the public switched telephone network, it shall be permitted to be tested weekly to confirm its operation to each public fire service communications center.

Subsections 7-4.1 and 7-4.2 require that periodic maintenance be performed. The emphasis is on cleaning, which should be done in strict accordance with the manufacturer's instructions and as frequently as the ambient conditions of the placement area necessitate. Subsection 7-4.4 of the code defines the testing frequency of retransmission means between the supervising station and the public fire service communications center.

7-5 Records

7-5.1* Permanent Records.

After successful completion of acceptance tests approved by the authority having jurisdiction, a set of reproducible as-built installation drawings, operation and maintenance manuals, and a written sequence of operation shall be provided to the building owner or the owner's designated representative. The owner shall be responsible for maintaining these records for the life of the system for examination by any authority having jurisdiction. Paper or electronic media shall be permitted.

A historic record of the system installation that includes the information required by 7-5.1 gives the technician valuable assistance in promptly diagnosing and repairing system faults.

A-7-5.1 For final determination of record retention, refer to 7-3.2.1 for sensitivity options.

7-5.2 Maintenance, Inspection, and Testing Records.

7-5.2.1 Records shall be retained until the next test and for 1 year thereafter.

7-5.2.2 A permanent record of all inspections, testing, and maintenance shall be provided that includes the following information regarding tests and all the applicable information requested in Figure 7-5.2.2.

(1) Date
(2) Test frequency
(3) Name of property
(4) Address

INSPECTION AND TESTING FORM

DATE: _____

TIME: _____

SERVICE ORGANIZATION

Name: _____

Address: _____

Representative: _____

License No.: _____

Telephone: _____

PROPERTY NAME (USER)

Name: _____

Address: _____

Owner Contact: _____

Telephone: _____

MONITORING ENTITY

Contact: _____

Telephone: _____

Monitoring Account Ref. No.: _____

APPROVING AGENCY

Contact: _____

Telephone: _____

TYPE TRANSMISSION
- ❏ McCulloh
- ❏ Multiplex
- ❏ Digital
- ❏ Reverse Priority
- ❏ RF
- ❏ Other (Specify) _____

_____

SERVICE
- ❏ Weekly
- ❏ Monthly
- ❏ Quarterly
- ❏ Semiannually
- ❏ Annually
- ❏ Other (Specify) _____

_____

Panel Manufacturer: _____   Model No.: _____

Circuit Styles: _____

Number of Circuits: _____

Software Rev.: _____

Last Date System Had Any Service Performed: _____

Last Date that Any Software or Configuration Was Revised: _____

ALARM-INITIATING DEVICES AND CIRCUIT INFORMATION

Quantity	Circuit Style	
_____	_____	Manual Stations
_____	_____	Ion Detectors
_____	_____	Photo Detectors
_____	_____	Duct Detectors
_____	_____	Heat Detectors
_____	_____	Waterflow Switches
_____	_____	Supervisory Switches
_____	_____	Other (Specify): _____

(NFPA Inspection and Testing 1 of 4)

Figure 7-5.2.2 Example of an inspection and testing form.

ALARM NOTIFICATION APPLIANCES AND CIRCUIT INFORMATION

Quantity	Circuit Style	
_____	_____	Bells
_____	_____	Horns
_____	_____	Chimes
_____	_____	Strobes
_____	_____	Speakers
_____	_____	Other (Specify): _____

No. of alarm indicating circuits: _____

Are circuits supervised?   ❑ Yes   ❑ No

SUPERVISORY SIGNAL-INITIATING DEVICES AND CIRCUIT INFORMATION

Quantity	Circuit Style	
_____	_____	Building Temp.
_____	_____	Site Water Temp.
_____	_____	Site Water Level
_____	_____	Fire Pump Power
_____	_____	Fire Pump Running
_____	_____	Fire Pump Auto Position
_____	_____	Fire Pump or Pump Controller Trouble
_____	_____	Fire Pump Running
_____	_____	Generator In Auto Position
_____	_____	Generator or Controller Trouble
_____	_____	Switch Transfer
_____	_____	Generator Engine Running
_____	_____	Other: _____
		_____

SIGNALING LINE CIRCUITS

Quantity and style (See NFPA 72, Table 3-6) of signaling line circuits connected to system:

   Quantity _____   Style(s) _____

SYSTEM POWER SUPPLIES

   a. Primary (Main):   Nominal Voltage _____ , Amps _____

   Overcurrent Protection: Type _____ , Amps _____

   Location (Panel Number): _____

   Disconnecting Means Location: _____

   b. Secondary (Standby):

   _____ Storage Battery: Amp-Hr. Rating _____

   Calculated capacity to operate system, in hours: _____ 24 _____ 60 _____

   _____ Engine-driven generator dedicated to fire alarm system:

   Location of fuel storage: _____

TYPE BATTERY

   ❑ Dry Cell
   ❑ Nickel-Cadmium
   ❑ Sealed Lead-Acid
   ❑ Lead-Acid
   ❑ Other (Specify): _____

   c. Emergency or standby system used as a backup to primary power supply, instead of using a secondary power supply:

   _____ Emergency system described in NFPA 70, Article 700

   _____ Legally required standby described in NFPA 70, Article 701

   _____ Optional standby system described in NFPA 70, Article 702, which also meets the performance requirements of Article 700 or 701.

(NFPA Inspection and Testing 2 of 1)

Figure 7-5.2.2 Continued.

(continues)

PRIOR TO ANY TESTING

NOTIFICATIONS ARE MADE	Yes	No	Who	Time
Monitoring Entity	❏	❏	_____	_____
Building Occupants	❏	❏	_____	_____
Building Management	❏	❏	_____	_____
Other (Specify)	❏	❏	_____	_____
AHJ (Notified) of Any Impairments	❏	❏	_____	_____

SYSTEM TESTS AND INSPECTIONS

TYPE	Visual	Functional	Comments
Control Panel	❏	❏	_____
Interface Eq.	❏	❏	_____
Lamps/LEDS	❏	❏	_____
Fuses	❏	❏	_____
Primary Power Supply	❏	❏	_____
Trouble Signals	❏	❏	_____
Disconnect Switches	❏	❏	_____
Ground-Fault Monitoring	❏	❏	_____

SECONDARY POWER

TYPE	Visual	Functional	Comments
Battery Condition	❏		_____
Load Voltage		❏	_____
Discharge Test		❏	_____
Charger Test		❏	_____
Specific Gravity		❏	_____
TRANSIENT SUPPRESSORS	❏		_____
REMOTE ANNUNCIATORS	❏	❏	_____

NOTIFICATION APPLIANCES

	Visual	Functional	Comments
Audible	❏	❏	_____
Visual	❏	❏	_____
Speakers	❏	❏	_____
Voice Clarity		❏	_____

INITIATING AND SUPERVISORY DEVICE TESTS AND INSPECTIONS

Loc. & S/N	Device Type	Visual Check	Functional Test	Factory Setting	Meas. Setting	Pass	Fail
_____	_____	❏	❏	_____	_____	❏	❏
_____	_____	❏	❏	_____	_____	❏	❏
_____	_____	❏	❏	_____	_____	❏	❏
_____	_____	❏	❏	_____	_____	❏	❏
_____	_____	❏	❏	_____	_____	❏	❏
_____	_____	❏	❏	_____	_____	❏	

Comments: _____

_____

_____

_____

(NFPA Inspection and Testing 3 of 4)

Figure 7-5.2.2 Continued.

EMERGENCY COMMUNICATIONS EQUIPMENT	Visual	Functional	Comments
Phone Set	❑	❑	_____
Phone Jacks	❑	❑	_____
Off-Hook Indicator	❑	❑	_____
Amplifier(s)	❑	❑	_____
Tone Generator(s)	❑	❑	_____
Call-in Signal	❑	❑	_____
System Performance	❑	❑	_____

	Visual	Device Operation	Simulated Operation
INTERFACE EQUIPMENT			
(Specify) _____	❑	❑	❑
(Specify) _____	❑	❑	❑
(Specify) _____	❑	❑	❑
SPECIAL HAZARD SYSTEMS			
(Specify) _____	❑	❑	❑
(Specify) _____	❑	❑	❑
(Specify) _____	❑	❑	❑

Special Procedures: _____
_____
_____
_____

Comments: _____
_____
_____
_____

ON/OFF PREMISES MONITORING	Yes	No	Time	Comments
Alarm Signal	❑	❑	_____	_____
Alarm Restoral	❑	❑	_____	_____
Trouble Signal	❑	❑	_____	_____
Supervisory Signal	❑	❑	_____	_____
Supervisory Restoral	❑	❑	_____	_____

NOTIFICATIONS THAT TESTING IS COMPLETE	Yes	No	Who	Time
Building Management	❑	❑	_____	_____
Monitoring Agency	❑	❑	_____	_____
Building Occupants	❑	❑	_____	_____
Other (Specify)	❑	❑	_____	_____

The following did not operate correctly: _____
_____
_____

System restored to normal operation: Date: _____   Time: _____

THIS TESTING WAS PERFORMED IN ACCORDANCE WITH APPLICABLE NFPA STANDARDS.

Name of Inspector: _____   Date: _____   Time: _____

Signature: _____

Name of Owner or Representative: _____

Date: _____   Time: _____

Signature: _____

(NFPA Inspection and Testing 1 of 4)

Figure 7-5.2.2 *Continued.*

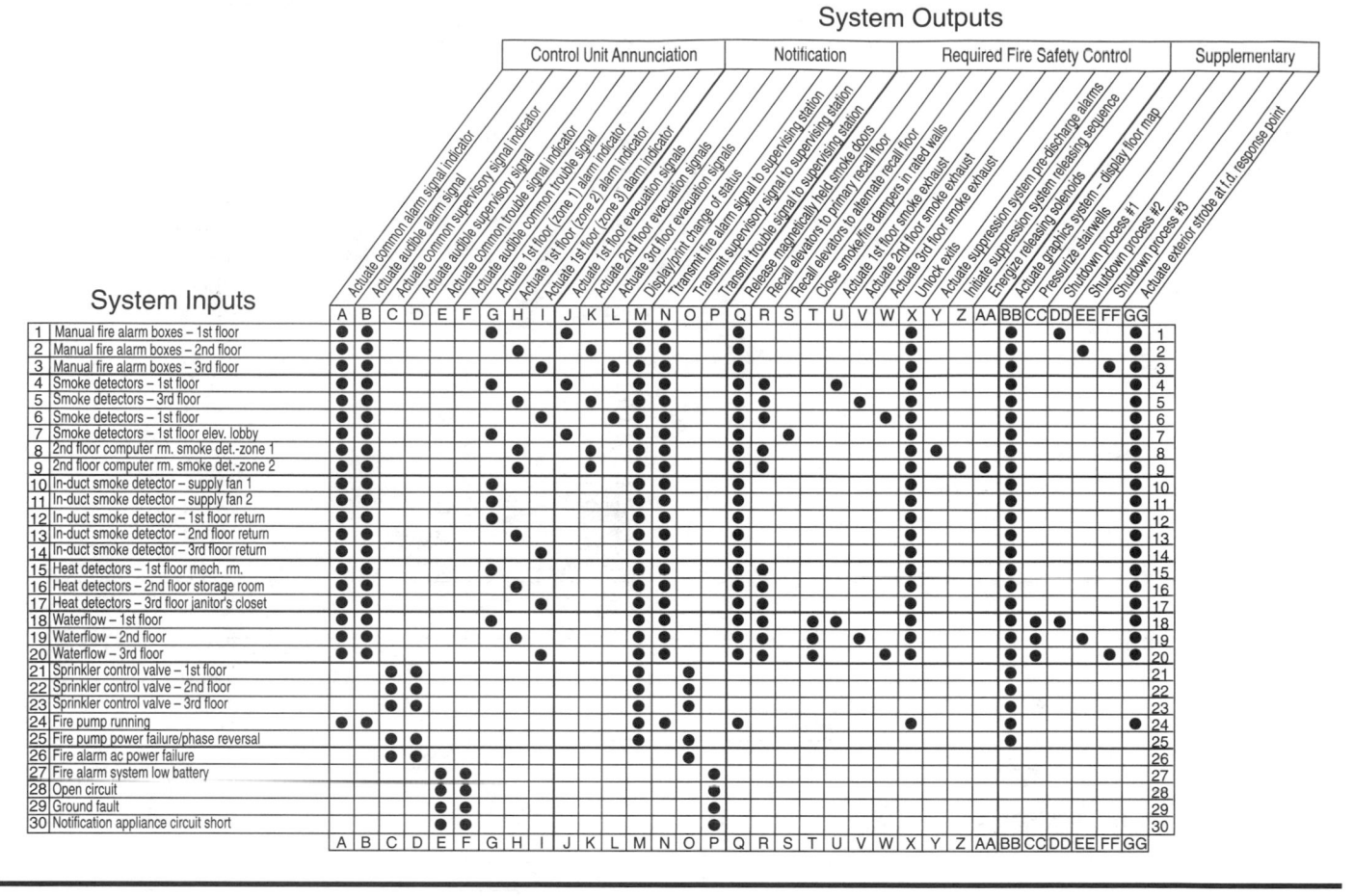

Figure A-7-5-2.2(9) *Typical input/output matrix.*

(5) Name of person performing inspection, maintenance, tests, or combination thereof, and affiliation, business address, and telephone number

(6) Name, address, and representative of approving agency(ies)

(7) Designation of the detector(s) tested, for example, "Tests performed in accordance with Section _____."

(8) Functional test of detectors

(9)*Functional test of required sequence of operations

A-7-5.2.2(9) One method used to define the required sequence of operations and to document the actual sequence of operations is an input/output matrix. *[Refer to Figure A-7-5.2.2(9).]*

(10) Check of all smoke detectors

(11) Loop resistance for all fixed-temperature, line-type heat detectors

(12) Other tests as required by equipment manufacturers

(13) Other tests as required by the authority having jurisdiction

(14) Signatures of tester and approved authority representative

(15) Disposition of problems identified during test (for example, owner notified, problem corrected/successfully retested, device abandoned in place

7-5.3 For supervising station fire alarm systems, records pertaining to signals received at the supervising station that result from maintenance, inspection, and testing, shall be maintained for not less than 12 months. Upon request, a hard copy record shall be provided to the authority having jurisdiction. Paper or electronic media shall be permitted.

7-5.4 If the operation of a device, circuit, control panel function, or special hazard system interface is simulated, it shall be noted on the certificate that the operation was simulated, and the certificate shall indicate by whom it was simulated.

For future reference in determining overall system reliability, it is important to document whether the interfaced system operation was fully tested by actual operation of the interfaced system or if its operation was simulated.

Reference Cited in Commentary

NFPA 12, *Standard on Carbon Dioxide Extinguishing Systems*, National Fire Protection Association, Quincy, MA, 1998.

CHAPTER 8

Fire Warning Equipment for Dwelling Units

Chapter 8 deals solely with fire detection and warning equipment and fire alarm systems used in dwelling units. (See definition of Dwelling Unit in Section 1-4.) Chapter 8 was relocated from Chapter 2 of the 1996 edition of the code.

Chapter 8 was completely revised for the 1999 code to include performance-based design criteria. However, Chapter 8 still contains prescriptive requirements that are similar to past editions of the code. The prescriptive requirements provide alternative methods to the performance-based design criteria.

A performance-based approach requires the designer to consider the conditions of the occupant(s), the time necessary for escape, the reliability of the alarm equipment used, suppression systems used, and other mitigating factors. See Supplement 1.

8-1* Primary Function

Fire warning equipment for dwelling units shall provide a reliable means to notify the occupants of a dwelling unit of the presence of a threatening fire and the need to escape to a place of safety before such escape might be impeded by untenable conditions in the normal path of egress.

A-8-1 *Household Fire Warning Protection.*

(a) *Fire Danger in the Home.* Fire is the third leading cause of accidental death. Residential occupancies account for most fire fatalities, and most of these deaths occur at night during the sleeping hours.

Most fire injuries also occur in the home. It is estimated that 1.5 million Americans are injured by fire each year. Many never resume normal lives.

It is estimated that each household will experience three (usually unreported) fires per decade and two fires serious enough to report to a fire department per lifetime.

(b) *Fire Safety in the Home.* NFPA 72 is intended to provide reasonable fire safety for persons in family living units. Reasonable fire safety can be produced through the following three-point program:

(1) Minimizing fire hazards
(2) Providing a fire warning system
(3) Having and practicing an escape plan

(c) *Minimizing Fire Hazards.* This code cannot protect all persons at all times. For instance, the application of this code might not provide protection against the following three traditional fatal fire scenarios:

(1) Smoking in bed
(2) Leaving children home alone
(3) Cleaning with flammable liquids such as gasoline

However, Chapter 8 can lead to reasonable safety from fire when the three points under A-8-1(b) are observed.

(d) *Fire Warning System.* There are two types of fire to which household fire warning equipment needs to respond. One is a rapidly developing, high-heat fire. The other is a slow, smoldering fire. Either can produce smoke and toxic gases.

(e) Household fires are especially dangerous at night when the occupants are asleep. Fires produce smoke and deadly gases that can overcome occupants while they are asleep. Furthermore, dense smoke reduces visibility. Most fire casualties are victims of smoke and gas inhalation rather than burns. To warn against a fire, Chapter 8 requires smoke detectors in accordance with 8-1.4.1 and recommends heat or smoke detectors in all other major areas. *[Refer to A-8-1.2.1(c).]*

(f) *Family Escape Plan.* There often is very little time between the detection of a fire and the time it becomes deadly. This interval can be as little as 1 or 2 minutes. Thus, this code requires detection means to give a family some advance warning of the development of conditions that

become dangerous to life within a short period of time. Such warning, however, could be wasted unless the family has planned in advance for rapid exit from their residence. Therefore, in addition to the fire warning system, this code requires exit plan information to be furnished.

Planning and practicing for fire conditions with a focus on rapid exit from the residence are important. Drills should be held so that all family members know the action to be taken. Each person should plan for the possibility that exit out of a bedroom window could be necessary. An exit out of the residence without the need to open a bedroom is essential.

(g) *Special Provisions for the Disabled.* For special circumstances where the life safety of an occupant(s) depends on prompt rescue by others, the fire warning system should include means of prompt automatic notification to those who are to be depended on for rescue.

8-1.1* Limitations.

A-8-1.1 Chapter 8 does not attempt to cover all equipment, methods, and requirements that might be necessary or advantageous for the protection of lives and property from fire.

NFPA 72 is a "minimum code." It provides a number of requirements related to household fire-warning equipment that are deemed to be the practical and necessary minimum for average conditions at the present state of the art.

Family living units lead all other types of occupancies as the site of fire-related deaths in the United States. Smoke alarms required by the code and other means of detection recommended in Chapter 8 provide warning, but do not extinguish the fire. It is the responsibility of the occupants to follow their emergency exit plan when the alarm signal sounds. Home fire statistics, recommendations for fire safety and life safety, fire warning system capabilities, a more detailed explanation of an escape plan, and special provisions for people with disabilities are described in Section A-8-1 of the code.

Additional safety equipment such as residential fire sprinklers can provide additional escape time and may limit damage to the premises.

8-1.1.1 Life safety from fire in residential occupancies shall be based primarily on early notification to occupants of the need to escape, followed by the appropriate egress actions by those occupants. Fire warning systems for dwelling units are capable of protecting about half of the occupants in potentially fatal fires. Victims are often intimate with the fire, too old or too young, or physically or mentally impaired such that they cannot escape even when warned early enough that escape should be possible. For these people, other strategies such as protection-in-place or assisted escape or rescue shall be necessary.

8-1.1.2* The performance of fire warning equipment for dwelling units discussed in Chapter 8 shall depend on such equipment being properly selected, installed, operated, tested, and maintained in accordance with the provisions of this code and with the manufacturer's instructions provided with the equipment.

A-8-1.1.2 Good fire protection requires that the equipment periodically be maintained. If the householder is unable to perform the required maintenance, a maintenance agreement should be considered.

Proper installation, maintenance, and testing of smoke alarms and detectors are essential to proper operation.

8-1.2 Performance Criteria.

8-1.2.1* Sufficient initiating devices shall be installed within the dwelling unit so that their operation provides adequate egress time before the occurrence of untenable conditions at any point along the normal path of egress for all design fire scenarios specified in applicable codes and any supplementary fire scenarios specified by the authority having jurisdiction.

A-8-1.2.1 One of the most critical factors of any fire alarm system is the location of the fire-detecting devices. This appendix is not a technical study. It is an attempt to provide some fundamentals on detector location. For simplicity, only those types of detectors recognized by Chapter 8 (that is, smoke and heat detectors) are discussed. In addition, special problems requiring engineering judgment, such as locations in attics and in rooms with high ceilings, are not covered.

Smoke detectors and smoke alarms must be properly located to avoid nuisance alarms. Dust, grease, dirt, insects, and other contaminants can make smoke alarms and smoke detectors more sensitive to stimuli such as cooking vapors and bathroom moisture. Higher sensitivities frequently cause nuisance alarms and subsequent disconnection of the device, leaving the occupants without adequate protection. See 8-1.4.2 for additional information on required locations of smoke alarms and smoke detectors.

(a) *Where to Locate the Required Smoke Detectors in Existing Construction.* The major threat from fire in a family living unit occurs at night when everyone is asleep. The principal threat to persons in sleeping areas comes from fires in the remainder of the unit. Therefore, a smoke detector(s) is best located between the bedroom areas and the rest of the unit. In units with only one bedroom area on one floor, the smoke detector(s) should be located as shown in Figure A-8-1.2.1(a).

In family living units with more than one bedroom area or with bedrooms on more than one floor, more than one smoke detector is required, as shown in Figure A-8-1.2.1(b).

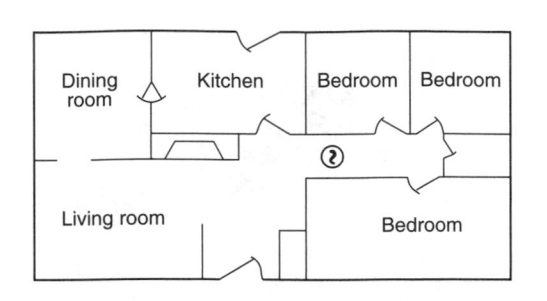

Figure A-8-1.2.1(a) *A smoke detector should be located between the sleeping area and the rest of the family living unit.*

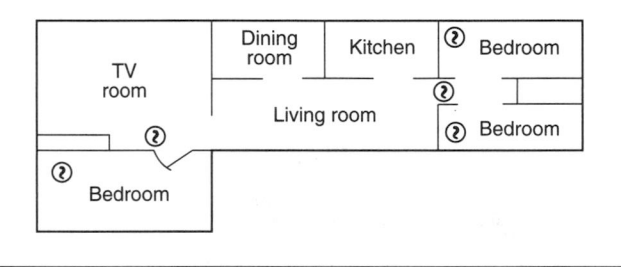

Figure A-8-1.2.1(b) *In family living units with more than one sleeping area, a smoke detector should be provided to protect each sleeping area in addition to detectors required in bedrooms.*

In addition to smoke detectors outside of the sleeping areas, Chapter 8 requires the installation of a smoke detector on each additional story of the family living unit, including the basement. These installations are shown in Figure A-8-1.2.1(c). The living area smoke detector should be installed in the living room or near the stairway to the upper level, or in both locations. The basement smoke detector should be installed in close proximity to the stairway leading to the floor above. Where installed on an open-joisted ceiling, the detector should be placed on the bottom of the joists. The detector should be positioned relative to the stairway so as to intercept smoke coming from a fire in the basement before the smoke enters the stairway.

(b) *Where to Locate the Required Smoke Detectors in New Construction.* All of the smoke detectors specified in A-8-1.2.1(a) for existing construction are required, and, in addition, a smoke detector is required in each bedroom.

(c) *Are More Smoke Detectors Desirable?* The required number of smoke detectors might not provide reliable early warning protection for those areas separated by a door from the areas protected by the required smoke detectors. For this reason, it is recommended that the householder

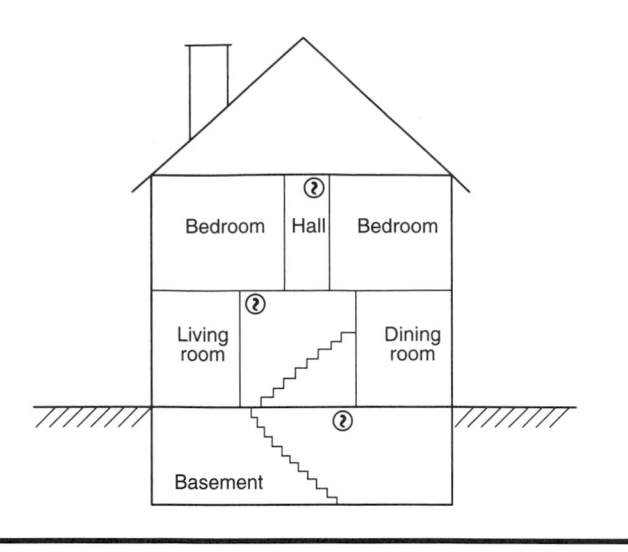

Figure A-8-1.2.1(c) *A smoke detector should be located on each story.*

consider the use of additional smoke detectors for those areas for increased protection. The additional areas include the basement, bedrooms, dining room, furnace room, utility room, and hallways not protected by the required smoke detectors. The installation of smoke detectors in kitchens, attics (finished or unfinished), or garages is not normally recommended, as these locations occasionally experience conditions that can results in improper operation.

8-1.2.2* Fire warning equipment for dwelling units shall provide a sound that is audible in all occupiable dwelling areas. Audible fire alarm signals shall meet the performance requirements of 4-3.4. If the dwelling unit is occupied by people with hearing deficiencies, visible appliances shall be provided in all dwelling areas to meet the requirements of Section 4-4.

Because many more residential fire deaths occur as a result of smoke inhalation rather than burns, the National Fire Protection Association recommends sleeping with bedroom doors closed to limit the spread of smoke in the event of a fire at night. To ensure that smoke alarms or notification appliances can be heard by occupants inside closed sleeping areas, detectors must be tested with bedroom doors shut and in conditions as similar as possible to those conditions normally occurring in the home at night. For example, if air conditioning units or humidifiers are routinely used in a home, those appliances should be operating when the detectors are tested. Chapter 4 requires a minimum of 15 dBA above the average ambient sound pressure level (SPL) or a minimum of 5 dBA above the maximum sound pressure level that lasts 1 minute or longer, or 70 dBA in the bedroom at pillow level, whichever is greater. These measurements

must be made using a sound pressure level meter as described in Chapter 7.

Paragraph 8-1.2.2 of the code also addresses the needs of hearing impaired individuals who cannot respond to audible alarms alone. Visible appliances enhance the ability of the system or equipment to alert hearing impaired persons of a fire condition. However, Chapter 4 of the code permits the use of a visible signal of lower intensity for non-sleeping areas. See Section 4-4 for requirements to visible notification appliances.

Additional notification appliances connected to, and powered through, the dry contacts of a single-station or multiple-station smoke alarm relay may also be used to comply with this requirement. See Exhibits 8.1 and 8.2 for examples of equipment used for notification.

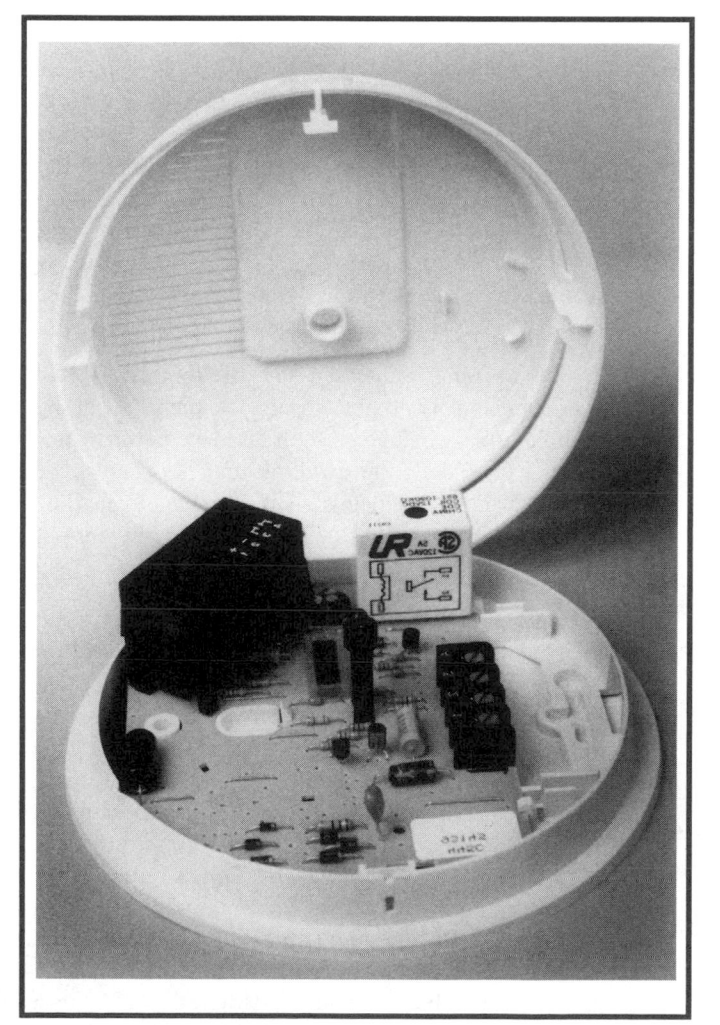

Exhibit 8.1 *Single/multiple-station smoke alarm with relay contacts for remote notification appliance. (Source: Mammoth Fire Alarms, Inc., Lowell MA)*

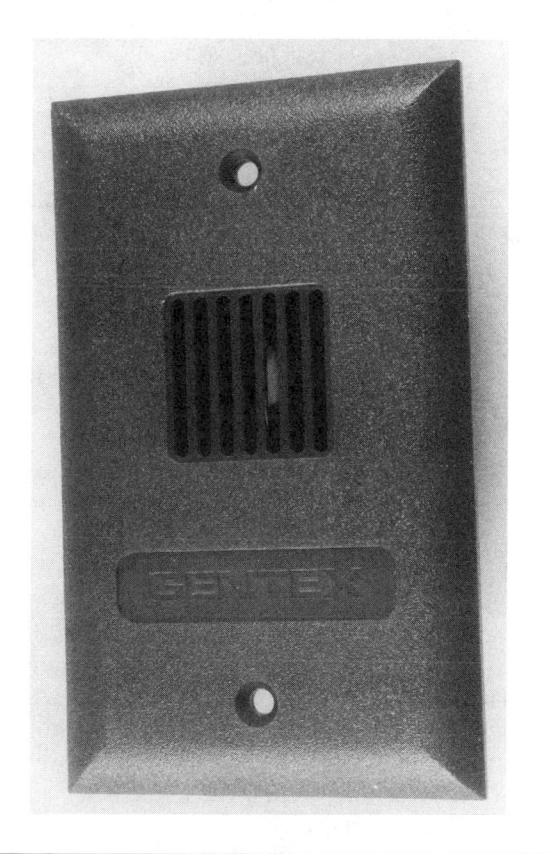

Exhibit 8.2 *Typical notification appliances that could be used in a household system. Top: Bell. (Source: Wheelock, Long Branch, NJ) Bottom: Mini-horn. (Source: Gentex Corp., Zeeland, MI)*

There are federal laws and regulations that affect the placement and intensity of visible signals. The intensity levels required by 8-1.2.2 were established through testing conducted by Underwriters Laboratories Inc. These levels have

been accepted by the Technical Committee on Fire Warning Equipment for Dwelling Units and the engineering community as sufficient for use in dwelling fire alarm systems.

A 110-cd visible notification appliance, installed at least 24 in. (610 mm) below the ceiling, generally provides sufficient light intensity to awaken a sleeping person. A 177-cd appliance is required when mounted within 24 in. (610 mm) of the ceiling because during a fire the light signal may be attenuated by the smoke layer. See Exhibits 8.3 and 8.4 for examples of smoke detectors with integral and remote notification appliances.

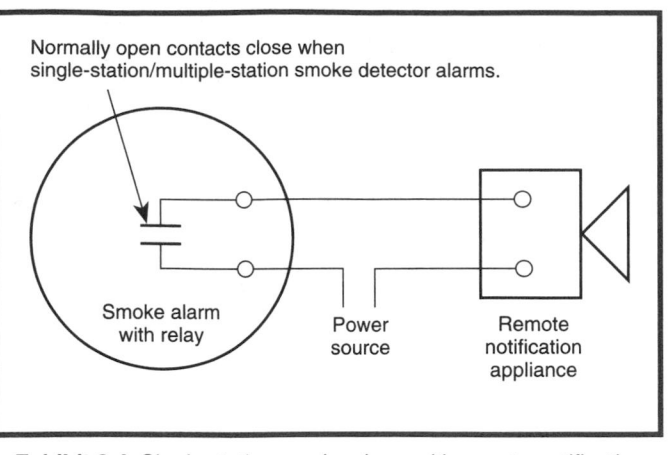

Exhibit 8.4 *Single station smoke alarm with remote notification appliance. (Source: Hughes Associates, Inc., Warwick, RI)*

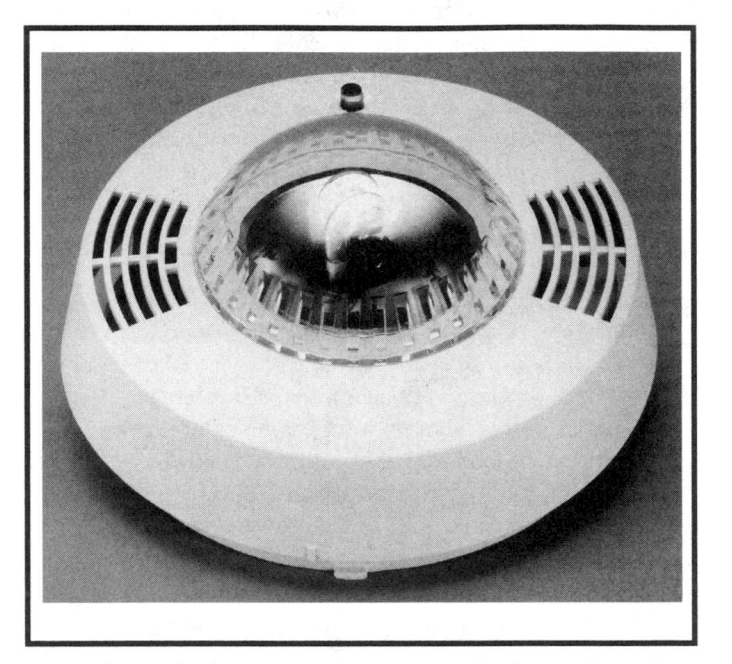

Exhibit 8.3 *Smoke alarm with integral notification appliance for the hearing impaired. (Source: BRK/First Alert, Aurora, IL)*

A-8-1.2.2 At times, depending on conditions, the audibility of detection devices could be seriously impaired when occupants are within the bedroom area. For instance, there might be a noisy window air conditioner or room humidifier generating an ambient noise level of 55 dBA or higher. The detection device alarms need to penetrate through the closed doors and be heard over the bedroom's noise levels with sufficient intensity to awaken sleeping occupants therein. Test data indicate that detection devices that have sound pressure ratings of 85 dBA of 10 ft (3 m) and that are installed outside the bedrooms can produce about 15 dBA over ambient noise levels of 55 dBA in the bedrooms. This is likely to be sufficient to awaken the average sleeping person.

Detectors located remote from the bedroom area might not be loud enough to awaken the average person. In such cases, it is recommended that detectors be interconnected in

such a way that the operation of the remote detector causes an alarm of sufficient intensity to penetrate the bedrooms. The interconnection can be accomplished by the installation of a fire detection system, by the wiring together of multiple station alarm devices, or by the use of line carrier or radio frequency transmitters/receivers.

8-1.2.3* Newly installed fire warning equipment for dwelling units (including self-contained devices) shall produce the audible emergency signal described in ANSI S3.41, *Audible Emergency Evacuation Signal*, whenever the intended response is to evacuate the building. The same audible signal shall be permitted to be used for other devices as long as the desired response is immediate evacuation.

Exception: Fire warning equipment that is installed in accordance with provisions of this chapter in other than dwelling units, unless required by another chapter.

A-8-1.2.3 The use of the distinctive three-pulse temporal pattern fire alarm evacuation signal required by 3-8.4.1.2 had been previously recommended by this code since 1979. It has since been adopted as both an American National Standard (ANSI S3.41, *Audible Emergency Evacuation Signal*) and an International Standard (ISO 8201, *Audible Emergency Evacuation Signal*).

Copies of both of these standards are available from the Standards Secretariat, Acoustical Society of America, 335 East 45th Street, New York, NY 10017-3483.

The standard fire alarm evacuation signal is a three-pulse temporal pattern using any appropriate sound. The pattern consists of the following in this order:

(1) An on phase lasting 0.5 second ± 10 percent.
(2) An off phase lasting 0.5 second ± 10 percent for three successive on periods.

(3) An off phase lasting 1.5 seconds ± 10 percent *[refer to Figures A-8-1.2.3(a) and (b)]*. The signal should be repeated for a period appropriate for the purposes of evacuation of the building, but for not less than 180 seconds. A single-stroke bell or chime sounded at on intervals lasting 1 second ± 10 percent, with a 2-second ± 10 percent off interval after each third on stroke, may be permitted *[refer to Figure A-8-1.2.3(c)]*.

The minimum repetition time may be permitted to be manually interrupted.

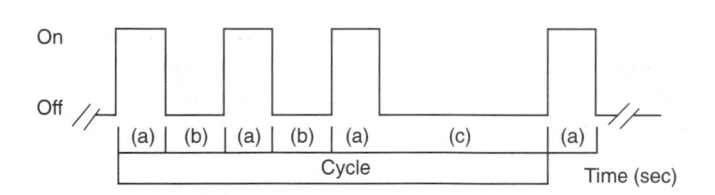

Key:
Phase (a) signal is on for 0.5 sec ± 10%
Phase (b) signal is off for 0.5 sec ± 10%
Phase (c) signal is off for 1.5 sec ± 10% [(c) = (a) + 2(b)]
Total cycle lasts for 4 sec ± 10%

Figure A-8-1.2.3(a) *Temporal pattern parameters.*

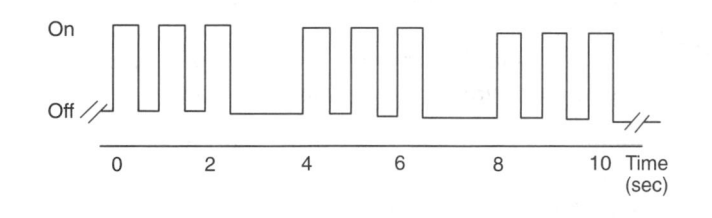

Figure A-8-1.2.3(b) *Temporal pattern imposed on signaling appliances that emit a continuous signal while energized.*

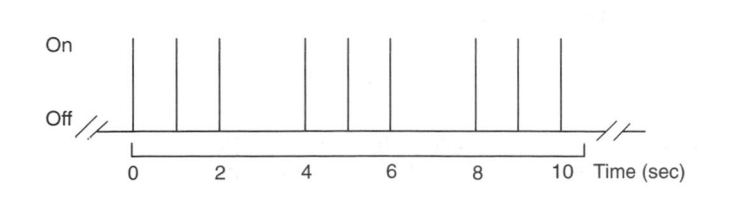

Figure A-8-1.2.3(c) *Temporal pattern imposed on a single-stroke bell or chime.*

8-1.2.4* Location of initiating devices shall consider conditions such as high or low temperature, humidity, or sources of smoke that can lead to nuisance alarms.

Smoke detectors that produce nuisance alarms should be relocated away from areas that produce excessive moisture, cooking vapors, or exhaust. Photoelectric-type smoke detection may provide some additional relief. These devices are more resistant to nuisance alarms caused by moisture in bathrooms as well as moisture and cooking vapors in kitchen areas.

To gain the full benefit of smoke detection, it is imperative that detectors be located as outlined in Chapter 8. However, repeated nuisance alarms, triggered by common household conditions, undermine the response of family members to the alarm. These nuisance alarms are most often experienced when smoke alarms are improperly placed in kitchens, bathrooms, garages, attics, and basements with dirt floors or moisture problems. In locations with conditions likely to cause frequent nuisance alarms, heat detectors or other fire detection devices less likely to be affected by false triggering conditions may offer additional protection. In any case, the requirements of the code regarding placement of smoke alarms near sleeping areas (or inside the bedrooms in new construction) should be met and these other devices installed as additional means of protection.

See Exhibits 8.5 and 8.6 for examples of alarms. Heat detectors are not considered life safety devices because they actuate after the fire has created untenable conditions in the space being protected. They should be used only to provide additional detection in areas not suitable for smoke alarms.

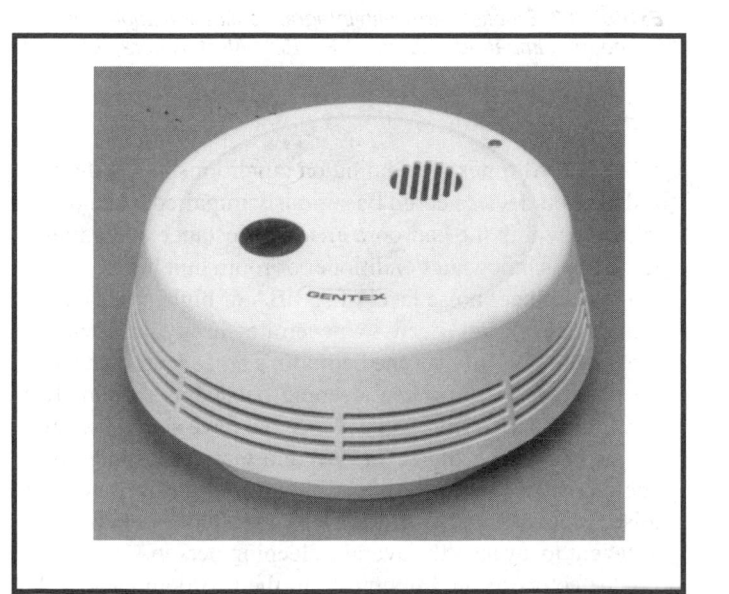

Exhibit 8.5 *Battery-operated smoke alarm. (Source: Gentex, Inc., Zeeland, MI)*

Exhibit 8.6 *Combination smoke alarm and heat alarm. (Source: Gentex, Inc., Zeeland, MI)*

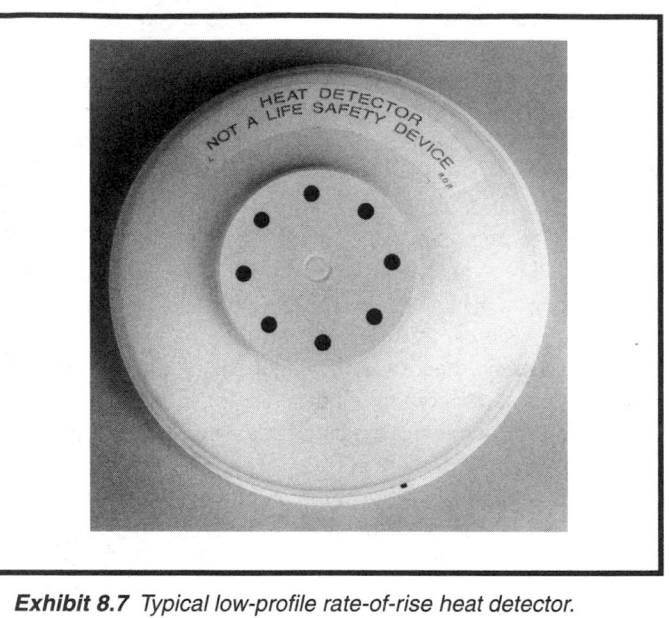

Exhibit 8.7 *Typical low-profile rate-of-rise heat detector. (Source: Mammoth Fire Alarms, Inc., Lowell, MA)*

See A-2-3.6.1.2 (a) for a list of sources of airborne contaminants that can cause nuisance alarms. Smoke alarms or detectors should not be located in garages, unfinished attics, crawl spaces, or other areas where conditions can cause nuisance alarms. These areas are best suited to heat detection.

The heat detector types that are allowed by the requirement of 8-1.2.4 are fixed-temperature, rate-of-rise detectors, or rate compensation type.

Multiple station applications where smoke alarms and heat detectors are connected together on the same circuit must use equipment that is listed as compatible.

Rate-of-rise heat detectors respond to rapid temperature increases. The designer should consider the environment in which rate-of-rise heat detectors are to be installed. Areas near dishwashers, hot air vents, and ovens are examples of areas to be avoided. See Exhibit 8.7 for an example of a rate-of-rise heat detector.

A-8-1.2.4 Location and Type of Devices.

(a) *Smoke Detector Mounting—Dead Air Space.* The smoke from a fire generally rises to the ceiling, spreads out across the ceiling surface, and begins to bank down from the ceiling. The corner where the ceiling and wall meet is an air space into which the smoke could have difficulty penetrating. In most fires, this dead air space measures about 4 in. (0.1 m) along the ceiling from the corner and about 4 in. (0.1 m) down the wall, as shown in Figure A-8-1.2.4(a). Detectors should not be placed in this dead air space.

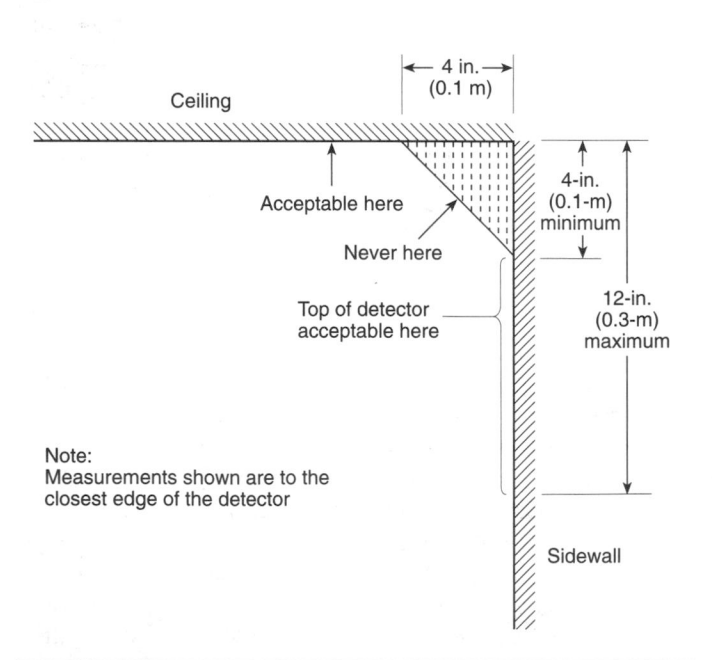

Figure A-8-1.2.4(a) *Example of proper mounting of detectors.*

Smoke and heat detectors should be installed in those locations recommended by the manufacturer, except in those cases where the space above the ceiling is open to the outside and little or no insulation is present over the ceiling. Such cases result in the ceiling being excessively cold in the winter

or excessively hot in the summer. Where the ceiling is significantly different in temperature from the air space below, smoke and heat have difficulty reaching the ceiling and a detector that is located on that ceiling. In this situation, placement of the detector on a sidewall, with the top 4 in. to 12 in. (0.1 m to 0.3 m) from the ceiling is recommended.

The situation described above for uninsulated or poorly insulated ceilings can also exist, to a lesser extent, in outside walls. The recommendation is to place the smoke detector on a sidewall. However, where the sidewall is an exterior wall with little or no insulation, an interior wall should be selected. It should be recognized that the condition of inadequately insulated ceilings and walls can exist in multiple dwelling unit housing (apartments), single dwelling unit housing, and mobile homes.

In those family living units employing radiant heating in the ceiling, the wall location is the recommended location. Radiant heating in the ceiling can create a hot-air, boundary layer along the ceiling surface, which can seriously restrict the movement of smoke and heat to a ceiling-mounted detector.

A 1992 survey of more than 1000 homes conducted by the U. S. Consumer Product Safety Commission found that 25 percent of the smoke alarms failed to operate when tested with smoke and the integral test button. Of these, 60 percent were restored to operation simply by installing a new battery or reconnecting the ac power. Most owners reported that they had disconnected power because of nuisance alarms, overwhelmingly related to cooking. Many of these problem smoke alarms were located within a few feet of the cooking facilities.

Locating smoke alarms away from sources of nuisance alarms can minimize these problems. Photoelectric-type smoke alarms are less likely to alarm from normal cooking than are the ionization type and should be considered for installation on the floor containing the cooking facilities. Some smoke alarms have a temporary silencing means that can be activated during cooking to reduce nuisance alarms. The two types of smoke alarms are equally susceptible to steam from bathrooms but locating them at least 3 ft from the bathroom door should minimize such problems. Having both types of smoke alarms in a home is a good idea from the viewpoint of response to different types of fires, but where interconnected it is important to make sure that the interconnect circuits are compatible.

(b) *Heat Detection.*

(1) *General.* While Chapter 8 does not require heat detectors as part of the basic protection scheme, it is recommended that the householder consider the use of additional heat detectors for the same reasons presented under A-8-1.2.1(c). The additional areas lending themselves to protection with heat detectors are the kitchen, dining room, attic, (finished or unfinished), furnace room, utility room, basement, and integral or attached garage. For bedrooms, the installation of a smoke detector is recommended over the installation of a heat detector for protection of the occupants from fires in their bedrooms.

(2) *Heat Detector Mounting—Dead Air Space.* Heat from a fire rises to the ceiling, spreads out across the ceiling surface, and begins to bank down from the ceiling. The corner where the ceiling and the wall meet is an air space into which heat has difficulty penetrating. In most fires, this dead air space measures about 4 in. (0.1 m) along the ceiling from the corner and 4 in. (0.1 m) down the wall as shown in Figure A-8-1.2.4(a). Heat detectors should not be placed in this dead air space.

The placement of the heat detector is critical where maximum speed of fire detection is desired. Thus, a logical location for a detector is the center of the ceiling. At this location, the detector is closest to all areas of the room.

If the heat detector cannot be located in the center of the ceiling, an off-center location on the ceiling can be used.

The next logical location for mounting heat detectors is on the sidewall. Any detector mounted on the sidewall should be located as near as possible to the ceiling. A detector mounted on the sidewall should have the top of the detector between 4 in. and 12 in. (0.1 m and 0.3 m) from the ceiling. *[See Figure A-8-1.2.4(a).]*

(3) *Spacing of Heat Detectors.* If a room is too large for protection by a single heat detector, several detectors should be used. It is important that they be properly located so all parts of the room are covered. *(For further information on the spacing of detectors refer to Chapter 2.)*

(4) *Where the Distance Between Heat Detectors Should be Further Reduced.* The distance between heat detectors is based on data obtained from the spread of heat across a smooth ceiling. If the ceiling is not smooth, the placement of the detector should be tailored to the situation.

For instance, with open wood joists, heat travels freely down the joist channels so that the maximum distance between detectors [50 ft (15 m)] can be used. However, heat has trouble spreading across the joists, so the distance in this direction should be one-half the distance allowed between detectors, as shown in Figure A-8-1.2.4(b), and the distance to the wall is reduced to $12\frac{1}{2}$ ft (3.8 m). Since one-half of 50 ft (15 m) is 25 ft (7.6 m), the distance between detectors across open wood joists should not exceed 25 ft (7.6 m), as shown in Figure A-8-1.2.4(b), and the distance to the wall is reduced [$\frac{1}{2}$ × 25 ft (7.6 m)] to 12.5 ft (3.8 m). Detectors should be mounted on the bottom of the joists and not up in joist channels.

Walls, partitions, doorways, ceiling beams, and open joists interrupt the normal flow of heat, thus creating new areas to be protected.

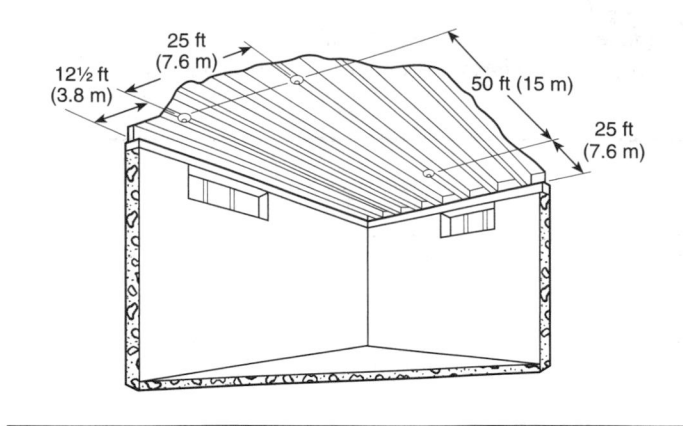

Figure A-8-1.2.4(b) Open joists, attics, and extra-high ceilings are some of the areas that require special knowledge for installation.

In addition to the special requirements for heat detectors installed on ceilings with exposed joists, reduced spacing also might be required due to other structural characteristics of the protected area, possible drafts, or other conditions that could affect detector operation.

8-1.3 Verification Methods.

The following methods of verifying compliance with the performance criteria of 8-1.2 shall be permitted. Other methods permitted by the authority having jurisdiction shall be accepted if appropriate.

The designer must provide verification of compliance with the performance-based requirements using calculations and modeling.

8-1.3.1 Calculation of detector activation times for all fire scenarios shall demonstrate that sufficient time is provided to allow for successful escape of all occupants.

Time available to escape can vary significantly, depending on the fuels present, the condition of the occupant(s), and other factors. Many residential fires occur at night when occupants are asleep, which requires more escape time. Fires can expand quickly from incipient stages to room flashover in a matter of minutes under the right conditions. Fast detector/alarm responses are necessary to ensure safe egress of occupants. See Supplement 1 and Supplement 4.

8-1.3.2 Calculation of sound distribution or light intensity levels, or both, shall demonstrate compliance with the performance criteria of 8-1.2.2.

See 4-3.4 and 4-4.4.3 for audible and visible notification signaling equivalent requirements.

8-1.3.3 Timing the signal to determine that it meets the required temporal pattern shall be permitted.

8-1.4 Acceptable Solutions.

Equipment and installations that comply with 8-1.4.1 through 8-1.4.4 shall be considered one way of meeting the performance criteria of 8-1.2. Other means judged by the authority having jurisdiction as providing equivalent performance shall be permitted.

Acceptable solutions are the prescriptive requirements that serve as alternatives to the performance-based requirements of 8-1.2. Either the performance-based approach or the prescriptive approach must be followed, but not both.

8-1.4.1* Smoke detectors and smoke alarms located in accordance with the requirements of NFPA *101®*, *Life Safety Code®*, for the applicable occupancy classification of the building shall be considered in compliance with these requirements.

The requirements of NFPA *101*, *Life Safety Code*, have been extracted and placed into Chapter 8 for the 1999 code. Chapter 3 describes the requirements for protected premises systems in common areas. Requirements for transmission of signals to a supervising station are covered in Chapter 5.

A-8-1.4.1 Refer to Figure A-8-1.4.1 where required smoke detectors are shown. Smoke detectors are optional where a door is not provided between a living room and a recreation room.

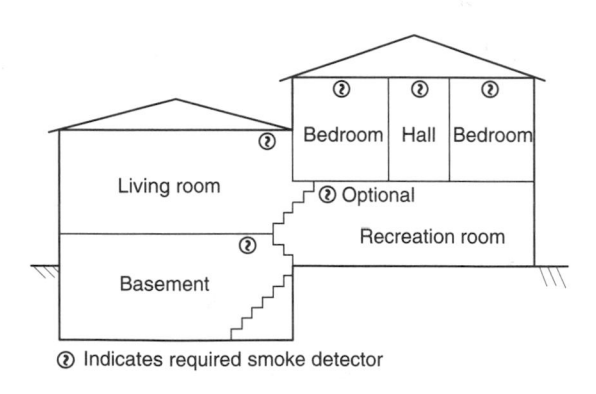

Ⓢ Indicates required smoke detector

Figure A-8-1.4.1 Split level arrangement.

8-1.4.1.1 New Hotels and Dormitories.

8-1.4.1.1.1 Detection.

8-1.4.1.1.2 A corridor smoke detection system in accordance with Section 7-6 of NFPA *101* shall be provided. (**101**:16-3.4.4.1)

This corridor detection system must comply with Chapters 1–7 of this code.

Exception: Buildings protected throughout by an approved, supervised automatic sprinkler system installed in accordance with 16-3.5.1 of NFPA 101. (101:16-3.4.4.1)

8-1.4.1.1.3 An approved, single-station smoke alarm shall be installed in accordance with 7-6.2.10 of NFPA *101* in every guest room and every living area and sleeping room within a guest suite. *(101:16-3.4.4.2)*

Subparagraph 8-1.4.1.1.3 requires single-station smoke alarms in each guest room or each room in a suite. Paragraph 7-6.2.10 of NFPA *101* requires the smoke alarms to be powered by 120-V ac power from the building electrical system. In new construction, and where more than one smoke alarm is used, interconnection of smoke alarms within a suite is required. Interconnection of smoke alarms between different dormitory rooms, hotel rooms, or guest suites is not permitted. Additionally, actuation of smoke alarms within a dormitory room, hotel room, or guest suite is not permitted to actuate the building fire alarm system.

8-1.4.1.2 Existing Hotels and Dormitories.

8-1.4.1.2.1 Detection.

8-1.4.1.2.2 An approved, single-station smoke alarm shall be installed in accordance with 7-6.2.10 of NFPA *101* in every guest room and every living area and sleeping room within a guest suite. *(101:17-3.4.4.2)*

Exception 1: These alarms shall not be required to be interconnected. (101:17-3.4.4.2)

Exception 2: Single-station smoke alarms without a secondary (standby) power source shall be permitted. (101:17-3.4.4.2)

The smoke alarms required by 8-1.4.1.2.2 are required to be powered by 120-V ac power from the building electrical system. However, the exceptions permit single-station alarms to be used without a secondary power supply. Battery-powered smoke alarms are not permitted to be used in existing hotels and dormitories.

8-1.4.1.3 New Apartment Buildings.

8-1.4.1.3.1 Detection.

8-1.4.1.3.2 Approved, single-station smoke alarms shall be installed in accordance with 7-6.2.10 of NFPA *101* outside every sleeping area in the immediate vicinity of the bedrooms and on all levels of the dwelling unit including basements. *(101:18-3.4.4.1)*

The minimum requirement is for a smoke alarm outside each separate sleeping area and in each bedroom of an apartment. Multiple-story apartments are required to have an additional smoke alarm on every level. This requirement can be met using multiple-station alarm devices or a system using a control panel. Interconnection of multiple-station smoke alarms within a new apartment is required where more than one

smoke alarm is used. See Exhibit 8.8 for a schematic of a typical multiple-station smoke alarm arrangement.

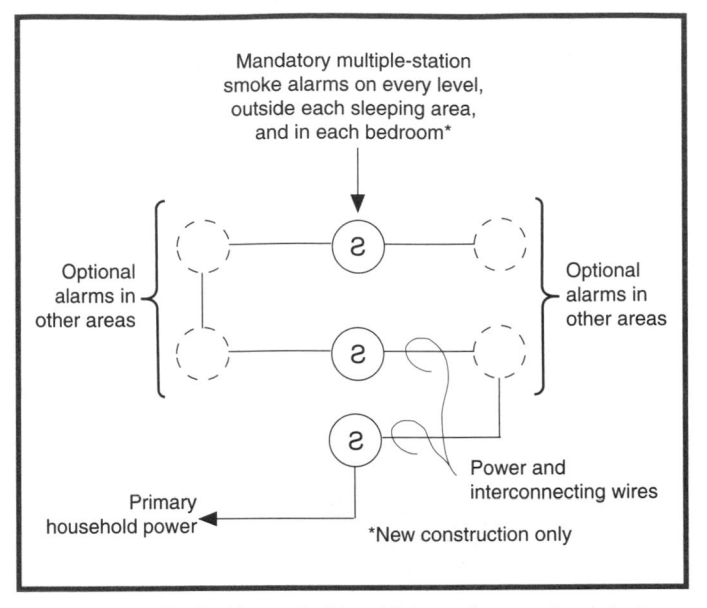

Exhibit 8.8 Typical household multiple-station smoke detector system.

The code also allows the use of a complete household fire alarm system that contains system-type smoke detectors connected to a listed fire alarm system control unit along with other devices, such as heat detectors and alarm notification appliances. The complete system would be used in place of, and not in addition to, the interconnected 120-V ac multiple-station smoke alarm devices.

Doors at the tops of stairwells prevent smoke from flowing upward. The stairwell acts as a dead air space, trapping smoke below and preventing smoke from reaching a detector located in the stairwell.

Detectors should always be mounted on the bottom of floor joists in unfinished construction. Placing the detector in the pockets between joists may prevent or delay detector response.

8-1.4.1.3.3 Approved, single-station smoke alarms shall be installed in accordance with 7-6.2.10 of NFPA *101* in every sleeping room. *(101:18-3.4.4.2)*

Exception: In buildings protected throughout by an approved, supervised automatic sprinkler system installed in accordance with 18-3.5 of NFPA 101. (101:18-3.4.4.2)

Smoke alarms are not required in individual sleeping rooms of fully sprinklered buildings.

8-1.4.1.4 Existing Apartment Buildings.

8-1.4.1.4.1 Detection.

8-1.4.1.4.2 Approved, single-station smoke alarms shall be installed in accordance with 7-6.2.10 of NFPA *101* outside

every sleeping area in the immediate vicinity of the bedrooms and on all levels of the dwelling unit including basements. (*101:19-3.4.4.1*)

Exception 1: The single-station smoke alarm shall not be required where the building is equipped throughout with an existing total automatic smoke detection system. (101:19-3.4.4.1)

Exception 2: Single-station smoke alarms without a secondary (standby) power source shall be permitted. (101:19-3.4.4.1)

Single-station, battery-powered smoke alarms may be used to meet these requirements.

8-1.4.1.4.3 In buildings using Option 2, a complete automatic fire detection system in accordance with 7-6.1.4 of NFPA *101* shall be required. (*101:19-3.4.4.2*)

Option 2 refers to buildings provided with a complete automatic fire detection and notification system as covered by Chapters 1 through 7 of the code.

8-1.4.1.5 Lodging or Rooming Houses.

8-1.4.1.5.1 Detection.

8-1.4.1.5.2 Approved single-station smoke alarms shall be installed in accordance with 7-6.2.10 of NFPA *101* in every sleeping room.

Exception 1: These detectors shall not be required to be interconnected. (101:20-3.3.4)

The smoke alarms required by 8-1.4.1.5.1 are not required to be interconnected because a nuisance alarm in one guest room would actuate all interconnected smoke alarms.

Exception 2: Existing battery-powered smoke alarms, rather than house electric-powered smoke alarms, shall be permitted where the facility has demonstrated to the authority having jurisdiction that the testing, maintenance, and battery replacement programs will ensure reliability of power to the smoke alarms. (101:30-3.3.4)

Batteries must be changed at least annually. The owner of the premises is responsible for maintenance of smoke alarms.

8-1.4.1.6 One- and Two-Family Dwellings.

8-1.4.1.6.1 Detection, Alarm, and Communications Systems.

8-1.4.1.6.2 Approved, single-station smoke alarms shall be installed in accordance with 7-6.2.10 of NFPA *101* in the following locations:

(1) All sleeping rooms

Exception: Smoke alarms shall not be required in sleeping rooms in existing construction.

(2) Outside of each separate sleeping area, in the immediate vicinity of the sleeping rooms
(3) On each additional story of the dwelling unit including basements (*101:21-3.3.1*)

Exception 1: Dwelling units protected by an approved smoke detection system installed in accordance with Section 7-6 of NFPA 101, having an approved means of occupant notification.

Exception 2: In existing construction, approved smoke alarms powered by batteries shall be permitted.

The minimum requirement is for a smoke alarm outside each separate sleeping area and in each sleeping room of a dwelling. Multiple-story dwellings are required to have an additional smoke alarm on every level. This requirement can be met using multiple-station smoke alarm devices. See Exhibit 8.8 for a typical multiple-station smoke alarm arrangement. In new construction, smoke alarms are required in each sleeping room. Interconnection of multiple-station smoke alarms within a dwelling is required where more than one smoke alarm is used. However, smoke alarms should not be interconnected between separate dwellings, such as duplex arrangements.

The code also allows the use of a complete household fire alarm system that contains system-type smoke detectors connected to a listed fire alarm system control unit along with other devices, such as heat detectors and alarm notification appliances. See Exhibit 8.9 for an illustration of a household fire alarm system. The complete system would be used in place of, and not in addition to, the smoke alarms required by 8-1.4.1.6.2.

Doors at the tops of stairwells prevent smoke flow in the upward direction. The stairwell acts as a dead air space and traps smoke below. This can prevent smoke from reaching a detector located in the stairwell. Detectors should always be mounted on the bottom of floor joists in basements with unfinished construction. Placing the detector in the pockets between joists may prevent or delay detector response.

8-1.4.2* Initiating devices shall be located in areas where ambient conditions are within the limits specified by the manufacturer, and smoke alarms or smoke detectors shall not be closer than 3 ft (1 m) from the door to a bathroom or kitchen. Smoke alarms or smoke detectors that are located within 20 ft (6.1 m) of a cooking appliance and are equipped with an alarm silencing means or are of the photoelectric type shall be considered acceptable.

Photoelectric-type smoke alarms may provide slightly more resistance to nuisance alarms caused by cooking vapors or bathroom moisture. However, every effort should be made to locate smoke alarms away from potential sources of nuisance alarms. Smoke alarms with silencing means or "hush buttons" are required to be listed for such use. See commentary following 8-1.2.4.

Exhibit 8.9 *Typical combination household fire alarm/burglary system. (Source: Radionics Inc., Salinas, CA)*

A-8-1.4.2 One of the common problems associated with residential smoke detectors is the nuisance alarms that are usually triggered by products of combustion from cooking, smoking, or other household particulates. While an alarm for such a condition is anticipated and tolerated by the occupant of a dwelling unit through routine living experience, the alarm is not permitted where it also sounds alarms in other dwelling units or in common use spaces. Nuisance alarms caused by cooking are a very common occurrence, and inspection authorities should be aware of the possible ramifications where the coverage is extended beyond the limits of the dwelling unit.

8-1.4.3 Smoke alarms conforming to the requirements of ANSI/UL 217, *Standard for Safety Single and Multiple Station Smoke Alarms,* or smoke detectors conforming to ANSI/UL 268, *Standard for Safety Smoke Detectors for Fire Protective Signaling Systems,* shall be considered acceptable. Visible signaling appliances shall conform to the requirements of ANSI/UL 1971, *Signaling Devices for*

Hearing Impaired. Smoke alarms and smoke detectors shall be capable of detecting abnormal quantities of smoke and shall alarm prior to a gray smoke level of 4 percent per ft (0.58 dB/m optical density) that can occur in a dwelling unit. Visible notification appliances located on the ceiling over the bed and within 16 ft of a sleeping occupant shall have a light output rating of at least 177 cd. Where a visible notification appliance in a sleeping room is mounted more than 24 in. (610 mm) below the ceiling and within 16 ft of the pillow, a minimum rating of 110 cd shall be permitted.

ANSI/UL 268, *Standard for Safety Smoke Detectors for Fire Protective Signaling Systems,* is the standard for system smoke detectors. These detectors are connected to and powered by a fire alarm system control unit and may also have integral notification appliances, depending on the model and manufacturer. All ac-powered, battery-powered, or combination ac, battery-powered single-station and multiple-station smoke alarms must comply with ANSI/UL 217, *Standard*

for Safety Single and Multiple Station Smoke Alarms, in order to be listed. Smoke alarms are not permitted to be connected to a control panel unless they have been specifically listed for that purpose. Both the ionization spot-type and the photoelectric spot-type smoke detectors are available under ANSI/UL 268 or ANSI/UL 217. See Chapter 2 for more details on these detectors.

Formal Interpretation 89-1 further clarifies this section.

Formal Interpretation 89-1

Reference: 1-4, 8-1.4.3

Question: Is it the intent of the Committee that only devices which comply with the definition of a smoke detector in 1-4 meet the intent of 8-1.4.3?

Answer: No. Paragraph 1-3.2 allows for any device or system that can demonstrate equivalent performance to be "approved." Data currently exists that could be used by any interested party to demonstrate such equivalency for a device sensing carbon monoxide. If an analysis of this data shows an equivalent performance, carbon monoxide sensing devices could be listed or approved as meeting the requirements of NFPA 72.

Issue Edition: 1989 of NFPA 74

Reference: 1-4, 4-2.1

Issue Date: January 2, 1990

Effective Date: January 22, 1990

8-1.4.4 Smoke detectors shall be connected to central controls for power, signal processing, and activation of notification appliances. Smoke alarms shall not be interconnected in numbers that exceed the manufacturer's recommendations, but in no case shall more than 18 initiating devices be interconnected (of which 12 can be smoke alarms) where the interconnecting means is not supervised, nor more than 64 initiating devices be interconnected (of which 42 can be smoke alarms) where the interconnecting means is supervised.

Manufacturer's instructions also provide guidance on the compatibility of smoke detectors with control units and on the maximum number of smoke alarms that may be interconnected. Because wiring that interconnects multiple station smoke alarms is not monitored for integrity, the number of devices permitted to be interconnected is limited. For applications that require more than 12 smoke alarms, a system must be used.

8-2 Optional Functions

The following optional functions of fire warning equipment for dwelling units shall be permitted:

(1) Notification of the fire department, either directly or through an alarm monitoring service

See Chapter 5 for off-premises connections using supervising stations.

(2) Monitoring of other safety systems, such as fire sprinklers for proper operating conditions

Connection of an automatic sprinkler system waterflow alarm-initiating device to the dwelling fire warning equipment or alarm system should be approved by the authority having jurisdiction. Where used, such a connection must use a multiple-station smoke alarm that has been specifically listed for this purpose. The smoke alarm must have the necessary terminals to allow connection of the waterflow alarm-initiating device. A listed fire alarm control unit or combination burglar/fire alarm control unit already serving the residence could be used to accept a connection from the waterflow alarm-initiating device. Additionally, a waterflow alarm-initiating device could be connected to a separate fire alarm system control unit to sound separate notification appliances throughout the family living unit.

(3) Notification of occupants or others of potentially dangerous conditions, such as the presence of fuel gases or toxic gases such as carbon monoxide

Carbon monoxide (CO) warning systems are covered by NFPA 720, *Recommended Practice for the Installation of Household Carbon Monoxide Warning Equipment.* NFPA 720 provides the recommendations for the location and quantity of CO detectors, and the means of interconnecting with fire warning or alarm equipment.

(4) Notification of occupants or others of the activation of intrusion (burglar alarm) sensors

Article 725 of NFPA 70, *National Electrical Code®* covers wiring requirements for security systems.

(5) Any other function, safety related or not, that may share components or wiring

8-2.1 Performance Criteria.

8-2.1.1* If designed and installed to perform additional functions, fire warning equipment for dwelling units shall operate reliably and without compromise to its primary functions.

A-8-2.1.1 Experience has shown that all hostile fires in dwelling units generate smoke to some degree. This is also true with respect to heat buildup from fires. However, the

results of full-scale experiments conducted over the past several years in the United States, using typical fires in dwelling units, indicate that detectable quantities of smoke precede detectable levels of heat in nearly all cases. In addition, slowly developing, smoldering fires can produce smoke and toxic gases without a significant increase in the room's temperature. Again, the results of experiments indicate that detectable quantities of smoke precede the development of hazardous atmospheres in nearly all cases.

For the above reasons, the required protection in this code uses smoke detectors as the primary life safety equipment for providing a reasonable level of protection against fire.

Of course, it is possible to install fewer detectors than required in this code. It could be argued that the installation of only one fire detector, whether a smoke or heat detector, offers some lifesaving potential. While this is true, it is the recommendation of NFPA 72 that the smoke detector requirements as stated in 8-2.1.1 are the minimum that should be considered.

The installation of additional detectors of either the smoke or heat type should result in a higher degree of protection. Adding detectors to rooms that are normally closed off from the required detectors increases the escape time because the fire does not need to build to the higher level necessary to force smoke out of the closed room to the required detector. As a consequence, it is recommended that the householder consider the installation of additional fire protection devices. However, it should be understood that Chapter 8 does not require additional detectors over and above those required in 8-1.4.1.

8-2.1.2 Fire signals shall take precedence over any other signal or functions, even if a non-fire signal is activated first.

8-2.1.3 Signals shall be distinctive so that a fire signal can be easily distinguished from signals that require different actions by the occupants.

Unless the fire alarm signals are unique, they could be confused with security, carbon monoxide alarms, or other signals in the home. Where the intended response is to evacuate, 8-1.2.3 requires new systems to produce the audible emergency evacuation signal described in ANSI S3.41-1990 (R1996), *Audible Emergency Evacuation Signal*. The requirement for a unique signal does not mean that two separate notification appliances must be used. A single notification appliance may be used provided it can supply different, distinctive signals. For example, a fully integrated system might sound the National Standard Audible Emergency Evacuation Signal for a fire alarm, sound a different signal for a security alarm, and sound a third signal to indicate detection of excessive levels of carbon monoxide.

8-2.1.4 Faults in other systems or components shall not affect the operation of the fire warning system.

Fire warning systems in dwellings are permitted to be combination systems. Equipment not required for the operation of the fire alarm system that is modified, removed, or is malfunctioning in any way must not impair the operation of the fire alarm system.

In order to comply with the requirement of 8-2.1.4, it may be necessary to provide additional detection beyond the basic requirements of 8-2.1.4. With some combination fire/security alarm control panels, a fault on the security system wiring may affect the operation of the entire control panel. If this situation might occur, protection of the wiring and control panel accessories (such as keypad controls) should be considered. This protection could take the form of additional detectors (heat or smoke detectors, depending on the location) or physical protection of the wiring using one or more metal conduits or raceways. (See Exhibit 8.9 for a typical listed combination fire/burglar alarm control unit.)

8-2.2 Verification Methods.

8-2.2.1 The following methods of verifying compliance with the performance criteria of 8-2.1 shall be permitted. Other methods permitted by the authority having jurisdiction shall be accepted if appropriate.

8-2.2.2 Equipment designed and installed such that the required operation is obtained shall be considered acceptable.

8-2.3 Acceptable Solutions.

Acceptable solutions are the alternative prescriptive approach to a performance-based design approach. The authority having jurisdiction has the sole responsibility for accepting a proposed system, method, or device as equivalent.

The authority having jurisdiction is responsible for approving or disapproving the placement of devices and equipment used in fire alarm systems based on the information presented for the equipment and systems proposed. The authority having jurisdiction may also request tests of equipment, appliances, or devices to determine equivalency.

8-2.3.1 Equipment and installations that comply with the following shall be considered one way of meeting the performance criteria of 8-2.1. Other means judged by the authority having jurisdiction as providing equivalent performance shall be permitted.

8-2.3.2 Fire warning equipment for dwelling units listed for the functions and combinations stated shall be considered acceptable when judged by the authority having jurisdiction as providing equivalent performance.

8-3 Reliability

8-3.1 Fire warning systems located in dwelling units and having all of the following features shall be considered to have a functional reliability of 95 percent:

(1) Utilizes a control panel

A listed commercial fire alarm system, installed in accordance with Chapter 3 of the code, that meets the basic detection and warning requirements of Chapter 8 would be an acceptable alternative to the more typical systems listed for dwelling use described in 8-3.3.

The authority having jurisdiction should require the submission of battery calculations for dwelling fire alarm systems. These calculations are used to determine the capacity required for standby and alarm time requirements for the dwelling fire warning system when the standby power is supplied by rechargeable batteries. See Exhibit 8.10 for a schematic of a household fire alarm system.

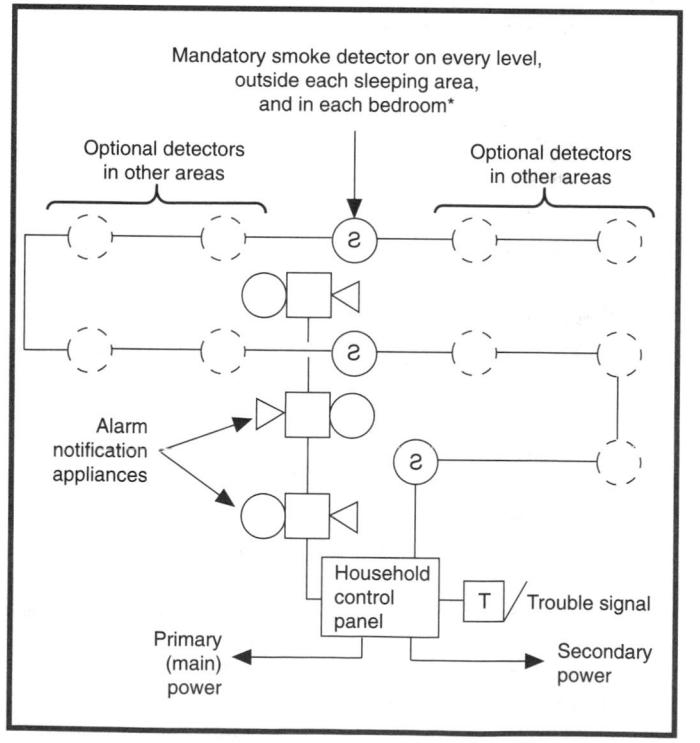

Exhibit 8.10 *Typical household fire alarm system with separate control panel.*

(2) Has at least two independent sources of operating power

The requirement of 8-3.1(2) addresses the increased need to have a functioning system during a power outage. Periodic

testing of batteries is vital to ensure that this secondary power supply to the detector will function.

During a power outage, occupants may greatly increase the risk of fire by using candles, gas lanterns, space heaters, and other equipment unusual in the home environment. Therefore, the smoke detector plays an even more important role in warning residents of a fire in these conditions during a power outage. Two sources of operating power significantly increase the likelihood that the system will operate during power outages.

(3) Monitors all initiating and notification circuits for integrity

Monitoring circuit integrity significantly increases the system reliability. Faults such as power failure, opens, and shorts are annunciated.

(4) Transmits alarm signals to a constantly attended, remote monitoring location

The primary objective of a dwelling fire warning system is life safety. Although some jurisdictions require a dwelling fire warning system (using a control panel) to be connected to a supervising station, the basic requirements of the occupant warning features in Chapter 8 must be followed regardless of any supervising station connection. Also see Chapter 5.

(5) Is tested regularly by the homeowner and at least every 3 years by a qualified service technician

Testing of equipment is vital to ensure that it is functioning properly. System owners should seriously consider a maintenance contract with a qualified service organization.

8-3.2 Fire warning systems for dwelling units with all of the features of 8-3.1 except (4) or systems that use low-power wireless transmission from initiating devices within the dwelling units shall be considered to have a functional reliability of 90 percent.

Low-power wireless systems specifically listed for fire alarm service provide a convenient way of installing a monitored system without extensive wiring.

8-3.3 Fire warning systems for dwelling units comprised of interconnected smoke alarms where the interconnecting means is monitored for integrity shall be considered to have a functional reliability of 88 percent. If the interconnecting means is not supervised or the alarms are not interconnected, such systems shall be considered to have a functional reliability of 85 percent.

8-3.4* Fire warning equipment for dwelling units shall be provided with a convenient means for testing its operability by the homeowner.

Generally, testing of single-station and multiple-station smoke alarms involves pushing a test button. Open flames should never be used to test smoke alarms because of the obvious fire hazard. Listed test aerosol can be used, provided that the smoke alarm manufacturer does not prohibit its use. Homeowners who are unable to perform these tests should consider a service contract.

A-8-3.4 If homeowner inspection, testing, and maintenance are required or assumed, the equipment should be installed in an accessible manner. It is a good practice to establish a specific schedule for tests.

8-3.5 Unless otherwise recommended by the manufacturer, smoke alarms installed in accordance with Chapters 18, 19, or 21 of NFPA *101, Life Safety Code,* shall be replaced when they fail to respond to tests conducted in accordance with 8-3.4 but shall not remain in service longer than 10 years from the date of installation.

Early reliability studies, most notably Canada's Ontario Housing Corporation studies (1980), indicate that most electronics products, including smoke alarms, fail at approximately 3 percent per year. This suggests that after 10 years, approximately one-third of the installed population of smoke alarms would be inoperable. However, thorough testing of smoke alarms helps discover units that have failed. To preserve the desired level of protection, failed units must be immediately replaced.

8-4 Performance Criteria

8-4.1 The two independent power sources shall consist of a primary source that uses the commercial light and power source and a secondary source that consists of a rechargeable battery or standby generator that can operate the system for at least 24 hours in the normal condition followed by 4 minutes of alarm.

8-4.2 All monitored circuits shall indicate with a distinctive trouble signal the occurrence of a single fault (open or unintentional ground), whether or not such fault affects the system's operation. Circuits interconnecting multiple station alarms shall not be required to be monitored provided a fault does not prevent an alarm signal from the detector that senses the abnormal condition.

8-4.3 Supervising Stations.

8-4.3.1 Means to transmit alarm signals to a constantly attended, remote monitoring location shall perform as described in Chapter 5 except that the DACT serving the protected premises only requires a single telephone line and only requires a call to a single DACR number. DACT test

signals shall be transmitted at least monthly. Such systems shall not be required to be certificated or placarded.

Paragraph 8-4.3.1 permits the dwelling fire alarm system to be connected to a supervising station using a digital alarm communicator transmitter (DACT), as defined in Section 1-4 of the code. In addition, paragraph 8-4.3.1 provides exceptions to some of the requirements of Chapter 5. Specifically, a single telephone line may be used. Also, the frequency of the daily test signal may be extended to monthly. However, these exceptions do not prevent a homeowner from installing and using a system that meets the requirements of Chapter 5, if desired. Such a system would also have to meet the requirements for dwelling units.

8-4.3.2* Remote monitoring locations shall be permitted to verify residential alarm signals prior to reporting them to the fire service provided that the verification process does not delay the reporting by more than 90 seconds.

Paragraph 8-4.3.2 permits supervising station personnel to place a verification call before retransmitting the alarm signal. The homeowner should use a pre-assigned personal identification code or password to verify that the source of an alarm signal does not require emergency response by the fire department.

A-8-4.3.2 If screening alarm signals to minimize response to false alarms is to be implemented, the following should be considered.

(1) Was the verification call answered at the protected premises?
(2) Did the respondent provide proper identification?
(3) Is it necessary for the respondent to identify the cause of the alarm signal?
(4) Should the public service fire communications center be notified and advised that an alarm signal was received, including the response to the verification call, when an authorized respondent states that the fire service response is not desired?
(5) Should the public service fire communications center be notified and advised that an alarm signal was received, including the response to the verification call, for all other situations, including both a hostile fire and no answer verification call?
(6) What other actions should be required by a standard operating procedure?

8-4.4 Low-power wireless systems shall comply with the performance criteria of Section 3-10.

8-4.5 Smoke alarms shall be powered from the commercial light and power source along with a secondary battery source that is capable of operating the device for at least 7 days in the normal condition followed by 4 minutes of alarm. Alternatively, smoke alarms shall be powered by a

non-replaceable primary battery that is capable of operating the device for at least 10 years followed by 4 minutes of alarm, followed by 7 days of trouble.

Where installing single-station or multiple-station smoke alarms, it is good practice to connect the detector's power to a branch circuit serving electrical outlets or lighting in a habitable area such as the living room or family room. This practice ensures that if for any reason the circuit breaker is in the off position, the condition will be noticed more quickly because lights and other appliances used frequently in the home will not operate. The power connection to a household control panel can be connected in the same way. Where connecting to a power circuit that serves other appliances, the installer must ensure that the circuit is not overloaded, causing the circuit breaker to frequently trip. Some state and local codes may require this branch circuit connection to be made to a dedicated circuit breaker. Consult with the authority having jurisdiction to determine if local codes or regulations differ from code requirements in 8-4.3.2.

Smoke alarms powered by a 10-year battery must also be of the multiple-station type in new construction, which requires interconnection of the alarms. Smoke alarms in existing construction may be of the single station type.

8-4.6 Smoke alarms powered by a primary battery capable of operating the device for at least 1 year in the normal condition, followed by 4 minutes of alarm, followed by at least 7 days of trouble, shall be used only if specifically permitted.

Battery-powered smoke alarms, once installed, are often not maintained by household occupants. NFPA studies (M. Ahrens 1998) indicate that nearly 20 percent of installed detectors do not operate, primarily because of dead or missing batteries. The requirements of 8-4.6 address the maintenance steps necessary for the smoke detector to function reliably. The trouble signal requirement was added to allow occupants to be alerted to an imminent battery failure. However, many homeowners or tenants do not recognize the trouble signal and may think it is a nuisance alarm. Establishing a routine battery replacement program is important to keep smoke alarms functioning. A popular program sponsored by the International Association of Fire Chiefs in conjunction with a battery manufacturer is the "Change Your Clocks, Change Your Batteries" campaign. This program reminds people living where the time changes to and from daylight savings time occurs to also change the batteries in their smoke alarms. In those few areas where time changes are not observed, some other means of public awareness should be devised.

8-4.7 Verification Methods.

The following methods of verifying compliance with the performance criteria of Section 8-4 shall be permitted.

Other methods permitted by the authority having jurisdiction shall be accepted if appropriate.

8-4.7.1 Engineering analysis such as fault tree analysis (FTA) or failure modes and effects criticality analysis (FMECA) shall be acceptable methods of determining equipment reliability.

When conducting analysis of this nature, all assumptions must be clearly stated and provided with the calculations.

8-4.7.2 Equipment designed and installed such that the required operation is obtained shall be considered acceptable.

8-4.8 Acceptable Solutions.

8-4.8.1 Equipment and installations that comply with 8-4.8.2 shall be considered one way of meeting the performance criteria of Section 8-4. Other means judged by the authority having jurisdiction as providing equivalent performance shall be permitted.

8-4.8.2 Equipment listed and approved to standards that verify the required performance such as ANSI/UL 985, *Standard for Safety Household Fire Warning Control Units;* ANSI/UL 217, *Standard for Safety Single and Multiple Station Smoke Alarms;* or ANSI/UL 268, *Standard for Safety Smoke Detectors for Fire Protective Signaling Systems,* shall be considered acceptable.

The approval or listing of fire alarm system devices is typically conducted by a qualified testing laboratory. It is the responsibility of the authority having jurisdiction to either accept or reject the laboratory's approval or listing. The authority having jurisdiction may require a product to be listed or labeled, but the listing or label alone does not constitute approval. The authority having jurisdiction has the right to approve products or systems that are not labeled or listed.

References Cited in Commentary

M. Ahrens, *1998, U.S. Experience with Smoke Detectors. Who Has Them, and How Well Do They Work,* National Fire Protection Association, Quincy, MA.

ANSI/UL 217, *Standard for Safety Single and Multiple Station Smoke Alarms,* Underwriters Laboratories Inc., Northbrook, IL.

ANSI/UL 268, *Standard for Safety Smoke Detectors for Fire Protective Signaling Systems,* Underwriters Laboratories Inc., Northbrook, IL.

ANSI S3.41, *Audible Emergency Evacuation Signal,* 1990 (R 1196), American National Standards Institute, New York.

NFPA 70, *National Electrical Code®,* National Fire Protection Association, Quincy, MA, 1999.

NFPA *101*®, *Life Safety Code*®, National Fire Protection Association, Quincy, MA, 2000.

NFPA 720, *Recommended Practice for the Installation of Household Carbon Monoxide (CO) Warning Equipment,* National Fire Protection Association, Quincy, MA, 1998.

Ontario Housing Corporation Studies, *Residential Smoke Detectors "In-Use" Reliability,* 1980, Ministry of Municipal Affairs and Housing, Toronto, ON.

CHAPTER 9

Referenced Publications

9-1 The following documents or portions thereof are referenced within this code as mandatory requirements and shall be considered part of the requirements of this code. The edition indicated for each referenced mandatory document is the current edition as of the date of the NFPA issuance of this code. Some of these mandatory documents might also be referenced in this code for specific informational purposes and, therefore, are also listed in Appendix C.

9-1.1 NFPA Publications.

National Fire Protection Association, 1 Batterymarch Park, P.O. Box 9101, Quincy, MA 02269-9101.

NFPA 10, *Standard for Portable Fire Extinguishers,* 1998 edition.

NFPA 13, *Standard for the Installation of Sprinkler Systems,* 1999 edition.

NFPA 13D, *Standard for the Installation of Sprinkler Systems in One- and Two-Family Dwellings and Manufactured Homes,* 1999 edition.

NFPA 13R, *Standard for the Installation of Sprinkler Systems in Residential Occupancies up to and Including Four Stories in Height,* 1999 edition.

NFPA 20, *Standard for the Installation of Stationary Pumps for Fire Protection,* 1999 edition.

NFPA 25, *Standard for the Inspection, Testing, and Maintenance of Water-Based Fire Protection Systems,* 1998 edition.

NFPA 37, *Standard for the Installation and Use of Stationary Combustion Engines and Gas Turbines,* 1998 edition.

NFPA 54, *National Fuel Gas Code,* 1999 edition.

NFPA 58, *Liquefied Petroleum Gas Code,* 1998 edition.

NFPA 70, *National Electrical Code®,* 1999 edition.

NFPA 75, *Standard for the Protection of Electronic Computer/Data Processing Equipment,* 1999 edition.

NFPA 90A, *Standard for the Installation of Air-Conditioning and Ventilating Systems,* 1999 edition.

NFPA *101®, Life Safety Code®,* 1997 edition.

NFPA 110, *Standard for Emergency and Standby Power Systems,* 1999 edition.

NFPA 111, *Standard on Stored Electrical Energy Emergency and Standby Power Systems,* 1996 edition.

NFPA 601, *Standard for Security Services in Fire Loss Prevention,* 1996 edition.

NFPA 780, *Standard for the Installation of Lightning Protection Systems,* 1997 edition.

NFPA 1221, *Standard for the Installation, Maintenance, and Use of Emergency Services Communications Systems,* 1999 edition.

9-1.2 Other Publications.

9-1.2.1 ANSI Publications. American National Standards Institute, Inc., 11 West 42nd Street, 13th floor, New York, NY 10036.

ANSI A-58.1, *Building Code Requirements for Minimum Design Loads in Buildings and Other Structures.*

ANSI S1.4a, *Specifications for Sound Level Meters,* 1985.

ANSI S3.41, *Audible Emergency Evacuation Signal,* 1996.

ANSI/ASME A17.1, *Safety Code for Elevators and Escalators,* 1998.

ANSI/IEEE C2, *National Electrical Safety Code,* 1997.

ANSI/UL 217, *Standard for Safety Single and Multiple Station Smoke Alarms,* 1997.

ANSI/UL 268, *Standard for Safety Smoke Detectors for Fire Protective Signaling Systems,* 1999.

ANSI/UL 827, *Standard for Safety Central-Station for Watchman, Fire-Alarm and Supervisory Services,* 1997.

ANSI/UL 985, *Standard for Safety Household Fire Warning Control Units,* 1994.

ANSI/UL 1971, *Signaling Devices for Hearing Impaired,* 1995.

9-1.2.2 EIA Publication. Electronic Industries Alliance, 2500 Wilson Boulevard, Arlington, VA 22201-3834.

EIA Tr 41.3, *Telephones.*

9-1.3 Additional Reference.

International Municipal Signal Association, P.O. Box 539, Newark, NY 14513.

National Institute for Certification in Engineering Technologies, 1420 King Street, Alexandria, VA 22314-2794.

APPENDIX A

Explanatory Material

The material contained in Appendix A of the 1999 edition of the *National Fire Alarm Code*® is not a part of the requirements of the code but is included with the code for informational purposes only. For the convenience of readers, in this handbook the Appendix A material is interspersed among the verbiage of Chapters 1 through 8 and, therefore, not repeated here.

APPENDIX B

Engineering Guide for Automatic Fire Detector Spacing

B-1 Introduction

B-1.1 Scope.

Appendix B provides information intended to supplement Chapter 2. It includes a procedure for determining detector spacing based on the objectives set for the system, size, and rate of growth of fire to be detected, various ceiling heights, ambient temperatures, and response characteristics of the detectors. In addition to providing an engineering method for the design of detection systems using plume-dependant detectors, heat detectors, and smoke detectors, this appendix also provides guidance on the use of radiant energy-sensing detectors.

Many jurisdictions permit technologists to design fire alarm systems following the prescriptive requirements found in the code. Designers using a performance-based approach must review the relevant engineering licensure laws in the jurisdictions in which they practice. It is very likely that performance-based designs of fire alarm systems will be deemed engineering of the type requiring licensure as a professional engineer. The designer should also be knowledgeable in the principles of fire protection engineering and apply these principles judiciously.

B-1.2 General.

Appendix B has been revised in its entirety from previous editions. The correlations originally used to develop the tables and graphs for heat and smoke detector spacings in the earlier editions have been updated to be consistent with current research. These revisions correct the errors in the original correlations. In earlier editions the tables and graphs were based on an assumed heat of combustion of 20,900 kJ/kg. The effective heat of combustion for common cellulosic materials is usually taken to be approximately 12,500 kJ/kg. The equations in this appendix were produced using test data and data correlations for cellulosic (wood) fuels that have a total heat of combustion of about 12,500 kJ/kg.

For the technical basis for the changes to Appendix B, see Reference 11 in Appendix C.

B-1.2.1 For the purposes of this appendix, the heat produced by a fire is manifested either as convective heat or radiant heat. It is assumed that conductive heat transfer is of little consequent during the early stages of the development of a fire, where this appendix is relevant. A convective heat release rate fraction equal to 75 percent of the total heat release rate has been used in this appendix. Users should refer to references 12 and 13 in C-2 for fuels or burning conditions that are substantially different from these conditions.

B-1.2.2 The design methods for plume-dependant fire detectors provided in this appendix are based on full-scale fire tests funded by the Fire Detection Institute in which all fires were geometrically growing flaming fires. (See *Environments of Fire Detectors—Phase 1: Effect of Fire Size, Ceiling Height and Material;* Measurements Vol. I and Analysis Vol. II.)

B-1.2.3 The guidance applicable to smoke detectors is limited to a theoretical analysis based on the flaming fire test data and is not intended to address the detection of smoldering fires.

The design methods in Appendix B rely on the presence of a buoyant plume. The relatively large heat release rate from flaming fires produces the plume. A smoldering fire essen-

tially has no buoyant plume. The pre-existing, ambient air currents provide the dominant smoke transport mechanism. Unfortunately, a designer cannot use the design methods in Appendix B to analyze the ambient air currents. Consequently, the designer can only use the design methods outlined in Appendix B for cases in which a flaming fire produces a buoyant plume.

B-1.2.4 The design methods for plume-dependant fire detectors do not address the detection of steady-state fires.

B-1.2.5 The design methods for plume-dependant fire detectors used in this appendix are only applicable when employed in the context of applications where the ceiling is smooth and level. It cannot be used for ceilings where there are beams, joists, or bays formed by beams and purlins. The research upon which the following methods have been based did not consider the effect of beams, joists, and bays in sufficient detail to justify the use of this appendix to those applications.

B-1.3 Purpose.

The purpose of Appendix B is to provide a performance basis for the location and spacing of fire detection initiating devices. The sections for heat and smoke detectors provide an alternative design method to the prescriptive approach presented in Chapter 2 (that is, based on their listed spacings). The section on radiant energy-sensing detectors elaborates on the performance-based criteria already existing in Chapter 2. A performance-based approach allows one to consider potential fire growth rates and fire signatures, the individual compartment characteristics, and damageability characteristics of the targets (for example, occupants, equipment, contents, structures, and so on) in order to determine the location of a specific type of detector to meet the objectives established for the system.

B-1.3.1 Under the prescriptive approach, heat detectors are installed according to their listed spacing. The listed spacing is determined in a full-scale fire test room. The fire test room used for the determination of listed spacing for heat detectors has a ceiling height of 15 ft 9 in. (4.8 m). A steady-state, flammable liquid fire with a heat release rate of approximately 1200 Btu/sec (1137 kW), located 3 ft (0.9 m) above the floor is used as the test fire. Special, 160°F (71°C) test sprinklers are installed on a 10 ft (3 m) × 10 ft (3 m) spacing array such that the fire is in the center of the sprinkler array. The heat detectors being tested are installed in a square array with increasing spacing centered about the fire location. The elevation of the test fire is adjusted during the test to produce the temperature versus time curve at the test sprinkler heads to yield actuation of the heads in 2.0 minutes ± 10 seconds. The largest heat detector spacing that achieves alarm before the actuation of the sprinkler heads in

the test becomes the listed spacing for the heat detector. *See Figure A-2-2.4.1(c).* If the room dimensions, ambient conditions, and fire and response characteristics of the detector are different from above, the response of the heat detector must be expected to be different as well. Therefore, the use of an installed detector spacing that is different from the listed spacing might be warranted through the use of a performance-based approach if

(1) The design objectives are different from designing a system that operates at the same time as a sprinkler in the approval test.
(2) Faster response of the device is desired.
(3) A response of the device to a smaller fire than used in the approved test is required.
(4) Accommodation to room geometry that is different than used in the listing process.
(5) Other special considerations, such as ambient temperature, air movement, ceiling height, or other obstruction, are different from or are not considered in the approval tests.
(6) A fire other than a steady-state 1200 Btu/sec (1137 kW) fire is contemplated.

B-2 Performance-Based Approach to Designing and Analyzing Fire Detection Systems

Section B-2, and particularly B-2.2 and B-2.3, are largely based on the work of Custer and Meacham as found in "Performance-Based Fire Safety Engineering: An Introduction of Basic Concepts" (Meacham and Custer 1995) and *Introduction to Performance-Based Fire Safety* (Custer and Meacham 1997).

The National Fire Protection Association and the Technical Committee on Initiating Devices for Fire Alarm Systems gratefully acknowledge the technical contributions of the Society of Fire Protection Engineers, Richard Custer, and Brian Meacham to performance-based design and this appendix.

B-2.1 Overview.

Subsection B-2.1 provides an overview of a systematic approach to conducting a performance-based design or analysis of a fire detection system. The approach has been outlined by Custer and Meacham and the SFPE *Engineering Guide to Performance Based Fire Protection Analysis and Design* [40], and is summarized below in the context of design and analysis of fire detection systems. *(Refer to Figure B-2.1.)* This approach has been divided into two phases: defining goals and objectives and system design and evaluation.

B-2.2 Phase 1—Defining Goals and Objectives.

B-2.2.1 Define Scope of Project. The initial step of this approach is to identify information relative to the overall scope of work on the project, including characteristics of the building, design intent, design and construction team organization, constraints on design and project schedule, proposed building construction and features, relevant hazards, how the building functions, occupant characteristics, and so on.

At this time, the designer should also consider the following:

- Design limitations
- Fire service capabilities
- Historical preservation
- Building management
- Applicable regulations

B-2.2.1.1 While defining the project's scope, the designer will identify which of the three situations in Table B-2.2.1.1 best describes the project at hand (that is, a performance-

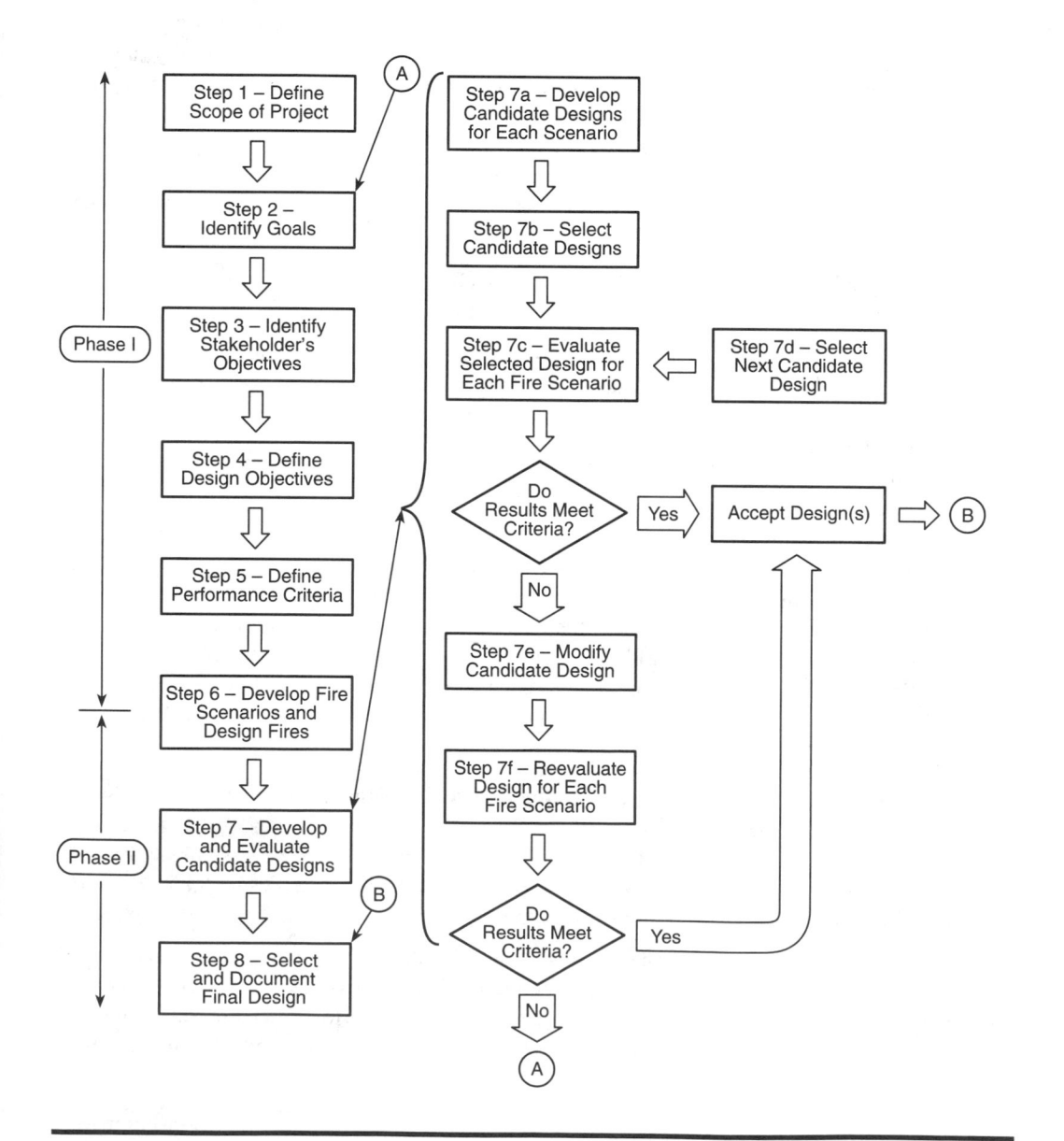

Figure B-2.1 *Overview of the performance-based design process. [25]*

based analysis of an existing detection system in an existing building).

Table B-2.2.1.1 Design/Analysis Situation

Building Type	System Type	Design/Analysis
New	New	Design
Existing	New	Design
Existing	Existing	Analysis

B-2.2.2 Identify Goals. Fire protection assets are acquired in order to attain one or more of the following four goals:

(1) To provide life safety (occupants, employees, fire fighters, and so on)
(2) To protect property and heritage (structure, contents, and so on)
(3) To provide for continuity of operations (protect stakeholder's mission, operating capability, and so on)
(4) To limit the environmental impact of fire (toxic products, fire-fighting water run-off, and so on)

B-2.2.2.1 Fire protection goals are like other goals in that they are generally easy to agree on, are qualitative, and are non-controversial in nature. They express the desired overall outcome to be achieved, that is, to provide life safety to the building occupants.

The design team may also have other goals that do not relate directly to fire safety. These goals may include specific design objectives (such as working with a large open space that has no compartment, minimizing damage to historic building fabric, reducing costs, etc.). The design team must consider these design parameters at the start of a project. See also Custer and Meacham, 1997, page 29.

B-2.2.2.2 When starting the performance-based process, the various parties including the stakeholders (that is, the architect, building owner, insurance carrier, building or fire officials, and so on), the authority having jurisdiction, and the design engineer work together to prioritize the basic fire protection goals. Prioritizing is based on the stakeholder's objective and the building and occupancy involved. For example, life safety is a high priority in a hospital or stadium, while property protection might have an equally high priority in a large warehouse or historic building.

Stakeholders can also include the following parties:

- Building manager
- Design team members
- Authorities having jurisdiction
- Tenants
- Neighbors
- Accreditation agencies

- Construction team
- Fire service

See also Custer and Meacham, 1995, page 40.

B-2.2.3 Identify Stakeholder's Objectives. Each stakeholder must explicitly state her or his objectives in terms of acceptable loss for the various goals previously stated.

B-2.2.3.1 Stakeholder objectives specify how much safety the stakeholder wants, needs, or can afford. "No loss of life within the room of origin" is a sample stakeholder objective or statement of the stakeholder's maximum acceptable loss.

Designers should note that most buildings contain ignition sources and some type of fuel. The chance that a fire could occur always exists. Similarly, some probability always exists that a fire in an occupied building may result in injury or death, some degree of property damage, or interruption to business (or all three). Therefore, the design engineer should ensure that the stakeholders know the impossibility of creating an entirely hazard-free or risk-free environment. (Custer and Meacham 1997)

B-2.2.3.2 The stakeholder's objectives are generally not stated in fire protection engineering terms.

See Tables B-2.2.5.4 (a) through (c) for additional examples of stakeholder's objectives.

B-2.2.3.3 Note that in a performance-based code environment, the code will most likely define a performance objective or stakeholder objective.

Stakeholders should define their objectives as clearly as possible. A code may establish other objectives. Each party or stakeholder must agree to the objectives for the system as early in the design phase as is practical. Without the stakeholders reaching early consensus on the objectives, the designer will have a far more difficult task developing quantitative engineering criteria with which to evaluate trial designs.

B-2.2.4 Define Design Objectives. The stakeholder's objective must then be explicitly stated and quantified in fire protection engineering terms that describe how the objective will be achieved. This demands that the design objectives be expressed quantitatively. See Tables B-2.2.5.4(a) through (c).

The designer uses the design objectives as a benchmark against which the predicted performance of a trial design is measured. He or she will use performance criteria expressed in engineering terms. See also Custer and Meacham, 1997, page 38.

B-2.2.4.1 The design objective provides a description of how the stakeholder's objective will be achieved in general fire protection engineering terms prior to this description being quantified. The general objective is then reduced to

explicit and quantitative fire protection engineering terms. The explicit fire protection engineering objectives provide a performance benchmark against which the predicted performance of a candidate design is evaluated.

The designer quantifies the design objectives in either deterministic or probabilistic terms. Deterministic methods consider any and all possible incident scenarios equally, regardless of how likely or unlikely a scenario might be. For example, the designer can translate the stakeholder objective of "no fire damage outside the compartment of origin" to "limiting the spread of flame to the compartment of origin." This translation of the stakeholder objective is a deterministic statement of the design objective. Probabilistic methods assign probabilities to incident scenarios, weighing the more likely ones higher than the less likely ones. The designer restates the stakeholder objective of "no fire damage outside the compartment of origin" as "limiting the probability of flame spreading to an adjacent compartment to a value that does not exceed a threshold value." (Custer and Meacham 1997)

B-2.2.5 Define Performance Criteria. Once the design objective has been established, specific, quantitatively expressed criteria that indicate attainment of the performance objective are developed.

B-2.2.5.1 Performance criteria provide a yardstick or threshold values that can measure a potential design's success in meeting stakeholder objectives and their associated design objectives. [25]

When defining performance criteria, a designer cannot achieve an environment totally free of risk or hazard. Also, the cost associated with an incremental reduction in risk typically increases as the intended risk or hazard level decreases.

B-2.2.5.2 Quantification of the design objectives into performance criteria involves determination of the various fire-induced stresses that are a reflection of the stated loss objectives. Performance criteria can be expressed in various terms, including temperature, radiant flux, a rate of heat release, or concentration of a toxic or corrosive species that must not be exceeded.

Other performance criteria include visibility, clear layer height, smoke concentration, ignition levels of adjacent fuel packages, and smoke product and toxic product damage. See also Custer and Meacham, 1995, page 41.

B-2.2.5.3 Once the design performance criteria are established, appropriate safety factors are applied to obtain the working design criteria. The working design criteria reflect the performance that must be achieved by the detection system. This performance level must allow appropriate actions to be undertaken (for example, activate suppression systems, occupants egress, notify fire department, and so on) to meet the objectives. An acceptable fire detection system

design provides the detection of the fire sufficiently early in its development to permit the other fire protective systems to meet or exceed the relevant performance criteria established for those systems.

B-2.2.5.4 Throughout the process identified as Phase I and II, communication should be maintained with the authorities having jurisdiction (AHJs) to review and develop consensus on the approach being taken. It is recommended that this communication commence as early in the design process as possible. The AHJ should also be involved in the development of performance criteria. Often the acceptance of a performance-based design in lieu of a design based on a prescriptive approach relies on demonstrating equivalence. This is called the comparative method, where the engineer demonstrates that the performance-based design responds at least as well, if not better than, a system designed using a prescriptive approach.

Tables B-2-2.5.4(a) through (c) present sample goals, objectives, and performance criteria. See also Custer and Meacham, 1997, page 42.

Table B-2.2.5.4(a) Defining Goals and Objectives — Life Safety

Fire Protection Goal	Provide life safety
Stakeholder's Objective	No loss of life within compartment of origin
Design Objective	Maintain tenable conditions within the compartment of origin
Performance Criteria	Maintain: Temperatures below xx °C Visibility above yy ft CO concentration below zz ppm for tt minutes

Table B-2.2.5.4(b) Defining Goals and Objectives— Property Protection

Fire Protection Goal	Provide protection of property
Stakeholder's Objective	No fire damage outside compartment of origin
Design Objective	Limit the spread of flame to the compartment of origin
Performance Criteria	Maintain upper layer temperature below xx °C and radiation level to the floor below yy kW/m^2 to prevent flashover.

Table B-2.2.5.4(c) Defining Goals and Objectives—Continuity of Operations

Fire Protection Goal	Provide continuity of operations
Stakeholder's Objective	Prevent any interruption to business operations in excess of 2 hours
Design Objective	Limit the temperature and the concentration of HCl to within acceptable levels for continued operation of the equipment
Performance Criteria	Provide detection such that operation of a gaseous suppression system will maintain temperatures below xx °C, and HCl levels below yy ppm

B-2.3 Phase II—System Design and Evaluation.

B-2.3.1 Develop Fire Scenarios. A fire scenario defines the development of a fire and the spread of combustion products throughout a compartment or building. A fire scenario represents a set of fire conditions that are deemed a threat to a building and its occupants and/or contents, and, therefore, should be addressed in the design of the fire protection features of the structure. [25]

B-2.3.1.1 The process of developing a fire scenario is a combination of hazard analysis and risk analysis. The hazard analysis identifies potential ignition sources, fuels, and fire development. Risk is the probability of occurrence multiplied by the consequences of that occurrence. The risk analysis looks at the impact of the fire to the surroundings or target items.

B-2.3.1.2 The fire scenario should include a description of various conditions, including building characteristics, occupant characteristics, and fire characteristics. [25, 40]

B-2.3.1.2.1 Building Characteristics. Building characteristics include the following:

Appendix B uses the term *building characteristics* to encompass the physical layout, ambient environment, structural features, fire hazards, and target locations within a compartment. Each of these affects fire initiation and growth, the spread of the products of combustion, and occupant evacuation. The designer must address these building characteristics when developing a design fire scenario.

(1) Configuration (area; ceiling height; ceiling configuration, such as flat, sloped beams; windows and doors, and thermodynamic properties)
(2) Environment (ambient temperature, humidity, background noise, and so on)
(3) Equipment (heat-producing equipment, HVAC, manufacturing equipment, and so on)

(4) Functioning characteristics (occupied, during times, days, and so on)
(5) Target locations

(Note target items, that is, areas associated with stakeholder objectives, along the expected route of spread for flame, heat, or other combustion products.)

The designer should also consider these additional building characteristics:

- Potential ignition sources
- Architectural details to be designed around (i.e., ornate ceilings)
- Concealed, enclosed spaces or voids

B-2.3.1.2.2 Occupant Characteristics. Occupant characteristics include the following:

(1) Alertness (sleeping, awake, and so on)
(2) Age
(3) Mobility
(4) Quantity and location within the building
(5) Sex
(6) Responsiveness
(7) Familiarity with the building
(8) Mental challenges

Human behavior plays a key role in life safety, as well as in other fire safety goals. The designer must consider the full range of possible actions the occupants can take after fire detection, such as how quickly occupants will respond once they hear an alarm and what they will do. Occupants can elect to alert and even rescue other family members, gather belongings, interpret or verify the message, or shut down processes. The designer should also consider the differences between the behavior of occupants as individuals and their behavior in the context of a larger group.

Once the designer analyzes occupant characteristics and behavior, he or she should compute realistic evacuation times for the occupants. Due to the nature of human behavior, a designer can never precisely quantify the movements and evacuation times of occupants from a building. A number of factors control the speed of evacuation. These factors include the number of occupants, their distribution throughout the building, their pre-movement times, their motivation, their state of wakefulness, their familiarity with the building, and the capacity and layout of the means of egress. Consequently, the designer must pay particular attention to assumptions and uncertainties assigned to these occupant characteristics. He or she must carefully weigh the sensitivity of the results to variations in the assumptions. (See also Supplement 4.)

B-2.3.1.2.3 Fire Characteristics. Fire characteristics include the following:

(1) Ignition sources—temperature, energy, time, and area of contact with potential fuels

(2) Initial fuels—state (solid, liquid, gas spray, vapor); type and quantity of fuel; fuel configuration; fuel location (against wall, in corner, in open); rate of heat release and fire growth; type and production rate of fire signatures (smoke, CO, CO$_2$, and so on)

(3) Secondary fuels—proximity to initial fuels; amount; distribution, ease of ignitibility *(see initial fuels);* and extension potential (beyond compartment, structure, area, if outside)

The following list elaborates on the fire characteristic of initial fuels.

(a) *State.* Fuels can come in various states (i.e., solid, liquid, or gas). Each state may have very different combustion characteristics (i.e., a solid block of wood versus wood shavings versus wood dust).

(b) *Type and quantity of fuel.* A fire's development and duration depends also on the type and quantity of the fuel. The burning properties of cellulosic materials differ markedly from those of plastics or flammable liquids. Each fuel produces different fire growth rates, heat release rates, and products of combustion.

(c) *Fuel configuration.* The geometrical arrangement of the fuel can also influence the fire growth rate and heat release rate. A wood block burns in a different manner from a wood crib. The crib has much more surface area and ventilation, which increases the radiation feedback between the combustible materials.

(d) *Fuel location.* The location of the fuel (i.e., against a wall, in a corner, in an open area, against the ceiling) influences the development of the fire. A fire in the corner of a room or against a wall typically grows faster than a fire located in the center of a room (Custer 1997).

(e) *Heat release rate.* The rate at which a fire releases heat depends on the fuel's heat of combustion, the mass loss rate, the combustion efficiency, and the amount of incident heat flux. The mass loss rate also directly influences the rate that a fire produces smoke, toxic gases, and other products of combustion.

(f) *Fire growth rate.* Fires grow at various rates. These rates depend on the type of fuel, the fuel's configuration, and the amount of ventilation. Some fires, such as confined flammable liquid fires, may not grow beyond their fixed burning area. These fires are referred to as steady-state fires. The faster a fire develops, the faster the temperature rises, and the faster the fire generates products of combustion.

(g) *Production rate of combustion products (smoke, CO, CO$_2$, etc.).* As the characteristics of various fuels vary, so do the type and quantity of materials generated during combustion. A designer can estimate species production rates with species yields. These production rates of combustion products represent the mass of species produced per mass of fuel loss.

Conduction, convection, radiation, or a combination of these can ignite secondary fuels. The designer must consider the issues itemized under item (b), when considering the participation of secondary fuels in a fire scenario.

See also Custer and Meacham, 1997, page 44.

B-2.3.1.2.4 An example of a fire scenario in a computer room might be as follows.

The computer room is 30 ft (9.1 m) × 20 ft (6 m) and 8 ft (2.8 m) high. It is occupied 12 hours a day, 5 days a week. The occupants are mobile and familiar with the building. There are no fixed fire suppression systems protecting this location. The fire department is capable of responding to the scene in 6 minutes, and an additional 15 minutes for fire ground evolution is needed.

Overheating of a resistor leads to the ignition of a printed circuit board and interconnecting cabling. This leads to a fire that quickly extends up into the above ceiling space containing power and communications cabling. The burning of this cabling produces large quantities of dense, acrid smoke and corrosive products of combustion that spread throughout the computer suite. This causes the loss of essential computer and telecommunications services for 2 months.

As mentioned in commentary to B-2.2.4.1, designers have a number of analytical techniques available to identify fire scenarios. Generally, these techniques can fall into one of two categories: probabilistic or deterministic.

Probabilistic approaches consider the probability or statistical likelihood that ignition and resulting fire will occur. When employing probabilistic approaches, the designer usually uses the following as sources of data:

- Fire statistics (ignition, first items ignited, etc.)
- Past history
- Hazard/Failure Analysis
- Failure Modes and Effects Analysis (FMEA)
- Event trees
- Fault trees
- HAZOP studies
- Cause–Consequence Analysis

Deterministic approaches assume that if an event can occur, it will occur, and that the design must address it. In this environment, the designer uses all of the previously listed sources of data. However, the designer treats every hazard scenario as equal. He or she applies equal resources to the management of the risk associated with the hazard.

The designer must exercise great care in choosing a design fire scenario. Addressing a fire scenario that seems very improbable can lead to an overly conservative and costly design. In contrast, a scenario developed using more liberal assumptions (i.e., short incipient phase, slow fire growth, etc.) will likely lead to a system design that will fail to meet the performance objectives. Such a design could present an unacceptably high risk to the people, contents, or processes to be protected.

B-2.3.2 Develop Design Fires. The design fire is the fire the system is intended to detect. When specifying a design fire, the specifics regarding the ignition, growth, steady-state output (if appropriate), and decay of the fire are expressed quantitatively.

B-2.3.2.1 Fire development varies depending on the combustion characteristics of the fuel or fuels involved, the physical configuration of the fuels, the availability of combustion air, and the influences due to the compartment. Once a stable flame is attained, most fires grow in an accelerating pattern *(see Figure B-2.3.2.1.2.8)*, reach a steady state characterized by a maximum heat release rate, and then enter into a decay period as either the availability of fuel or combustion air becomes limited. Fire growth and development are limited by factors such as quantity of fuel, arrangement of fuel, quantity of oxygen, and the effect of manual and automatic suppression systems.

Designers have very little data available that they can use to predict the behavior of smoldering fires. The designer should therefore carefully specify the duration of the smouldering period.

A variety of factors determine the fire growth rate of flaming fires. These factors include the following:

- Type of fuel and ease of ignition
- Fuel configuration and orientation
- Location of secondary fuel packages
- Proximity of fire to walls and corners
- Ceiling height
- Ventilation

When using heat release data, the designer must consider the fuel that is burning as well as the fire compartment. A couch might produce sufficient heat to cause flashover in a small compartment. The same couch placed in a large compartment with high ceilings might result in fire that is limited to a single fuel package that never reaches flashover.

B-2.3.2.1.1 Fires can be characterized by their rate of heat release, measured in terms of the number of Btus per second (kW) of heat liberated. Typical maximum heat release rates (Q_m) for a number of different fuels and fuel configurations are provided in Tables B-2.3.2.3.1(a) and (c). The heat release rate of a fire can be described as a product of a heat release density and fire area using the following equation:

$$Q_m = qA \tag{1}$$

where:

Q_m = maximum or peak heat release rate (Btu/sec)

q = heat release rate density per unit floor area (Btu/sec·ft^2)

A = floor area of the fuel (ft^2)

B-2.3.2.1.2 Example. A particular hazard analysis is to be based on a fire scenario involving a 10-ft × 10-ft (3-m × 3-m) stack of wood pallets stored 5 ft (1.5 m) high. Approximately what peak heat release rate can be expected?

B-2.3.2.1.2.1 From Table B-2.3.2.3.1(a), the heat release rate density (q) for 5-ft (1.5-m) high wood pallets is approximately 330 Btu/sec·ft^2.

B-2.3.2.1.2.2 The area is 10 ft × 10 ft = (3 m × 3 m), or 100 ft^2 (9 m^2). Using equation (1) to determine the heat release rate yields the following:

$$Q_m = qA$$

$$330 \times 100 = 33,000 \text{ Btu/sec}$$

B-2.3.2.1.2.3 As indicated in the Table B-2.3.2.3.1(a), this fire generally produces a medium to fast fire growth rate reaching 1000 Btu/sec (1055 kW) in approximately 90 to 190 seconds.

B-2.3.2.1.2.4 Fires can also be defined by their growth rate or the time (t_g) it takes for the fire to reach a given heat release rate. Previous research [16] has shown that most fires grow exponentially and can be expressed by what is termed the "power law fire growth model," which follows

$$Q \cong t^p \tag{2}$$

where:

Q = heat release rate (Btu/sec or kW)

p = 2

t = time (sec)

B-2.3.2.1.2.5 In fire protection, fuel packages are often described as having a growth time (t_g). This is the time necessary after the ignition with a stable flame for the fuel package to attain a heat release rate of 1000 Btu/sec (1055 kW). The following equations describe the growth of design fires:

$$Q = \frac{1000}{t_g^2} t^2 \text{ (Btu/sec)} \tag{3}$$

or

$$Q = \frac{1055}{t_g^2} t^2 \text{ (Kw)}$$

and thus

$$Q = \alpha t^2$$

where:

α = fire growth rate [$1000/t_g^2$ (Btu/sec^3) or $1055/t_g^2$ (kW/sec^2)]

Q = heat release rate (Btu/sec)

t_g = fire growth time to reach 1000 Btu/sec (1055 kW) after established burning

t = time after established burning occurs (sec)

B-2.3.2.1.2.6 Tables B-2.3.2.3.1(a) and (e) provide values for t_g, the time necessary to reach a heat release rate of 1000 Btu/sec (1055 kW), for a variety of materials in various configurations.

B-2.3.2.1.2.7 Test data from 40 furniture calorimeter tests, as indicated in Table B-2.3.2.3.1(e), have been used to independently verify the power law fire growth model, $Q = \alpha t^2$. [14] For reference, the table contains the test numbers used in the original NIST reports.

The virtual time of origin (t_v) is the time at which a stable flame had appeared and the fires began to obey the power law fire growth model. Prior to t_v, the fuels might have smoldered but did not burn vigorously with an open flame. The model curves are then predicted by the following equations:

$$Q = \alpha(t - t_v)^2 \tag{4}$$

or

$$Q = \left(\frac{1000}{t_g^2}\right)(t - t_v)^2 \text{ (Btu/sec)}$$

$$Q = \left(\frac{1055}{t_g^2}\right)(t - t_v)^2 \text{ (Kw)}$$

where:

α = fire growth rate [$1000/t_g^2$ (Btu/sec^3) or $1055/t_g^2$ (kW/sec^2)]

Q = heat release rate (Btu/sec)

t_g = fire growth time to reach 1000 Btu/sec (1055 kW)

t = time after established burning occurs (sec)

t_v = virtual time of origin (sec)

B-2.3.2.1.2.8 Figure B-2.3.2.1.2.8 is an example of an actual test data with a power law curve superimposed.

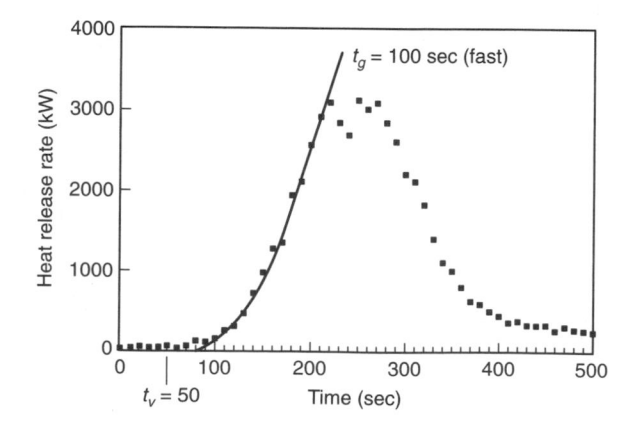

Figure B-2.3.2.1.2.8 *Test 38, foam sofa.*

B-2.3.2.1.2.9 For purposes of this appendix, fires are classified as being either slow-, medium-, or fast-developing from the time that established burning occurs until the fire reaches a heat release rate of 1000 Btu/sec (1055 kW). Table B-2.3.2.1.2.9 results from using the relationships discussed above. *[See also Table B-2.3.2.3.1(a).]*

Table B-2.3.2.1.2.9 *Power Law Heat Release Rates*

Fire Growth Rate	Growth Time (t_g)	α (kW/sec^2)
Slow	$t_g \geq 400$ sec	$\alpha \leq 0.0066$
Medium	$150 \leq t_g < 400$ sec	$0.0066 < \alpha \leq 0.0469$
Fast	$t_g < 150$ sec	$\alpha > 0.0469$

B-2.3.2.1.2.10 The correlation between flame height and heat release rate can be used to assist in deciding on an appropriate design fire. As shown in Figure B-2.3.2.1.2.10, flame height and fire size are directly related. [2] The lines in Figure B-2.3.2.1.2.10 were derived from the following equation:

$$h_f = 0.584(kQ)^{2/5} \tag{5}$$

where:

h_f = flame height (ft)

k = wall effect factor

Q = heat release rate (Btu/sec)

Where there are no nearby walls, use $k = 1$.

Where the fuel package is near a wall, use $k = 2$.

Where the fuel package is in a corner, use $k = 4$.

B-2.3.2.1.3 Example. What is the average flame height of a fire with a heat release rate of 1000 Btu/sec located in the middle of a compartment?

From Figure B-2.3.2.1.2.10, find the heat release rate on the abscissa and read estimated flame height from the ordinate, or use equation (5).

$$h_f = 0.584(kQ)^{2/5}$$

$$h_f = 0.584(1 \times 1000 \text{ Btu/sec})^{2/5}$$

$$h_f = 9.25 \text{ ft (2.8 m)}$$

B-2.3.2.2 Selection of Critical Fire Size. Because all fire control means require a finite operation time, there is a critical difference between the time at which the fire must be detected and the time at which it achieves the magnitude of the design fire. Even though a fire has been detected, this does not mean that it stops growing. Fires typically grow exponentially until they become ventilation controlled, and limited by the availability of fuel, or until

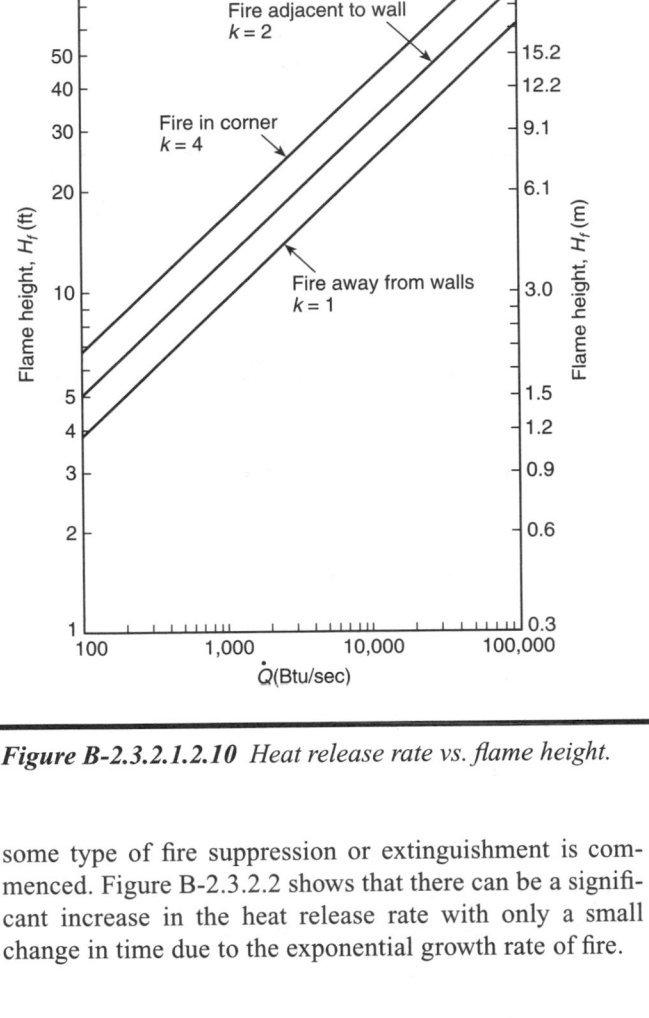

Figure B-2.3.2.1.2.10 *Heat release rate vs. flame height.*

some type of fire suppression or extinguishment is commenced. Figure B-2.3.2.2 shows that there can be a significant increase in the heat release rate with only a small change in time due to the exponential growth rate of fire.

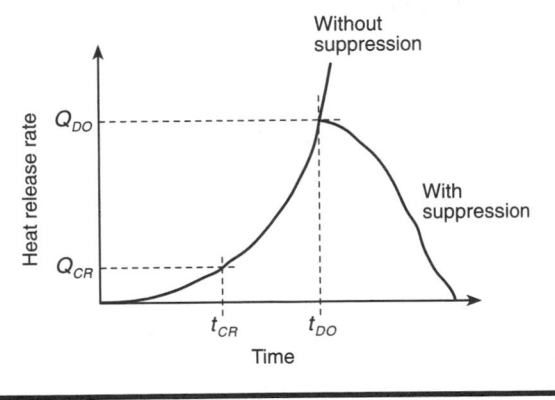

Figure B-2.3.2.2 *Critical and design objective heat release rates vs. time.*

B-2.3.2.2.1 Once the design objectives and the design fire have been established, the engineer will need to establish two points on the design fire curve: Q_{DO} and Q_{CR}.

B-2.3.2.2.2 Q_{DO} represents the heat release rate, or product release rate, which produces conditions representative of the design objective. This is the "design fire." However, Q_{DO} does not represent the point in time at which detection is needed. Detection must occur sufficiently early in the development of the fire to allow for any intrinsic reaction time of the detection as well as the operation time for fire suppression or extinguishing systems. There will be delays in both detection of the fire as well as the response of equipment, or persons, to the alarm.

B-2.3.2.2.3 A critical fire size (Q_{CR}) is identified on the curve that accounts for the delays in detection and response. This point represents the maximum permissible fire size at which detection must occur that allows appropriate actions to be taken to keep the fire from exceeding the design objective (Q_{DO}).

B-2.3.2.2.4 Delays are inherent in both the detection system as well as in the response of the equipment or people that need to react once a fire is detected. Delays associated with the detection system include a lag in the transport of combustion products from the fire to the detector and response time lag of the detector, alarm verification time, processing time of the detector, and processing time of the control panel. Delays are also possible with an automatic fire extinguishing system(s) or suppression system(s). Delay can be introduced by alarm verification or crossed zone detection systems, filling and discharge times of pre-action systems, delays in agent release required for occupant evacuation (for example, CO_2 systems), and the time required to achieve extinguishment.

B-2.3.2.2.5 Occupants do not always respond immediately to a fire alarm. The following must be accounted for when evaluating occupant safety issues:

(1) Time expected for occupants to hear the alarm (due to sleeping or manufacturing equipment noise)
(2) Time to decipher the message (for example, voice alarm system)
(3) Time to decide whether to leave (get dressed, gather belongings, call security)
(4) Time to travel to an exit

B-2.3.2.2.6 Response of the fire department or fire brigade to a fire incident involves several different actions that need to occur sequentially before containment and extinguishment efforts of the fire can even begin. These actions should also be taken into account to properly design detection systems that meet the design objectives. These actions typically include the following:

(1) Detection (detector delays, control panel delays, and so on)
(2) Notification to the monitoring station (remote, central station, proprietary, and so on)

(3) Notification of the fire department
(4) Alarm handling time at the fire department
(5) Turnout time at the station
(6) Travel time to the incident
(7) Access to the site
(8) Set-up time on site
(9) Access to building
(10) Access to fire floor
(11) Access to area of involvement
(12) Application of extinguishant on the fire

B-2.3.2.2.7 Unless conditions that limit the availability of combustion air or fuel exist, neither the growth of the fire nor the resultant damage stop until fire suppression begins. The time needed to execute each step of the fire response sequence of actions must be quantified and documented. When designing a detection system, the sum of the time needed for each step in the response sequence (t_{delay}) must be subtracted from the time at which the fire attains the design objective (t_{DO}) in order to determine the latest time and fire size (Q_{CR}) in the fire development at which detection can occur and still achieve the system design objective.

B-2.3.2.2.8 The fire scenarios and design fires selected should include analysis of best and worst case conditions and their likelihood of occurring. It is important to look at different conditions and situations and their effects on response.

B-2.3.2.3 Data Sources. To produce a design fire curve, information is needed regarding the burning characteristics of the object(s) involved. Data can be obtained from either technical literature or by conducting small or large scale calorimeter tests.

B-2.3.2.3.1 Some information is contained here in Figure B-2.3.2.3.1 and Tables B-2.3.2.3.1(a) through (e).

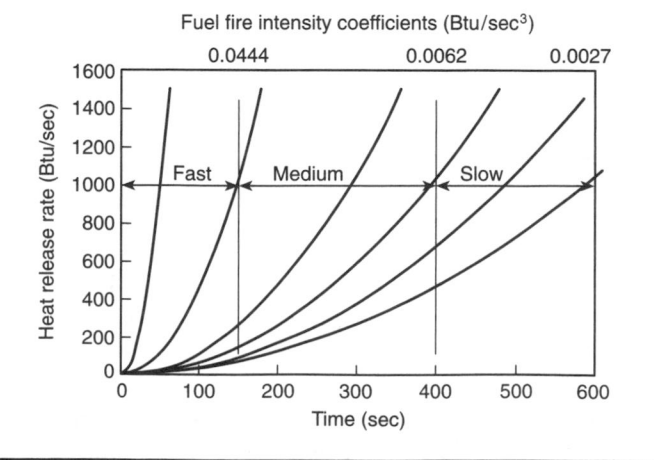

Figure B-2.3.2.3.1 *Power law heat release rates.*

B-2.3.2.3.2 Graphs of heat release data from the 40 furniture calorimeter tests can be found in *Investigation of a New Sprinkler Sensitivity Approval Test: The Plunge Test.* Best fit power law fire growth curves have been superimposed on the graphs. Data from these curves can be used with this guide to design or analyze fire detection systems that are intended to respond to similar items burning under a flat ceiling. Table B-2.3.2.3.1(e) is a summary of the data.

B-2.3.2.3.3 In addition to heat release rate data, the original NIST reports [3] contain data on particulate conversion and radiation from the test specimens. These data can be used to determine the threshold fire size (heat release rate) at which tenability becomes endangered or the point at which additional fuel packages might become involved in the fire.

B-2.3.2.3.4 The *NFPA Fire Protection Handbook* [22], SFPE *Handbook of Fire Protection Engineering,* and *Upholstered Furniture Heat Release Rates Measured with a Furniture Calorimeter* contain further information on heat release rates and fire growth rates.

B-2.3.2.3.5 Technical literature searches can be performed using a number of resources including FIREDOC, a document base of fire literature that is maintained by NIST.

B-2.3.2.3.6 A series of design fire curves are included as part of the "Fastlite" computer program available from NIST.

Additional sources of information for developing design fire curves include the following:

- Heat release rate data determined from experimental work (i.e., using a furniture calorimeter)
- Full scale tests of actual contents
- Generic curves
- Fire growth model that automatically generates heat release rate data

B-2.3.3 Develop and Evaluate Candidate Fire Detection Systems. Once the design objectives, the potential fire scenarios, and the room characteristics are well understood, the designer can select an appropriate detection strategy to detect the fire before its critical fire size (Q_{CR}) is reached. Important factors to consider include the type of detector, its sensitivity to expected fire signatures, its alarm threshold level and required duration at that threshold, expected installed location (for example, distance from fire, or below ceiling), and freedom from nuisance response to expected ambient conditions. *(See Chapter 2 and Appendix A.)*

Candidate designs usually differ from the design outlined in the prescriptive code or standard. Usually a designer compares the response of the detection system designed using a performance-based approach to the response of a system designed following prescriptive requirements. A designer also

Table B-2.3.2.3.1(a) Maximum Heat Release Rates—Warehouse Materials

Warehouse Materials	Growth Time (t_g) (sec)	Heat Release Density (Q) $(Btu/sec·ft^2)$	Classification
1. Wood pallets, stack, $1^1/_2$ ft high (6%–12% moisture)	150–310	110	fast–medium
2. Wood pallets, stack, 5 ft high (6%–12% moisture)	90–190	330	fast
3. Wood pallets, stack, 10 ft high (6%–12% moisture)	80–110	600	fast
4. Wood pallets, stack, 16 ft high (6%–12% moisture)	75–105	900	fast
5. Mail bags, filled, stored 5 ft high	190	35	medium
6. Cartons, compartmented, stacked 15 ft high	60	200	fast
7. Paper, vertical rolls, stacked 20 ft high	15–28	—	†
8. Cotton (also PE, PE/cot, acrylic/nylon/PE), garments in 12-ft high racks	20–42	—	†
9. Cartons on pallets, rack storage, 15 ft–30 ft high	40–280	—	fast–medium
10. Paper products, densely packed in cartons, rack storage, 20 ft high	470	—	slow
11. PE letter trays, filled, stacked 5 ft high on cart	190	750	medium
12. PE trash barrels in cartons, stacked 15 ft high	55	250	fast
13. FRP shower stalls in cartons, stacked 15 ft high	85	110	fast
14. PE bottles, packed in item 6	85	550	fast
15. PE bottles in cartons, stacked 15 ft high	75	170	fast
16. PE pallets, stacked 3 ft high	130	—	fast
17. PE pallets, stacked 6 ft–8 ft high	30–55	—	fast
18. PU mattress, single, horizontal	110	—	fast
19. PE insulation board, rigid foam, stacked 15 ft high	8	170	†
20. PS jars, packed in item 6	55	1200	fast
21. PS tubs nested in cartons, stacked 14 ft high	105	450	fast
22. PS toy parts in cartons, stacked 15 ft high	110	180	fast
23. PS insulation board, rigid, stacked 14 ft high	7	290	†
24. PVC bottles, packed in item 6	9	300	†
25. PP tubs, packed in item 6	10	390	†
26. PP and PE film in rolls, stacked 14 ft high	40	350	†
27. Distilled spirits in barrels, stacked 20 ft high	23–40	—	†
28. Methyl alcohol	—	65	—
29. Gasoline	—	200	—
30. Kerosene	—	200	—
31. Diesel oil	—	180	—

For SI units: 1 ft = 0.305 m.
Notes:
1. The heat release rates per unit floor area are for fully involved combustibles, assuming 100 percent combustion efficiency. The growth times shown are those required to exceed 1000 Btu/sec heat release rate for developing fires, assuming 100 percent combustion efficiency.
2. PE = polyethylene, PS = polystyrene, PVC = polyvinyl chloride, PP = polypropylene, PU = polyurethane, and FRP = fiberglass-reinforced polyester.
†Fire growth rate exceeds design data.

evaluates candidate designs against acceptance criteria previously established by a consensus of the relevant stakeholders.

In addition to the operational and response characteristics, stakeholders often set limits on the amount of disruption, visibility, or other ways the fire alarm system will impact the intended use of the protected space. For example, in historic structures the intended use must preserve the history and heritage by preserving the appearance of the structure. In such cases, the visibility of the fire protection system must be limited; the less visible the fire alarm system the better.

B-2.3.3.1 Reliability of the detection system and individual components should be computed and included in the selection and evaluation of the candidate fire detection system. A performance-based alternative design cannot be deemed

Table B-2.3.2.3.1(b) Maximum Heat Release Rates from Fire Detection Institute Analysis

Materials	Approximate Values (Btu/sec)
Medium wastebasket with milk cartons	100
Large barrel with milk cartons	140
Upholstered chair with polyurethane foam	350
Latex foam mattress (heat at room door)	1200
Furnished living room (heat at open door)	4000–8000

Table B-2.3.2.3.1(c) Unit Heat Release Rates for Fuels Burning in the Open (NFPA 92B)

Commodity	Heat Release Rate (Btu/sec)
Flammable liquid pool	$290/\text{ft}^2$ of surface
Flammable liquid spray	2000/gpm of flow
Pallet stack	1000/ft of height
Wood or PMMA* (vertical)	
2-ft height	30/ft of width
6-ft height	70/ft of width
8-ft height	180/ft of width
12-ft height	300/ft of width
Wood or PMMA*	
Top of horizontal surface	$63/\text{ft}^2$ of surface
Solid polystyrene (vertical)	
2-ft height	63/ft of width
6-ft height	130/ft of width
8-ft height	400/ft of width
12-ft height	680/ft of width
Solid polystyrene (horizontal)	$120/\text{ft}^2$ of surface
Solid polypropylene (vertical)	
2-ft height	63/ft of width
6-ft height	100/ft of width
8-ft height	280/ft of width
12-ft height	470/ft of width
Solid polypropylene (horizontal)	$70/\text{ft}^2$ of surface

*Polymethyl Methacrylate (Plexiglas™, Lucite™, Acrylic)

performance-equivalent unless the alternative design provides comparable reliability to the prescriptive design it is intended to replace.

RAMS includes reliability, availability, maintainability, and safety studies. Designers use a RAMS study as an assess-ment method to manage dependability in mission-critical systems. The designer has to consider each of these factors to ensure that the system will continue to operate as designed, as well as to ensure ease of maintenance and safety during maintenance.

A RAMS study is a systematic process, based on the system life cycle and tasks within it, including:

(1) Assistance to the client specifying system requirements, in terms of dependability, from a general mission statement to availability targets for systems and sub-systems, and components (including software)
(2) Assessment of proposed designs, using formal RAMS techniques, to see how targets are met and where objectives are not achieved
(3) Provision of a means to make recommendations to designers and a system of hazard logging, to record and eventually check off identified necessary actions

The technical concepts of availability and reliability are based on a knowledge of and means to assess the following:

(1) All possible system failure modes in the specified application environment
(2) Probability (or rate) of occurrence of a system failure mode
(3) Cause and effect of each failure mode on the functionality of the system
(4) Efficient failure detection and location
(5) Efficient restorability of the failed system
(6) Economic maintenance over the required life cycle of the system
(7) Human factors issues regarding safety during inspection, testing, and maintenance

Fire alarm systems typically have high levels of supervision and fault-tolerant designs. Consequently, designers usually use mission effectiveness to evaluate fire alarm systems rather than strict reliability. The equipment, the system design, the installation, and the maintenance all contribute to the inherent mission effectiveness of a fire alarm system.

B-2.3.3.2 Various methods are available to evaluate whether a candidate design will achieve the previously established performance criteria. Some methods are presented in Section B-3.

Section B-6 discusses some additional modeling methods.

B-2.3.4 Select and Document Final Design. The last step in the process is the preparation of design documentation and equipment and installation specifications.

The designer must be sure to properly document each design decision. Proper documentation establishes the reasoning behind the design decisions, minimizing the opportunity for error. Proper documentation also ensures that all of the involved parties understand the steps needed to imple-

Table B-2.3.2.3.1(d) Characteristics of Ignition Sources (NFPA 92B)

	Typical Heat Output (W)	Burn Time[a] (sec)	Maximum Flame Height (mm)	Flame Width (mm)	Maximum Heat Flux (kW/m²)
Cigarette 1.1 g (not puffed, laid on solid sur face), bone dry, conditioned to 50% relative humidity	5 5	1200 1200	— —	— —	42 35
Methenamine pill, 0.15 g	45	90	—	—	4
Match, wooden, laid on solid surface	80	20–30	30	14	18–20
Wood cribs, BS 5852 Part 2					
No. 4 crib, 8.5 g	1,000	190	—	—	15[d]
No. 5 crib, 17 g	1,900	200	—	—	17[d]
No. 6 crib, 60 g	2,600	190	—	—	20[d]
No. 7 crib, 126 g	6,400	350	—	—	25[d]
Crumpled brown lunch bag, 6 g	1,200	80	—	—	—
Crumpled wax paper, 4.5 g (tight)	1,800	25	—	—	—
Crumpled wax paper, 4.5 g (loose)	5,300	20	—	—	—
Folded double-sheet newspaper, 22 g (bottom ignition)	4,000	100	—	—	—
Crumpled double-sheet newspaper, 22 g (top ignition)	7,400	40	—	—	—
Crumpled double-sheet newspaper, 22 g (bottom ignition)	17,000	20	—	—	—
Polyethylene wastebasket, 285 g, filled with 12 milk cartons (390 g)	50,000	200[b]	550	200	35[c]
Plastic trash bags, filled with cellulosic trash (1.2–14 kg)[e]	120,000– 350,000	200[b]	—	—	—

For SI Units: 1 in. = 25.4 mm; 1 Btu/sec = 1.055 W; 1 oz = 0.02835 kg = 28.35 g; 1 Btu/ft²·sec = 11.35 kW/m²

[a] Time duration of significant flaming
[b] Total burn time in excess of 1800 seconds
[c] As measured on simulation burner
[d] Measured from 25 mm away
[e] Results vary greatly with packing density

ment the design. Such steps include the selection of equipment, the methods of installation, and the maintenance program.

B-2.3.4.1 These documents should encompass the following information [25]:

(1) Participants in the process—persons involved, their qualifications, function, responsibility, interest, and contributions
(2) Scope of work—purpose of conducting the analysis or design, part of the building evaluated, assumptions, and so on.
(3) Design approach—approach taken, where and why assumptions were made, and engineering tools and methodologies applied
(4) Project information—hazards, risks, construction type, materials, building use, layout, existing systems, occupant characteristics, and so on.

(5) Goals and objectives—agreed upon goals and objectives, how they were developed, who agreed to them and when
(6) Performance criteria—clearly identify performance criteria and related objective(s), including any safety, reliability, or uncertainty factors applied, and support for these factors where necessary
(7) Fire scenarios and design fires—description of fire scenarios used, bases for selecting and rejecting fire scenarios, assumptions, and restrictions.
(8) Design alternative(s)—describe design alternative(s) chosen, basis for selecting and rejecting design alternative(s), heat release rate, assumptions, and limitations. [This step should include the specific design objective (Q_{DO}) and the critical heat release rate (Q_{CR}) used, comparison of results with the performance criteria and design objectives, and a discussion of the sensitivity of the selected design alternative to changes in the

Table B-2.3.2.3.1(e) Furniture Heat Release Rates [3, 14, 16]

Test No.	Item/Description/Mass	Growth Time (t_g) (sec)	Classification	Fuel Fire Intensity Coefficient (α) (kW/sec^2)	Virtual Time (t_v) (sec)	Maximum Heat Release Rates (kW)
15	Metal wardrobe, 41.4 kg (total)	50	fast	0.4220	10	750
18	Chair F33 (trial love seat), 29.2 kg	400	slow	0.0066	140	950
19	Chair F21, 28.15 kg (initial)	175	medium	0.0344	110	350
19	Chair F21, 28.15 kg (later)	50	fast	0.4220	190	2000
21	Metal wardrobe, 40.8 kg (total) (initial)	250	medium	0.0169	10	250
21	Metal wardrobe, 40.8 kg (total) (average)	120	fast	0.0733	60	250
21	Metal wardrobe, 40.8 kg (total) (later)	100	fast	0.1055	30	140
22	Chair F24, 28.3 kg	350	medium	0.0086	400	700
23	Chair F23, 31.2 kg	400	slow	0.0066	100	700
24	Chair F22, 31.9 kg	2000	slow	0.0003	150	300
25	Chair F26, 19.2 kg	200	medium	0.0264	90	800
26	Chair F27, 29.0 kg	200	medium	0.0264	360	900
27	Chair F29, 14.0 kg	100	fast	0.1055	70	1850
28	Chair F28, 29.2 kg	425	slow	0.0058	90	700
29	Chair F25, 27.8 kg (later)	60	fast	0.2931	175	700
29	Chair F25, 27.8 kg (initial)	100	fast	0.1055	100	2000
30	Chair F30, 25.2 kg	60	fast	0.2931	70	950
31	Chair F31 (love seat), 39.6 kg	60	fast	0.2931	145	2600
37	Chair F31 (love seat), 40.4 kg	80	fast	0.1648	100	2750
38	Chair F32 (sofa), 51.5 kg	100	fast	0.1055	50	3000
39	$^1/_2$-in. plywood wardrobe with fabrics, 68.5 kg	35	†	0.8612	20	3250
40	$^1/_2$-in. plywood wardrobe with fabrics, 68.32 kg	35	†	0.8612	40	3500
41	$^1/_8$-in. plywood wardrobe with fabrics, 36.0 kg	40	†	0.6594	40	6000
42	$^1/_8$-in. plywood wardrobe with fire-retardant interior finish (initial growth)	70	fast	0.2153	50	2000
42	$^1/_8$-in. plywood wardrobe with fire-retardant interior finish (later growth)	30	†	1.1722	100	5000
43	Repeat of $^1/_2$-in. plywood wardrobe, 67.62 kg	30	†	1.1722	50	3000
44	$^1/_8$-in. plywood wardrobe with fire-retardant latex paint, 37.26 kg	90	fast	0.1302	30	2900
45	Chair F21, 28.34 kg	100	†	0.1055	120	2100
46	Chair F21, 28.34 kg	45	†	0.5210	130	2600
47	Chair, adj. back metal frame, foam cushions, 20.82 kg	170	medium	0.0365	30	250
48	Easy chair CO7, 11.52 kg	175	medium	0.0344	90	950
49	Easy chair F34, 15.68 kg	200	medium	0.0264	50	200
50	Chair, metal frame, minimum cushion, 16.52 kg	200	medium	0.0264	120	3000
51	Chair, molded fiberglass, no cushion, 5.28 kg	120	fast	0.0733	20	35
52	Molded plastic patient chair, 11.26 kg	275	medium	0.0140	2090	700

(continues)

Table B-2.3.2.3.1(e) Continued

Test No.	Item/Description/Mass	Growth Time (t_g) (sec)	Classification	Fuel Fire Intensity Coefficient (α) (kW/sec^2)	Virtual Time (t_v) (sec)	Maximum Heat Release Rates (kW)
53	Chair, metal frame, padded seat and back, 15.54 kg	350	medium	0.0086	50	280
54	Love seat, metal frame, foam cushions, 27.26 kg	500	slow	0.0042	210	300
56	Chair, wood frame, latex foam cushions, 11.2 kg	500	slow	0.0042	50	85
57	Love seat, wood frame, foam cushions, 54.6 kg	350	medium	0.0086	500	1000
61	Wardrobe, $^3/_4$-in. particleboard, 120.33 kg	150	medium	0.0469	0	1200
62	Bookcase, plywood with aluminum frame, 30.39 kg	65	fast	0.2497	40	25
64	Easy chair, molded flexible urethane frame, 15.98 kg	1000	slow	0.0011	750	450
66	Easy chair, 23.02 kg	76	fast	0.1827	3700	600
67	Mattress and box spring, 62.36 kg (later)	350	medium	0.0086	400	500
67	Mattress and box spring, 62.36 kg (initial)	1100	slow	0.0009	90	400

For SI units: 1 ft = 0.305 m; 1000 Btu/sec = 1055 kW; 1 lb = 0.435 kg.

Note: For tests 19, 21, 29, 42, and 67, different power law curves were used to model the initial and the latter realms of burning. In examples such as these, engineers should choose the fire growth parameter that best describes the realm of burning to which the detection systems is being designed to respond.

[†] Fire growth exceeds design data.

building use, contents, fire characteristics, occupants, and so on.]

(9) Engineering tools and methods used—description of engineering tools and methods used in the analysis or design, including appropriate references (literature, date, software version, and so on), assumptions, limitations, engineering judgments, input data, validation data or procedures, and sensitivity analyses

(10) Drawings and specifications—detailed design and installation drawings and specification

(11) Test, inspection, and maintenance requirements *(see Chapter 7)*

(12) Fire safety management concerns—allowed contents and materials in the space in order for the design to function properly, training, education, and so on.

(13) References—software documentation, technical literature, reports, technical data sheets, fire test results, and so on.

Critical design assumptions include all assumptions on which the design relies. These assumptions must be maintained throughout the life cycle of the building. Maintaining the assumption ensures that the design will function as intended.

Critical design features include the design features and parameters that must be maintained throughout the life of the building so that the design functions as intended.

Specifications and drawings result from a properly conducted performance-based design process.

An operations and maintenance manual is an essential part of the design. The manual should clearly state the requirements for ensuring that the components of the performance-based design are placed correctly and functioning as intended. The manual must identify all subsystems, as well as their operation and interaction with the fire detection system. The operations and maintenance manual should also include maintenance and testing frequencies, and methods and forms for recording the maintenance activities. The manual must also detail the importance of testing interconnected systems (i.e., elevator recall, suppression systems, HVAC shutdown, etc.).

B-2.3.5 Management. It is important to ensure that the systems are designed, installed, commissioned, maintained,

and tested on regular intervals as indicated in Chapter 7. In addition, the person conducting the testing and inspections should be aware of the background of the design and the need to evaluate not only the detector and whether it operates, but also to be aware of changing conditions including the following:

(1) Hazard being protected changes
(2) Location of the hazard changes
(3) Other hazards are introduced into the area
(4) Ambient environment
(5) Invalidity of any of the design assumptions

B-3 Evaluation of Heat Detection System Performance

B-3.1 General.

Section B-3 provides a method for determining the application spacing for both fixed-temperature heat detectors (including sprinklers) and rate-of-rise heat detectors. This method is only valid for use when detectors are to be placed on a large, flat ceiling. It predicts detector response to a geometrically growing flaming fire at a specific fire size. This method takes into account the effects of ceiling height, radial distance between the detector and the fire, threshold fire size [critical heat release rate (Q_{CR})], rate of fire development, and detector response time index. For fixed temperature detectors, the ambient temperature and the temperature rating of the detector are also considered. This method also allows for the adjustment of the application spacing for fixed-temperature heat detectors to account for variations in ambient temperature (T_a) from standard test conditions.

B-3.1.1 This method can also be used to estimate the fire size at which detection will occur, an existing array of listed heat detectors installed at a known spacing, ceiling height, and ambient conditions.

To analyze the response of an existing fire detection system, the designer must also quantify the fire growth rate and detector temperature rating.

B-3.1.2 The effect of rate of fire growth and fire size of a flaming fire, as well as the effect of ceiling height on the spacing and response of smoke detectors can also be determined using this method.

A designer can predict the response of a smoke detector by modeling the smoke detector as a very sensitive heat detector. Engineers have used this model for many years. The model relies on the premise that the ceiling jet, formed by the buoyant plume as it collides with the ceiling, provides the force to move smoke horizontally beneath the ceiling from the fire to the detector. Consequently, the smoke and the heat are conveyed together. This model supports the notion that a correlation exists between temperature rise and smoke density. However, the prudent designer exercises caution because he or she knows that this correlation is loose at best.

B-3.1.3 The methodology contained herein uses theories of fire development, fire plume dynamics, and detector performance. These are considered the major factors influencing detector response. This methodology does not address several lesser phenomena that, in general, are considered unlikely to have a significant influence. A discussion of ceiling drag, heat loss to the ceiling, radiation to the detector from a fire, re-radiation of heat from a detector to its surroundings, and the heat of fusion of eutectic materials in fusible elements of heat detectors and their possible limitations on the design method are provided in References 4, 11, 16, and 18 in Appendix C.

B-3.1.4 The methodology in Section B-3 does not address the effects of ceiling projections, such as beams and joists, on detector response. While it has been shown that these components of a ceiling have a significant effect on the response of heat detectors, research has not yet resulted in a method for quantifying this effect. The prescriptive adjustments to detector spacing in Chapter 5 should be applied to application spacings derived from this methodology.

This method does not consider the response of heat detectors on sloped ceilings. The designer should adjust detector spacings that have been determined using this method, by applying the spacing adjustments outlined in Chapter 2.

B-3.2 Considerations Regarding Input Data.

B-3.2.1 Required Data. The following data are necessary in order to use the methods in this appendix for either design or analysis.

(a) *Design.* Data required to determine design include the following:

H = ceiling height or clearance above fuel

Q_d or t_d = threshold fire size at which response must occur or the time t_d to detector response

RTI = response time index for the detector (heat detectors only) or its listed spacing

T_a = ambient temperature

T_s = detector operating temperature (heat detectors only)

T_s/min = rate of temperature change set point for rate-of-rise heat detectors

α or t_g = fuel fire intensity coefficient (α) or the fire growth time (t_g)

(b) *Analysis.* Data required to determine analysis include the following:

H = ceiling height or clearance above fuel

RTI = response time index for the detector (heat detectors only) or its listed spacing

S = actual installed spacing of the existing detectors

T_a = ambient temperature

T_s = detector operating temperature (heat detectors only)

T_s/min = rate of temperature change set point for rate-of-rise heat detectors

α or t_g = fuel fire intensity coefficient (α) or the fire growth time (t_g)

The following are the units for variables listed in B-3.2.1:

$$
\begin{aligned}
T &= {}^\circ\text{C} \\
H &= \text{m} \\
\text{RTI} &= \text{m}_{1/2}\text{s}_{1/2} \\
\alpha &= \text{kW/s}^2 \\
t_g &= \text{sec} \\
S &= \text{m} \\
Q &= \text{kW}
\end{aligned}
$$

B-3.2.2 Ambient Temperature Considerations.

B-3.2.2.1 The maximum ambient temperature expected to occur at the ceiling will directly effect the choice of temperature rating for a fixed-temperature heat detector application. However, the minimum ambient temperature likely to be present at the ceiling is also very important. When ambient temperature at the ceiling decreases, more heat from a fire is needed to bring the air surrounding the detector's sensing element up to its rated (operating) temperature. This results in slower response when the ambient temperature is lower. In the case of a fire that is growing over time, lower ambient temperatures result in a larger fire size at the time of detection.

B-3.2.2.2 Selection of the minimum ambient temperature can therefore have a significant effect on the calculations. The engineer must decide what temperature to use for these calculations and document why that temperature was chosen. Because the response time of a given detector to a given fire is dependent only on the detector's time constant and the temperature difference between ambient and the detector rating, the use of the lowest anticipated ambient temperature for the space results in the most conservative design. For unheated spaces, a review of historical weather data would be appropriate. However, such data might show extremely low temperatures that occur relatively infrequently, such as every 100 years. Depending on actual design considerations, it might be more appropriate to use an average minimum

ambient temperature. In any case, a sensitivity analysis should be performed to determine the effect of changing the ambient temperature on the design results.

The National Oceanic and Atmospheric Administration (NOAA) provides one frequently used source for weather data in the United States.

B-3.2.2.3 In a room or work area that has central heating, the minimum ambient temperature would usually be about 68°F (20°C). On the other hand, certain warehouse occupancies might be heated only enough to prevent water pipes from freezing and, in this case, the minimum ambient temperature may be considered to be 35°F (2°C), even though, during many months of the year, the actual ambient temperature may be much higher.

B-3.2.3 Ceiling Height Considerations.

B-3.2.3.1 A detector ordinarily operates sooner if it is nearer to the fire. Where ceiling heights exceed 16 ft (4.9 m), ceiling height is the dominant factor in the detection system response.

When the calculations show that the fire size at the time of response exceeds the design objective (design fire), the designer may reduce the spacing of detectors. He or she can move detectors closer to the fire plume centerline. In some circumstances, particularly with high ceilings and small design fires, the designer must understand that further reductions in the detector spacing cannot improve the system response. When the detector spacing is reduced to less that 0.4 times the ceiling height, a detector is in the plume regardless of the fire location. Further spacing reductions will not enhance system performance. If the design goals for the hazard area require a faster response, the designer should consider other types of detection.

B-3.2.3.2 As flaming combustion commences, a buoyant plume forms. The plume is comprised of the heated gases and smoke rising from the fire. The plume assumes the general shape of an inverted cone. The smoke concentration and temperature within the cone varies inversely as a variable exponential function of the distance from the source. This effect is very significant in the early stages of a fire, because the angle of the cone is wide. As a fire intensifies, the angle of the cone narrows and the significance of the effect of height is lessened.

B-3.2.3.3 As the ceiling height increases, a larger-size fire is necessary to actuate the same detector in the same length of time. In view of this, it is very important that the designer consider the size of the fire and rate of heat release that might develop before detection is ultimately obtained.

B-3.2.3.4 The procedures presented in this section are based on analysis of data for ceiling heights up to 30 ft (9.1 m). No data was analyzed for ceiling heights greater than 30

ft (9.1 m). In spaces where the ceiling heights exceed this limit, this section offers no guidance. [40]

B-3.2.3.5 The relationships presented here are based on the difference between the ceiling height and the height of the fuel item involved in the fire. It is recommended that the designer assume the fire is at floor level and use the actual distance from floor to ceiling for the calculations. This will yield a design that is conservative, and actual detector response can be expected to exceed the needed speed of response in those cases where the fire begins above floor level.

When analyzing an existing detection system, if the designer assigns a value for H that represents the distance from the floor to the ceiling, then the assumption will lead to a maximum predicted detection time. In such a case, the predicted detection time will exceed the actual detection time during a fire. Detection time occurs because in many cases a building occupant has located the fuel load a significant distance above the floor. For a more accurate analysis, the designer should select a value for H that represents a reasonable worst case.

B-3.2.3.6 Where the designer desires to consider the height of the potential fuel in the room, the distance between the base of the fuel and the ceiling should be used in place of the ceiling height. This design option is only appropriate if the minimum height of the potential fuel is always constant and the concept is approved by the authority having jurisdiction.

B-3.2.4 Operating Temperature.

B-3.2.4.1 The operating temperature, or rate of temperature change, of the detector required for response is obtained from the manufacturer's data and is determined during the listing process.

B-3.2.4.2 The difference between the rated temperature of a fixed temperature detector (T_s) and the maximum ambient temperature (T_a) at the ceiling should be as small as possible. To reduce unwanted alarms, the difference between operating temperature and the maximum ambient temperature should be not less than 20°F (11°C). *(See Chapter 2.)*

The designer should thoroughly analyze the location of heat detectors to ensure that there are no non-fire sources of heat (i.e., equipment, vehicles, heaters, etc.) in the vicinity of the detector. These types of heat sources cause local or intermittent hot spots and lead to unwanted alarms.

B-3.2.4.3 If using combination detectors incorporating both fixed temperature and rate-of-rise heat detection principles to detect a geometrically growing fire, the data contained herein for rate-of-rise detectors should be used in selecting an installed spacing, because the rate-of-rise principle controls the response. The fixed temperature set point is determined from the maximum anticipated ambient temperature.

B-3.2.5 Time Constant and Response Time Index (RTI).

B-3.2.5.1 The flow of heat from the ceiling jet into a heat detector sensing element is not instantaneous. It occurs over a period of time. A measure of the speed with which heat transfer occurs, the thermal response coefficient is needed to accurately predict heat detector response. This is currently called the detector time constant (τ_0). The time constant is a measure of the detector's sensitivity. In theory, the sensitivity of a heat detector, τ_0 or RTI, should be determined by validated test. [8] Currently, such a test is not available. Given the detector's listed spacing and the detector's rated temperature (T_s), Table B-3.2.5.1, developed in part by Heskestad and Delichatsios [10], can be used to find the detector time constant.

As a detector's time constant increases, the response time also increases. Currently, the only available measure of heat detector thermal response coefficient is the Response Time Index (RTI). The designer should note that UL-listed and FMRC-approved detectors of the same temperature rating and listed spacing have different values for RTI. This difference in RTI values occurs because UL and FMRC use different methods to test detectors.

See B-3.3.3 for additional information regarding time constants.

B-3.2.6 Fire Growth Rate.

B-3.2.6.1 Fire growth varies depending on the combustion characteristics and the physical configuration of the fuels involved. After ignition, most fires grow in an accelerating pattern. Some information regarding the fire growth rate for various fuels have been provided previously in this appendix.

See discussion on fire characteristics and growth rates under B-2.3.1.2.3.

B-3.2.6.2 If the heat release history for a particular fire is known, the α or t_g can be calculated using curve fitting techniques for implementation into the method detailed herein.[16]

B-3.2.6.3 In most cases, the exact fuel(s) and growth rates will not be known. Engineering judgment should therefore be used to select α or t_g that is expected to approximate the fire. Sensitivity analysis should also be performed to determine the effect on response from changes in the expected fire growth rate. In some analyses the effect on response will be negligible. Other cases might show that a more thorough analysis of potential fuels and fire scenarios is necessary.

B-3.2.7 Threshold Fire Size. The user should refer to previous sections regarding discussions on determining threshold fire sizes (Q_{DO} and Q_{CR}) to meet the design objectives.

Table B-3.2.5.1 Time Constants (τ_0) for Any Listed Heat Detector

Listed Spacing (ft)	Underwriters Laboratories Inc.						Factory Mutual Research Corporation (All Temperatures)
	128°	135°	145°	160°	170°	196°	
10	400	330	262	195	160	97	196
15	250	190	156	110	89	45	110
20	165	135	105	70	52	17	70
25	124	100	78	48	32	—	48
30	95	80	61	36	22	—	36
40	71	57	41	18	—	—	—
50	59	44	30	—	—	—	—
70	36	24	9	—	—	—	—

For SI units: 1 ft = 0.305 m.

Notes:

1. These time constants are based on an analysis [10] of the Underwriters Laboratories Inc. and Factory Mutual listing test procedures.

2. These time constants can be converted to response time index (RTI) values by using the following equation:

RTI = τ_0 (5.0 ft/sec)$^{1/2}$ *(Refer also to B-3.3.)*

*At a reference velocity of 5 ft/sec (1.5 m/sec).

The designer should select threshold fire sizes carefully. He or she should perform a sensitivity analysis to quantify the effect that variations in threshold fire size have on the system response.

B-3.3 Heat Detector Spacing.

B-3.3.1 Fixed-Temperature Heat Detector Spacing. The following method can be used to determine the response of fixed-temperature heat detectors for designing or analyzing heat detection systems.

B-3.3.1.1 The objective of designing a detection system is to determine the spacing of detectors required to respond to a given set of conditions and goals. In order to achieve the objectives, detector response must occur when the fire reaches a critical heat release rate, or in a specified time.

B-3.3.1.2 When analyzing an existing detection system, the engineer is looking to determine the size of the fire at the time that the detector responds.

B-3.3.2 Theoretical Background. [26, 28]

B-3.3.2.1 The design and analysis methods contained in Appendix B are the joint result of extensive experimental work and of mathematical modeling of the heat and mass transfer processes involved. The original method was developed by Heskestad and Delichatsios [9, 10], Beyler [4], and Schifiliti [16]. It was recently updated by Marrion [28] to reflect changes in the original correlations as discussed in work by Heskestad and Delichatsios [11] and Marrion [27]. Subsection B-3.3.2 outlines methods and data correlations used to model the heat transfer to a heat detector, as well as velocity and temperature correlations for growing fires at

the location of the detector. Only the general principles are described here. More detailed information is available in References 4, 9, 10, 16, and 28 in Appendix C.

B-3.3.3 Heat Detector Correlations. [26, 28] The heat transfer to a detector can be described by the following equation:

$$Q_{total} = Q_{cond} + Q_{conv} + Q_{rad} \text{ (kW or Btu/sec)} \quad (6)$$

where:

Q_{total} = total heat transfer to a detector

Q_{cond} = conductive heat transfer

Q_{conv} = convective heat transfer

Q_{rad} = radiative heat transfer

B-3.3.3.1 Because detection typically occurs during the initial stages of a fire, the radiant heat transfer component (Q_{rad}) can be considered negligible. In addition, because the heat sensing elements of most of the heat detectors are thermally isolated from the rest of the detection unit, as well as from the ceiling, it can be assumed that the conductive portion of the heat release rate (Q_{cond}) is also negligible, especially when compared to the convective heat transfer rate. Because the majority of the heat transfer to the detection element is via convection, the following equation can be used to calculate the total heat transfer:

$$Q = Q_{cond} = H_c A(T_g - T_d) \text{ (kW or Btu/sec)} \quad (7)$$

where:

Q_{conv} = convective heat transfer

H_c = convective heat transfer coefficient for the detector in kW/m^2·°C or in Btu/ft^2·sec·°F

A = surface area of the detector's element

T_g = temperature of fire gases at the detector

T_d = temperature rating, or set point, of the detector

B-3.3.3.2 Assuming the detection element can be treated as a lumped mass (m) (kg or lbm) its temperature change can be defined as follows:

$$\frac{dT_d}{dt} = \frac{Q}{mc} \quad \text{(deg/sec)} \tag{8}$$

where:

Q = heat release rate (Btu/sec or kW)

m = detector element's mass

c = detector element's specific heat (kJ/kg·°C or Btu/lbm·°F)

B-3.3.3.3 Substituting this into the previous equation, the change in temperature of the detection element over time can be expressed as follows:

$$\frac{dT_d}{dt} = \frac{H_c A (T_g - T_d)}{mc} \tag{9}$$

Note that the variables are identified in Section B-7.

B-3.3.3.4 The use of a time constant (τ) was proposed by Heskestad and Smith [8] in order to define the convective heat transfer to a specific detector's heat sensing element. This time constant is a function of the mass, specific heat, convective heat transfer coefficient, and area of the element and can be expressed as follows:

$$\tau = \frac{mc}{H_c A} \quad \text{(sec)} \tag{10}$$

where:

m = detector element's mass

c = detector element's specific heat (kJ/kg·°C or Btu/lbm·°F)

H_c = convective heat transfer coefficient for the detector (kW/m^2·°C or Btu/ft^2·sec·°F)

A = surface area of the detector's element

τ = detector time constant

B-3.3.3.5 As seen in the equation (10), τ is a measure of the detector's sensitivity. By increasing the mass of the detection element, the time constant, and thus the response time, increases.

B-3.3.3.6 Substituting into equation (9) produces the following:

$$\frac{dT_d}{dt} = \frac{T_g - T_d}{\tau}$$

Note that the variables are identified in Section B-7.

B-3.3.3.7 Research has shown [24] that the convective heat transfer coefficient for sprinklers and heat detection elements are similar to that of spheres, cylinders, and so on, and is thus approximately proportional to the square root of the velocity of the gases passing the detector. As the mass, thermal capacity, and area of the detection element remain constant, the following relationship can be expressed as the response time index (RTI) for an individual detector:

$$\tau u^{1/2} \sim \tau_0 u_0^{1/2} = \text{RTI} \tag{11}$$

where:

τ = detector time constant

u = velocity of fire gases (m/sec)

u_0 = instantaneous velocity of fire gases (m/sec or ft/sec)

RTI = response time index

B-3.3.3.8 If τ_0 is measured at a given reference velocity (u_0), τ can be determined for any other gas velocity (u) for that detector. A plunge test is the easiest way to measure τ_0. It has also been related to the listed spacing of a detector through a calculation. Table B-3.2.5.1 presents results from these calculations [10]. The RTI value can then be obtained by multiplying τ_0 values by $u_0^{1/2}$.

B-3.3.3.9 It has become customary to refer to the time constant using a reference velocity of u_0 = 5 ft/sec (1.5 m/sec). For example, where u_0 = 5 ft/sec (1.5 m/sec), a τ_0 of 30 seconds corresponds to an RTI of 67 sec$^{1/2}$/ft$^{1/2}$ (or 36 sec$^{1/2}$/m$^{1/2}$). On the other hand, a detector that has an RTI of 67 sec$^{1/2}$/ft$^{1/2}$ would have a τ_0 of 23.7 sec, if measured in an air velocity of 8 ft/sec (2.4 m/sec).

B-3.3.3.10 The following equation can therefore be used to calculate the heat transfer to the detection element, and thus determine its temperature from its local fire-induced environment.

$$\frac{dT_d}{dt} = \frac{u^{1/2}(T_g - T_d)}{\text{RTI}} \tag{12}$$

Note that the variables are identified in Section B-7.

B-3.3.4 Temperature and Velocity Correlations. [26, 28] In order to predict the operation of any detector, it is necessary to characterize the local environment created by the fire at the location of the detector. For a heat detector, the important variables are the temperature and velocity of the gases at the detector. Through a program of full-scale tests and the use of mathematical modeling techniques, general expressions for temperature and velocity at a detector location have been developed by Heskestad and Delichatsios (refer to references 4, 9, 10, and 16 in Section C-2). These expressions are valid for fires that grow according to the following power law relationship:

$$Q(\text{kW}) = \alpha(\text{Btu/sec}^2)t^p \tag{13}$$

where:

Q = theoretical convective fire heat release rate

α = fire growth rate

t = time

p = positive exponent

B-3.3.4.1 Relationships have been developed by Heskestad and Delichatsios [9] for temperature and velocity of fire gases in a ceiling jet. These have been expressed as follows [26]:

$$U_p{}^* = \frac{u}{A^{1/(3+p)}u^{1/(3+p)}H^{(p-1)/(3+p)}} = f\left(t_p{}^*, \frac{r}{H}\right) \tag{14}$$

$$\Delta T_p{}^* = g\left(t_p{}^*, \frac{r}{H}\right) = \frac{\Delta T}{A^{2/(3+p)}\left(\dfrac{T_a}{g}\right)\alpha^{2/(3+p)}H^{-(5-p)/(3+p)}}$$

where

$$t_p{}^* = \frac{t}{A^{-1/(3+p)}\alpha^{-1/(3+p)}H^{4/(3+p)}}$$

and

$$A = \frac{g}{C_p T_a \rho_0}$$

Note that the variables are identified in Section B-7.

B-3.3.4.2 Using the above correlations, Heskestad and Delichatsios [9], and with later updates from another paper by Heskestad [11], the following correlations were presented for fires that had heat release rates that grew according to the power law equation, with $p = 2$. As previously discussed [10, 18], the $p = 2$ power law fire growth model can be used to model the heat release rate of a wide range of fuels. These fires are therefore referred to as *t-squared* fires.

$$t_{2f}^* = 0.861\left(1 + \frac{r}{H}\right) \tag{15}$$

$$\Delta T_2^* = 0 \text{ for } t_2^* < t_{2f}^*$$

$$\Delta T_2^* = \left[\frac{t_2^* - t_{2f}^*}{0.146 + 0.242r/H}\right]^{4/3} \text{ for } t_2^* \geq t_{2f}^*$$

$$\frac{u_2^*}{(\Delta T_2^*)^{1/2}} = 0.59\left(\frac{r}{H}\right)^{-0.63}$$

Note that the variables are identified in Section B-7.

B-3.3.4.3 Work by Beyler [4] determined that the above temperature and velocity correlations could be substituted

into the heat transfer equation for the detector and integrated. His analytical solution is as follows:

$$T_d(t) - T_d(0) = \left(\frac{\Delta T}{\Delta T_2^*}\right)\Delta T_2^*\left[\frac{1 - (1 - e^{-Y})}{Y}\right]$$

$$\frac{dT_d(t)}{dt} = \frac{\left(\dfrac{4}{3}\right)\left(\dfrac{\Delta T}{\Delta T_2^*}\right)(\Delta T_2^*)^{1/4}(1 - e^{-Y})}{\left(\dfrac{t}{t_2^*}\right)D} \tag{16}$$

where

$$Y = \left(\frac{3}{4}\right)\left(\frac{u}{u_2^*}\right)^{1/2}\left[\frac{u_2^*}{\Delta T_2^{*1/2}}\right]^{1/2}\left(\frac{\Delta T_2^*}{\text{RTI}}\right)\left(\frac{t}{t_2^*}\right)D \tag{17}$$

and

$$D = 0.146 + 0.242r/H$$

Note that variables are identified in Section B-7.

B-3.3.4.4 The steps involved in solving these equations for either a design or analysis situation are presented in Figure B-3.3.4.4 [28].

B-3.3.5 Limitations. [26] If velocity and temperature of the fire gases flowing past a detector cannot be accurately determined, errors will be introduced when calculating the response of a detector. The graphs presented by Heskestad and Delichatsios indicate the errors in the calculated fire gas temperatures and velocities [10]. A detailed analysis of these errors is beyond the scope of this appendix, however, some discussion is warranted. In using the method described above, the user should be aware of the limitations of these correlations, as outlined in Reference 26. The designer should also refer back to the original reports.

Graphs of actual and calculated data show that errors in ΔT_2^* can be as high as 50 percent, although generally there appears to be much better agreement. The maximum errors occur at r/H values of about 0.37. All other plots of actual and calculated data, for various r/H, show much smaller errors. In terms of the actual change in temperature over ambient, the maximum errors are on the order of 5°C to 10°C. The larger errors occur with faster fires and lower ceilings.

At $r/H = 0.37$, the errors are conservative when the equations are used in a design problem. That is, the equations predicted lower temperatures. Plots of data for other values of r/H indicate that the equations predict slightly higher temperatures.

Errors in fire–gas velocities are related to errors in temperatures. The equations show that the velocity of the fire

Fire Detection Design and Analysis Worksheet [28]
Design Example

1.	Determine ambient temperature (T_a) ceiling height or height above fuel (H).	$T_a = $ _____ °C + 273 = _____ K $H = $ _____ m
2.	Determine the fire growth characteristic (α or t_g) for the expected design fire.	$\alpha = $ _____ kW/s^2 $t_g = $ _____ sec
3a.	Define the characteristics of the detectors.	$T_s = $ _____ C RTI = _____ m$^{1/2}$s$^{1/2}$ $\dfrac{dT_d}{dt} = $ _____ °C/min $\tau_0 = $ _____ sec
3b. or 3b.	*Design* — Establish system goals (t_{CR} or Q_{CR}) and make a first estimate of the distance (r) from the fire to the detector.	$t_{CR} = $ _____ sec $r = $ _____ m $Q_{CR} = $ _____ kW
	Analysis — Determine spacing of existing detectors and make a first estimate of the response time or the fire size at detector response ($Q = \alpha t^2$).	$r = $ _____ *1.41 = _____ = S (m) $Q = $ _____ kW $t_d = $ _____ sec
4.	Using equation (15), calculate the nondimensional time $\left(t^*_{2f}\right)$ at which the initial heat front reaches the detector.	$t^*_{2f} = 0.861 \left(1 + \dfrac{r}{H}\right)$ $t^*_{2f} = $
5.	Calculate the factor A defined in equation (14).	$A = \dfrac{g}{C_p T_a \rho 0}$ $A = $
6.	Use the required response time along with equation (14) and $p = 2$ to calculate the corresponding value of t^*_2.	$t^*_2 = \dfrac{t}{A^{-1/(3+p)}\, \alpha^{-1/(3+p)}\, H^{4/(3+p)}}$ $t^*_2 = $
7.	If $t^*_2 > t^*_{2f}$ continue to step 8. If not, try a new detector position (r) and return to step 4.	
8.	Calculate the ratio $\dfrac{u}{u^*_p}$ using equation (18).	$\dfrac{u}{u^*_p} = A^{1/(3+p)}\, a^{1/(3+p)}\, H^{(p-1)/(3+p)}$ $\dfrac{u}{u^*_p} = $
9.	Calculate the ratio $\dfrac{\Delta T}{\Delta T^*_2}$ using equation (19).	$\dfrac{\Delta T}{\Delta T^*_2} = A^{2/(3+p)}(T_a/g)\, a^{2/(3+p)}\, H^{-(5-p)/(3+p)}$ $\dfrac{\Delta T}{\Delta T^*_2} = $
10.	Use equation (15) to calculate ΔT^*_2.	$\Delta T^*_2 = \left[\dfrac{t^*_2 - t^*_{2f}}{(0.146 + 0.242 r/H)}\right]^{4/3}$ $\Delta T^*_2 = $
11.	Use equation (17) to calculate the ratio $\dfrac{u^*_2}{(\Delta T^*_2)^{1/2}}$.	$\dfrac{u^*_2}{(\Delta T^*_2)^{1/2}} = 0.59\left(\dfrac{r}{H}\right)^{-0.63}$ $\dfrac{u^*_2}{(\Delta T^*_2)^{1/2}} = $
12.	Use equations (16) and (17) to calculate Y.	$Y = \left(\dfrac{3}{4}\right)\left(\dfrac{u}{u^*_2}\right)^{1/2}\left[\dfrac{u^*_2}{\Delta T^{*\,1/2}_2}\right]^{1/2}\left(\dfrac{\Delta T^*_2}{RTI}\right)\left(\dfrac{t}{t^*_2}\right)D$ $Y = $
13.	*Fixed Temperature HD* — Use equation (16) to calculate the resulting temperature of the detector.	$T_d(t) = \left(\dfrac{\Delta T}{\Delta T^*_2}\right)\Delta T^*_2\left[1 - \dfrac{(1 - e^{-Y})}{Y}\right] + T_d(0)$ $T_d(t) = $
14.	*Rate of Rise HD* — Use equation (16).	$dT_d = \left[\left(\dfrac{4}{3}\right)\left(\dfrac{\Delta T}{\Delta T^*_2}\right)(\Delta T^*_2)^{1/4}\dfrac{(1 - e^{-y})}{[(t/t^*_2)D]}\right]dT_d$ $dT_d = $
15.	If: 1. $T_d < T_s$ 2. $T_d > T_s$ 3. $T_d = T$	Repeat Procedure Using Design Analysis 1. a larger r 1. a larger t_r 2. a smaller r 2. a smaller t_r 3. s = 1.41 × r = _____ m 3. $t_r = $ _____ sec

Figure B-3.3.4.4 *Fire detection design and analysis worksheet.* *[28]*

gases is proportional to the square root of the change in temperatures of the fire gases. In terms of heat transfer to a detector, the detector's change in temperature is proportional to the change in gas temperature and the square root of the fire–gas velocity. Hence, the expected errors bear the same relationships.

Based on the above, errors in predicted temperatures and velocities of fire gases will be greatest for fast fires and low ceilings. Sample calculations simulating these conditions show errors in calculated detector spacings on the order of plus or minus one meter, or less.

B-3.3.5.1 The procedures presented in this appendix are based on an analysis of test data for ceiling heights up to 30 ft (9.1 m). No data was analyzed for ceilings greater than 30 ft (9.1 m). The reader should refer to Reference 40 for additional insight.

B-3.3.6 Design Examples.

B-3.3.6.1 Define Project Scope. A fire detection system is to be designed for installation in an unsprinklered warehouse building. The building has a large, flat ceiling that is approximately 4 m (13.1 ft) high. The ambient temperature inside is normally 10°C (50°F). The municipal fire service has indicated that it can begin putting water on the fire within 5.25 minutes of receiving the alarm.

B-3.3.6.2 Identify Goals. Provide protection of property.

B-3.3.6.3 Define Stakeholder's Objective. No fire spread from initial fuel package.

B-3.3.6.4 Define Design Objective. Prevent radiant ignition of adjacent fuel package.

B-3.3.6.5 Develop Performance Criteria. After discussions with the plant fire brigade with regard to their capability and analyzing the radiant energy levels necessary to ignite adjacent fuel packages it was determined that the fire should be detected and suppression activities started prior to it reaching 10,000 kW.

B-3.3.6.6 Develop Fire Scenarios and the Design Fire. Evaluation of the potential contents to be warehoused identified the areas where wood pallets are stored to be one of the highest fire hazards.

The designer cannot come to the conclusion in B-3.3.6.6 unless he or she reviews all of the combustibles, their heat release rates, and their orientation. With this analysis, the designer can then determine if the stack of wood pallets indeed presents the worst case. The designer must analyze other scenarios using fire loads in other areas to verify that this represents the worst case scenario. Furthermore, if stakeholders desire future flexibility in the use of the space, the designer must use additional factors of safety.

B-3.3.6.6.1 The fire scenario involving the ignition of a stack of wood pallets will therefore be evaluated. The pallets are stored 1.5 m (5 ft) high. Fire test data [see Table B-2.3.2.3.1(a)] indicate that this type of fire follows the t^2 power law equation with a t_g equal to approximately 150 to 310 seconds. In order to be conservative, the faster fire growth rate will be used. Thus,

$$Q(\text{kW}) = \alpha(\text{kW/sec}^2)t^p$$

$$1055(\text{kW}) = \alpha(\text{kW/sec}^2)(150 \text{ sec})^2$$

$$\alpha = 0.047(\text{kW/sec}^2)$$

Note that variables are identified in Section B-7.

B-3.3.6.7 Using the power law growth equation with $p = 2$, the time after open flaming until the fire grows to 10,000 kW can be calculated as follows:

$$Q = \left| \frac{1055}{t^2} \right| \quad t_{DO}^2 = \alpha t^2 \tag{18}$$

$$t_{DO} = 461 \text{ seconds}$$

Note that variables are identified in Section B-7.

As part of this analysis, the designer should verify that sufficient fuel exists in the initial fuel package to allow the fire to sustain the continued growth rate over this length of time. Insufficient fuel or a change in fire growth rate will affect the detector response.

B-3.3.6.8 The critical heat release rate and time to detection can therefore be calculated as follows, assuming t_{respond} equals the 1 minute necessary for the fire brigade to respond to the alarm and begin discharging water.

$$t_{CR} = t_{DO} - t_{\text{respond}} \tag{19}$$

$$t_{CR} = 461 - 315 = 146 \text{ seconds}$$

and thus

$$Q_{CR} = \alpha t_{CR}^2$$

$$Q_{CR} = 1000 \text{ kW}$$

Note that variables are identified in Section B-7.

B-3.3.7 Develop Candidate Designs. Fixed-temperature heat detectors have been selected for installation in the warehouse with a 57°C (135°F) operating temperature and a UL-listed spacing of 30 ft (9.1 m). From Table B-3.2.5.1, the time constant is determined to be 80 seconds when referenced to a gas velocity of 1.5 m/sec (5 ft/sec). When used with equation (11), the detector's RTI can be calculated as follows:

$$\text{RTI} = t_0 u_0^{1/2}$$

$$\text{RTI} = 98 \text{ m}^{1/2} \text{ sec}^{1/2} \tag{20}$$

B-3.3.7.1 In order to begin calculations, it will be necessary to make a first guess at the required detector spacing. For

this example, a first estimate of 4.7 m (15.3 ft) is used. This correlates to a radial distance of 3.3 m (10.8 ft).

B-3.3.8 Evaluate Candidate Designs. These values can then be entered into the design and analysis worksheet shown in Figure B-3.3.8 in order to evaluate the candidate design.

B-3.3.8.1 After 146 seconds, when the fire has grown to 1000 kW and at a radial distance of 3.3 m (10.8 ft) from the center of the fire, the detector temperature is calculated to be 57°C. This is the detector actuation temperature. If the calculated temperature of the detector were higher than the actuation temperature, then the radial distance could be increased. The calculation would then be repeated until the calculated detector temperature is approximately equal to the actuation temperature.

If, for some reason, the designer cannot change the detector spacing, the designer can select another type of heat detector with a different listed spacing and, hence, a different RTI. He or she can then repeat the design calculation to determine if a system using the second type of detector meets the performance criteria.

B-3.3.8.2 The last step is to use the final calculated value of r with the equation relating spacing to radial distance. This will determine the maximum installed detector spacing that will result in detector response within the established goals.

$$S = 2^{1/2} r \qquad (21)$$
$$S = 4.7 \text{ m}$$

where:

S = spacing of detectors
r = radial distance from fire plume axis (m or ft)

At this point, the designer should perform a sensitivity analysis for each of the variables to determine if the design inordinately relies on an assumed value for one or more parameters.

B-3.3.8.3 Example of Analysis.

B-3.3.8.3.1 The following example shows how an existing heat detection system or a proposed design can be analyzed to determine the response time or fire size at response. The scenario that was analyzed in the previous example will be used again, with the exception that the warehouse building has existing heat detectors. The fire, building, and detectors have the same characteristics as the previous example with the exception of spacing. The detectors are spaced evenly on the ceiling at 9.2 m (30 ft) intervals.

B-3.3.8.3.2 The following equation is used to determine the maximum radial distance from the fire axis to a detector:

$$S = 1.414r \qquad (22)$$

or

$$r = \frac{S}{1.414}$$
$$r = 6.5 \text{ m}$$

where:

S = spacing of detectors
r = radial distance from fire plume axis (m or ft)

B-3.3.8.3.3 Next, the response time of the detector or the fire size at response is estimated. In the design above, the fire grew to 1000 kW in 146 seconds when the detector located at a distance of 3.3 m (10.8 ft) responded. As the radial distance in this example is larger, a slower response time and thus a larger fire size at response is expected. A first approximation at the response time is made at 3 minutes. The corresponding fire size is found using the following p = 2 power law fire growth equation:

$$Q(\text{kW}) = \alpha(\text{kW/sec}^2)t^p$$
$$Q = (1055/150^2)(280 \text{ sec})^2$$
$$Q = 1519 \text{ kW}$$

B-3.3.8.3.4 This data can be incorporated into the fire detection design and analysis worksheet shown in Figure B-3.3.8.3.4 in order to carry out the remainder of the calculations.

B-3.3.8.3.5 Using a radial distance of 6.5 m (21 ft) from the axis of this fire, the temperature of the detector is calculated to be 41°C (106°F) after 3 minutes of exposure. The detector actuation temperature is 57°C (135°F). Thus, the detector response time is more than the estimated 3 minutes. If the calculated temperature were more than the actuation temperature, then a smaller t would be used. As in the previous example, calculations should be repeated varying the time to response until the calculated detector temperature is approximately equal to the actuation temperature. For this example, the response time is estimated to be 213 seconds. This corresponds to a fire size at response of 2132 kW.

B-3.3.8.4 The above examples assume that the fire continues to follow the t-squared fire growth relationship up to detector activation. These calculations do not check whether this will happen, nor do they show how the detector temperature varies once the fire stops following the power law relationship. The user should therefore determine that there will be sufficient fuel, as the above correlations do not perform this analysis. If there is not a sufficient amount of fuel, then there is the possibility that the heat release rate curve will flatten out or decline before the heat release rate needed for actuation is reached.

The use of the t-square model presumes sufficient fuel to permit the fire to grow according to the t-square model. The

Fire Detection Design and Analysis Worksheet [28]
Design Example

1.	Determine ambient temperature (T_a) ceiling height or height above fuel (H).	$T_a = $ ____10____ °C + 273 = ____283____ K $H = $ ____4____ m
2.	Determine the fire growth characteristic (α or t_g) for the expected design fire.	$\alpha = $ ____0.047____ kW/s^2 $t_g = $ ____150____ sec
3a.	Define the characteristics of the detectors.	$T_s = $ ____57____ C RTI = ____98____ m$^{1/2}$s$^{1/2}$ $\dfrac{dT_d}{dt} = $ _____ °C/min $\tau_0 = $ _____ sec
3b. or	*Design* — Establish system goals (t_{CR} or Q_{CR}) and make a first estimate of the distance (r) from the fire to the detector.	$t_{CR} = $ ____146____ sec $r = $ ____3.3____ m $Q_{CR} = $ ____1000____ kW
3b.	*Analysis* — Determine spacing of existing detectors and make a first estimate of the response time or the fire size at detector response ($Q = \alpha t^2$).	$r = $ _____ *1.41 = _____ = S (m) $Q = $ _____ kW $t_d = $ _____ sec
4.	Using equation (15), calculate the nondimensional time (t_{2f}^*) at which the initial heat front reaches the detector.	$t_{2f}^* = 0.861 \left(1 + \dfrac{r}{H}\right)$ $t_{2f}^* = $ **1.57**
5.	Calculate the factor A defined in equation (14).	$A = \dfrac{g}{C_p T_a \rho_0}$ $A = $ **0.030**
6.	Use the required response time along with equation (14) and $p = 2$ to calculate the corresponding value of t_2^*.	$t_2^* = \dfrac{t}{A^{-1/(3+p)} \alpha^{-1/(3+p)} H^{4/(3+p)}}$ $t_2^* = $ **12.98**
7.	If $t_2^* > t_{2f}^*$ continue to step 8. If not, try a new detector position (r) and return to step 4.	
8.	Calculate the ratio $\dfrac{u}{u_p^*}$ using equation (18).	$\dfrac{u}{u_p^*} = A^{1/(3+p)} \alpha^{1/(3+p)} H^{(p-1)/(3+p)}$ $\dfrac{u}{u_p^*} = $ **0.356**
9.	Calculate the ratio $\dfrac{\Delta T}{\Delta T_2^*}$ using equation (19).	$\dfrac{\Delta T}{\Delta T_2^*} = A^{2/(3+p)} (T_a/g) \alpha^{2/(3+p)} H^{-(5-p)/(3+p)}$ $\dfrac{\Delta T}{\Delta T_2^*} = $ **0.913**
10.	Use equation (15) to calculate ΔT_2^*.	$\Delta T_2^* = \left[\dfrac{t_2^* - t_{2f}^*}{(0.146 + 0.242 r/H)}\right]^{4/3}$ $\Delta T_2^* = $ **105.89**
11.	Use equation (17) to calculate the ratio $\dfrac{u_2^*}{(\Delta T_2^*)^{1/2}}$.	$\dfrac{u_2^*}{(\Delta T_2^*)^{1/2}} = 0.59 \left(\dfrac{r}{H}\right)^{-0.63}$ $\dfrac{u_2^*}{(\Delta T_2^*)^{1/2}} = $ **0.66**
12.	Use equations (16) and (17) to calculate Y.	$Y = \left(\dfrac{3}{4}\right)\left(\dfrac{u}{u_2^*}\right)^{1/2}\left[\dfrac{u_2^*}{\Delta T_2^{*\,1/2}}\right]^{1/2}\left(\dfrac{\Delta T_2^*}{RTI}\right)\left(\dfrac{t}{t_2^*}\right)D$ $Y = $ **1.533**
13.	*Fixed Temperature HD* — Use equation (16) to calculate the resulting temperature of the detector.	$T_d(t) = \left(\dfrac{\Delta T}{\Delta T_2^*}\right)\Delta T_2^*\left[1 - \dfrac{(1-e^{-Y})}{Y}\right] + T_d(0)$ $T_d(t) = $ **57.25**
14.	*Rate of Rise HD* — Use equation (16).	$dT_d = \left[\left(\dfrac{4}{3}\right)\left(\dfrac{\Delta T}{\Delta T_2^*}\right)(\Delta T_2^*)^{1/4}\dfrac{(1-e^{-y})}{[(t/t_2^*)D]}\right]dT_d$ $dT_d = $
15.	If: 1. $T_d < T_s$ 2. $T_d > T_s$ 3. $T_d = T$	Repeat Procedure Using Design Analysis 1. a larger r 1. a larger t_r 2. a smaller r 2. a smaller t_r 3. $s = 1.41 \times r = $ ____4.7__ m 3. $t_r = $ _____ sec

Figure B-3.3.8 *Fire detection design and analysis worksheet [28]—design example.*

Fire Detection Design and Analysis Worksheet [28]
Analysis Example 2

1.	Determine ambient temperature (T_a) ceiling height or height above fuel (H).	$T_a = \underline{\quad 10 \quad}$ °C + 273 = $\underline{\quad 283 \quad}$ K $H = \underline{\quad 4 \quad}$ m
2.	Determine the fire growth characteristic (α or t_g) for the expected design fire.	$\alpha = \underline{\quad 0.047 \quad}$ kW/s^2 $t_g = \underline{\quad 150 \quad}$ sec
3a.	Define the characteristics of the detectors.	$T_s = \underline{\quad 57 \quad}$ C RTI = $\underline{\quad 98 \quad}$ m$^{1/2}$s$^{1/2}$ $\dfrac{dT_d}{dt} = \underline{\qquad}$ °C/min $\tau_0 = \underline{\qquad}$ sec
3b. or	*Design* — Establish system goals (t_{CR} or Q_{CR}) and make a first estimate of the distance (r) from the fire to the detector.	$t_{CR} = \underline{\qquad}$ sec $r = \underline{\qquad}$ m $Q_{CR} = \underline{\qquad}$ kW
3b.	*Analysis* — Determine spacing of existing detectors and make a first estimate of the response time or the fire size at detector response ($Q = \alpha t^2$).	$r = \underline{\quad 6.5 \quad}$ *1.41 = $\underline{\quad 9.2 \quad}$ = S (m) $Q = \underline{\quad 1,523 \quad}$ kW $t_d = \underline{\quad 180 \quad}$ sec
4.	Using equation (15), calculate the nondimensional time $\left(t^*_{2f}\right)$ at which the initial heat front reaches the detector.	$t^*_{2f} = 0.861 \left(1 + \dfrac{r}{H}\right)$ $t^*_{2f} = 2.26$
5.	Calculate the factor A defined in equation (14).	$A = \dfrac{g}{C_p T_a \rho 0}$ $A = 0.030$
6.	Use the required response time along with equation (14) and $p = 2$ to calculate the corresponding value of t^*_2.	$t^*_2 = \dfrac{t}{A^{-1/(3+p)}\,\alpha^{-1/(3+p)}\,H^{4/(3+p)}}$ $t^*_2 = 16$
7.	If $t^*_2 > t^*_{2f}$ continue to step 8. If not, try a new detector position (r) and return to step 4.	
8.	Calculate the ratio $\dfrac{u}{u^*_p}$ using equation (18).	$\dfrac{u}{u^*_p} = A^{1/(3+p)}\,\alpha^{1/(3+p)}\,H^{(p-1)/(3+p)}$ $\dfrac{u}{u^*_p} = 0.356$
9.	Calculate the ratio $\dfrac{\Delta T}{\Delta T^*_2}$ using equation (19).	$\dfrac{\Delta T}{\Delta T^*_2} = A^{2/(3+p)}(T_a/g)\,\alpha^{2/(3+p)}\,H^{-(5-p)/(3+p)}$ $\dfrac{\Delta T}{\Delta T^*_2} = 0.913$
10.	Use equation (15) to calculate ΔT^*_2.	$\Delta T^*_2 = \left[\dfrac{t^*_2 - t^*_{2f}}{(0.146 + 0.242 r/H)}\right]^{4/3}$ $\Delta T^*_2 = 75.01$
11.	Use equation (17) to calculate the ratio $\dfrac{u^*_2}{(\Delta T^*_2)^{1/2}}$.	$\dfrac{u^*_2}{(\Delta T^*_2)^{1/2}} = 0.59\left(\dfrac{r}{H}\right)^{-0.63}$ $\dfrac{u^*_2}{(\Delta T^*_2)^{1/2}} = 0.435$
12.	Use equations (16) and (17) to calculate Y.	$Y = \left(\dfrac{3}{4}\right)\left(\dfrac{u}{u^*_2}\right)^{1/2}\left[\dfrac{u^*_2}{\Delta T^{*1/2}_2}\right]^{1/2}\left(\dfrac{\Delta T^*_2}{RTI}\right)\left(\dfrac{t}{t^*_2}\right)D$ $Y = 1.37$
13.	*Fixed Temperature HD* — Use equation (16) to calculate the resulting temperature of the detector.	$T_d(t) = \left(\dfrac{\Delta T}{\Delta T^*_2}\right)\Delta T^*_2\left[1 - \dfrac{(1-e^{-Y})}{Y}\right] + T_d(0)$ $T_d(t) = 41$
14.	*Rate of Rise HD* — Use equation (16).	$dT_d = \left[\left(\dfrac{4}{3}\right)\left(\dfrac{\Delta T}{\Delta T^*_2}\right)(\Delta T^*_2)^{1/4}\dfrac{(1-e^{-y})}{(t/t^*_2)D}\right]dT_d$ $dT_d =$
15.	If: 1. $T_d < T_s$ 2. $T_d > T_s$ 3. $T_d = T$	Repeat Procedure Using Design 1. a larger r 2. a smaller r 3. $s = 1.41 \times r = \underline{\qquad}$ m Analysis 1. a larger t_r 2. a smaller t_r 3. $t_r = \underline{\qquad}$ sec

Figure B-3.3.8.3.4 *Fire detection design and analysis worksheet [28]—analysis example 2.*

designer should verify that the hazard provides sufficient fuel. Since $Q = mH_c$, the designer can determine the required fuel load by integrating this relation over the time period in question and then dividing by H_c.

B-3.3.8.5 Tables B-3.3.8.5(a) through (k) provide a comparison of heat release rates, response times, and spacings when variables characteristic of the fires, detectors, and room are changed from the analysis example.

Table B-3.3.8.5(a) Operating Temperature Versus Heat Transfer Rate (S = 30 ft)

| Operating Temperature | | Heat Release Rate (kW)/Response Time (sec) |
°C	°F	
57	135	2132/213
74	165	2798/244
93	200	3554/275

Table B-3.3.8.5(b) Operating Temperature Versus Spacing (Q_d = 1000 kW)

| Operating Temperature | | Spacing (m) |
°C	°F	
57	135	4.7
74	165	3.5
93	200	2.5

Table B-3.3.8.5(c) RTI Versus Heat Release Rate (S = 30 ft)

| RTI | | Heat Release Rate (kW)/Response Time (sec) |
$m^{1/2} sec^{1/2}$	$ft^{1/2} sec^{1/2}$	
50	93	1609/185
150	280	2640/237
300	560	3898/288

Table B-3.3.8.5(d) RTI Versus Spacing (Q_d = 1000 kW)

| RTI | | Spacing (m) |
$m^{1/2} sec^{1/2}$	$ft^{1/2} sec^{1/2}$	
50	93	6.1
150	280	3.7
300	560	2.3

Table B-3.3.8.5(e) Ambient Temperature Versus Heat Release Rate (S = 30 ft)

| Ambient Temperature | | Heat Release Rate (kW)/Response Time (sec) |
°C	°F	
0	32	2552/233
20	68	1751/193
38	100	1058/150

Table B-3.3.8.5(f) Ambient Temperature Versus Spacing (Q_d = 1000 kW)

| Ambient Temperature | | Spacing (m) |
°C	°F	
0	32	3.8
20	68	5.7
38	100	8.8

Table B-3.3.8.5(g) Ceiling Height Versus Heat Release Rate (S = 30 ft)

| Ceiling Height | | Heat Release Rate (kW)/Response Time (sec) |
m	ft	
2.4	8	1787/195
4.9	16	2358/224
7.3	24	3056/255

Table B-3.3.8.5(h) Ceiling Height Versus Spacing (Q_d = 1000 kW)

| Ceiling Height | | Spacing (m) |
m	ft	
2.4	8	5.8
4.9	16	4.0
7.3	24	2.1

Table B-3.3.8.5(i) Detector Spacing Versus Heat Release Rate (S = 30 ft)

| Detector Spacing | | Heat Release Rate (kW)/Response Time (sec) |
m	ft	
4.6	15	1000/146
9.1	30	2132/213
15.2	50	4146/297

Table B-3.3.8.5(j) Fire Growth Rate Versus Heat Release Rate (S = 30 ft)

Fire Growth Rate	Heat Release Rate (kW)/ Response Time (sec)
Slow t_g = 400 sec	1250/435
Medium t_g = 250 sec	1582/306
Fast t_g = 100 sec	2769/162

Table B-3.3.8.5(k) Fire Growth Rate Versus Spacing (Q_d = 1000 kW)

Fire Growth Rate	Spacing (m)
Slow t_g = 400 sec	8.2
Medium t_g = 250 sec	6.5
Fast t_g = 100 sec	3.7

B-3.3.9 Rate-of-Rise Heat Detector Spacing. The procedure presented above can be used to estimate the response of rate-of-rise heat detectors for either design or analysis purposes. In this case, it is necessary to assume that the heat detector response can be modeled using a lumped mass heat transfer model.

B-3.3.9.1 The user must determine the rate of temperature rise at which the detector will respond from the manufacturer's data. [Note that listed rate-of-rise heat detectors are designed to activate at a nominal rate of temperature rise of 15°F (8°C) per minute.] The user must use equation (17) instead of equation (16) in order to calculate the rate of change of the detector temperature. This value is then compared to the rate of change at which the chosen detector is designed to respond.

> Note: The assumption that heat transfer to a detector can be modeled as a lumped mass might not hold for rate-of-rise heat detectors. This is due to the operating principle of this type of detector, in that most rate-of-rise detectors operate when the expansion of air in a chamber expands at a rate faster than it can vent through an opening. To accurately model the response of a rate-of-rise detector would require modeling the heat transfer from the detector body to the air in the chamber, as well as the air venting through the hole.

B-3.3.9.2 Rate Compensation-Type Heat Detectors. Rate-compensated detectors are not specifically covered by Appendix B. However, a conservative approach to predicting their performance is to use the fixed-temperature heat detector guidance contained herein.

B-4 Smoke Detector Spacing for Flaming Fire

B-4.1 Introduction.

B-4.1.1 The listing investigation for smoke detectors does not yield a "listed spacing" as it does for heat detectors. Instead, the manufacturers recommend a spacing. Because the largest spacing that can be evaluated in the full-scale fire test room is 30 ft (9.1 m), it has become common practice to recommend this 30-ft (9.1-m) spacing for smoke detectors when they are installed on flat, smooth ceilings. Reductions in smoke detector spacing are made empirically to address factors that can affect response, including ceiling height, beamed or joisted ceilings, and areas that have high rates of air movement.

Chapter 2 addresses the effects of exposed joists and beams, and the effect of ceiling slope, on a qualitative basis. No one has yet performed the research necessary to provide quantitative measures of these effects.

B-4.1.1.1 The placement of smoke detectors, however, should be based on an understanding of fire plume and ceiling jet flows, smoke production rates, particulate changes due to aging, and the operating characteristics of the particular detector being used. The heat detector spacing information presented in Section B-3 is based on knowledge of plume and jet flows. An understanding of smoke production and aging lags considerably behind an understanding of heat production. In addition, the operating characteristics of smoke detectors in specific fire environments are not often measured or made generally available for other than a very few number of combustible materials. Therefore, the existing knowledge base precludes the development of complete engineering design information for smoke detector location and spacing.

Designers should proceed with great caution when designing or analyzing the response of smoke detection systems. Currently available analytical methods do not account for variations in the composition of smoke, the effects of smoke aging, or the detection mechanism appropriate to different detection technologies.

B-4.1.1.2 In design applications where predicting the response of smoke detectors is not critical, the spacing criteria presented in Chapter 2 should provide sufficient information to design a very basic smoke detection system. However, if the goals and objectives established for the detection system require detector response within a certain amount of time, optical density, heat release rate, or temperature rise, then additional analysis might be needed. For these situations, information regarding the expected fire characteristics (fuel and its fire growth rate), transport characteristics, detector characteristics, and compartment char-

acteristics is required. The following information regarding smoke detector response and various performance-based approaches to evaluating smoke detector response is therefore provided.

B-4.1.2 Response Characteristics of Smoke Detectors.

B-4.1.2.1 General.

B-4.1.2.1.1 In order to determine whether a smoke detector will respond to a given Q_{CR}, a number of factors need to be evaluated. These factors include smoke characteristics, smoke transport, and detector characteristics.

B-4.1.2.2 Smoke Characteristics.

B-4.1.2.2.1 Smoke characteristics are a function of the fuel composition, the mode of combustion (smoldering or flaming), and the amount of mixing with the ambient air (dilution). These factors are important for determining the characteristics of the products of combustion, such as particle size, distribution, composition, concentration, refractive index, etc. The significance of these features with regard to smoke detector response are well documented. [29,30]

B-4.1.2.2.2 Whether smoke detectors detect by sensing scattered light, loss of light transmission (light extinction), or reduction of ion current, they are particle detectors. Thus, particle concentration, size, color, size distribution, and so on, affect each sensing technology differently. It is generally accepted that a flaming, well-ventilated, energetic fire produces smoke having a larger proportion of the sub-micron diameter particulates as opposed to a smoldering fire that produces smoke with a predominance of large, super-micron particulates. It is also known that as the smoke cools the smaller particles agglomerate, forming larger ones as they age, and are carried away from the fire source. More research is necessary to provide sufficient data to allow the prediction of smoke characteristics at the source, as well as during transport. Furthermore, response models must be developed that can predict the response of a particular detector to different kinds of smoke as well as smoke that has aged during the flow from the fire to the detector location.

B-4.1.2.3 Transport Considerations.

B-4.1.2.3.1 General. All smoke detection relies on the plume and ceiling jet flows to transport the smoke from the locus of the fire to the detector. Various considerations must be addressed during this transport time, including changes to the characteristics of the smoke that occur with time and distance from the source, and transport time of smoke from the source to the detector.

B-4.1.2.3.2 The smoke characteristic changes that occur during transport relate mainly to the particle size distribu-

tion. Particle size changes during transport occur mainly as a result of sedimentation and agglomeration.

B-4.1.2.3.3 Transport time is a function of the characteristics of the travel path from the source to the detector. Important characteristics that should be considered include ceiling height and configuration (for example, sloped, beamed), intervening barriers such as doors and beams, as well as dilution and buoyancy effects such as stratification that might delay or prevent smoke in being transported to the detector.

B-4.1.2.3.4 In smoldering fires, thermal energy provides a force for transporting smoke particles to the smoke sensor. However, usually in the context of smoke detection, the rate of energy (heat) release is small and the rate of growth of the fire is slow. Consequently, other factors such as ambient airflow from HVAC systems, differential solar heating of the structure, and wind cooling of the structure can have a dominant influence on the transport of smoke particles to the smoke sensor when low-output fires are considered.

B-4.1.2.3.5 In the early stages of development of a growing fire, the same interior environmental effects, including ambient airflow from HVAC systems, differential solar heating of the structure, and wind cooling of the structure, can have a dominant influence on the transport of smoke. This is particularly important in spaces having high ceilings. Greater thermal energy release from the fire is necessary to overcome these interior environmental effects. Because the fire must attain a sufficiently high level of heat release before it can overcome the interior environmental airflows and drive the smoke to the ceiling-mounted detectors, the use of closer spacing of smoke detectors on the ceiling might not significantly improve the response of the detectors to the fire. Therefore, when considering ceiling height alone, smoke detector spacing closer than 30 ft (9.1 m) might not be warranted, except in instances where an engineering analysis indicates additional benefit will result. Other construction characteristics also should be considered. (Refer to the appropriate sections of Chapter 2 dealing with smoke detectors and their use for the control of smoke spread.)

B-4.1.2.3.6 Smoke Dilution. Smoke dilution causes a reduction in the quantity of smoke per unit of air volume of smoke reaching the detector. Dilution typically occurs either by entrainment of air in the plume or the ceiling jet or by effects of HVAC systems. Forced ventilation systems with high air change rates typically cause the most concern, particularly in the early stages of fire development, when smoke production rate and plume velocity are both low. Airflows from supply as well as return vents can create defined air movement patterns within a compartment, which can either keep smoke away from detectors that are located outside of

these paths or can inhibit smoke from entering a detector that is located directly in the airflow path. [26]

The National Institute of Standards and Technology (NIST) first investigated these issues using the Harwell Flow 3D Computational Fluid Dynamics (CFD) model as part of the International Fire Detection Research Project. NIST conducted this research under the auspices of the National Fire Protection Research Foundation with technical advice and support from the Fire Detection Institute.

B-4.1.2.3.7 There currently are no quantitative methods for estimating either smoke dilution or airflow effects on locating smoke detectors. These factors should therefore be considered qualitatively. The designer should understand that the effects of airflow become larger as the fire size at detection (Q_{CR}) gets smaller. Depending on the application, the designer might find it useful to obtain airflow and velocity profiles within the room or to even conduct small-scale smoke tests under various conditions to assist in the design of the system.

B-4.1.2.3.8 Stratification.

B-4.1.2.3.8.1 The potential for the stratification of smoke is another concern in designing and analyzing the response of detectors. This is of particular concern with the detection of low-energy fires and fires in compartments with high ceilings.

B-4.1.2.3.8.2 The upward movement of smoke in the plume depends on the smoke being buoyant relative to the surrounding air. Stratification occurs when the smoke or hot gases flowing from the fire fail to ascend to the smoke detectors mounted at a particular level (usually on the ceiling) above the fire due to the loss of buoyancy. This phenomenon occurs due to the continuous entrainment of cooler air into the fire plume as it rises, resulting in cooling of the smoke and fire plume gases. The cooling of the plume results in a reduction in buoyancy. Eventually the plume cools to a point where its temperature equals that of the surrounding air and its buoyancy diminishes to zero. Once this point of equilibrium is reached, the smoke will cease its upward flow and form a layer, maintaining its height above the fire, regardless of the ceiling height, unless and until sufficient additional thermal energy is provided from the fire to raise the layer due to its increased buoyancy. The maximum height to which plume fluid (smoke) will ascend, especially early in the development of a fire, depends on the convective heat release rate of the fire and the ambient temperature in the compartment.

B-4.1.2.3.8.3 Because warm air rises, there will usually be a temperature gradient in the compartment. Of particular interest are those cases where the temperature of the air in the upper portion of the compartment is greater than at the lower level before the ignition. This can occur as a result of solar load where ceilings contain glazing materials. Computational methods are available to assess the potential for intermediate stratification for the following two cases, depicted in Figure B-4.1.2.3.8.3(a).

Case 1. The temperature of the ambient is relatively constant up to a height above which there is a layer of warm air at uniform temperature. This situation can occur if the upper portion of a mall, atrium, or other large space is unoccupied and the air is left unconditioned.

Case 2. The ambient interior air of the compartment has a constant and uniform temperature gradient (temperature change per unit height) from floor to ceiling. This case is generally encountered in industrial and storage facilities that are normally unoccupied.

The analysis of intermediate stratification is presented in Figure B-4.1.2.3.8.3(b). Plume centerline temperatures from two fires, 1000 kW and 2000 kW, are graphed based on estimates from correlations presented in this section. In Case 1, a step function is assumed to indicate a 30°C/m change in temperature 15 m above the floor due to the upper portion of the atrium being unconditioned. For Case 2, a temperature gradient of 1.5°C/m is arbitrarily assumed in an atrium that has a ceiling height of 20 m.

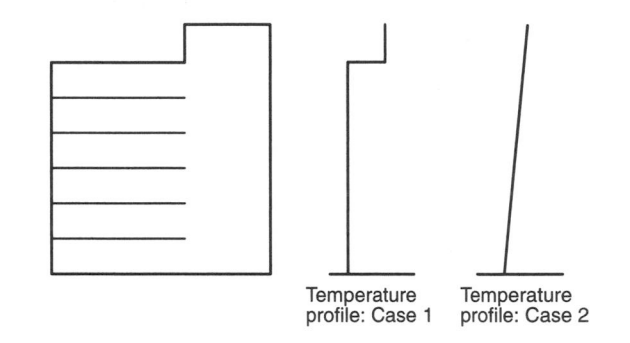

Temperature profile: Case 1   Temperature profile: Case 2

Figure B-4.1.2.3.8.3(a) Pre-fire temperature profiles.

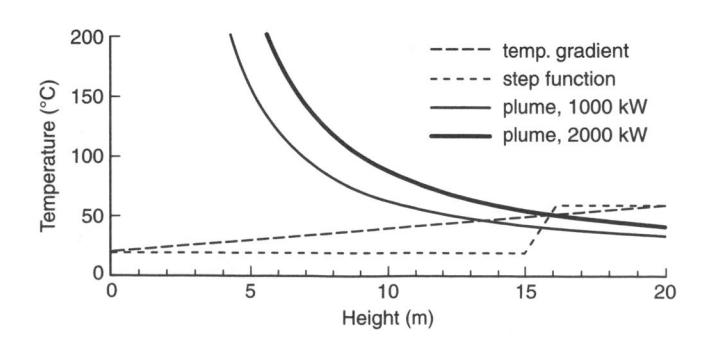

Figure B-4.1.2.3.8.3(b) Indoor air and plume temperature profiles with potential for intermediate stratification.

B-4.1.2.3.8.4 Step Function Temperature Gradient Spaces. If the interior air temperature exhibits a discrete change at some elevation above the floor, the potential for stratification can be assessed by applying the plume centerline temperature correlation. If the plume centerline temperature is equal to the ambient temperature, the plume is no longer buoyant, loses its upward momentum, and stratifies at that height. The plume centerline temperature can be calculated by using the following equation:

$$T_c = 316Q_c^{2/3}z^{-5/3} + 70 \ (^\circ\text{F}) \tag{23}$$

or

$$T_c = 25Q_c^{2/3}z^{-5/3} + 20 \ (^\circ\text{C})$$

where:

T_c = plume centerline temperature (°F or °C)

Q_c = convective portion of fire heat release rate (Btu/sec or kW)

z = height above the top of the fuel package involved (ft or m)

B-4.1.2.3.8.5 Linear Temperature Gradient Spaces. To determine whether or not the rising smoke or heat from an axisymmetric fire plume will stratify below detectors, the following equation can be applied where the ambient temperature increases linearly with increasing elevation:

$$Z_m = 14.7Q_c^{1/4}\left(\frac{\Delta T_0}{dZ}\right)^{-3/8} \text{ (ft)} \tag{24}$$

or

$$Z_m = 5.54Q_c^{1/4}\left(\frac{\Delta T_0}{dZ}\right)^{-3/8} \text{ (m)}$$

where:

Z_m = maximum height of smoke rise above the fire surface (ft or m)

ΔT_0 = difference between the ambient temperature at the location of detectors and the ambient temperature at the level of the fire surface (°F or °C)

Q_c = convective portion of the heat release rate (Btu/sec or kW)

B-4.1.2.3.8.6 The convective portion of the heat release rate (Q_c) can be estimated as 70 percent of the heat release rate.

B-4.1.2.3.8.7 As an alternative to using the noted expression to directly calculate the maximum height to which the smoke or heat will rise, Figure B-4.1.2.3.8.7 can be used to determine Z_m for given fires. Where Z_m, as calculated or determined graphically, is greater than the installed height of detectors, smoke or heat from a rising fire plume is predicted to reach the detectors. Where the compared values of Z_m and the installed height of detectors are comparable

heights, the prediction that smoke or heat will reach the detectors might not be a reliable expectation.

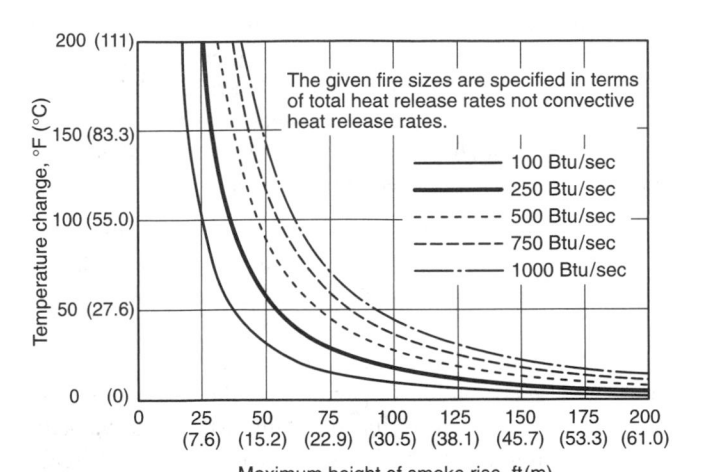

Figure B-4.1.2.3.8.7 Temperature change and maximum height of smoke rise for given fire sizes.

B-4.1.2.3.8.8 Assuming the ambient temperature varies linearly with the height, the minimum Q_c required to overcome the ambient temperature difference and drive the smoke to the ceiling $(Z_m = H)$ may be determined from the following equation:

$$Q_c = 2.39 \times 10^{-5} H^{5/2} \Delta T_0^{3/2} \text{ (Btu/sec)} \tag{25}$$

or

$$Q_c = 0.0018 H^{5/2} \Delta T_0^{3/2} \text{ (kW)}$$

Note that variables are identified in Section B-7.

B-4.1.2.3.8.9 The theoretical basis for the stratification calculation is based on the works of Morton, Taylor, and Turner [15] and Heskestad [9]. For further information regarding the derivation of the expression defining Z_m, the user is referred to the work of Klote and Milke [13] and NFPA 92B, *Guide for Smoke Management Systems in Malls, Atria, and Large Areas.*

B-4.1.2.4 Detector Characteristics.

The following discussion applies primarily to spot-type smoke detectors. Some of the comments may apply to projected beam and air sampling–type smoke detectors. Nevertheless, B-4.1.2.4 does not provide a detailed discussion of these types of detectors. The original Fire Detection Institute research project did not include projected beam and air sampling type smoke detectors.

B-4.1.2.4.1 General. Once smoke is transported to the detector, additional factors become important in determin-

ing whether response will occur. These include the aerodynamic characteristics of the detector and the type of sensor within the detector. The aerodynamics of the detector relate to how easily smoke can pass through the detector housing and enter the sensor portion of the unit. Additionally, the location of the entry portion to the sensor with respect to the velocity profile of the ceiling jet is also an important factor. Finally, different sensing methods (for example, ionization or photoelectric) will respond differently, depending on the smoke characteristics (smoke color, particle size, optical density, and so on). Within the family of photoelectric devices, there will be variations depending on the wavelengths of light and the scattering angles employed. The following paragraphs discuss some of these issues and various calculation methods.

B-4.1.2.4.2 Resistance to Smoke Entry.

B-4.1.2.4.3 All spot-type smoke detectors require smoke to enter the detection chamber in order to be sensed. This requires additional factors to be taken into consideration when attempting to estimate smoke detector response, as smoke entry into the detection chamber can be affected in several ways, for example, insect screens, sensing chamber configuration, and location of the detector with respect to the ceiling.

B-4.1.2.4.4 In trying to quantify this, Heskestad [32] developed the idea of smoke detector lag to explain the difference in optical density outside (D_{ur}) versus inside (D_{uo}) of a detector when the detector activates. It was demonstrated that this difference could be explained by the use of a correction factor D_{uc} using the following relationship:

$$D_{uc} = \frac{L \dfrac{d(D_u)}{dt}}{V} \qquad (26)$$

where:

L = characteristic length for a given detector design, represents the ease of smoke entry into the sensing chamber

$\dfrac{d(D_u)}{dt}$ = rate of increase of optical density outside the detector

V = velocity of the smoke at the detector

B-4.1.2.4.5 Various studies regarding this correlation have provided additional insight regarding smoke entry and associated lags [33,34]; however, the difficulty in quantifying L for different detectors and relating it to spacing requirements may have limited usefulness, and the concept of critical velocity (u_c) may be more applicable. [21]

B-4.1.2.4.6 Critical Velocity. [26] A smoke detector's critical velocity refers to the minimum velocity of the smoke

necessary to enter the sensing chamber to cause an alarm. Flow across a detector causes a pressure differential between the up-stream and down-stream sides of the detector. This pressure differential is the principal driving force for the smoke entering the unit. Experimental work has indicated that this minimum velocity is approximately 0.15 m/sec for the detectors tested in one particular study. [21] Once velocities were reduced below this level, the smoke concentration level outside the detector before an alarm condition increased dramatically when compared to smoke concentration levels when the velocity was above the critical value. Estimating the critical velocity can therefore be useful for design and analysis. It is interesting to note that this critical velocity value (0.15 m/sec) is close to that at which a smoke detector must respond in the UL smoke detector sensitivity chamber in order to become listed. [35] The location in the ceiling jet where this velocity occurs for a given fire and ceiling height might therefore be considered as a first approximation for locating detectors. This again assumes a horizontal, smooth ceiling.

B-4.1.2.4.7 Response to Smoke Color. Smoke detectors that use an optical means to detect smoke respond differently to smokes of different colors.

B-4.1.2.4.8 Manufacturers currently provide limited information regarding the response of smoke detectors in their specifications as well as in the information contained on the labels on the backs of the detectors. This response information indicates only their nominal response values with respect to gray smoke, not to black, and is often provided with a response range instead of an exact response value. This range is in accordance with UL 268, *Standard for Safety, Smoke Detectors for Fire Protective Signalling Systems.*

B-4.1.2.4.9 The response ranges allowable by UL for gray versus black smoke are shown in Table B-4.1.2.4.9.

Table B-4.1.2.4.9 UL 268, Smoke Detector Test Acceptance Criteria for Different Colored Smoke [35]

Color of Smoke	Acceptable Response Range		Maximum: Minimum
	%/m	%/ft	
Gray	1.6 – 12.5	0.5 – 4.0	7:8
Black	1.6 – 29.2	0.5 – 10.0	18:25

B-4.1.2.4.10 As seen in Table B-4.1.2.4.9 response levels are different for black and gray smoke. This is due to the fact that detectors respond at different optical density levels to different fuels and different types of smoke. Examples of this are shown by Heskestad and Delichatsios [10] in tests

they performed as shown in Table B-4.1.2.4.10. This is why it is critical that when analyzing or designing a detection system one knows how the individual detector will respond to various types of smoke.

Table B-4.1.2.4.10 Values of Optical Density at Response (For Flaming Fires Only) [18]

| Material | $10^2 D_{ur}$ | | Relative Smoke Color |
	Ionization	Scattering	
Wood	0.5	1.5	Light
Cotton	0.05	0.8	Light
Polyurethane	5.0	5.0	Dark
PVC	10.0	10.0	Dark
Variation	200:1	12.5:1	

Note the large variations in response not only to materials producing relatively the same color of smoke, but also to smoke of different color, which is much more pronounced.

B-4.1.2.4.11 Optical Density and Temperature. During a flaming fire, smoke detector response is affected by ceiling height and the size and rate of fire growth in much the same way as heat detector response. The thermal energy of the flaming fire transports smoke particles to the sensing chamber just as it does heat to a heat sensor. While the relationship between the amount of smoke and the amount of heat produced by a fire is highly dependent on the fuel and the way it is burning, research has shown that the relationship between temperature and the optical density of smoke remains somewhat constant within the fire plume and on the ceiling in the proximity of the plume.

B-4.1.2.4.12 These results were based on the work by Heskestad and Delichatsios [10] and are indicated in Table B-4.1.2.4.12. Note that for a given fuel, the optical density to temperature rise ratio between the maximum and minimum levels is 10 or less.

B-4.1.2.4.13 In situations where the optical density at detector response is known and is independent of particle size distribution, the detector response can be approximated as a function of the heat release rate of the burning fuel, the fire growth rate, and the ceiling height, assuming that the above correlation exists.

B-4.1.2.4.14 When Appendix C of NFPA 72E was first published in 1984, a 20°F (13°C) temperature rise was used to indicate detector response. Schifiliti and Pucci [8] have combined some of the data from Heskestad and Delichatsios to produce Table B-4.1.2.4.14 showing the temperature rise at detector response. Note that the temperature rise required for detector response varies significantly depending on the detector type and fuel.

Table B-4.1.2.4.12 Ratio of Optical Density to Temperature Rise for Various Fuels [18]

Material	$10^2 D_u$ $(1/ft°F)$ (T)	Value Range	Maximum: Minimum
Wood	0.02	0.015–0.055	3:6
Cotton	0.01/0.02	0.005–0.03	6:0
Paper	0.03	Data not available	—
Polyurethane	0.4	0.2–0.55	2:8
Polyester	0.3	Data not available	—
PVC	0.5/1.0	0.1–1.0	10
Foam rubber PU	1.3	Data not available	—
Average	0.4	0.005–1.3	260

Table B-4.1.2.4.14 Temperature Rise for Detector Response [18]

| Material | Temperature Rise (°F) | |
	Ionization	Scattering
Wood	25	75
Cotton	3	50
Polyurethane	13	13
PVC	13	13
Average	14	38

B-4.1.3 Methods for Predicting Smoke Detector Response.

B-4.1.3.1 Method 1—Optical Density Versus Temperature.

B-4.1.3.1.1 It is intended to determine whether an existing fire detection system can detect a fire in part of a warehouse used to store wardrobes in sufficient time to prevent radiant ignition of adjacent wardrobes. The area under review has a large, flat ceiling, 16.5 ft (5 m) high. The ambient temperature within the compartment is 68°F (20°C). The compartment is not sprinklered. The wardrobes are constructed mainly of particleboard. The detectors are ionization-type smoke detectors spaced 20 ft (6.1 m) on center. The design objective is to keep the maximum heat release rate (Q_{DO}) below 2 MW in order to ensure that radiant ignition of the wardrobes in the adjacent aisle will not occur. There is an on-site fire brigade that can respond to and begin discharging water on the fire within 90 seconds of receiving the alarm. It can be assumed that there are no other delays between the time the detector reaches its operating threshold and the time to notification of the fire brigade. Given the above, would the existing system be sufficient?

B-4.1.3.1.2 Assumptions. The following assumptions are made for this example:

$\alpha = 0.047$

RTI $= 25 \text{ m}^{1/2}\text{sec}^{1/2}$

Temperature rise for response $= 14°C$ (25°F)

Refer to Table B-4.1.2.4.14 for temperature rise to response of an ionization smoke detector for a wood fire.

B-4.1.3.1.3 Using the power law equation, the design objective response time is calculated as follows:

$$Q_{DO} = \alpha t_{DO}^2$$

$$2000 \text{ kW} = 0.047 t_{DO}^2 \qquad (27)$$

$$t_{DO} = 210 \text{ sec}$$

B-4.1.3.1.4 Next, subtract the time for the fire brigade to respond to determine what time after ignition that detection should occur. Note that a 30-second safety factor has been added to the fire brigade's response time.

$$t_{CR} = 210 \text{ sec} - 120 \text{ sec} = 90 \text{ sec} \qquad (28)$$

B-4.1.3.1.5 Then, calculate the critical heat release rate at which detection should occur as follows:

$$Q_{CR} = \alpha t_{CR}^2$$

$$Q_{CR} = 0.047(90)^2 = 380 \text{ kW} \qquad (29)$$

B-4.1.3.1.6 Using the numbers in the fire detection design and analysis worksheet at 90 seconds into the fire when the heat release rate is 380 kW, the temperature rise at the detector is calculated to be approximately 17°C. This, therefore, might be a reasonable approximation to show that the detector may respond.

Once again, the designer should perform a sensitivity analysis on pertinent variables used in this analysis to determine if the results inordinately rely on specific parameter values. For example, considerable variation exists in the response range of the detectors. The Value Range column in Table B-4.1.2.4.12 illustrates this variation.

B-4.1.3.2 Method 2—Mass Optical Density.

B-4.1.3.2.1 Data regarding smoke characteristics for given fuels can be used as another method to evaluate detector response.

B-4.1.3.2.2 Example. The design objective established for this scenario is to detect the smoke from a flaming 400-g (1.0-lb) polyurethane chair cushion in less than 2 minutes. The chair is placed in a compartment that is 40 m² (431 ft²). The ceiling height is 3.0 m (10 ft). It has been determined that the burning rate of the cushion is a steady rate of 50 g/min. Determine if the design objective will be met.

B-4.1.3.2.3 The total mass loss of the cushion due to combustion at 2 minutes is 100 g. Therefore, the optical density in the room produced by the burning cushion can be calculated from the following equation. [5]

$$D = \frac{D_m M}{V_c} \qquad (30)$$

where:

D_m = mass optical density (m²/g) [26]

M = mass

V_c = volume of the compartment

D = $[(0.22 \text{ m}^2/\text{g})(100\text{g})]/(40 \text{ m}^2)(3 \text{ m}) = 0.183 \text{ m}^{-1}$

B-4.1.3.2.4 If it is assumed that the detector responds at the UL upper sensitivity limit of 0.14 m^{-1} for black smoke [35], it can be assumed that the detector will respond within 2 minutes.

B-4.1.3.2.5 It should be noted that this method presents a very simplified approach, and that various assumptions would need to be made including that the smoke is confined to the room, is well mixed, can reach the ceiling, and can enter the detector.

B-5 Radiant Energy Detection

The Technical Committee on Initiating Devices introduced performance-based design criteria in the section on Radiant Energy-Sensing Fire Detectors in the 1990 edition of NFPA 72E. The authors added this commentary on Appendix B to provide the designer with more specific guidance on how to design consistent with the performance criteria in the code. A designer may NOT use Section B-5 as an alternate to the requirements in the body of the code. Instead, Section B-5 outlines how to meet the current requirements of the code.

B-5.1 General.

B-5.1.1 Electromagnetic radiation is emitted over a broad range of the spectrum during the combustion process. The portion of the spectrum in which radiant energy-sensing detectors operate has been divided into three bands: ultraviolet (UV), visible, or infrared (IR). These wavelengths are defined with the following wavelength ranges: [3]

(1) Ultraviolet 0.1—0.35 microns
(2) Visible 0.35—0.75 microns
(3) Infrared 0.75—220 microns

B-5.1.2 These wavelength ranges correspond to the quantum-mechanical interaction between matter and energy. Photonic interactions with matter can be characterized by wavelength as shown in Table B-5.1.2:

Table B-5.1.2 Wavelength Ranges

Wavelength	Photonic Interaction
$\lambda < 50$ micron	Gross molecular translations
$50 \, \mu < \lambda < 1.0 \, \mu$	Molecular vibrations and rotations
$1.0 \, \mu < \lambda < 0.05 \, \mu$	Valence electron bond vibrations
$0.3 \, \mu < \lambda < 0.05 \, \mu$	Electron stripping and recombinations

B-5.1.3 When a fuel molecule is oxidized in the combustion process, the combustion intermediate molecule must lose energy to become a stable molecular species. This energy is emitted as a photon with a unique wavelength determined by the following equation:

$$e = \frac{hc}{\lambda} \qquad (31)$$

where:

e = energy (joules)

h = Plank's constant (6.63E-23 joule-sec)

c = speed of light (m/sec)

λ = wavelength (microns)

[1.0 joule = 5.0345E+18(λ), where λ is measured in microns.]

B-5.1.4 The choice of the type of radiant energy-sensing detector to use is determined by the type of emissions that are expected from the fire radiator.

B-5.1.4.1 Fuels that produce a flame, a stream of combustible or flammable gases involved in the combustion reaction with a gaseous oxidizer, radiate quantum emissions. These fuels include flammable gases, flammable liquids, combustible liquids, and solids that are burning with a flame.

B-5.1.4.2 Fuels that are oxidized in the solid phase or radiators that are emitting due to their internal temperature (sparks and embers) radiate Plankian emissions. These fuels include carbonacious fuels such as coal, charcoal, wood, and cellulosic fibers that are burning without an established flame, as well as metals that have been heated due to mechanical impacts and friction.

B-5.1.4.3 Almost all combustion events produce Plankian emissions, emissions that are the result of the thermal energy in the fuel mass. Therefore, spark/ember detectors that are designed to detect these emissions are not fuel specific. Flame detectors detect quantum emissions that are the result of changes in molecular structure and energy state in the gas phase. These emissions are uniquely associated with particular molecular structures. This can result in a flame detector that is very fuel specific.

If a photon could be held in the hand, it could not be determined whether it was a Plankian photon or a quantum photon. The distinction between the two merely alludes to the theory of physics that explains the mechanism of their formation. The designer should note this distinction because it helps in the detector selection process. The designer must understand what emits photons and why. Only then, can he or she select the appropriate type of detection device.

B-5.1.5 Affects of Ambient. The choice of radiant energy-sensing detector is also limited by the affect of ambient conditions. The design must take into account the radiant energy absorption of the atmosphere, presence of nonfire-related radiation sources that might cause nuisance alarms, the electromagnetic energy of the spark, ember, or fire to be detected, the distance from the fire source to the sensor, and characteristics of the sensor.

B-5.1.5.1 Ambient Non-Fire Radiators. Most ambients contain non-fire radiators that can emit at wavelengths used by radiant energy-sensing detectors for fire detection. The designer should make a thorough evaluation of the ambient to identify radiators that have the potential for producing unwarranted alarm response from radiant energy-sensing detectors. Since radiant energy-sensing detectors use electronic components that can act as antennas, the evaluation should include radio band, microwave, infrared, visible, and ultraviolet sources.

B-5.1.5.2 Ambient Radiant Absorbance. The medium through which radiant energy passes from fire source to detector has a finite transmittance. Transmittance is usually quantified by its reciprocal, absorbance. Absorbance by atmospheric species varies with wavelength. Gaseous species absorb at the same wavelengths that they emit. Particulate species can transmit, reflect, or absorb radiant emission and the proportion that is absorbed is expressed as the reciprocal of its emissivity, ε.

B-5.1.5.3 Contamination of Optical Surfaces. Radiant energy can be absorbed or reflected by materials contaminating the optical surfaces of radiant energy-sensing detectors. The designer should evaluate the potential for surface contamination and implement provisions for keeping these surfaces clean. Extreme caution must be employed when considering the use of surrogate windows. Common glass, acrylic, and other glazing materials are opaque at the wavelengths used by most flame detectors and some spark/ember detectors. Placing a window between the detector and the hazard area that has not been listed by a nationally recognized testing laboratory (NRTL) for use with the detector in question is a violation of the detector listing and will usually result in a system that is incapable of detecting a fire in the hazard area.

B-5.1.5.4 These factors are important for several reasons. First, a radiation sensor is primarily a line-of-sight device, and must "see" the fire source. If there are other radiation sources in the area, or if atmospheric conditions are such

that a large fraction of the radiation could be absorbed in the atmosphere, the type, location, and spacing of the sensors could be affected. In addition, the sensors react to specific wavelengths, and the fuel must emit radiation in the sensor's bandwidth. For example, an infrared detection device with a single sensor tuned to 4.3 microns (the CO_2 emission peak) cannot be expected to detect a non-carbon-based fire. Furthermore, the sensor needs to be able to respond reliably within the required time, especially when activating an explosion suppression system or similar fast-response extinguishing or control system.

B-5.1.6 Detector Response Model. The response of radiant energy-sensing detectors is modeled with a modified inverse square relationship as shown in the following equation [5]:

$$S = \frac{kPe^{-\zeta d}}{d^2} \tag{32}$$

where:

S = radiant power reaching the detector (W) sufficient to produce alarm response

k = proportionality constant for the detector

P = radiant power emitted by the fire (W)

ζ = extinction coefficient of air at detector operating wavelengths

d = distance between the fire and the detector

This relationship models the fire as a point source radiator, of uniform radiant output per steradian, some distance *(d)* from the detector. This relationship also models the effect of absorbance by the air between the fire and the detector as being a uniform extinction function. The designer must verify that these modeling assumptions are valid for the application in question.

B-5.2 Design of Flame Detection Systems.

B-5.2.1 Detector Sensitivity. Flame detector sensitivity is traditionally quantified as the distance at which the unit can detect a fire of given size. The fire most commonly used by the NRTLs in North America is a 1.0 ft^2 (0.9 m^2) fire fueled with regular grade, unleaded gasoline. Some special purpose detectors are evaluated using 6.0-in.(0.015-m) diameter fires fueled with isopropanol.

B-5.2.1.1 This means of sensitivity determination does not take into account that flames can best be modeled as an optically dense radiator in which radiant emissions radiated from the far side of the flame toward the detector are reabsorbed by the flame. Consequently, the radiated power from a flame is not proportional to the area of the fire but to the flame silhouette, and hence to the height and width of the fire.

B-5.2.1.2 Because flame detectors detect the radiant emissions produced during the formation of flame intermediates and products, the radiant intensity produced by a flame at a given wavelength is proportional to the relative concentration of the specific intermediate or product in the flame and that portion of the total heat release rate of the fire resulting from the formation of that specific intermediate or product. This means that the response of a detector may vary widely as different fuels are used to produce a fire of the same surface area and flame width.

The designer must verify that the fuels present in the hazard area match those used in the listing evaluation of the flame detector. Relatively small variations in chemical composition can have profound effects on the response of the detector. For instance a detector might detect a gasoline fire (C6 to C9 fraction) at 80 ft (24 m) but detect the same size fire fueled with #2 Fuel Oil (C9 to C12 fraction) at 40 ft (12 m). This variation represents a fourfold difference in sensitivity.

B-5.2.1.3 Many flame detectors are designed to detect specific products such as water (2.5 microns) and CO_2 (4.35 microns). These detectors cannot be used for fires that do not produce these products as a result of the combustion process.

A designer could expect a flame detector that uses 2.5 micron (water emission) photocells to promptly detect a methane fire. Detection occurs because the hydrogen of the methane molecule combines with oxygen to produce two water molecules. However, a designer could not expect such a detector to detect burning sulfur or metals. Likewise, a designer cannot expect a detector using the 4.35 micron (CO_2 emission) photocell to respond to methane fires.

B-5.2.1.4 Many flame detectors use time variance of the radiant emissions of a flame to distinguish between non-fire radiators and a flame. Where a deflagration hazard exists, the designer must determine the sample time period for such flame detectors and how such detectors will operate in the event of a deflagration of fuel vapor or fuel gases.

The organization and design of the electronics in a flame detector can have unanticipated effects on its performance as a fire detection device. Many flame detectors require a time-variant, repetitive radiant signal before the radiation is interpreted as a flame emission. Ideally, this type of circuit detects growing pool fires. However, this type of circuitry might not detect a deflagration of a fuel vapor and air mixture. The designer should verify that a nationally recognized testing laboratory has tested the detector for the fire scenarios appropriate to the hazard area.

B-5.2.2 Design Fire. Using the process outlined in Section B-2, determine the fire size (kW or Btu/sec) at which detection must be achieved.

B-5.2.2.1 Compute the surface area the design fire is expected to occupy from the correlations in Table B-2.3.2.3.1(a) or other sources. Use the flame height correlation to determine the height of the flame plume:

$$h_f = 0.584 \, (kQ)^{2/5} \tag{33}$$

where:

h_f = flame height (ft)

Q = heat release rate (Btu/sec)

k = wall effect factor

Where there are no nearby walls, use $k = 1$

Where the fuel package is near a wall, use $k = 2$

Where the fuel package is in a corner, use $k = 4$

Determine the minimum anticipated flame area width (w_f). Where flammable or combustible liquids are the fuel load and are unconfined, model the fuel as a circular pool. Compute the radiating area (A_r) using the following equation:

$$A_r = 1/2 h_f w_f \tag{34}$$

where:

A_r = radiating area (ft^2)

h_f = flame height (ft)

w_f = flame width (ft)

This design fire computation models the fire as an optically dense radiator of uniform radiant intensity per unit area. It should be noted that this method does NOT employ the commonly used ratio of 35 percent radiant heat release and 65 percent heat release. This method cannot use these ratios because the test methods employed by nationally recognized testing laboratories in the listing process do not quantify the test fires. The method in Appendix B uses a consistent approach to estimate the radiating area of the design fire and the test fire. It then compares the design fire to the test fire on the basis of radiating area and power per unit area.

B-5.2.2.2 The radiant power output of the fire to the detector can be approximated as being proportional to the radiating area (A_r) of the flame.

$$P = cA_r \tag{35}$$

where:

A_r = radiating area (ft^2)

c = power per unit area proportionality constant

P = radiated power (W)

B-5.2.3 Calculate Detector Sensitivity. Using equation (33) compute the radiating area of the test fire used by the NRTL in the listing process (A_t). The radiant power output of the test fire to the detector in the listing process is proportional to the radiating area (A_t) of the listing test flame.

B-5.2.4 Calculate Detector Response to Design Fire. Because the sensitivity of a flame detector is fixed during the manufacturing process, the following is the relationship that determines the radiant power reaching the detector sufficient to produce an alarm response.

$$S = \frac{kcA_t^{-\zeta d}}{d^2} \tag{36}$$

where:

S = radiant power reaching the detector (W) sufficient to produce alarm response

k = proportionality constant for the detector

A_t = radiant area of the listing test fire (W)

ζ = extinction coefficient of air at detector operating wavelengths

d' = distance between the fire and the detector during the listing fire test

c = emitted power per unit flame radiating area correlation

Because the sensitivity of the detector is constant over the range of ambients for which it is listed

$$S = \frac{kcA_r^{-\zeta d}}{d^2}$$

where:

S = radiant power reaching the detector (W) sufficient to produce alarm response

k = proportionality constant for the detector

A_r = radiant area of the design fire (W)

ζ = extinction coefficient of air at detector operating wavelengths

d' = distance between the design fire and the detector

c = emitted power per unit flame radiating area correlation

Therefore, use the following equation to determine the following:

$$\frac{kcA_r^{-\zeta d}}{d^2} = \frac{kcA_t^{-\zeta d'}}{d'^2} \tag{37}$$

To solve for d' use the following equation:

$$\frac{(d^2)A_t^{-\zeta d'}}{A_t^{(-\zeta d)^{1/2}}} = d' \tag{38}$$

This relation is solved iteratively for d', the distance at which the detector can detect the design fire.

This method relies on several important assumptions. First, it assumes that the design fire has the same fuel as the fire in

the listing evaluation. This assumed scenario allows the emitted power per unit flame area correlation parameter (c) to cancel out in the final equation. Second, this method assumes that the fire can be modeled as a point source radiator. This assumption becomes invalid when the flame area occupies a substantial fraction of the total field of view of the detector. Finally, it demands that the data generated in the listing evaluation must include a numerical value for the atmospheric extinction coefficient (ξ).

B-5.2.5 Correction for Angular Displacement.

B-5.2.5.1 Most flame detectors exhibit a loss of sensitivity as the fire is displaced from the optical axis of the detector. This correction to the detector sensitivity is shown as a polar graph in Figure A-2-4.3.2.3.

B-5.2.5.2 When the correction for angular displacement is expressed as a reduction of normalized detection distance, the correction is made to detection distance (d').

B-5.2.5.3 When the correction for angular displacement is expressed as a normalized sensitivity (fire size increment), the correction must be made to A_r prior to calculating response distance (d').

B-5.2.6 Corrections for Fuel. Most flame detectors exhibit some level of fuel specificity. Some manufacturers provide "fuel factors" that relate detector response performance to a fire of one fuel to the response performance of a benchmark fuel. Other manufacturers provide performance criteria for a list of specific fuels. Unless the manufacturer's manual, bearing the listing mark, contains explicit instructions for the application of the detector for fuels other than those used in the listing process, the unit cannot be deemed listed for use in hazard areas containing fuels different than those employed in the listing process.

B-5.2.6.1 When the fuel factor correction is expressed as a detection distance reduction, the correction should be applied after the detection distance has been computed.

B-5.2.6.2 When the fuel factor correction is expressed as a function of normalized fire size, the correction must be made prior to calculating detection distance.

B-5.2.7 Atmospheric Extinction Factors. Because the atmosphere is not infinitely transmittant at any wavelength, all flame detectors are affected by atmospheric absorption to some degree. The effect of atmospheric extinction on the performance of flame detectors is determined to some degree by the wavelengths used for sensing and the detector electronic architecture. Values for the atmospheric extinction coefficient (ζ) should be obtained from the detector manufacturer.

B-5.3 Design of Spark/Ember Detection Systems.

The similarity between the method for the design of flame detection systems and spark/ember detection systems is not accidental. Each method employs the same physics, but different chemistry. Because spark/ember detectors are designed to detect the Plankian emissions emanating from an ember due to its temperature, the designer does not have to deal with fuel specificity as when designing with flame detectors. It is important to note that all flames emit radiation over the range of wavelengths normally used for spark/ember detectors. However, normal ambient light as well as light from artificial light sources are also rich in near infrared light. This fact prevents the use of most spark/ember detectors in normally lit ambient environments.

B-5.3.1 Design Fire. Using the process outlined in Section B-2 determine the fire size (kW or Btu/sec) at which detection must be achieved.

B-5.3.1.1 The quantification of the fire is generally derived from the energy investment per unit time sufficient to propagate combustion of the combustible particulate solids in the fuel stream. Because energy per unit time is power, expressed in watts, the fire size criterion is generally expressed in watts or milliwatts.

B-5.3.1.2 The radiant emissions, integrated over all wavelengths, from a non-ideal Plankian radiator is expressed with the following form of the Steffan-Boltzmann equation:

$$P = \varepsilon A \sigma T^4 \tag{39}$$

where:

P = radiant power (W)

ε = emissivity, a material property expressed as a fraction between 0 and 1.0

A = area of radiator (m^2)

σ = Steffan-Boltzmann constant 5.67E-8W/m^2K^4

T = temperature (K)

B-5.3.1.3 This models the spark or ember as a point source radiator.

B-5.3.2 Fire Environment. Spark/ember detectors are usually used on pneumatic conveyance system ducts to monitor combustible particulate solids as they flow past the detector(s). This environment puts large concentrations of combustible particulate solids between the fire and the detector. A value for ζ must be computed for the monitored environment. The simplifying assumption that absorbance at visible levels is equal to or greater than that at infrared wavelengths yields conservative designs and is used.

B-5.3.3 Calculate Detector Response to Design Fire. Because the sensitivity of a spark/ember detector is fixed during the manufacturing process,

$$S = \frac{kPe^{-\zeta d}}{d^2}$$

where:

S = radiant power reaching the detector (W) sufficient to produce alarm response

k = proportionality constant for the detector

P = radiant power emitted by test spark (W)

ζ = extinction coefficient of air at detector operating wavelengths

d = distance between the fire and the detector during the listing fire test

Because the sensitivity of the detector is constant over the range of ambients for which it is listed

$$S = \frac{kP'e^{-\zeta d'}}{d'^2} \qquad (40)$$

where:

S = radiant power reaching the detector (W) sufficient to produce alarm response

k = proportionality constant for the detector

P' = radiant power from the design fire

ζ = the extinction coefficient of air at detector operating wavelengths

d' = the distance between the design fire and the detector

Therefore, use the following equation to solve for

$$\frac{kPe^{-\zeta d}}{d^2} = \frac{kP'e^{-\zeta d'}}{d'^2}$$

To solve for d',

$$\left[\frac{(d^2)P'^{-\zeta d'}}{P^{-\zeta d}}\right]^{1/2} = d'$$

This relation is solved iteratively for d', the distance at which the detector can detect the design fire.

Because the spark is essentially a point source radiator of measurable radiant power, the designer does not need to perform a flame area calculation for spark/ember detectors. However, the designer should keep in mind that spark/ember detectors generally respond only to a step-function increase in radiant power.

B-5.3.4 Correction for Angular Displacement.

B-5.3.4.1 Most spark/ember detectors exhibit a loss of sensitivity as the fire is displaced from the optical axis of the detector. This correction to the detector sensitivity is shown as a polar graph in Figure A-2-4.3.2.3.

B-5.3.4.2 When the correction for angular displacement is expressed as a reduction of normalized detection distance the correction is made to detection distance (d').

B-5.3.4.3 When the correction for angular displacement is expressed as a normalized sensitivity (fire size increment) the correction must be made to P' prior to calculating response distance (d').

B-5.3.5 Corrections for Fuel. Because spark/ember detectors respond to Plankian emission in the near infrared portion of the spectrum, corrections for fuels are rarely necessary.

B-6 Computer Fire Models

Several special application computer models are available to assist in the design and analysis of both heat detectors (for example, fixed-temperature, rate-of-rise, sprinklers, fusible links) and smoke detectors. These computer models typically run on personal computers and are available from the NIST Center for Fire Research computer bulletin board.

B-6.1 DETACT—T2.

DETACT—T2 (DETector ACTuation—time squared) calculates the actuation time of heat detectors (fixed-temperature and rate-of-rise) and sprinklers to user-specified fires that grow with the square of time. DETACT—T2 assumes the detector is located in a large compartment with an unconfined ceiling, where there is no accumulation of hot gases at the ceiling. Thus, heating of the detector is only from the flow of hot gases along the ceiling. Input data includes H, τ_0, RTI, T_s, S, and α. The program calculates the heat release rate at detector activation, as well as the time to activation. The response of a smoke detector can also be modeled by assuming the smoke detector to be a low-temperature, zero lag-time heat detector.

B-6.2 DETACT—QS.

DETACT—QS (DETector ACTuation—quasi-steady) calculates the actuation time of heat detectors and sprinklers in response to fires that grow according to a user-defined fire. DETACT—QS assumes the detector is located in a large compartment with unconfined ceilings, where there is no accumulation of hot gases at the ceiling. Thus, heating of the detector is only from the flow of hot gases along the ceiling. Input data includes H, τ_0, RTI, T_s, the distance of the detec-

tor from the fire's axis, and heat release rates at user-specified times. The program calculates the heat release rate at detector activation, the time to activation, and the ceiling jet temperature. The response of a smoke detector can also be modeled by assuming the smoke detector to be a low-temperature, zero lag-time heat detector.

B-6.2.1 DETACT—QS can also be found in HAZARD I, FIREFORM, FPETOOL.

B-6.3 LAVENT.

LAVENT (Link Actuated VENT) calculates the actuation time of sprinklers and fusible link-actuated ceiling vents in compartment fires with draft curtains. Inputs include the ambient temperature, compartment size, thermophysical properties of the ceiling, fire location, size and growth rate, ceiling vent area and location, RTI, and temperature rating of the fusible links. Outputs of the model include the temperatures and release times of the links, the areas of the vents that have opened, the radial temperature distribution at the ceiling, and the temperature and height of the upper layer.

B-7 Nomenclature

The nomenclature used in Appendix B is defined as follows:

α	=	fire intensity coefficient (Btu/sec³ or kW/sec²)
A	=	area (m² or ft²)
A_0	=	$g/(C_p T_a \rho_0)$ [m⁴/sec²kJ) or ft⁴/(sec²Btu)]
A_r	=	radiating area (ft²)
A_t	=	radiating area of test fire
C	=	specific heat of detector element (Btu/lbm·°F or kJ/kg·°C)
c	=	speed of light (m/sec)
C_p	=	specific heat of air [Btu/lbm R or kJ/(kg K) (1.040 kJ/kg K)]
D_m	=	mass optical density (m²/g)
d	=	distance between fire and radiant energy-sensing detector
d'	=	distance between fire and detector
$\frac{d(D_u)}{dt}$	=	rate of increase of optical density outside the detector
D	=	$0.146 + 0.242 r/H$
Δt	=	change in time (sec)
ΔT	=	increase above ambient in temperature of gas surrounding a detector (°C or °F)
Δt_d	=	increase above ambient in temperature of a detector (°C or °F)
Δt_p^*	=	change in reduced gas temperature
e	=	energy (joules)
f	=	functional relationship

g	=	gravitational constant [m/sec² or ft/sec² (9.81 m/sec²)]
h	=	Plank's constant (6.63E-23 joule-sec)
H	=	ceiling height or height above fire (m or ft)
H_c	=	convective heat transfer coefficient (kW/m²·°C or Btu/ft²·sec·°F)
ΔH_c	=	heat of combustion (kJ/mol)
h_f	=	flame height (ft)
H_f	=	heat of formation (kJ/mol)
L	=	characteristic length for a given detector design
k	=	detector constant, dimensionless
m	=	mass (lbm or kg)
p	=	positive exponent
P	=	radiant power (watts)
q	=	heat release rate density per unit floor area (Btu/sec·ft²)
Q	=	heat release rate (Btu/sec or kW)
Q_c	=	convection portion of fire heat release rate (Btu/sec)
Q_{cond}	=	heat transferred by conduction (Btu/sec or kW)
Q_{conv}	=	heat transferred by convection (Btu/sec or kW)
Q_d	=	threshold fire size at which response must occur
Q_{rad}	=	heat transferred by radiation (Btu/sec or kW)
Q_{total}	=	total heat transfer (Btu/sec or kW)
Q_{CR}	=	critical heat release rate (Btu/sec or kW)
Q_{D0}	=	design heat release rate (Btu/sec or kW)
Q_m	=	maximum heat release rate (Btu/sec or kW)
Q_p	=	predicted heat release rate (Btu/sec or kW)
Q_T	=	threshold heat release rate at response (Btu/sec or kW)
r	=	radial distance from fire plume axis (m or ft)
ρ_0	=	density of ambient air [kg/m³ or lb/ft³ (1.1 kg/m³)]
RTI	=	response time index (m$^{1/2}$sec$^{1/2}$ or ft$^{1/2}$sec$^{1/2}$)
S	=	spacing of detectors or sprinkler heads (m or ft)
S	=	radiant energy
t_{DO}	=	time at which the design objective heat release rate (Q_{DO}) is reached (sec)
t_{CR}	=	time at which the critical heat release rate (Q_{CR}) is reached (sec)
t	=	time (sec)
t_c	=	critical time—time at which fire would reach a heat release rate of 1000 Btu/sec (1055 kW) (sec)
t_d	=	time to detector response
t_g	=	fire growth time to reach 1000 Btu/sec (1055 kW) (sec)
t_r	=	response time (sec)
$t_{respond}$	=	time available, or needed, for response to an alarm condition (sec)
t_v	=	virtual time of origin (sec)
t_{2f}	=	arrival time of heat front (for $p = 2$ power law fire) at a point r/H (sec)
t_{2f}^*	=	reduced arrival time of heat front (for $p = 2$ power law fire) at a point r/H (sec)
t_p^*	=	reduced time

T	=	temperature (°C or °F)
T_a	=	ambient temperature (°C or °F)
T_c	=	plume centerline temperature (°F)
T_d	=	detector temperature (°C or °F)
T_g	=	temperature of fire gases (°C or °F)
T_s	=	rated operating temperature of a detector or sprinkler (°C or °F)
u_0	=	instantaneous velocity of fire gases (m/sec or ft/sec)
u	=	velocity (m/sec)
u_c	=	critical velocity
$U_p{}^*$	=	reduced gas velocity
V	=	velocity of smoke at detector
w_f	=	flame width (ft)
Y	=	defined in equation (16)
z	=	height above top of fuel package involved (ft)
λ	=	wavelength (microns)
Z_m	=	maximum height of smoke rise above fire surface (ft or m)
τ	=	detector time constant mc/H_cA (sec)
τ_0	=	detector time constant measured at reference velocity u_0 (sec)

ε	=	emissivity, a material property expressed as a fraction between 0 and 1.0

References Cited in Commentary

1. Meacham, Brian J., and Custer, Richard L.P., 1995, "Performance-Based Fire Safety Engineering: An Introduction to Basic Concepts," *Journal of Fire Protection Engineering*, vol. 7, no. 2.
2. Custer, Richard L.P., and Meacham, Brian J., *Introduction to Performance-Based Fire Safety*, National Fire Protection Association, Quincy, MA, 1997.
3. Custer, Richard L. P., 1997, "Dynamics of Compartment Fire Growth," *Fire Protection Handbook*, 18th ed., pp. 1-84 through 1–91, National Fire Protection Association, Quincy, MA.
4. *Fire Protection Handbook*, 1997, 18th ed., National Fire Protection Association, Quincy, MA.

APPENDIX C

Referenced Publications

C-1 The following documents or portions thereof are referenced within this code for informational purposes only and are thus not considered part of the requirements of this code unless also listed in Chapter 9. The edition indicated here for each reference is the current edition as of the date of the NFPA issuance of this code.

C-1.1 NFPA Publications.

National Fire Protection Association, 1 Batterymarch Park, P.O. Box 9101, Quincy, MA 02269-9101.

NFPA 10, *Standard for Portable Fire Extinguishers*, 1998 edition.

NFPA 11, *Standard for Low-Expansion Foam*, 1998 edition.

NFPA 11A, *Standard for Medium- and High-Expansion Foam Systems*, 1999 edition.

NFPA 12, *Standard on Carbon Dioxide Extinguishing Systems*, 1998 edition.

NFPA 12A, *Standard on Halon 1301 Fire Extinguishing Systems*, 1997 edition.

NFPA 13, *Standard for the Installation of Sprinkler Systems*, 1999 edition.

NFPA 14, *Standard for the Installation of Standpipe and Hose Systems*, 1996 edition.

NFPA 15, *Standard for Water Spray Fixed Systems for Fire Protection*, 1996 edition.

NFPA 17, *Standard for Dry Chemical Extinguishing Systems*, 1998 edition.

NFPA 70, *National Electrical Code®*, 1999 edition.

NFPA 80, *Standard for Fire Doors and Fire Windows*, 1999 edition.

NFPA 90A, *Standard for the Installation of Air-Conditioning and Ventilating Systems, 1999 edition.*

NFPA 90B, *Standard for the Installation of Warm Air Heating and Air Conditioning Systems*, 1999 edition.

NFPA 92A, *Recommended Practice for Smoke-Control Systems*, 1996 edition.

NFPA 92B, *Guide for Smoke Management Systems in Malls, Atria, and Large Areas*, 1995 edition.

NFPA *101®*, *Life Safety Code®*, 1997 edition.

NFPA 170, *Standard for Fire Safety Symbols*, 1999 edition.

NFPA 1221, *Standard for the Installation, Maintenance, and Use of Emergency Services Communication Systems*, 1999 edition.

C-1.2 Other Publications.

C-1.2.1 ANSI Publications. American National Standards Institute, Inc., 11 West 42nd Street, 13th floor, New York, NY 10036.

ANSI A17.1, *Safety Code for Elevators and Escalators*, 1998.

ANSI S3.2, *Method for Measuring the Intelligibility of Speech Over Communications Systems,* 1989.

ANSI S3.41, *Audible Emergency Evacuation Signal*, 1990.

C-1.2.2 IEC Publications. International Electrotechnical Commission; 3 rue de Varembé, P.O. Box 131, Geneva 1, Switzerland. IEC documents are available through ANSI.

IEC 60849, *Sound systems for emergency purposes,* Second Edition: 1998.

IEC 60268, Part 16, *The objective rating of speech intelligibility by speech transmission index,* Second Edition: 1998.

C-1.2.3 IES Publication. Illuminating Engineering Society of North America, 120 Wall Street, 17th floor, New York, NY 10005.

Lighting Handbook Reference and Application, 1993.

C-1.2.4 ISO Publication. Standards Secretariat, Acoustical Society of America, 335 East 45th Street, New York, NY 10017-3483.

ISO 8201, *Audible Emergency Evacuation Signal*, 1990.

C-1.2.5 UL Publications. Underwriters Laboratories Inc., 333 Pfingsten Road, Northbrook, IL 60062.

UL 268, *Standard for Safety, Smoke Detectors for Fire Protective Signaling Systems,* 1999.
Visual Signaling Appliances—Private Mode Emergency and General Utility Signaling, UL 1638, 1995.
UL 1971, *Standard for Safety Signaling Devices for the Hearing Impaired,* 1992.

C-1.2.6 U.S. Government Publications. U.S. Government Printing Office, Superintendent of Documents, Washington, DC 20402.

Title 47, *Code of Federal Regulations,* Part 15.
FCC Rules and Regulations, Volume V, Part 90, March 1979.

C-2 Bibliography

This part of the appendix lists other publications pertinent to the subject of this NFPA document that might or might not be referenced.

1. Alpert, R. "Ceiling Jets," *Fire Technology,* Aug. 1972.

2. Evaluating Unsprinklered Fire Hazards, SFPE Technology Report 83-2.

3. Babrauskas, V.; Lawson, J. R.; Walton, W. D.; and Twilley, W. H. "Upholstered Furniture Heat Release Rates Measured with a Furniture Calorimeter" (NBSIR 82-2604) (Dec. 1982). National Institute of Standards and Technology (formerly National Bureau of Standards), Center for Fire Research, Gaithersburg, MD 20889.

4. Beyler, C. "A Design Method for Flaming Fire Detection" *Fire Technology,* vol. 20, No. 4, Nov. 1984.

5. DiNenno, P., ed. Chapter 31, *SFPE Handbook of Fire Protection Engineering,* by R. Schifiliti, Sept. 1988.

6. Evans, D. D. and Stroup, D. W. "Methods to Calculate Response Time of Heat and Smoke Detectors Installed Below Large Unobstructed Ceilings" (NBSIR 85-3167) (Feb. 1985, issued Jul. 1986). National Institute of Standards and Technology (formerly National Bureau of Standards), Center for Fire Research, Gaithersburg, MD 20889.

7. Heskestad, G. "Characterization of Smoke Entry and Response for Products-of-Combustion Detectors" Proceedings, 7th International Conference on Problems of Automatic Fire Detection, Rheinish-Westfalischen Technischen Hochschule Aachen (Mar. 1975).

8. Heskestad, G. "Investigation of a New Sprinkler Sensitivity Approval Test: The Plunge Test," FMRC Tech. Report 22485, Factory Mutual Research Corporation, 1151 Providence Turnpike, Norwood, MA 02062.

9. Heskestad, G. and Delichatsios, M. "The Initial Convective Flow in Fire: Seventeenth Symposium on Combustion," The Combustion Institute, Pittsburgh, PA (1979).

10. Heskestad, G. and Delichatsios, M. A. "Environments of Fire Detectors - Phase 1: Effect of Fire Size, Ceiling Height and Material," Measurements vol. I (NBS-GCR-77-86), Analysis vol. II (NBS-GCR-77-95). National Technical Information Service (NTIS), Springfield, VA 22151.

11. Heskestad, G. and Delichatsios, M. A. "Update: The Initial Convective Flow in Fire," *Fire Safety Journal,* vol. 15, No. 5, 1989.

12. International Organization for Standardization, *Audible Emergency Evacuation Signal,* ISO 8201,1987.

13. Klote, J. and Milke, J. "Design of Smoke Management Systems," American Society of Heating, Refrigerating and Air Conditioning Engineers, Atlanta, GA 1992.

14. Lawson, J. R.; Walton, W. D.; and Twilley, W. H. "Fire Performance of Furnishings as Measured in the NBS Furniture Calorimeter, Part 1" (NBSIR 83-2787) (Aug. 1983). National Institute of Standards and Technology (formerly National Bureau of Standards), Center for Fire Research, Gaithersburg, MD 20889.

15. Morton, B. R.; Taylor, Sir Geoffrey; and Turner, J.S. "Turbulent Gravitational Convection from Maintained and Instantaneous Sources," Proc. Royal Society A, 234, 1-23, 1956.

16. Schifiliti, R. "Use of Fire Plume Theory in the Design and Analysis of Fire Detector and Sprinkler Response," Master's thesis, Worcester Polytechnic Institute, Center for Firesafety Studies, Worcester, MA, 1986.

17. Title 47, *Code of Federal Regulations,* Communications Act of 1934 Amended.

18. R. Schifiliti, P.E. and W. Pucci, "Fire Detection Modelling, State of the Art," 6 May, 1996, sponsored by the Fire Detection Institute, Bloomfield, CT

19. G. Forney, R. Bukowski, W. Davis, "Field Modelling: Effects of Flat Beamed Ceilings on Detector and Sprinkler Response," Technical Report, Year 1, International Fire Detection Research Project, National Fire Protection Research Foundation, Quincy, MA, October, 1993

20. W. Davis, G. Forney, R. Bukowski, "Field Modelling: Simulating the Effect of Sloped Beamed Ceilings on Detector and Sprinkler Response," Year 1. International Fire Detection Research Project Technical Report, National Fire Protection Research Foundation, Quincy, MA, October, 1994

21. E. Brozovski, "A Preliminary Approach to Siting Smoke Detectors Based on Design Fire Size and Detector Aerosol Entry Lag Time," Master's Thesis, Worcester Polytechnic, Worcester, MA, USA, 1989.

22. A. Cote, *Fire Protection Handbook*, 17th Edition, National Fire Protection Association, Quincy, MA. USA, 1992.

23. Tewarson, A., "Generation of Heat and Chemical Compounds in Fires," *SFPE Handbook of Fire Protection Engineering*, Second Edition, NFPA and SFPE, 1995.

24. J.P. Hollman, *Heat Transfer*, McGraw-Hill, New York (1976).

25. Custer, R.L.P and Meacham B., *Introduction to Performance Based Fire Safety*, SFPE, 1997.

26. Schifiliti, R.P., Meacham B., Custer, R.L.P., "Design of Detection Systems," *SFPE Handbook of Fire Protection Engineering*.

27. Marrion, C., "Correction Factors for the Heat of Combustion in NFPA 72, Appendix B," *SFPE Handbook of Fire Protection Engineering*, NFPA, 1998.

28. Marrion, C. "Designing and Analysing the Response of Detection Systems: An Update to Previous Correlations," 1998. Unpublished paper.

29. R. Custer and R. Bright, "Fire Detection: The State-of-the-Art," NBS Tech. Note 839, National Bureau of Standards, Washington (1974).

30. Brian J. Meacham, "Characterization of Smoke from Burning Materials for the Evaluation of Light Scattering-Type Smoke Detector Response," MS Thesis, WPI Center for Firesafety Studies, Worcester, MA (1991).

31. M.A. Delichatsios, "Categorization of Cable Flammability, Detection of Smoldering, and Flaming Cable Fires," Interim Report, Factory Mutual Research Corporation, Norwood, MA NP-1630, Nov. 1980.

32. G. Heskestad, FMRC Serial Number 21017, Factory Mutual Research Corp., Norwood, MA (1974).

33. C.E. Marrion, "Lag Time Modeling and Effects of Ceiling Jet Velocity on the Placement of Optical Smoke Detectors," MS Thesis, WPI Center for Firesafety Studies, Worcester, MA (1989).

34. M. Kokkala et al., "Measurements of the Characteristic Lengths of Smoke Detectors," *Fire Technology*, Vol. 28, No. 2, National Fire Protection Association, Quincy, MA (1992).

35. UL 268, *Standard for Safety, Smoke Detectors for Fire Protective Signaling Systems*, Underwriters Laboratories, Inc., Northbrook, IL (1989).

36. Scott Deal, "Technical Reference Guide for FPEtool Version 3.2," NISTIR 5486, National Institute for Standards and Technology, U.S. Department of Commerce, Gaithersburg, MD, Aug. (1994).

37. F.W. Mowrer, "Lag Times Associated with Detection and Suppression," *Fire Technology*, Vol. 26, No. 3, pp. 244\N265 (1990).

38. J.S. Newman, "Principles for Fire Detection," *Fire Technology*, Vol. 24, No. 2, pp. 116\N127 (1988).

39. Custer, R., Meacham, B., Wood, C. "Performance Based Design Techniques for Detection and Special Suppression Applications," Proceedings of the SFPE Engineering Seminars on Advances in Detection and Suppression Technology, 1994.

40. SFPE *Engineering Guide to Performance Based Fire Protection Analysis and Design*, National Fire Protection Association, Quincy, MA, 2000.

PART TWO

Supplements

In addition to the code text and commentary presented in Part One, the *National Fire Alarm Code Handbook* includes supplements. They are not part of the code but are included as additional information for handbook users. In the following four supplements, Part Two explores the background of four selected topics related to NFPA 72 in more detail than the commentary:

1 Performance-Based Design and Fire Alarm Systems

2 Fire Alarm Systems Providing Central Station Service: Applications and Advancements

3 Integrating Fire Alarm Systems with Other Building Systems

4 Occupant Response to Fire Alarm Signals

Performance-Based Design and Fire Alarm Systems

John M. Cholin, P.E.
J.M. Cholin Consultants, Inc.

Wayne D. Moore, P.E.
Hughes Associates, Inc.

Editors' Note: Chapter 8 of the 1999 edition of the National Fire Alarm Code includes performance-based language for the first time. Long-time users of the code, however, will recognize many tools of performance-based design in Appendix B. This supplement introduces the concept of performance-based design of fire alarm systems.

John M., Cholin, P.E., is the president of J.M. Cholin Consultants, Inc., in Oakland, New Jersey. He is a member of the NFPA Technical Committees on Initiating Devices for Fire Alarm Systems and on Handling and Conveying of Dusts, Vapors, and Gases and Chairman of the Technical Committee on Wood and Cellulosic Materials Processing.

Wayne D. Moore, P.E., is the Director of Operations of the New England office of Hughes Associates, Inc., in Warwick, Rhode Island. He has served as an editor of the first, second, and third editions of the National Fire Alarm Code Handbook. Mr. Moore is a former member of the NFPA Standards Council, chairs the Technical Correlating Committee on the National Fire Alarm Code, and holds membership on the NFPA Technical Committees on Cultural Resources, Fire Prevention Code, and Protected Premises Fire Alarm Systems.

INTRODUCTION

Over the past decade, engineers have introduced the term *performance-based design* into the language of the fire protection community. To many, it is a new concept; to others it is the long overdue recognition of how most engineering has been performed since the birth of the engineering profession. As designers employ performance-based design methods in the design of buildings, fire alarm systems will become far more important. The fire alarm systems will provide an alternative means of achieving the overall design objectives.

The National Fire Protection Association has assumed a leadership role in formulating the concepts that make the development of performance-based codes and standards possible. As part of that larger effort, Supplement 1 provides a review of the performance-based design concepts for the fire alarm system designer and technician. All too often, no one makes the fire alarm technologist and technician aware of how the fire alarm system fits into the overall fire protec-

tion strategy for a given building. When performance-based design methods are used, an understanding of the interdependency of all the fire protection features is critical.

In fire protection engineering, the building design, type of construction, water supplies, fixed suppression systems, fire alarm systems, fixed extinguishing systems, off-site reporting, and fire service capabilities all become essential components in the fire safety plan for the building. Each element interacts with, and often relies on, the others. When a change is made to one of those elements, changes must often be made to one or more of the others in order to maintain an equivalent level of fire safety. The interrelation and interdependency between the various elements of the facility fire safety assets becomes far more apparent when an engineer considers performance-based design.

Although many jurisdictions permit technologists to design fire alarm systems pursuant to the prescriptive criteria found in the code, designers using a performance-based approach must review the relevant engineering licensure law

in the jurisdiction where the designer plans to practice. Performance-based designs of fire alarm systems are likely to be deemed engineering of the type requiring licensure as a professional engineer. The designer should also be knowledgeable in the principles of fire protection engineering, and apply these principles judiciously.

DEFINING PERFORMANCE-BASED DESIGN

Performance-based design derives the design from the performance objectives for what is being designed, whether it is a building, fire protection system, or bridge. Rather than relying upon past experience, the designer develops and evaluates the design in terms of its ability to achieve the performance objective through calculations. This method is not a new approach. In other engineering disciplines this method is the norm rather than an exception. For example, an engineer designs a bridge to support a design load under the range of conditions expected at the contemplated location. Engineers use a similar design process for automobiles, computers, or aircraft. In the case of an aircraft, the design process begins when someone defines the payload, range, and cost of the contemplated airplane. The objectives for the finished product then lead to specific design and performance criteria. These criteria relate to the engine thrust, wing dimensions, cruising altitude, required runway, and so forth. The designer benchmarks each of these criteria against the overall objectives of the contemplated aircraft. Where unavoidable, the designer makes concessions. Eventually a design emerges. If the designer has done his or her job properly, the prototype airplane flies and fulfills the objectives of the purchaser.

Currently, the design of the fire protection for a building is less flexible and follows a profoundly different process. The building code and fire code dictate that the design must include a collection of specific features. Traditionally, building codes explicitly prescribe the design features a building must have based on its size and contemplated use. The prescribed building features stated in the locally adopted building codes and fire codes generally result from building failures and fire experience throughout history. The codes reflect a consensus among the building community on how to prevent the result of a specific fire. The building community has a collective understanding of the outcome they wish to prevent and adopt a prescribed means to prevent that outcome. However, these objectives are generally not explicitly stated nor do the prescribed means represent the only means by which the objective can be achieved.

Consequently, when using prescriptive codes, the design process is reduced to identifying the type of intended use (occupancy classification) for the structure and then applying the prescribed design features without determining whether those features are necessary or sufficient to attain the intended level of fire safety for the particular structure.

Codes and standards that are written in prescriptive terms, like NFPA 72, *National Fire Alarm Code*, are often referred to as "cookbook codes." If the user follows the rules as outlined in the code, he or she needs to give very little thought to the design basis for the fire alarm system. Most fire alarm system installers and many fire alarm system designers often do not consider design parameters such as fire department response time, egress capacity of existing buildings, fire protection water supply availability, or construction type.

Users should consider many of these parameters in the design of a fire alarm system, even when using the prescriptive approach. However, when the designer is required to observe minimum compliance requirements of a prescriptive code there is often insufficient latitude in the code to allow the designer to address the specific needs of the building owner and occupants.

In addition, few understand the limitations of a fire alarm system. Most assume that if the fire alarm system complies with the building code or *Life Safety Code*® and if the installation complies with NFPA 72, it will provide early warning of a fire regardless of where the fire originates. The purpose of NFPA 72 states: "It is the intent of this code to establish the required levels of performance, extent of redundancy, and quality of installation but not to establish the methods by which these requirements are to be achieved." Even so, the required levels are implied minimums. There are many design issues that are not addressed in the regional building codes or *National Fire Alarm Code* because they are relevant only for a subset of facilities and are not appropriate for inclusion into a minimum compliance consensus standard. Designing to a minimum prescriptive code does not guarantee early warning of a fire regardless of where the fire originates. Essentially, if the fire is remote from the detection devices, notification will be delayed until the fire grows larger. Owners who accept the minimums prescribed by these codes do not have the benefit of increased performance for a fire alarm system that would meet their goals and objectives.

The SFPE *Engineering Guide to Performance-Based Fire Protection Analysis and Design* (2000) defines *performance-based design* as "an engineering approach to fire protection design based on (1) agreed upon fire safety goals and objectives, (2) deterministic and/or probabilistic analysis of fire scenarios, and (3) quantitative assessment of design alternatives against the fire safety goals and objectives using accepted engineering tools, methodologies, and performance criteria." The *Guide* also defines *stakeholder* as "one who has a share or an interest, as in an enterprise. Specifically, an individual (or representative of same) having an interest in the successful completion of a project. The reason for having an interest in the successful completion of a project may be financial, safety related, etc." It is necessary to identify all of the stakeholders at the beginning of the project. They will

help to establish goals and objectives for the project, as well as approve the methods that are used to achieve them. The *Guide* identifies the following possible stakeholders:

Building owner

Building manager

Jurisdictional authorities

- Fire
- Building
- Electrical
- Insurance
- Accreditation agencies

Construction team

Construction manager

General contractor

Subcontractors

Tenants

Building operations and maintenance

Emergency responders / Fire service

Each of the individuals on the list will have similar, different, or additional goals and objectives that they contribute during the design process. Again to quote the *Guide*, "It is imperative for the engineer to identify the stakeholders in order to obtain acceptance of the performance-based strategies used in the [design] process." A stakeholder cannot reduce or rescind a goal or objective of some other stakeholder, but the objectives can be strengthened to achieve a higher level of safety. Consequently, the goals and objectives become the minimum compliance criteria rather than the specific design features in the prescriptive code.

In summary, performance-based design is a method in which the design features are derived from the explicitly stated goals and objectives established by either the locally adopted building codes or relevant stakeholders. These goals and objectives become the minimum compliance criteria for the design rather than a set of prescribed design features, as is found in our current codes.

THE PRESCRIPTIVE CODE CONTEXT

In the current environment of prescriptive building and fire codes, the process of design is largely a process of complying with a law. Generally, a jurisdiction adopts one of the model building codes, often with local amendments. This tradition started a long time ago with the Code of Hammurabi. In ancient Mesopotamia the King Hammurabi erected an obelisk. The king ordered his minions to write the laws of his kingdom on the faces of the obelisk. One law stated: "Who ever takes commission to build a house and that house falleth down, killing a subject of Hammurabi, the builder shall be put to death."

Although the severity of the penalty Hammurabi administered can be questioned, he did establish the concept that holds a builder accountable for failing to construct a sound building. That concept has carried through to the present day; the owner of a building has an obligation to society to construct a safe building.

The design process in the current prescriptive environment generally begins by defining the use group or occupancy type of the contemplated structure. When an architect categorizes a building or space as a particular use group, he or she is accepting a set of assumptions regarding the anticipated type and quantity of combustibles, the probable sources of ignition, the number of occupants, and the capabilities of the occupants for that building or space. None of these assumptions are explicitly stated but they are implicit in the prescriptive requirements as are the expectations regarding the performance of the structure under fire conditions. The performance-based design method is radically different from the prescriptive design method.

REASONS FOR A PERFORMANCE-BASED DESIGN METHOD

The current system of prescriptive codes and standards provides increased levels of fire safety in the new built environment. Most newly constructed buildings provide more safety from fire than ever before. Despite this positive trend, the ever increasing need to use resources more wisely dictates change.

In an increasingly competitive world environment the need to use resources wisely has become the fundamental driving force behind the transition toward a performance-based design environment. The inherent flexibility in the performance-based design method permits the selection of fire protection features based on the design goals and objectives. The design flexibility results in a greater efficiency in the use of fire protection resources. Performance-based design also allows objectives to be met in unusual environments.

In some instances, the routine application of the requirements of a prescriptive code results in a design that commits fire protection resources to systems that are not likely to make significant contributions to the fire safety of the structure. For example, general purpose sprinkler heads are often installed in an atrium space at ordinary hazard spacing, 70 feet or 80 feet above the floor. In the past engineers have specified spot-type smoke detectors in that same location.

A quick calculation shows that a fire would probably have to reach a heat release rate of approximately 15 MW (approximately equivalent to a 15 m^2 pool of gasoline in free burn) before the first sprinkler head would actuate. Yet, under these conditions the water discharging from the sprinkler head might not penetrate the fire plume to the flame surface where it could contribute to controlling the fire.

Relying on experience rather than calculation, it would be expected that a smoke detector installed at such a height would need a similar size fire before an alarm response would be attained. The smoke detectors at those heights would certainly not meet the implied goal of early warning. Finally, it is possible that the available fuel load in the space could not produce the needed 15-MW fire. So, neither the sprinkler system nor the smoke detection system can be expected to achieve the implied design objectives in the event of a fire in this example. The objectives, which have not changed, demand a design appropriate to the space. The designer must compute the impact the fire will have in the space before a design can be selected. In this case the expenditure of fire safety resources could have been spent on different fire protection assets that make a more compelling contribution to the fire safety of the facility.

This is a classic case of designing a building space by including the design features prescribed by the relevant building code and fire code, rather than designing fire protective features that provide for a given level of fire safety. Numerous historical fire incidents have shown that sprinklers prove extremely effective in controlling fire, by holding the fire to the compartment of origin, protecting the compartmentation, and maintaining structural integrity until the fire is extinguished. Accordingly, the model building codes have adopted a requirement for most occupancies to equip all such compartments with sprinkler systems. Undeniably, smoke detectors provide early warning in most fires, so the model building codes have adopted requirements for smoke detection for many occupancies. Numerous other historical fire incidents have shown that limiting the quantity and type of combustibles available in compartments enhances the survival prospects of the occupants and the protection of the structure. Therefore, requirements relating to the flame spread rating of wall coverings and interior furnishings for public places have become incorporated into the model building codes. Lastly, the model building codes also might require 2-hour rated compartment construction under some circumstances. Required construction provides a passive compartmentation of the fire, limiting the probability of fire spread to adjacent areas.

Separately, each of these building code requirements offers valid strategies for limiting the hazard of a fire in the particular compartment. To a significant degree in the preceding example, the requirements for sprinklers, smoke detectors, and fuel load limitations address the same general goal: to limit the size of a fire. Considering the fire resistance rating required of the compartment in the atrium example, at least four required fire protection strategies address the same fundamental objective: to contain the fire to the compartment of ignition. Is it necessary to provide all of these required features for the same compartment? The current prescriptive building and fire codes don't explicitly address this question nor do they provide a means to quan-

tify the contribution each feature makes to the overall fire safety of the facility. The result is required redundancies.

There is nothing wrong with redundancies as long as they are intentional and based upon a rational analysis. The logic usually used to justify redundancy is if strategy A doesn't work, then strategy B is there to fall back on. It is essentially a "belt and suspenders" argument. A fire protection strategy that uses redundant features is more reliable than a strategy that relies on a single feature, assuming that all of the fire protection features have equivalent reliability. If the fire compromises one feature, then the other still maintains the safety of the structure. But, doesn't this mean that redundant requirements work their way into the prescriptive building and fire codes because of a tacit concern for the fundamental reliability of the "required" building features? How can reliability be quantified? How many redundancies are enough? By the sixth or seventh set of "suspenders" is it fair to ask how much incremental benefit is accrued from the last set?

In a performance-based design environment the fire safety objective remains the same: to limit the spread of a fire to the compartment of fire origin. The engineer is not limited to the choice of means adopted by a building code. He or she is free to develop any means that can be shown to accomplish the fundamental performance objective, regardless of whether the means has been prescribed by the locally adopted code. Appropriate construction of the walls and doors can contain a fire involving the worst-case fire load. The fire load can be limited. The fire can be suppressed. The fire can be detected early in its development and manually extinguished before it breaches the compartment. Any of these approaches can achieve the basic objective of confining the fire to the compartment of origin. In the performance-based environment, the engineer responsible for designing the fire protective features is free to select one or more means as part of an integrated fire protection strategy that best fulfills the objective at an acceptable level of reliability, even if that strategy is not the one prescribed by law.

The use of a performance-based design method does not necessarily mean that the engineer won't employ intentional redundancies to ensure that he or she meets the objective in the event of a failure of one of the fire protection features. However, the engineer will base the selection of fire protection features on a rational analysis of the possible fire scenarios for a particular compartment and the computed mission effectiveness of the fire protection features. Then, he or she will select the most efficient means to accomplish the objectives in the event of that fire. Thus, performance-based design provides the engineer with the flexibility to address unique structures and sets of performance objectives that might not have been contemplated during the drafting of a model building or fire code.

The locally adopted prescriptive code doesn't explicitly state what level of functional performance the prescribed

design attains. How can the designer determine the types and extents of necessary and sufficient changes to the structure in order to achieve the objectives of society? Answers to these questions emerge through the use of performance-based codes.

In a performance-based code environment, the code writers must qualify and, whenever possible, quantify the level of risk acceptable to the community served by the code. These levels of risk translate into broad fire safety goals for each type of occupancy. Additionally, users must develop detailed performance objectives as a means to meet the established fire safety goals. The inherent flexibility of the performance-based design environment also enables the designer to satisfy conflicting social values through the use of creative design. The inherent flexibility of performance-based design permits the selection of fire protection features based on the design goals and objectives, allowing the engineer to tailor the fire protection features to the specific structure and circumstances.

For example, an eighteenth century house that has been restored to its original, eighteenth century condition and is now used as a site for education and lectures on the local history. Conflicting goals immediately surface. The social value of the structure relies on its being preserved exactly as it existed in the eighteenth century. This preservation connects us to our history. Yet one of the means to preserve this heritage will change the structure to use modern fire detection and suppression techniques to limit the damage to the building if a fire occurs. Furthermore, because the public assembles in this structure, social value exists to provide for a level of occupant life safety consistent with the expectations of the community in general. This life safety objective also might necessitate changes to the structure. Certainly, the historical value will erode the least with minimized changes. Here is where the problem arises. With a prescriptive-based code, the designer of the fire protection systems only has a few alternatives. The authority having jurisdiction (AHJ) may or may not allow non-compliance with the prescriptive requirements of the code because it would not want to assume the inherent responsibility. With a performance-based environment, the solution becomes a "reasonable engineering approach" to finding acceptable methods to meet the fire safety goal.

Obviously, only qualified designers working with the building design team should undertake the design of engineered fire protection systems in the performance-based context. Because many of the fire protection features of the structure are interdependent, a thorough understanding of each of those features and their interdependencies is critical. In the case of fire alarm systems, the designer should possess substantial design experience not only with fire alarm systems but also with the other interdependent fire protection systems including the passive fire resistance of the structure, sprinkler fire suppression systems, fire service

response capabilities, special extinguishing systems, and fire protection in general, before undertaking a performance-based design project.

In addition, the designer should be included in the building design team beginning with the feasibility stage. The designer of the fire alarm system addresses a wide range of fire protection issues that impact the fire alarm system design. As the owner and the AHJ establish their fire safety goals and objectives, the fire alarm system designer must be able to advise them of the feasibility and prerequisite conditions for attaining those goals. Clearly, this design environment places far greater demands upon the designer. This increased demand is the price paid for the design freedom, improved cost efficiency, and design flexibility that performance-based design provides.

EXPLICIT STATEMENT OF THE FIRE PROTECTION GOALS

The performance-based design method begins with explicitly stated fire safety goals. Basically, the engineer asks the question: "What level of fire safety does society expect from the building?" The goals generally reflect a consensus of social values. The basic goals relating to life safety, protection of adjacent properties, environmental impact, and fire fighter safety will be part of a locally adopted performance-based code. Other fire safety goals such as those relating to mission continuity will be established by the owner and insurance authorities. For example, a building should maintain its structural integrity during certain situations such as a fire. This goal could include maintaining the structural integrity even after all of the combustible material within it has burned. The goal may also require that a structure maintains its integrity during, and in spite of, fire-fighting activities.

Another fire safety goal might be to warn the occupants of a fire in sufficient time to allow them to escape from imminent danger without harm. Some will complain that these goals seem so general that they add nothing to the design process. However, they are important to ensure that all of the relevant social values are reflected in the design of the building and its fire protection systems.

Fire protection goals generally address a range of issues relating to occupant life safety, fire-fighter life safety, citizen life safety, citizen property rights, property loss limitation, the continuity of the facility mission, the preservation of heritage, and the environmental impact from the fire. Society puts these issues in a hierarchy, reflecting consensus social values. Avoiding a long-term environmental impact might be considered more important than limiting property damage. Therefore, the use of a very effective extinguishing agent for the protection of property that also produces a long-term environmental degradation would not be permitted. However, society will

permit an airline to use the same extinguishing agent to extinguish engine fires on their commercial aircraft. In this case, society values the obvious life safety goal for the occupants of an airplane at a higher level than the environmental goal.

Further examples of fire safety goals are listed in the proposed *SFPE Guide*, including efforts to

- Minimize fire-related injuries and prevent undue loss of life
- Minimize fire-related damage to the building, its contents, and its historical features and attributes
- Minimize undue loss of operations and business-related revenue due to fire-related damage
- Limit environmental impact of fire and fire protection measures

EXPLICIT STATEMENT OF FIRE PROTECTION OBJECTIVES

Once the fire safety goals have been agreed upon for a building or a design, specific objectives are developed that reflect how to meet those goals. [See NFPA *Primer #1: Performance-Based Goals, Objectives and Criteria* (1997).] Although the objectives are derived from the goals, they are more specific and quantifiable. If the goal is to ensure that there is no loss of life due to fire in the building, then the objective derived from this goal can be that the building should provide sufficient warning to all occupants in sufficient time to permit them to escape or relocate without loss of life or injury. By refining the goal into an objective, the engineer is beginning the process of developing potential strategies. The use of warning as the means to attain the goal implies the use of a fire alarm system as part of the occupant protection strategy. This objective also presumes that the design of the building provides a means of egress to the building exterior or to an area of refuge. All of these fire protective features must be in place in order to attain the life safety objective. Because performance-based design permits flexibility, the designer can freely consider alternatives to the traditional approach if these alternatives provide equivalent or superior performance.

QUANTITATIVELY EXPRESSED PERFORMANCE CRITERIA

Once an engineer establishes explicit objectives, he or she can develop *quantitative* performance criteria that provide the yardstick for measuring performance and attainment of the design objective. An earlier example used an objective of "providing warning in sufficient time to permit the occupants to escape without injury." An engineer will measure this objective using performance criteria that relates specifically to the following:

- Warning of the occupants
- Size of the fire at the moment of detection
- Rate of fire growth
- Rate of deterioration of the tenability in each of the compartments, as well as along the egress route
- Number of occupants
- Condition of the occupants
- Egress speed of the occupants along the egress path
- Length of the egress path, and others

The designer must quantify each of these criteria before the process of outlining possible designs can begin.

For example, a quantitative criterion for the warning of the occupants must exist. Research has shown that audible notification with sound pressure levels of 15 dBA above average ambient or 5 dBA above momentary ambient peaks effectively warns occupants with normal hearing. For conscious, hearing-impaired occupants, visible notification intensities of 0.405 lm/m^2 (0.0375 lm/ft^2) have been shown to be sufficient to warn such occupants. Consequently, the designer could formulate the performance criteria for occupant notification to read as follows:

> Occupant notification shall be deemed to have been provided when the following conditions occur:
>
> 1. Attainment in all occupiable portions of the compartment, the Temporal Code-3 audible notification having a sound pressure of at least 15 dBA above average ambient or 5 dBA above momentary maximum (greater than 60 seconds duration) ambient and,
>
> 2. Attainment in all occupiable portions of the compartment, visible notification producing an effective illuminance of 0.405 lm/m^2 (0.0375 lm/ft^2) with an integrated flash rate no greater than 2 Hz and no less than 1 Hz.

One way of achieving these performance criteria is to install audible and visible notification appliances according to the prescriptive criteria in the current edition of *National Fire Alarm Code*.

However, unoccupiable compartments might exist within the building. A prescriptive code would require notification appliances throughout the building. Under the performance-based approach, if portions of the building are inherently unoccupiable, the designer can make a case to omit notification for those portions of the structure.

The Technical Committee on Notification Appliances for Fire Alarm Systems has framed the prescriptive criteria with parallel performance criteria in the appendix. However, the prescriptive requirements of other chapters in the *National Fire Alarm Code* are not as easily restated as performance-based criteria. In this case, a validated performance demonstration method must be used to determine the implied performance criteria of a design using the pre-

scriptive criteria. Then the designer must use that same method to demonstrate that an alternative design meets or exceeds the performance implied by the prescriptive design.

Once again, the SFPE *Engineering Guide to Performance-Based Fire Protection Analysis and Design* designates areas in which performance criteria may be needed:

- *Life safety criteria.* These criteria address the survivability of persons exposed to fire and fire products
- *Thermal effects.* These effects include both the effects on the occupants and on materials and equipment
- *Toxicity effects.* These effects, primarily on humans, consist of reduced decision-making capability and impaired motor activity leading to incapacity or death
- *Visibility.* This criterion affects the ability of occupants to safely exit from a fire
- *Non-life safety criteria.* These criteria address issues relating to acceptable damage levels to property
- *Ignition of objects.* These effects include the source of energy and what can be expected to ignite
- *Flame spread.* These effects assess the propagation of flame once ignition has occurred
- *Smoke damage.* This criterion includes smoke aerosols and particulates as well as corrosive combustion products
- *Fire barrier damage and structural integrity.* This criterion addresses the loss of fire barriers resulting in fire extension, increased damage, and structural collapse
- *Damage to exposed properties.* The engineer may need to develop this criterion in order to measure the potential for fire spread or damage to exposed properties

The *SFPE Guide* provides references to assist the designer in determining how to account for the list of performance criteria given.

VERIFICATION METHOD FOR DEMONSTRATING PERFORMANCE

Once the parties have agreed on the performance criteria, the designer must identify and develop methods for demonstrating the performance of the design. Designers often use computer modeling to develop solutions to performance-based requirements. Computer modeling programs exist that iteratively solve the equations that describe fire plume dynamics, fluid flow, heat transfer, and other physical phenomena involved in a fire and their impact on the building compartment. These programs require detailed input data about the particular compartment, fire, and ambient conditions. The programs account for all of the heat and mass evolved from the fire in order to predict the impact the fire has on the compartment and fire protection equipment.

Even though computer fire modeling is often a fundamental part of the process of demonstrating performance,

engineers often address many issues in a performance-based design with algebraic formulas. Appendix B of NFPA 72 provides numerous algebraic formulas for solving specific aspects of the performance prediction of a fire alarm system. The *SFPE Handbook of Fire Protection Engineering* (1995) provides additional formulas.

It is critical that the method used to demonstrate the performance of the fire alarm system be validated to the greatest extent possible using documented research. First principles of physics or the reduction of experimental data to engineering correlations generally form the basis for performance prediction methods used by the designer. The performance-based design must document the source of the correlations or physical relation used to demonstrate performance for review by the AHJ.

The sources, methodologies, and data used in performance-based designs must be based upon technical references that are widely accepted and used by the fire protection community. As advised by the *SFPE Guide,* "The engineer and other stakeholders should determine the acceptability of the sources and methodologies for the particular applications in which they are used." The *Guide* provides guidance as to what constitutes a valid technical reference.

COMPARISON OF PREDICTED PERFORMANCE WITH CRITERIA

Once the designer has established specific criteria and has adopted a verification method for demonstrating performance, he or she proposes a trial system design as an hypothesis and uses the performance demonstration methods to determine if the proposed design accomplishes the objectives.

The performance demonstration starts with describing the fire scenarios. The choice of fire scenarios establishes the severity of the fire challenge the facility is expected to handle. Therefore, the choice implies value judgments on relative probabilities of occurrence and acceptable losses. Usually, an analysis of the range of types of combustibles, the extremes of combustible quantity, extremes of ambient conditions, extremes of asset vulnerability, and other circumstances lead to a limited number of worst-case scenarios. The engineer presumes that if the proposed system can achieve the design objectives under worst-case conditions, it will achieve the objectives under less arduous conditions. Profound errors can occur during this phase when the engineer defines "reasonable" worst-case conditions so caution must be used.

Once the engineer defines the scenario, he or she usually uses a fire model to predict the impact of the fire on the compartment. The rate of heat release and rate of fire growth are used in the computer fire model to predict the development of the ceiling jet, its velocity, and temperature. The

model uses the ceiling jet dynamics to predict the rate of formation of a ceiling layer and, hence, the rate of interface descent. The computer fire model provides estimates of smoke and heat detector response time as well as a determination of the response time for the first sprinkler head. The engineer uses the predicted time for the upper layer to descend to a level that impedes egress to infer the time available for escape. At the same time, he or she uses computational methods for modeling the response of occupants to predict the response to the notification and the rate of occupant egress. Other issues such as the rate of heat transfer through a fire barrier are addressed with other models or algebraic relations.

Ultimately, the designer compares the performance of the proposed design with the objectives established for the building to determine whether the proposed design passes or fails. If the design passes the evaluation of the first scenario, the design moves to the next scenario and the engineer repeats the process. If the design fails to meet these conditions, the engineer must develop a modified design and put that design to the same test. The designer repeats this evaluation process until a design emerges that achieves the design objectives for all contemplated fire scenarios.

CAUTIONS AND CAVEATS FOR THE DESIGNER

The analysis of fire alarm systems and the development of a system design predicated upon performance objectives and criteria are one step in the process of developing a fire protection strategy for the building as a whole. Although some very compelling advantages of performance-based design exist, some very important disadvantages also exist. Performance-based design does provide a method that enables the designer to tackle difficult and unique hazards where either consensus standards do not outline an accepted protection method or where no consensus standard exists. However, performance-based design relies entirely on the designer's understanding of the hazard area, as well as an understanding of the process and progress of the potential fires in that context. Failure to consider material aspects of the hazard area can lead to fire protection systems that are doomed to failure.

Performance-based designs of fire alarm systems are likely to be deemed engineering of the type requiring licensure as a professional engineer. The designer should be knowledgeable in the principles of fire protection engineering and apply these principles judiciously.

The process of performance-based design often relies on the use of computer fire models. Some of these models are no longer actively supported and have been released to the public domain by the developer. In some cases, the validation of the computational routines nested within the software is tenuous or entirely lacking. In other cases, the software has minimal documentation. Consequently, the

designer must take care that he or she only uses the model within the range of parameters over which the developer has validated the model.

The level of precision of the available performance measurements for fire alarm initiating devices is not equivalent to the level of precision generally implied by the results of the computer modeling techniques.

The most critical issue in evaluating the performance of a fire alarm system in a performance-based environment is the prediction of the time at which the system responds to the design fire. How big, or small, a fire will be detected by the detection portion of the fire alarm system. Sound validated performance metrics for heat and smoke detectors do not currently exist. No credible measurement for detector performance has yet been developed for use in the performance-based environment. At present, the design engineer may have to make do with the listed spacing for heat detectors and the recommended spacing for smoke detectors derived from the UL listing evaluation. The intent behind the UL test protocol for determining the listed spacing of heat detectors was to compare the response of these devices to that of automatic sprinklers and, subsequently, to compare one heat detector to another. The listed spacing is essentially a measure of the detector response in a particular test designed for this comparison. UL never intended for anyone to use the data from the listed spacing test outside the context of the listing investigation. The listed spacing provides only a rough estimate of detector response. Unfortunately, the listed spacing is the only guideline currently available.

On the other hand, engineers can predict the response of sprinkler heads when the temperature rating, Response Time Index (RTI) for the particular model of head, ceiling height, ambient temperature, and fire heat release rate are known. Appendix B of NFPA 72, presents a rough correlation that converts listed spacing to an estimated RTI. When a designer designs a fire alarm system in a performance-based environment, he or she has no alternative but to rely on this rough correlation until manufacturers provide thermal response coefficients for the heat detectors they supply. This requirement also illustrates the need for continuous education and staying up-to-date with current research and new technological developments in the field. The designer must always use the most recent validated data for inclusion in a performance-based design of any fire protection system.

The situation becomes even more difficult in the context of smoke detectors. The detector sensitivity measured in the UL 268 laboratory smoke box is intended to serve only for manufacturer quality control and is not applicable outside the context of that test. Consequently, the only performance metric available for smoke detectors is the smoke levels attained during the full-scale room fire tests conducted during the listing evaluation. These tests produce maximum optical obscurations that range between 10 percent per foot and 37 percent per foot at the detector loca-

tions, depending on the fuel and the individual test run. (Refer to Sections 39 and 40 in U.L. *Standard for Safety for Smoke Detectors for Fire Protective Signaling*, UL 268, 4th ed. 1996.)

Because a credible performance metric for smoke detectors is lacking, one way of predicting the response of a smoke detector is to adopt the simplifying assumption that in a flaming fire the plume's buoyancy serves as the driving force that conveys the smoke to the detector. This assumption allows a designer to model the smoke detector as a very sensitive heat detector using the iterative method outlined in Appendix B of NFPA 72. The second method is a mass density approximation, also outlined in Appendix B. Both of these methods tend to predict performance that is slower than that observed of smoke detectors in actual fire tests. Consequently, the predictions are conservative. However, the inaccuracy and lack of precision in these methods can lead to *predictions* of failure to achieve design objectives when, in reality, the system will meet the performance objectives. This lack of precision can deprive the designer of the option to use smoke detection where it could provide a viable solution.

Very little credible data exist regarding the relative reliability of various fire protection strategies. Clearly, unless the engineer can compare the mission effectiveness of a proposed fire protection strategy with the alternatives, he or she cannot make a legitimate decision between them.

Fire alarm systems equipment, being assembled from electronic components with documented failure rates, can be assessed for equipment reliability using the methods outlined in the *Military Handbook for Reliability Prediction of Electronic Equipment*, MIL HDBK 217X ("X" stands for the revision letter). However, contributions to the mission effectiveness of the fire alarm system are also made by the design, installation, and maintenance elements of the system. These elements are more difficult to assess.

When comparing a design reliant upon a fire alarm system to some other strategy the designer must compute the mission effectiveness of that other strategy using the same method he or she uses for the fire alarm system. In general, little information exists regarding the failure rates of system components. Consequently, estimates of the reliability of these systems must be used. In addition, the quality of the fire alarm system installation, testing, and maintenance has a large impact on the mission effectiveness of all active fire protection systems. A performance-based design environment both permits and demands that the designer evaluate these factors and incorporate them into the overall design of the building protection scheme.

Lastly, the performance-based design is far more reliant upon a complete documentary trail of the entire decision-making process. There are no prescriptive requirements that can be used as a reference years after the project has been completed. Because any change in the facility can trigger the need for a reassessment of the design, the design must be thoroughly documented and the documentation must be maintained for the life of the structure. The documentation of the entire basis for developing the performance-based alternatives must include the following items in order to be considered complete:

(1) *Project scope.* Includes the extent of the fire alarm system design and issues such as occupant characteristics, building characteristics, location of the property, fire service capabilities, utilities, environmental considerations, heritage preservation, building management, security, economic and social value of the building, the project delivery process, and the applicable regulations

(2) *Goals and Objectives.* Includes those agreed upon by the owner, AHJ, and so on

(3) *Performance criteria.* Must include how the engineer developed the criteria, and how he or she included safety factors

(4) *Fire scenarios and design fires.* Discusses the expected conditions under which the design will be valid, including the following:

- Form of ignition source
- Different items first ignited
- Ignition in different rooms of a building
- Effects of compartment geometry
- Whether doors and windows are open or closed, and at what time in the fire scenario they are open or closed
- Ventilation, whether natural (doors and windows) or mechanical (HVAC, etc.)
- Form of intervention (occupants, automatic sprinklers, fire department, etc.)

(5) *Final design.* Discusses how the design meets the performance criteria

(6) *Evaluation.* Discusses how to evaluate the design. What are the uncertainty factors? What are the safety factors?

(7) *Critical design assumptions.* Asks how the system will hold up and what the maintenance schedule is

(8) *Critical design features.* Discusses what must stay in place from a building design scenario to ensure the fire alarm system will continue to operate

(9) References

A review of Chapter 8 of NFPA 72 reveals that the chapter on fire warning equipment for dwelling units has included both prescriptive- and performance-based approaches to design and compliance. The applications used in Chapter 8 point out the fact that after all is said and done, the original prescriptive approach to code compliance continues to serve as one of the acceptable methods to meet the performance objectives. When use of the prescriptive approach is chosen, the resulting system can be expected to

achieve the minimum performance criteria established by the performance-based code.

CONCLUSION

As the fire protection community moves toward a performance-based code, an engineer could encounter performance requirements something like the following: "Fire detection systems shall be designed to activate before a fire reaches a size that represents an unreasonable hazard to the building occupants or to the building itself." How an engineer approaches that requirement will depend on his or her understanding of basic fire principles and the accepted procedure for developing a performance-based design.

The recommended steps in the process of developing performance-based approaches to a design problem can be recapped as follows:

(1) Define the project scope.
(2) Identify goals.
(3) Define objectives.
(4) Develop performance criteria.
(5) Develop fire scenarios and design fire scenarios.
(6) Develop trial designs.
(7) Develop a design brief.
(8) Evaluate trial designs.
(9) Select the final design.
(10) Document the design.

In conclusion, Supplement 1 of the *National Fire Alarm Code Handbook* has discussed the concept of performance-based design and what it is, and it has described the design process. Even after performance-based design has become adopted into code, most fire protection features for most facilities will be designed using the prescriptive criteria that exist in the current codes and standards. As the need for greater design flexibility and efficiency increases however, the trend toward the use of performance-based design methods will continue.

References

U.S. Dept. of Defense, *Military Handbook for Reliability Prediction of Electronic Equipment*, MIL HDBK 217.

NFPA *101*®, *Life Safety Code*®, National Fire Protection Association. Quincy, MA, 2000.

Society of Fire Protection Engineers and National Fire Protection Association, 2000, *The SFPE Engineering Guide to Performance-Based Fire Protection Analysis and Design*.

SFPE Handbook of Fire Protection Engineering, National Fire Protection Association, Quincy, MA, 1995.

Underwriters Laboratories Inc., UL 268, *Standard for Safety for Smoke Detectors for Fire Protective Signaling Systems*, 4th ed., 1996.

Bibliography

Custer, R.L.P., Meacham, B. J. *Introduction to Performance-Based Fire Safety*. Society of Fire Protection Engineers and National Fire Protection Association. June 1997.

National Fire Protection Association. *Primer #1: Performance-Based Goals, Objectives and Criteria*. Revision 1.1.19, September 1997. Quincy, MA.

National Fire Protection Association. *Primer #3: Performance-Based Fire Scenarios*. Revision 1.1, September 11, 1998. Quincy, MA.

National Fire Protection Association. *Primer #4: Performance-Based Verification Methods*. Draft 9, January 5, 1999. Quincy, MA.

National Fire Protection Association. *NFPA's Future in Performance-Based Codes and Standards*. Report of the NFPA In-House Task Group, July 1995. Quincy, MA.

Fire Alarm Systems Providing Central Station Service: Applications and Advancements

Dean K. Wilson, P.E.
Hughes Associates, Inc.

Editors' Note: In some cases, fire alarm systems provide dual service: the traditional role of detection and signaling and the additional function of integrating various fire protection functions. This supplement explores how central stations integrate fire protection functions.

Dean K. Wilson, P.E., is with the Windsor, Connecticut, office of Hughes Associates, Inc. He is the former chairman and a member emeritus of the Technical Correlating Committee for the National Fire Alarm Code.

INTRODUCTION

The purpose of central station fire alarm systems is to offer service that integrates the overall fire protection design of a facility. Many times throughout the course of a week, in locations around the world, an architect carefully explains each page of a set of blueprints. During each of these meetings, at some point, someone asks a question that invites evaluation of the project. Most often the answer lies in how much background work has contributed to the completeness of the design.

Fire protection, at both the most complex facility and the most simple facility, must include a carefully developed strategy. A holistic approach usually proves best. This approach emphasizes the organic or functional relationship between parts and wholes. In other words, a holistic approach to fire protection asserts that the effectiveness of the strategy depends on a series of interconnected and interrelated protection features. These features must function as a complete entity. They cannot work effectively when only applied as individual components.

Some of these features provide active physical protection, such as automatic sprinkler systems, fire extinguishers, and special hazard fire extinguishing or suppression systems. Other features provide passive protection, such as fire walls, fire barriers, fire doors, or other construction features. Still other features provide supervision and feedback by sensing conditions and reporting those conditions, such as

the central station fire alarm system. Lastly, some features provide management control of the human response to fire. Every element is critical. Leaving out an element significantly reduces the overall effectiveness of the fire protection for the facility.

When selecting the elements to include, the design professional must begin by conducting a needs assessment to determine the overall site-specific fire protection goals for the facility. The designer must analyze and define goals for life safety, property protection, mission continuity, heritage preservation, and environmental protection.

Once the designer has defined the goals, then he or she must determine the objectives of each element of the overall fire protection system. For example, if the occupants of a building cannot move freely on their own to escape a fire, a central station fire alarm system can summon aid. This aid could come from the public fire department or from a private fire brigade.

At another facility with complex property protection issues, the fire protection must provide a means of preserving the value that the physical property represents. The overall system must meet objectives that provide the necessary level of fire protection.

Finally, the design professional must choose some way to oversee or manage the interrelationship between the individual elements of fire protection for the facility. He or she must choose the tool that management will use to help ensure the fire protection systems will work as intended.

DEFINING CENTRAL STATION SERVICE

A fire alarm system installed throughout a facility and connected to a supervising station operated by a listed central station operating company provides one of the most effective tools for managing the fire protection at a facility. For example, property insurance companies have long required high-value industrial and commercial facilities to have at least one of the following: continuous occupancy in all areas of the facility; recorded guard patrol tours in all unoccupied areas; or a complete central station fire alarm system. See Exhibit S2.1 for a photo of a typical central station operating company.

Section 1-4 of NFPA 72, *National Fire Alarm Code* (1999), provides the following definitions:

Central Station. A supervising station that is listed for central station service.

Central Station Fire Alarm System. A system or group of systems in which the operations of circuits and devices are transmitted automatically to, recorded in, maintained by, and supervised from a listed central station having competent and experienced servers and operators who, upon receipt of a signal, take such action as required by this code. Such service is to be controlled and operated by a person, firm, or corporation whose business is the furnishing, maintaining, or monitoring of supervised fire alarm systems.

Central Station Service. The use of a system or a group of systems in which the operations of circuits and devices at a protected property are signaled to, recorded in, and supervised from a listed central station having competent and experienced operators who, upon receipt of a signal, take such action as required by this code. Related activities at the protected property such as equipment installation, inspection, testing, maintenance, and runner service are the responsibility of the central station or a listed fire alarm service–local company. Central station service is controlled and operated by a person, firm, or corporation whose business is the furnishing of such contracted services or whose properties are the protected premises.

Because of unique requirements, central station fire alarm systems that comply with the code offer seven important advantages over other types of fire alarm systems.

(1) The code requires the central station operating company to obtain listing by one of the qualified testing laboratories, Underwriters Laboratories Inc. or Factory Mutual Research Corporation.
(2) The code requires very tight control over the manner in which a central station operating company provides service.
(3) The code requires the central station operating company to verify compliance.
(4) The code requires the central station operating company to automatically record signals received from a protected premises.

Exhibit S2.1 *A typical central station operating company. (Source: Property Protection Monitoring, Lowell, MA.)*

(5) The code has carefully stated the procedures for handling the various types of signals.

(6) The code has spelled out the manner in which central station operating companies may use the various transmission technologies to receive signals from a protected premises.

(7) The code requires tight control over who conducts the testing and maintenance of the system.

In examining each of these advantages, the design professional will recognize that the significant challenge to life safety, the high value of the property, the critical nature of the mission, the importance of preserving the heritage, or the crucial necessity of protecting the environment drives these requirements. Although every protected property could benefit from having central station service, not every property can justify the additional cost. Thus, owners normally purchase central station service only when the fire protection goals demand it.

LISTING OF THE CENTRAL STATION

To assure the baseline level of quality for a central station fire alarm system, the code requires a qualified testing laboratory to list both the equipment and the operating company providing the service.

NFPA 72, *National Fire Alarm Code*, states the following regarding equipment:

> **1-5.1.2** *Equipment.* Equipment constructed and installed in conformity with this code shall be listed for the purpose for which it is used.

It further defines *listed* as follows:

> *Listed.** Equipment, materials, or services included in a list published by an organization acceptable to the authority having jurisdiction and concerned with evaluation of products or services that maintains periodic inspection of production of listed equipment or materials or periodic evaluation of services and whose listing states either that the equipment, material, or service meets identified standards or has been tested and found suitable for a specified purpose.

From the outset, the code requires that fire alarm system service providers only use listed equipment. The listing process involves not only testing the equipment to make certain it performs properly. The listing process also involves inspecting the production of listed equipment to make certain the manufacturer has not changed the product after the laboratory has tested it.

Listing can also apply to a material or a service, which is important to note when considering the third distinct advantage of central station service.

Two qualified testing laboratories, Underwriters Laboratories Inc. (UL) and Factory Mutual Research Corporation (FMRC), both rely on the requirements of the code to guide their testing requirements. In addition, each laboratory has developed a performance standard for central station service. UL 827, *Standard for Central Station Alarm Services* (1996), serves this purpose. Factory Mutual Standard 3011, *Approval Standard for Central Station Service for Fire Alarms and Protective Equipment Supervision (1999)*, also serves this purpose.

Representatives of the laboratory visit each central station operating company to review records of signals and to audit the personnel performing operations and service. The representatives verify the construction of the physical central station and check the equipment and the power supplies.

Both Underwriters Laboratories Inc. and Factory Mutual Research Corporation also provide for the listing of fire alarm service–local companies. UL does so under the category "Protective Signaling Services-Local, Auxiliary, Remote Station, and Proprietary (UUJS)." FMRC offers listing under the category "Fire Alarm Service Local Company (FIRE)."

UL publishes the results of the listing process in the *UL Fire Protection Equipment Directory* (1999). Factory Mutual Research Corporation publishes the results of its listing process in the *Factory Mutual Approval Guide* (1999).

Public and private authorities having jurisdiction can use these publications to determine whether or not a central station operating company has obtained listing.

PROVIDING CENTRAL STATION SERVICE

Paragraph 5-2.2.2 of the code provides a six-element description of central station service. Three elements relate to the protected premises. The other three elements relate to the central station.

The three elements at the protected premises include installation, testing and maintenance, and runner service. The three elements at the central station include monitoring of signals, retransmission of signals to appropriate authorities, and record keeping. These very distinct and descriptive items help define the high level of quality that central station service provides.

The code also permits fire alarm service providers to offer central station service in one of three contractual methods. Each method gives a central station service subscriber a single point of contact. Using the first method, a single listed central station alarm service provider delivers all six elements of central station service.

Using the second method, a single listed central station alarm service provider subcontracts one or more of the three elements at the protected premises. For example, a listed central station operating company may subcontract all or

part of the installation at the protected premises. For example, where an installation requires supervisory initiating devices on various parts of a protected premises automatic sprinkler system, the central station alarm service provider subcontracts the installation of those devices to a sprinkler contractor.

Using the third method, a listed fire alarm service–local company provides installation, testing, and maintenance at the protected premises. The listed local company then subcontracts the three elements at the central station to a listed central station operating company. The runner service at the protected premises comes from either the listed local company or the listed central station operating company.

VERIFYING COMPLIANCE WITH THE CODE

Paragraph 5-2.2.3 of the code requires that the prime contractor must conspicuously indicate that the fire alarm system providing service at a protected premises complies with all the requirements of the code, by providing a means of third-party verification.

As part of a process that will maintain a very high level of system performance, this particular requirement helps ensure that a central station fire alarm system has built-in procedures for quality assurance. Note that the code gives the responsibility for complying with these requirements to the prime contractor. In Section 1-4, the code defines the term *prime contractor* as follows:

> *Prime Contractor.* The one company contractually responsible for providing central station services to a subscriber as required by this code. This can be either a listed central station or a listed fire alarm service–local company.

The code states two methods of conspicuously indicating that the fire alarm system providing service at a protected premises complies with all the requirements of the code: the posting of a certificate or the posting of a placard. See Exhibit 5.4 for samples of those two methods.

The option of posting a certificate applies where a fire alarm system complying with all requirements of the code receives certification from the organization that has listed the prime contractor. The prime contractor must then post a document attesting to this certification on or near the fire alarm system control unit, or a system component if no control unit exists. Further, the organization that has listed the prime contractor must maintain a central repository of the certification documents it has issued.

Alternatively, the prime contractor may conspicuously mark a fire alarm system that complies with all requirements of the code by posting one or more placards that meet the requirements of the organization that has listed the prime contractor. The size of the placard must cover an area of at least 20 in.^{2} or 130 cm^2. The prime contractor must post the placard on or near the fire alarm system control unit, or a system component if no control unit exists.

Note that these requirements do not dictate that every installed system must have its code compliance verified by a third party. Rather, the requirements direct that a means to provide third-party verification must exist. Both UL and FMRC have such programs. UL offers a certificate program. FMRC directs the use of a placard. The authority having jurisdiction has the opportunity to choose which method he or she will require a building owner to use.

In the case of the certificate, when an installation nears completion, the prime contractor fills out a detailed form and requests that UL issue a certificate for that installation. This application form contains sufficient detail to allow UL to review the nature and character of the extent of protection provided by the system. The prime contractor must sign a statement that the installation complies with the code. The form also has a place where the prime contractor must indicate any deviations from the requirements of the code.

UL receives the application and reviews it. If the information appears valid, UL issues a certificate with a serial number. The certificate contains many of the details stated in the application. UL then mails copies of the certificate to the prime contractor for presentation to the authorities having jurisdiction and the subscriber.

During follow-up visits to the prime contractor, a UL representative will choose to visit a statistically significant number of certified installations. These visits will serve to verify the correctness of certification practices.

In the case of the placard, when an installation nears completion, the prime contractor fills out a copy of the placard described in 5-2.2.3.2.2. He or she then posts this placard near the fire alarm equipment at the protected premises.

During follow-up visits to the prime contractor, an FMRC representative will choose to visit a statistically significant number of placarded installations. These visits will serve to verify the correctness of certification practices.

AUTOMATICALLY RECORDING SIGNALS

When signals transmit to a central station from a protected premises, the central station must automatically record those signals. Over time, the central station accumulates a detailed history of the performance of the fire protection systems and equipment at the protected premises. This written record helps management of a facility to properly supervise the interaction of the fire protection systems with the ongoing operations of the facility.

Several examples of this supervision exist. A central station receives numerous supervisory off-normal signals from an air pressure supervisory switch mounted on an automatic sprinkler system's dry pipe valve. These off-normal signals might indicate that air is leaking from the system. A proper investigation may find the problem and

provide a solution. This solution, in turn, could prevent a premature tripping of the dry pipe valve.

For example, every morning at 10:05 a.m. a central station receives a low water pressure supervisory signal followed almost immediately by a sprinkler system waterflow alarm signal from Building 422. Investigators are stumped. They arrange to view Building 422 from a nearby rooftop. Imagine their surprise when precisely at 10:05 a.m. a worker emerges from a Building 422 rooftop door and places his coffee mug underneath the discharge nozzle of an automatic sprinkler system inspector's test connection.

When a fire does occur, the record of signals received at the central station assists investigators tremendously. The date-stamped and time-stamped automatic record of signals received helps the investigators develop a step-by-step sequence of events for the fire. Investigators can piece together the direction of fire and smoke travel based on patterns of which initiating devices operated at which particular points in the fire development timeline. Sometimes the fire alarm system control unit at the protected premises has a memory that records system function and operation in a log. Comparing the record from the protected premises with the record of signals received at the central station can further clarify details regarding the fire development.

HANDLING VARIOUS SIGNALS

Paragraph 5-3.5.1 of the code requires that the central station operating company must have sufficient personnel on duty at the central station at all times to ensure attention to signals received. The code further defines *sufficient* to mean a minimum of two persons. The operation and supervision of the central station fire alarm system must serve as the primary function of the operators. No other interest or activity can take precedence over the handling of incoming signals.

The central station operators play a key role in the effective use of central station service. The code emphasizes that the servers and operators must be competent and experienced. Proper education and training coupled with on-the-job experience under the supervision of a competent mentor can develop individuals with the necessary skills.

The code does not leave the operators on their own regarding how to handle signals, however. In 5-3.6.6, the code specifies exactly what action should be taken for each type of signal received: fire alarm signals, guard patrol tour supervisory signals, supervisory signals, trouble signals, and test signals.

Fire Alarm Signals

Unless the central station operator has received notice of a prearranged test, if a manual fire alarm box, automatic fire detector, waterflow from an automatic sprinkler system, or actuation of any other fire suppression or extinguishing sys-

tem initiates the signal, the operator must treat the signal as a fire alarm signal. Upon receipt of such a signal, the central station operator must immediately retransmit the alarm signal to the public fire service communications center. In this context, immediately means without unreasonable delay. Routine handling should take a maximum of 90 seconds from receipt of an alarm signal by the central station until the initiation of retransmission to the public fire service communications center.

Note that the code does not permit the operator to verify whether or not the fire alarm signal comes from the scene of a real fire. The operator must immediately retransmit the signal. An exception to this requirement to immediately retransmit does exist. Upon receipt of a fire alarm signal from a household fire warning system, the supervising station must immediately (within 90 seconds) retransmit the alarm to the public fire communications center. However, the supervising station may contact the residence for verification of an alarm condition and, where, within 90 seconds, the operator receives acceptable indication that the fire department does not need to respond, the operator may withhold the retransmission to the public service fire communications center.

Some proponents of quick-fix solutions to the problem of false fire alarm signals urge that central station operating companies verify all signals including fire alarm signals. The code does not permit this verification for other than household systems.

In addition to the very specific responsibilities outlined for operators, for central station service the code requires that upon receipt of a fire alarm signal the operator must dispatch a runner or technician to the protected premises whenever the prime contractor must reset the system. In Section 1-4, the code defines the terms *runner* and *runner* service as follows:

> *Runner.* A person other than the required number of operators on duty at central, supervising, or runner stations (or otherwise in contact with these stations) available for prompt dispatching, when necessary, to the protected premises.
>
> *Runner Service.* The service provided by a runner at the protected premises, including resetting and silencing of all equipment transmitting fire alarm or supervisory signals to an off-premises location.

In some cases, the prime contractor can install the system at the protected premises so that it does not require manual reset by the prime contractor. The equipment could automatically reset once the device that initiated the signal restores to normal. Alternatively, the equipment could permit a designated representative of the subscriber to operate a manual reset. In any case, if the prime contractor does not have to manually reset the system, then a runner does not

need to respond when the central station receives a fire alarm signal from a protected premises.

Upon receipt of a fire alarm signal, the operator must also notify the subscriber by the quickest available method. Also, where the authorities having jurisdiction require it, the operator must provide written notice to the subscriber, to the authorities having jurisdiction, or to both.

Guard's Patrol Tour Supervisory Signal

Unless the central station operator has received notice of a prearranged test, if a guard's patrol tour supervisory device either initiates a signal or fails to initiate a signal within a 15-minute grace period following a scheduled time, the operator must treat either the signal, or the failure to receive the signal, as a guard's patrol tour supervisory signal. Upon receipt of such a signal, the central station operator must communicate without unreasonable delay with personnel at the protected premises. If the operator cannot establish communications with the protected premises, the operator must dispatch a runner to the protected premises to arrive within 30 minutes.

Upon receipt of a guard's patrol tour supervisory signal, where the authorities having jurisdiction require it, the operators must also report all delinquencies to the subscriber, to the authorities having jurisdiction, or to both.

Supervisory Signal

Unless the central station operator has received notice of a prearranged test, if the equipment monitoring the availability or operational integrity of an automatic sprinkler system, or if another automatic fire suppression or extinguishing system initiates the signal, the operator must treat the signal as a supervisory signal. Upon receipt of such a signal, the central station operator must immediately communicate with the person designated by the subscriber. The term *immediately* in this context means without unreasonable delay. Routine handling should take a maximum of 4 minutes from receipt of a supervisory signal by the central station until the initiation of communications with the person designated by the subscriber.

If that communication does not resolve the restoration of the affected fire extinguishing or fire suppression system, then the operator must dispatch a runner to the protected premises. The runner must arrive at the protected premises within 1 hour.

Where the authorities having jurisdiction require it, the operator must notify the fire department or law enforcement agency, or both. First, the central station operator should attempt to notify designated personnel at the protected premises. If he or she cannot make such notification, it might be appropriate to notify law enforcement or the fire department, or both. For example, if the central station receives a valve supervisory signal from an unoccupied protected premises, the operator may decide to notify the police.

In addition, the operator must notify the authorities having jurisdiction when sprinkler systems or other fire suppression or extinguishing system(s) or equipment have been wholly or partially out of service for 8 hours.

Upon restoration of equipment that has been out of service for 8 hours or more, and where the authorities having jurisdiction require it, the operator must also provide notice to the subscriber, to the authorities having jurisdiction, or to both as to the nature of the signal, time of occurrence, and time of restoration.

Trouble Signal

Unless the central station operator has received notice of a prearranged test, if the fire alarm system initiates a signal that indicates a need for maintenance or repair, the operator must treat the signal as a trouble signal. Upon receipt of such a signal, the central station operator must immediately communicate with the person designated by the subscriber. The term *immediately* in this context means without unreasonable delay. Routine handling should take a maximum of 4 minutes from receipt of a supervisory signal by the central station until the initiation of communications with the person designated by the subscriber.

The operator must also dispatch to the protected premises a technician who can initiate maintenance or repair. The maintenance or repair technician must arrive at the protected premises within 4 hours.

In addition, when the equipment has been out of service for 8 hours or more, and where the authorities having jurisdiction require it, the operator must notify the subscriber, or the authorities having jurisdiction, or both as to the nature of the interruption, time of occurrence, and time of restoration.

The operator must record the date, time, and type of all test signals received. Whenever the subscriber or an authority having jurisdiction inquires, the operator must acknowledge all test signals initiated by the subscriber, including those initiated for the benefit of an authority having jurisdiction.

If the operator does not receive an initiated test signal, he or she must immediately investigate and take appropriate action to reestablish system integrity. The term *immediately* in this context means without unreasonable delay. Routine handling should take a maximum of 4 minutes from the failure to receive a test signal by the central station until the initiation of communications with the person designated by the subscriber.

The operator must also dispatch a runner to arrive at the protected premises within 1 hour if equipment must be manually reset.

As another important function of his or her job, the central station operator oversees record keeping and reporting. The code requires the central station operating company to retain records of all signals received for at least one year. If any of the authorities having jurisdiction wish to see reports

of those signals, the central station operating company must furnish the reports in a form acceptable to the particular authority having jurisdiction.

The presence of competent and experienced servers and operators at the central station helps create an important distinction between central station fire alarm systems and other fire alarm systems.

USING TRANSMISSION TECHNOLOGIES

Over the years, the greatest advancement in central station service has come through a burgeoning of transmission technologies. As recently as thirty years ago, virtually all central station signals used a single transmission technology.

When connecting a building fire alarm system to a central station, a prime contractor currently has six distinct transmission technologies available. These include the following:

(1) McCulloh system
(2) Directly connected noncoded system
(3) Active multiplex system, including derived local channel systems and private microwave radio systems
(4) Two-way radio frequency (RF) multiplex systems
(5) Digital alarm communicator systems, including digital alarm radio systems
(6) One-way private radio alarm systems

The requirements for each transmission technology help ensure that no matter which technology a prime contractor may choose to use, the signals from the protected premises will reach the central station. Though the technologies have different techniques and different requirements, they represent a common effort to address reliability of signal pathways.

In some cases, the central station operating company may locate an unstaffed subsidiary central station in a specific geographic area. The signals from that geographic area will first transmit to the subsidiary station. The subsidiary station then relays the signals to the central station. This arrangement permits a single central station to cover a vast number of geographic areas.

McCulloh System

The oldest form of transmission technology is the McCulloh system. Named for its inventor, this system evolved out of Samuel F. B. Morse's invention of the telegraph. A similar transmission technology has long been used for public fire alarm reporting system street fire alarm boxes.

The McCulloh system uses a mechanical spring-wound or electric motor-driven transmitter to alternately open and ground a series direct current circuit. A code wheel produces a pattern of signals that can be interpreted at the central station as a distinctly coded number. An operator can

cross-reference that number to identify the particular type of signal and which protected premises originated the signal.

If a single open fault or single ground fault occurs on a McCulloh circuit, an operator can turn a switch on the receiving equipment that allows the signal to find an alternate path using a connection to earth ground.

To help maintain system quality and reliability, each McCulloh circuit can serve up to 25 protected premises with a total of no more than 250 code wheels.

A central station operating company leases the actual wires used for the McCulloh circuit from the public telephone utility. However, the telephone company does not supply power for these circuits. The equipment at the central station applies operating power for each circuit.

Where McCulloh circuits have to traverse long distances, too great a loop resistance prevents the use of metallic circuits. In such a case, the telephone company can offer a carrier system that will encode the McCulloh signals as a multiplex signal at the first telephone company wire center downstream from the protected premises. The multiplex signal can then transmit through the public telephone network as an ordinary digital signal until it reaches the telephone company wire center that connects to the central station. There the carrier system decodes the McCulloh signals and passes them on to the McCulloh receiving equipment in the central station. The code also contains specific additional requirements for the nonmetallic portion of this type of transmission channel.

Directly Connected Noncoded Systems

In rare cases, each protected premises may transmit to the central station using a directly connected noncoded system. A central station operating company leases the actual wires used for each directly connected noncoded circuit from the public telephone utility. However, the telephone company does not supply power for these circuits, either. Instead, the equipment at the protected premises applies operating power for the circuit extending from that premises.

When the fire alarm system control unit at the protected premises transmits a signal to the central station over a directly connected noncoded circuit, either a change in current flow, a reversal of direct current polarity, or a frequency shift of an audible tone causes the receiving equipment to notify operators of the nature of an incoming signal.

Active Multiplex Systems

An active multiplex transmission method, including telephone company supplied derived local channel, offers another transmission method between the protected premises and the central station. Digital signals transmit from the central station to the protected premises and back again. This interrogation and response sequence repeats at least

once every 90 seconds. The signal traffic between the central station and the several protected premises monitors the interconnecting pathway for integrity. The pathway may consist of a metallic circuit, a nonmetallic circuit such as optical fiber cable, or private microwave radio.

A derived local channel places the digital multiplex signal in tandem with the normal telephone signals on a single regular telephone circuit. The multiplex signals piggyback on the regular voice telephone transmission.

The code identifies three types of active multiplex transmission systems, Type 1, Type 2, and Type 3. Types 1 and 2 use a closed window bridge that scans one protected premises leg facility at a time. This separate scanning prevents a fault on one protected premises leg facility (communications channel) from degrading signals transmitted over another protected premises leg facility. A Type 3 multiplex transmission system uses an open window bridge. This open window bridge makes the leg facilities more vulnerable to fault conditions on other leg facilities.

In addition, Type 1 active multiplex systems provide for an alternate trunk facility in case the main trunk facility becomes disarranged. Providing an alternate trunk helps safeguard the continuity of signals. Telephone company supplied equipment, either DataPhone Select-A-Station or "Dial up make good," enables the alternate trunk facility to function when the main trunk facility fails.

Loading tables allow Type 1 active multiplex circuits to serve a greater number of protected premises initiating devices and leg facilities than Type 2. Likewise, Type 2 may serve a greater number of protected premises initiating devices and leg facilities than Type 3.

A central station operating company leases the actual wires or communications channel used for the active multiplex circuit from the public telephone utility. In this case, the telephone company does supply power for these circuits or channels.

Two-Way Radio Frequency (RF) Multiplex Systems

These systems use licensed two-way radio frequency channels to carry multiplexed signals from protected premises fire alarm systems. These systems operate the same as active multiplex systems, except they do not use circuits supplied by the telephone company. Signals transmit over licensed radio channels only. Loading tables allow a Type 4 two-way RF multiplex system with two or more control sites to serve a greater number of protected premises initiating devices and leg facilities than a Type 5 system with a single control site.

Digital Alarm Communicator System

The digital alarm communicator system (DACS) provides the most common type of central station transmission method currently used. In its simplest form, DACS connects a digital alarm communicator transmitter (DACT) to two communications channels or pathways. One pathway consists of an ordinary loop start telephone circuit. The other pathway may consist of another ordinary loop start telephone circuit or one of several other alternate communications systems.

Other telephone instruments, called subsets, may be connected to the loop start telephone circuit. Occupants of the protected premises may use those subsets for ordinary telephone communications. However, the subsets must connect downstream of the DACT. This positioning allows the DACT to seize the circuit when transmitting a signal to the central station.

The loop start telephone circuit has a direct current voltage present at all times. The DACT can measure this voltage and monitor the connected integrity of the telephone circuit between the protected premises and the telephone company wire center.

When a protected premises fire alarm system initiates a signal to the central station, the DACT seizes the telephone circuit and dials a preprogrammed telephone number. At the central station, a digital alarm communicator receiver (DACR) answers the call. The DACT and DACR exchange a "handshake" recognition signal. Then the DACT transmits signals to the DACR, and the DACR displays and records those signals. In most cases, the DACRs at a central station provide input into a central station automation system that assists the operators in efficiently handling a large amount of signal traffic.

If the DACT does not successfully connect to the DACR, it goes on-hook (hangs up) and redials over the alternate pathway. This process continues until the DACT successfully connects to the DACR, or until the DACT has completed between five and ten attempts. If the DACT fails to successfully connect to the DACR, it initiates a local trouble signal at the protected premises.

The DACT tests the integrity of the communications pathways by transmitting a test signal once every 24 hours.

The code permits the pathways serving a DACT to consist of the following:

- Two metallic telephone circuits
- One telephone line (number) and one cellular telephone connection
- One telephone line (number) and a one-way radio system
- One telephone line (number) equipped with a derived local channel
- One telephone line (number) and a one-way private radio alarm system
- One telephone line (number) and a private microwave radio system
- One telephone line (number) and a two-way RF multiplex system

- A single integrated services digital network (ISDN) telephone line using a specifically listed terminal adapter

Where a central station incorporates a one-way private radio alarm system receiver into a digital alarm communicator receiver, the code designates the receiver as a digital alarm radio receiver (DARR).

One-Way Private Radio Alarm Systems

In some communities, a service provider has erected a series of transmitter/receiver locations and has offered a one-way private radio alarm system to the central station operating companies serving the community. The fire alarm system control unit at the protected premises connects to a one-way private radio alarm transmitter (RAT). Radio propagation tests confirm that under all conditions at least two, one-way private radio alarm repeater system receivers (RARSR) will receive the signals from each RAT. These RARSRs will repeat the transmitted signal to either another RARSR or to one or more one-way private radio alarm supervising station receivers (RASSR). This multiple-path transmission helps ensure the integrity of the transmitted signal.

No matter which transmission technology the central station operating company chooses to use, the operating company emphasizes the integrity of the signal transmission. This emphasis helps ensure the highest possible cost-effective level of system reliability.

TESTING AND MAINTAINING THE SYSTEM

The code provides for acceptance, reacceptance, and periodic testing, and both preventive and emergency maintenance for central station service in accordance with the requirements of Chapter 7. The prime contractor must provide each of its representatives and each alarm system user with a unique personal identification code. In order to authorize the placing of an alarm system into test status, a representative of the prime contractor or an alarm system user must first provide the central station with his or her personal identification code.

Although the owner of the protected premises must bear the ultimate responsibility for the testing and maintenance of the fire alarm system, testing and maintenance of central station service systems must be performed under the contractual arrangements specified in 5-2.2.2. In other words, the prime contractor is responsible for the testing and maintenance. However, the owner can subcontract maintenance of his or her system to the prime contractor at no charge.

Tables 7-2.2, 7-3.1, and 7-3.2 of the code give details regarding the test methods, frequency of visual inspection, and frequency of testing, respectively. The prime contractor must perform maintenance required by the code and by the manufacturer's instructions for each piece of equipment.

The rigorous schedule of testing and maintenance ensures that the quality and reliability of the installed system will continue throughout the life of the system.

Based on these seven advantages, one can clearly understand why a fire alarm system installed throughout a facility and connected to a supervising station operated by a listed central station operating company provides one of the most effective tools for managing the fire protection at a facility.

References

FM 3011, *Approval Standard for Central Station Service for Fire Alarms and Protective Equipment Supervision,* 1999, Factory Mutual Research Corporation, Norwood, MA.

Factory Mutual Approval Guide, 1999, Factory Mutual Research Corporation, Norwood, MA.

UL 827, *Standard for Central Station Alarm Services,* 1996, Underwriters Laboratories Inc., Northbrook, IL.

UL Fire Protection Equipment Directory, Underwriters Laboratories Inc., Northbrook, IL.

Integrating Fire Alarm Systems with Other Building Systems

Ronald H. Kirby
Simplex Time Recorder Co.

Dean K. Wilson, P.E.
Hughes Associates, Inc.

Editors' Note: Fire alarm systems not only detect and signal the presence of fire or other products of combustion, but may also manage non-fire building systems. This supplement introduces the emerging field of integrated systems.

Ronald H. Kirby is Vice President, Industry Relations at Simplex Time Recorder Co. in Gardner, Massachusetts. He is a member of the Technical Correlating Committee on the National Fire Alarm Code and serves as a member of the NFPA Technical Committees on Building Service and Fire Protection Equipment; Fundamentals of Fire Alarm Systems; and Residential Occupancies (Safety to Life).

Dean K. Wilson, P.E., is senior engineer with the Windsor, Connecticut, office of Hughes Associates, Inc. He is former chairman and a member emeritus of the Technical Correlating Committee for the National Fire Alarm Code.

INTRODUCTION—WHY INTEGRATE?

In the simplest form, a fire alarm system notifies occupants of the presence of a fire. This notification allows the occupants sufficient time to escape. In other cases, the fire alarm system connects to an off-premises supervising station that alerts the public fire department so that search, rescue, and fire extinguishing can begin. But, in still other cases, the fire alarm system provides an even greater degree of fire protection for the facility.

Building owners or the authority having jurisdiction may wish to preserve not only life safety, but also the physical property, the mission of the facility, the heritage of the contents, and the environment. To do this, they depend on the fire alarm system to help them manage the overall fire protection for the facility.

When the fire alarm system serves as a management tool, it adds complexity to the system to match the complexity of the facility it protects. The system already has fire alarm-initiating devices such as manual fire alarm boxes, heat, smoke, or flame detectors, automatic sprinkler water-flow switches, or switches that indicate the discharge of other fire extinguishing or fire suppression systems.

To these basic features, an installer may add a variety of supervisory initiating devices. These devices include valve tamper switches for all sprinkler system control valves that are $2\frac{1}{2}$-in. or larger and switches that detect high or low air pressure in a dry pipe sprinkler system or pressure tank. Other features include switches to monitor the level in any fire protection water storage tank and switches that monitor low temperature in areas protected by wet pipe sprinkler systems, water storage tanks, fire pump rooms or houses, and for dry pipe valve or deluge valve closets where the facility is located in a climate that is subject to freezing temperatures. The installer can also include connections to monitor the running of electric motor-driven fire pumps, power failure, and phase reversal; and for monitoring the running of diesel engine-driven fire pumps, pump or controller trouble, and the position of controller main switches anywhere other than automatic.

Collecting information through the fire alarm system may not offer enough help to building owners and authorities having jurisdiction. They may also use the fire alarm system to provide fire fighters with information about building conditions that will support the fire fighters' decision-making

process in the midst of working a fire. Such information may include information regarding the operation of building heating, ventilating, and air-conditioning equipment; the status of fire and smoke doors; the presence of occupants in normally unoccupied rooms; the status of all active fire protection systems; the progress of the fire as it moves through the building; and so forth. The fire alarm system may also be used to control other building equipment and systems. This control will help to make the building safer during a fire.

Wherever the fire alarm system controls other building systems, the question arises: "Which system will maintain ultimate control?" Will the fire alarm system take over control from the normal building operating system? Or will the building operating system receive commands from the fire alarm system, take action, and then simply report that action to the fire alarm system? Because two or more separate vendors often supply the fire alarm system and building controls, a lot of finger-pointing can occur.

So, what about an integrated system? A system where somehow the fire alarm system and the building control systems, sometimes collectively called the building management systems, communicate in an orderly manner to help meet the overall fire protection goals for the facility.

Traditionally fire alarm systems have helped provide building fire safety by recalling elevators; shutting off power to elevators before automatic sprinklers operate; shutting down air-handling systems; closing smoke or fire doors; operating smoke control systems, including stairwell pressurization; unlocking doors; and actuating fire extinguishing or suppression systems. In fact, Section 3-9 of NFPA 72, *National Fire Alarm Code* (1999), offers very specific requirements as to how the fire alarm system can control these functions.

Some of the requirements have been correlated with related codes or standards. For example, the Technical Committee on Protected Premises Fire Alarm Systems has correlated the Chapter 3 requirements for elevator recall and elevator shutdown with the committee of the American Society of Mechanical Engineers that writes ANSI/ASME A17.1, *Safety Code for Elevators and Escalators*. Likewise, the Technical Committee on Protected Premises Fire Alarm Systems has correlated the requirements on the shutdown of air-handling systems with the NFPA Technical Committee on Air-Conditioning that writes NFPA 90A, *Standard for the Installation of Air-Conditioning and Ventilating Systems*, and NFPA 90B, *Standard for the Installation of Warm Air Heating and Air-Conditioning Systems*.

The Technical Committee on Protected Premises Fire Alarm Systems has correlated the requirements for smoke control systems with the NFPA Technical Committee on Smoke Management Systems that writes NFPA 92A, *Recommended Practice for Smoke-Control Systems*, and NFPA 92B, *Guide for Smoke Management Systems in Malls, Atria and Large Areas*. And, the Technical Committee on Pro-

tected Premises Fire Alarm Systems has correlated the requirements for door release and door unlocking with the Technical Committee on the Safety to Life that writes NFPA 101®, *Life Safety Code®*.

The code provides a philosophical context for initiating these building fire safety functions. Paragraph 1-5.4.1 gives permission to automatically perform these functions. The paragraph also states that the operation of these functions must not interfere with power for lighting or for operating elevators. Paragraph 1-5.4.1 also makes provision for combining fire alarm services with other services requiring monitoring of operations. So, the code permits the combining of fire alarm systems with other building management systems.

The requirements of 3-8.2 and 3-9.2.5 permit fire alarm systems to share components, equipment, circuitry, and installation wiring with non-fire alarm systems. However, short circuits, open circuits, or grounds in the non-fire equipment, or between the non-fire equipment and the fire alarm system wiring, must not interfere with the integrity of the fire alarm system. Such faults on the non-fire equipment must not prevent alarm or supervisory signal transmission on the fire alarm system.

In addition to these requirements, the code also insists that the integrity of the fire alarm system functions must be maintained. Thus, the removal of, replacement of, failure of, or maintenance procedure on any hardware, software, or circuit from the part of the combination system that does not perform any of the fire alarm system functions, must not cause loss of any of the fire alarm system functions. Also, the combination system must provide distinctive, clearly recognizable fire alarm signals that take precedence over any other signal, even when a non-fire alarm signal is initiated first.

The authority having jurisdiction may also determine that the signal traffic on the combination system has reached an excessive level that causes confusion. In this case, the authority having jurisdiction may require that the system annunciate fire alarm signals separately, and on a priority basis.

These requirements place a significantly restrained framework around the integration of a fire alarm system with other building management systems. Yet, the need to have various building systems communicate with other building systems has increased in recent years.

Some applications where a fire alarm system communicates with other building systems have included the following:

(a) *Environmental Monitoring.* The heating, ventilating, and air-conditioning system (HVAC) uses sensors to monitor temperature or to monitor carbon monoxide or carbon dioxide levels. Theoretically, the fire alarm system could use these same sensors to detect specific fire signatures.

(b) *Door Monitoring.* Door switches used for security might also provide information on whether smoke doors and

fire doors are open or closed. This feature could give fire fighters valuable information regarding the integrity of the opening protection for fire walls, fire barriers, and smoke barriers while working a fire.

(c) *Selective Door Unlocking.* Security concerns dictate the locking of doors. Fire safety concerns dictate readily accessible means of egress. These needs frequently conflict. Upon receipt of a fire alarm signal, the fire alarm system could automatically unlock all doors. But, this action may not satisfy the needs of a comprehensive security plan. A thief could merely operate a manual fire alarm box to unlock a secured door. However, selectively unlocking doors from both the fire alarm system and the security/access control system can satisfy both applications.

(d) *Motion Detection.* Building energy conservation systems often use passive infrared motion sensors (PIR) to detect human activity in a space. The control system then turns on lighting, raises the heat, or starts the air conditioning. When the space empties, the PIR sends a signal that allows the control system to turn out the lights, reduce the heat, or shut off the air conditioning. A security system may use a PIR to indicate unauthorized entry into secured spaces for security management. These same sensors might well integrate into the fire alarm system to provide fire fighters with real-time information on which rooms are occupied. In a typical working fire, fire fighters often must devote as much as half of their time to search and rescue. Knowing which rooms are occupied, and which are not, can significantly enhance fire-fighter efficiency.

(e) *Fire Alarm/CCTV Interface.* Fire alarm systems can integrate with closed circuit television systems (CCTV) to automatically switch on cameras at the actual location of the origin of a fire alarm signal. These cameras can then provide fire fighters with a real-time view of conditions in the area of the fire.

(f) *Elevator Control.* The building codes require that elevators be "captured and recalled" by the fire alarm system when smoke detectors sense smoke in an elevator lobby or elevator machine room. Many elevators also interface with a security/access control system to manage access to specific floors at predetermined times.

(g) *HVAC Override by Fire Alarm Systems and Security Systems.* Security/access control systems often interact with HVAC control systems to place spaces into an occupied mode on nights and weekends. Fire alarm systems frequently interface with HVAC systems to open and close dampers and turn fans on and off for smoke control.

WHAT DOES INTEGRATION MEAN?

In trying to describe a smoothly operating interface between a fire alarm system and other building management systems, fire safety professionals use such terms as *interconnected*,

interoperable, and *integrated*. Some examples may best define these terms.

In many applications, fire alarm systems do not connect to other building systems. Exhibit S3.1 clearly shows two separate, discrete and distinct systems. On the left, a smoke detector connects to a fire alarm system. On the right, a security system connects to an electric door lock.

HOW DOES INTEGRATION WORK?

At a rather simplistic level, a fire alarm-initiating device may connect to an appliance from another building system. In Exhibit S3.2, an installer has connected the smoke detector to the electric door lock. A signal from the smoke detector causes the electric door lock to react. To comply with 3-9.2.1, the circuit between the smoke detector and the lock must fail safe. Fail safe means that a power failure or an open circuit causes the door to unlock. Of course, the lock is also controlled from the security system.

Moving up one level higher than a device-to-appliance connection, a fire alarm system control unit can connect with another building system control panel. Exhibit S3.3 shows a circuit that extends from the fire alarm system control unit and connects to the security system control panel. When the smoke detector initiates a fire alarm signal, the fire alarm control unit signals the security system control panel. The security system control panel then signals the door to unlock. Even though the signal passes through two control units, this path is clearly unidirectional. The signal proceeds from the smoke detector to the door lock. Again, compliance with 3-9.2.1 would require the circuit between the two panels to fail safe or be monitored for integrity so that a fault would cause a trouble signal at the fire alarm control unit.

However, the connection does not need to limit communication to a single direction. Exhibit S3.4 adds a switch inside the lock pocket. This switch permits two-way communication. The smoke detector, through the fire alarm system and the security system, tells the door to unlock. The pocket switch tells the fire alarm system that the door has unlocked.

The two-way communication can take place at a control unit level. Exhibit S3.5 shows a two-way communication between the fire alarm control unit and the security control panel. An actuated smoke detector causes the fire alarm control unit to signal the security control panel to unlock the door. The lock pocket switch signals the security control panel to confirm to the fire alarm control unit that the door has unlocked. The circuit from the fire alarm control unit that initiates the unlocking of the door must be fail safe or be monitored for integrity. The circuit from the security control panel that confirms the unlocking must either be monitored for integrity or fail safe.

These illustrations explore the connection possibilities between a fire alarm system and a security system. However,

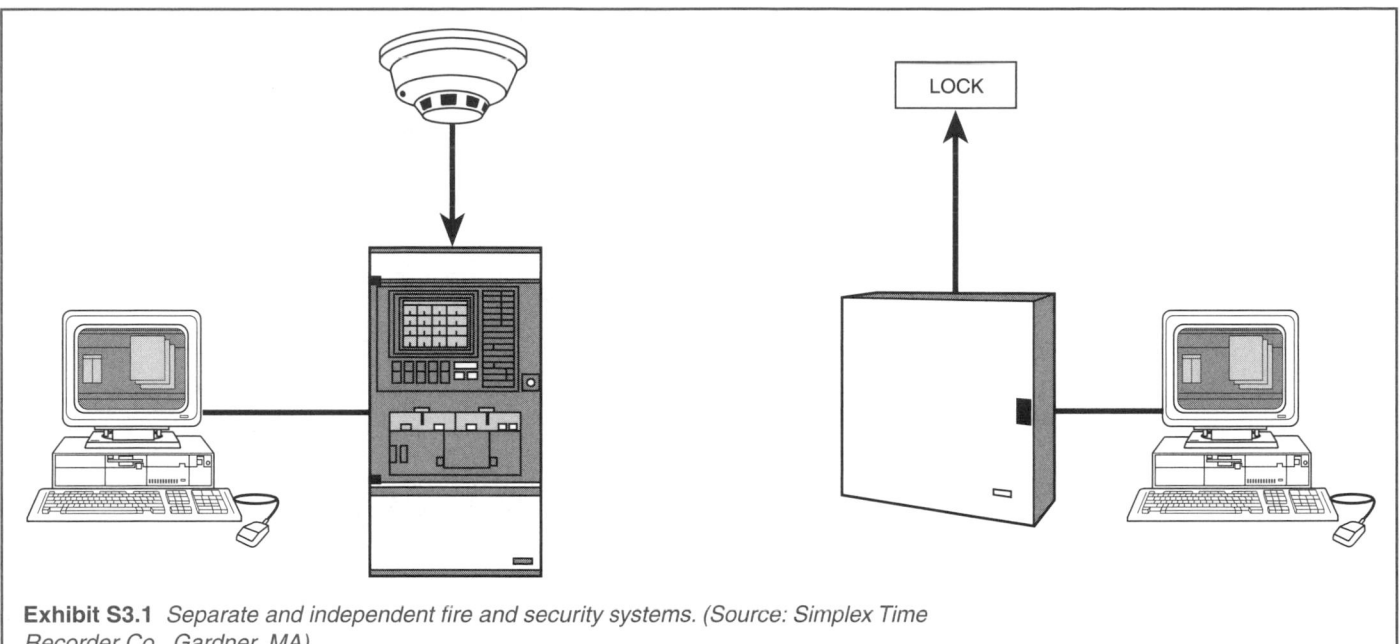

Exhibit S3.1 *Separate and independent fire and security systems. (Source: Simplex Time Recorder Co., Gardner, MA)*

the same concepts could easily apply to a fire alarm system and an HVAC control system.

In Exhibit S3.6 the smoke detector initiates an alarm signal to the fire alarm system control unit. The fire alarm system control unit sends a signal to the HVAC control panel. The HVAC control panel turns a fan off (or on). A sail switch or pressure differential switch provides positive confirmation of the fan's status. The HVAC control panel sends this confirmation back to the fire alarm system control unit. In this case, the circuit in both directions between the fire alarm system control unit and the HVAC control panel must either be monitored for integrity or must fail safe.

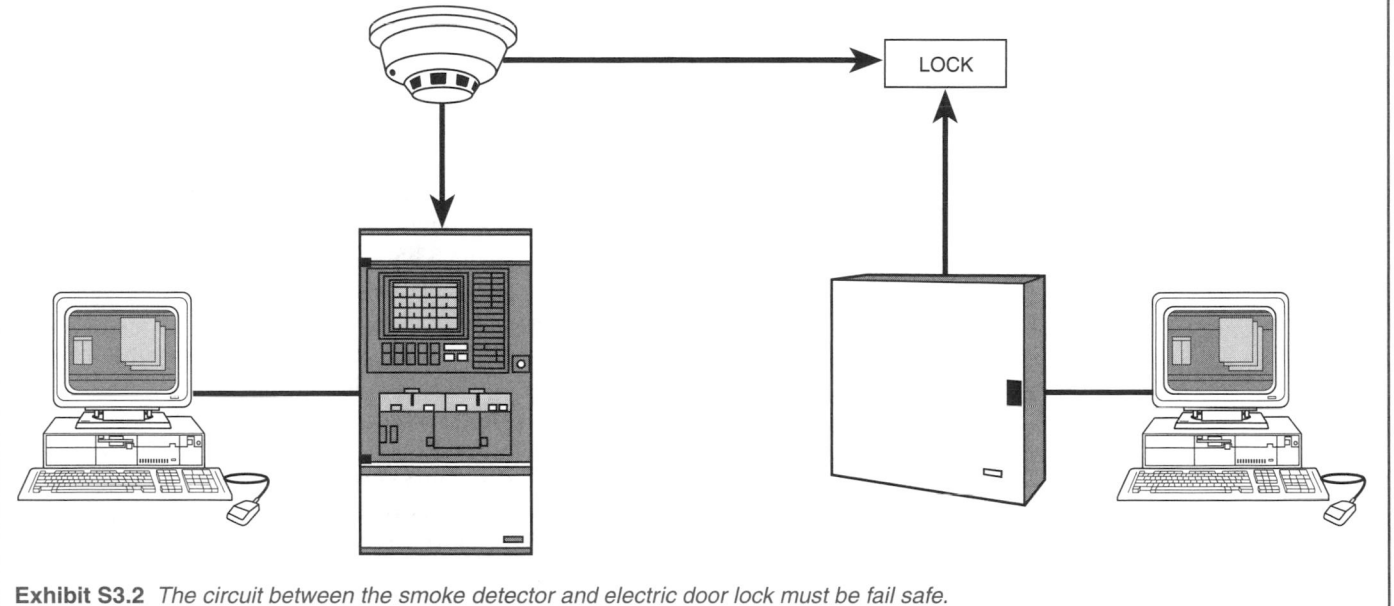

Exhibit S3.2 *The circuit between the smoke detector and electric door lock must be fail safe. (Source: Simplex Time Recorder Co., Gardner, MA)*

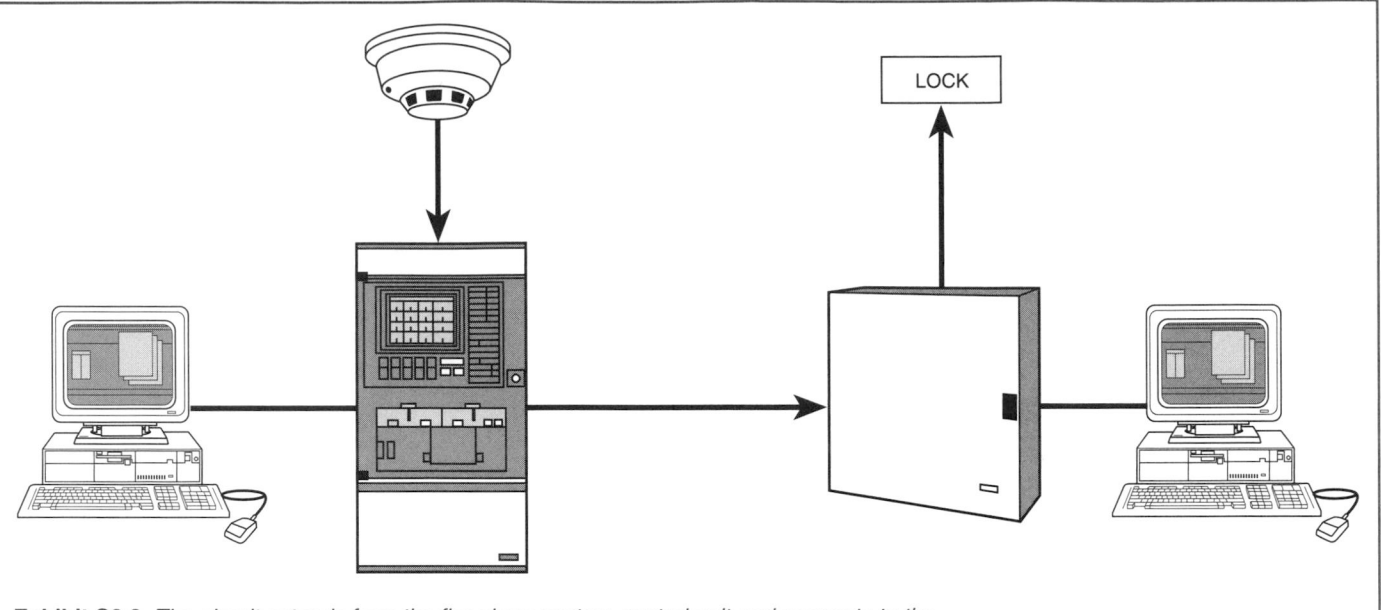

Exhibit S3.3 *The circuit extends from the fire alarm system control unit and connects to the security system control panel. (Source: Simplex Time Recorder Co., Gardner, MA)*

Once an overall building management design establishes two-way communication between different building systems, the design can also establish a single point of control. Exhibit S3.7 shows a single proprietary supervising station system control unit exercising control over both the fire alarm system and the security system. To meet the requirements of the code, the system design must provide some protection for the integrity of the fire alarm system functions. This protection,

termed a *fire wall*, can be created in hardware, in software, or in a combination of both hardware and software. This fire wall can prevent operators at the supervising station from changing the fire alarm system software unless they have the proper priority level. For example, a proper fire wall might authorize a non-fire alarm operator to interrogate a smoke detector to determine the detector's sensitivity level, but would prohibit that operator from changing that level.

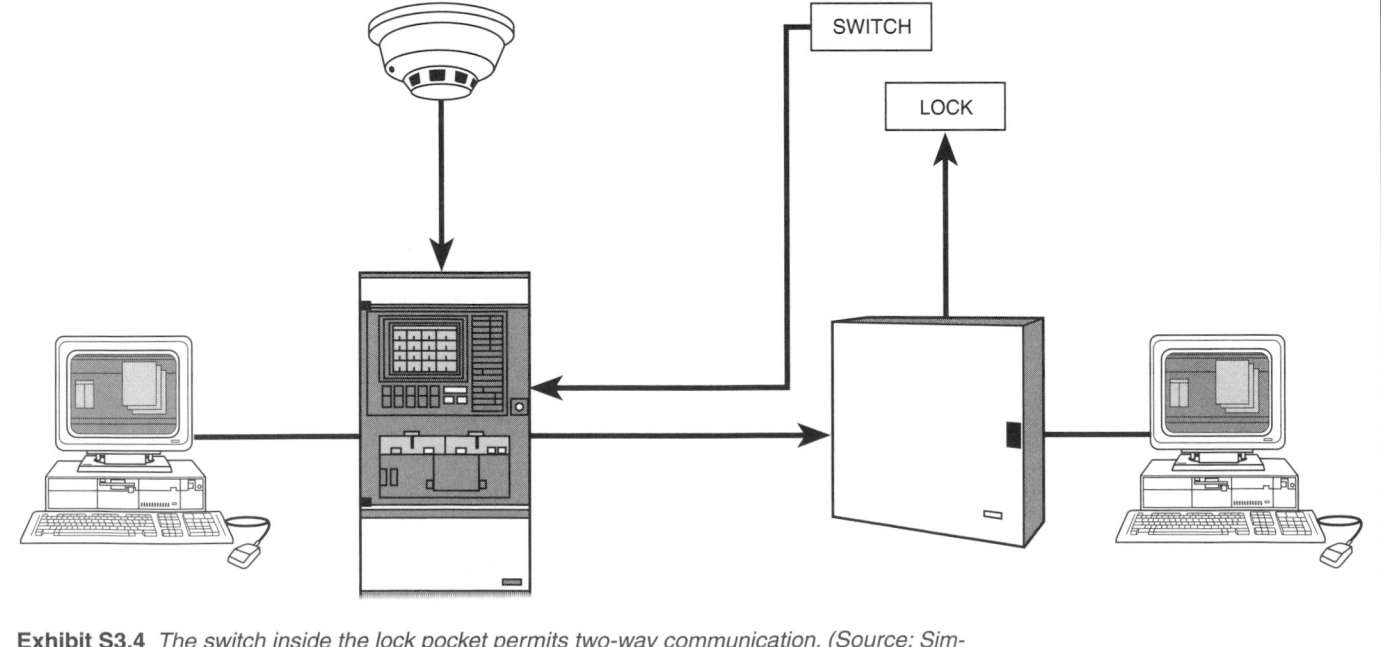

Exhibit S3.4 *The switch inside the lock pocket permits two-way communication. (Source: Simplex Time Recorder Co., Gardner, MA)*

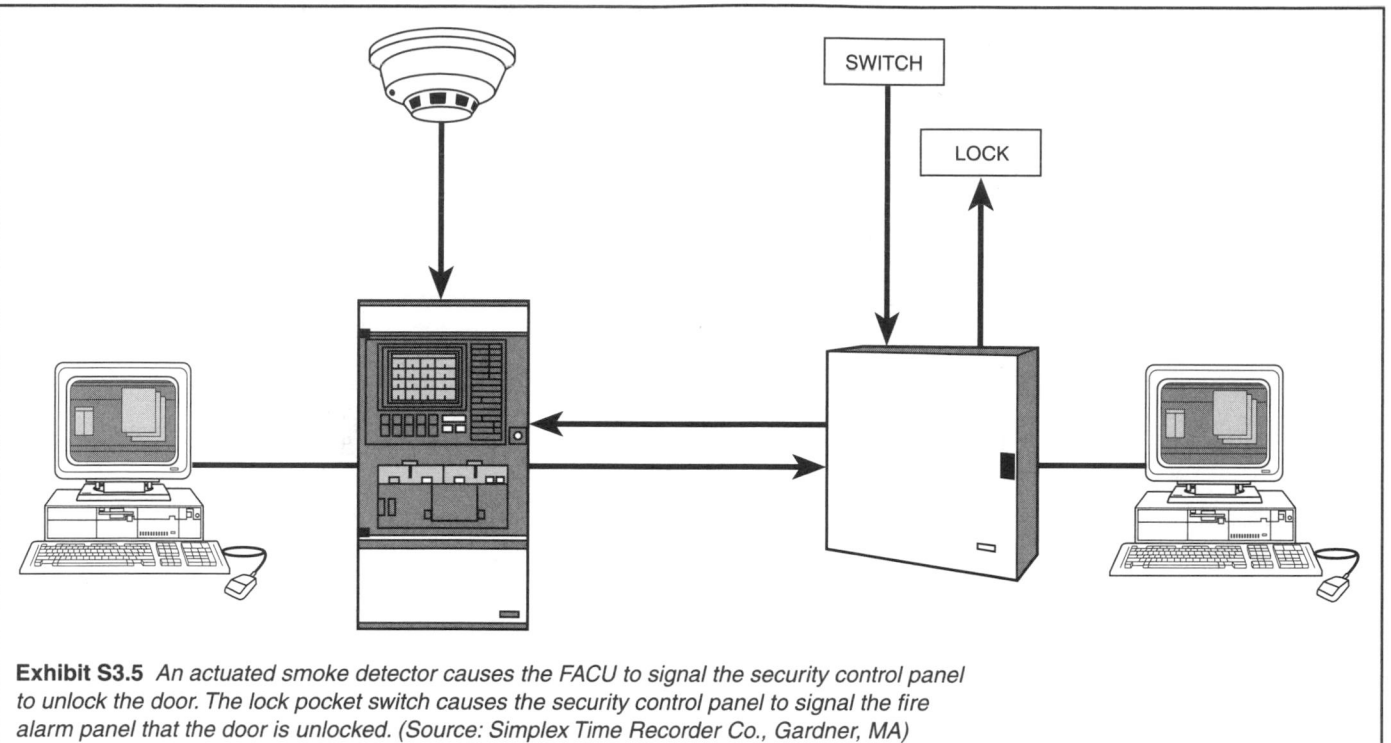

Exhibit S3.5 *An actuated smoke detector causes the FACU to signal the security control panel to unlock the door. The lock pocket switch causes the security control panel to signal the fire alarm panel that the door is unlocked. (Source: Simplex Time Recorder Co., Gardner, MA)*

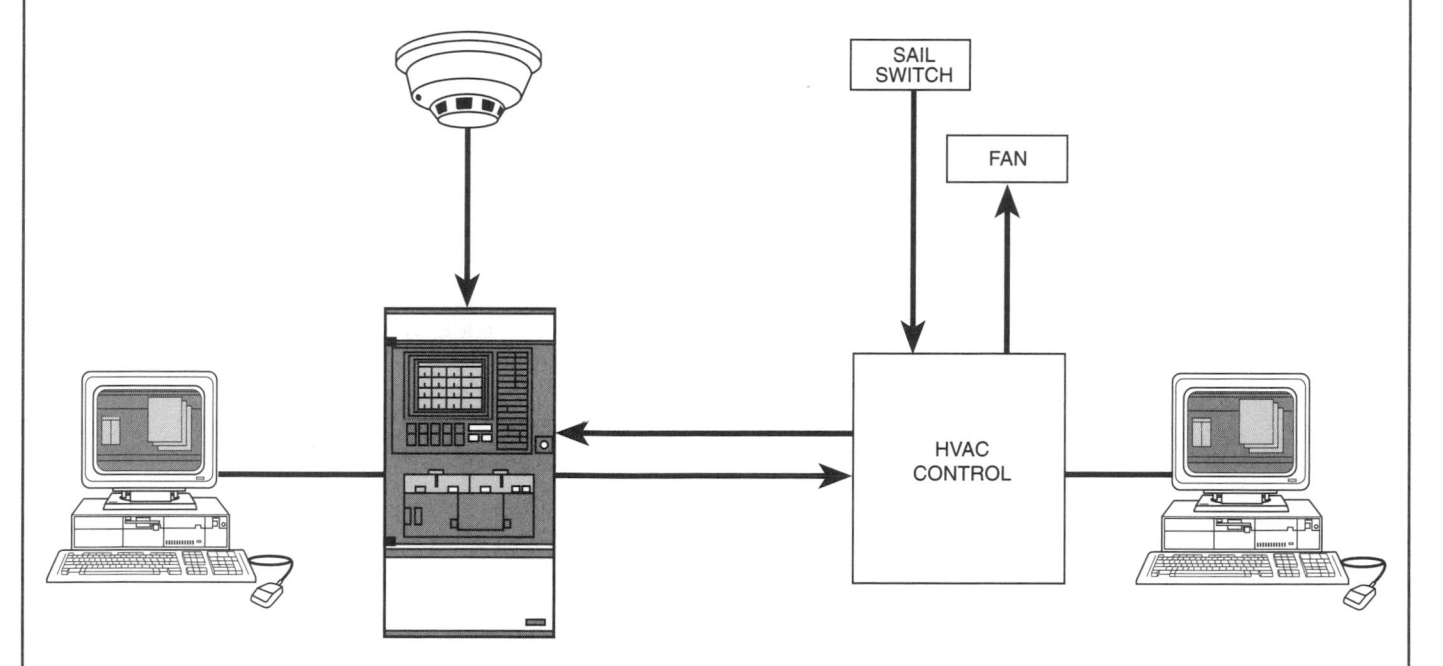

Exhibit S3.6 *The circuit in both directions must be monitored or fail safe. (Source: Simplex Time Recorder Co., Gardner, MA)*

In the previous seven exhibits, the connections between the fire alarm system and the other building management system have used circuits connected to either open or closed relay contacts. Modern computer-based systems can also communicate digitally. Exhibit S3.8 shows a serial connection between the fire alarm system control unit and the security control panel. Even though the two systems exchange information digitally, the circuit still must be either moni-

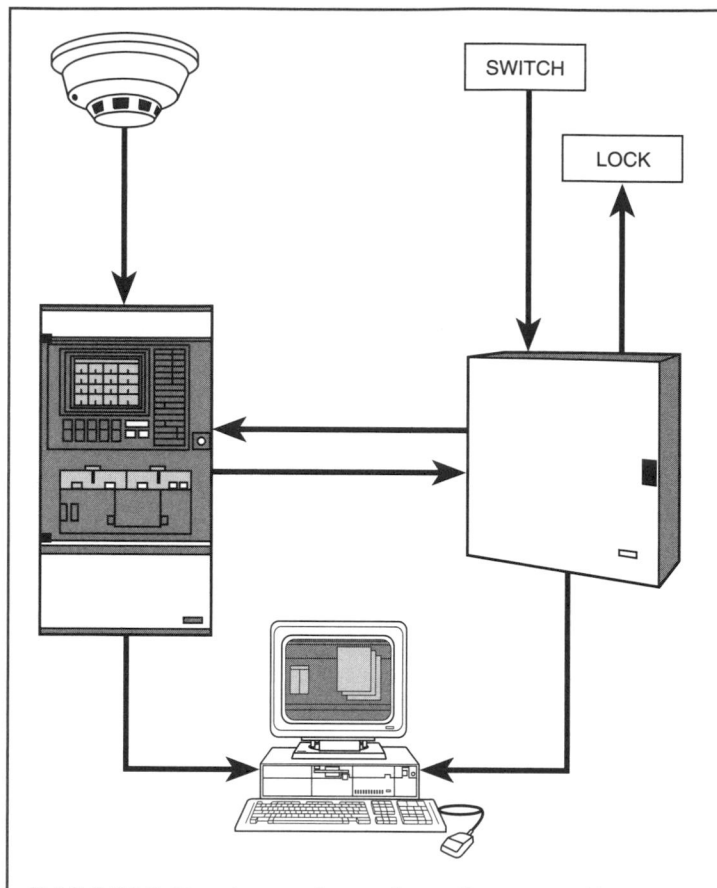

Exhibit S3.7 *Fire alarm system and security system with shared interface. (Source: Simplex Time Recorder Co., Gardner, MA)*

tored for integrity or must fail safe. Likewise, some method of interrogation and response, such as a "handshake" or a "ping," must ensure the integrity of the data.

Once these systems begin to exchange data by means of digital technology, they must use the same data protocols. Otherwise, while one system attempts to "speak" in a particular digital "language" the other system may "listen" in another digital "language."

Using some type of industry standard communications protocol can break through this language barrier. Fortunately, building system management experts, including the American Society of Heating, Refrigeration, and Air Conditioning Engineers (ASHRAE) have developed just such a communications protocol. The protocol, called BACnet, allows any building system designed to use the protocol to communicate with any other building system so equipped. Echleon Corporation has developed another protocol, known as LONworks. Communications between separate building systems must use such protocols whenever the building systems do not originate from the same manufacturer.

Of course, if a building management system has only moderate complexity, a manufacturer could build all the building system functions into a single system. Exhibit S3.9 shows a single system control unit performing both fire alarm functions and security functions. In order to meet the requirements of 1-5.1.2 of the code, a qualified testing laboratory would have to list the equipment for the particular purposes. This arrangement truly represents an integrated system.

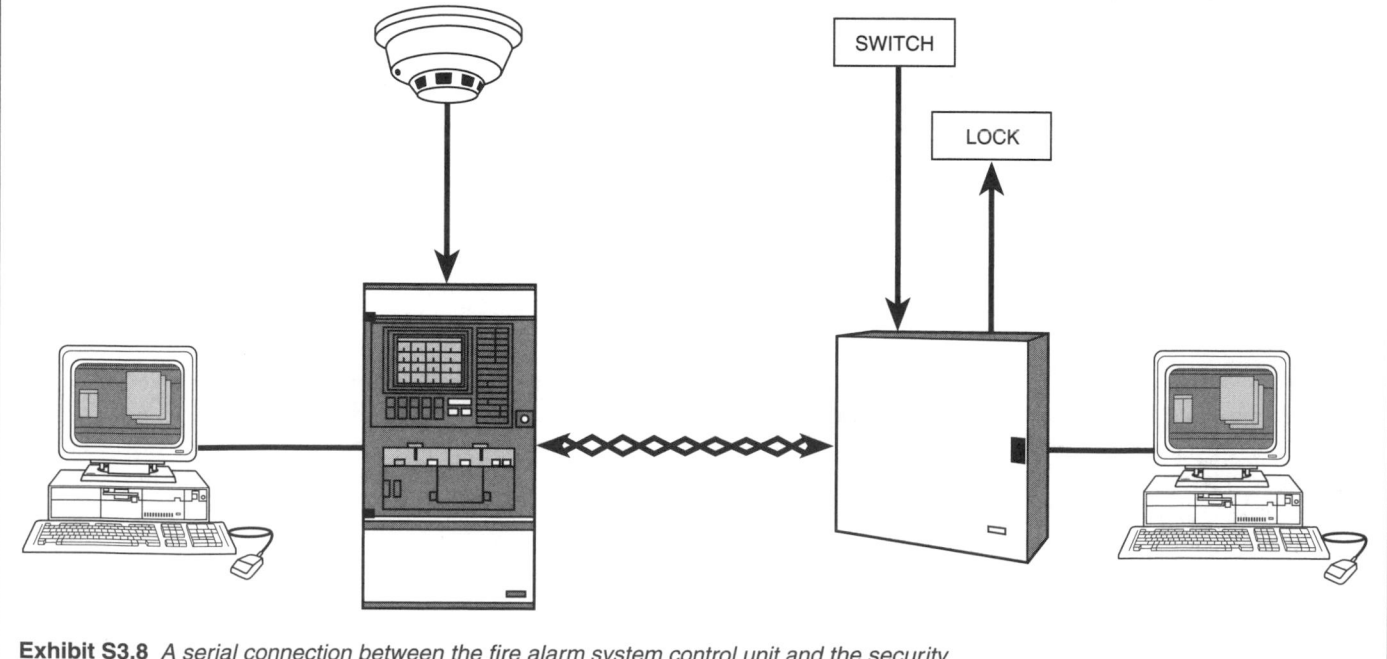

Exhibit S3.8 *A serial connection between the fire alarm system control unit and the security control panel. (Source: Simplex Time Recorder Co., Gardner, MA)*

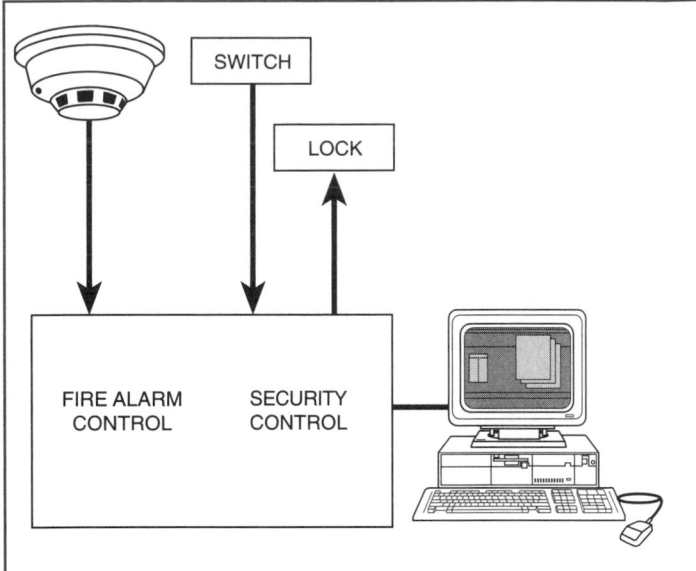

Exhibit S3.9 *In this truly integrated system, a single control unit performs fire alarm and security functions. (Source: Simplex Time Recorder Co., Gardner, MA)*

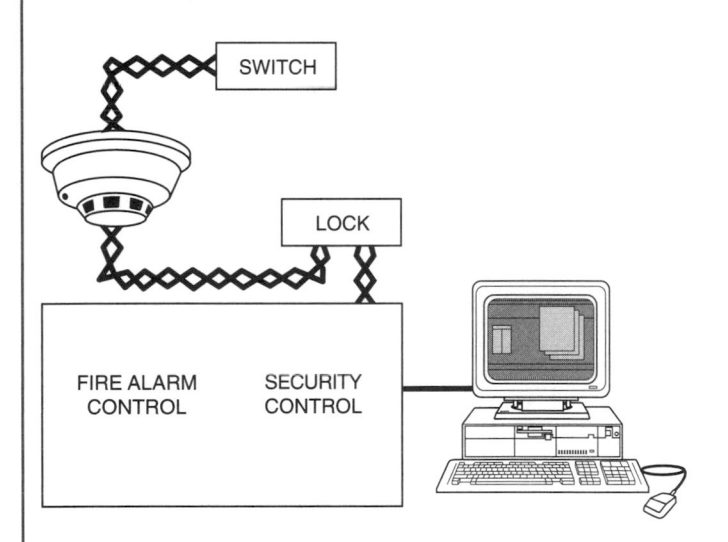

Exhibit S3.10 *Devices connect on a common circuit that must be monitored. (Source: Simplex Time Recorder Co., Gardner, MA)*

Handling multiple systems integrations at high speed may call for a peer-to-peer communication scheme. Exhibit S3.10 shows this even higher level of integration. Here the devices for both fire alarm and security connect on a common circuit. Once again, because this circuit contains a fire alarm-initiating device, it must be monitored for integrity. A digital interrogation and response, at a frequency prescribed by the code, could provide the monitoring for integrity.

The design of the building systems may extend peer-to-peer communication to the systems level. Exhibit S3.11

shows the fire alarm system control unit connected to and operating on a local area network (LAN). In this scenario, the LAN must be monitored for integrity and fire alarm signals must have the highest priority. The system administrator must configure the LAN so that no other traffic on the network will impede the fire alarm. Problems on non-fire alarm systems must not adversely affect the fire alarm system.

TO INTEGRATE OR NOT TO INTEGRATE?

Author Stephen Covey suggests, "Begin with the end in mind." This adage proves especially true when an owner or design professional considers how to allow the several building management systems to communicate with each other. Each design goal must have a list of expected outcomes that define exactly how each system will communicate with and respond to the other systems in the building. Once the owner or design professional determines these outcomes, then he or she can more easily decide if integrating the building systems will facilitate the outcomes.

The owner or design professional must clearly understand the cost implications. Integrated systems may cost more than separate, discreet systems. Soliciting separate bids for discreet systems and for integrated systems discloses those costs.

The owner or design professional must determine whether the integrated system will truly be easier to operate than separate systems. Sometimes looking at two separate screens will be much less confusing to an operator than toggling back and forth between multiple systems displays on one screen.

The owner or design professional must determine whether the integrated system will be easier and less expensive to maintain. Including a request for a quotation on the price of future service and maintenance as a part of the original bid documents provides a mechanism to confirm and control life cycle costs.

The owner or design professional must also determine whether the integrated systems provider has expertise across all the disciplines reflected in the several applications. Lower cost must not compromise fire protection.

And, equally important, the owner or design professional must make certain the authority having jurisdiction will permit integration. Many authorities having jurisdiction do not want all the eggs placed in a single basket. Such authorities having jurisdiction often require that the fire alarm system must stand alone as a separate and distinct system.

CONCLUSION

The building owner or the design professional must define what the building systems must accomplish. He or she must determine the potential level of disruption that the protec-

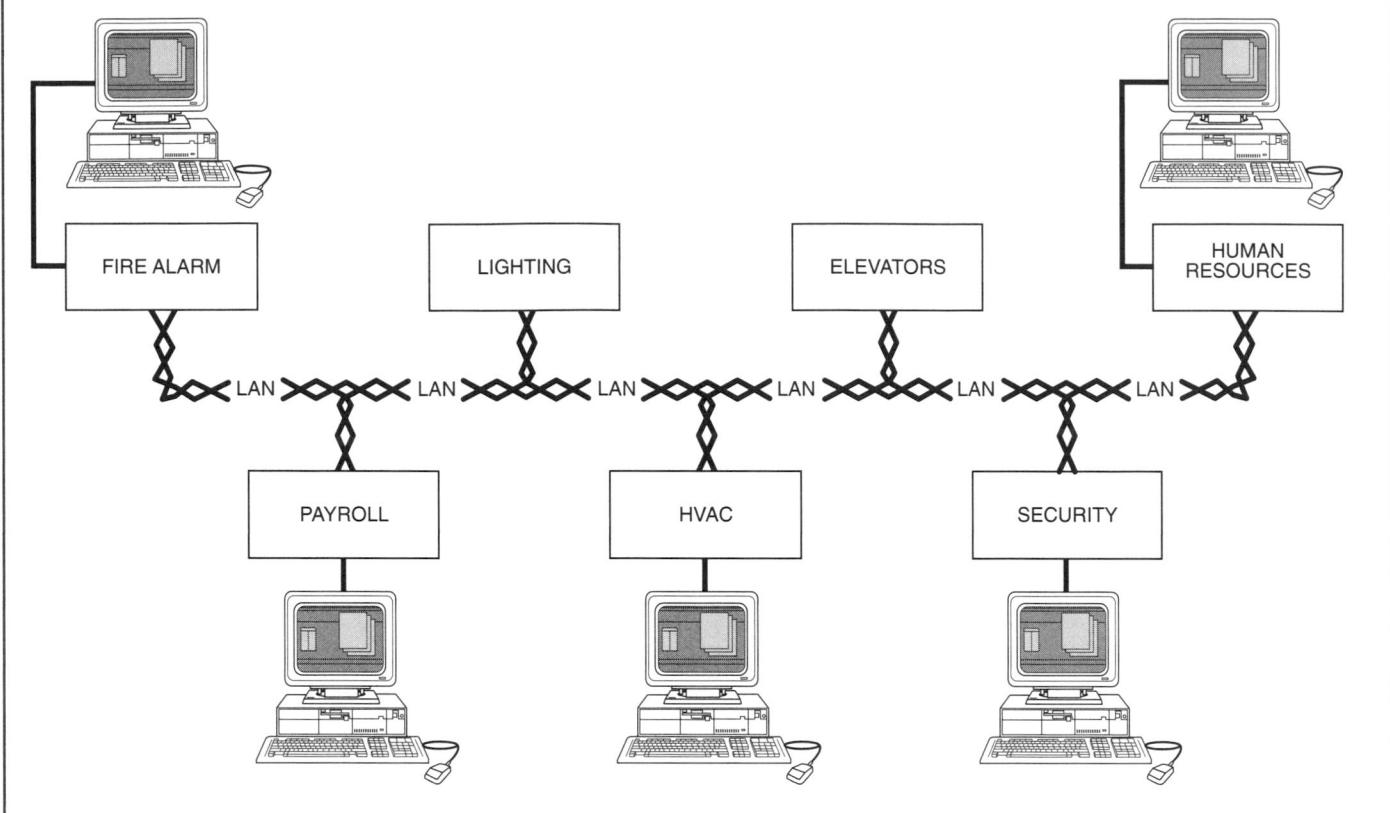

Exhibit S3.11 *This fire alarm control unit is part of a LAN that must be monitored for integrity. (Source: Simplex Time Recorder Co., Gardner, MA)*

tion design can tolerate. Then, he or she must determine if integrating the fire alarm system into other building systems helps further those objectives.

Above all, by following the requirements of the code, the fire alarm system can integrate with other building management systems and still preserve the integrity of the fire alarm system.

References

ANSI/ASME A17.1, *Safety Code for Elevators and Escalators,* 1998, American National Standards Institute, Inc., New York.

NFPA 90A, *Standard for the Installation of Air-Conditioning and Ventilating Systems,* 1999, National Fire Protection Association, Quincy, MA.

NFPA 90B, *Standard for the Installation of Warm Air Heating and Air-Conditioning Systems,* 1999, National Fire Protection Association, Quincy, MA.

NFPA 92A, *Recommended Practice for Smoke-Control Systems,* 1999, National Fire Protection Association, Quincy, MA.

NFPA 92B, *Guide for Smoke Management Systems in Malls, Atria, and Large Areas,* 1995, National Fire Protection Association, Quincy, MA.

NFPA *101*®, *Life Safety Code*®, 2000, National Fire Protection Association, Quincy, MA.

Occupant Response to Fire Alarm Signals

Guylène Proulx, Ph.D.
National Research Council of Canada

Editors' Note: The National Fire Alarm Code is based on the premise that fire alarm systems will detect and signal the presence of fire or other products of combustion and that building occupants will respond to alarm signals.

Guylène Proulx earned a Ph.D. in Architectural Planning from the University of Montreal. She has been a research officer of the Fire Risk Management Program at the National Research Council of Canada since 1992. She focuses her research on investigating human response to alarms, evacuation movement, typical actions taken, timing of escape, time-to-start evacuation, and social interaction during a fire.

INTRODUCTION

Fire alarms go off from time to time in most buildings. Users of NFPA 72, *National Fire Alarm Code* expect that the fire alarm system will improve the level of fire safety. It is generally assumed that the signal transmitted by the fire alarm system notifying occupants about a fire requires their immediate action. The most common action expected of building occupants upon hearing a fire alarm, is that they will immediately start to evacuate the building. However, it has been observed that occupants are slow in taking action when hearing a fire alarm, especially in deciding to evacuate. In fact, research shows that in some buildings, occupants tend to completely ignore the fire alarm signal. Supplement 4 will attempt to explain such situations and identify the means to change occupants' indifference.

WHY INSTALL A FIRE ALARM SYSTEM?

Questioning the intentions behind the installation of fire alarm systems in buildings that are required by NFPA *101®, Life Safety Code®*, and many other codes is a good beginning. Why install fire alarm notification appliances that will transmit a signal to the building occupants? This question may appear strange in the code but is worthwhile asking to demystify occupants' reactions. The first objective for installing fire alarm systems is stated in the first chapter of the code under Section 1-4. According to the code, an *alarm* is defined as a warning of fire danger. The first objective of a fire alarm signal is to notify occupants of a fire. Another objective is implicit in the definition of the *alarm signal*: a signal indicating an emergency requiring immediate action. The second objective for installing fire alarms is the expectation that occupants will immediately react to the alarm signal. It is usually expected that the fire alarm signal will be understood by occupants as the *evacuation signal,* which is defined as a distinctive signal intended to be recognized by the occupants as requiring evacuation of the building. The third objective of fire alarms is that, upon hearing its signal, occupants will start evacuation. A final objective is that the fire alarm activation will allow sufficient time for the occupants to escape.

In summary, there are four occupant behavior objectives intended to result from the activation of the fire alarm signal:

(1) Warn occupants of a fire
(2) Prompt immediate action
(3) Initiate evacuation movement
(4) Allow sufficient time to escape

Are these rather demanding objectives met when the fire alarm signal is activated? In some buildings, they are met, but in some others they are not. For example, when the fire alarm goes off in elementary schools, all pupils leave in ranks with their teachers and gather on the playground. In such situations, it is concluded that the four objectives of the fire alarm signal are met. In comparison, when a fire alarm

goes off in a shopping center or a high-rise office building, fire fighters often observe upon arriving on location that most, if not all, occupants are still in the building continuing their activities and ignoring the fire alarm. In such cases, the objectives of the fire alarm signal are not met. This lack of occupant response could have tragic consequences in the event of an actual fire.

THE MEANING OF THE FIRE ALARM SIGNAL

One explanation for the occupants' lack of reaction to the fire alarm signal is the occupants' failure to recognize the signal as the fire alarm signal. Fire alarm signals can be delivered through appliances such as bells, horns, chimes, or electronic appliances. In turn, these appliances can emit either a continuous, pulsating, slow whoop, or voice instruction. This situation means that a large variety of sounds can be used as a fire alarm signal. Furthermore, there are other types of alarms that can be activated in buildings, such as burglar alarms, elevator fault alarms, security door alarms, and so forth. Consequently, the public may have difficulty recognizing the sound of the fire alarm signal. This problem in identifying the fire alarm signal could explain, in part, the public's indifference when a fire alarm actuates.

The need to identify a unique fire alarm signal that could become universal was acknowledged a long time ago. Since the 1970s, numerous discussions to develop a standard fire alarm signal have taken place (Mande 1975; CHABA 1975). Experts finally agreed not to limit the fire alarm signal to any one sound, but instead to support universal identification through the use of a consistent sound pattern. The temporal three pattern, described in ISO 8201, *Acoustics—Audible Emergency Evacuation Signal,* is expected to become the standard fire alarm signal to be used everywhere to warn occupants of a fire danger. The temporal three signal has been required by NFPA 72 since 1996 and, in Canada, since 1995 in the *National Building Code* of Canada (NBC 1995). (See Exhibit S4.1.)

It is expected that it may take 15 years to 20 years before most buildings in Canada and the United States have the temporal three fire alarm signal installed. With time, it is hoped that more countries around the world will adopt this standard signal. The objective of the implementation of the temporal three alarm signal is to facilitate occupants' recognition of the fire alarm signal by using a standardized sound pattern. This sound pattern allows for cultural and language differences, and it can also facilitate perception by people with minor hearing limitations.

The introduction of the temporal three alarm signal is aimed at meeting the first objective of the fire alarm, which is to warn occupants of a fire. The question remains, however, whether the implementation of the new temporal three alarm signal will solve the problem of occupants ignoring

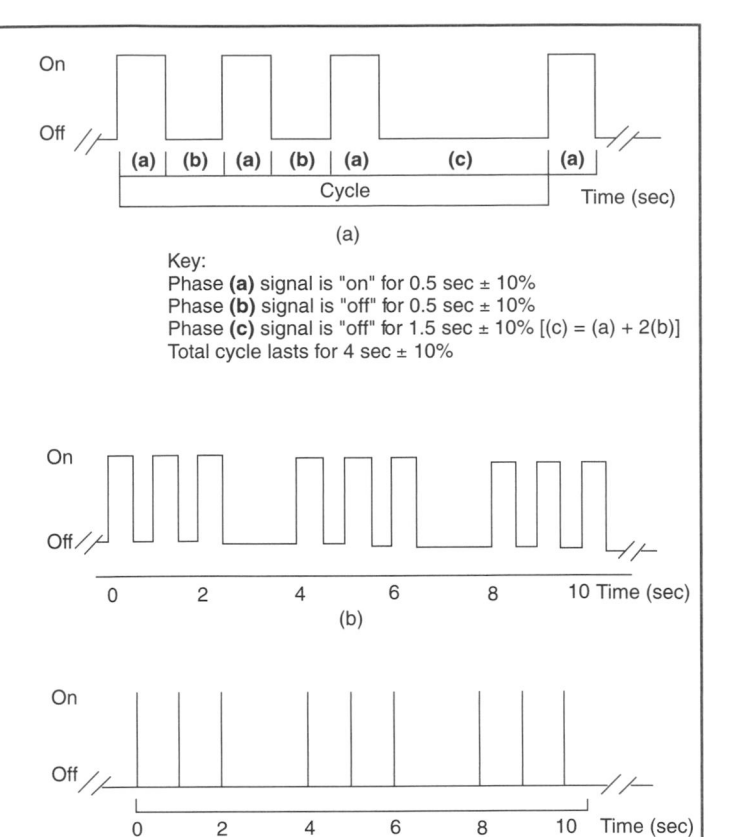

Key:
Phase **(a)** signal is "on" for 0.5 sec ± 10%
Phase **(b)** signal is "off" for 0.5 sec ± 10%
Phase **(c)** signal is "off" for 1.5 sec ± 10% [(c) = (a) + 2(b)]
Total cycle lasts for 4 sec ± 10%

Exhibit S4.1 *Temporal three pattern for fire alarm signal. (Source: Life Safety Code® Handbook, 7th ed., National Fire Protection Association, Quincy, MA)*

the fire alarm. It is one thing to recognize a signal, it is another thing to act upon perceiving a signal.

Assuming that the fire alarm signal is recognized by occupants, is this signal sufficient to trigger evacuation movement? After studying numerous actual fires, false alarms, and evacuation drills, it appears that a fire alarm signal alone is often not sufficient to prompt all occupants to leave a building. The intention behind the installation of the temporal three signal is that occupants will recognize this signal and will know that they should immediately evacuate the building. Will the new temporal three alarm signal, by itself, meet this intention? Probably not. For the temporal three signal to be understood by occupants as a signal indicating that they should immediately evacuate the building would require considerable public education and occupant training in each specific building. The implementation of the temporal three alarm signal is a big step in the right direction but it cannot solve all the problems. In the long run, it should solve the problem of recognition but, to ensure occupant safety, the problem of proper response must also be handled. Proper response can be achieved by means complementary to the fire alarm signal. Meanwhile, fire alarm devices emitting the temporal three pattern signal should

become the universal means to notify occupants of a fire, no matter what behavior is expected from the occupants.

WHY DO PEOPLE TEND TO IGNORE THE FIRE ALARM?

A first element of the explanation as to why occupants tend to ignore fire alarms was previously discussed: the occupants do not recognize the meaning of the fire alarm signal. But there is more. It seems that the large number of false alarms, test alarms, and drills can also explain the fact that occupants are reluctant to act when the fire alarm sounds. The problem with nuisance or false alarms is that, after a time, occupants lose confidence in the system, assuming that when the fire alarm goes off it is only another false alarm. The number of nuisance alarms and their deterring effects have to be studied over a period of time. Three nuisance alarms in the course of the same week will not have the same impact as three nuisance alarms over a year. The time of occurrence and the type of building might also play an important role in the impact of such nuisance alarms. If the nuisance alarm occurs in the middle of the night in a high-rise residential building, it may have a more negative lasting effect on occupants than a nuisance alarm happening in an office building on a nice, sunny day.

How many nuisance alarms are too many? How many will make people lose faith in this signal? Over the period of a year, are 3, 5, or 10 nuisance alarms the maximum? No research data has been found on that question. What is known, though, is that nuisance alarms tend to wear out the meaning of danger or urgency that should be associated with the fire alarm signal. It has been documented by the NFPA that in households, 25 percent of single-station smoke alarms failed to activate when tested because the power source had been removed or disconnected. When asked why the power source was missing, one-third of the respondents cited too many nuisance alarms (Smith et al. 1997).

Confronted with many nuisance alarms, people are likely to ignore the signal or attempt to disconnect the system if they can. This type of behavior is similar to that shown in the great movie with Audrey Hepburn and Peter O'Toole, entitled "How to Steal a Million" in which the couple wants to steal a statue in a Paris museum. The statue is surrounded by a sophisticated burglar alarm, and the couple manage to trigger the alarm from a distance with a boomerang. After the alarm has been triggered twice in the middle of the night for no apparent reason, the chief guard, who has become annoyed with the alarm system, disconnects it. So the thieves can run away with the statue without any problem.

One way to mitigate the impact of nuisance alarms is to reduce them to a minimum. Reducing nuisance alarms is not easy in some buildings where the fire alarm system is antiquated, receives poor maintenance, or is badly installed. Nonetheless, it is necessary to investigate each nuisance alarm to try to solve the problem. A code-compliant fire alarm system will produce fewer false alarms.

The public often assumes that false alarms are due to teenagers' mischief. Teenagers are blamed for many things. However, it was observed on surveillance cameras that pranksters could be teenagers, but were also children, adults, or even elderly people. Further, the assumption that nuisance alarms are usually prank alarms is not founded. In fact, most nuisance alarms are usually due to system malfunctions. In 1997 fire departments in the United States received close to 2 million false alarm calls. As presented in Table S4.1 (Karter 1998), 45 percent of these calls were systems malfunctions, 27 percent were well-intentioned calls that turned out not to be fires, 16 percent were mischievous false calls, and 12 percent were other tpes of false alarms (such as bomb scares).

Table S4.1 Estimated False Alarms Received by U.S. Fire Departments in 1997

Reasons for Alarm	Estimated Number	Percent (%)
System malfunctions	816,500	45.0
Unintentional calls	490,000	27.0
Malicious, mischievous false calls	286,500	15.8
Other false alarms (bomb scares, etc.)	221,500	12.2
Total	1,814,500	100

The important thing to remember is to attempt to reduce the number of nuisance alarms to a minimum. If nuisance alarms still occur, it is essential to inform occupants of the situation. After a nuisance alarm, occupants should be informed of the cause of the problem so they will know that the building owners and managers are aware of the situation and are doing something about it. Informing the occupants will help maintain a certain level of confidence about the fire alarm system and the building management.

Due to the deterring effect of actuating the alarm for non-fire reasons, should conducting system tests and evacuation drills with occupants in the building be avoided? Yes and no. System tests should be conducted when there is a limited number of occupants in the building, because the objective of such tests is to assess whether the system is functioning properly or not. Walkthrough tests conducted on the input side without any output are not sufficient. Tests of the notification appliances must be conducted by actuating the devices. If such tests are conducted when occupants are in the building, it is important to inform the occupants before the fire alarm activation that a test will be conducted, and after the test, that the test is completed. Silent, walkthrough tests are frequently conducted on the input side (i.e., for pull boxes and smoke/heat detectors) without any output. This test procedure minimizes the disturbance. Tests of the

notification appliances must be done by actuating the devices.

EVACUATION DRILLS

Evacuation drills are a different issue. There is some discussion in the field whether or not building management should tell occupants when drills are expected, so they are not perceived as nuisance alarms. Often if occupants are told in advance, they tend not to participate or avoid being in the building during a drill. After the drill, however, occupants should be told that "this was a drill," that "the drill was conducted to improve safety," and "if you have comments or concerns, please contact so and so."

Exercises or drills are held to assess if staff and occupants can apply the fire safety plan and if the evacuation procedure is appropriate. Consequently, evacuation drills should be conducted at least annually in all buildings (regardless of occupancy), with the full participation of every occupant. Each drill may add to the number of perceived nuisance alarms, but drills are essential. It is the best way to train occupants. During a drill, occupants learn to recognize the sound of the fire alarm signal; they also learn and practice the actions expected from them. Usually, drills are the only opportunity for occupants to experience alternative ways out, such as stairwells and emergency exits not usually used or even prohibited from normal use. A drill allows management to assess the evacuation procedure that is in place but, more importantly, it allows occupants to become familiar with the procedure. Through a drill, occupants can assess their own capacity in carrying out the evacuation procedure. For instance, it might be the only opportunity for an occupant to evaluate how long it takes to evacuate the building using the stairwell from the 45th floor. Drill participation allows an occupant to confirm that the stairwell is accessible, that the door at the bottom is indeed unlocked, and that the exit leads to a safe outside area. A person who has never experienced a route is very unlikely to give it a first try during an emergency. People tend to go toward the familiar (Sime 1980). Gaining experience with a non-familiar route to evacuate a building is the best way to ensure that occupants are likely to use this means of egress during an actual fire.

AUDIBILITY OF FIRE ALARM SIGNALS

The problem of recognition of the fire alarm signal and the number of nuisance alarms are two phenomena that can explain occupants' tendency to ignore fire alarm signals. A third explanation is also possible: the audibility problem of the fire alarm signal. Studies in mid-rise and high-rise residential buildings have shown that, in some buildings, occupants could not hear the fire alarm signal from inside their apartments (Proulx et al. 1995a). This audibility problem was typically observed in apartment buildings where the fire alarm appliances were located in the common corridors. Even though the fire alarm signal was very loud in the corridor, sound attenuation was such that the signal was not audible inside dwelling units, especially in rooms located farthest from the corridor. Ambient sound created by television, radio, air conditioning, or human activities can easily mask the sound of the fire alarm signal.

The audibility problem is very important because people cannot be expected to do the right things if they are not notified of the fire in the first place. To ensure alarm audibility, Chapter 4 of the code requires certain sound pressure levels. In most multi-dwelling buildings, the required levels can be met only by locating the appliance inside the dwelling unit. In fact, locating a notification appliance in each unit is probably the best way to ensure alarm audibility.

The traditional location of fire alarm appliances in corridors and stairwells can create areas where the alarm is not audible. Further, locating appliances in common areas can be counter-productive. It was observed during evacuation drills, and was reported after fires, that once occupants were notified of the fire and decided to leave their unit, they often went to their neighbors or discussed the best course of action with others in the corridor. Communication among members of a group from the same dwelling or with neighbors becomes paramount to ensure that everybody is accounted for, to decide what to do, where to go, to confirm decisions with others, and so forth. Very loud alarm signals in corridors and stairwells can prevent these essential exchanges among people. Once in the corridor or stairwell, occupants no longer need to be notified of the fire; what they need is the possibility to obtain and exchange information. Corridors and stairwells are also locations where occupants might receive instructions from wardens, staff, or fire fighters, so the volume of the fire alarm should be low enough to allow for efficient communication in these locations. Designers, installers, and authorities having jurisdiction should consider these human reactions when locating, installing, or inspecting the notification appliances in corridors and stairwells.

Despite the fact that the volume of the fire alarm signal should be low enough to allow verbal exchanges, it should not be so low that occupants might think that the signal has been switched off. During an evacuation study, the fire alarm signal was turned off after 5 minutes to facilitate walkie-talkie communication among fire fighters (Proulx et al. 1995b). It was observed, on camera, that most occupants who had started to evacuate stopped and returned home when the alarm signal was deactivated. Because the fire alarm was switched off, occupants assumed the emergency was over and they could return home. This reaction from the occupants explains why it is very important to maintain the actuation of the alarm signal as long as the situation is not totally resolved. For the occupants, interruption of the fire

alarm signal is the sign that the situation is over. As long as there are occupants in the building, the fire alarm signal should be functioning to maintain awareness of the state of emergency.

THE BUILDING, THE OCCUPANT, AND THE FIRE

When a fire alarm actuates in a building, what occupants will do about it—assuming they heard the signal, recognized it as the fire alarm, and have not been completely desensitized by too many nuisance alarms—depends on a number of complex factors. These complex factors can be organized around three major headings: the building characteristics, the occupant characteristics, and the fire characteristics.

There are a number of characteristics that could simultaneously have an impact on occupant behavior during a fire. The characteristics to be presented are not an exhaustive list. Furthermore, some characteristics can have a greater impact than others.

Among the building characteristics, a few types of occupancies can be identified to illustrate the importance of looking at the essence of each building and building area. The traditional way to approach occupancy classification is sometimes too broad to support predictions relative to occupant behavior in fire. For example, it cannot be expected that occupants in a church, a cinema, or a skating rink will react the same way in the event of a fire even though these buildings are all assembly-type occupancies. Each of these locations presents a specific challenge. The architecture of the space is another important building characteristic. If the space is complex, it can have a major impact on occupant movement and on the possibility of finding an alternative way out if the familiar route is blocked. At the time of the fire, the activities happening in the protected premises will have a major impact on occupants' response and reaction time. For example in a hotel, whether the location of the guests, are in their rooms, at the swimming pool or on the casino floor, will have an impact on their reactions. Finally, the building fire safety features will also play a key role in informing the occupants of the situation.

The following represents a partial list of building characteristics that can impact occupants' response and reaction time to a fire alarm signal.

(1) Type of occupancy.

    a. Residential (low-rise, mid-rise, high-rise)
    b. Office
    c. Factory
    d. Hospital
    e. Hotel
    f. Cinema
    g. College and university
    h. Shopping center

(2) Architecture.

    a. Number of floors
    b. Floor area
    c. Location of exits
    d. Location of stairwells
    e. Complexity of space
    f. Building shape
    g. Visual access

(3) Activities in the building.

    a. Working
    b. Sleeping
    c. Eating
    d. Shopping
    e. Watching a show, a play, a film, etc.

(4) Fire safety features.

    a. Fire alarm signal (type, audibility, location, number of nuisance alarms)
    b. Voice communication system
    c. Fire safety plan
    d. Trained staff
    e. Refuge area

The occupant characteristics will be paramount in explaining and predicting potential occupant behavior. Occupant characteristics include the occupants' profile, which represents a grouping of important parameters that can be influential in predicting occupants' response to a fire such as the occupants' age and mobility. Occupants' knowledge and experience are also important factors, because occupants who have, or don't have, training can react very differently. The condition of the occupants at the time of the event can also determine their potential to react promptly and appropriately. Personality and decision-making styles of each occupant can be influential; some occupants copy the reactions of others, whereas other occupants are prepared to take on a leadership role. Finally, the occupant's role in the building can explain different responses, for example, in a restaurant, the owner might be more likely than a client to fight a kitchen fire.

The following represents a partial list of occupant characteristics that can impact occupants' response and reaction time to a fire alarm signal.

(1) Profile.

    a. Gender
    b. Age
    c. Ability
    d. Limitation

(2) Knowledge and experience.

    a. Familiarity with the building
    b. Past fire experience

c. Fire safety training
d. Other emergency training

(3) Condition at the time of event.

a. Alone vs. with others
b. Active vs. passive
c. Alert
d. Under influence of drug, alcohol, medication

(4) Personality.

a. Influenced by others
b. Leadership qualities
c. Attitude toward authority
d. Anxiety level

(5) Role.

a. Visitor
b. Employee
c. Owner

Fire characteristics also can play an important role in the occupant response. During a fire, people perceive different cues from the fire and their interpretation of the situation changes rapidly, influencing their behavior. Perceiving a smell of smoke initiates a different response than directly seeing the fire.

The following represents a partial list of fire characteristics that can impact occupants' response and reaction time to a fire alarm signal.

(1) Visual cues.

a. Flame
b. Smoke (color, thickness)
c. Deflection of wall, ceiling, floor

(2) Olfactory cues.

a. Burn smell
b. Acrid smell

(3) Audible cues.

a. Cracking
b. Broken glass
c. Object falling

(4) Other cues.

a. Heat
b. Air draft

The difficulty with attempting to predict the occupants' behavior is that a number of the characteristics previously listed are mixed in different patterns according to each situation. There are a few concepts that can help explain and predict some of the occupant behavior, however. The concept of commitment is one of them. For example, imagine occu-

pants of a cinema watching a suspense movie. The fire alarm signal sounds, the sound level of the alarm is audible above the sound track, and occupants recognize the signal. According to the objectives of the fire alarm system, the signal should prompt immediate action and initiate evacuation movement. Unfortunately, these reactions are unlikely to happen. It should be expected that most occupants will stay in their seat hoping that the alarm signal will shut up soon. Such a response could be explained using the concept of commitment. Occupants who have paid good money to watch a movie are not prepared to leave while they are engrossed in the story. They are committed to the activity of watching this movie, and the fire alarm signal by itself is probably insufficient to make them leave. Being committed to an activity such as eating a meal, waiting in line for a ticket, or watching a show, is very powerful. People have a decision plan to carry out a specific activity and are reluctant to switch their attention to something unrelated.

As another example, the concept of role can explain the lack of response of some occupants in public buildings. In a museum or a department store, most occupants play the role of visitors and as such, they expect to be taken care of. When the fire alarm signal is actuated, there are social interactions taking place: people will be looking at what others are doing. Therefore, if others are not paying attention to the fire alarm signal, occupants become reluctant to take any action that would make them appear to be out of place or over-reacting to an insignificant situation. The role of visitors is usually to conform to the general behavior of others. Furthermore, visitors feel that it is their role to wait for instructions, even if they have recognized the signal as a fire alarm signal. They expect that someone will tell them what to do if something serious is really happening.

Despite constant efforts to educate the public as to the meaning of the fire alarm signal, that is, "that a fire alarm signal means leave immediately," this association is not automatic for every situation. For instance, in most public buildings. such as airports, occupants' interpretation of the fire alarm signal is that something is happening, which is unlikely to be a fire, so they should stay put and wait to see what happens.

FIRE SAFETY PLAN

It is reasonable to expect occupants to be warned of a fire by the alarm signal, but the related response expected from the occupants may not be known by them unless it is detailed in a document such as the fire safety plan. Different names are used such as "fire safety information," "emergency instructions," or "escape plan" but these are all fire safety plans. Every building should have a fire safety plan from the single family home, to airports, shopping centers, hospitals, and warehouses. A fire safety plan should contain all the fire safety features of the building (including how the fire

department is called), instructions regarding the actions expected by the occupants (whether they are staff or visitors, and including people with disabilities), number and frequency of drills, and so forth. This fire safety plan should be available to everybody; it should be posted in the building, updated regularly, and used during training and drills.

In more and more large buildings, the fire alarm signal does not indicate that occupants should evacuate the building. Instead, occupants are expected to remain on location, move to another area, move to an area of refuge, or implement any other plan of action that is the most appropriate for the building or some specific locations of a building. A massive evacuation movement could bring tragic outcomes in many large buildings. In some cases, such as high-rise hotels protected by a complete automatic sprinkler system, it might be safer for occupants to stay in their rooms and start protect-in-place activities such as sealing doors and cracks to prevent smoke from entering, waiting for the situation to be controlled, or waiting to be rescued.

The fire safety plan is an essential tool to ensure occupant fire safety. The fire alarm signal may not make occupants start to move, but if the intention and occupant response expected are clearly stated in the fire safety plan, and this plan is known and available, it might help to make occupants respond the way it has been planned.

Change the Environment

Once occupants have been warned of the fire danger through the fire alarm signal, the second step is to initiate action. Because it cannot be expected that the alarm signal, by itself, will initiate action, some other means must come into play. In many buildings, what is needed to make occupants move is to stop the current activities and change the environment.

In a shopping center, such change of the environment would be turning off the music; in a cinema, the movie should be stopped and the lights should be turned on as soon as the fire alarm actuates. In a discotheque or restaurant, the music should be stopped and full lighting should flood the space. Initial protestation from the crowd will diminish as occupants perceive new information. This kind of atmosphere change will help occupants to understand that something is going on and will facilitate the shift of occupants' attention from their current activities to the emergency situation.

This ambiance change is essential to ensure that occupants pay attention to the fire alarm signal and the emergency situation. Occupants are usually committed to specific activities such as eating, shopping, watching a show, or participating in a sporting event. As long as the show continues, people focus their attention on this activity and are very reluctant to shift their attention to an unexpected or ambiguous event. It is, therefore, critical to have a sharp change in the environment to alter the behavior of occupants.

After the fire alarm signal actuation to warn occupants, additional means to convey information will help initiate evacuation movement, such as using a voice communication system.

Use of a Voice Communication System

Most modern buildings are equipped with a voice communication system that is used to broadcast music and specific messages directed to the occupants or the staff on location. In the past, this means was rarely used to provide information to the occupants during fire emergencies. This is unfortunate because a voice communication system is probably one of the best ways to provide essential information to the occupants.

The reluctance to use the voice communication system to provide information was mainly due to the false idea that occupants will panic if they are told that there is a fire (Sime 1980; Keating 1982). In fact, being told the truth is more likely to trigger the appropriate reaction than to trigger dysfunctional behavior. Research and actual fires demonstrate that receiving information through a voice communication system is one of the best ways to ensure that occupants will react immediately. Telling occupants that "there is a fire on the 3rd floor, please leave immediately" makes it easier for occupants to decide what to do. Contrary to some beliefs, occupants tend to immediately obey instructions given through the voice communication system (Proulx and Sime 1991; Proulx 1998).

As soon as the situation has been assessed as a fire emergency, there should be no delay in using the voice communication system to deliver messages to occupants. Voice messages will confirm the meaning of the fire alarm signal and instruct occupants on the best course of action. On-site management should be prepared to rapidly make the decision to evacuate the building or to direct occupants to a safe location in accordance with the fire safety plan. Waiting for the arrival of the fire department and for their assessment of the situation to deliver messages could be counter-productive. In fact, the fire department's first priority upon arrival at a fire scene is the location of occupants. This important activity occurs more quickly if occupants have gathered in a meeting place. Furthermore, when the fire department arrives on location, 5 minutes to 10 minutes, if not more, have passed since the fire was first detected. By that time, the situation could be lethal in some locations; if occupants are required to evacuate at that time, they may have to move through smoke-filled areas to reach the outside (Proulx 1998).

Some buildings are equipped with an emergency voice alarm communication system that delivers prerecorded messages. Although such a system saves staff time, the use of prerecorded messages has proven ineffective and even dangerous in some situations. A field study demonstrated that a prerecorded message could not be precise enough to help

occupants locate the nearest exit. During the evacuation of an underground station where the main escalator was blocked, occupants did not know where to go because the prerecorded message could not provide information as to the location of an alternative way out (Proulx and Sime 1991).

The information content of a prerecorded message is always limited because it needs to be general enough to cover all situations of an alarm actuation. There are some new systems that can deliver different messages according to the location of the actuated detectors, but this technology has not yet proven totally efficient. During the Düsseldorf Airport fire in 1996, prerecorded messages in different languages were transmitted; unfortunately, the information delivered during the initial 10 minutes was erroneous, directing passengers toward the most dangerous areas of the airport (NFPA 1998).

On two recent transatlantic British Airways' flights, a prerecorded emergency evacuation message was broadcast mistakenly. The crew could not stop the messages for approximately 30 minutes. During that time, passengers were crying, praying, and putting on life safety vests for emergency landing at sea. The distressful moments these passengers experienced were terrible. Issuing a prerecorded evacuation message by mistake is similar to a false alarm, in that it discredits a means of providing emergency information.

Evidently, the best approach is to broadcast live messages. Live messages allow the flexibility of altering the messages as new information is relayed to the person delivering the information. The tone of live messages can convey the urgency and importance of the information. Occupants are more receptive to live messages because they consider this information more likely to be genuine.

Messages delivered to the public during a fire emergency should contain three essential types of information:

(1) Identification of the problem
(2) Location of the problem
(3) Instructions

If occupants are expected to react correctly, it is imperative for them to understand the situation. Attempting to minimize the danger or using technical jargon to disguise the real situation could confuse occupants and prevent them from reacting appropriately. Instead, it is important to identify the problem in common terms such as "we suspect a fire" or "there is a fire." The second important type of information is the location of the problem. The occupants will want to know if they are at immediate risk; knowing the location of the problem will help them in their decision-making process. Finally, the message should clearly explain what is expected from the occupants. In some cases, it might be best for the occupants to remain where they are when the alarm sounds; in others, they can be directed through a specific route and to a specific exit with the aid of live messages.

The availability of closed-circuit televisions (CCTVs) to broadcast useful and precise messages becomes a must for the person issuing messages. Since many buildings are now equipped with CCTVs for security purposes, these tools are an incomparable source of information to deliver the most precise messages during an emergency. Strategically placed CCTVs allow the person behind the microphone an overview of conditions in different areas of the premises. Messages can then be tailored to the crowd movement and the developing situation.

Well-Trained Occupants

If the occupants are expected to do the right things during a fire emergency, the best way to meet this intention is to train them. The public, in general, do not have the knowledge of those who deal with fire safety issues day after day. Occupants' knowledge and assumptions regarding the development of a fire are often wrong. The literature is full of anecdotes reporting about people not doing what they were expected to do or worse, doing things that endangered their lives. For example, occupants broke windows during the World Trade Center fire to vent the smoke, which made the situation worse, as fire professionals would predict (Fahy and Proulx 1995). In a high-rise residential fire, occupants did not close the main door upon leaving, judging that a wood door would burn right through (Proulx 1996). Some people poured water on burning oil (read in the newspaper); other people have attempted to hold their breath moving long distances through smoke (Proulx 1998); and some people entered a subway station and went down an escalator next to the fire, focused on their journey back home (Donald and Canter 1990). More horror stories are not needed.

The public, in general, should be educated about fire, how it can start, how it develops, and what impact it has on people. Most fire safety education programs are targeted toward children, which is excellent and should continue, but other groups are at risk and need to be educated as well. Everybody needs general education about fire safety. Further, occupants need to be trained in the fire safety plan for buildings they are visiting. This is easier for buildings that people visit frequently such as a place of work or where they stay for a period of time, such as a cinema or theater, where a short message could inform occupants of the fire safety plan for that location before the performance. But there are other types of buildings, such as airports or sport centers, where training occupants is not practical. In such locations, a large part of the responsibility for occupants' safety rests with staff. Consequently, staff training is paramount.

In public buildings, occupants are unlikely to initiate evacuation movement by hearing the fire alarm alone, but they are very likely to respond to members of staff. Staff members are regarded as knowledgeable; they are expected

to know what is going on, what is the best course of action, and where the closest exit is. In uniform or wearing a name tag, staff are likely to be listened to. Evacuations of Marks & Spencer's department stores in the United Kingdom demonstrated that customers, even though the fire alarm had been ringing for some time, were prompted to evacuate only when requested to do so by the staff; then they complied right away to instructions (Shields et al. 1998). In another situation, during an evacuation drill in an underground subway station, passengers waited three minutes on the platform under the ringing fire alarm signal. They complied immediately with the instruction to reboard the train when the uniformed guard arrived.

Proper staff training should include regular classroom training sessions as well as evacuation drills. An evacuation drill is a valuable occasion for staff to put into practice ideas learned in the training class, and for management to assess the application of the fire safety plan. Changes and other adjustments might be required after an evacuation drill; feedback from staff can help to identify areas for improvement. An assessment is also advisable after false alarms or actual fires in order to improve the fire safety plan.

If staff are expected to play an important role during the evacuation of the public, it becomes essential to train them. Each staff member, whether part-time or permanent, should be educated about the content of the fire safety plan. Staff should not be allowed to begin work before having received proper fire safety training. The lives of hundreds of people could be in the hands of a few staff members. Employees need to be made aware of the importance of their role and of their responsibility to look after the public in case of an emergency.

When dealing with large spaces or with large crowds, it is not practical to rely entirely on staff to direct occupants to safety, as the number of employees required might be very large. For such situations, it is more efficient to rely on a few well-trained staff members, the emergency voice alarm communication system, and CCTVs. With CCTVs, the person issuing information will have an overview of the situation that will contribute to the delivery of precise messages. Staff on location will then be able to assist with the evacuation in conjunction with the instructions being delivered through the voice communication system.

Time to Escape

When the fire alarm is actuated, it should provide enough time for occupants to move to a safe location before conditions become dangerous. However, if the occupants do not start to move immediately after perceiving the fire alarm signal, the time available for safe escape becomes shorter. In an effort to reduce the delay between the time of alarm activation and the time at which people start to move, information should be provided to the occupants to prompt their

movement. Movement can be prompted through a dramatic change in the environment (people are dancing the night away and suddenly the music stops and full lighting floods the space; scary but effective to make people pay attention), voice communication messages, staff instructions, and so forth. These means to inform occupants of the emergency should come into play as soon as possible after the alarm actuation to provide sufficient time for occupants to leave safely.

In residential evacuations, the delay time to start evacuation after hearing the fire alarm signal was three-fourths of the whole evacuation time (Proulx et al. 1995b). In other words, if the total evacuation time was 4 minutes, 3 minutes were spent in delay time (investigating the situation, gathering family members and pets, finding wallet and keys, etc.), then 1 minute was used to move to safety. The initial delay time could be dramatically shortened if additional means to inform occupants were used. In an underground station, not all occupants managed to be evacuated; 15 minutes after the fire alarm signal activation, some passengers were still patiently waiting for their train, reading their newspaper. When the fire alarm signal was paired with emergency voice alarm communication messages or staff, the space was cleared in just over 5 minutes (Proulx and Sime 1991).

Consequently, even a good fire alarm system that can issue early warnings to occupants might not ensure that occupants have sufficient time to escape safely if they do not respond to the alarm rapidly. Complementary means to provide information to occupants will help shorten the delay in response, providing enough time for occupants to reach safety.

CONCLUSION

It was demonstrated that it might be overly optimistic to expect that the fire alarm signal alone will warn occupants, prompt immediate action, initiate evacuation movement, and allow sufficient time to escape safely. A fire alarm signal emitting the temporal three pattern should be excellent at warning occupants of a fire danger. To meet the other objectives of the fire alarm signal regarding rapid appropriate response from occupants, other means to convey information are required.

The actuation of a fire alarm signal is unlikely to trigger a massive evacuation movement. This observation does not imply that fire alarms should be removed, because fire alarms are indispensable in warning occupants of an imminent danger. Occupants want and need to be warned of an occurring fire. With the recognizable temporal three pattern, good audibility, and minimum nuisance alarms, warning occupants of a fire danger with the fire alarm signal is a realistic objective. Obtaining a specific occupant response can be achieved through complementary means including voice

communication messages, staff–warden instruction, training, drills and a well-devised fire safety plan. It is the combination of the fire alarm signal with other means to convey information that will ensure occupants' safety.

References

CHABA, 1975, "A Proposed Standard Fire Alarm Signal," *Fire Journal* 69:4:24–27.

Donald, I., and D. Canter, 1990, "Behavioural Aspects of the King's Cross Disaster," *Fires & Human Behaviour*, pp. 15–30. ed., D. Canter, 2nd ed, David Fulton Publishers, London.

Fahy, R.F., and G. Proulx, 1995, "Collective Common Sense: A Study of Human Behavior During the World Trade Center Evacuation," *NFPA Journal* 9:2:59–67, National Fire Protection Association, Quincy MA.

Karter, J. Michael, 1998, "Fire Loss in the United States During 1997," *NFPA Journal* 5:72–82, National Fire Protection Association, Quincy, MA.

Keating, P.J., 1982, "The Myth of Panic," *Fire Journal* 5:57–61.

Mande, I., 1975, "A Standard Fire Alarm Signal Temporal or 'Slow Whoop,'" *Fire Journal* 69:6:25–28.

NBC, 1995, *National Building Code of Canada*, Institute for Research in Construction, National Research Council Canada, Ottawa.

NFPA 1998, "Hard Lessons Learned from the Düsseldorf Fire," *Fire Prevention* 312:32–33, Fire Protection Association, UK.

Proulx, G., 1996, "Critical Factors in High-Rise Evacuations," *Fire Prevention* 291:24–27, Borehamwood, England.

Proulx, G., 1998, "The Impact of Voice Communication Messages During a Residential High-Rise Fire," *Human Behaviour in Fire—Proceedings of the First International Symposium*. ed. J. Shields, pp. 265–274, University of Ulster, UK.

Proulx, G., C. Laroche, and J.C. Latour, 1995a, "Audibility Problems with Fire Alarms in Apartment Buildings," *Proceedings of the Human Factors and Ergonomics Society,* 39th Annual Meeting, 2:989–993.

Proulx, G., J.C. Latour, J.W. MacLaurin, J. Pineau, L.E. Hoffman, and C. Laroche, 1995b, "Housing Evacuation of Mixed Abilities Occupants in High-Rise Buildings," IRC-IR 706, Internal Report, 92, Institute for Research in Construction, National Research Council of Canada, Ottawa.

Proulx, G., and J.D. Sime, 1991, "To Prevent 'Panic' in an Underground Emergency: Why Not Tell People the Truth?" *Fire Safety Science—Proceedings of the Third International Symposium*, pp. 843–852, Elsevier Applied Science, London.

Shields, T.J., K.E. Boyce, and G.W.H. Silcock, 1998, "Towards the Characterization of Large Retail Stores," *Human Behaviour in Fire—Proceedings of the First International Symposium*, ed. J. Shields, pp. 277–289, University of Ulster, UK.

Sime, J.D.,1980, "The Concept of Panic," *Fires and Human Behaviour*, ed. D. Canter, pp. 63–81, John Wiley & Sons, Chichester, England.

Smith, E.L., L.C. Smith, and L.J. Ayres, 1997, "When Detectors Don't Work," *NFPA Journal*, pp. 40–56, National Fire Protection Association, Quincy, MA.

NFPA 101®, 2000, *Life Safety Code*®, National Fire Protection Association, Quincy, MA.

ISO 8201, *Acoustics—Audible Emergency Evacuation Signal*, 1987, International Standards Organization, Geneva, Switzerland.

Index